THE Regional Geography OF CANADA

Fourth Edition

THE Regional Geography OF CANADA

Robert M. Bone

OXFORD
UNIVERSITY PRESS

70 Wynford Drive, Don Mills, Ontario M3C 1J9
www.oup.com/ca

Oxford University Press is a department of the University of Oxford.
It furthers the University's objective of excellence in research, scholarship,
and education by publishing worldwide in

Oxford New York
Auckland Cape Town Dar es Salaam Hong Kong Karachi
Kuala Lumpur Madrid Melbourne Mexico City Nairobi
New Delhi Shanghai Taipei Toronto

With offices in
Argentina Austria Brazil Chile Czech Republic France Greece
Guatemala Hungary Italy Japan Poland Portugal Singapore
South Korea Switzerland Thailand Turkey Ukraine Vietnam

Published in Canada
by Oxford University Press

First published 2008

Library and Archives Canada Cataloguing in Publication

Bone, Robert M.
The regional geography of Canada / Robert M. Bone. — 4th ed.

Includes bibliographical references and index.
ISBN 978-0-19-542536-9

1. Canada—Geography—Textbooks. I. Title.

FC76.B66 2007 917.1 C2007-904823-4

1 2 3 4 – 11 10 09 08

Cover image: Alain Juteau/iStockPhoto

This book is printed on permanent (acid-free) paper ♾.
Printed in Canada

Contents Overview

Detailed Contents

List of Figures

List of Tables

Preface

The purpose of this book is to introduce university students to Canada's regional geography. In studying the regional geography of Canada, the student not only gains an appreciation of the country's amazing diversity but also learns how its regions interact with one another. By developing the central theme that Canada is a country of regions, this text presents a number of images of Canada, revealing its physical, cultural, and economic diversity as well as its regional complexity. A number of features such as photos, key terms, maps, vignettes, tables, graphs, and references and further readings are designed to facilitate and enrich student learning.

The Regional Geography of Canada divides Canada into six geographic regions: Ontario, Québec, British Columbia, Western Canada, Atlantic Canada, and the Territorial North. Each region has a particular regional geography, history, population, and a unique location. These factors have determined each region's character, set the direction for its development, and created a sense of place. In examining these themes, this book underscores the dynamic nature of Canada's regional geography. Part of the dynamism of Canada's regional geography is the changing relationship between Canada's six regions. Trade liberalization provides one element of change while regional tensions provide another element of change.

Following World War II, the liberalization of world trade and the creation of the North American trading bloc exposed Canada's economy to new challenges and dangers. The consequences of trade liberalization were sixfold:

- Canada's old spatial economic structure was transformed from one that was national to one that is continental.
- Canada's economy was integrated into the North American market.
- Canada's manufacturing sector was restructured.
- Rates of unemployment first increased and then declined.
- Canada's regional economies were reoriented to adjust to the larger North American market.
- Canada's economic growth and regional development have become tied more and more to trade with the United States.

This book explores both the national and regional implications of these economic changes and proposes scenarios for greater stability in the new global economic order.

Regional tensions have always existed in Canada. These tensions often place pressure on the existing social and political systems. This pressure for change reflects the 'potential' for a redress of the division of power within Canada's federation. From this perspective, regional tensions can lead to a shift of power within a region and may even alter relations between regions and/or the federal government. In this book, four tensions or faultlines are examined. These tensions occur between Aboriginal and non-Aboriginal Canadians, French and English Canadians, centralist and decentralist forces, and recent immigrants (newcomers) and those born in Canada (oldtimers). Each of these tensions exists in all parts of Canada, but because of spatial variations in Canada's human and/or physical geography, each appears more prominently in a particular region or regions. For instance, the Aboriginal/non-Aboriginal faultline is most deeply felt in the Territorial North, where a larger proportion of the population is Aboriginal. The French/English faultline manifests itself most commonly in Québec. The centralist/decentralist faultline is frequently played out between the federal government and Western Canada, often over resources. The impact of immigration is heavily concentrated in the three major cities of Toronto, Vancouver, and Montréal. This book explores the nature of these faultlines, the need to reach compromises, and the fact that these faultlines provide the country with its greatest strength—diversity. Ultimately, these faultlines are shown to be not divisive forces but forces of change that ensure Canada's existence as a country of regions.

Organization of the Text

This book consists of 11 chapters. Chapters 1 through 4 deal with general topics related to Canada's national and regional geographies—Canada's physical, historical, and human geography—thereby setting the stage for a discussion of the six main geographic regions of Canada. Chapters 5 through 10 focus on these six geographic regions. The core/periphery model provides a guide for the ordering of these six regions. The regional discussion begins with Ontario and Québec, which represent the traditional demographic, economic, and political core of Canada. The core regions are followed by British Columbia, Western Canada, Atlantic Canada, and the Territorial North. Chapter 11 provides a conclusion.

Chapter 1 discusses the nature of regions and regional geography, including the core/periphery model and its applications. Chapter 2 introduces the major physiographic regions of Canada and other elements of physical geography that affect Canada and its regions. Chapter 3 is devoted to Canada's historical geography, such as its territorial evolution and the emergence of regional tensions and regionalism. This discussion is followed, in Chapter 4, by an examination of the basic demographic, economic, and social factors that influence both Canada and its regions. This chapter explores the national and global forces that have shaped Canada's regions as well as the features of its population (size, urbanization, etc.). To sharpen our awareness of how economic forces affect local and regional developments, four major themes running throughout this text are introduced in these first four chapters. The primary theme is that Canada is a country of regions. Two secondary themes—the integration of the North American economy and the changing world economy—reflect the recent shift in economic circumstances and its effects on regional geography. These two economic forces, described as continentalism and globalization, exert both positive and negative impacts on Canada and its regions, and are explored through the core/periphery model.

In Chapters 5 to 10, the text moves from a broad, national overview to a more regional focus. Each of these six chapters profiles one of Canada's large geographic regions: Ontario, Québec, British Columbia, Western Canada, Atlantic Canada, and the Territorial North. These regional chapters explore the physical and human characteristics that distinguish each region from the others and that give each region its special sense of place. To emphasize the economic specialty of each region, a predominant economic activity is identified and explored through a 'Key Topic'. They are: the automobile industry in Ontario; Hydro-Québec in Québec; forestry in British Columbia; agricultural transition in Western Canada; the fishing industry in Atlantic Canada; and megaprojects in the Territorial North. From this presentation, the unique character of each region emerges. The book concludes with Chapter 11, which discusses the future of Canada as a country of regions.

Fourth Edition

Canada's regional geography is not static. In the relatively short period of four years since the third edition was published, dramatic developments have reshaped Canada's regional geography. This new edition, the fourth, focuses on these developments.

The first development is the rising demand for resources, especially energy resources, which has caused prices to reach record heights and created an economic boom for resource-rich regions. Of course, resource price cycles have traditionally risen to boom levels and then collapsed in accordance with the global business cycle. However, the global economy may have set a new basic price level for certain resources because of the strong and seemingly steady demand from the new economic giants of Asia, especially China and India.

The second development is that Asian countries, led by China, Japan, and South Korea, are producing an ever-widening range of manufactured goods that are finding a ready market in Western countries, including Canada.

The consequences of these global economic developments for Canada's regions are twofold. First, Western Canada (led by Alberta) and British Columbia are enjoying record prices for their resources. The resulting boom has meant that the red-hot economies of Alberta and British Columbia are attracting record numbers of newcomers. With the spillover of Alberta's economic boom into Saskatchewan, this traditionally have-not province is undergoing its own mini-boom. High oil prices have also benefited Newfoundland and Labrador, though its population continues to decline.

Second, while Ontario and Québec remain the economic and population pillars of Canada, a shift of regional power seems to be in the wind. Over the last four years, Ontario and Québec have suffered a decline in their manufacturing sectors. This decline began with the loss of most of their textile firms at the beginning of trade liberalization. Often these firms relocated offshore where labour costs were significantly lower. Now many other manufacturing firms are following the same path—moving offshore but selling their goods in the Canadian market. Canadian manufactured goods also are troubled by the so-called 'Dutch disease'—a combination of high energy prices and a rising Canadian dollar—which has made their production and export more difficult, thus magnifying the problems facing the industrial heartland of Canada.

Both the text and the accompanying statistics in this fourth edition have been revised extensively to reflect Canada in 2007. More maps, graphs, and photographs have been added to enrich the text. At the end of each chapter, I have added a set of 'Challenge Questions'. These questions are designed to provoke class discussion. In this way, the students can enter the debate of what makes Canada's regions so stimulating and intriguing.

Acknowledgements Past and Present

I have continued to rely on the help of colleagues across Canada. With each edition, I have benefited from the constructive comments of anonymous reviewers selected by Oxford University Press. I especially owe a debt of thanks to one of those anonymous reviewers who spiced his critical comments with words of encouragement that kept me going.

Professors at the University of Saskatchewan were particularly helpful—several provided me with their comments on specific matters that fall under their areas of specialization while some were pressed into reviewing draft versions of regional chapters. Most contributed photographs. These colleagues and friends include Alec Aitken, Bill Archibold, William Barr, Dirk de Boer, K.I. Fung, the late Walter Kupsch, Lawrence Martz, the late John McConnell, John Newell, Jim Randall, Jack Stabler, and Mike Wilson. Professors from other institutions who played a role in shaping this book include Jim Miller, Robert MacKinnon, and Gilles Viaud from Cariboo College; Hugh Gayler and Dan McCarthy from Brock University; Ben Moffat from Medicine Hat College; and Keith Storey from Memorial University.

Thanks to a sabbatical leave granted to me by the University of Saskatchewan, I was able to spend the fall term of 1996 at the Département de géographie at Université Laval. My generous hosts not only put up with my limited command of French but, more importantly, contributed significantly to the design and content of my chapter on Québec. Among those contributing to my education were Louis-Edmond Hamelin, who must be considered the Dean of Canadian geographers; Jean-Jacques Simard, a well-known sociologist who has written extensively on Québec's north; and Benoit Robitaille, the Chair of the Département de géographie at Université Laval. Discussions of critical issues with Jacques Bernier, Marc St-Hilaire, and Eric Waddell proved extremely fruitful. For me, one of the highlights at Laval was the opportunity to present my ideas on Québec at one of the Friday afternoon seminars organized by Dean Louder. The constructive and stimulating responses from the cultural geographers attending that seminar were especially helpful to me.

Maps are a critical element in geography. I must thank Keith Bigelow, the cartographer in the Department of Geography at the University of Saskatchewan, for the help he provided on the maps. The library staff in Government Documents at the University of Saskatchewan was particularly helpful in my search for detailed information on Canada's regions.

The National Atlas of Canada and Statistics Canada have created important websites for geography students. These provide access to a wide range of geographic data and maps. The National Atlas of Canada has placed many of its printed maps on its website <www.atlas.gc.ca>. Statistics Canada has published key elements of its 2006 census data by provinces and territories on its website <www.statcan.ca>. Statistics Canada has also published these key statistics for all urban places on another website <www2.statcan.ca>. Students can also access annually published statistical data and population estimates from the Statistics Canada site under the heading CANSIM (Canadian Socio-economic Information Management System). Joel Yan of Statistics Canada was particularly helpful in identifying suitable data from CANSIM used in this book.

The staff at Oxford University Press, but particularly Phyllis Wilson, made the preparation of the fourth edition a pleasant and rewarding task. Janna Green deserves special thanks for obtaining many of the new photographs. Richard Tallman, who diligently and skilfully has edited my manuscripts into polished finished products for each of the last three editions, deserves special mention. He has become an old friend who often pushes me to expand my ideas.

Finally, a special note of appreciation to my wife, Karen.

Note To Students and Instructors

A package of Student's Resources to accompany this text is available at:

www.oup.com/ca/geogcanada
Login: Regional
Password: hudsonbay

Instructors, please contact your local Oxford University Press sales representative for further information regarding the instructor's manual and test item file.

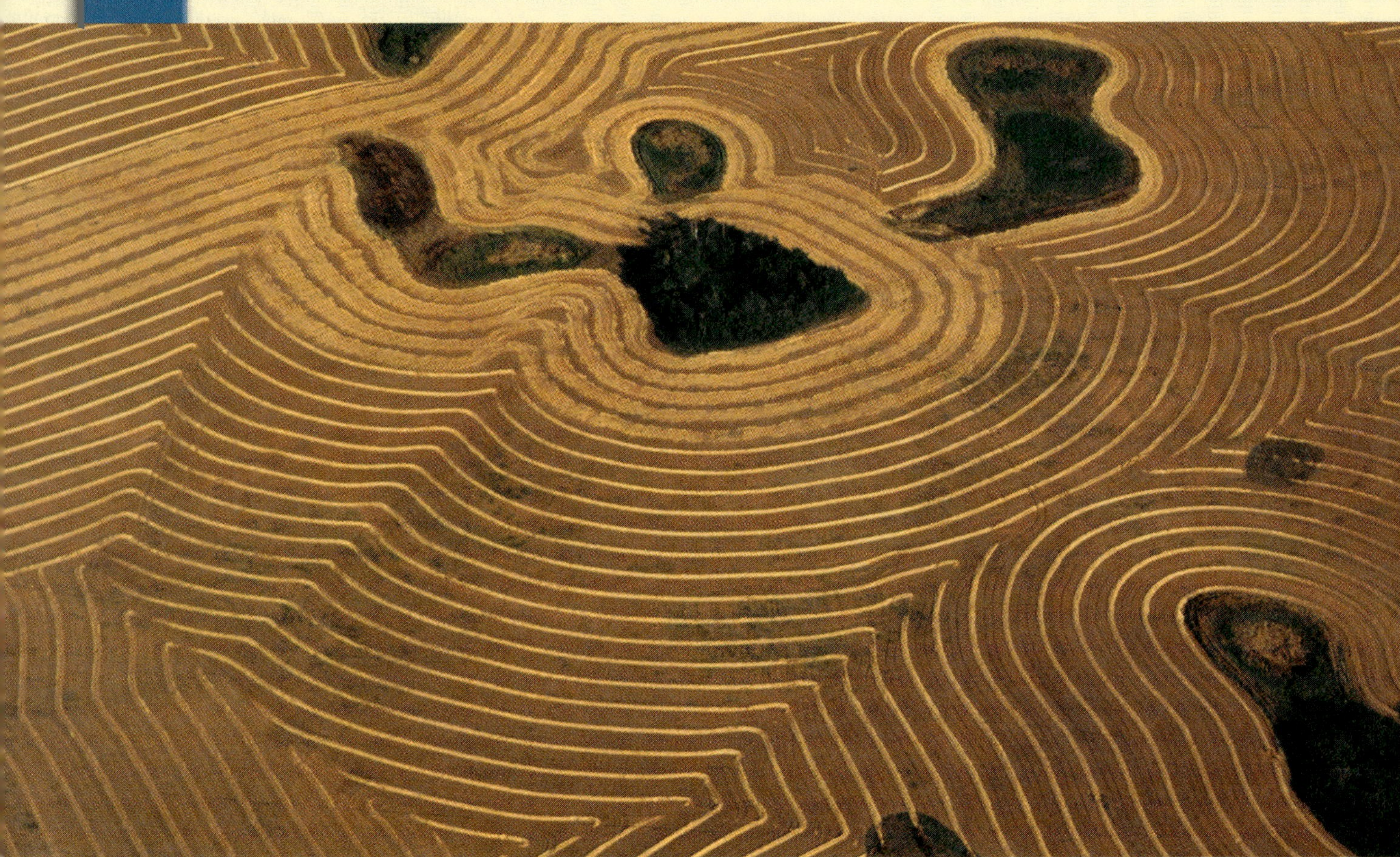

North of Regina, Saskatchewan (Al Harvey/The Slide Farm)

Overview and Objectives

Canada is a country of regions. As the second largest country in the world, regional geography presents a geographic perspective of this complex and diverse country, reveals relationships between its regions, and identifies Canada's place in the North American and global economies. Canada has six major geographic regions. Each major geographic region is defined by its particular geography and historic evolution. Over time, each region has evolved—and continues to evolve—its own sense of regional consciousness and its place within Confederation. Part of this evolution has seen the emergence of a strong north–south economic axis—the outcome of the North American Free Trade Agreement—to complement the traditional east–west axis dominated by the manufacturing industries in Ontario and Québec. Canada, as a federal state, is not without internal tensions and international disputes. Regional tensions often spring from sensitive issues buried in Canada's past or from a geography that has favoured some regions over others. The challenge of interpreting Canada's diversity is not an easy task and a spatial conceptual framework helps us to understand Canada's regional geography. The major topics discussed in Chapter 1 are:

- Geography as a discipline.
- Regional geography as a subject.
- The origins of regional thought.
- Canada's geographic regions.
- The four major faultlines within Canada.
- The power of place.
- The sense of place.
- The dynamic nature of regions.
- The core/periphery model.
- Three stages in Canada's regional development.
- Canada within North America.

CHAPTER 1

Regions of Canada

INTRODUCTION

Geography helps us understand our world. Since Canada is such a huge and diverse country, it is best understood from a regional perspective. Such a perspective comes from dividing Canada into six regions. Each region has a distinct location, physical geography, and historical development. Each also has a strong sense of regional consciousness that developed over time as people reacted to challenges presented by their economic, physical, and social environments. Geographic analysis and synthesis of our world require an intuitive grasp of its regional nature and spatial relationships. Such a grasp is provided by a conceptual framework based on the core/periphery theory. Within that framework social tensions exist, and these tensions are called faultlines.

Geography as a Discipline

Geography provides a description and explanation of lands, places, and peoples beyond our personal experience. Geography also determines life's opportunities. De Blij and Murphy (2006: 3) believe that 'Geography is destiny', meaning that for most people, place is the most powerful determinant of their life chances, experiences, and opportunities. In this instance, place refers to the community where one was born and raised. Not surprisingly, then, geographers are interested in cultural and physical characteristics of places and how they influence local, regional, and national societies. Sharing of a common space is a critical factor in this spatial differentiation because it inevitably leads to a regional consciousness. Regional consciousness is a dynamic product of the region's physical geography, its historical events, and economic situation. As such, regional consciousness forms one of the cornerstones of regional geography. Regional self-interest, a logical outcome of regional consciousness, shows its face in the centralist/decentralist faultline.

Regional Geography as a Subject

The geographic study of a particular part of the world is called **regional geography**. In such studies, people, interacting with their economic, physical, and social environments, place their imprint on the landscape. In layman's terms, the goal of regional geography is to find out what makes a region 'tick'. By achieving such an understanding, we gain a fuller appreciation of the complexity, diversity, and interconnectivity of our world.

Regional geography has evolved over time.[1] Traditionally, geographers focused their attention on the physical aspects of a **region** that affected and shaped the people and their institutions. Today, geographers place more emphasis on the human side because the physical environment is largely mediated through culture, economy, and technology. Over time, a multitude of profound and often repeated extreme experiences, whether they are economic, natural, or political events, mark the people, forcing them to respond. In turn, these responses help create a common sense of regional belonging and regional consciousness. Other expressions of regional belonging and consciousness are sense of place and power of place.

Regional self-interest often results in conflicts with other regions. The current brawl over equalization payments is part of Canada's political life, and different positions on equalization payments make up part of the regional consciousness of each region (see Vignette 3.5 for a fuller discussion of equalization payments). Back in 1995, similar heated disagreements between premiers and Ottawa took place. The political cartoonist Brian Gable captured these regional expressions of self-interest (Figure 1.1).

The Origins of Geographic Thought

Curiosity forms the starting point for geography. Curiosity about distant places is not a new phenomenon. The ancient Greeks were curious about the world around them. From reports of travellers, they recognized that the earth varied from place to place and that different peoples inhabited each place. Stimulated by the travels, writings, and map-making

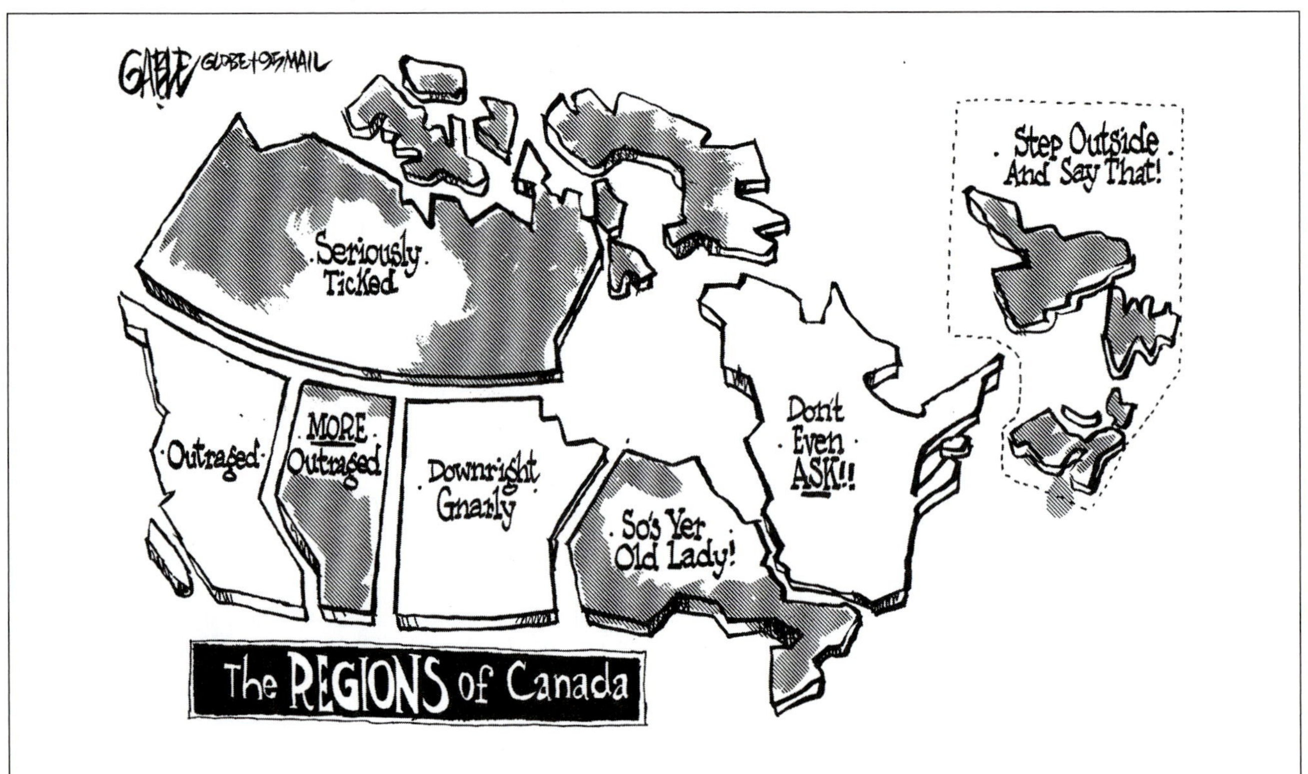

Figure 1.1 Gable's regions of Canada. Political cartoonist Brian Gable has aptly depicted the regionalized tensions in Canada. (*Globe and Mail*, 11 Dec. 1995, A14. Reprinted with permission from *The Globe and Mail*.)

of scholars such as Herodotus (484–*c.* 425 BC), Aristotle (384–332 BC), Thales (*c.* 625–*c.* 547 BC), Ptolemy (AD 90–168), and Eratosthenes (*c.* 276–*c.* 192 BC), the ancient Greeks coined the word 'geography' and mapped their known world. By considering both human and physical aspects of a region, geographers have developed an integrative approach to the study of our world. This approach, which is the essence of geography, separates geography from other disciplines. The richness and excitement of geography are revealed in Canada's six regions—each region is the product of past events and contemporary issues that result in a set of regional identities. Geographers often refer to such regional identities as a **sense of place**. Within Canada, there is no stronger sense of place than that exhibited by the Québécois who translate their social identity into political aspirations. While Québec is one of 10 provinces, most Québécois perceive their place within Canada as more of a partnership with the rest of Canada while others seek independence from Canada. Consequently, this divide between English and French-speaking Canadians on the nature of Confederation is a recurring political issue. This divide represents the most critical social/political faultline and its origins reach back to 1759. The competing concepts of Canada as two nations or as 10 provinces are discussed more fully in Chapter 4 ('The French/English Faultline') and in Chapter 6 ('The French/English Faultline in Québec').

Canada's Geographic Regions

The geographer's challenge is to divide a large spatial unit like Canada into a series of 'like places'. To do so, a regional geographer selects the critical physical and human characteristics that logically divide a large spatial unit into a series of regions and that distinguish each region from adjacent ones. Towards the margins of a region, its core characteristics become less distinct and merge with those characteristics of a neighbouring region. In that sense, boundaries separating regions are best considered transition zones rather than finite limits.

In this book, we examine Canada as composed of six geographic regions (Figure 1.2):

- Atlantic Canada
- Québec
- Ontario
- Western Canada
- British Columbia
- Territorial North

Six regions are identified as a basis of study for several reasons. First, a huge Canada needs to be divided into a set of manageable segments. Too many regions would distract the reader from the goal of easily grasping the basic nature of Canada's regional geography. Six regions allow us to readily comprehend Canada's regional geography and to place these regions within a conceptual framework based on the core/periphery model, discussed later in this chapter. This is not to say that there are not internal regions or subregions. In Chapter 5, Ontario provides such an example. Ontario is subdivided into southern Ontario (the industrial core of Canada) and northern Ontario (a resource **hinterland**). Southern Ontario is Canada's most densely populated area and contains the bulk of the nation's manufacturing industries. Northern Ontario, on the other hand, is sparsely populated and is losing population because of the decline of its mining and forestry activities.

Second, an effort has been made to balance these regions by their geographic size, economic importance, and population size, thus allowing for comparisons (see Table 1.1). For this reason, Alberta was combined with Saskatchewan and Manitoba to form Western Canada, while Newfoundland and Labrador along with Prince Edward Island, New Brunswick, and Nova Scotia comprise Atlantic Canada. The Territorial North, consisting of three territories, makes up a single region. Three provinces, Ontario, Québec, and British Columbia, have the geographic size, economic importance, and population size to form sepa-

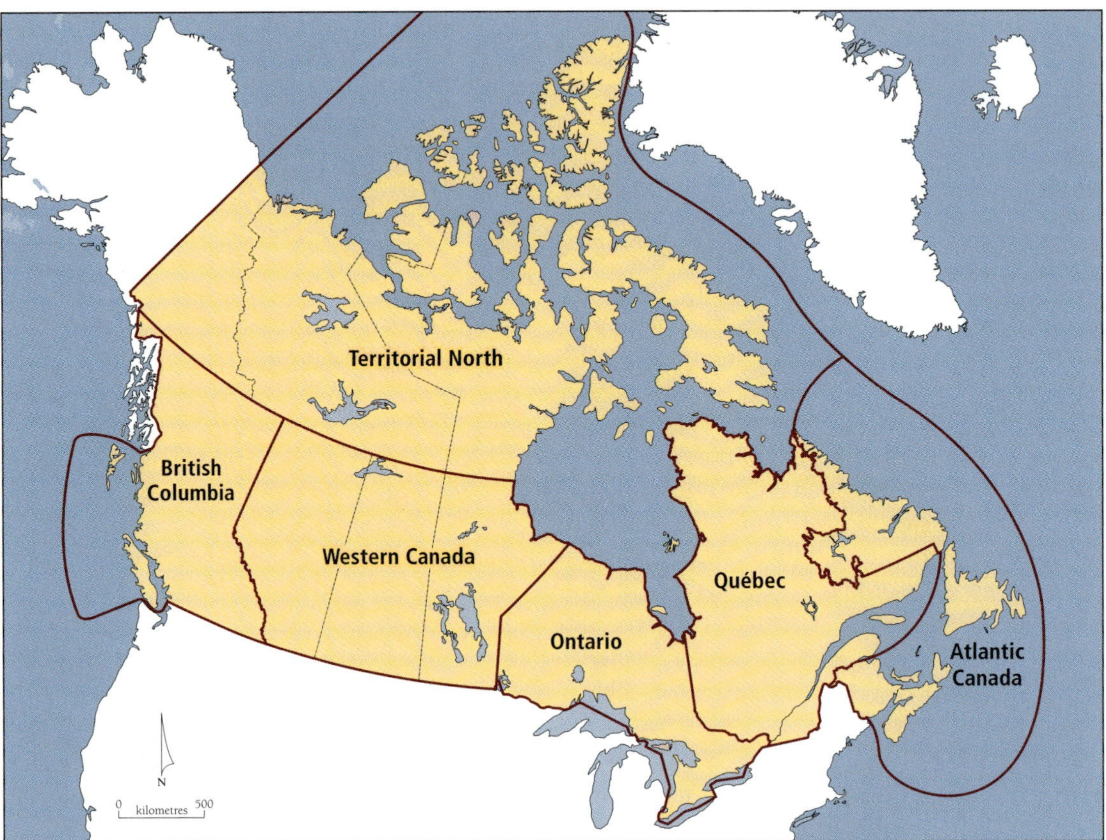

Figure 1.2 The six geographic regions of Canada. The coastal boundaries of Canada are recognized by other nations except for the 'sector' boundary in the Arctic Ocean. For a fuller discussion, see the subsection 'Arctic Sovereignty' in Chapter 10.

rate geographic regions. In addition, these regions are:

- readily understood by Canadians;
- associated with distinctive physical features, natural resources, and economic activities;
- reflect the political structure of Canada;
- facilitate the use of statistical data;
- contain a sense of regional identity;
- reveal regional economic strengths and cultural presence.

Faultlines within Canada

Geologists refer to cracks in the earth's crust as faultlines. In this text, 'faultlines' serves as a metaphor for stresses within Canadian society. Since Canada became a nation, four major faultlines—English and French, centralist and decentralist, old and new Canadians, and Aboriginal and non-Aboriginal—have played a key role in its evolution. These faultlines are often dormant or 'just below the surface' for long periods of time, but they inevitably flare up first as local tensions, then as regional ones, and sometimes they become national issues. The cracks within Canadian society can and have threatened the regional cohesiveness of Canada. By seeking a compromise between these differing positions, Canada, over time, has become what John Ralston Saul (1997: 8–9) describes as a 'soft' country, meaning a society where conflicts are, more often than not, resolved through discussion and negotiations.[2] Disagreement over the nature of Canada—is it a partnership between the two so-called founding societies of the French and

Table 1.1 General Characteristics of the Six Canadian Regions, 2006

Geographic Region	Area* (000km^2)	Area (%)	Population (000s)	Population (%)	GDP (%)
Ontario	1,076.4	10.8	12,160.3	38.5	38.7
Québec	1,542.1	15.4	7,546.1	23.9	19.7
British Columbia	944.7	9.5	4,113.5	13.0	12.5
Western Canada	1,960.7	19.6	5,406.9	17.1	22.6
Atlantic Canada	539.1	5.4	2,278.3	7.2	6.0
Territorial North	3,909.8	39.3	101.3	0.3	0.5
Canada	9,972.8	100.0	31,612.9	100.0	100.0

*Includes freshwater bodies such as the Canadian portion of the Great Lakes.
Sources: Statistics Canada (2006c, 2007a, 2007b).

the English or is it composed of 10 equal provinces?—has troubled the country since Confederation in 1867 and in recent years has come to a head twice with the sovereignty-association and independence referendums in Quebec in 1980 and 1995, the latter of which was won by the federalist side by the narrowest margin—a mere percentage point. At the height of the referendum campaigns, uneasy relationships tore at the very fabric of Canada. But the ensuing dialogue and goodwill between Québec and the rest of the country have led to an unspoken compromise that cements the country and its regions together.

How can we conceptualize such complex events? As Canadian society moves through time, history provides a view of the past, but the future is much less certain. While all societies change, Canada is evolving at a more rapid rate. Four factors contribute to this rapid change. First, population and economic power are shifting from Atlantic Canada, Québec, and even Ontario to Alberta and British Columbia. The shift is related to a slowdown in manufacturing centred in Ontario and Québec and to resource industries in BC and especially in Alberta that are benefiting from higher world prices. Second, the pluralistic nature of Canadian society represents another significant factor. Pluralism is a relatively new phenomenon in Canada, superseding the biculturalism represented by the French and English. Pluralism is an expanding phenomenon because of the flood of non-Western immigrants into Canada. Adjustments and accommodation of the newcomers and their cultural/religious practices are necessary for continued social harmony. Aboriginal peoples form the third factor. Outstanding land claims, the rapidly growing Aboriginal population, and the ever-increasing numbers of Native peoples in Canadian cities are three issues that require attention. The last factor is the place of Québec within Canada. With its distinct culture and French language, Québec, as a people and as a government, views its place within Canada and North America through a different lens from that used in other regions of Canada. For the first time, the House of Commons in 2006 declared that the Québécois are a nation within Canada. Does this declaration support the concept of two founding nations? For more on this subject, see the subsection 'One Country, Two Visions' in Chapter 3.

Continuity with the past is essential for successfully navigating into the future. For Canada, this desire for continuity must be tempered by accommodating the many new cultures arriving in Canada. In simple terms, Canada must grip tightly its core values while necessarily reinventing itself to meet new circumstances. This situation is of special significance in the discussion of these four faultlines.

English-speaking/ French-speaking Canadians

Québec, with its Québécois culture and French language, forms a distinct cultural region within Canada. Tensions between English-speaking Canada and Québec have often erupted over language issues. Within Québec, an internal faultline exists between separatists and federalists. Separatists argue that Québec cannot fulfill its aspirations unless it becomes a nation. Federalists take the position that Québec can best flourish within the Canadian federation. This argument was played out in the 1980 and 1995 referendums. The struggle to maintain its culture and language has been long and hard, going back to the defeat of the French army on the Plains of Abraham in 1759. Until 1969, English was Canada's only official language. This fact underscores the long and difficult struggle francophones have had in finding a place within Canada.

A key to that struggle is the French language. French as a viable language in a continent that speaks mainly English resulted in the controversial Bill 101 in 1977. The bill proclaimed French as the official language in Quebec for just about every facet of life: government, the judicial system, education, advertising, and business. In 1982, the Supreme Court of Canada struck down Bill 101. However, the Liberal government of Robert Bourassa overturned this rule by employing the 'notwithstanding clause' (section 33 of the 1982 Charter of Rights and Freedoms). While language is much less of an issue in Québec today, language disputes remain a hot button between English and French Canadians residing in Québec.

Centralist/Decentralist Faultline

Of the over 30 million Canadians, most live and work in Ontario and Québec. Adding to their demographic superiority, Ontario and Québec house most manufacturing activities in Canada. These demographic and economic advantages illustrate the dominant position of these provinces in the nation's economic and political matters and they support the notion of a core region. Not surprisingly, other regions of Canada over the years have perceived Ottawa as favouring these two provinces. While Ottawa may consider its considerable subsidies to manufacturing industries in Ontario and Québec in the national interest, the rest of Canada sees these subsidies as 'special treatment' designed to gain the favour of voters in Ontario and Québec. Because no political party can form a majority government without strong support from these two provinces, the regions in the rest of Canada feel powerless to get their voice heard in Ottawa. This feeling of powerlessness is the source of regional alienation. Western alienation in Alberta is fuelled by Alberta's fear that Ottawa will once again intrude into provincial power over natural resources with a new version of the 1980 National Energy Program. Royalties from oil and gas have already made Alberta extremely rich, allowing the province to eliminate its debt and to maintain lower tax rates. Now other provinces—Saskatchewan, British Columbia, Newfoundland and Labrador, and Nova Scotia—hope to profit from their petroleum deposits. As the economies of petroleum-producing provinces gain new strength, will Ottawa again seek a share of their natural wealth?

Aboriginal Peoples and the Non-Aboriginal Majority

As the first people to occupy the territory now called Canada, Aboriginal peoples have had to deal with the settlement of their lands, with subjugation as colonized peoples, and with Ottawa's restrictive Indian Act. The Métis mounted two rebellions to defend their land (see the subsection 'Manitoba: The Métis against the Newcomers' in Chapter 3 for a more complete discussion). While Aboriginal Canadians reside in all provinces, they form the largest proportion of the population in the

Territorial North and in the northern areas of the other geographic regions. While treaty Indians were assigned to reserves, many have relocated to cities where more opportunities are available.

Modern land-claim agreements have led to the possibility of self-government on reserves.[3] But what does self-government mean? Would it take the form of ethnic government or public government? Could it contain provincial-like powers or just municipal powers? Some First Nations have obtained self-government in the form of an ethnic government with powers at least at the municipal level but not at the provincial level. Public government within an area dominated by Aboriginal peoples provides another approach. Nunavut is one example while Nunavik in Arctic Québec may provide another example. In Nunavik, the Inuit population constitutes over 85 per cent of the total population and almost 100 per cent of residents born and raised there. The case for self-government is strong—as it was in Nunavut—but the political ramifications of creating a new level of public government within a province are huge. Negotiations among the federal government, the Quebec government, and Makivik Corporation (which represents the Inuit of Nunavik) are well underway and may result in an innovative political development—a semi-autonomous region within a province. The concept of 'nested federalism' would break new ground in Canada's political realm and could have ramifications for other provinces.

Yellowknife is no longer a rough-and-tumble mining town. This city has evolved into a sophisticated centre with expensive housing along its waterfront. (Terry Parker/NWTT)

Newcomers and Old-timers

Canada is a land of immigrants. Early French and English settlers had a very difficult time adjusting to the New World. In the early twentieth century, immigrants came to settle the prairie lands of Western Canada. Finding a place—a comfort level that would allow them to be themselves and gain a sense of rootedness and 'home'—was not as easy for the many non-English speaking immigrants to Canada from Central Europe and czarist Russia as it was for those from Great Britain who were English-speaking and who found strong similarities in Canadian society and its institutions to those in their home country. The Doukhobors, for example, who practised collective farming, failed to see their lifestyle gain acceptance (for a fuller discussion, see the subsection 'The Doukhobors' in Chapter 3).

The rubbing and bumping between those whose cultural roots are in the Old World and those whose roots developed in the New World necessarily require adjustments and compromises. Old-timers are not always prepared to give ground and sometimes, as in the case of the code of standards created by the municipal council in Hérouxville, Québec, they react strongly to protect their status quo (Vignette 1.1). Yet, these sometimes unpleasant interactions between newcomers and old-timers are a necessary fact of Canadian life and signal that a dialogue is taking place. Generally speaking, the second generation of immigrant groups, by being born and raised in Canada, had a much easier time feeling connected to Canada. Yet, as with those nicknamed CBCs (Chinese born in Canada), a cultural gap emerges within the diasporic community. Now the same challenge faces immigrants from Asia, Africa, and the Middle East.

Unlike before the 1960s, when the vast majority of immigrants to Canada came from the British Isles and from Northern Europe, most immigrants in more recent years have come from non-European countries. This has made integration into Canadian society more difficult, especially for visible minorities who may face both open and subtle forms of racial discrimination. In the current socio-political climate, for example, Muslim Canadians must deal with the fallout from the 2001 terrorist attacks on New York and Washington, the broader struggles in the Muslim world, and Canada's role in Afghanistan. Racial profiling by the American 'No Fly' policy poses another restriction for Canadian citizens who were born in the Middle East and who hold dual passports.

The concentration of new immigrants in Canada's major cities has both advantages and disadvantages. One advantage is that newcomers have more cultural anchors to support them, such as family and friends who speak their native tongue, restaurants that serve traditional foods, and religious institutions that provide both spiritual and community support. The disadvantage may be a sense of isolation from other Canadians. For example, recent research suggests that visible minorities in Toronto are feeling less connected to Canada, and the next generation seems to feel even less of a bond with the country (Lowington, 2007). On the other hand, the Muslim Canadian TV sitcom, *Little Mosque on the Prairies*, may help expose Canadians to the reality of its pluralistic society as well as make a connection for Muslim Canadians with their adopted country.

The Power of Place

Power of place, a concept from economic geography, is a product of natural wealth, geographic location, and global forces, and reveals regional economic strengths and weaknesses. For resource-rich Canadian regions, world prices dictate whether a resource is a strength or a weakness. Two examples are found in grain and oil prices. For many years, the price for most resources was low, especially for agricultural products. The price for wheat, the heart of Prairie agriculture, has remained stalled at very low levels, forcing many farmers into bankruptcy. Quite the opposite has been the case for oil and many other resources. Since 2000, global demand for energy resources increased sharply, pushing prices to record levels. Oil and natural gas were the first to experience a sharp rise in prices, but other commodities—coal, copper, nickel, potash, and uranium—soon experienced an abrupt rise in demand and prices. Not only did companies profit from this resource boom, but provinces, especially Alberta, Saskatchewan, and British Columbia, and, to a lesser degree, Nova Scotia and Newfoundland and Labrador, saw their royalties from natural resources rise to new heights. Two other regions recognize an opportunity to benefit from this situation. Québec sees high energy prices as an incentive to begin another phase of the James Bay Project by diverting water from three rivers into La Grande River, where hydroelectric stations are located. Surplus power would be sold in energy-deficient places like New England and Ontario. The Northwest Territories government strongly supports the construction of the Mackenzie Delta Pipeline, which will deliver natural gas to the Chicago region of the United States. The territorial government would like to obtain the royalties from the production and transport of this natural gas, which would otherwise mainly go to Ottawa. Ontario is less well blessed with energy resources. As an energy-deficient region, Ontario is struggling to find new sources of energy so necessary to fuel its manufacturing-based economy.

World prices for energy and resources tend to rise and fall in accordance with the global economic cycle. However, the world may have reached the tipping point where the combined demand from 'old' industrial nations found in the European Union, Japan, and North America and from 'new' industrial nations such as China, India, and other Asian

Vignette 1.1 Hérouxville's Code of Standards

Canadians who were born and raised in Canada watch changes to Canadian society with both pleasure and trepidation. Pleasure comes from the enrichment of Canadian society while trepidation comes from fear of losing control of their world, whether it is the so-called 'monster houses' built by wealthy immigrant Chinese in Vancouver or the call for space for Muslim prayers in Montréal universities. Because of its concern over culture, Québec is at the forefront of this debate about what is considered reasonable accommodation of racial, ethnic, and religious minorities. In 2007, matters came to a boil in Hérouxville, Québec, when the municipal council announced its code of standards, which it sent to the federal and provincial governments with the expectation that these senior governments will distribute their code to potential immigrants. The purpose of the code is 'to inform the new arrivals that the lifestyle that they left behind in their birth country cannot be brought here with them and they would have to adapt to their new social identity' (Hamilton, 2007: A7).

Shortly after this declaration, a group of Muslim women visited the town to present a fuller picture of Islam and its customs. 'There was a real exchange', Najat Boughaba told Radio-Canada. 'There were people who reached out to us. I really think our visit to Hérouxville benefited both sides' (CBC News, 2007). Najat Boughaba was struck by how little the townspeople knew about Islam and its customs. Following this exchange, the town council in Hérouxville amended its immigrant code of conduct to remove references to 'no stoning of women in public' and 'no female circumcision'.

countries will push the cycle of energy and resource prices to higher levels. Such a scenario will not stop prices from rising and falling in accordance with the global economic cycle, but it will keep commodity prices at a higher level. If true, then a long-term higher cycle of energy and resource prices will continue to shift Canada's economic power from Ontario and Québec to the three resource-rich provinces of Alberta, British Columbia, and Saskatchewan.

The Sense of Place

Sense of place is based in cultural geography, from which we learn that local circumstances bond people to their environment and institutions. As such, sense of place provides some protection from the economic and cultural globalization that drives many international companies. Sense of place contains within it a powerful psychological bond between people and a region. The physical nature, human activities, and institutional bodies of a region combine to form a powerful psychological bond between people and a region.

Canada, with its vast spaces and magnificent diversity of physical settings, lends itself to strong regional bonds. Atlantic Canada and Western Canada provide two examples. The sea exerts a profound influence on all who live in Atlantic Canada. The fishery sparked early settlement in Atlantic Canada and, until recently, provided the economic basis for many coastal communities. The tragic loss of the northern cod fishery in the 1990s brought great economic, social, and personal dislocation to the small fishing communities in Atlantic Canada, but especially to the outports in Newfoundland (for more on this subject, see Chapter 9). The vast open expanses of the Prairie region also exert a powerful influence on its inhabitants. Its dry continental climate, however, has left agriculture on the margins.

Climatic events often have made the memories of farmers bitter ones. For example, the Dust Bowl of the 1930s deeply marked Prairie farmers, forcing many to abandon their homesteads. While the Dust Bowl has faded into past collective memory, this semi-arid environment continues to influence life on the Prairies, with drought a real threat to Prairie agriculture. Indeed, two years of arid growing conditions (the summers of 2001 and 2002) followed by three years of untimely frosts and wet falls (2003, 2004, and 2005) have left many western grain farmers deep in debt or forced them to declare bankruptcy (for more on this topic, see Chapter 8). In both regions, poor economic circumstances have forced families to relocate to major cities, especially those in Alberta and British Columbia, or have resulted in heads of households effectively commuting across the country, as many Newfoundlanders do in working in the Alberta oil patch.

The concept of a sense of place, then, recognizes that people living in a region have undergone a collective experience that leads to shared aspirations, concerns, goals, and values. Over time, such experiences develop into a social cohesiveness among those people living within a spatial unit. Often these experiences are coloured by extreme weather events in a region. The rainstorm and subsequent flooding in the Saguenay Valley of Québec in July 1996 was such an experience. Ice storms often have devastating effects by knocking out regional electrical systems. In early February 2003, New Brunswick was hit with a severe ice storm that dropped 60 millimetres of frozen rain, leaving some residents from Saint John to Moncton without power for several days while temperatures were hovering around –10°C.

A sense of place can also evolve from a region's history and geography. Early in its history, British Columbia was isolated from the rest of Canada by the Rocky Mountains, which form the largest mountain system in Canada, extending 1,200 km from the US–Canada border northward towards 60° N. While this formidable physical barrier, considered by many to be the backbone of Canada, has been surmounted by modern transportation systems, it remains an important physical and cultural feature unique to British Columbia, and the promise of a trans-Canada rail line, the Canadian Pacific Railway, was instrumental in BC joining Confederation in 1871. In *The Unknown Country*, Bruce Hutchison (1942: 266) described the effect of the Rocky Mountains on people as he observed changes in the landscape from a passenger car on the Canadian Pacific train: 'Unlike the life of the Prairies or the East—so unlike it that, crossing the Rockies, you are in a new country, as if you had crossed a national frontier. Everyone feels it, even the stranger feels the change of outlook, tempo and attitude. What makes it so, I do not know.' Feelings of alienation run deep in British Columbia. This theme reverberates in Philip Resnick's *The Politics of Resentment: British Columbia, Regionalism and Canadian Unity* (2000). Resnick supports Hutchison's contention that BC is a distinct region, but he also stresses that a sense of alienation from Ottawa is widespread within the province.

An equally powerful expression of place and alienation exists in Québec, where history and geography have had four centuries to nurture a strong sense of place and to give birth to a nationalist movement that has sought to sep-

In 1811 Nova Scotia granted six families 800 acres of land, and the land was given the name Peggy's Cove. Originally a tiny fishing village, Peggy's Cove has become a major tourist attraction. (© Peter Burian/Corbis)

arate Québec from the rest of Canada. New France was born and grew within the confines of the St Lawrence Lowlands. Since Québec's early days, the St Lawrence River has played a key role in its settlement and economic development. Prior to the British Conquest of New France, formalized in 1763 by the Treaty of Paris, this mighty river promoted the interests of the French colony by providing a supply route to France, allowing the fur traders (coureurs du bois) to penetrate far inland to trade with distant Indian tribes, and giving French explorers access into the heart of North America

A region, then, is a synthesis of physical and human characteristics that, combined with its distinctiveness from surrounding regions, produce a regional character that includes a sense of place and power. People living and working in a region are conscious of belonging to that place and frequently demonstrate an attachment and commitment to their 'home' region. A sense of place, falling somewhere between regional tribalism and commitment to Canada, unites people on common issues and challenges, and compels them to seek solutions to these issues and challenges. For the Inuit, a sense of place is embodied in their natural homeland of the Arctic and in their search for a place within Canadian society, while for new Canadians, a sense of place may seem elusive at first, but, with time, they too will develop such feelings as they (or more likely their children) sink their roots into Canadian society. Indeed, the theme of this book is that Canada is a country of regions, each of which has a strong sense of regional pride, but also a commitment to Canadian federalism.

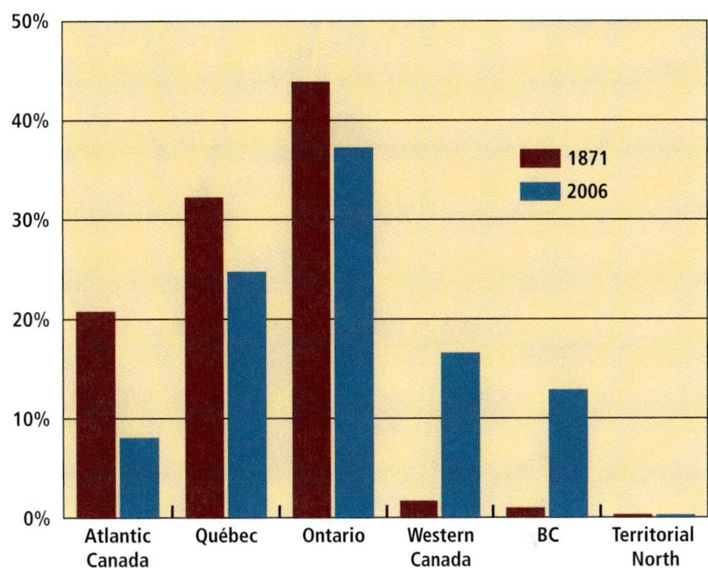

Figure 1.3 Regional populations by percentage, 1871 and 2006.
Sources: Statistics Canada (1997, 2007a).

The Dynamic Nature of Regions

Regions are not static entities. Regions change over time and those changes alter relationships between regions. One measure of such change is revealed in regional population shifts from 1871 to 2006 (Figure 1.3). The most dramatic changes occurred in Atlantic Canada and Western Canada. In 1871, Atlantic Canada contained 21 per cent of Canada's population while Western Canada's figure was less than 3 per cent. By 2006, the population figures were almost reversed with Western Canada accounting for 17 per cent and Atlantic Canada for 7 per cent.

Rapid structural change has taken place over the last 25 years. Canada and its regions have transformed themselves by leaving behind a closed economy for a much more open North American/world economy. Much of this transformation can be attributed to free trade agreements and to rising global prices for energy and resources. Ottawa has entered into trade agreements that have moved Canada from a closed economy with its east/west trade axis and high tariffs to a more open free trade economy with a north/south axis (Table 1.2). Except for agricultural commodities but especially for grains, commodity prices have increased in the first years of the twenty-first century. As we have seen, western provinces have benefited greatly from higher energy and resource prices.

Canada's society has changed in many ways. Much of this transformation came following modifications in 1967 to Canada's Immigration Act that resulted in large numbers of new Canadians coming from Asia,

Table 1.2 Key Steps Altering Canada's Economy

Date	Event
1947	The General Agreement on Tariffs and Trade (GATT) was signed by Ottawa and 22 other governments. These countries agreed to reduce tariffs and eliminate import quotas among members through multilateral negotiations and agreements.
1965	The Auto Pact between Canada and the United States created a continental market for automobiles. An important provision for Canada was that its share of automobile production be maintained. The three large American car manufacturers (General Motors, Ford, and Chrysler) could now expand their production to a continental scale. In this process, Canadian branch plants ceased to produce a wide variety of automobiles for the Canadian market and began to produce only a few models (but more of them) for the continental market.
1985	The Royal Commission on the Economic Union and Development Prospects for Canada examined the question of Canada–United States free trade and concluded that such an agreement would greatly benefit Canada.
1989	The Canada–US Free Trade Agreement (FTA) was approved. The aim of the FTA was to integrate the two economies. Unlike the Auto Pact, however, there were no assurances that Canada's share of economic activities would be maintained. Instead, the North American marketplace would determine the type and amount of economic activity in Canada.
1994	Superseding the FTA, the North American Free Trade Agreement (NAFTA) broadened the geographic area of the FTA to include Mexico. The North American economy now had a member with much lower wage rates, causing many labour-intensive industries to close their operations in Canada and reopen in Mexico.
1995	The final GATT treaty (the so-called Uruguay Round) was ratified by 124 governments on 1 January 1995. This agreement, which established the World Trade Organization (WTO), a permanent institution with powers to enforce trade rules and assess penalties against members, is the most far-reaching trade pact in world history and moves closer to the idea of global free trade.
2000	The WTO rules that the Auto Pact discriminates against foreign companies. In 2001, the Auto Pact is dissolved.
2001	Following the suicide mission of al-Qaeda territorists who commandeered and then crashed two passenger planes into the twin towers of the World Trade Center in New York City and another one into the Pentagon in Washington, DC, the US government created a Department of Homeland Security, promoted a North American Security Perimeter, and announced the Enhanced Border Security and Visa Entry Reform Act that, together, had enormous implications for Canadian access to the United States.
2002	Canada signed the Kyoto Protocol that committed 38 industrialized countries to cut their emissions of greenhouse gases between 2008 and 2012 to levels 5.2 per cent below 1990 levels.
2006	The United States continues with its North American Security Perimeter, which will require all cross-border travellers to carry passports by 2008. Delays and modifications have taken place, largely because of strong opposition from border provinces and states. Consequently, the US continues to search for a balance between security and trade.

Africa, and South America rather than from Europe. Since most have settled in Canada's three largest cities, Toronto, Montreal, and Vancouver, the spatial impact is highly concentrated. Efforts to encourage a more widespread distribution of newcomers have not been particularly successful. This important subject is explored more fully in Chapter 4.

The changing nature of Canada's regions is also reflected in major events, some positive and some negative. They include:

- The loss of the northern cod stocks on the Grand Banks.
- The controversial James Bay hydroelectric project.

- The Auto Pact between Canada and the United States.
- The US–Canada softwood lumber dispute and the conditions imposed by the 2006 agreement.
- The end of the Crow Benefit for Prairie grain farmers.
- The creation of the Territory of Nunavut.

The Core/ Periphery Model

The **core/periphery model** is an abstract theory devised and refined by scholars that explains how the world economic system evolved into abstract spatial units. This evolution began with the start of the Industrial Revolution, thus giving economic advantage to England and then other European states. In the most simplistic terms, industrialization takes place in a few favoured areas known as cores. As these industrial cores expand, so does their need for raw materials, energy, and foodstuffs, so they form colonies (peripheries) that serve as a source of raw materials and a market for surplus manufactured goods. In this way, the world is divided into two interrelated functional units: industrial cores and resource peripheries. In theory, the **core** controls the direction and rate of development of its **periphery**.

By their very nature, however, grand theories must simplify the real world. In doing so, when applied to actual regions, they often lose touch with the complexity and variety of economic, political, and social realities. The model may even suggest that regions are not dynamic places whereas change is constantly taking place. On the world stage, for example, European nations are no longer the only 'industrial regions' in the world. Japan and now China are considered leading industrial countries.

Applying the Core/Periphery Model to Canada

A national version of the core/periphery model as it related to Venezuela was devised by John Friedmann (Table 1.3). His spatial model provides a framework for classifying Canada's six geographic regions as well as describing the fundamental relationships and tensions between Canada's regions (Vignette 1.2).

Core and periphery regions have different characteristics. In Table 1.4, the geographic, economic, cultural, and political differences between the core and periphery regions are outlined. Applying one of the four characteristics to Canada's Territorial North, geography through space (distance) and physical barriers (river and sea ice) continue to hinder its economic development. Geography can change over time. For instance, the construction of the CPR in 1885 overcame the distance and physical barriers between the original four provinces of Confederation and British Columbia.

The distinctive economic character of Canada's six geographic regions is revealed in Table 1.5. Each region has a different mix of primary, secondary, and tertiary activities As well, Table 1.5 provides empirical support for the classification of the six regions into core, upward transition, downward transition, and resource frontier. For example, Ontario has the highest proportion of its labour force engaged in secondary activities (22.5 per cent in 2006). At the other end of the scale, the Territorial North has the lowest proportion of its labour force involved in the secondary sector (an estimated 2 per cent in 2006).

An examination of Table 1.5 leads us to ask why there is such a spatial variation in Canada's industrial structure. An equally important question: Is this spatial arrangement subject to change over time? History and geography have played a role in determining the existing spatial pattern of economic development. On the one hand, the Maritimes, Québec, and Ontario had a head start in the process of development. On the other hand, geography, in the form of the great Western Sedimentary Basin, has provided enormous petroleum wealth for Alberta, making this province the fastest growing in Canada. Ontario, with by far the largest economy, lacks similar oil-rich sedimentary strata as well vast hydroelectric resources like those found in northern Québec. Today, Ontario faces a serious energy crisis. To better understand these spatial patterns and geographic realities, we can com-

Vignette 1.2 Friedmann's Version of the Core/Periphery Model

In the 1960s, John Friedmann, an American scholar, was working as a regional planner in Venezuela where he devised a version of the core/periphery model. He applied this model to regions within a single country. He viewed core regions as territorially organized subsystems of society that have high capacity for generating and absorbing innovative change, and peripheral regions as subsystems whose development is determined chiefly by core region institutions on which they depend. From a modernization perspective, Friedmann conceived an industrial core and three types of peripheral regions, all of which, in different ways, are dominated by the core.

In applying Friedmann's model to Canada, a core region and three periphery regions can be identified. They are:

- core region (Ontario and Québec);
- periphery region: upward transitional region (British Columbia and Western Canada);
- periphery region: downward transitional region (Atlantic Canada);
- periphery region: resource frontier (the Territorial North).

The broad brush of history provides an explanation for the current classification of Ontario and Québec as core regions—their favourable resource base coupled with their head start in settlement, rapid population growth, and development of a manufacturing base. For those reasons, plus high tariffs on imported manufactured goods, Ontario and Québec formed the industrial and population core in Canada.

But such a model is not set in stone. What about current developments with British Columbia and Western Canada outpacing Québec and challenging Ontario in terms of economic and population growth? Will these two so-called periphery regions continue at that fast pace? Another key question is: will Ontario's economy slow because of rising energy costs, the higher Canadian dollar, and the loss of manufacturing jobs to developing countries? In other words, Canada's six regions can change their position within this theoretical classification scheme.

bine the abstract notion of the core/periphery model with Canada's spatial distribution of resources, the historical imbalances of population size, economic power, political control, and social innovations, and the resulting tensions that existed between the major regions of Canada at Confederation. To be sure, these tensions have modified and altered somewhat over the course of nearly a century and a half.[4]

Three Stages in Canada's Regional Development

Three stages of regional development can be identified for Canada—from shortly after Confederation until the 1960s, when an east–west trade pattern driven by a central government held sway; from the 1965 Auto Pact to the formation of the WTO; and from the 1995 creation of the WTO to the present. Within this temporal framework, spatial changes have and are taking place. As history has revealed, Canada's regional development has demonstrated a capacity to transfer economic and demographic power from one region to another. In part, this transfer process is a product of the historical geography of Canada: British and French colonies were established in the Maritimes and Central Canada, giving those places a head start. After 1867, Canada acquired vast territories in the West and North. Finally, in 1949, Newfoundland joined Confederation. The transfer process also has been affected by location,

Table 1.3 National Version of the Core/Periphery Model

Regions	Characteristics
Core Region	The core region is the focus of economic, political, and social activity. Most people live in the core, which is highly urbanized and industrialized. The core has a high capacity for innovation and economic change. Innovations and economic advances are disseminated downward through the national urban hierarchical system to the periphery. Ontario and Québec form Canada's core region.
Upward Transitional Region	The upward transitional region's economy and population are growing as both capital and labour flow into this rapidly developing area. While initial development occurred in the primary sector, there is now a greater emphasis on manufacturing and service activities. British Columbia and Western Canada are examples of upward transitional regions.
Downward Transitional Region	In a downward transitional region, the economy is declining, unemployment is rising, and out-migration is occurring. Often this is an 'old' region dependent on resource development for its economic growth. Now that these resources have passed their prime or have been exhausted, the regional economy has stalled. Atlantic Canada is an example of a downward transitional region.
Resource Frontier	Located far from the core region, few people live in this frontier and little development has taken place. Resource companies are just beginning to penetrate into this remote area. As energy and mineral deposits are discovered, the prospects for economic growth are enhanced. The Territorial North represents Canada's last resource frontier.

Sources: John Friedmann, *Regional Development Policy: A Case Study of Venezuela* (Cambridge, Mass.: MIT Press, 1966); Immanuel Wallerstein, *The Capitalist World-Economy* (Cambridge: Cambridge University Press, 1979).

with southern Ontario benefiting from close proximity to what traditionally had been the American manufacturing belt in the Upper Midwest and Northeast states. Lastly, the uneven distribution of natural wealth has blessed British Columbia with a vast forest, Québec, British Columbia, and Manitoba with significant hydroelectric potential, and Alberta with huge petroleum deposits.

The first stage began soon after Confederation. This stage can be described as a national version of the core/periphery model with a dominant east–west economic axis, especially in manufactured goods. In 1879, Ottawa enacted the National Policy, which, by raising tariffs on foreign manufactured goods, encouraged manufacturing in the Windsor–Québec City industrial and population core of Ontario

Table 1.4 Basic Characteristics of a Core and a Periphery

	Core	Periphery
Geographic	Space and physical barriers overcome by a modern transportation system	Space and physical barriers continue to hinder economic development
Economic	Industrial production emphasizes manufacturing	Primary production dominates
Cultural	Attitudes, language, social customs, and values prevail	Forced to accept the core's culture
Political	Controls the periphery	Is subservient to the core

Source: Adapted from Edward J. Malecki, *Technology and Economic Development* (New York: Wiley, 1991). Reprinted by permission of Pearson Education Limited.

Table 1.5 Industrial Structure of Canada's Six Geographic Regions, 2006 (per cent)

Industrial Structure	Ontario	Québec	British Columbia	Western Canada	Atlantic Canada	Territorial North
Primary	2.2	2.8	3.6	10.1	5.4	13.0
Secondary	22.5	21.2	17.7	16.7	16.0	2.0
Tertiary	75.3	76.0	78.7	73.2	78.6	85.0
Total	100.0	100.0	100.0	100.0	100.0	100.0

Sources: Statistics Canada (2007c); for Territorial North, author's estimate based on Government of Yukon (2006) and Government of the Northwest Territories (2006). Part of data is adapted from Statistics Canada website Distribution of Employed People, by Industry, by Province [on-line database], Searched 12 September 2006: http://www40.statcan.ca/101/cst01/labor21c.htm.

and Québec. For the rest of Canada, the downside was higher prices for manufactured goods. In this way, a political decision expanded the economic power base of Central Canada (Ontario and Québec) where the major market was found; weakened manufacturing in Atlantic Canada (then the Maritime provinces), which could not export to Central Canada because of high transportation costs or to New England because of high tariffs; and left the rest of Canada heavily dependent on resource development and the sale of its staples in foreign markets. The completion of the Canadian Pacific Railway in 1885 provides a visible sign of this east–west economic axis. Along with economic power, Central Canada accounted for approximately 75 per cent of Canada's population in 1871 (Figure 1.3). At the same time, Atlantic Canada (then consisting of Nova Scotia and New Brunswick) made up 21 per cent of the national population, with the rest of Canada (sparsely populated British Columbia and the North-West Territories) comprising only 4 per cent.

The second stage began in 1965 when the Canada–US Automotive Products Agreement—the Auto Pact—created a North America market for the Big Three automakers. Suddenly, a north–south economic axis emerged, first in the automobile industry, as Canada's automobile manufacturers moved from supplying a national market to a North American one. The result was twofold: a shift from producing an entire range of vehicles to a more specialized form of production that led to **economies of scale** and a reorientation of trade from an east–west axis to a north–south one. With the 1989 Canada–US Free Trade Agreement all manufacturing firms had to compete in the North American market. Canada now had two economic axes. By 1998, Canadian National Railway had become a North American railway by purchasing American railway companies, notably Illinois Central, thus providing concrete evidence of a north–south trade axis (see Figure 1.4). Canada's manufacturing industry began to change. Branch plants of American firms, for instance, lost their cost advantage to serve the Canadian market. Many closed as their larger American-based companies had the advantage of economies of scale and therefore could offer lower prices in the Canadian market.

The third stage took Canada into the global economy. In 1995, with the conclusion of the Uruguay Round of the GATT negotiations that created the World Trade Organization, Canada and 123 other nations became committed to global free trade. While the elimination of all trade barriers between countries remains a goal, most manufacturing trade barriers have disappeared. Many labour-intensive firms in the industrialized West, such as the textile industry, have been unable to compete

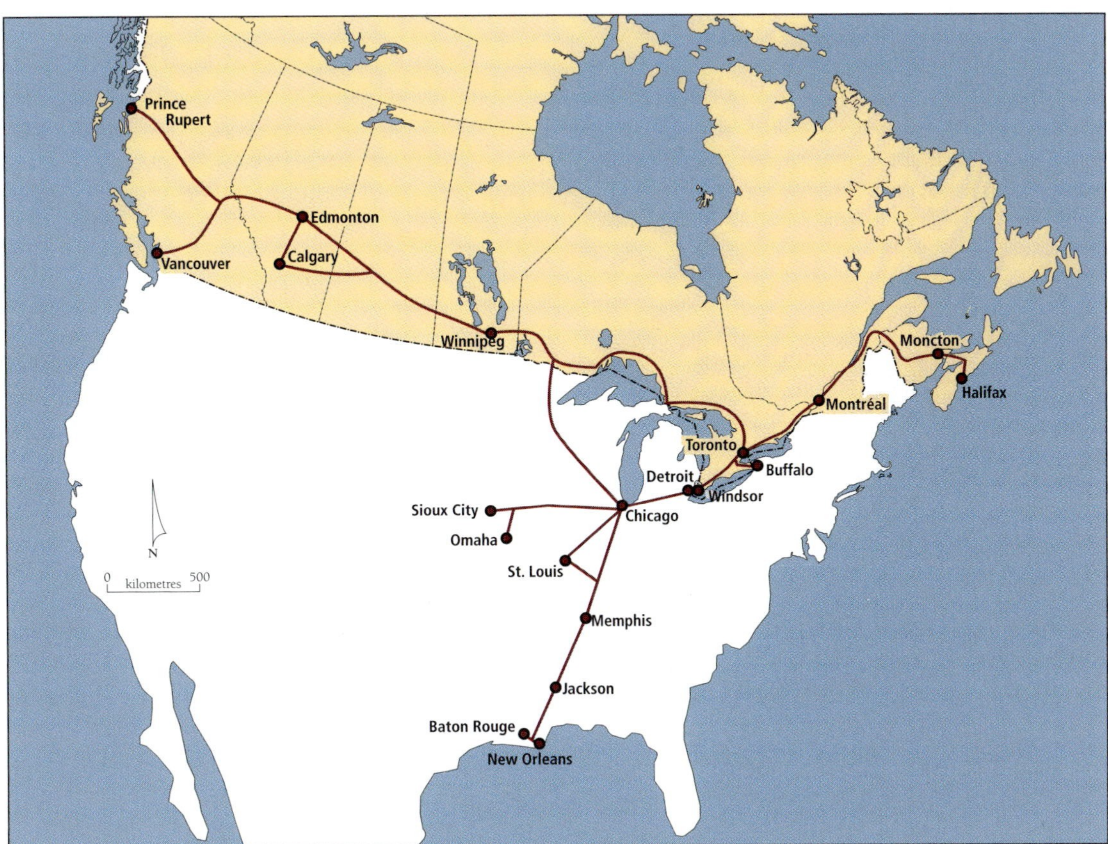

Figure 1.4 The CN North American rail system. With its purchase in 1998 of Illinois Central, Canadian National Railway extended its market access to the populous American Midwest and the American South.

against companies based in low-wage countries. China, for example, produces most of the textiles, shoes, and appliances found in retail stores in Canada. With China, India, and other Asian countries undergoing rapid industrialization, the demand for raw materials and energy rose. By the end of the twentieth century, world prices for raw materials and energy reached record levels. Alberta, with its vast petroleum and coal reserves, benefited most. Not surprisingly, Alberta has become the wealthiest as well as the fastest-growing province in Canada. Other provinces have benefited, too. Saskatchewan and British Columbia, and, to a lesser degree, Newfoundland and Labrador plus Nova Scotia, have enjoyed an economic boost from the oil and gas industry. Since energy prices tend to rise in tandem, Québec, with its vast hydroelectric resources, and Saskatchewan, with most of the world's uranium deposits, have profited from high world prices. Natural gas prices have reached levels that make the tapping of the Mackenzie Delta gas deposits and the construction of the world's longest natural gas pipeline from Inuvik to Chicago a reality. Of Canada's six geographic regions, only Ontario is left out of this energy resource boom. Instead, Ontario is searching for a solution to its energy shortage.

Canadian Trade within North America and the World

With the acceleration of globalization, Canada's trade has become more international in focus. Although the value of interprovincial trade grew by 40 per cent from 1994 to 2004, international trade nearly doubled. In 1984, international and interprovincial trade were nearly equal, but by 2004 the volume of international trade was more than twice that of interprovincial trade (Statistics Canada, 2006f. The only province to have a trade surplus in interprovincial trade in 2004 was Ontario, with automotive products being far and away its most valuable domestic export.

The United States is Canada's largest trading partner, followed by China. Trade between the two North American countries has boomed since the signing of the FTA in 1989. More than $620 billion worth of merchandise was traded between Canada and the United States in 2006—an average of $1.7 billion crossing the Canada–US border every day (ibid.).

With a continental economy in place, a North American core/periphery spatial framework has been superimposed on the traditional Canadian version. The increasing integration of the North American economy has had both positive and negative impacts. Improved access to the American market has boosted Canadian production and trade and, consequently, Canadian living standards. One telling measure of the degree of that integration is revealed in Canada's international merchandise trade. In 2006, 79 per cent of Canadian exports went to the United States while 66 per cent of its imports came from the US (Statistics Canada, 2007d). From 2005 to 2006, exports to the United States declined by 2.3 per cent (Table 1.6). The main reasons were the slowdown in the US economy and the high Canadian dollar. The three export sectors affected were automotive, forestry, and natural gas (Statistic Canada, 2007e). Table 1.6 outlines Canadian exports.

Yet, trade disputes have hurt Canada's economy and certain regions have felt the full weight of these trade restrictions. The ongoing softwood lumber dispute, which was officially resolved in 2006 but still continues to fester, has affected the forest industry, especially small producers in British Columbia and Québec. For that reason, tariff and non-tariff barriers still have a substantial effect on Canada–US trade in certain industrial sectors.

Canada shares the continent with the United States and Mexico. As we have seen, three trade agreements—the Auto Pact (1965), the FTA (1989), and **NAFTA** (1994)—led to a realignment of the Canadian economy so that it was more thoroughly integrated with the North American economy. These trade agreements saw more and more manufactured goods exported to the United States, thus breaking the old pattern of exporting primarily low-value unprocessed or semi-processed resource products, and the volume of exports to the United States grew dramatically. In spite of these trade agreements, however, Washington is prepared to defend US business interests by imposing trade barriers, as has been the case with duties on softwood lumber and grain, to restrict the natural flow of certain Canadian goods into the United States. The purpose of these duties is twofold: (1) to protect US farmers and forest companies in the short run by imposing duties on targeted Canadian exports, and (2) to force Canada to accept a long-term agreement that will limit its grain and lumber exports.

While Canadian relations with the rest of the world are important, the long-term trend is clear—Canada's geopolitical interests, cultural development, and economic priorities are, for better or worse, linked to the United States. One advantage of sharing the North American continent with our powerful American neighbour is proximity to the American market, while many would argue that another is the shared defence of the continent. Economic and military agreements such as NAFTA and the North American Aerospace Defence Command (**NORAD**) have, for the most part, worked well for Canada. Both of these agreements foster a closer co-operation between the two countries. While common interests have led Canada and the US to sign numerous agreements and to undertake joint ventures, Canada remains a distinct society (Adams, 2003). Evidence of such independence was revealed in Ottawa's decision not to join the US-led invasion of Iraq but to participate in

Table 1.6 Canadian Exports, 2006

Country	$ millions	%	% Change since 2005
United States	361,310	78.9	–2.0
European Union	33,556	7.3	16.2
Other OECD countries	18,379	4.0	20.6
Japan	10,760	2.3	2.8
All other countries	34,161	7.5	14.3
Total Trade	458,166	100.0	

Source: Statistics Canada (2007d). Adapted from Statistics Canada publication *The Daily*, Catalogue 11-001, Canada's International Merchandise Trade, 13 February 2007, URL: http://www.statcan.ca/Daily/English/070213/td070213.htm.

the UN-sanctioned and NATO-led military intervention in Afghanistan. Yet, living next door to the most powerful country in the world caused Prime Minister Pierre Trudeau to remark in 1969 that 'living next to you is in some ways like sleeping with an elephant: No matter how friendly and even-tempered the beast, one is affected by every twitch and grunt' (Colombo, 1987: 54).

Following the al-Qaeda attacks of 11 September 2001, America declared war on terrorists and countries that harboured terrorists. At the same time, the US took unprecedented measures to protect itself against future attacks. While its Department of Homeland Security is charged with preventing terrorists from slipping into the United States, the US government reaction to the events of 9/11 and its aftermath has reshaped the architecture of Canada–US relations, transformed the 'unguarded' Canada–US border into a quasi-military checkpoint, and raised the issue of a North American security perimeter. Since Canada's economic well-being depends on American willingness to maintain an open border, the consequences of border restrictions are potentially crippling. A slowing of the flow of goods and people across the border would have a negative impact on Canada's economic well-being and tourist industry. Ontario's automobile industry, which is based on the just-in-time delivery of auto parts, would be devastated. In 2007, the Department of Homeland Security tightened the requirements for Canadians travelling to the United States by requiring Canadians to have a passport. Passports are required for air travel to the United States, but not yet for land travel. Border provinces and states have demanded a more flexible approach to land travel, forcing the US to review the balance between security and trade. With cross-border tourist business so important, the substitution of a 'security-designed' driver's licence for a passport is under review. Already one concession has been made—children up to age 15 who are Canadian citizens and have parental consent need only show a birth certificate to enter the US. As well, an unknown number of Canadians are on the US Homeland Security's 'no-fly' list.

What is at stake? The North American economy is based on the free flow of goods and services. Delays in cross-border shipments create uncertainty about maintaining production and such uncertainty makes it less likely that foreign companies will choose to locate in Canada.

In 2006 the United States announced plans for a virtual fence stretching nearly 9,000 km along the Canadian–US border. The proposed virtual fence would consist of high-

Figure 1.5 You've changed, Sam. Following al-Qaeda suicide attacks on the World Trade Center in New York and the Pentagon in Washington, the United States declared an all-out war on terrorists and increased its efforts to prevent future attacks by creating the Department of Homeland Security. One consequence has been to make the flow of goods and people from Canada to the US more complicated and even unpleasant, especially for Canadian citizens born in the Middle East. (CAM/Regina Leader-Post)

tech monitoring equipment designed to spot illegal crossings. More important, Washington has called for a North American security perimeter that includes a common position on immigration, military, and trade policies. All of these measures are designed to reduce the risk of terrorist attacks. The forces of continental integration have gained strength under the guise of North American security. Canada faces a difficult decision—staying outside of the North American security perimeter and running the risk of greater delays at the US border, or subscribing to a North American security perimeter and harmonizing its border and immigration regulations with the United States. Dancing with the American elephant may facilitate commercial traffic crossing into the United States, but it may also diminish Canada's sovereignty. Is this a concern? According to John Manley, former Deputy Prime Minister in the Jean Chrétien government:

> Canada has no choice but to co-operate with American anti-terrorism efforts to keep the border open and the economy running smoothly. I think we have to be . . . a little bit less religious about the issue of sovereignty and be a little bit more practical about it. (Blackwell, 2006: A7)

Yet, compare this 2006 view with the 1988 position of John Turner, who, in the leaders' debate, declared:

> We built a country east and west and north. We built it on an infrastructure that deliberately resisted the continental pressure of the United States. For 120 years, we've done it. With one signature of a pen, you've [Prime Minister Mulroney] reversed that, thrown us into the north–south influence of the United States and will reduce us, I am sure, to a colony of the United States, because when the economic levers go the political independence is sure to follow. (Azzi, 2006: A18)

Regional Implications of the Canada–US Relationship

Twitches and grunts from the American elephant often resonate more strongly in one or more regions of Canada. The explanation is simple: each region of Canada has a different set of natural resource and manufacturing activities. Consequently, the type of trade with the United States varies from one region to the next. Exports to the United States reveal that primary and semi-processed products dominate trade from British Columbia, Western Canada, Atlantic Canada, and the Territorial North, while manufactured goods are key exports from Ontario and Québec. For example, increased oil, potash, and uranium exports to the US market will have a positive impact on Western Canada, but will have no direct effect on the rest of Canada.

Trade disputes affect regions differently. Washington may wish to punish certain Canadian industries, but the impacts of American-erected trade barriers take on a highly regional character. Forest industries across Canada suffered from the US duties on softwood lumber, an issue that has impacted trade in wood products for the past generation, but British Columbia, Québec, and Ontario were most severely affected, and the 2006 agreement remains less than satisfactory at least for some producers. Western Canada is just recovering from a US ban on cattle. The blow to the cattle industry was triggered by a case of bovine spongiform encephalopathy (BSE) in an Alberta steer. Exports were halted, causing the industry to lose over 60 per cent of its markets and to create a glut of cattle in Canada. Prices fell dramatically for cattle producers, causing some to declare bankruptcy. In July 2005, the US border reopened to Canadian cattle exports, causing prices for Canadian cattle to increase.

Key Topics

The heart of this book is discussion of the six geographic regions of Canada (Chapters 5–10). Broad profiles of these regions in regard to area, population, and economic strength are in Table 1.1; their social characteristics are outlined in Table 1.7. These basic geographic elements provide a start to understanding the six regions. Further understanding is provided by analysis of key economic activities found in each region. Extensive description of a dominant economic activity for each of Canada's six regions is presented in Chapters 5–10. These key topics offer fuller and more detailed insights into the nature and strength of the regional economies as well as the challenges faced by these economies. These key economy activities are:

- Ontario: manufacturing of automobiles
- Québec: production of hydroelectric power
- British Columbia: forest industry
- Western Canada: agriculture
- Atlantic Canada: fisheries
- The Territorial North: megaprojects

Understanding Canada's Regions

As a prelude to understanding the many interrelated physical and human dimensions of Canada's geographic regions, the next three chapters focus on Canada's physical geography, history, and human geography. Chapter 2 ('Canada's Physical Base') explores the wide range of natural and geomorphic processes that account for wide variations in landforms across Canada. Seven physiographic regions are described with brief accounts of their geology, soils, natural vegetation, and climate. A major theme is the physical environment's influence on human occupancy of the land and the effects of human occupancy on the physical environment. Chapter 2, therefore, is more than a mere backdrop for the remainder of this text; rather, it illuminates how human activities and human decision-making both shape and are shaped by the physical environment in which they occur, and it introduces the extent of environmental problems resulting from industrial development, as described in each of the regional chapters.

Chapter 3 ('Canada's Historical Geography') focuses on the arrival of Canada's First Peoples, early French and British settlements, and the territorial evolution of Canada. Four faultlines affect Canada and its six geographic regions differently both in time and location. The tensions surrounding the four faultlines outlined in this chapter often lie dormant only to flare up unexpectedly, and these tensions provide a background for the subject of Chapter 4 ('Canada's Human Face'), which is directed to contemporary issues dealing with demography (baby boom and bust), Canada's pluralistic society (the impact of a more open immigration policy), and a continental economy (the effect of the Canada–US Free Trade Agreement). This chapter also looks at the population trends

Table 1.7 Social Characteristics of the Six Canadian Regions, 2001

Geographic Region	French* (000s)	French (% by region)	Aboriginal Peoples** (000s)	Aboriginal Peoples (% by region)
Ontario	307,295	2.7	188,315	1.6
Québec	5,918,390	83.4	79,400	1.1
British Columbia	16,905	0.4	170,025	4.4
Western Canada	46,375	0.9	436,450	8.6
Atlantic Canada	241,375	10.6	37,785	2.4
Territorial North	1,040	1.1	45,865	51.7
Canada	6,531,375	22.0	976,310	3.3

*French includes those who speak French most often at home.
**In 2001 the census asked two questions to determine the number of Aboriginal peoples. The 2001 Aboriginal population, as determined by a question about identity, was 976,310. The three census categories for Aboriginal population were North American Indian (608,850), Métis (292,310), and Inuit (45,070). The 2001 Aboriginal population, as determined by a question based on ethnic origin, was 1,319,890. The explanation for the difference in the number of Aboriginal peoples may be one of 'multiple identity'.
Sources: Statistics Canada (2003a, 2003b).

associated with the tensions that arise from centralist and decentralist visions for Canada, French and English views of 'nation', immigration and Canada's pluralistic society, and claims to the land and their place on it from Aboriginal and non-Aboriginal populations.

Summary

Canada's six regions provide a vehicle to explore the geographic essence of Canada. To simplify the complexities of space, regional geographers divide the world and countries into regions, which vary by scale but often are interrelated in a hierarchical order. A regional geographer selects critical physical, historic, and human characteristics that logically divide a large spatial unit into a series of regions. Towards the margins of a region, its main characteristics become less distinct and merge with those of a neighbouring region. For that reason, boundaries are best considered as transition zones. Canada is a country of regions. Shaped by its history and physical geography, Canada is distinguished by six geographic regions: Ontario, Québec, British Columbia, Western Canada, Atlantic Canada, and the Territorial North. Within Canada and its regions, four key tensions exist that, through their interaction, demonstrate the very essence of Canada as a 'soft' nation, where conflicts are usually resolved or ameliorated through compromise rather than by political or military power.

The essential foundation for studying regional geography is to conceptualize places and regions as components of a constantly changing global system. The core/periphery model provides an abstract spatial framework for understanding the general workings of the modern capitalist system. It consists of an interlocking set of industrial cores and resource peripheries. This model can function at different geographic scales and serves as an economic framework for interpreting Canada's regional nature. In addition to the core region, three types of regions devised by Friedmann extend our appreciation of the diversity of the Canadian periphery. They are: (1) an upward transitional region, (2) a downward transitional region, and (3) a resource frontier.

Time affects geographic regions. Perhaps the most dramatic events to impact on Canada and its regions were (1) changes to immigration regulations and (2) the shift over time from a closed economy to an open one. Liberalization of immigration policy began in 1967 when Canada no longer favoured European applicants by deciding to accept applicants on merit from around the world. Although Canada was one of the original signatories to the General Agreement on Tariffs and Trade in 1947, which worked towards freer trade over the course of eight rounds of negotiations spanning nearly a half-century, Canada truly opened its economy to greater international influence with the 1989 Canada–US Free Trade Agreement, the signing of NAFTA in 1994, and the establishment of the World Trade Organization in 1995. Canada's place in that world has changed. Trade is one measure of those changes. International trade clearly dominates the regional economies in Canada while interprovincial trade remains an important element, especially for Ontario and Québec. Canadian regions reflect these changes, with resource-rich regions now benefiting from high prices due to growing world demand for **primary products**. On the other hand, Québec and Ontario manufacturers are facing stiff competition from other countries where costs of labour are low.

Notes

1. In the late nineteenth century, geographers believed that the physical environment determined human affairs. Geographers soon rejected that position, although they recognize that the environment does exert a strong influence on the nature of human activities in various regions of the world. Students can find a more complete discussion of environmental determinism and other philosophical options in geography, including possibilism, positivism, humanism, and Marxism, in William Norton, *Human Geography*, 6th edn (Toronto: Oxford University Press, 2007). The writings of most geographers reflect either one of these philosophical positions or some combination. The core/periphery model, for example, sprang from Marxist scholarship. Because of the model's powerful spatial implications, other scholars holding different philosophical positions have modified this theory by removing its economically deterministic character. Instead, they accept that external forces (such as global institutions like the WTO) and internal forces (such as federal–provincial agreements like equalization payments) can modify the impact of the physical environment on regional development. Hinterlands, therefore, are not locked into a single outcome because of their physical geography.
2. While Canadians and Americans occupy the same continent, historic events and geographic differences laid the foundation for the emergence of two different societies within one continent. These differences are found in many aspects of the two societies. Each country has a different approach to gun control legislation, multiculturalism, and a national health-care system. As a result of these and other differences, some scholars believe that Canadians are more trusting of their governments and more tolerant of social diversity than are Americans (Hartz, 1955; Lipset, 1990; Lemon, 1996; Saul, 1997).
3. With the federal government's recognition of Aboriginal title to ancestral lands in 1973 following the judgement by the Supreme Court of Canada in *Calder v. Attorney General*, comprehensive land-claim settlements have followed this watershed decision, and these agreements have transformed the cultural, economic, and political landscape of the Territorial North and the northern areas of many provinces. Already the James Bay Cree, Inuit, and Innu of Québec, the Inuvialuit, Gwich'in, Sahtu Dene, and Dogrib of the Northwest Territories, and the First Nations in Yukon, the Inuit of Nunavut, the Nisga'a of BC, and the Inuit of Labrador have obtained land-claim settlements and received land and cash in exchange for surrendering their Aboriginal rights to vast areas of land. In 1975, the James Bay Cree and Inuit achieved a regional form of self-government in northern Québec. Following 1995, self-government became a reality in land claim negotiations and those groups who settled before 1995 can return to the negotiation table on the issue of self-government. While the North remains a resource hinter-

land, land-claim settlements ensure that those Aboriginal peoples will have more control over resource development and will therefore be better able to protect the wildlife and their hunting lifestyle.

4. The **staples thesis**, developed by Harold Innis (1930), is a Canadian variant of the core/periphery model. Within this theoretical context, Innis conceived of Canada's early history in terms of its regional geography but with the country functioning as a resource economy driven by foreign demands. From this core/periphery perspective, Innis projected a new vision of Canada's economic history, asserting that the export of Canada's natural resources to Great Britain, the United States, and other advanced industrial countries affected the national and regional economies of Canada and their social and political systems. Innis insisted that Canada is a nation because of its geography; that is, Canada's history was triggered by the exploitation of a major resource (staple) found in a region and shaped by its trade relation with external markets. In some cases resource development led to economic diversification; in other cases it did not. In several books, Innis provided detailed historical accounts of staples development in several regions of Canada. For instance, the fur trade in early colonial Canada did not result in the diversification of the northern economy but, rather, in the enrichment of England and France (Innis, 1930). The wheat trade, on the other hand, led to the diversification of the economy of Western Canada. The staples thesis, then, provided Innis with a Canadian version of the core/periphery model (Barnes, 1993; Hayter and Barnes, 2001).

Challenge Questions

1. From a geographer's perspective, what is the basis of the statement that 'Canada is a country of regions'?
2. Does the Friedmann version of the core/periphery model correspond to Canada's six regions?
3. What is the argument that Canada consists of six geographic regions?
4. Are regions static or dynamic?
5. Does Hérouxville's code of standards fit into one of the four faultlines.
6. Can you envisage another faultline? What would it be?
7. What is the difference between 'power of place' and 'sense of place'?
8. Do you agree with John Manley's statement that 'I think we have to be . . . a little bit less religious about the issue of sovereignty and be a little bit more practical about it.'

Key Terms

core
An abstract area or real place where economic power, population, and wealth are concentrated; sometimes described as an industrial core, heartland, or metropolitan centre.

core/periphery model
A theoretical concept based on a dual spatial structure of the capitalist world and a mutually beneficial relationship between its two parts, which are known as the core and the periphery. While both parts are dependent on each other, the core (industrial heartland) dominates the economic relationship with its periphery (resource hinterland) and thereby benefits the most from this relationship. The core/periphery model can be applied at several geographical levels, including international, national, and regional.

economies of scale
A reduction in unit costs of production that results from an increase in output.

gross domestic product (GDP)
An estimate of the total value of all materials, foodstuffs, goods, and services produced by a country or province in a particular year.

heartland
A geographic area in which a nation's industry, population, and political power are concentrated; also known as a core.

hinterland
A geographic area based on resource development that supplies the heartland with many of its primary products; also known as a periphery.

NAFTA
In January 1994, Canada, the United States, and Mexico launched the North American Free Trade Agreement and formed the world's largest free trade area. The Agreement has increased trade among the three countries and rearranged the location of labour-intensive manufacturing firms to Mexico where wages are much lower than in either the United States or Canada.

NORAD
In September 1957, Canada and the United States agreed to create the North American Air Defence Command (NORAD) headquartered in Colorado Springs, Colorado, as a binational command, centralizing operational control of continental air defenses against the threat of Soviet bombers. In March 1981, the name was changed to North American Aerospace Defence Command but the acronym NORAD continued to be used. Since 11 September 2001, NORAD is responsible for protecting North America from domestic as well as foreign air attacks.

periphery
The weakly developed area surrounding an industrial core; also known as a hinterland.

primary products
Goods derived from agriculture, fishing, logging, mining, and trapping; non-processed products.

region
An area of the earth's surface defined by its distinctive human or natural characteristics. Boundaries between regions are often transition zones where the main characteristics of one region merge into those of a neighbouring region. Geographers use the concept of regions to study parts of the world.

regional geography
The study of the geography of regions and the interplay between physical and human geography, which results in an understanding of human society, its physical geographical underpinnings, and a sense of place.

sense of place
The special and often intense feelings that people have for the region in which they live. These feelings are derived from a variety of experiences; some are due to natural factors such as climate, while others are from cultural factors such as language. Whatever its origin, a sense of place is a powerful psychological bond between people and their region.

staples thesis
The idea that the history of Canada, especially its regional economic and institutional development, was linked to the discovery, utilization, and export of key resources in Canada's vast frontier. Harold Innis devised this thesis in the early 1930s, and his ideas continue to influence Canadian scholars.

Bibliography

Adams, Michael. 2003. *Fire and Ice: The United States, Canada and the Myth of Converging Values*. Toronto: Penguin.

Alexandroff, Alan S., and Don Guy. 2003. 'What Canadians Have to Say about Relations with the United States', *Border Papers No. 73*. Toronto: C.D. Howe Institute.

Ali, Anar. 2006. 'Life in the Hyphenated Spaces', *National Post*, 14 June, A21.

———. 2006. *Baby Khaki's Wings: Stories*. Toronto: Viking Canada.

Allan, John, Doreen Massey, and Alan Cochrane. 1998. *Rethinking the Region*. London: Routledge.

Azzi, Stephen. 2006. 'Debating Free Trade', *National Post*, 14 Jan., A18.

Barnes, Trevor, ed. 1993. 'Focus: A Geographical Appreciation of Harold A. Innis', *Canadian Geographer* 37, 4: 352–64.

Blackwell, Tom. 2006. 'Be "Less Religious" about Sovereignty, Manley Urges', *National Post*, 7 Feb., A7.

Brethour, Patrick. 2004. 'NAFTA Needs Fixing, Martin Says', *Globe and Mail*, 7 July, B1.

Bunting, Trudi, and Pierre Filion, eds. 2006. *Canadian Cities in Transition: Local Through Global Perspectives*, 3rd edn. Toronto: Oxford University Press.

CBC News. 2007. 'Hérouxville Drops Some Rules from Controversial Code', 13 Feb. At: <www.cbc.ca/canada/montreal/story/2007/02/13/qc-herouxville20070213.html>.

Columbo, John Robert, ed. 1987. *New Canadian Quotations*. Edmonton: Hurtig.

De Blij, J.H., and Alexander B. Murphy. 2006. *Human Geography: Culture, Society, and Space*, 8th edn. Toronto: John Wiley.

d'Haenens, Leen, ed. 1998. *Images of Canadianness: Visions on Canada's Politics, Culture, Economics*. Ottawa: University of Ottawa Press.

Frank, A.G. 1969. *Capitalism and Underdevelopment in Latin America*. New York: Monthly Review Press.

Friedmann, John. 1966. *Regional Development Policy: A Case Study of Venezuela*. Cambridge, Mass.: MIT Press

Government of the Northwest Territories, Bureau of Statistics. 2006. *Northwest Territories—2005 . . . by the Numbers*. At: <www.stats.gov.nt.ca/Statinfo/Generalstats/bythenumbers/BTNhome(dvo).html>.

Government of Yukon. 2006. *Yukon Labour Force Survey Review, 2005*. At: <www.eco.gov.yk.ca/stats/onetime/lforcerev05.pdf>.

Hamilton, Graeme. 2007. 'Welcome! Leave Your Customs at the Door', *National Post*, 30 Jan., A1, A7.

Hartz, Louis. 1995. *The Liberal Tradition in America*. New York: Harcourt Brace Jovanovich.

Hayter, Roger, and Trevor J. Barnes. 2001. 'Canada's Resource Economy', *Canadian Geographer* 45, 1: 36–41.

Hutchison, Bruce. 1942. *The Unknown Country: Canada and Her People*. Toronto: Longmans, Green and Company.

Innis, Harold. 1930. *The Fur Trade in Canada: An Introduction to Canadian Economic History*. New Haven: Yale University Press.

James, William. 1981 [1890]. *The Principles of Psychology*

Lipset, Seymour M. 1990. *Continental Divide: The Values and Institutions of the United States and Canada*. New York: Routledge.

Lewington, Jennifer. 2007. 'Immigrants and Integration—Is the City Ready to Listen?', *Globe and Mail*, 29 Jan. At: <www.theglobeandmail.com/servlet/story/RTGAM.20070116.wimmig16home/BNStory/National/home>.

Matthews, Ralph. 1983. *The Creation of Regional Dependency*. Toronto: University of Toronto Press.

Norton, William. 2007. *Human Geography*, 6th edn. Toronto: Oxford University Press.

Resnick, Philip. 2000. *The Politics of Resentment: British Columbia Regionalism and Canadian Unity*. Vancouver: University of British Columbia Press.

Saul, John Ralston. 1997. *Reflections of a Siamese Twin: Canada at the End of the Twentieth Century*. Toronto: Viking.

Savoie, Donald J. 1986. *The Canadian Economy: A Regional Perspective*. Toronto: Methuen.

Scott, Allen J. 1998. *The Coming Shape of Global Production, Competition, and Political Order*. Oxford: Oxford University Press.

Simpson, Jeffrey. 2000. *Star-Spangled Canadians*. Toronto: HarperCollins Canada.

Stanford, Quentin H., ed. 2006. *Canadian Oxford World Atlas*, 6th edn. Toronto: Oxford University Press.

Statistics Canada. 1997. *A National Overview: Population and Dwelling Counts*. 1996 Census of Canada. Catalogue no. 93–357–XPB. Ottawa: Industry Canada.

———. 2002. Census of Canada 2001—Census Geography. Highlights and Analysis: Canada's 2001 population. At: <www12.statcan.ca/English/census01/>.

———. 2003a. Aboriginal Peoples of Canada: A Demographic Profile. At: <www12.statcan.ca/english/census01/products/analytic/companion/abor/contents.cfm>.

———. 2003b. Census of Population: Language, Mobility and Migration. At: <www12.statcan.ca/english/census01/products/analytic/companion/abor/contents.cfm>.

———. 2006a. Population by Year, by Province and Territory. At: <www.statcan.ca/Daily/English/021210/d021210a.htm>.

———. 2006b. Gross Domestic Product, Expenditure-based, by Province and Territory. At: <www40.statcan.ca/l01/cst01/econ15.htm>.

———. 2006c. Land and Freshwater Area, by Province and Territory, 2006. At: <www40.statcan.ca/l01/cst01/phys01.htm>.

———. 2006d. 'Canada's International Merchandise Trade', *The Daily*, 9 Mar. At: <www.statcan.ca/Daily/English/060309/d060309a.htm>.

———. 2006e. Canadian Statistics: Distribution of Employed People, by Industry, by Province. At: <www40.statcan.ca/101/cst01/labor21c.htm>.

———. 2006f. Trade. At: <www41.statcan.ca/1130/ceb1130_000_e.htm>.

———. 2007a. Population and Dwelling Counts for Canada, Provinces and Territories, 2006 and 2001 Censuses—100% data. At: <www12.statcan.ca/english/census06/data/popdwell/Table.cfm?T=101>.

———. 2007b. Gross Domestic Product, Expenditure-based, by Province and Territory. At: <www40.statcan.ca/l01/cst01/econ15.htm>.

———. 2007c. Canadian Statistics: Distribution of Employed People, by Industry, by Province. At: <www40.statcan.ca/l01/cst01/labor21c.htm>.

———. 2007d. 'Canada's International Merchandise Trade', *The Daily*, 13 Feb. At: <www.statcan.ca/Daily/English/070213/td070213.htm>.

———. 2007e. Export of Goods on a Balance-of-Payments Basis, by Product. At: <www40.statcan.ca/l01/cst01/gblec04.htm>.

Telford, Hamish, and Harvey Lazar. 2003. *Canada: The State of the Federation 2001: Canadian Political Culture(s) in Transition*. Montréal and Kingston: McGill-Queen's University Press.

Thomas, David M., ed. 2000. *Canada and the United States: Differences that Count*, 2nd edn. Peterborough, Ont.: Broadview Press.

Wallerstein, Immanuel. 1979. *The Capitalist World Economy*. Cambridge: Cambridge University Press.

Further Reading

Hare, Kenneth F. 1968. 'Canada', in John Warkinton, ed., *Canada: A Geographical Interpretation*. Toronto: Methuen, 3–12.

Professor Hare had an illustrious career as a geographer, climatologist, and senior university administrator. In 1968, he wrote the lead chapter in a major geography book on Canada, in which he described a Canada that had just emerged from 20 years of rapid economic expansion and population growth. The Auto Pact, signed in 1965, had already had an impact on Ontario's economic growth and its success foretold more economic gains from trade with the United States. The baby boom had just peaked, marking the end of Canada's high rate of natural increase. Yet, Canada was about to change, and change radically, from the impact of three critical social events: the emergence of a powerful separatist party, the Parti Québécois, which formed the government of Québec in 1976; the increasing assertion of Aboriginal rights following the Liberal government's 1969 White Paper on Aboriginal peoples; and the impact of the 1967 immigration policy, which began to shift the main flow of new Canadians from Europe to the entire world. For students, Professor Hare's version of Canada in the 1960s serves as a useful historical picture just prior to these critical changes.

Bourne, Larry S. and Damaris Rose. 2001. 'The Changing Face of Canada: The uneven geographies of population and social change', *Canadian Geographer* 45, 1: 105–19.

Over the last 50 years, Canada and its regional geography have seen dramatic changes, as noted above. Larry Bourne and Damaris Rose, two prominent Canadian geographers, have identified the demographic and social forces that caused these changes. Bourne and Rose argue that these forces are not yet spent. In fact, they predict that these demographic and social forces will continue to shape the country's future more so than economic and political forces. However, their impact is not evenly spread across Canada, but, in a synergetic manner, they come together in particular places at particular times to reshape the regional and local rural/urban landscape. In their analysis, Bourne and Rose identify processes of change that, in combination, have recast Canadian society into a different format. To prove their case, the authors focus their attention on four critical forces:

- population and spatial demography, including the changing components of population growth;
- lifestyles, families, and living arrangements, including changes in family structure, domestic relations, and household composition;
- social diversity in the urban system caused by immigration and migration;
- the labour market nexus, including shifts in the linkages between the domestic or household sphere, the sphere of work and production, and the changing nature of the state and civil society.

Cape Breton, Nova Scotia (Ivy Images)

Overview and Objectives

Physical geography has a powerful influence on the pattern of human settlement. Climate and physiography represent two key elements of physical geography that affect the human occupation of the land, including resource development. Then, too, physical geography's link to the core/periphery model is obvious: areas with advantageous locations, climates, and natural resources attract the majority of settlers and gradually form population and economic cores, leaving less advantageous areas with fewer people and weaker economies. Different physiographic regions and climatic zones have different soil, natural vegetation, and wildlife zones. This chapter provides a basic introduction, emphasizing the interrelationships between the influence of physical geography on human occupation and humans' impact on the land. This chapter will:

- Show how physical geography has shaped the regional nature of Canada.
- Examine the geological structure, origins, and characteristics of Canada's physical base and seven physiographic regions.
- Describe the main global factors that influence climate.
- Explore the interrelationships among environmental factors.
- Discuss how physical conditions influence human occupation of the land, with reference to the core/periphery model.
- Examine how human activities and the physical environment are interrelated.
- Consider the impact of human activities on Canada's natural environment and the potential threat of global warming.

CHAPTER 2

Canada's Physical Base

INTRODUCTION

The earth provides a wide variety of natural settings for human beings. For that reason, physical geography helps us understand the regional nature of our world. The basic question posed in this chapter is: Why is Canada's physical geography so essential to an understanding of its regional geography? The answer lies in the regional character of Canada's physical geography and in the interrelationships between physical geography and human settlement and activity. Physical geography is an underlying factor in shaping Canada's national and regional character, and it provides a fundamental explanation for the distribution of population within Canada.[1] In fact, population differences between Canada and the United States can be, in part, attributed to physical geography (Vignette 2.1). In this text, physical geography also provides the raison d'être for the basis of the core/periphery model. The argument is a simple one: regions with a more favourable physical base are more likely to develop into core regions, while regions with less favourable physical conditions have less opportunity to encourage settlement and economic development.

Physical Variations within Canada

Geographers recognize that the physical nature of Canada varies in a number of ways. For example, climate varies from place to place. The Maritimes have a mild, wet climate, while the Arctic has a cold, dry climate. Climate also affects the shape of landforms (mountains, plateaus, and lowlands) through a variety of weathering and erosional processes. Major landforms illustrate the regional distinctiveness of Canada's physical geography. For instance, topography of the Canadian Prairies is totally different from that found in the Canadian Shield. The Prairies have a flat to gently rolling landscape, while the adjacent Canadian Shield consists of rugged, rocky, hilly terrain.

Geographers perceive an interaction between people and the physical world. This interactive two-way relationship is a fundamental component of regional geography. Favourable physical conditions can make a region more attractive for human settlement. The combination of a mild climate and fertile soils in the Great Lakes–St Lawrence Lowlands encourages agricultural settlement, while the St Lawrence River and the Great Lakes provide low-cost water transportation to local and world markets. The favourable physical features of this region have allowed it to become Canada's industrial heartland.

Vignette 2.1 Two Different Geographies

Canada and the United States occupy the northern and central parts of North America, yet the two countries have strikingly different geographies. Canada, while much larger in geographic area, has a much smaller area suitable for agriculture and settlement. Much of Canada lies in high latitudes where polar climates and permafrost place these lands far beyond the limits of commercial agriculture and settlement. Consequently, most Canadians live in a narrow zone close to the border with the United States (see Figure 4.1). Here, more temperate climates prevail. Geography, therefore, has been kinder to the United States, giving it more suitable physical space for settlement and allowing its population to reach 300 million in 2006, compared to 33 million for Canada. Geography is partly responsible for Canada's much smaller population. Canada, for instance, has a population density of 3 people per square kilometre compared to 29 in the United States. This physical reality, best described as Canada's northern handicap, limits the areas suitable for settlement. Recent immigrants to Canada have recognized this geographic fact and most take up residence in one of Canada's three largest cities.

As scientists who study the spatial aspects of nature and the processes that shape nature, physical geographers are concerned with all aspects of the physical world: **physiography** (landforms), bodies of water, climate, soils, and natural vegetation. Regional geographers, however, are more interested in how physical geography varies and subsequently influences human settlement of the land. The Rocky Mountains, for instance, offer few opportunities for agricultural settlement, but the spectacular scenery has led to the emergence of an economy based on tourism. Nature often works slowly and change may take centuries, but nature can work quickly. Floods and storms have had sudden and dramatic impacts on human settlements.

Regional geographers are also concerned about the effect of human activities on the natural environment. In most cases, humans have a negative impact on the environment. For example, within the Bow Valley of the Rocky Mountains, extensive land developments have reduced the size of the natural habitat of wild animals such as bears and elk. Ironically, if more land is converted into golf courses, resort facilities, and housing developments, the animals that make this wilderness region so unique and attractive to tourists may no longer be able to survive. Another example is urban sprawl, which has gobbled up some of Canada's best farm land in the Niagara Peninsula, the Fraser Valley, and the Okanagan Valley. In our contemporary world, therefore, humans are the most active and, some would say, the most dangerous agents of environmental change.

The discussion of physical geography in this chapter and in the six regional chapters is designed to provide basic information about the natural environment and its essential role in the regional geography of Canada. To that end, the following points are emphasized:

- While physical geography varies across Canada, it has distinct and unique regional patterns.
- Landforms are one aspect of this physical diversity.
- Climate, soils, and natural vegetation are another aspect and provide the basis for biodiversity.
- The impact of human activity is changing the natural environment and, in the case of air, soil, and water pollution, there are long-term negative implications for all life forms.

- Physical geography has a powerful impact on Canadians by making certain areas more attractive for settlement and urban/industrial development. This relationship between the natural environment and the human world forms the basis of the core/periphery model.

We begin our discussion of physical geography by examining the nature and origin of landforms.

The Nature of Landforms

The earth's surface features a variety of landforms: mountains, plateaus, and lowlands. These landforms are subject to change by various physical processes. Some processes create new landforms while others reduce them. The earth, then, is a dynamic planet, and its surface is actively shaped and reshaped over very long periods of time. For instance, the process known as **denudation** gradually wears down mountains by erosion and weathering. The Appalachian Uplands represents a prime example of the denudation of an ancient mountain chain. How did this happen? First, over millions of years, **weathering** broke down the solid rock of these ancient mountains into smaller particles. Second, **erosion** transported these smaller particles by means of air, ice, and water to lower locations where they were deposited. The result was a much subdued mountain chain from what once resembled the Rocky Mountains. Denudation and deposition, then, are constantly at work and, over long periods of time, dramatically reshape the earth's surface.

The earth's crust, which forms less than 0.01 per cent of the earth and is its thin solidified shell, consists of three types of rocks: igneous, sedimentary, and metamorphic. When the earth's crust cooled about 3.5 billion years ago, **igneous rocks** were formed from molten rock known as magma. Nearly 3 billion years later, **sedimentary rocks** were formed from particles derived from previously existing rock. Through denudation (weathering and erosion), rocks are broken down and transported by water, wind, or ice and then deposited in a lake or sea. At the bottom of a water body, these sediments form a soft substance or mud. In geological time, they harden into rocks. Hardening occurs because of the pressure exerted by the weight of additional layers of sediments and because of chemical action that cements the particles together. Since only sedimentary rocks are formed in layers (called **strata**), this feature is unique to this type of rock. **Metamorphic rocks** are distinguished from the other two types of rock by their origin: they are igneous or sedimentary rocks that have been transformed into metamorphic rocks by the tremendous pressures and high temperatures beneath the earth's surface. Metamorphic rocks are often produced when the earth's crust is subjected to folding and faulting. **Faulting** is a process that fractures the earth's crust, while **folding** bends and deforms the earth's crust.

The earth's crust is broken into at least 14 huge slabs or plates, each moving in response to the currents of molten material just below the crust. This motion, originally described as **continental drift**, is now known as plate tec-

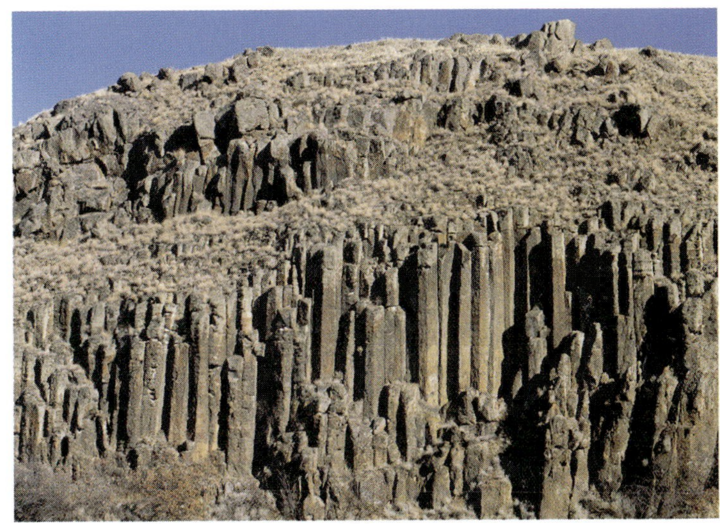

Basalt is a hard, black volcanic rock that, when cooled, can form various shapes, including tabular columns. These weathered basalt columns are found on Axel Heiberg Island, Nunavut. (David Nunuk/Science Photo Library)

tonics. It can result in the folding and faulting of the earth's crust.[2] For example, as these huge plates drift, they may collide with one another, thus causing earthquakes. Over sufficient geological time, these plates have compressed parts of the earth's crust into mountain chains.

Physiographic Regions

The earth's surface can be classified into a series of physiographic regions. A **physiographic region** is a large area of the earth's crust that has distinct characteristics. There are three key characteristics of a physiographic region:

- It extends over a large, contiguous area with similar relief features.
- Its landform has been shaped by a common set of geomorphic processes.
- It possesses a common geological structure and history.

Canada has seven physiographic regions (Figure 2.1). The Canadian Shield is by far the largest region, while the Great Lakes–St Lawrence Lowlands is the smallest. Perhaps the most spectacular and varied topography (i.e., the landforms of the earth's surface) occurs in the Cordillera, while the Hudson Bay Lowland has the most uniform relief. The remaining three regions are the Interior Plains, Arctic Lands, and Appalachian Uplands.

Each physiographic region has a different geological structure. These structural differ-

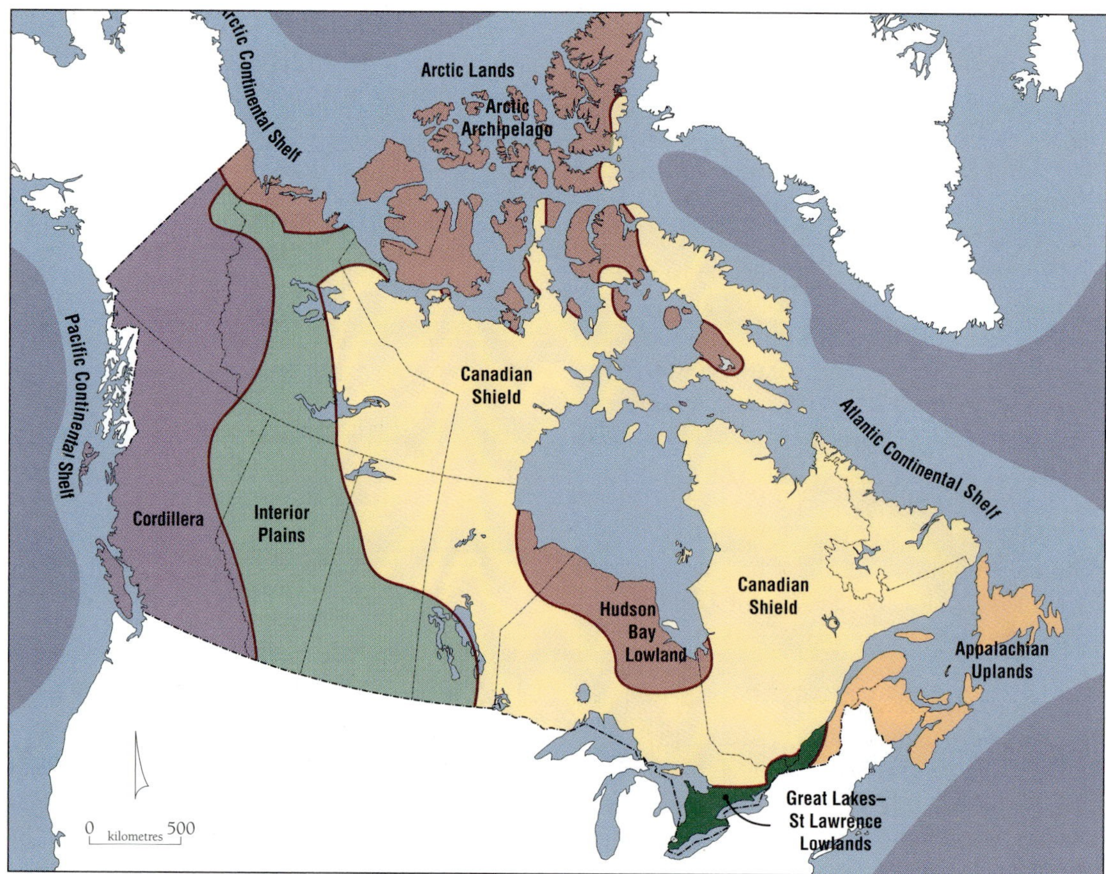

Figure 2.1 Physiographic regions and continental shelves in Canada. The Arctic Lands consist of a dozen large islands and numerous small islands that together are known as the Arctic Archipelago. The Canadian Shield is the largest physiographic region and extends beneath the Interior Plains, the Hudson Bay Lowland, and the Great Lakes–St Lawrence Lowlands. (Further resources: Student website, National Atlas section, Maps 2 and 3. Website instructions are found on p. xx.)

ences have produced a particular set of mineral resources in each physiographic region. For example, formed from the solidification of the earth's crust about 3.5 billion years ago, the Precambrian crystalline rock that makes up the Canadian Shield contains deposits of copper, diamonds, gold, nickel, iron, and uranium. The famous Sudbury nickel mines are in the Canadian Shield. Other physiographic regions were formed much later, as shown in the geological time chart (Table 2.1). The formation of the Interior Plains began about 500 million years ago when rivers deposited sediment in a shallow sea that existed in this area. Over a period of about 300 million years, more and more material was deposited into this inland sea, including massive amounts of vegetation and the remains of dinosaurs and other creatures. Eventually, these deposits were solidified into layers of sedimentary rocks 1 to 3 km thick. As a result, the Interior Plains have a sedimentary structure that contains oil and gas deposits. Such variation in the geological structure of each physiographic region has produced unique mineral resources for the human occupants of these regions to extract. Furthermore, as these regions developed their various resources, differences in regional economies began to take shape.

The Canadian Shield

As noted previously, the Canadian Shield is the largest physiographic region in Canada. It extends over nearly half of the country's land mass. The Canadian Shield forms the ancient geological core of North America. More than 3 billion years ago, molten rock solidified into the Canadian Shield (Table 2.1). Today, these ancient Precambrian rocks are not only exposed at the surface of the Shield but also underlie many of Canada's other physiographic regions.

During the last ice advance, the surfaces of the Canadian Shield and those of other physiographic regions were subjected to glacial erosion and deposition (Vignette 2.2). **Glacial erosion** and deposition are caused by giant ice sheets slowly grinding over the earth's surface. As the ice sheet moved over the surface of the Canadian Shield, the ice scraped, scoured, and scratched its massive rock surface. During the movement of ice sheets, a variety of loose materials such as sand, gravel, and boulders were trapped within the ice sheet. As the ice sheet reached its maximum extent, the edge of the ice sheet melted, depositing rocks, soil, and other debris. This debris is called **till**. Towards the end of the ice age, these ice sheets melted in situ, depositing whatever debris they contained. Sometimes the water from the melting ice was blocked from reaching the sea by the retreating ice sheet. These waters then formed temporary lakes. Once this ice was removed, these waters surged towards the sea.

Evidence of the impact of these processes on the surface of the Canadian Shield is widespread. Drumlins and eskers, both depositional

Table 2.1 Geological Time Chart

Geological Era	Geological Time (millions of years ago)	Physiographic Region(s) Formed
Precambrian	600 to 3,500	Canadian Shield
Paleozoic	250 to 600	Appalachian Uplands, Interior Plains, and Arctic Lands
Mesozoic	100 to 250	Interior Plains
Cenozoic	0 to 100	Cordillera
Quaternary		The Quaternary Period is divided into the Pleistocene (ice ages) and the Holocene (the present warm period).

Vignette 2.2 The Last Ice Age

During the earth's geological history, climatic cooling has produced a number of ice ages. The last ice age took place during the **Pleistocene Epoch**, which began nearly 2 million years ago (Table 2.1). Over the last 10,000 years, world temperatures increased and this geological time period is called the **Holocene Epoch**. Climatic cooling is not fully understood, but scientists have two theories. One is that climate cooling occurs when the distance between the earth and the sun is at its maximum every 21,000 years, due to variations in the earth's orbit. The other theory postulates that an increase in the amount of dust in the atmosphere from volcanoes known as supervolcanoes, the eruptions of which radically alter the amount of sunlight that reaches the earth, has a global cooling effect. Both theories offer explanations for a reduction in the amount of sunshine (solar energy) reaching the surface of the earth, which would cool the planet enough to trigger an **ice age**. While the average world temperature has increased over the last few decades—thus supporting the notion of global warming, which is attributed to the burning of fossil fuels—geologists believe that we are living in an interglacial period and that, within the next 100,000 years, the climate will again cool, resulting in another ice age.

At least 20 times during the last two million years (Pleistocene Epoch), huge sheets of ice, perhaps over 5 km thick, spread over Canada and the northern edge of the United States and then quickly retreated. The most recent glacial advance is called the late Wisconsin, reaching as far south as the state of Wisconsin (hence the name of this particular glacial advance). The late Wisconsin ice sheet reached its maximum extent 18,000 years ago.

The late Wisconsin ice advance consisted of two major ice sheets, the Laurentide and the Cordillera. The Laurentide ice sheet was centred in the Hudson Bay area. As its mass increased, the sheer weight of the ice sheet caused it to move, eventually covering much of Canada east of the Rocky Mountains. In the Cordillera, a series of alpine glaciers coalesced into the Cordillera ice sheet, which spread westward into the continental shelf off the Pacific coast and eastward, eventually merging with the Laurentide ice sheet.

Roughly 15,000 years ago, the climate began to warm, causing these ice sheets to retreat. Seven thousand years ago, the last main remnants of these ice sheets were in the Rocky Mountains and in the uplands of the Canadian Shield in northern Québec–Labrador and on Baffin Island. Today, the largest glaciers in Canada are in the mountains of Ellesmere Island. These ancient ice sheets and alpine glaciers have had a lasting impact on Canada's landforms.

Source: <en.wikipedia.org>.

landforms, are common to this region. **Drumlins** are long, low hills composed of till (material deposited and shaped by the movement of an ice sheet), while **eskers** are long, narrow mounds of sand and gravel deposited by melt water streams found under a glacier. There are also **glacial striations**, which are scratches in the rock surface caused by large rocks embedded in the slowly moving ice sheet.

The Canadian Shield consists mainly of a rugged, rolling upland. Shaped like an inverted saucer, the region's lowest elevations are along the shoreline of Hudson Bay, while its highest elevations occur in Labrador and Baffin Island,

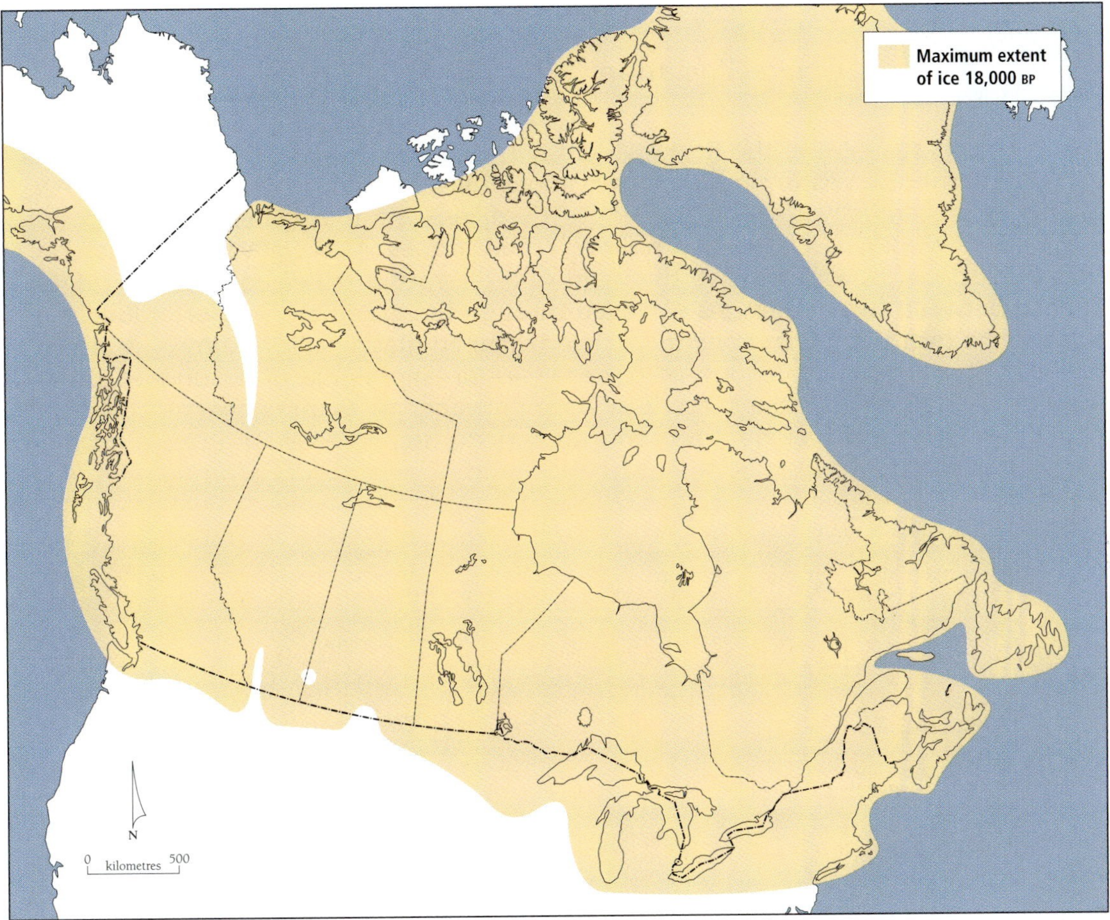

Figure 2.2 Maximum extent of ice, 18,000 BP. The last advance of the Wisconsin ice sheet covered almost all of Canada and extended into the northern part of the United States. Geologists believe that the present 'warm' climate is an interlude before the next ice advance. (Further resources: Student website, National Atlas section, Map 4. Website instructions are found on p. xx.)

where the most rugged and scenic landforms in the Canadian Shield are found. The Torngat Mountains, for instance, provide spectacular scenery with a coastline of fjords. These mountains reach elevations of 1,600 m, making them the highest land east of the Rocky Mountains. They also form the boundary between northern Québec and Labrador.

Another area of the Canadian Shield, known as the Laurentides, is located just north of Montréal. It has many lakes and hills that are the summer and winter playground for local residents as well as for tourists from Ontario and New England.

Other areas of the Canadian Shield are dotted by large communities that operate as single-industry mining towns, such as the iron-mining town of Labrador City in Newfoundland and Labrador, the nickel centre of Sudbury in Ontario, and the copper mining and smelter town of Flin Flon in Manitoba. Often located in remote places, resource towns such as these are vulnerable to closure if the mining operation ceases.

The Cordillera

The Cordillera, a complex region of mountains, plateaus, and valleys, occupies over 16 per cent of Canada's territory. With its north–south alignment, the Cordillera extends from southern British Columbia to Yukon. The

Vignette 2.3 Alpine Glaciation

While glaciers still exist in the Rocky Mountains, they are slowly melting and retreating. During the late Wisconsin ice advance about 18,000 years ago, these glaciers grew in size and eventually covered the entire Cordillera. At that time, alpine glaciers advanced down slopes, carving out hollows called **cirques**. As the glaciers increased in size, they spread downward into the main valley, creating **arêtes**, steep-sided ridges formed between two cirques. As it moved through the valley, the glacier eroded the sides of the river valleys, creating distinctive U-shaped glacial valleys known as **glacial troughs**. The Bow Valley is one of Canada's most famous glacial troughs. Cutting through the Rocky Mountains, the Bow Valley now serves as a major transportation corridor. It has also developed into an international tourist area. The centre of this tourist trade is the world-famous ski resort of Banff.

Located along the Continental Divide between British Columbia and Alberta, the Athabasca Glacier forms part of the massive Columbia Icefield. Known as the 'mother of rivers', the meltwaters from the Columbia Icefield nourish four river systems (the Saskatchewan, Columbia, Athabasca, and Fraser river systems) whose waters empty into three oceans—the Atlantic, Arctic, and Pacific oceans. (Barrett & MacKay Photography, Inc.)

Cordillera, classified as a young geological structure, was formed about 40 to 80 million years ago (Table 2.1) when the North American tectonic plate slowly moved westward, eventually colliding with the Pacific plate. The collision compressed sedimentary rocks into a series of mountains and plateaus now known as the Cordillera. These ancient sedimentary rock strata can be seen on exposed mountain sides in the Rockies. Along the Pacific coast, tectonic plate movement continues, making the coast of British Columbia vulnerable to both earthquakes and volcanic activity. With the vast majority of the population of this region clustered along the coast in the cities of Vancouver, Victoria, New Westminster, and Nanaimo, the potential damage and loss of life from a major earthquake (measuring 7.0 or greater on the Richter scale) might be the worst natural disaster to strike Canada. The strongest earthquake ever recorded in Canada shook the sparsely populated Queen Charlotte Islands in August 1949. This earthquake measured 8.1 on the Richter scale.

In more recent geological times, the Cordillera ice sheet altered the landforms of the region. Over the last 20,000 years, alpine glaciation has sharpened the features of the mountain ranges in the Cordillera and broadened its many river valleys (Vignette 2.3). The Rocky Mountains are the best known of these mountain ranges. Most have elevations between 3,000 and 4,000 m. Their sharp, jagged peaks create some of the most striking landscape in North America. The highest mountain in Canada—at nearly 6,000 m—is Mount Logan, part of the St Elias Mountain Range in southwest Yukon.

The Interior Plains

The Interior Plains region is a vast sedimentary plain that covers nearly 20 per cent of Canada's land mass. The Interior Plains are wedged between the Canadian Shield and the Cordillera, extending from the Canada–US border to the Arctic Ocean. Within the Interior Plains, most of the population lives in the southern area where a longer growing season permits grain farming.

Millions of years ago, a huge shallow inland sea occupied the Interior Plains. Over the course of time, sediments were deposited into this sea. Eventually, the sheer weight of these deposits produced sufficient heat and pressure to transform these sediments into sedimentary rocks. The oldest sedimentary rocks were formed during the Paleozoic era, about 500 million years ago (Table 2.1). Since then, other sedimentary deposits have settled on top of them, including those associated with the Mesozoic era when dinosaurs roamed the earth.

Tectonic forces have had little effect on the geology of this region. For that reason, the Interior Plains are described as a stable geological region. For example, sedimentary rocks formed millions of years ago remain as a series of flat rock layers within the earth's crust.

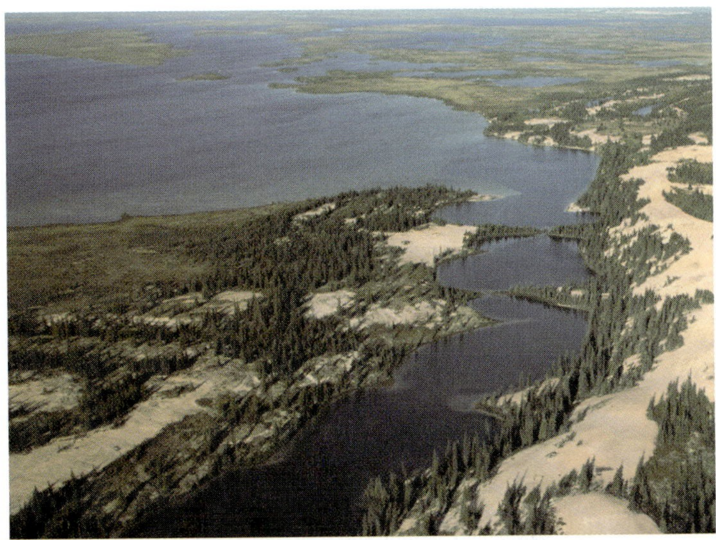

Eskers take the form of long, narrow mounds of sand and gravel that wind their way across glaciated landscapes. This esker is located in the Northwest Territories and was likely formed some 13,000 years ago when a meltwater stream flowed in an ice-walled tunnel inside the Laurentide Ice Sheet. Eskers are found in many parts of Canada. In Ontario, the Boulter esker extends approximately 160 km from near North Bay (Mattawa) to near Orillia (Washago). (© George D. Lepp/Corbis)

The stunning St Elias Mountain Range in Yukon Territory was subjected to intense erosion from alpine glaciers during the last ice age. Its sharp peaks and arêtes are a result of the powerful impact of mountain glaciers. (Al Harvey/The Slide Farm)

The Alberta Badlands were formed at the end of the last ice age when vast amounts of meltwater flowed through the Red Deer River and its tributaries. These quick-moving waters easily cut through soft sedimentary rocks to reach rock layers that date back to the late Cretaceous Period. The Dinosaur Trail that explores these badlands and the Royal Tyrrell Museum of Paleontology are located near Drumheller, Alberta. (© Paul A. Souders/Corbis)

Geologists have used such sedimentary structures as geological time charts. In Alberta and Saskatchewan, rivers have cut deeply into these soft rocks, exposing Mesozoic strata. The Alberta Badlands provide an example of this rough terrain. Archaeologists have discovered many dinosaur fossils within these Mesozoic rocks in southern Alberta and Saskatchewan.

Beneath the Interior Plains, valuable deposits of oil and gas are in sedimentary structures called **basins**. Known as fossil fuels, oil and gas deposits are the result of the capture of the sun's energy by plants and animals in earlier geologic time. The storage of this energy in the form of hydrocarbon compounds takes place in sedimentary basins. The Western Sedimentary Basin is the largest such basin. Most oil and gas production in Alberta comes from this basin. Fossil fuels are non-renewable resources, meaning that they cannot regenerate themselves. Renewable resources, such as trees, can reproduce themselves.

As the Laurentide ice sheet melted and began to retreat from the Interior Plains about 12,000 years ago, the surface of the region was covered with as much as 300 m of debris deposited by the ice sheet. Huge glacial lakes were formed in a few places. Later, the melt water from these lakes drained to the sea, leaving behind an exposed lakebed. Lake Agassiz, for example, was once the largest glacial lake in North America and covered much of Manitoba, northwestern Ontario, and eastern Saskatchewan—its lakebed is now flat and fertile land that provides some of the best farmland in Manitoba. When glacial waters escaped into the existing drainage system, they cut deeply into the glacial till and sedimentary rocks, creating huge river valleys known as **glacial spillways**. The geological history of the Interior Plains accounts for the great variety of landforms found within the area.

Just north of Edmonton, the Interior Plains slope towards the Arctic Ocean. The Athabasca River marks this northward course as it eventually enters the Mackenzie River. Edmonton lies on the banks of the North Saskatchewan River, which flows eastward into Lake Winnipeg. These waters then enter the Nelson River, which empties into Hudson Bay. Across this section of the Interior Plains, the land slopes towards Hudson Bay. Elevations decline from 1,200 m just east of the Rocky Mountains to about 200 m near Lake Winnipeg. These changes in elevation create three subregions within the Prairies: the Manitoba Lowland, the Saskatchewan Plain, and the Alberta Plateau. Typical elevations are 250 m in the Manitoba Lowland, 550 m in the Saskatchewan Plain, and 900 m in the Alberta Plateau. The Cypress Hills, however, provide a sharp contrast to the flat or rolling terrain of the Canadian Prairies (Vignette 2.4), as do the deeply incised river valleys of the Peace River country.

Vignette 2.4 Cypress Hills

The Cypress Hills, a subregion of the Interior Plains, consist of a rolling plateau-like upland that is deeply incised by fast-flowing streams. Situated in southern Alberta and Saskatchewan, this subregion is the highest point in Canada between the Rocky Mountains and Labrador. These hills are an erosion-produced remnant of an ancient higher-level plain formed in the Cenozoic era (see Table 2.1). With an elevation of over 1,400 m, these hills rise 600 m above the surrounding plain formed about 50 million years ago from materials borne eastward by rivers originating in the Rocky Mountains. During the maximum extent of the Laurentide ice sheet about 18,000 years ago, the higher parts of the Cypress Hills remained above the Laurentide ice sheet. Known as **nunataks**, these areas served as refuge for animals and plants. As the alpine glacier melted, streams flowing from the Rocky Mountains deposited a layer of gravel up to 100 m thick on these hills.

Today, the Cypress Hills area is a humid 'island' surrounded by a semi-arid environment and has an entirely different natural vegetation compared to the area surrounding it. Unlike the grasslands, the Cypress Hills have a mixed forest of lodgepole pine, white spruce, balsam poplar, and aspen. The Cypress Hills also contain many varieties of plants and animals found in the Rocky Mountains.

The Hudson Bay Lowland

The Hudson Bay Lowland comprises about 3.5 per cent of the area of Canada. It lies mainly in northern Ontario, though small portions stretch into Manitoba and Québec. This region extends from James Bay along the west coast of Hudson Bay to just north of the Churchill River. Much of the ground's surface consists of wet peatland known as **muskeg**. Low ridges of sand and gravel are interspersed between these extensive areas of muskeg. These ridges are the remnants of former beaches of the Tyrrell Sea. Because of its almost level surface, much of the land is poorly drained. Underneath the peatland are recently deposited marine sediments mixed with glacial till. With few resources to support human activities, the region has only a handful of tiny settlements. From this perspective, the Hudson Bay Lowland is one of the least favourably endowed physiographic regions of Canada. Moosonee (at the mouth of the Moose River in northern Ontario) and Churchill (at the mouth of the Churchill River in northern Manitoba) are the largest settlements in the region. Each has a population of just over 1,000 people. These two settlements, formerly fur-trading posts, are now the termini of two northern railways (the Ontario Northland Railway and the Hudson Bay Railway, respectively).

The Hudson Bay Lowland was formed by two events. First, a warmer climate appeared some 15,000 years ago, causing the Laurentide ice sheet to retreat. By 12,000 years ago, the ice-free coastal plain now known as the Hudson Bay Lowland was submerged by the waters from the Atlantic Ocean. Second, with the huge weight of the ice removed, the earth's crust began to rise, thus causing the coastal plain to emerge from beneath the waters. This inland extension of the Atlantic Ocean began some 12,000 years ago and has been named the Tyrrell Sea.

This inland saltwater sea reached its maximum extent about 7,000 years ago, extending over much of the lowlands surrounding Hudson and James bays. With the huge weight of the ice sheet removed, the earth's crust began to rise, forcing the Tyrrell Sea to retreat. This process is called **isostatic rebound** (Vignette 2.5). Slowly the isostatic rebound caused the seabed of the Tyrrell Sea to rise above sea level, thus exposing a low, poorly drained coastal plain (most of

Vignette 2.5 Isostatic Rebound

At its maximum extent about 18,000 years ago, the weight of the huge Laurentide ice sheet caused a depression in the earth's crust. When the ice sheet covering northern Canada melted, this enormous weight was removed, and the elastic nature of the earth's crust allowed it to return to its original shape. This process, known as isostatic rebound or uplift, follows a specific cycle. As the ice mass slowly diminishes, the isostatic recovery begins. This phase is called a **restrained rebound**. Once the ice mass is gone, the rate of uplift reaches a maximum. This phase is called a **postglacial uplift**. It is followed by a period of final adjustment called the **residual uplift**. Eventually the earth's crust reaches an equilibrium point and this isostatic process ceases. In the Canadian North, this process began about 12,000 years ago and has not yet completed its cycle.

which is called the Hudson Bay Lowland). This process of isostatic rebound has slowed over time from 600 cm per century to 100 cm per century. This relatively recent geomorphic process makes the Hudson Bay Lowland the youngest of the physiographic regions in Canada (Table 2.1).

Arctic Lands

The Arctic Lands region stretches over nearly 10 per cent of the area of Canada. Centred in the Canadian Arctic Archipelago, this region lies north of the Arctic Circle. It is a complex composite of coastal plains, plateaus, and mountains. The Arctic Platform, the Arctic Coastal Plain, and the Innuitian Mountain Complex are the three principal physiographic subregions. The Arctic Platform consists of a series of plateaus composed of sedimentary rocks. This subregion is in the western half of the Arctic Archipelago around Victoria Island. The Arctic Coastal Plain extends from the Yukon coast and the adjacent area of the Northwest Territories into the islands located in the western part of the Beaufort Sea. The third subregion, the Innuitian Mountain Complex, is located in the eastern half of the Arctic Archipelago. It is composed of ancient sedimentary rocks. Like the Rocky Mountains, its sedimentary rocks were folded and faulted. However, unlike the Rocky Mountains, the plateaus and mountains in the Innuitian subregion were formed in the early Paleozoic era (Table 2.1). At 2,616 m, Mount Barbeau on Ellesmere Island is the highest point in the Arctic Lands region.

Across these lands, the ground is permanently frozen to great depths, never thawing even in the short summer. This cold thermal condition is called **permafrost**. Physical weathering, consisting mainly of differential heating and frost action, shatters bedrock and produces various forms of patterned ground. **Patterned ground** consists of rocks arranged in polygonal forms by minute movements of the ground caused by repeated freezing and thawing. Patterned ground and **pingos** (ice-cored mounds or hills) give the Arctic Lands a unique landscape.

The climate in this region is cold and dry. In the mountainous zone of Ellesmere Island, glaciers are still active. That is, as these alpine glaciers advance from the land into the sea, the ice is 'calved' or broken from the glacier, forming icebergs. On the plains and plateaus, it is a polar desert environment. The term 'polar desert' describes barren areas of bare rock, shattered bedrock, and sterile gravel. Except for primitive plants known as lichens, no vegetation grows. Aside from frost action, there are no other geomorphic processes, such as water erosion, to disturb the patterned ground.

Most people live in the coastal plain in the western part of this physiographic region. The three largest settlements are situated at the mouth of the Mackenzie River. Inuvik has a population of almost 3,000, while Aklavik and Tuktoyaktuk are smaller communities.

The Appalachian Uplands

The Appalachian Uplands region represents only about 2 per cent of Canada's land mass. Sometimes known as Appalachia, this physiographic region consists of the northern section of the Appalachian Mountains (stretching south to the eastern United States), though few mountains are found in the Canadian section. With the exception of Prince Edward Island (Vignette 2.6), its terrain is a mosaic of rounded uplands and narrow river valleys. These weathered uplands are the remnants of ancient mountains that underwent a variety of weathering and erosional processes over a period of almost 500 million years. Together, weathering and erosion (including transportation of loose material by water, wind, and ice to lower elevations) has worn down these mountains, creating a much subdued mountain landscape. The highest elevations are in the Gaspé Peninsula. Here, Mount Jacques Cartier, at an elevation of 1,268 m, is found. The coastal area has been slightly submerged; consequently, ocean waters have invaded the lower valleys, creating bays or estuaries. The result is a number of excellent small harbours and a few large ones, such as Halifax harbour. The island of Newfoundland consists of a rocky upland with only pockets of soil found in valleys. Like the Maritimes, it has an indented coastline where small harbours abound. The nature of this physiographic region favoured early European settlement along the heavily indented coastline where there was easy access to the vast cod stocks. With the demise of the cod stocks, these tiny settlements are declining or, like Great Harbour Deep, have been abandoned (see Vignette 9.6).

The Great Lakes–St Lawrence Lowlands

The Great Lakes–St Lawrence Lowlands physiographic region is small but important. Extending from the St Lawrence River near

Tundra polygons are a type of patterned-ground found in Arctic Canada. Tundra polygons are formed by the repeated freezing of water in cracks in the ground and take centuries to form. (© Peter Dunwiddle/Visuals Unlimited)

Some 400 icebergs reach the coast of Newfoundland and Labrador each spring and summer. Almost all come from the Greenland Ice Sheet. Icebergs consist of a floating mass of freshwater ice and are a hazard to ships and offshore oil rigs. (Search4Stock Inc.)

Vignette 2.6 Prince Edward Island

Unlike other areas of Appalachia, Prince Edward Island has a flat to rolling landscape. While the island is underlain by sedimentary strata, these rocks consist of relatively soft, red-coloured sandstone that is quickly broken down by weathering and erosional processes. Occasionally, outcrops of this sandstone are exposed, but for the most part the surface is covered by reddish soil that contains a large amount of sand and clay. The heavy concentrations of iron oxides in the rock and soil give the island its distinctive reddish-brown hue. Prince Edward Island, unlike the other provinces in this region, has an abundance of arable land.

Québec City to Windsor, this narrow strip of land is wedged between the Appalachians, the Canadian Shield, and the Great Lakes.[3] Near the eastern end of Lake Ontario, the Canadian Shield extends across this region into the United States where it forms the Adirondack Mountains in New York state. Known as the Frontenac Axis, this part of the Canadian Shield divides the Great Lakes–St Lawrence Lowlands into two distinct subregions.

As the smallest physiographic region in Canada, the Great Lakes–St Lawrence Lowlands comprises less than 2 per cent of the area of Canada. As its name suggests, the landscape is flat to rolling. This topography reflects the underlying sedimentary strata and its thin cover of glacial deposits. In the Great Lakes subregion, flat sedimentary rocks are found just below the surface. This slightly tilted sedimentary rock, which consists of limestone, is exposed at the surface in southern Ontario, forming the Niagara Escarpment. A thin layer of glacial and lacustrine (i.e., lake) material, deposited after the melting of the Laurentide ice sheet in this area about 12,000 years ago, forms the surface, covering the sedimentary rocks.

In the St Lawrence subregion, the landscape was shaped by the Champlain Sea, which occupied this area for about 2,000 years. It retreated about 10,000 years ago and left marine materials, which are now broad terraces that slope gently towards the river (Vignette 2.7). The sandy to clay surface materials are a mixture of recently deposited sea, river, or glacial materials. For the most part, this subregion's soils are fertile, which, when combined with a long growing season, allow agricultural activities to flourish.

The physiographic region lies well south of 49° parallel, which forms the US–Canada

Vignette 2.7 Champlain Sea

About 12,000 years ago, vast quantities of glacial water from the melting ice sheets around the world drained into the world's oceans. Sea levels rose because of this additional water, causing the Atlantic Ocean to surge into the St Lawrence and Ottawa valleys, perhaps as far west as the edge of Lake Ontario. Known as the Champlain Sea, this body of water occupied the depressed land between Québec City and Cornwall and extended up the Ottawa River Valley to Pembroke. These lands had been depressed earlier by the weight of the Laurentide ice sheet. About 10,000 years ago, the earth's crust rebounded sufficiently to cause the Champlain Sea to retreat. However, the sea left behind marine deposits, which today form the basis of the fertile soils in the St Lawrence Lowlands.

border west of Ontario. The Great Lakes subregion extends from 42° N to 45° N, while the St Lawrence subregion lies somewhat further north, reaching towards 47° N. As a result of its southerly location, its proximity to the industrial heartland of the United States, and its favourable physical setting, the Great Lakes–St Lawrence region is home to Canada's main ecumene and manufacturing core.

The Impact of Physiography on Human Activity

Not only do physiographic regions provide a basic understanding of the physical shape and geological structure of Canada, they have also exerted a powerful influence over the geographic pattern of early settlers' land selection. In some instances, settlers were attracted to certain types of land while they avoided other types. Two examples illustrate this point. In the seventeenth century, the St Lawrence Lowlands was an attractive area for the establishment of a French colony because of its agricultural lands and its accessibility by water to France. Few settlers ventured beyond this favoured area. To the north was the rocky Canadian Shield, while to the south was the Appalachian Uplands. Neither of these surrounding physiographic regions offered attractive land for farming. Instead, Indians occupied these lands where they hunted game and trapped furs in order to barter with French traders for European goods.

Physical features can also create barriers to settlement. The Rocky Mountains were such a barrier in the nineteenth century. In 1867, when the Dominion of Canada was formed, the Rocky Mountains isolated the small British colony on the southern tip of Vancouver Island from the settled area of Canada. At that time, communications and trade with adjacent American settlements along the Pacific coast proved much easier and quicker than the overland route used by fur traders to reach Montréal. Until the completion of the CPR in 1885, the Rocky Mountains were such an imposing physical barrier that many residents of Vancouver Island favoured joining the United States rather than the Dominion of Canada.

In our contemporary world, such physical barriers are no longer the obstacles they once were. Technological advances in transportation and communications have greatly reduced the **friction of distance**, the term used to describe how interactions between two points decrease as the distance between them increases. The obstacles presented by physical barriers and distance have been greatly diminished.

Geographic Location

Canada's location in the northern half of North America is a critical climatic factor. For instance, Canada has a much cooler climate than the country to its south, the United States of America (Vignette 2.1). As well, Canada is bounded by the Arctic, Atlantic, and Pacific oceans, making it a marine nation (Vignette 2.8). Shipping routes are crucial for a trading country like Canada. Ocean shipping is not possible, however, in the Arctic Ocean because much of the Arctic Ocean is covered by a slow-moving permanent ice pack, ice floe, and ice fields. In the long winter, fast ice (**sea ice** that has frozen along coasts and extends out from land to the ice pack) completes the ice cover of the Arctic Ocean. Only small areas of open water, called **polynyas**, occur in the winter. In the late summer, fast ice disappears, leaving a narrow stretch of open water that makes coastal shipping possible. All this could change if predictions of global warming, resulting in more open water and perhaps even an ice-free Arctic Ocean, are accurate. Under these conditions, the Arctic Ocean would become the world's most important ocean route between Asia and Europe.

A measure of geographic location on the earth's surface is provided by latitude and longitude. Because the earth is a spherical body, this measure is given in degrees. By **latitude**, we mean the measure of distance north and south of the equator. For example, Ottawa is 45 degrees 24 minutes north of the equator. Degrees and minutes are expressed as ° and ', respectively. Since the distance between each degree of latitude is about 110 km, Ottawa is

Vignette 2.8 **Facts about Canada as a Maritime Nation**

- At 243,792 km, Canada has the longest coastline in the world, forming 25 per cent of the world's coastline.
- With an offshore economic zone stretching seaward some 200 nautical miles and comprising 3.7 million km^2, Canada has the largest offshore zone in the world.
- Canada, with two million lakes and rivers covering 755,000 km^2 or 7.6 per cent of the country's landmass, has the largest freshwater system in the world.
- From the Gulf of St Lawrence to Lake Superior, Canada's inland waterway extends over 3,700 km, making it the longest in the world.
- The Arctic Archipelago covers 1.4 million km^2, making it the largest archipelago in the world.
- Approximately 7 million Canadians live along its coastal literal.

Source: Adapted from Fisheries and Oceans Canada: <www.dfo-mpo.gc.ca/communic/facts-info/facts-info_e.html>.

about 5,000 km north of the equator. By **longitude**, we mean the distance east or west of the prime meridian. As the equator represents zero latitude, the prime meridian represents zero longitude. It is an imaginary line that runs from the North Pole to the South Pole and passes through the Royal Observatory at Greenwich, England. Canada lies entirely in the area of west longitude. Ottawa, for example, is 75°28' west of the prime meridian. Within Canada, latitude and longitude vary enormously. The variation in latitude has considerable implications for climate, which in turn affects the types of soils, natural vegetation, and wildlife found in each climatic zone. Examples of the range of latitudes and longitudes found in Canada are shown in Table 2.2.

Climate

Our physical world encompasses more than just landforms, physiographic regions, and geographic location. Climate, for instance, plays a key role in our physical world. **Climate** describes average weather conditions for a specific place or region over a very long period of time while weather refers to the current state of the atmosphere with a focus on weather conditions that affect people living in a particular place. In short, climate is what we can expect while weather is what we get. Extreme weather events—such as blizzards, droughts, and ice storms—are also part of climate and often have very powerful impacts on humans. Extreme weather events have brought people together to combat natural disasters, and, as indicated in Chapter 1, have contributed to their sense of belonging to a region.

Floods provide such an example. Often, they reoccur. As de Loë (2000: 357) explains, 'Floods are considered hazards only in cases where human beings occupy floodplains and shoreland.' What weather conditions provide conditions for flooding of such landforms? Often heavy rainfall combined with rapid snowmelt triggers catastrophic floods. An excellent example is found in the flat Manitoba Lowland where the normally benign Red River winds its way from North Dakota in the United States northward to Lake Winnipeg. Since 1950, residents of Winnipeg and other communities along the Red River have suffered through five spring floods, in 1950, 1979, 1996, 1997, 2001, and 2006. In 1950, the Red River flood drove over 100,000 people from their homes. Following that disaster, the Red River Floodway, a wide channel nearly 50 km long, was constructed. Its purpose was

Table 2.2 Latitude and Longitude of Selected Centres

Centre	Latitude	Longitude
Windsor, Ontario	42°18′ N	83° W
Alert, Northwest Territories	83°63′ N	60°05′ W
St John's, Newfoundland	47°34′ N	52°43′ W
Victoria, British Columbia	48°26′ N	123°20′ W
Whitehorse, Yukon	60°41′ N	135°08′ W

to divert the flood waters around the city of Winnipeg. However, small communities in the Red River Basin remained vulnerable to flooding. In 1997, the largest flood in the twentieth century occurred (Rasid et al., 2000). While the Red River Floodway saved Winnipeg, the towns of Emerson, Morris, Ste Agathe, and St Adolphe and the surrounding farm buildings and lands were less fortunate.

Since climate is relatively stable over a long period of time, it plays a key role in the formation of soils and natural vegetation. One outcome is the emergence of global patterns of soils and natural vegetation. Climatic conditions vary around the world and within Canada. In the Köppen climatic classification scheme for the world, for example, there are 25 climate types that reflect different temperature patterns and seasonal precipitation patterns. Seven of Köppen's climatic types are found in Canada and the equivalent Canadian climatic type is shown in Table 2.3.

Table 2.3 Climatic Types

Köppen Classification	Canadian Climatic Zone	General Characteristics
Marine West Coast	Pacific	Warm to cool summers; mild winters Precipitation throughout the year with a maximum in winter
Highland	Cordillera	Cooler temperature at similar latitudes because of higher elevations
Steppe	Prairies	Hot, dry summers and long cold winters Low annual precipitation
Humid continental	Great Lakes–St Lawrence Lowlands	Hot, humid summers and short, cold winters Moderate annual precipitation with little seasonal variation
Humid continental, cool	Atlantic Canada	Cool to warm, humid summers and short, cool winters
Subarctic	Subarctic	Short, cool summers and long, cold winters Low annual precipitation
Tundra	Arctic	Extremely cool and very short summers; long, cold winters Very low annual precipitation

Sources: Adapted from Robert W. Christopherson, *Geosystems: An Introduction to Physical Geography*, 3rd edn (Upper Saddle River, NJ: Prentice-Hall, 1998); F. Kenneth Hare and Morley K. Thomas, *Climate Canada* (Toronto: Wiley, 1974).

Climatic Controls

Three dominant climatic controls affect Canada's weather and climate, and these are related to the global atmospheric and oceanic circulation system:

- Variations in the amount of solar energy reaching different parts of the earth's surface correspond with latitude and temperature. That is, lower latitudes receive more solar energy and therefore have higher temperatures than higher latitudes.
- The global circulation of air masses causes a westerly flow of air across Canada, although invasions of air masses from the Arctic and the Gulf of Mexico can temporarily disrupt this general pattern of air circulation.
- Distance from oceans plays an important role in temperature and precipitation. That is, as distance from oceans increases, the annual temperature range increases and the annual amount of precipitation decreases.

Global Circulation System

Regional climates are controlled by the amount of solar energy absorbed by the earth and its atmosphere and then converted into heat. The amount of energy received at the earth's surface varies by latitude. In low latitudes around the equator, there is a net surplus of energy (and therefore high temperatures), but in high latitudes around the North and South poles, more energy is lost through re-radiation than is received, and therefore annual average temperatures are extremely low. Canada, where settlements extend from 42° N (Windsor) to 83° N (Alert), experiences great variation in the amount of solar energy received (and therefore great variation in temperatures).

The **global circulation system** redistributes this energy (i.e., energy transfers) from low latitudes to high latitudes through circulation in the atmosphere (system of winds and air masses) and the oceans (system of ocean currents). For example, the Japan Current warms the Pacific Ocean, bringing milder weather to British Columbia. On Canada's east coast, the opposite process occurs as the Labrador Current brings Arctic waters to Atlantic Canada. While Halifax, at 44°40' N, lies about 500 km closer to the equator than Victoria, at 48°26' N, Halifax's winter temperatures, on average, are much lower than those experienced in Victoria.

The atmospheric circulation system travels in a west-to-east direction in the higher latitudes of the northern hemisphere, causing air masses that develop over large water bodies to bring mild and moist weather to adjacent land masses. Such air masses are known as **marine air masses**. In this way, energy transfers ultimately determine regional patterns of global weather and climate (see Tables 2.3 and 2.4). Air masses originating over large land masses are known as **continental air masses**. These air masses are normally very dry and vary in temperature depending on the season. In the winter continental air masses are cold, while in the summer they are associated with hot weather.

Canada experiences warmer and moister weather in its lower latitudes and colder and drier conditions in its higher latitudes. However, Canada's coastal areas (particularly its Pacific coast) experience smaller ranges of seasonal temperatures and more annual precipitation than do inland or continental areas at the same latitude (Figures 2.3, 2.4, and 2.5). Winnipeg, for example, experiences a much greater daily and annual range in temperature than does Vancouver, even though both lie near 49° N. The principal reason is Vancouver's greater proximity to the ameliorating effects of the Pacific Ocean.

Air Masses

Air masses are large bodies of air with similar temperature and humidity characteristics. They form over large areas that have uniform surface features and relatively consistent temperatures. Such areas are known as source regions. The Pacific Ocean is a marine source region, while the interior of North America is a continental source region. During a period of about a week or so, an air mass may form over

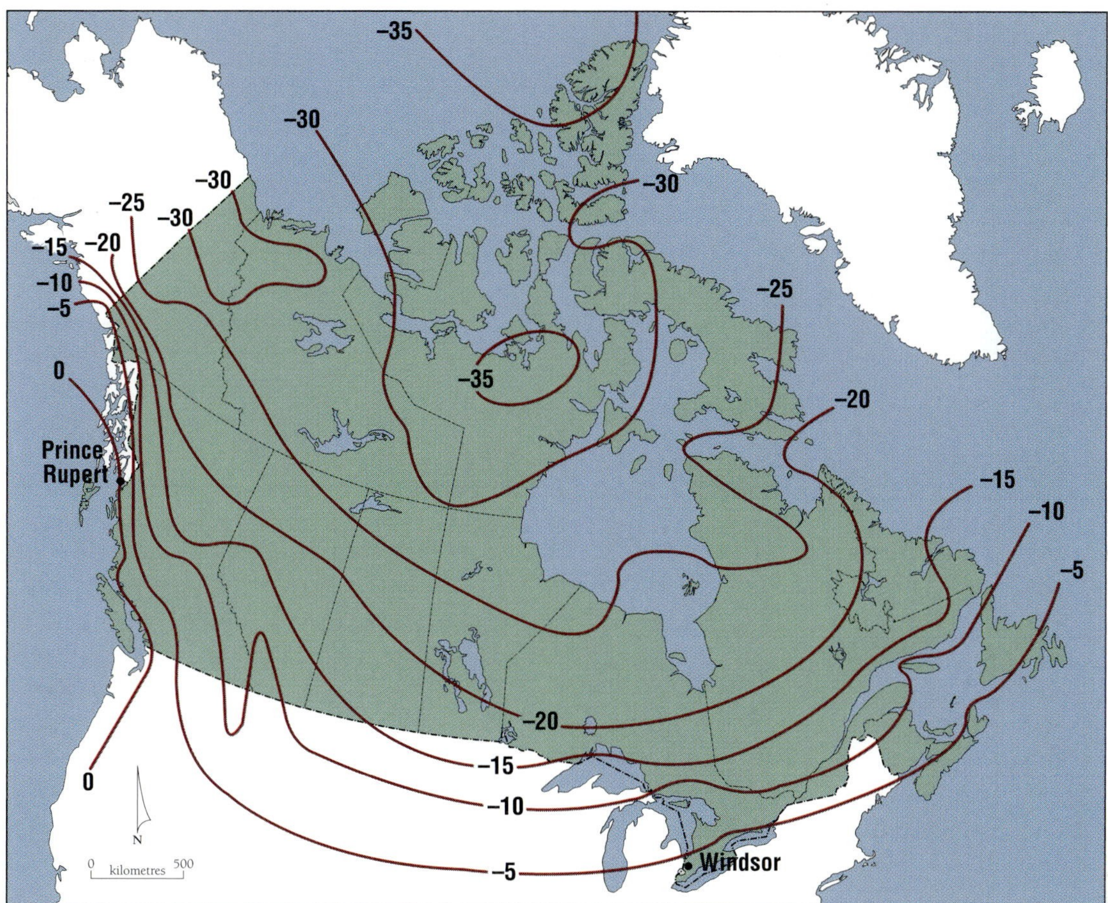

Figure 2.3 Seasonal temperatures in Celsius, January. The moderating influence of the Pacific Ocean and its warm air masses is readily apparent in the 0 to –5° C January isotherm. For example, Prince Rupert, located near 55° N, has a warmer January average temperature (0° C) than Windsor (–2° C), which is located near 42° N. (Further resources: Student website, National Atlas section, Map 5. Website instructions are found on p. xx.)

a source region, taking on the temperature and humidity characteristics of that source region. Canada's weather is affected by five air masses associated with the northern hemisphere. For example, Pacific air masses bring mild, wet weather to British Columbia's coast for most of the year. These air masses are much stronger in the winter, so British Columbia normally experiences greater precipitation in winter than in summer. In some years, British Columbia can have a relatively dry summer. The general characteristics of the five major air masses affecting Canada's weather are shown in Table 2.4.

Air masses bring moisture from oceans to land bodies. Across Canada, precipitation is unevenly distributed (Figure 2.5). The lowest average annual precipitation occurs in the Territorial North, indicating the dry nature of the Arctic air masses that originate over the ice-covered Arctic Ocean. The highest average annual precipitation takes place along the coast of British Columbia. Here, much falls as frontal and orographic rainfall due to two factors: (1) the warm Pacific Ocean serves as a source for eastward-moving Pacific air masses that often contain large quantities of water vapour, and (2) along the British Columbia coast, precipitation occurs either as the warm Pacific air mass rises over a colder one or as this same air mass must rise over the coastal mountain ranges. In both cases, the water vapour condenses and falls as rain or, at higher

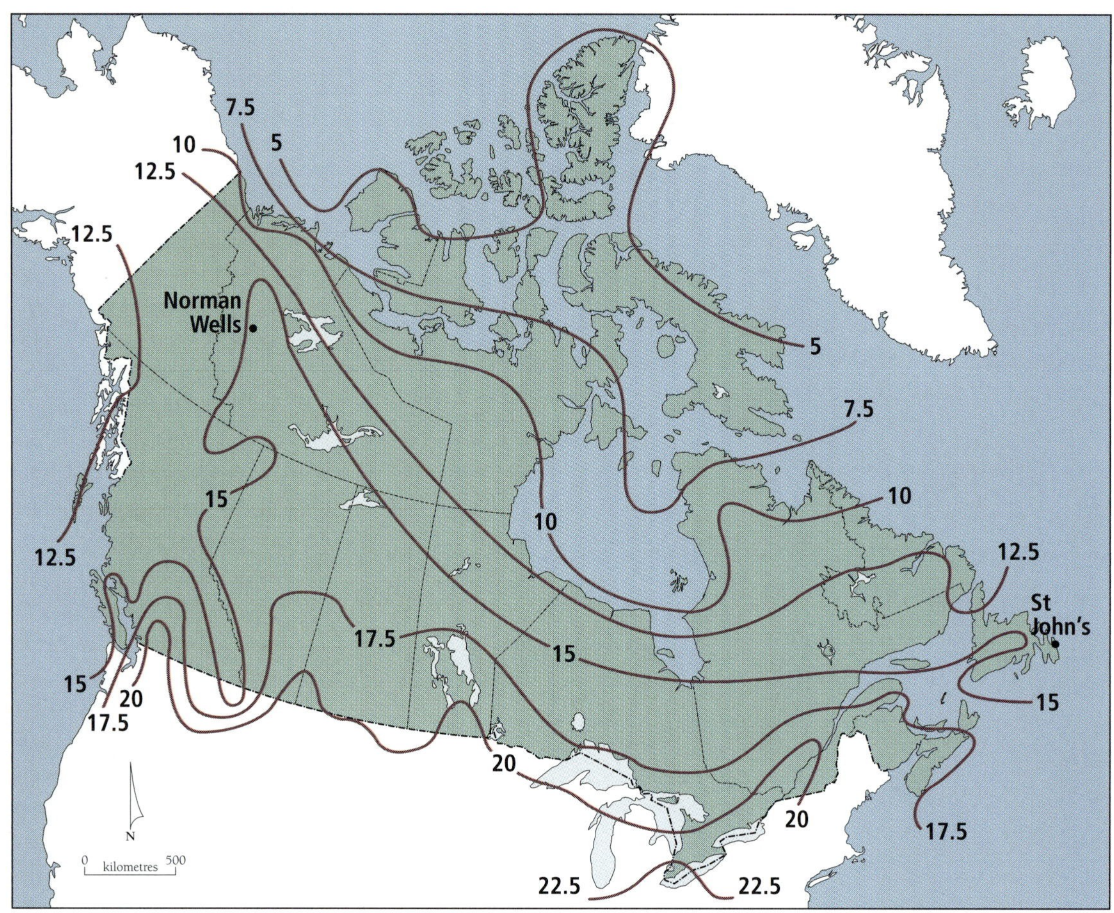

Figure 2.4 Seasonal temperatures in Celsius, July. The continental effect results in very warm summer temperatures that extend into high latitudes, as illustrated by the 15° C July isotherm. For example, Norman Wells, located near the Arctic Circle, has warmer July temperatures than St John's. (Further resources: Student website, National Atlas section, Map 5. Website instructions are found on p. xx.)

Table 2.4 Air Masses Affecting Canada

Air Mass	Type	Characteristics	Season
Pacific	Marine	Mild and wet	All
Atlantic	Marine	Cool and wet	All
Gulf of Mexico	Marine	Hot and wet	Summer
Southwest US	Continental	Hot and dry	Summer
Arctic	Continental	Cold and dry	Winter

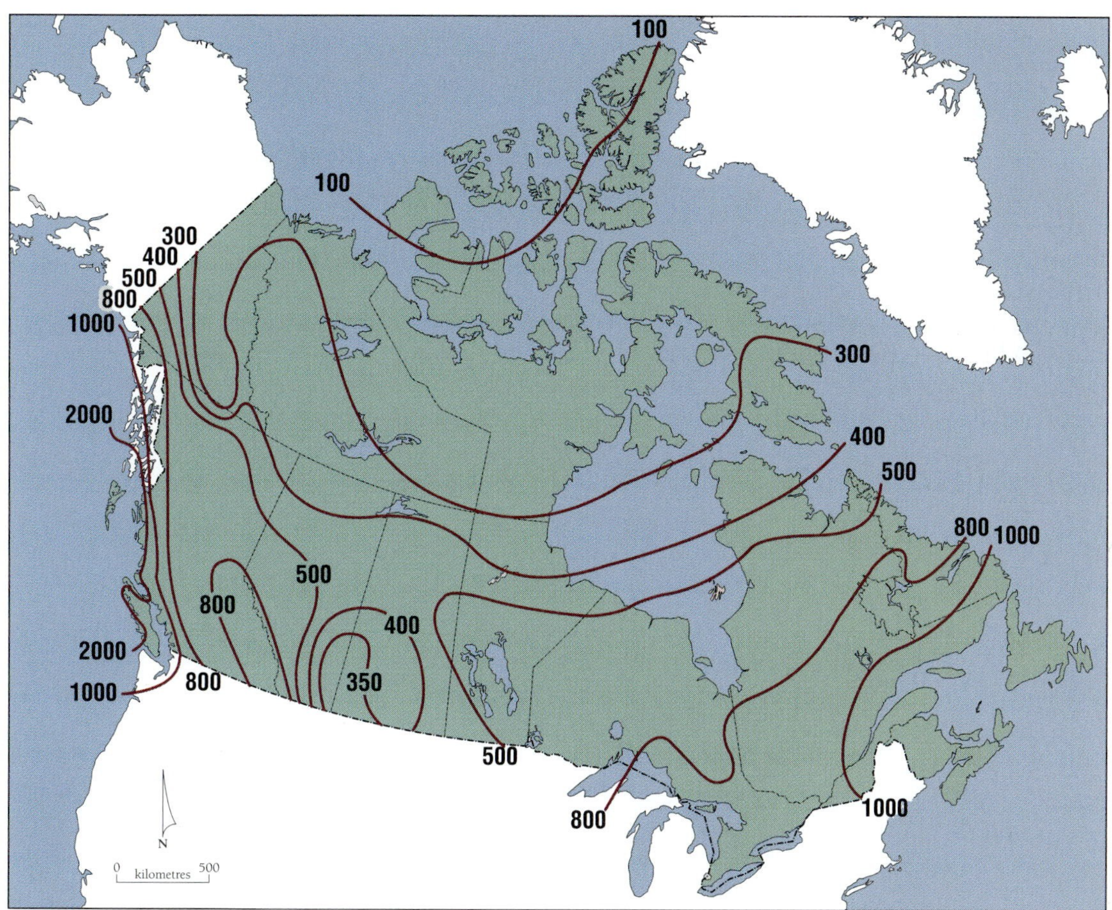

Figure 2.5 Annual precipitation in millimetres. The lowest average annual precipitation occurs in the Territorial North, indicating the dry nature of the Arctic air masses that originate over the ice-covered Arctic Ocean. The highest average annual precipitation occurs along the coast of British Columbia due to the moist marine air masses and the coastal mountains. (Further resources: Student website, National Atlas section, Map 6. Website instructions are found on p. xx.)

elevations, as snow. The three principal types of precipitation are discussed in Vignette 2.9.

Climate, Soils, and Natural Vegetation

As noted earlier, climate affects the development of soils and the growth of natural vegetation. In fact, the interdependency of climate, soils, and natural vegetation is so strong that physical geographers have identified an orderly and interrelated global pattern of climatic, soil, and natural vegetation zones. This relationship is revealed in the three maps indicating climatic, natural vegetation, and soil zones (Figures 2.6, 2.7, and 2.8). As shown in Table 2.5, climate determines to a large extent the **soil order** and native vegetation that exist in a given region and hence influences land use, such as crop cultivation, forestry, or grazing. Together with topography, climate determines the land's suitability for human settlement.

Climate has a direct impact on many economic activities. Long, cold winters cause people to use more energy to heat their homes; Canadians are among the highest consumers

Vignette 2.9 Types of Precipitation

As air masses rise, their temperature drops. This cooling process triggers condensation of water vapour contained in the air masses. With sufficient cooling, water droplets are formed. When these droplets reach a sufficient size, precipitation begins. Precipitation refers to rainfall, snow, and hail. There are three types of precipitation. **Convectional precipitation** results when moist air is forced to rise because the ground has become particularly warm. Often this form of precipitation is associated with thunderstorms. **Frontal precipitation** occurs when warm air masses are forced to rise over colder (and denser) air masses. **Orographic precipitation** results when air masses are forced to rise over high mountains. However, as those same air masses descend along the leeward slopes of those mountains (that is, the slopes that lie on the east side of the mountains), their temperature rises and precipitation is less likely to occur. This phenomenon is known as the **rain shadow effect**.

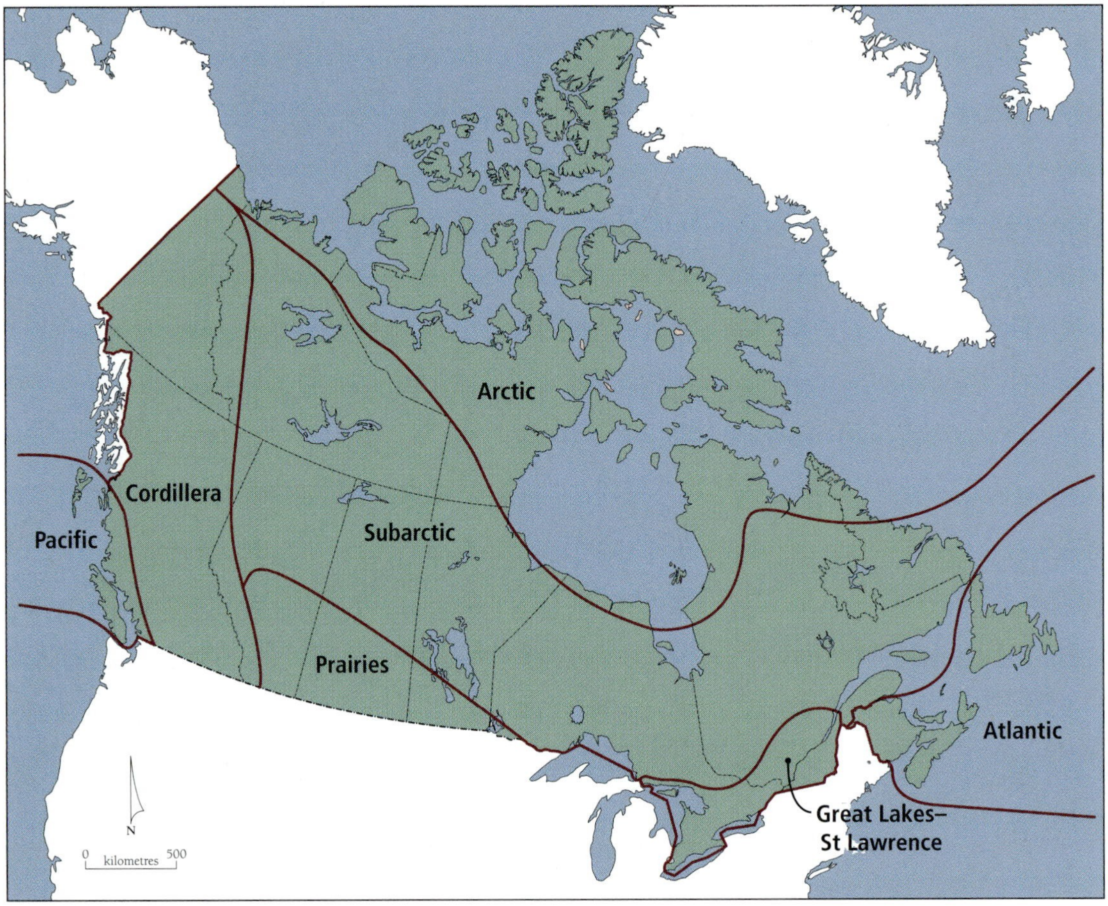

Figure 2.6 Climatic zones of Canada. Each climatic zone represents average climatic conditions in that area. Canada's most extensive climatic zone, the Subarctic, is associated with the boreal forest and podzolic soils. (Further resources: Student website, National Atlas section, Map 7. Website instructions are found on p. xx.)

Table 2.5 Canadian Climatic Zones

Canadian Climatic Zone	Natural Vegetation Type	Soil Order
Pacific	Coastal rain forest	Podzolic
Cordillera	Montane and boreal forests	Mountain complex
Prairies	Grassland and parkland	Chernozemic
Great Lakes–St Lawrence	Broadleaf and mixed forests	Luvisolic
Atlantic	Mixed and boreal forests	Podzolic
Subarctic	Boreal forest	Podzolic
Arctic	Tundra and polar desert	Cryosolic

Note: See Figures 2.6, 2.7, and 2.8. Also see Key Terms at end of chapter for definitions of soil orders.

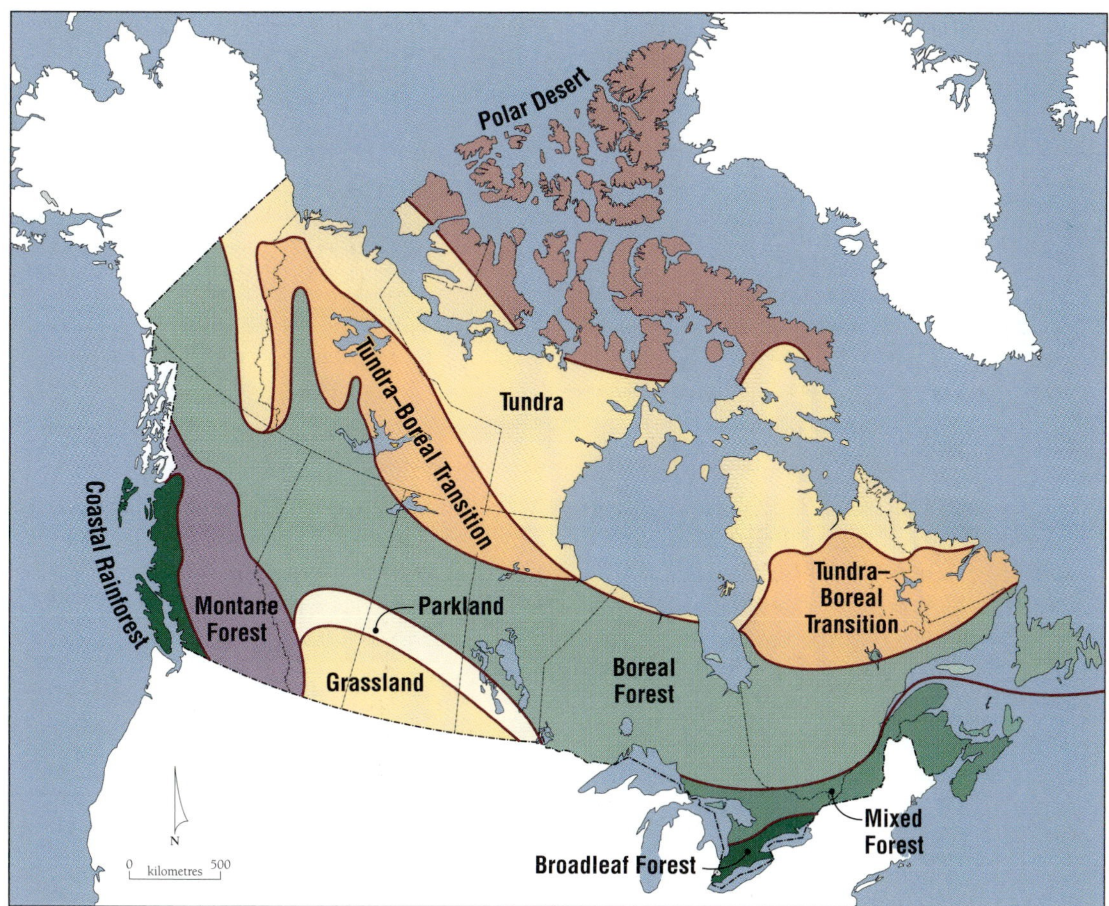

Figure 2.7 Natural vegetation zones. These natural vegetation zones have 'core' characteristics, which diminish towards their edges. Transitions exist between natural vegetation zones. Two major transition zones shown here are the Tundra–Boreal Transition and the Parkland. (Further resources: Student website, National Atlas section, Map 8. Website instructions are found on p. xx.)

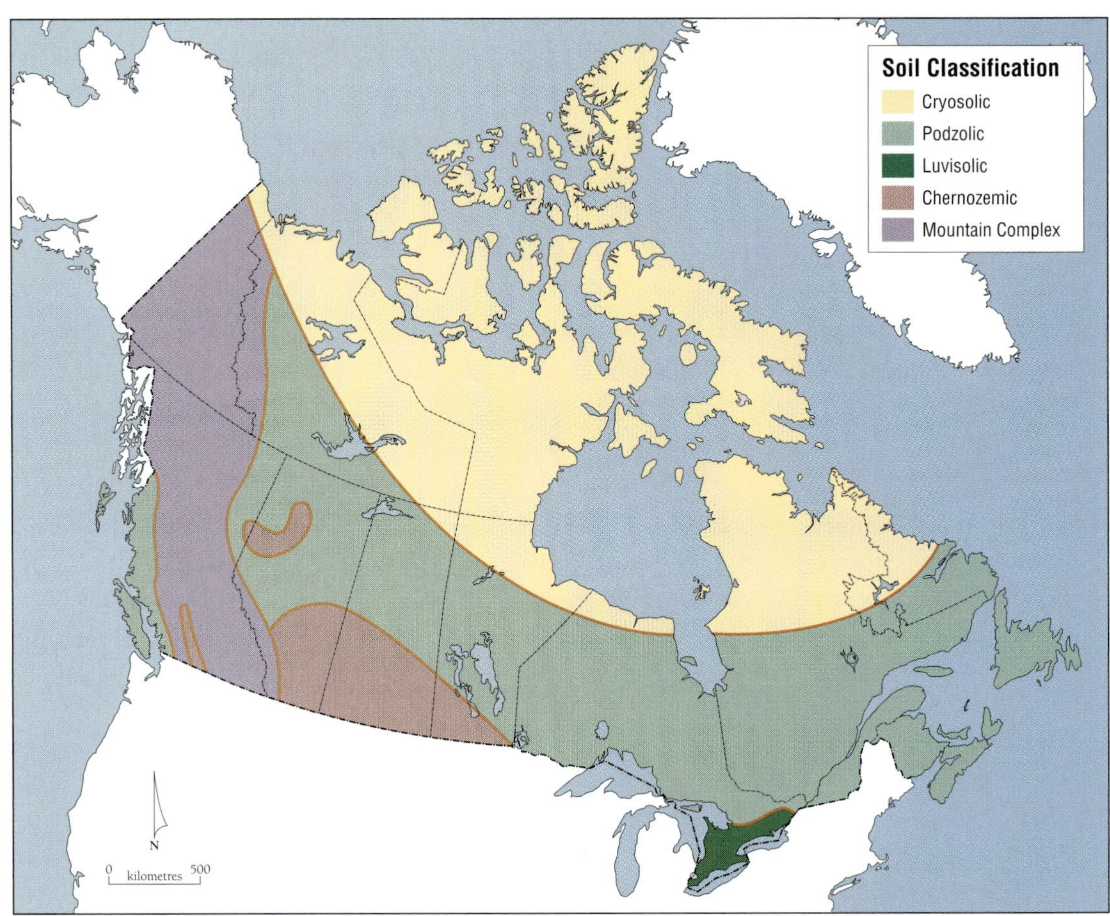

Figure 2.8 Soil zones. Most agricultural land is in luvisolic and chernozemic soil zones that together comprise about 5 per cent of Canada's land base. (Further resources: Student website, National Atlas section, Map 9. Website instructions are found on p. xx.)

of energy in the world. Variations in precipitation affect economic activities. Several years of below-normal precipitation often have a negative effect on agriculture, forestry, and hydroelectric production. For example, the 1988 drought in Canada cost the national economy approximately $1.8 billion in decreased agricultural and hydroelectric output, increased costs of fighting forest fires, and a loss of commercial timber and wildlife habitat (Shabbar et al., 1997: 3016). Certainly, the drought of 2001 and 2002 wreaked havoc on farmers and ranchers in Alberta and Saskatchewan. The lack of rainfall made pastures next to useless and grain farmers faced crop failure. The bountiful hay harvest in Ontario, Québec, and the Maritimes demonstrated the regional nature of this recent Prairie drought and, with the voluntary shipment of hay from these provinces to Alberta and Saskatchewan in the fall of 2002, indicated the cohesion of Canada's rural society.

Climatic Zones

The earth's atmosphere is a perpetually moving global system of air circulation that works to adjust the differences in pressure and temperature over different parts of the globe. This global circulation system, together with ocean bodies and major topographic features, affects Canada's weather. There is a climatic order within our complex and dynamic atmosphere. This order is expressed in several ways, including climatic zones. A **climatic zone** is an area of the earth's surface where similar weather conditions occur. Long-term data describing annual, seasonal, and daily temperatures and precipitation are used to define the extent of a climatic zone. Similar weather conditions occur in a particular area for complex reasons. For example, land near large water bodies usually receives more precipitation than land far from large water bodies. As well, coastal settlements have a small range of seasonal temperatures due to the sea's cooling effect in the summer and its warming effect in the winter. Land-locked places, however, do not benefit from the influence of the sea and have a much wider range of annual, seasonal, and daily temperatures.

Canada lies in the northern half of North America. This geographical location has several consequences for Canadians:

- Canada, located in the middle and high latitudes, receives much less solar energy than the continental United States and Mexico. It therefore has shorter summers and longer winters.
- Canada is noted for its long, cold winters, which affect Canadians in many ways. 'Coldness', wrote French and Slaymaker, 'is a pervasive Canadian characteristic, part of the nation's culture and history' (1993: i). They go on to state that winter's effects include not only low absolute temperatures but also exposure to wind chill, snow, ice, and permafrost.
- Continental climates are widespread in Canada's interior. These climates, but notably the Subarctic climate, are formed over large areas of the interior of Canada and are characterized by cold, dry winters and warm, dry summers. Except for the Pacific and Atlantic climates, Canadians live in areas with continental-type climates.
- Marine climates are limited in their geographic extent to the Pacific coast of British Columbia and to Atlantic Canada.

Canada has seven climatic zones (Figure 2.6): the Pacific, Cordillera, Prairies, Great Lakes–St Lawrence, Atlantic, Subarctic, and Arctic. The Subarctic zone is the largest climatic zone. It extends over much of the interior of Canada and is found in each geographic region. Though the Arctic climate exists along the Labrador coast and in the extreme northern reaches of Québec, the Subarctic climate prevails in the northern areas of Atlantic Canada, Québec, Ontario, and Western Canada, and it is present in northeast British Columbia. As well, the Subarctic climate is found in the Territorial North and is the principal climate in the Northwest Territories. The Subarctic climatic zone extends into much higher latitudes in northwest Canada than in northeast Canada because of warmer temperatures in the northwest. In northwest Canada, the average July temperature often reaches or exceeds 10° C, thus permitting the growth of trees. In similar latitudes of northeast Canada, summer temperatures are much lower. In the extreme north of Québec, for example, the average July temperature is below 10°C, thus resulting in tundra rather than a tree vegetation cover. The Subarctic climatic zone therefore has a southeast to northwest alignment (Figure 2.6). This alignment, somewhat modified in Québec, is caused by three factors:

- A continental effect means that land heats up more rapidly than water in the summer, resulting in higher summer temperatures in continental areas, such as Yukon and the Mackenzie Valley, compared to coastal areas, such as the coast of Hudson Bay and Baffin Island, at the same latitude. Also, oceans warm up more slowly than land because oceans reflect more solar energy and because solar energy is distributed throughout the water body.
- The snow cover in the western section of the Subarctic is much thinner than in the eastern half. Pacific air masses dominate the weather pattern in this area in the spring and bring warmer weather. A thinner snow cover and

warmer spring temperatures cause snow to disappear more quickly in the western Subarctic. Once the snow is gone, temperatures rise sharply.

- The Atlantic Ocean (including Hudson Bay) cools northern Québec and Labrador (thus breaking the pronounced southeast to northwest alignment of the Subarctic climatic zone). Part of that cooling effect is due to the Labrador Current, which brings Arctic waters to the middle latitudes of Atlantic Canada, and to the marine air masses that originate over the Atlantic Ocean. Combined with the deeper snow pack, this keeps spring temperatures low in the eastern subregion.

Each climatic zone has a particular natural vegetation type and soil (Table 2.5; Figures 2.7 and 2.8). The Subarctic climate, for instance, is associated with the boreal forest and podzolic soils (Table 2.5 and Vignette 2.10). The core characteristics of each of these climatic zones are presented in the appropriate regional chapter, e.g., the Pacific and Cordillera climatic zones are presented in the chapter on British Columbia.

Permafrost

A particularly distinctive feature of Canada's physical geography is permafrost. As noted earlier in this chapter, permafrost is permanently frozen ground with temperatures at or below zero for at least two years. The vast extent of permafrost in Canada provides a measure of the size of the country's cold environment (Figure 2.9). Permafrost exists in the Arctic and Subarctic climatic zones and occurs

Vignette 2.10 Subarctic Climate Type

The Subarctic climatic type has the greatest seasonal variation in temperatures of all the climatic types in Canada, with long, cold winters and short, warm summers. As is typical of continental climates, extremely cold winter temperatures occur. January minimum daily temperatures often drop to –40° C and sometimes even to –50° C. Winters are influenced by Arctic air masses and are therefore extremely dry. In the short summer period, daily temperatures often exceed 20° C and occasionally reach 30° C. During the summer, Pacific air masses usually dominate this weather pattern, providing most of the precipitation in the Subarctic zone. Under these air mass conditions, the annual temperature range is quite broad, perhaps reaching 80° C.

There are also important variations in annual precipitation within the Subarctic zone. In the western subzone of the Subarctic, annual precipitation is low—about 40 cm—due to the rain shadow effect of the Cordillera. In the eastern subzone, annual precipitation is much higher, sometimes exceeding 80 cm, most of which is provided by the Atlantic and Gulf of Mexico air masses.

The warm but short summers provide adequate growing conditions for coniferous trees. For example, average monthly summer temperatures exceed 10° C, thereby promoting tree growth. Black and white spruce are the most common species in the Canadian boreal forest. Birch and poplar also occur, especially along the southern edge of the boreal forest. Stands of Jack pine trees indicate an area that is recovering from a forest fire. Beneath this coniferous forest, there are podzolic soils. Wetlands are widespread: much of the land is poorly drained due to the disrupted drainage pattern caused by glaciation and permafrost. Wetlands contain numerous lakes, peat bogs, and marshes. Canadians often refer to this type of poorly drained land as muskeg.

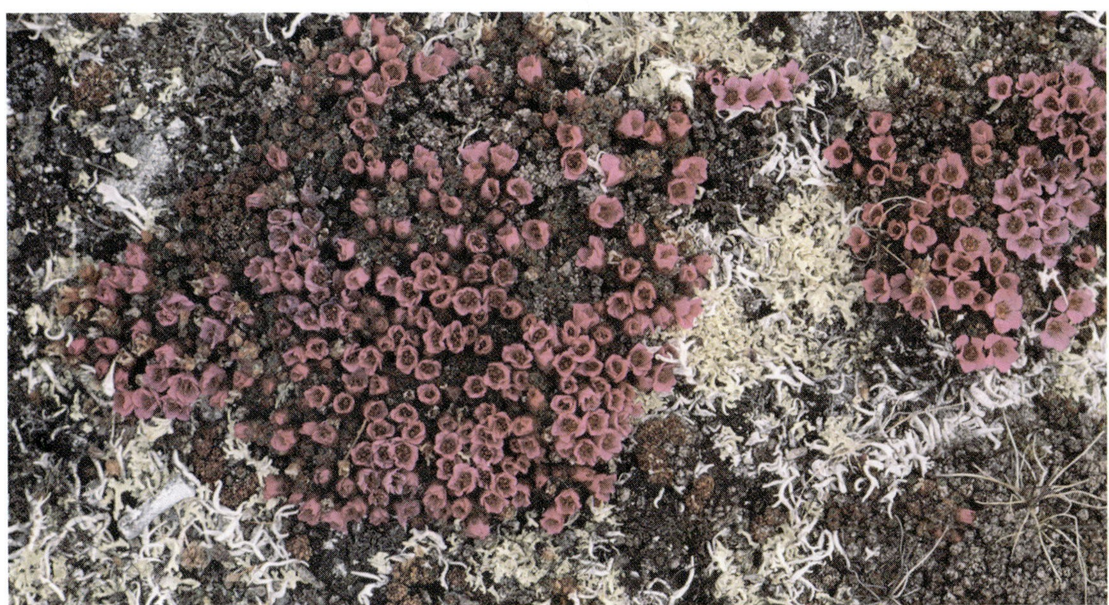

Tundra vegetation, such as lichens, mosses, and sedges, grow in sheltered areas within the Canadian Arctic in order to minimize exposure to arctic winds. Arctic plants have a short reproductive cycle that can be completed in the brief summer. This red flowering plant is purple saxifrage (*Saxifraga oppositifolia*). (Al Harvey/The Slide Farm)

at higher elevations in the Cordillera zone. Overall, permafrost is found in just over two-thirds of Canada's land mass.

Permafrost extends deeply into the ground. North of the Arctic Circle, permafrost may extend more than several hundred metres into the ground. Further south, permafrost is less frequent and where it occurs, it rarely penetrates more than 10 m into the ground. Permafrost is found in all six of Canada's geographic regions and reaches its most southerly position along 50° N in Ontario and Québec. Along the southern edge of permafrost, there is a transition zone where small pockets of frozen ground have a depth of less than 1 m. Further south, these pockets of permafrost disappear.

Permafrost is divided into four types. **Alpine permafrost** is found in mountainous areas and takes on a vertical pattern as elevations of a mountain increase. Over most of Canada, however, permafrost follows a zonal pattern, which does not correspond to latitude but rather to the annual mean temperatures that fall below zero.[4] The zonal pattern has a northwest to southeast alignment, that is, from Yukon to central Québec (see Figure 2.9).

As the mean annual temperature varies, the type of permafrost also changes. **Continuous permafrost** occurs in the higher latitudes of the Arctic climatic zone, where at least 80 per cent of the ground is permanently frozen, although it also extends into northern Québec. Continuous permafrost is associated with very low mean annual air temperatures of –15° C or less. **Discontinuous permafrost** occurs when 30 to 80 per cent of the ground is permanently frozen. It is found in the Subarctic climatic zone where mean annual air temperature ranges from –5° C in the south to –15° C in the north. **Sporadic permafrost** is found mainly in the northern parts of the provinces, where less than 30 per cent of the area is permanently frozen. Sporadic permafrost is associated with mean annual temperatures of zero to –5° C.

Major Drainage Basins

Facing three oceans, Canada is clearly a maritime nation (Vignette 2.8). Canada has four

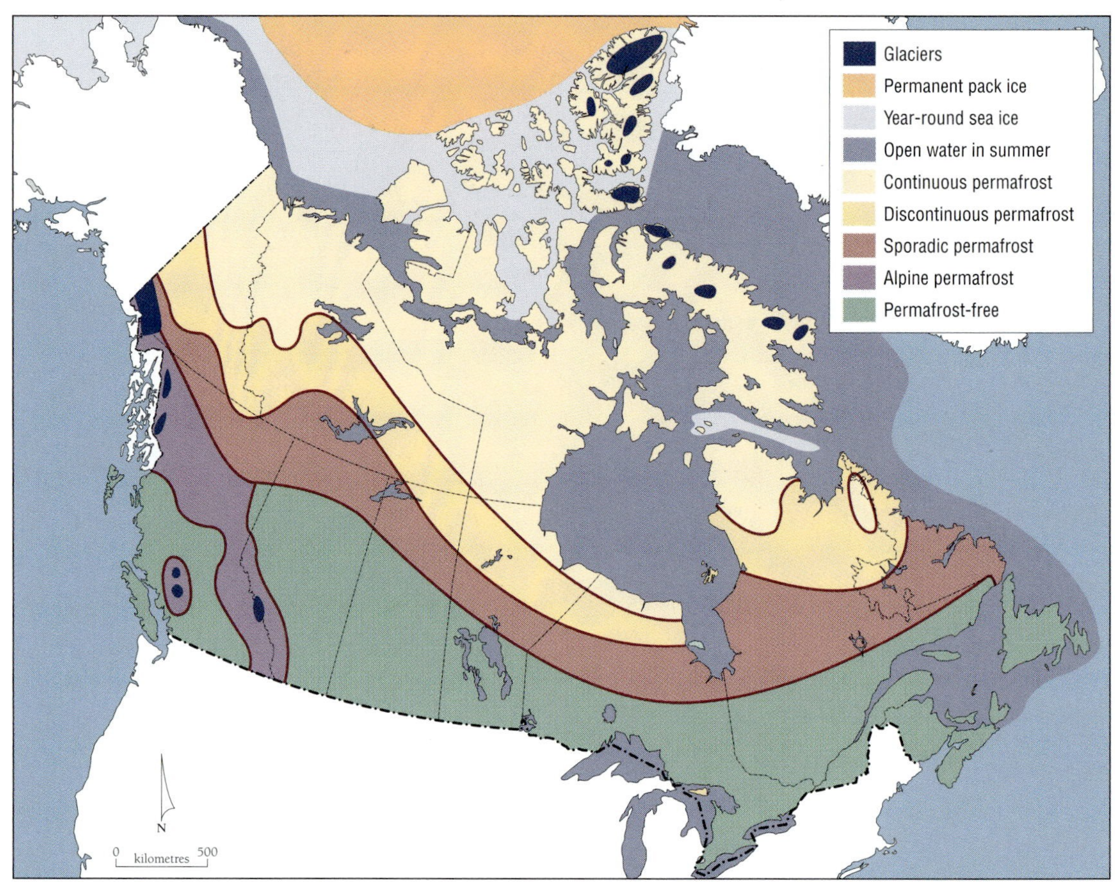

Figure 2.9 Permafrost zones. Canada's cold environment is demonstrated by the permanently frozen ground that extends over two-thirds of the country. Sea ice varies in thickness and duration. The most durable and thickest ice is found in the permanent ice pack. In the area of open water in summer, sea ice disappears first in the Great Lakes and in the offshore waters of Atlantic Canada, and last in the Arctic Ocean. In September 2007, satellite imagery indicated that the extent of open water in the Arctic Ocean was greater than in previous decades. (Further resources: Student website, National Atlas section, Map 10. Website instructions are found on p. xx.)

major drainage basins (Figure 2.10 and Table 2.6): the Atlantic Basin, the Hudson Bay Basin, the Arctic Basin, and the Pacific Basin. In addition, a small portion of southern Alberta and Saskatchewan forms part of the Mississippi River system, which drains into the Gulf of Mexico. A **drainage basin** is land that slopes towards the sea and is separated from other lands by topographic ridges. These ridges form drainage divides. The Continental Divide of the Rocky Mountains, for example, separates those streams flowing to the Pacific Ocean from those flowing to the Arctic and Atlantic oceans. While each stream has a drainage basin, those streams flowing to the same ocean combine their basins to form a major drainage basin.

Historically, the rivers in each basin have played major roles in the development of the country. For example, Aboriginal peoples and Europeans both used the St Lawrence and Mackenzie rivers as transportation routes during the fur trade. These rivers remain important waterways today.

Atlantic Basin

The Atlantic Basin is centred on the Great Lakes and the St Lawrence River and its tributaries, but the basin also includes Labrador.

Figure 2.10 Drainage basins of Canada. The Hudson Bay drainage basin is by far the largest of the five basins in Canada. It also serves as a boundary between southern Alberta and British Columbia, and between northern Québec and Labrador. (Further resources: Student website, National Atlas section, Map 11. Website instructions are found on p. xx.)

The Atlantic Basin has the third largest drainage area and also the third greatest stream-flow. The largest hydroelectric development in this drainage basin is located at Churchill Falls in Labrador. Development of the lower Churchill River has reached the discussion stage between the governments of Newfoundland/Labrador and Québec and three Aboriginal groups, the Labrador Inuit, the Innu First Nation, and the Labrador Métis. Earlier hydroelectric developments took place along the St Lawrence River in southern Québec and along its tributary rivers that flow out of the Laurentide Upland of the Canadian Shield. Rivers such as the Manicouagan River originate in the higher elevation of the Laurentide Upland. Here, the combination of abundant precipitation, natural lakes, and a sharp increase in elevation provides ideal natural conditions for the generation of hydroelectric power. Because there is a large market for electrical power in the St Lawrence Lowlands, virtually all potential sites in the Laurentides have been developed.

The Hudson Bay Basin

The Hudson Bay Basin is the largest drainage basin in Canada (Table 2.6). It occupies about 3.8 million km^2. In the West, its rivers originate in the Rocky Mountains and flow to Hudson Bay. In the East, the headwaters of its rivers in the uplands of northern Québec flow westward into James Bay. In northern Ontario and Manitoba, rivers drain into James and Hudson bays.

Niagara Falls, one of the most spectacular waterfalls in the world, lies along the border with the United States. On the Canadian side, the Horseshoe Falls are 54 metres high and 675 metres wide. Niagara Falls was formed some 10,000 years ago when an ice front, which separated lakes Erie and Ontario, melted. Water now flows from Lake Erie northward across the Niagara Escarpment into Lake Ontario. (© Charles Smith/Corbis)

The combination of large rivers and sudden drops in elevation that occur in the Canadian Shield makes this part of the basin ideal for developing hydroelectric power stations. In fact, most of Canada's hydroelectric power is generated in the Canadian Shield area of the Hudson Bay Basin—the largest installations are on La Grande Rivière in northern Québec and on the Nelson River in northern Manitoba. La Grande Rivière's hydroelectric developments are the first stage in the James Bay Project. The Great Whale River Project was to follow the completion of the hydroelectric projects on La Grande Rivière, but a variety of circumstances (low energy demand, low prices in New England, and strong opposition from environmental groups and the Cree Indians of northern Québec) stalled its development.

The Arctic Basin

The Arctic Basin is Canada's second-largest drainage basin. The Mackenzie River dominates the drainage system in this basin. Along with its major tributaries (the Athabasca, Liard, and Peace rivers), the Mackenzie River is the second-longest river in North America. However, because of low precipitation in the Arctic, this basin has only the fourth-largest stream-flow. There are few hydroelectric projects in the Arctic Basin because of the long distance to markets, with the exception of the hydroelectric development on the Peace River in British Columbia. Here, power from the Gordon M. Shrum generating facility is transmitted to the population centres in southern British Columbia and to the United States, pri-

Table 2.6 Canada's Drainage Basins

Drainage Basin	Area (million km^2)	Stream-flow (m^3/second)
Hudson Bay	3.8	30,594
Arctic	3.6	20,491
Atlantic	1.6	21,890
Pacific	1.0	24,951
Gulf of Mexico	<0.1	—
Total	10.0	105,135

Sources: A.H. Laycock, 'The Amount of Canadian Water and Its Distribution', in M.C. Healey and R.R. Wallace, eds, *Canadian Aquatic Resources* (Ottawa: Department of Fisheries and Oceans, 1987), 32; Philip Dearden and Bruce Mitchell, *Environmental Change and Challenge: A Canadian Perspective*, 2nd edn (Toronto: Oxford University Press, 2005), 124.

marily to the states of Washington, Oregon, and California.

The Pacific Basin

The Pacific Basin is the smallest basin. However, it has the second-highest volume of water draining into the sea. Heavy precipitation along the coastal mountains of British Columbia accounts for this unusually high streamflow. As a result, the Pacific Basin is the site of one of Canada's largest hydroelectric projects. Located at Kemano, this facility is owned and operated by Alcan, which uses the electrical generating station to supply power to its aluminum smelter at Kitimat. The ice-free, deepwater harbour at Kitimat and low-cost electric power generated at Kemano make Kitimat an ideal location for an aluminum smelter.

Environmental Challenges

Canadians once believed that their resources were infinite so that human activities could never harm the environment. While Canadians have traditionally exploited the country's natural wealth for their benefit, the current extent and pace of human impacts on the environment have reached and in some locations exceeded the danger point. Many serious environmental problems facing Canada today are the consequence of such human actions (Dearden and Mitchell, 2005: ch. 1) and geographers are calling for more protected areas and parks (Slocombe and Dearden, 2002: ch. 12). The rush to develop our resources has placed greater and greater demands on the natural environment. The discharge of toxic chemicals and raw sewage into our oceans, rivers, and lakes has fouled the waters, sometimes with disastrous consequences; car exhaust is a major source of air pollution in cities and a leading contributor of carbon dioxide to the atmosphere; the loss of forests has reduced the habitat for many plants and animals. Canada's air, land, and water are subject to pollution from a variety of sources. Air pollution, for instance, concerns geographers at three levels:

- *Global scale*. Air pollution contributes to global warming and, through the process of the global atmospheric system, adds minute toxic particles to the Canadian Arctic.
- *Regional scale*. Air pollution has damaged the forests and lakes of Ontario, Québec, and Atlantic Canada through acid rain.

- *Local scale*. Air pollution known as smog in Canada's major cities now poses a serious health hazard.

Air pollution is associated with industrial emissions and automobile exhaust. Alberta and Ontario account for two-thirds of industrial emissions in Canada (Table 2.7). Automobiles generate most other emissions shown in Table 2.7. In 2006, the minority Conservative government announced the Clear Air Act. Opposition parties proposed a number of amendments to strengthen the Act. Since the Conservative government objected strongly to many of these amendments the Act is unlikely to become law.

The environment that we want is slipping away as pollution of our air, land, and water continues. One sign of environmental degradation is the threat to Canada's biodiversity by a host of human-induced factors. Hydroelectric projects, often touted as a perfect means of harnessing rivers to produce a valuable product, can have serious environmental consequences for the local region. Sometimes, these consequences take the form of resource conflicts. The Pacific Basin provides such an example. This basin forms an integral part of the natural biological cycle for salmon (a renewable resource), which live most of their lives in the Pacific Ocean. When they reach maturity, however, they return to their spawning grounds in the headwaters of the various tributaries that flow into the major rivers of the Pacific Basin, including the Nechako River, which is an important tributary of the Fraser River. Alcan Canada's construction of a hydroelectric dam on this river flooded prime salmon spawning grounds and thus created a resource conflict.

Another sign, indeed one with much broader implications for the habitat in the Arctic, results from the dependency of our industrial world on obtaining energy from fossil

Table 2.7 Industrial Emissions by Provinces and Territories, 2004

Rank	Province/Territory	Million Tonnes CO_2	Percentage
1	Alberta	110.78	39.59
2	Ontario	78.39	28.02
3	Saskatchewan	22.87	8.17
4	Québec	22.10	7.90
5	New Brunswick	12.61	4.51
6	British Columbia	12.44	4.45
7	Nova Scotia	12.01	4.29
8	Newfoundland and Labrador	5.22	1.86
9	Manitoba	2.94	1.05
10	Northwest Territories	0.36	0.13
11	Prince Edward Island	0.10	0.04
12	Yukon	<0.10	<0.01
13	Nunavut	<0.10	<0.01
Total	Total Industrial Emissions	278.9	100.0
Total	All Emissions	758.0	

Source: www.PollutionWatch.org. PollutionWatch is a collaborative project of Environmental Defence and the Canadian Environmental Law Association.

fuels. The release of carbon dioxide into the atmosphere is the primary cause of the recent warming of the atmosphere. As more and more countries have become industrialized, most scientists believe that global temperatures have increased over the last 150 years due to the release of carbon dioxide and other greenhouse gases. This warming effect has been most noticeable over the last 30 years. In the 1990s, for instance, the retreat of glaciers in the Rocky Mountains was a clear sign of warmer temperatures. The Athabasca Glacier, which is part of the Columbia Icefield, has lost considerable ice mass. In the northern hemisphere, this phenomenon of glacier retreat is widespread (see Figure 2.11).

At the same time, average temperatures in the Arctic have increased with the result that the ice cover on the Arctic Ocean has thinned and the length of time that land-fast ice in Hudson Bay exists has decreased. Both these events have had a negative impact on the time available for seal hunting and on the size of the habitat for polar bears. A potential positive event would see the Northwest Passage become a commercial shipping route, although such an eventuality could have significant negative environmental impacts and would bring to the fore serious international issues surrounding Canadian claims to sovereignty in the High Arctic.

In the remainder of this chapter, two important results of industrial pollution—acid rain and global warming—are examined. In each regional chapter, a local environmental challenge is examined: for Ontario, the pollution of the Great Lakes; for Québec, toxic wastes in the St Lawrence; for British Columbia, clear-cut logging; for Western Canada, the threat to groundwater by wastes from large-scale hog barns; for Atlantic Canada, the toxic wastes at Cape Breton; and for the Territorial North, the conflict between placer gold mining and fish habitat.

Glacial retreat of the Athabasca Glacier, 1992 to 2005. (Copyright Hugh Saxby)

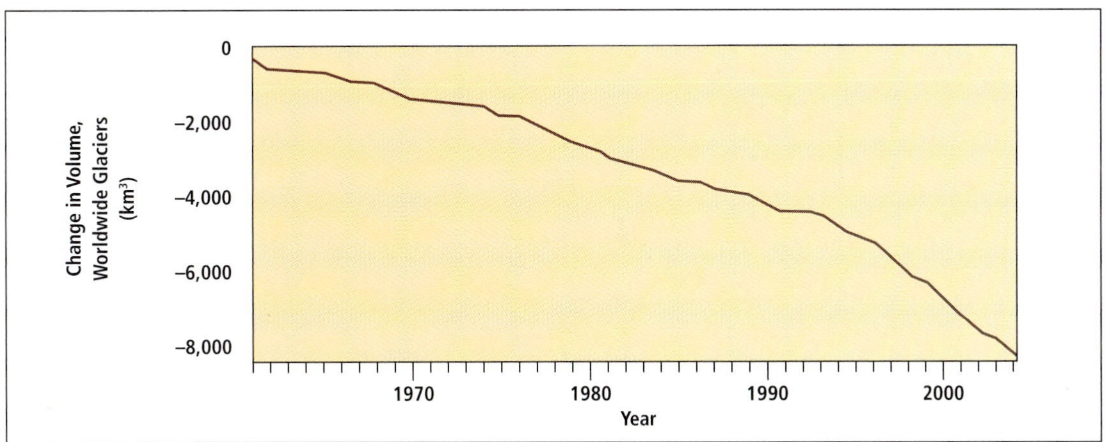

Figure 2.11 Worldwide change in volume (km^3) of glaciers, 1960–2004. (NASA, Earth Today. At: www.nasa.gov/vision/earth/features/index.html)

Air Pollution

Air pollution is the number-one problem facing urban residents of major cities. The source of air pollution (smog) comes from emissions from automobiles and coal-burning plants. Smog, a mixture of smoke, sulphur dioxide, and other contaminates, takes the form of a brownish haze over cities. Most contaminates are produced by automobiles (carbon dioxide) and thermal coal plants (sulphur dioxide). Vehicles account for over half of urban air pollution. Coal-burning plants are the second most serious source of urban air pollution. Densely populated areas like southern Ontario contain millions of automobiles and trucks. Furthermore, southern Ontario is an energy-deficient area. For years, coal thermal plants have produced much of their electricity at a relatively low cost. Now that urban air pollution has become a health problem, Ontario has plans to limit sulphur dioxide emissions by closings its coal-burning plants, replacing its electrical production with natural gas and nuclear power and by importing electricity from Québec, Michigan, New York, and Manitoba.

Acid Rain

Acid rain—precipitation that has an unusually acidic chemical composition—affects the forests and lakes of Ontario, Québec, and Atlantic Canada. Acidity levels are described in terms of the pH factor, which measures the acidity or alkalinity of a substance on a scale of 0 to 14, with 0 being extremely acidic and 14 being extremely alkaline. The average pH of normal rainfall is 5.6, making it slightly acidic. Acid rain, however, has a pH of 2.4 or less. At this pH level, acid rain is a serious environmental problem in many parts of the world, including eastern Canada.

Most acid rain results from the chemicals derived from the burning of fossil fuels in industrial plants and from automobile exhaust. Sulphur dioxide and nitrogen oxides are emitted in the smoke from coal-burning plants, while automobile engines are particularly prolific producers of nitrogen oxides. Acid rain is created when oxides of sulphur and nitrogen change chemically as they dissolve in water vapour in the atmosphere and then return to earth as tiny droplets with high concentrations of sulphuric acid and nitric acid.

The most visible impact of acid rain is in urban areas where it slowly corrodes limestone buildings and defaces marble sculptures. Less visible but just as destructive are the effects of acid rain on forests, lakes, and soils. In eastern Canada, acid rain has damaged many forests and caused a sharp decline in fish stocks in many lakes. The problem of acid rain is compounded by the fact that pollutants emitted from coal-burning electrical power plants in the Great Lakes area of the United States fall as acid rain in Ontario and Québec. Canada and the United States have turned their attention to this problem. In 1991, Canada and the United States signed a bilateral Air Quality Agreement, and a few years later, New England governors and eastern Canadian premiers agreed on an Acid Rain Action Plan to co-ordinate efforts to reduce acid rain. Progress has been made by reducing sulphur emissions at coal-burning power plants, substituting natural gas for coal at power plants, and introducing 'scrubbing' in existing coal-burning thermal facilities.

Global Warming

Is global warming underway? Most scientists think so. Their argument is based on anthropogenic-caused warming in combination with the greenhouse effect. As more carbon dioxide and other greenhouse gases are added to the atmosphere by the burning of fossil fuels, the **greenhouse effect** causes air temperatures to rise. Certainly the mean world air temperature has increased over the last 250 years, but is it due entirely to the burning of fossil fuels, or is something else, such as a natural solar cycle, a significant factor as well? For instance, the explanation for the world's temperature rise during the Medieval Warming Period, which was followed by a temperature decline during the Little Ice Age, is unknown (Vignette 2.11).

A prominent scientific body, the Intergovernmental Panel on Climatic Change (IPCC), has played a lead role in assessing global climate change, and the evidence amassed by this group is significant. The IPCC, established in 1988 by the World Meteorological Organization and the United Nations Environment Program, has produced a comprehensive assessment on climate change every five years beginning in 1991. In 2001, the authors of the Third Assessment Report of the IPCC concluded that new and stronger evidence indicates that most of the warming observed over the last 50 years is attributable to human activities. They also announced that the global average surface temperature has increased by 0.6° C since the late nineteenth century and that the 1990s were the warmest decade, with 1998 being the warmest year since 1861 (Houghton et al., 2001: 26). The IPCC also produced a computer-simulated graph of global temperature variations over the last 1,000 years. This graph gained notoriety as the 'hockey stick' with the long shaft representing most of the first 900 years, when annual temperature variation was slight, and the short blade representing the last 100 years, when temperatures increased rapidly. The hockey stick graph purports to show that the northern hemisphere (and probably the world) is now experiencing the warmest climate in a millennium, and that the earth, after remaining cool for centuries, suddenly began to heat up about 100 years ago—just at the time that the burning of coal and oil led to an increase in atmospheric levels of greenhouse gases, notably carbon dioxide. Some scientists have challenged the hockey stick graph (see Figure 2.12).

The same research group predicts rapid increases in world temperatures over the next 100 years, with increases in the range of 1.7° C to 4.2° C (ibid., 70). However, temperature increases have so far fallen within the so-called normal temperature variation associated with the earth's climate, that is, within plus or minus 1 per cent of the world mean air surface temperature. Until world temperatures exceed this limit, global warming has not 'officially' begun.

While the physics of global warming in the greenhouse model are elementary, the actual process of climate change is extremely complex and remains unclear. Global warming seems inevitable, but other events might diminish the amount of solar energy reaching the earth's surface and thereby reverse the temperature trend. For example, volcanic eruption, by releasing huge amounts of dust into the air, could reflect significant amounts of incoming solar energy to outer space and thereby chill the world's climate. But can we count on such natural events occurring to reverse the current warming trend? Probably not, and so efforts—such as more efficient automobile engines—are underway to reduce the release of carbon dioxide into the atmosphere. Then, too, there are other efforts such as the Weyburn Project, which involves injecting carbon dioxide into Saskatchewan's Weyburn oil field. The injection of gas accomplishes two goals. First, a liquefied version of carbon dioxide is stored underground, thereby reducing the release of this gas into the atmosphere. The origin of the gas is an oil refinery in the United States. By means of a 320-km pipeline, the liquefied carbon dioxide is transported to the Weyburn oil field. Second, the injection of carbon dioxide greatly increases the recovery of oil. Such tech-

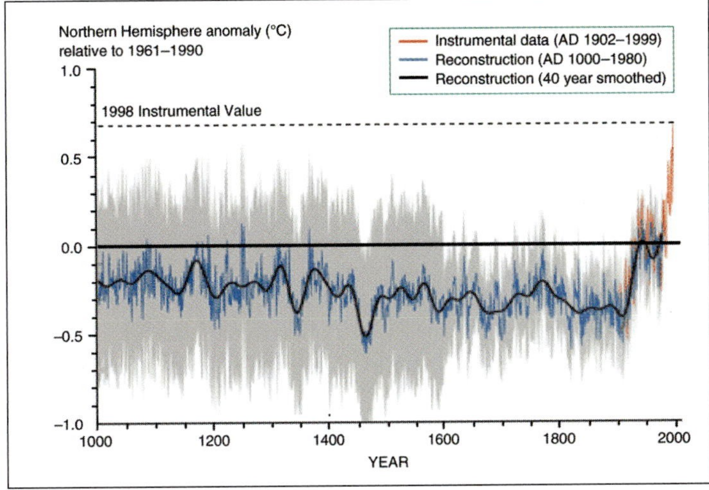

Figure 2.12 The hockey stick graph. The hockey stick estimates millennial northern hemisphere (NH) temperature reconstruction (blue: tree rings, corals, ice cores, historical records) and instrumental data (red) from AD 1000 to 1999. A smoother version (black) and two standard error limits (grey) are shown.
Source: Climate Change 2001: The Scientific Basis, Intergovernmental Panel on Climate Change.

nology, though costly, could be used by the companies producing heavy oil from the Alberta tar sands.

If global warming is for real, temperatures will rise significantly, especially in the Arctic. Such temperatures would alter Canada's environment (Bouchard, 2001). If world temperatures rose sufficiently, Canada's climatic zones, followed by its natural vegetation zones, would shift northward. Under this scenario, the geographic extent of the Arctic would be greatly reduced because the greatest increases in annual temperatures would take place in the Arctic due to the loss of snow and ice cover (the so-called **albedo effect**). With a reduced period of snow cover, solar energy would be able to warm the ground more effectively and thus cause temperatures in the atmosphere to increase. Similarly, as the ice cover on the Arctic Ocean disappears, more solar energy would be absorbed by the open water.

The economic consequences of global warming for Canada would be far-reaching. Some consequences would be favourable; many others would not. Global warming would affect the Arctic and Surbarctic more than the temperate areas of Canada. For example, the ice-free shipping season in Hudson Bay and the Arctic Ocean would be greatly extended but, with a rising sea level, at a cost of flooding coastal towns and cities. Then, too, global warming could thaw the permanently frozen ground known as permafrost. Melting of the ice in permafrost could cause massive ground **subsidence**, resulting in an irregular relief referred to by physical geographers as 'thermokarst topography'. This would disrupt transportation and pipeline systems, and play havoc with foundations for buildings, bridges, and other human-made structures (Bone et al., 1997: 265–74). In southern Canada, the impacts would see agricultural activities take place further north, but, on the negative side, grain agriculture in the Canadian Prairies might be subject to greater risk of drought. Water transportation would greatly benefit from longer navigation seasons. The Great Lakes and the St Lawrence River, for instance, would be navigable year-round.

Unlike past climatic changes (Vignette 2.11), global warming is caused by human actions. Hence, we have the power to prevent it. Although no one can predict the precise impacts of global warming on Canada, higher temperature would translate into longer, hotter summers and shorter, milder winters.[5] Changes to Canada's precipitation patterns are more difficult to predict, but with more open water in Hudson Bay and the Arctic Ocean, arctic air masses could bring more precipitation to parts of Canada.

Summary

Physical geography varies across Canada. This variation is critical in understanding Canada's regional character. At the macro level, physiographic regions represent large areas with similar landforms and geological structures. Climate creates a similar zonal arrangement of soils and natural vegetation. Climate therefore determines to a large extent the type of soil and native vegetation in a given region and hence influences land use. Together with topography, climate partly determines the land's ability to support a population. This link between the physical and human worlds identifies those regions having a more favourable mix of physical characteristics for economic development, and translates the abstract core/periphery model into a geographic reality. The Great Lakes–St Lawrence Lowlands region is the most favoured physical region in Canada. Physical barriers, such as the Rocky Mountains, and extreme climatic conditions, such as the very long and cold winters in northern Canada, have also affected the historical settlement of the country and continue to influence contemporary economic activities.

Canada has several physiographic regions and climatic zones. The seven physiographic regions are: the Canadian Shield, the Cordillera, the Interior Plains, the Hudson Bay Lowland, the Arctic Lands, the Appalachian Uplands, and the Great Lakes–St Lawrence Lowlands. The seven climatic zones are: the Pacific, Cordillera, Prairie, Great Lakes–St Lawrence, Atlantic, Subarctic, and Arctic.

Fluctuations in World Temperatures and the Warming of Mars

Both major and minor temperature changes have occurred during the earth's history. A major cooling of the world's climate took place over a million years ago. For example, the Pleistocene Epoch represents such a long-term cooling (Table 2.1). Recent minor temperature variations for periods of 500 years or so are associated with the Medieval Warming Period and the Little Ice Age. The scientific assumption for these changes in world temperatures is related to variation in energy emitted by the sun. The sunspot theory is based on the notion that sunspots indicated greater solar activity, i.e., the emitting of more solar energy. According to Dr Habibullo Abdussamatov, head of the Pulkovo Astronomical Observatory in St Petersburg, Russia, 'These parallel global warmings—observed simultaneously on Mars and on Earth—can only be a straightline consequence of the effect of the one same factor: a long-time change in solar irradiance' (Solomon, 2007).

However, minor cooling of the world's climate has also occurred. Minor cooling refers to relatively short periods of time when the global climate is slightly cooler than normal. A minor cooling, known as the Little Ice Age, took place between 1450 and 1850. The Little Ice Age had a dramatic impact on human beings living in the Arctic. During the Little Ice Age, the global climate was much cooler than it is today. This period was characterized by lower temperatures and longer winters in higher latitudes. In northern Canada, the ice cover over the Arctic Ocean was more extensive, which prevented bowhead whales from entering these waters. The consequences for the Thule inhabitants, who had developed a hunting economy based on the bowhead whale, were devastating. They were forced to hunt smaller game—seals and caribou. The results were twofold: the new hunting system could not support as many people and it required smaller, more mobile hunting groups. Archaeologists believe that the Inuit, who were established in the Arctic by the mid-sixteenth century, are the descendants of the Thule people.

Because Canada lies in the northern latitudes, permafrost is common in the northern areas of provinces (Newfoundland and Labrador, Québec, Ontario, Manitoba, Saskatchewan, Alberta, and British Columbia) and in the Territorial North. Within Canada, rivers flow into four major drainage basins: the Atlantic, Hudson Bay, Arctic, and Pacific. Only a tiny area of Alberta and Saskatchewan, where waters originating in Canada drain into the Gulf of Mexico, belongs to the Mississippi Basin.

In our contemporary world, humans are the most active and dangerous agents of environmental change. The cultivation of the land, exploitation of its renewable and non-renewable resources, and processing of its primary products have forever changed our natural environment in many ways. The construction of the Confederation Bridge, which joins Prince Edward Island to the mainland of Canada, is one example. Often, however, industrial activities have damaged the environment by causing air pollution, acid rain, and global warming. Both air pollution and acid rain represent serious environmental problems today, while global warming may prove to be the environmental challenge of the twenty-first century.

Notes

1. Canadians have various visions of themselves, their region, and their country. For the most part, these visions are rooted in the physical nature and historical experiences that affected Canada and its regions. For example, people see Canada as a northern country because of its location in the North America and because of its climates, which are often noted for long, cold winters. Hamelin's concept of nordicity (as discussed in Chapter 10) exemplifies this northern perspective. Artists, too, have found this northern theme appealing. Gilles Vigneault, one of Québec's best-known chansonniers, wrote the song 'Mon pays'. Though referring to Québec, the opening line, *Mon pays ce n'est pas un pays, c'est l'hiver* (My country is not a country, it is winter), resonates equally well in all regions of Canada.
2. Tectonic forces press, push, and drag portions of the earth's crust in a slow but steady movement. This process is the basis for the theory of continental drift. In 1912, Alfred Wegener suggested that long ago all the earth's continents formed one huge land mass. Tectonic action caused the breakup of this land mass into huge slabs. These slabs of the earth's crust drifted slowly on the molten mass (magma) beneath the earth's crust.
3. The Champlain Sea covered Anticosti Island and the northern tip of the island of Newfoundland. For the purposes of this text, the eastern extent of this physiographic region ends just east of Québec City.
4. The mean annual temperature of a location on the earth's surface is a measure of the energy balance at that point. Solar energy is the source of heat for the earth, and this energy is returned to the atmosphere in a variety of ways. Therefore, a global energy balance exists. However, there are regional energy surpluses and deficits in different parts of the world. For example, the Arctic has an energy deficit, while the tropics have a surplus. These energy differences drive the global atmospheric and oceanic circulation systems. When the mean annual temperature is below zero Celsius, it indicates that an energy deficit exists.
5. Slaymaker and French discuss the effects of global warming and climatic change on Canada's North. They maintain that climatic change caused by the greenhouse effect would alter Canada's cold environments more dramatically than the country's other environments. The authors describe possible changes to sea ice, permafrost, snow cover, sea level, and natural vegetation. There are four key factors for such a remarkable climatic change in northern Canada:

    - The percentage of carbon dioxide in the atmosphere over the Territorial North is much greater than that over southern Canada.
    - The thinning of the ozone layer over the Territorial North permits more solar radiation to enter the atmosphere.
    - The release of methane gases from the muskeg in northern Canada will add more greenhouse gases to the northern atmosphere.
    - The reduction in the duration of snow cover will expose the northern lands to solar radiation for a longer time.

    In the same book, Ledrew presents the nature of climatic change, while Smith describes the impact of climatic change on permafrost (see French and Slaymaker, 1993: chs 11–13).

Challenge Questions

1. Is there a link between physical geography and the core/periphery model?
2. What are the three main characteristics of physiographic regions?
3. What is the impact of isostatic rebound on Hudson and James bays?
4. How does the global circulation system redistribute solar energy from low to high latitudes?
5. Based on the information in Table 2.3, what are the temperature and precipitation characteristics in the climatic zone where you live?
6. What is the scientific basis for global warming?
7. Why do scientists believe that global warming would have its greatest impact in the Arctic?

Key Terms

albedo effect
Proportion of solar radiation reflected from the earth's surface.

alpine permafrost
Permanently frozen ground that is found at high elevations.

arêtes
Sharp mountain ridges that are formed between two cirques.

basins
Structural depressions in sedimentary rock that are caused by a bending of sedimentary strata into huge bowl-like shapes. Petroleum may accumulate in sedimentary basins.

chernozemic
A soil order identified by a well-drained soil that is often dark brown to black in colour; associated with the grassland and parkland natural vegetation types and located in the Prairies climatic zone.

cirques
Large, shallow depressions found in mountains caused by the plucking action of alpine glaciers.

climate
An average condition of weather in a particular area over a very long period of time.

climatic zone
A geographic area where similar types of weather occur.

continental air masses
Homogeneous bodies of air that have taken on moisture and temperature characteristics of the land mass of their origin. Continental air masses are normally dry and cold in the winter and dry and hot in the summer.

continental drift
The movement of the earth's crust; also known as plate tectonics.

continuous permafrost
Extensive areas of permanently frozen ground in the Arctic, where at least 80 per cent of the ground is permanently frozen.

convectional precipitation
An upward movement of moist air that causes the air to cool, resulting in condensation and then precipitation.

cryosolic
A soil order associated with permafrost and poorly drained land; soil is either lacking or extremely thin; associated with the tundra and polar desert vegetation types and located in the Arctic climatic zone.

denudation
The process of breaking down and removing loose material found at the surface of the earth. In this way, erosion and weathering lead to a reduction of elevation and relief in landforms.

deposition
The deposit of material on the earth's surface by various processes such as ice, water, and wind.

discontinuous permafrost
Permanently frozen ground mixed with unfrozen ground in the Subarctic. At its northern boundary about 80 per cent of the ground is permanently frozen, while at its southern boundary about 30 per cent of the ground is permanently frozen.

drainage basin
Land sloping towards the sea; an area drained by rivers and their tributaries.

drumlins
Landforms created by the deposit of glacial till and shaped by the movement of the ice sheet.

erosion
The displacement of loose material by geomorphic processes such as wind, water, and ice by downward movement in response to gravity.

Eskers
Long, sinuous mounds of sand and gravel that were deposited on the bottom of a stream flowing under a glacier.

faulting
The breaking of the earth's crust.

folding
The bending of the earth's crust.

friction of distance
The effect of distance on spatial interaction; that is, as distance increases, the number of spatial interactions (such as telephone calls) diminishes.

frontal precipitation
When a warm air mass is forced to rise over a colder air mass, condensation and then precipitation occur.

glacial erosion
The scraping and plucking action of moving ice on the surface of the land.

glacial spillways
Deep and wide valleys formed by the flow of massive amounts of water originating from a

melting ice sheet or from water escaping from glacial lakes.

glacial striations
Scratches or grooves in the bedrock caused by rocks embedded in the bottom of a moving ice sheet or glacier.

glacial troughs
U-shaped valleys carved by alpine glaciers.

global circulation system
The movement of ocean currents and wind systems that redistribute energy around the world.

greenhouse effect
The absorption of long-wave radiation from the earth's surface by the atmosphere.

Holocene Epoch
The current geological division of the Geological Time Chart. It began some 10,000 years ago and is associated with the warm climate following the last ice age.

ice age
A long cold period accompanied by the appearance of continental ice sheets. The most recent ice age is called the Pleistocene Ice Age, which began some 2 million years ago.

igneous rocks
Rock formed when the earth's surface first cooled or when magma or lava reached the earth's surface.

isostatic rebound
The uplifting process of the earth's crust following the retreat of an ice sheet that, because of its weight, depressed the earth's crust. Also known as *postglacial uplift*.

latitude
An imaginary line parallel to the equator that encircles the globe.

longitude
An imaginary line that runs through both the North and South poles.

luvisolic
A soil order identified by a well-drained soil that is often grey-brown in colour; associated with the broadleaf and mixed forest natural vegetation types in the Great Lakes–St Lawrence climatic zone.

marine air masses
Large homogeneous bodies of air with moisture and temperature characteristics similar to the ocean where they originated. Marine air masses are normally moist and mild in both winter and summer.

metamorphic rocks
Formed from igneous and sedimentary rocks by means of heat and pressure.

muskeg
A wet, marshy area found in areas of poor drainage, such as the Hudson Bay Lowland. Muskeg contains peat deposits.

nunataks
An unglaciated area of a mountain that stood above the surrounding ice sheet.

orographic precipitation
Rain or snow created when air is forced up the side of a mountain, thereby cooling the air and causing condensation followed by precipitation.

patterned ground
The arrangement of stones and pebbles in polygonal shapes. Patterned ground occurs in the Arctic where continuous permafrost exists and where frost shattering is the principal erosion process.

permafrost
Permanently frozen ground.

physiographic region
A large geographic area where a single landform, such as the Interior Plains, is found.

physiography
A study of landforms, their underlying geology, and the processes that shape these landforms; geomorphology.

pingos
Hills or mounds that have an ice core and that are found in areas of permafrost.

Pleistocene Epoch
A minor division of the Geological Time Chart beginning nearly 2 million years ago. It forms part of the Quaternary Period and is associated with some 20 ice ages.

podzolic
A soil order identified by a poorly drained soil that is often grey in colour; associated with the boreal forest and the coastal rain forest and with climates that have large amounts of precipitation, such as the Pacific, Atlantic, and Subarctic climatic zones.

polynyas
Areas of open water surrounded by sea ice.

postglacial uplift
The slow rising of the earth's crust following the retreat of an ice sheet that, because of its weight, depressed the earth's crust. Also known as *isostatic rebound*.

Quaternary Period
This geological period consisting of the Pleistocene and Holocene epochs.

rain shadow effect
Results in a dry area on the lee side of mountains where air masses descend, causing them to become warmer and drier.

residual uplift
The final stages of isostatic rebound.

restrained rebound
The first stage of isostatic rebound.

sea ice
Ice formed from ocean water that freezes. Forms of sea ice include: *fast ice* is ice that has frozen along coasts and extends out from land; *pack ice* is floating consolidated sea ice that is detached from land and freely floating; *ice floe* is a floating chunk of sea ice that is less than 10 km (less than six miles) in diameter; *ice fields* are larger chunks of sea ice more than 10 km (more than six miles) in diameter.

sedimentary rocks
Rocks formed from the accumulation, in a layered sequence, of sediment deposited in the bottom of an ocean.

soil order
Classes of soil based on observable soil properties and soil-forming processes. In Canada there are nine soil orders, including chernozemic, cryosolic, and podzolic.

sporadic permafrost
Pockets of permanently frozen ground mixed with large areas of unfrozen ground. Sporadic permafrost ranges from a trace of permanently frozen ground to an area having up to 30 per cent of its ground permanently frozen.

strata
Layers of sedimentary rock.

subsidence
A downward movement of the ground. In areas of permafrost, subsidence occurs when large blocks of ice within the ground melt, causing the material above to sink or collapse.

supervolcano
A volcano that produces an eruption whose sheer volume of ejected ash, dust, and other material significantly reduces the sun's rays from reaching the earth's surface and thus radically cools the global climate for years, with a potentially cataclysmic effect on life (en.wikipedia.org/wiki/Volcano).

till
Unsorted glacial deposits.

weathering
The decomposition of rock and particles in situ.

Bibliography

Bone, R.M. 2003. *The Geography of the Canadian North: Issues and Challenges*, 2nd edn. Toronto: Oxford University Press.

———, Shane Long, and Peter McPherson. 1997. 'Settlements in the Mackenzie Basin: Now and in the Future 2050', in Cohen (1997: 265–74).

Bouchard, Mireillle. 2001. 'Un défi environnemental complexe du XXIe siècle au Canada: L'identification et la compréhension de la réponse des environnements face aux changements climatiques globaux', *Canadian Geographer* 45, 1: 54–70.

Briggs, David, Peter Smithson, and Timothy Ball. 1993. *Fundamentals of Physical Geography*. Toronto: Copp Clark Pitman.

Canada. 1997. *Canada Year Book 1998*. Ottawa: Minister of Supply and Services.

Christopher, Robert W. 1998. *Geosystems: An Introduction to Physical Geography*, 3rd edn. Upper Saddle River, NJ: Prentice-Hall.

Cohen, Stewart J. 1990. 'Bringing Global Warming Issue Closer to Home: The Challenge of Regional Impact Studies', *American Meteorological Society* 71, 4: 520–6.

———, ed. 1997. *The Final Report of the Mackenzie Basin Impact Study*. Downsview, Ont.: Environment Canada.

De Loë, R. 2000. 'Floodplain Management in Canada: Overview and Prospects', *Canadian Geographer* 44, 4: 354–68.

Draper, Dianne, and Maureen Reed. 2005. *Our Environment: A Canadian Perspective*, 3rd edn. Toronto: Nelson.

Dyke, Arthur S., and Victor K. Prest. 1987. 'Late Wisconsinan and Holocene History of the Laurentide Ice Sheet', *Géographie physique et quaternaire* 41, 2: 237–64.

Fisheries and Oceans Canada. 2003. Fast Facts. At: <www.dfo-mpo.gc.ca/communic/facts-info/facts-info_c.htm>.

French, H.M., and O. Slaymaker, eds. 1993. *Canada's Cold Environments*. Montréal and Kingston: McGill-Queen's University Press.

Fulton, Robert J., and Victor K. Prest. 1987. 'The Laurentide Ice Sheet and Its Significance', *Géographie physique et quaternaire* 41, 2: 181–6.

Hare, F. Kenneth, and Morley K. Thomas. 1974. *Climate Canada*. Toronto: Wiley.

Houghton, J.T., Y. Ding, D.J. Griggs, M. Noguer, P.J. van der Linden, X. Dai, K. Maskell, and C.A. Johnson. 2001. *Climate Change 2001: The Scientific Basis*. Contribution of Working Group I to the Third Assessment Report of the Intergovernmental Panel on Climate Change. Cambridge: Cambridge University Press. At: <www.ipcc.ch/pub/reports.htm>.

Laycock, A.H. 1987. 'The Amount of Canadian Water and Its Distribution', in M.C. Healey and R.R. Wallace, eds, *Canadian Arctic Resources*. Ottawa: Department of Fisheries and Oceans, 13–42.

Marsh, James H., ed. 1988. *The Canadian Encyclopedia*, 2nd edn. Edmonton: Hurtig.

Mitchell, Bruce, ed. 2004. *Resource and Environmental Management in Canada: Addressing Conflict and Uncertainty*, 3rd edn. Toronto: Oxford University Press.

Norton, William. 2007. *Human Geography*, 6th edn. Toronto: Oxford University Press.

Nuttall, Mark, and Terry V. Callaghan, eds. 2000. *The Arctic: Environment, People, Policy*. Amsterdam: Harwood Academic Publishers.

PollutionWatch. 2007. Industrial Emissions, 2005. At: <www.pollutionwatch.org/rank.do?change=pwsource&year=2005&pwSource=GHG&provincesByList=TOTAL_ALL_TONE&provincesByButton=Rank&facilitiesByList=TOTAL_ALL_TONE&provincesListFac=all&companiesByList=TOTAL_ALL_TONE>.

Rasid, Harun, Wolfgang Haider, and Len Hunt. 2000. 'Post-flood Assessment of Emergency Evacuation Policies in the Red River Basin, Southern Manitoba', *Canadian Geographer* 44, 4: 369–86.

Shabbar, Amir, Barrie Bonsal, and Madhav Khandekar. 1997. 'Canadian Precipitation Patterns Associated with Southern Oscillation', *Journal of Climate* 10: 3016–27.

Slocombe, D. Scott, and Phillip Dearden. 2002. 'Protected Areas and Ecosystem-Based Management', in Dearden and Rick Rollins, eds, *Parks and Protected Areas in Canada: Planning and Management*, 2nd edn. Toronto: Oxford University Press, 295–320.

Solomon, Lawrence. 2007. 'Look to Mars for the Truth on Global Warming', *National Post*. At: <www.canada.com/nationalpost/story.html?id=eabbe10d-3891-41eb-9ee1-a59b71743bec>.

Trenhaile, Alan S. 2003. *Geomorphology: A Canadian Perspective*, 2nd edn. Toronto: Oxford University Press.

Young, Steven B. 1989. *To the Arctic: An Introduction to the Far Northern World*. New York: Wiley.

Further Reading

Dearden, Philip, and Bruce Mitchell. 2005. *Environmental Change and Challenge: A Canadian Perspective*, 2nd edn. Toronto: Oxford University Press.

Canada's vastness is often equated with unspoiled wilderness and unlimited resources. Perhaps there was some validity to these images in the distant past when our numbers were smaller and our technology more limiting, but not now. False and misleading images are firmly put to rest in *Environmental Change and Challenge*. Our growing population and technological advances have placed our fragile environment at risk. Already, we have stepped over the line by decimating northern cod stocks and adding toxic chemicals to the Arctic food chain so that the Inuit must limit their consumption of seals and other sea mammals.

The scope of this book is very broad, ranging from how ecosystems function to how humans can conserve energy, lands, and waters. For geography students, the case studies throughout the book provide a rich and varied record of how our physical environment has been abused and how Canadians are just learning how to manage the environment. Dearden and Mitchell expose a multitude of environmental problems caused by human activities in an important introduction that places environmental issues in the context of societal needs and resource use and exploitation. The authors then examine their vast subject in three substantive parts: 'The Ecosphere', which describes the science of energy flows, ecosystem change, and ecosystems and matter cycling; 'Planning and Management', which looks at both the philosophy and the process and methods of environmental management; and 'Resource and Environmental Management in Canada', which explores various subject areas of the field, such as climate change, oceans and fisheries, forests, agriculture, endangered species, water, and minerals and energy. A concluding part considers what individuals and groups must do now to preserve and conserve our natural heritage, both in Canada and globally.

Of particular interest are the 12 guest statements throughout the book written by environmental activists, managers, and scientists on issues of their experience and special concern. These guest statements discuss such topics as the Sydney Tar Ponds, traditional ecological knowledge, ocean management, forest tenure, global hunger in a world of plenty, water management and heritage rivers, and the Great Bear Rainforest on mainland British Columbia's central coast.

St Boniface in Winnipeg, Manitoba
(Kennon Cooke/Valan Photos)

Overview and Objectives

The first humans arrived in North America perhaps as long ago as 30,000 years before the present, marking the start of Canada's historical geography. Since then, of course, and up to the present day, more newcomers have arrived. These migrations are the making of Canada's history and, with geography, have combined to forge Canada's regional geography. In this chapter we look at three migrations: (1) from Asia to North America; (2) from France and Britain to the four original colonies; and (3) from Central Europe and czarist Russia to the Canadian West. By the end of World War II, these three groups had made Canada a multicultural and pluralistic society. They have also led to tensions or faultlines based on fundamental differences between English- and French-speakers, centralists and decentralists, 'new' and 'old' Canadians, and Aboriginal and non-Aboriginal Canadians. The search for solutions to these tensions is a dominant feature of Canadian society. In this chapter we will:

- Describe the arrival of Canada's first people.
- Examine the colonization of Canada by the French and the British.
- Account for the settlement of Canada's West by peoples from Central Europe and czarist Russia.
- Outline the territorial evolution of Canada.
- Look more closely at the four faultlines.
- Discuss the notion of 'One Country, Two Visions'.
- Suggest 'solutions' to complex problems.

CHAPTER 3

Canada's Historical Geography

INTRODUCTION

In one sense, Canada is a young country. Its formal history began in 1867 with the passing of the British North America Act by the British Parliament. In another sense, Canada is an old country whose human history goes back perhaps as far as 30,000 years. As an old country, its history has followed many twists and turns, but three events stand out because they continue to have a profound impact on the nature of Canadian society. These events are the arrival of the first people in North America, the colonization of North America by France and England, and the arrival of people from Central Europe and czarist Russia, many of whom settled the virgin prairie lands.

Canada began as a collection of four small British colonies—Upper and Lower Canada, New Brunswick, and Nova Scotia. After the Deed of Surrender in 1870, when Ottawa obtained Rupert's Land and the North-Western Territory, Canada's territory expanded by over 20 times, making it the second largest country in the world. This territorial expansion continued with the transfer of the Arctic Islands from Great Britain to Canada in 1880 and only ceased in 1949, when Newfoundland joined Canada. As the country was spread over such a huge territory, Canada developed a distinct regionalized character shaped by history and geography.

Within the federal system of government, political reality favours Ontario and Québec largely because the governing party is dependent on holding a majority of seats from the combined ridings of those two provinces. Provincial governments, while having defined powers, have insufficient tax revenue to meet all their responsibilities, such as health care and education.[1] Consequently, an ongoing battle between provincial governments and Ottawa takes place annually over transfer payments from the federal government to the provinces. With no federal body such as the Senate representing provincial/territorial interests, regionally based political parties, such as the Reform Party of Canada (which became the Canadian Alliance, which became the Conservative Party by amalgamating with the Progressive Conservative Party) and the Bloc Québécois, have emerged. Both add to the country's diverse character. Therefore, much of this great political experiment called Canada involves the difficult task of forging a unity among its various political members.

While Canada is admired by most peoples of the world, seeds of regional discontent exist in each province and territory. Regional discontent, often related to economic disparities and squabbling over royalties from natural resources, has generated ongoing tensions between the federal government and the provinces, as well as tensions among the provinces over issues such as language laws and constitutional amendments. To use *Globe and Mail* columnist Jeffrey Simpson's metaphor, these tensions can be likened to faultlines—large cracks in the earth's crust caused by tectonic forces. Like geological faultlines, Canada's economic, social, and political faultlines divide regions and people, and threaten to destabilize Canada's integrity as a nation. For long periods of time, these cracks in Canadian society remain dormant, but they can shift at any time, dividing the country into wrangling factions. A recent example took place in 2007 with the squabble over equalization payments.

Four key tensions are examined in this chapter: (1) English-speaking/French-speaking Canadians; (2) centralists and decentralists; (3) 'new' and 'old' Canadians; and (4) Aboriginal/non-Aboriginal Canadians. In each case, differences and disagreements occur between a core (for example, Central Canada) and a periphery (for example, the rest of Canada). Central Canada is often the symbol of the political power held by Ottawa, and at other times it represents the economic power held by Ontario and Québec as well as their ability to obtain large subsidies for the automobile and aeronautical industries. Although the periphery refers in general to the rest of Canada, it can also represent a particular geographic region or even a group of people with a common complaint against the core.

Of the four faultlines, the French/English rift is by far the most threatening to Canada. For example, if another Québec referendum at some time in the future points conclusively to separation and independent nationhood for that province, it would have a devastating impact by partitioning Atlantic Canada from the rest of Canada. At the same time, the historic and contemporary interaction between French- and English-speaking Canadians is the very process that makes Canada so unique. This give and take between the two so-called founding peoples often spills over to affect the Aboriginal, immigration, and regional issues that test Canada's unity. Oddly enough, because of the interaction between Aboriginal and non-Aboriginal Canadians, new and old Canadians, centralists and decentralists, and English-speaking and French-speaking Canadians, events affecting one faultline often have a positive or negative effect on the other three faultlines.

One aspect of the geographic realities that have resulted from Canada's history is the emergence of tensions between regions and/or groups that live in those regions. Often these disagreements pit Ottawa against the provinces. Frequently, these disagreements fall into three categories: (1) the issue of national interests overriding provincial/territorial ones; (2) the issue of federal funding for provincial programs; and (3) the issue of federal responsibilities for Aboriginal peoples, have-not provinces, and the three territorial governments. Examples for each category would be (1) the Kyoto Accord, (2) health care, and (3) First Nations and the territorial governments.

The First People

Around 30,000 years ago, according to some archaeological estimates, the first people to set foot on North American soil were Old World hunters who crossed a land bridge (known as Beringia) into Alaska and Yukon. Beringia was a product of the last ice advance when so much water was contained in the continental glaciers that the sea level dropped, thus exposing the ocean bottom between Siberia and Alaska. Scientists estimate that the sea level dropped by at least 100 metres, thus creating a land bridge between Asia and North America.

Some 15,000 years ago, the climate warmed and the ice sheets began to melt. Soon Beringia was again well below the surface of the Pacific Ocean. The Old World hunters were now in Alaska but were blocked from proceeding south by an ice sheet that was perhaps 4 km thick. When an ice-free corridor was formed just east of the Rocky Mountains, the ancient hunting tribes were able to migrate southward into the heart of North America (Figure 3.1). Archaeologists have estimated that this ice-free corridor appeared about 12,000 to 13,000 years ago. Just when the first people travelled along this narrow corridor is unknown, but it probably occurred after tundra vegetation had taken hold, supplying food for the animals that also migrated into the heart of North America. With Beringia once more below sea level, further migration from

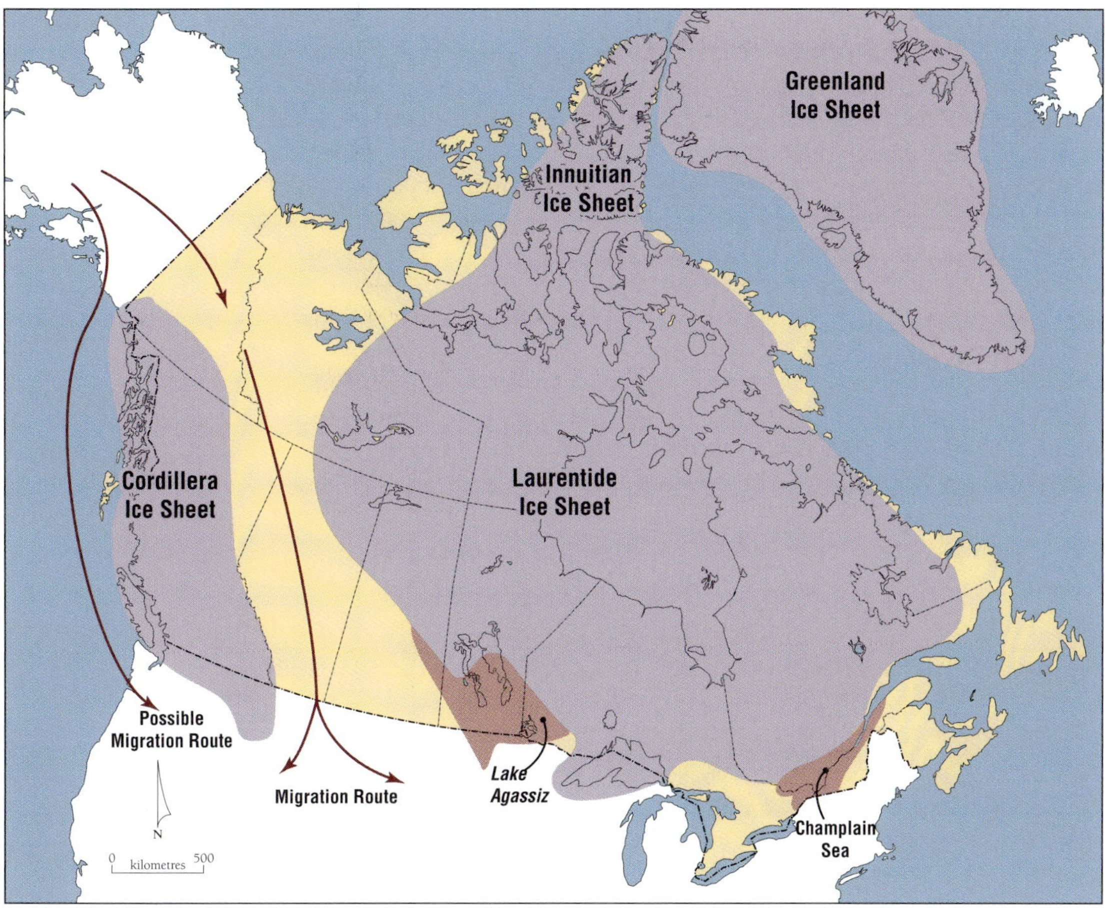

Figure 3.1 Migration routes into North America. The Wisconsin ice sheet had retreated by 12,000 BP (before present), leaving an ice-free corridor between its two remaining parts (the Cordillera and Laurentide ice sheets). In places where the meltwater could not flow to the sea, water collected in low-lying areas and created glacial lakes such as Lake Agassiz. As explained in Vignette 2.7, however, the Champlain Sea was not a glacial lake but an extension of the Atlantic Ocean.

Asia to North America was virtually impossible until the marine technology of the Paleo-Eskimos emerged about 5,000 years ago (Table 3.1).

Some archaeologists even speculate that Old World hunters may have penetrated into the lands south of the ice sheet much earlier. According to one theory, these early hunters made a coastal migration along the edge of the Cordillera ice sheet that covered all the mountains and valleys of British Columbia. By slowly moving southward along the unglaciated islands adjacent to the ice-covered British Columbia coast, these early people could have circumvented the Cordillera ice sheet by island-hopping, thereby reaching the unglaciated area of the US Pacific coast much earlier than 12,000 years ago. However, archaeological evidence of such a route is lacking because these ancient campsites, if they existed, are now well below the current sea level.

Although the question of when, precisely, humans first reached North America is still unsettled, archaeological evidence indicates that the Old World hunters had reached the southern United States by at least 11,000 years ago. Archaeologists have assigned the name Paleo-Indian to these first people of North America because they shared a common hunting culture, which was characterized by its uniquely designed fluted-point stone spearhead. In time, the Paleo-Indians developed more distinct hunting cultures that Thomas (1999: 10) divides into three groups:

- Clovis culture from 9500 to 8500 BC;
- Folsom culture from 9000 to 8200 BC;
- Plano culture from 8000 to 6000 BC.

Paleo-Indians

How did these Old World hunting cultures evolve into a New World hunting system? The Paleo-Indians, the people who devised the fluted points, were descendants of the Old World hunters. The oldest fluted points found in North America are about 11,500 years old. These spearheads, along with the bones of woolly mammoths, have been discovered in the southern part of the Canadian Prairies. By 9,000 years ago, many of the large species, such as the woolly mammoths and the mastodons, had become extinct, possibly as a result of excessive hunting and/or climatic change. After the extinction of the woolly mammoths and mastodons, later Paleo-Indian cultures, which archaeologists refer to collectively as Folsom and Plano, developed a variety of unfluted stone points with stems for attachment to spear shafts, which made weapons more suitable for hunting buffalo and caribou. About 8,000 years ago, hunters in the grasslands of the interior of Canada pursued

Table 3.1 Timeline: Old World Hunters to Contact with Europeans

Date (BP = before present)	Event
30,000–25,000 BP	Old World hunters of the woolly mammoth cross the Beringia land bridge into the unglaciated areas of Alaska and southern Yukon.
18,000 BP	The Wisconsin ice sheet reaches its maximum geographic extent, covering virtually all of Canada.
15,000 BP	The Wisconsin ice sheet retreats rapidly in western Canada, exposing a narrow ice-free area along the foothills of the Rocky Mountains. Ice persists in northern Québec until about 6,000 years ago.
13,000–12,000 BP	Descendants of the Old World hunters migrate southward through the ice-free corridor east of the Rocky Mountains.
9,000 BP	Mammoths and mastodons become extinct, forcing early inhabitants of North America to adjust their hunting practices and thereby become more mobile and less numerous.
5,000 BP	As the ice sheet disintegrated across the northern limits of Canada, Paleo-Eskimos (known as the Denbigh) were the first peoples to cross the Bering Strait to the Arctic Coast of Alaska. Later they move eastward along the Canadian section of the Arctic coast.
3,000 BP	The Dorset people, who were descendants of the Denbigh, develop a more advanced technology suited for an Arctic marine environment.
1,000 BP	A third wave of marine hunters, known as the Thule, migrate across the Arctic, eventually reaching the coast of Labrador. At the same time Vikings reach Greenland and North America, where they establish a settlement on the north coast of Newfoundland (L'Anse aux Meadows).
1497	John Cabot lands on the east coast (Newfoundland or Nova Scotia).
1534	Jacques Cartier plants the flag of France near Baie de Chaleur.

the buffalo, while those in the tundra and forest lands of northern and eastern Canada depended on caribou for most of their food. However, these smaller prey species could not support large numbers of people, so the Paleo-Indians had to develop new survival strategies. These strategies involved remaining in one area (and presumably keeping other peoples out of that area), developing effective hunting techniques for the local game, and making extensive use of fish and plants to supplement their principal diet of game.

This link between geographic territory and hunting societies marked the development of Paleo-Indian **culture areas**, which were geographic regions with the following two characteristics: (1) a common set of natural conditions that resulted in similar plants and animals, and (2) inhabitants who used a common set of hunting, fishing, and food-gathering techniques and tools. Under these conditions, Paleo-Indians formed more enduring social units that became the forerunners of the numerous North American Indian tribes at the time of contact with Europeans.

Indians

Climatic conditions vary across North America, ranging from a tropical climate in Mexico to an arctic climate just north of the Great Lakes–St Lawrence Lowlands. At that time, a great ice sheet still occupied most of Ontario and Québec. These varied conditions provided different opportunities for Indians, at least some of whom were descendants of the Paleo-Indians. Most archaeologists support the idea that Algonquians are direct descendants of Paleo-Indians, but they are less certain about Athapaskans, whose ancestors may have arrived from Asia some 10,000 years ago. Since the Paleo-Indian culture had emerged some 11,500 years ago, the Athapaskans represent a distinct Indian culture. But how did the early Athapaskans cross the Bering Strait? Archaeologists suggest that the ancient ancestors of the Athapaskans either walked across the frozen Bering Strait or crossed it in small, primitive boats.

About 5,000 years ago, Indians living in the tropical climate of Mexico began to domesticate plants and animals. This agricultural system and its people gradually spread northward into areas with more restrictive growing conditions. These climatic differences required Indians to adapt their agricultural system accordingly. About 3,000 years ago, Indians in what is today the eastern United States planted corn, beans, and squash, which supplemented their diet of game and fish.

Agriculture was not possible north of the Great Lakes–St Lawrence Lowlands because of its shorter growing season for corn and other crops. Algonquian-speaking Indians who lived north of the Great Lakes had to hunt big-game animals, particularly caribou, for sustenance. They also traded with those more sedentary Indians, such as the Huron and Iroquois, who practised a form of slash and burn agriculture in the Great Lakes–St Lawrence Lowlands and in the Ohio Valley. By the sixteenth century, the Huron controlled the agriculture lands between Lake Simcoe and the southeastern corner of Georgian Bay, where about 7,000 acres were under cultivation and where Indian

A tent ring representing the remains of a Thule house is located near Igloolik, Nunavut. Tent rings represent a site where a Thule family had located their tents. The stone rings mark the edges of the tent skin walls that were weighted down with rocks. (Alec Aitken)

villages, with populations as large as 5,000 persons, were commonplace (Dickason, 2002: 52). In western North America, agriculture spread northward along the valleys of the Mississippi River and its major tributary, the Missouri River. Tribes on the Canadian prairies continued to hunt buffalo but did engage in trade for agriculture products with tribes along the upper reaches of the Missouri River. In northwest Canada, Athapaskan-speaking Indians, whose ancestors probably came from Asia much later (perhaps between 7,000 and 10,000 years ago), continued to practise a nomadic lifestyle of hunting and gathering. They moved about in the forest lands stalking big-game animals and made summer hunting trips to the tundra where the caribou had their calving grounds.

Arctic Migration

Arctic Canada was settled much later than the forested lands of the Subarctic. Before people could occupy the Arctic Lands, two developments were necessary. The first was the melting of the ice sheets that covered Arctic Canada. About 8,000 years ago, only small remnants of the great Laurentide ice sheet remained in northeastern Canada, leaving the western Arctic free of the ice sheet. The second development was the emergence of a hunting technique that would enable people to live in an Arctic environment. About 5,000 years ago, the Paleo-Eskimos, who were living along the northeast coast of Asia and in Alaska, had developed an Arctic sea-based hunting technique, and they began to move eastward along the coast of Alaska and then along the Arctic coast of Canada. This hunting culture of Paleo-Eskimos is known as the Denbigh. Unlike previous marine hunting societies, these people invented a harpoon and other tools that enabled them to hunt seals and other marine mammals along the Arctic coast, though they also relied heavily on terrestrial game such as the caribou. About 3,000 years ago, a second migration from Alaska took place. Known as the Dorset culture, this culture replaced the Denbigh one. The third and final Arctic migration took place roughly 1,000 years ago, when the Thule people, who had developed a sophisticated sea-hunting culture, spread eastward from Alaska and gradually succeeded their predecessors. The Thule, the ancestors of the Inuit, hunted the bowhead whale and the walrus. However, the climate began to cool (known as the Little Ice Age) and whales no longer entered the Arctic Ocean in large numbers because now the ocean was covered by ice for most of the year. The Thule were forced to hunt smaller game such as the seal and the caribou.

Initial Contact

By the time of European contact, the descendants of Old World hunters occupied all of North and South America. Yet, Europeans consider the New World **terra nullius** or empty lands. North American Indian and Inuit tribes met the European explorers searching for a trade route to the Orient. While the total population of these tribes can only be estimated, many scholars now believe there may have been as many as 500,000 Indians and Inuit living in Canada at the time of contact. The greatest concentrations were found along the Pacific coast, where marine resources provide abundant food, and in the Great Lakes/St Lawrence Lowlands, where a primitive form of agriculture supported relatively large sedentary populations. Following contact, their numbers dropped sharply, perhaps declining to 100,000. Loss of hunting grounds to European settlers and the spread of new diseases by explorers, fur traders, and missionaries greatly contributed to this depopulation. In the early seventeenth century, the sudden collapse of Huronia, the most powerful Indian group in the region, took place shortly after contact with the French. Its numbers dropped from 21,000 to less than half this number within a decade. This demographic catastrophe was not unique to Huronia, but it does provide one example of the deadly impact of European diseases and colonial alliances on the Aboriginal peoples. French missionaries brought diseases to the Huron villages, while the Iroquois, who opposed the French/Huron alliance, attacked the Huron villages and eventually destroyed the Huron Confederacy in 1649. In little more than a generation, Huron villages were abandoned, the cornfields of Huronia reverted to forest, and its people were greatly reduced in numbers and scattered across the land, some finding their way to the north shore of the St Lawrence in Québec, others being captured and assimilated by the Iroquois, and another remnant joining related tribes in what are today Michigan and Ohio.

John Cabot, the first European explorer, reached the east coast in 1497. He was followed by others, including Jacques Cartier and Martin Frobisher. In 1534 Cartier made contact with two Indian tribes along the Gaspé coast, and in 1576 Frobisher encountered an Inuit encampment along the Arctic coast of southern Baffin Island. Both explorers were searching for a route to Asia, but instead discovered a new continent and peoples. In both instances, contact with North Americans ended badly. Lives were lost on both sides, and the ore that the explorers took back to Paris and London, respectively, proved worthless (Vignette 3.1). Instead of gold, they had found fool's gold (iron pyrites).

Relationships between explorers and Aboriginal peoples were not always peaceful. In this painting by John White, the artist records a fight between Frobisher's crew and Baffin Island Inuit. (© Copyright the Trustees of The British Museum)

Culture Regions

At the time of contact with Europeans, Aboriginal peoples occupied specific territories (cultural regions) where their hunting, fishing, and gathering activities took place. Within each cultural region, Aboriginal peoples developed distinct techniques suitable for the local environment and wildlife. Seven culture regions are Eastern Woodlands, Eastern Subarctic, Western Subarctic, Arctic, Plains, Plateau, and Northwest Coast (Figure 3.2). The Inuit occupied the Arctic cultural region. In the Eastern Subarctic, the Cree were the principal Algonquian tribe. The Cree in this region had developed a technology—snowshoes—to hunt moose in deep snow. In the Western Subarctic region, the Athapaskans, including a number of Dene tribes, hunted caribou and other big-game animals. Northwest Coast Indians harvested the rich marine life found along the Pacific coast. Tribes such as the Haida, Nootka (**Nuu-chah-nulth**), and Salish comprised the Northwest Coast cultural region. In the southern interior of British Columbia, the Plateau Indians—including the Carrier, Lillooet, Okanagan, and Shuswap—occupied the valleys of the Cordillera, forming the Plateau cultural region. Across the grasslands of the Canadian West, Plains Indians such as the Assiniboine, Blackfoot, Sarcee, and Plains Cree hunted bison. The Iroquois and Huron lived in the Eastern Woodlands of southern Ontario and Québec. Both groups combined agriculture with hunting. In the Maritimes, the Mi'kmaq and Maliseet also occupied the Eastern

Vignette 3.1 Contact between Cartier and Donnacona

In 1534, Jacques Cartier made contact at the Gaspé coast with Donnacona, the leader of the St Lawrence Iroquois. The Iroquois, whose village (Stadacona) was located at the site of Québec City, came to the Gaspé to fish for mackerel. The next year Cartier returned. After reaching the Iroquois village of Hochelaga, which later became the site of Montréal, Cartier realized that the St Lawrence did not provide a sea passage to China. Like the Spanish in Mexico and Peru who discovered gold and silver, Cartier hoped to find wealth in the New World. In 1541, on his third voyage to Stadacona, Cartier hoped to find diamonds and gold. He left France with several hundred colonists. By then, Donnacona and all but one of the Iroquois captives that Cartier had taken with him when he returned to France in 1536 were dead. The French authorities, who wished to conceal this fact, decided not to allow the surviving Indian woman to return to Stadacona with Cartier. The French arrived at Cap Rouge where they established a settlement just west of Stadacona. When the St Lawrence Iroquoians realized that Cartier was not going to return their chief and the other Iroquois captives, relationships quickly soured. Over the winter, the Iroquois killed at least 35 of the French settlers. Cartier and his surviving party left for France as soon as the river ice melted. He took along rocks that he believed contained gold and diamonds, but, like the ore mined on Baffin Island by Frobisher's men, Cartier brought back fool's gold and quartz. As there seemed to be no prospect for mineral wealth, the King of France lost interest in the New World. Sixty years went by before the French made another attempt at establishing a settlement at Stadacona. In 1608, Samuel de Champlain founded Québec City.

Source: Adapted from Ramsay Cook, *The Voyages of Jacques Cartier* (Toronto: University of Toronto Press, 1993).

Woodlands where they hunted and fished. The complexity and diversity of Aboriginal peoples can be gleaned from the spatial arrangement of their languages (Figure 3.3).

The Second People

The colonization of North America by the French and the British was the second major development in Canada's early history. France and England established colonies in North America in the seventeenth century. Québec City, founded in 1608 by Samuel de Champlain, was the first permanent settlement in Canada. By 1663, the French population in New France was about 3,000 compared to a population of about 10,000 Indians (mainly Huron and Iroquois), who lived in the same area of the St Lawrence Valley, the Great Lakes, and the Ohio Basin. By 1750, the French Canadians constituted most of the population in New France, where they numbered about 60,000, while the Indian population had dropped sharply because of disease and warfare. Following the British Conquest of New France in 1760, the flow of French colonists ceased and British immigrants began to move to what used to be New France. In doing so, they increased the number of British settlers. Meanwhile, the French-Canadian population now had to depend entirely on natural population increase to expand their numbers.

The first large wave of British immigrants consisted of refugees from the United States. These Loyalists had supported Britain during the American War of Independence (1775–83). After the defeat of the British army, they sought refuge in other parts of the British Empire,

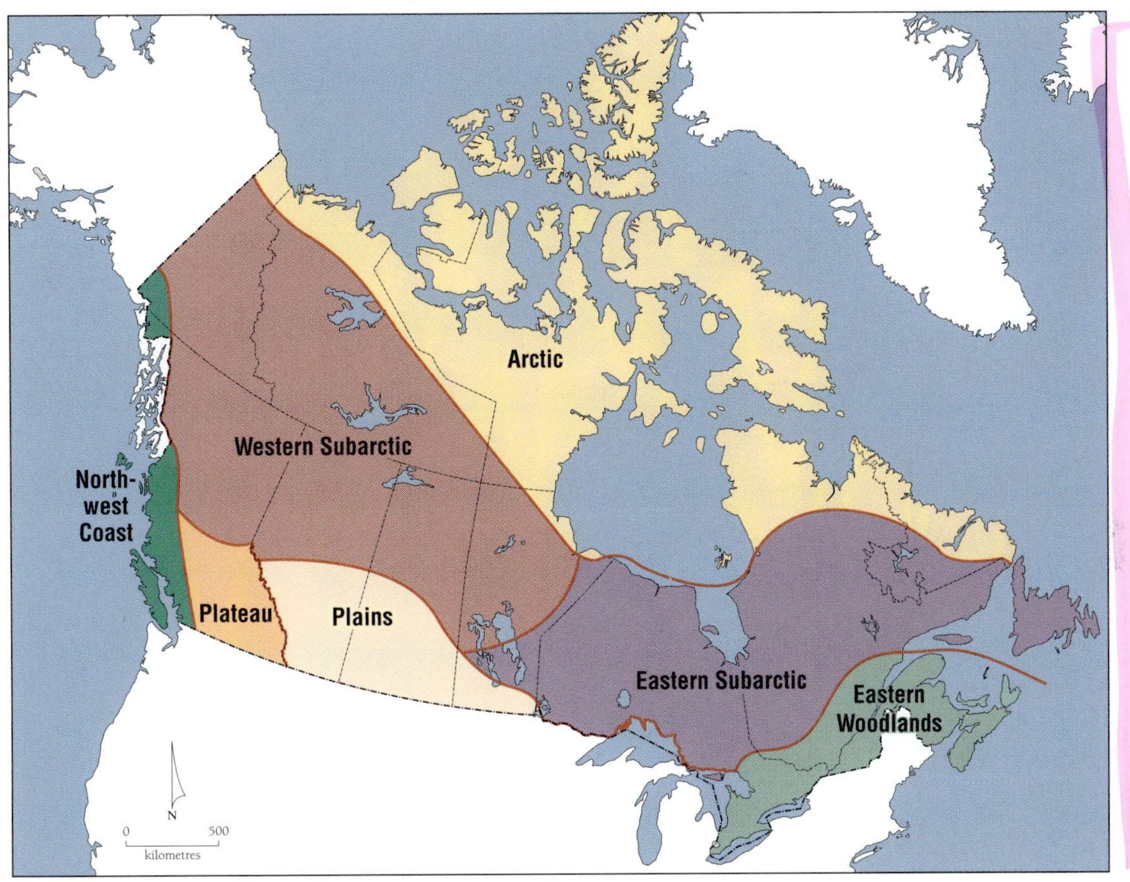

Figure 3.2 Culture regions of Aboriginal peoples. Common resources and natural conditions in certain geographic areas of North America were associated with similar ways of hunting wildlife and organizing economic and social activities.

including its North American possessions. In North America, most Loyalists settled in Nova Scotia, while a smaller number moved to the Eastern Townships of Québec and to Montréal. Others settled in what is now southern Ontario.

The second wave of immigrants came from the British Isles. From 1790 to 1860, almost a million people migrated from the British Isles to British North America, mostly because of deteriorating economic conditions in Great Britain and because of job opportunities in the merchant cod fishery. In the 1840s, the potato famines in Ireland added to the problems in the Old World, causing terrible hardships for the Irish people. Thousands fled the countryside and many left for the New World to settle in the towns and cities of British North America.

These two waves of migration greatly changed Canada. At the time of Confederation in 1867, the population of British North America had reached 3 million. Approximately 80 per cent of these British subjects lived in the Great Lakes–St Lawrence Lowlands, while about 20 per cent inhabited Atlantic Canada. Across the rest of British North America, Aboriginal peoples made up most of the population. In the Red River Colony, a new Aboriginal people, the Métis, who were of Native and European descent, had emerged. By 1869, the Métis, who were split between French- and English-speaking, greatly outnumbered white settlers and fur traders in the Red River Colony. The colony had a population of nearly 12,000 and the Métis formed over 80 per cent this population (Table 3.2).

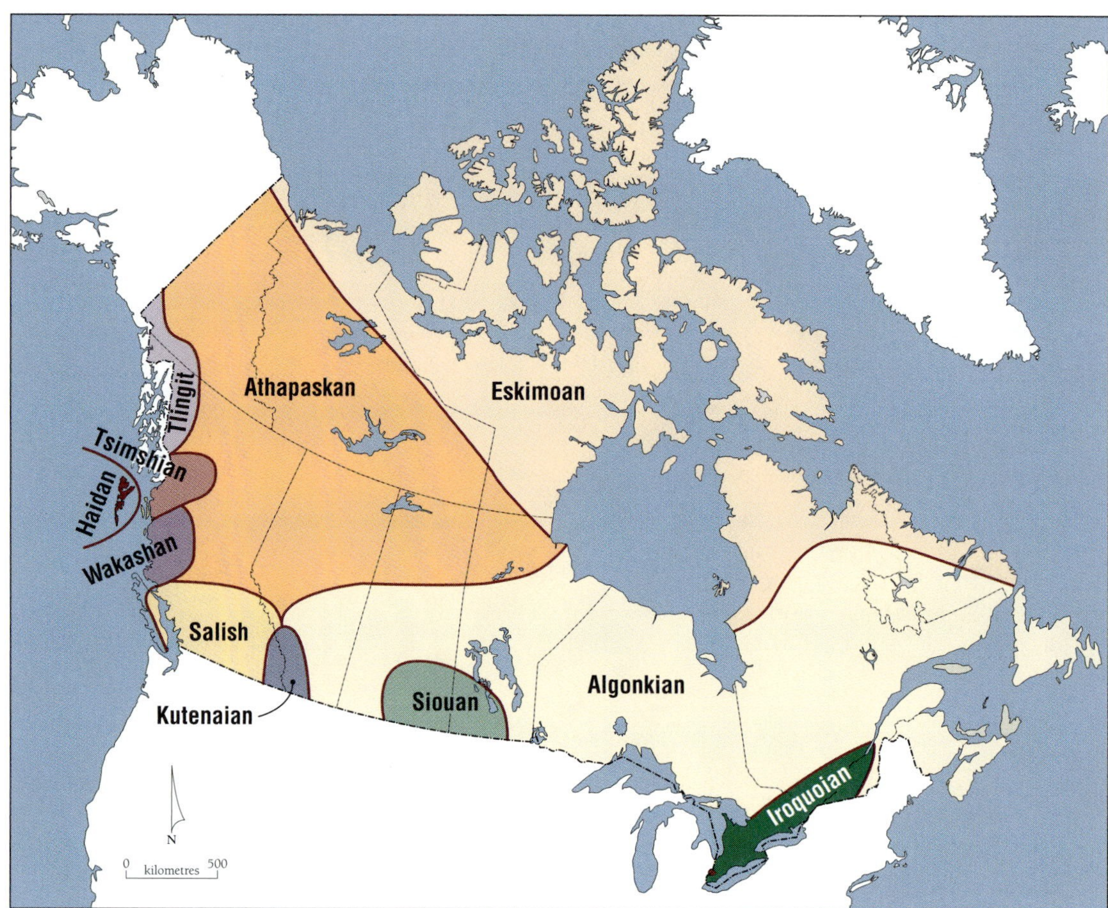

Figure 3.3 Aboriginal language families. Aboriginal peoples in Canada form a very diverse population. At the time of contact, there were over 50 distinct Aboriginal languages spoken. These languages formed 11 language families. Five of these language families are found in one natural cultural region, the Northwest Coast. Following contact, language loss was swift. By the end of the twentieth century, only three Aboriginal languages, Cree, Inuktitut, and Ojibwa, had over 20,000 speakers. (Further resources: Student website, National Atlas section, Map 12. Website instructions are found on p. xx.)

But more important, the old balance of demographic power between the French and English had been reversed. British migration to Canada had, over the course of 100 years, changed the demographic balance between French-speaking and English-speaking Canadians. Approximately 60 per cent of the European population was English-speaking. The new demographics also resulted in large English-speaking populations in the principal cities of Québec. Migration had also changed the ethnic composition of the English-speaking population from almost entirely English to a mixture of English, Irish, Scottish, and Welsh. Moreover, by the 1860s, Canada's ethnic character varied by region. In Atlantic Canada, the Scottish and the Irish outnumbered the English. In Québec, the English and Irish formed a sizable minority in the towns and cities, though rural Québec remained solidly French-speaking except in the Eastern Townships. Ontario, like Atlantic Canada, was decidedly British.

The separation of the British and French in different geographic parts of British North America ensured the continued existence of two visions of Canada. The greatest threat to Canadian unity stems from these two irrecon-

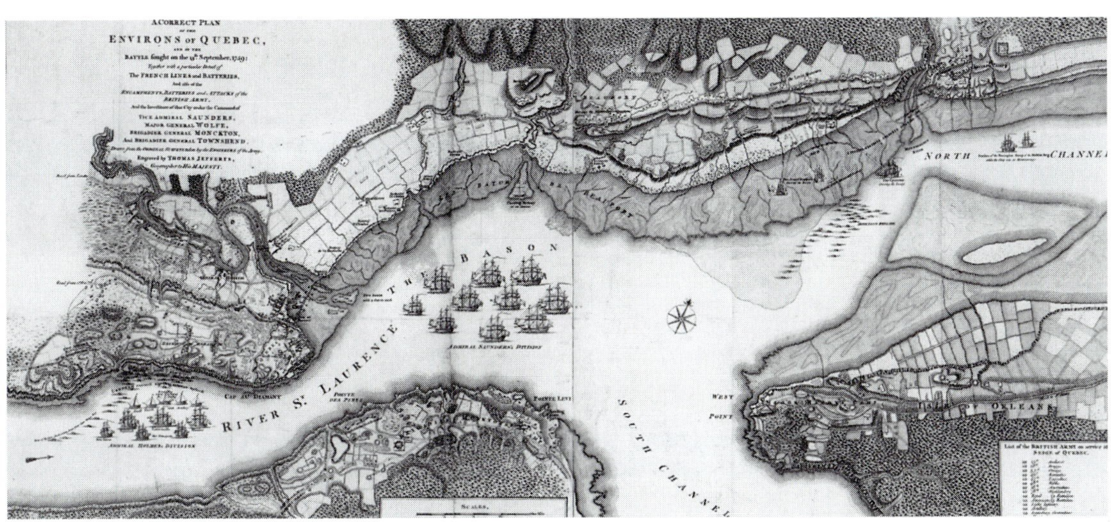

In 1759, British forces, under the command of General James Wolfe, defeated the French troops led by Louis-Joseph, Marquis de Montcalm, on the Plains of Abraham, which lay just west of the walled capital of New France, Québec City. (*Courtesy Library and Archives Canada 128079*)

The American Revolution (1775-83) divided the residents of the Thirteen British Colonies. With the defeat of British forces, those British subjects who did not support the rebel cause were forced to leave, losing their property and sometimes their lives. Considered traitors by Americans, Loyalists were often subjected to mob violence. (North Wind Picture Archives)

Table 3.2 Population of the Red River Colony, 1869

Ethnic Group	Population Size	Population Percentage
Whites born in Canada	294	2.5
Whites born in Britain or a foreign country	524	4.4
Indians	558	4.7
Whites born in Red River	747	6.2
English-speaking Métis	4,083	34.1
French-speaking Métis	5,757	48.1
Total Population	11,963	100.0

Source: Lower (1983: 96).

cilable visions (a further discussion is found in the subsection 'The French/English Faultline'). In 1867, 92 per cent of the population was either British or French. Each linguistic group developed its own vision of Canada: French Canadians saw Canada as two founding peoples, while English-speaking Canadians favoured the notion of equality among provinces. In English Canada, the notion of equal provinces grew out of the following factors:

- The nature of Confederation was such that provincial powers were shared equally.
- The British formed the majority of the population in three of the four provinces, thereby dominating the political affairs of those provinces.
- The British, while a minority within Québec, were the dominant business group.

The Third People

The hegemony of the British and French began to weaken in the early part of the twentieth century when the fertile lands of Western Canada were settled in large part by neither English-Canadian nor French-Canadian farmers. In 1870, Ottawa obtained the vast land of the Hudson's Bay Company and was faced with the question of settling this territory. Little progress was made until after the completion of the CPR railway in November 1885 and the signing of treaties (Figure 3.10). Now the time was right for massive immigration to settle these lands. For Ottawa, there were two key advantages in encouraging settlement. First, the threat of American settlers moving into the Canadian West and annexing these lands would be diminished. Second, the creation of a grain economy would provide freight for the Canadian Pacific Railway, thereby helping turn it into a viable operation. But where to find the people? Some came from Ontario, Québec, and Atlantic Canada to claim their 160 acres as homesteaders while others came from Britain and the United States (Vignette 3.2).

Still, much of Western Canada was unoccupied. Clifford Sifton, the Minister of the Interior, accepted the challenge to settle the West. By the beginning of the twentieth century, Sifton had launched an aggressive and innovative advertising campaign to lure people from Britain and the United States to 'The Last Best West', but this effort failed to bring sufficient immigrants. At that point, he recognized the need to go beyond these two countries. In a break with past immigration policy, Sifton turned his attention to the people of Central Europe, Scandinavia, and czarist Russia. Land-hungry peasants from Ukraine formed the largest single

Vignette 3.2 The Land Survey System and the Settling of the Prairies

In 1872, the federal government passed the Dominion Lands Act. This legislation established a system of survey that divided the land into square townships made up of 36 sections, each measuring 1 mile by 1 mile, with allowances for roads. Each section was further subdivided into four quarters, each quarter section measuring one-half mile by one-half mile and comprising 160 acres. This survey system gave the Canadian Prairies a distinctive 'checkerboard' pattern.

Ottawa sought to populate the arable lands of Western Canada by offering 'free' land to farmers. The free land took the form of a homestead (a quarter section or 160 acres). A free grant of one quarter section was available to persons 21 years or older with the payment of a $10 fee. Upon fulfilling cultivation and residency requirements within three years of acquiring the property, the homesteader would receive title to the land. Since Ottawa had made substantial land grants to the Canadian Pacific Railway and to the Hudson's Bay Company, not all the land was free. Both the Canadian Pacific Railway and the Hudson's Bay Company sold their land at market prices.

group of immigrants but Doukhobors and German-speaking Mennonites also came from czarist Russia, giving Western Canada a distinct and different mix of ethnic groups and landholdings from the rest of the homesteaders, with communal rather than individual landholdings the norm among some groups. Sifton recognized that peasants from czarist Russia were well suited for settling the Canadian Prairies, and Sifton was willing to accommodate their religious concerns and social customs, including communal farming practices. From 1901 to 1921, Western Canada's population increased from 400,000 to 2 million with Saskatchewan forming the third largest province (Table 3.3). As these ethnic groups and individuals spread across the Prairies, their impact was enormous on a landscape that only recently was populated largely by vast herds of buffalo and semi-nomadic Indian tribes (see Kerr and Holdsworth, 1990: Plate 17).

By opening the door for immigration from European countries without a French or British background, Sifton's immigration policy changed the face of Canada. His goal of settling the West was accomplished, and a new dimension had been added to Canada's social fabric—people with neither a French nor a British background. While the majority of these immigrants were homesteaders, some took jobs in Canada's booming mining and logging industries while others settled in Canada's major cities, especial Montréal and Toronto. While most soon learned English, this cultural/linguistic difference from the two founding peoples surfaced later in the twentieth century as a powerful force for multiculturalism and pluralism in English-speaking Canada. The immigration to the Canadian Prairies from 1896 to 1914 forms the topic for more detailed discussion later in this chapter under the subheading 'The Immigration Faultline'.

The Territorial Evolution of Canada

A brief presentation of the territorial evolution of Canada marks the start of our account of Canada's formal history as a nation. The British North America Act of 1867 united the colonies of New Brunswick, Nova Scotia, and the Province of Canada (formerly Upper Canada and Lower Canada) into the Dominion of Canada.[2] Canada soon acquired more territory. In 1870, the Deed of Surrender transferred Rupert's Land and the North-Western Territory to the federal government, at which time this

Vignette 3.3 America's Manifest Destiny

In the nineteenth century, many Americans believed in the doctrine of Manifest Destiny, which was based on the belief that the United States would eventually expand to all parts of North America, thus incorporating Canada and Mexico into the American republic. From its beginnings along the Atlantic seaboard, the United States had greatly increased its territory by a combination of force, negotiation, and purchase. In 1803, the United States purchased the Louisiana Territory (a vast land west of the Mississippi and east of the Rocky Mountains) from France, and in 1867 the country bought Alaska from Russia. To Americans, such an expansion was an expression of their right to North America. As well, it would rid North America of the much-hated European colonial powers.

Not surprisingly, the Fathers of Confederation were concerned about American designs on British North America. First, in 1866, the Fenians raided Upper Canada, Lower Canada, and New Brunswick with the intention of seizing British North America and holding it for ransom until Ireland was free of British rule. Second, in 1867, the US purchase of Alaska left British Columbia wedged between American territory to its north and south, and the exact boundary along the coastline south of 60° N was uncertain. Canada and the United States settled this final border dispute in 1903.

Table 3.3 Canada's Population by Provinces and Territories, 1901 and 1921

Political Unit	Population 1901	Per cent	Population 1921	Per cent
Ontario	2,182,947	40.6	2,933,662	33.4
Québec	1,648,898	30.7	2,360,510	26.9
Nova Scotia	459,574	8.6	523,837	6.0
New Brunswick	331,120	6.2	387,876	4.4
Manitoba	255,211	4.8	610,118	6.9
Northwest Territories*	21,064	0.4	8,143	0.1
Prince Edward Island	103,259	1.9	88,615	1.0
British Columbia	178,657	3.3	524,582	6.0
Yukon	27,219	0.5	4,147	>0.1
Saskatchewan	91,279	1.7	757,510	8.6
Alberta	73,022	1.3	588,454	6.7
Canada	5,371,315	100.0	8,787,949**	100.0

*Saskatchewan and Alberta did not become provinces until 1905 and were officially included in the population of the Northwest Territories.

**Includes 485 members of the armed forces.

Source: Adapted from Statistics Canada publication *Historical Statistics of Canada*, Catalogue 11–516, Searched 15 February 2003, http://www.statcan.ca/english/freepub/11-516-XIE/sectiona/sectiona.htm.

large expanse was renamed the North-West Territories. In 1871, British Columbia joined Confederation, and Prince Edward Island followed two years later. In 1880, Ottawa acquired the Arctic Islands from Great Britain. Thus, while Canada began as a small country, consisting of what is now known as southern Ontario, southern Québec, New Brunswick, and Nova Scotia, it quickly became one of the largest in the world (Figure 3.4).

What was behind these real estate deals? The British government was a strong advocate of uniting its colonies in North America. For Britain, the union of its North American colonies had three advantages: (1) a better chance for their political survival against the growing economic and military strength of the United States; (2) an improved environment for British investment, especially for the proposed trans-Canada railway; and (3) a reduction in British expenditures for the defence of its North American colonies. The British colonies perceived unification differently. The Province of Canada, led by John A. Macdonald, pushed hard for a united British North America because it would have a larger domestic market for its growing manufacturing industries and a stronger defensive position against a feared American invasion. The Atlantic colonies showed little interest in such a union. As they were part of the British Empire,

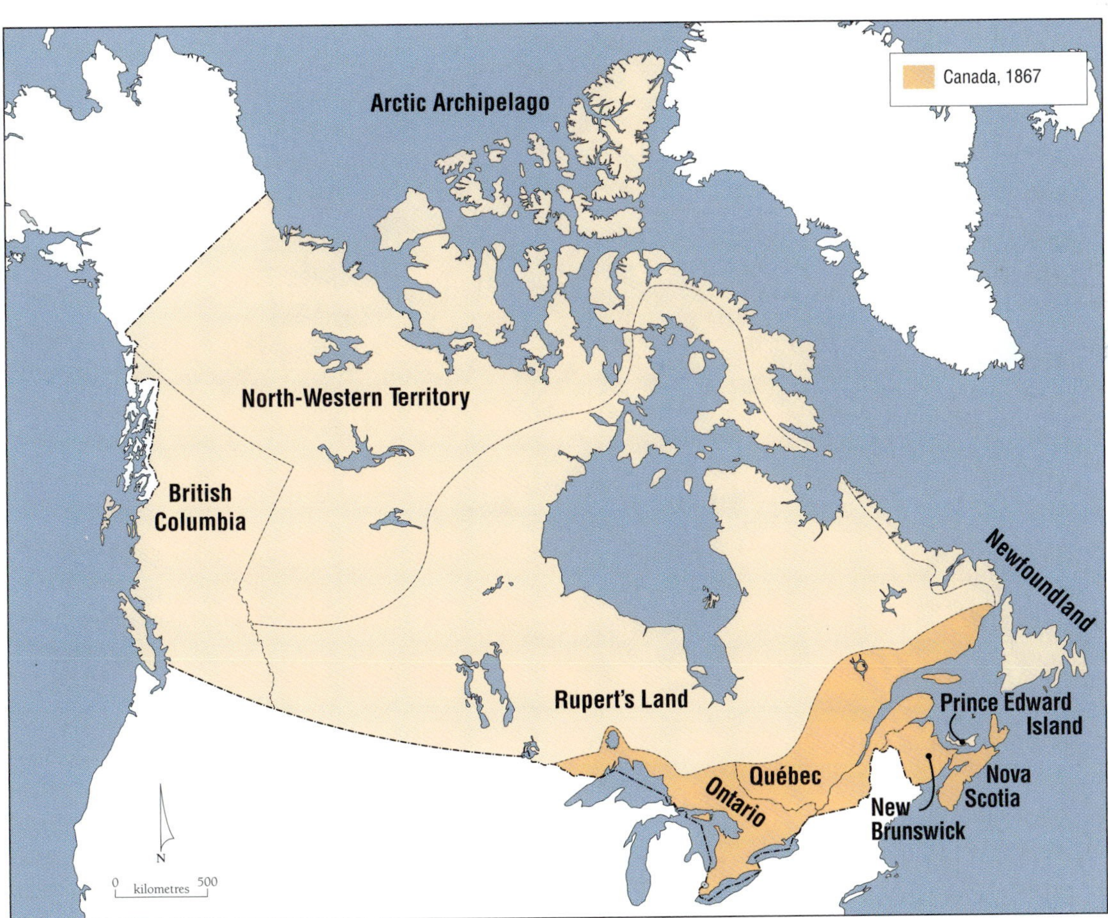

Figure 3.4 Canada, 1867. At Confederation, Canada was only a fraction of its current territorial extent. The Hudson's Bay Company controlled most of British North America, including Rupert's Land and the North-Western Territory. (Further resources: Student website, National Atlas section, Map 13. Website instructions are found on p. xx.)

the attraction of joining the Province of Canada had little appeal. Furthermore, unlike the Canadians, Maritimers continued to base their prosperity on a flourishing transatlantic trading economy with Caribbean countries and Great Britain. From 1840 to 1870, the backbone of the Maritime economy was the construction of wooden sailing-ships. Shipbuilding was so important that this period was known as the 'Golden Age of Sail' in the Maritimes. Even diplomatic pressure from Britain to join Confederation had little effect on Maritime politicians. But the Fenian raid into New Brunswick in 1866 and the termination of the Reciprocity Treaty with the United States quickly changed public opinion in the Maritimes (Vignette 3.3).[3] Shortly after the Fenian raids, the legislatures of both New Brunswick and Nova Scotia voted to join Confederation.

As we have seen, within a decade, the territorial extent of Canada expanded from four British colonies to the northern half of North America. The new Dominion grew in size with the addition of other British colonies and territories, and the creation of new political jurisdictions (Figure 3.5), while the British government transferred its claim to the Arctic Archipelago to Canada in 1880. Negotiations

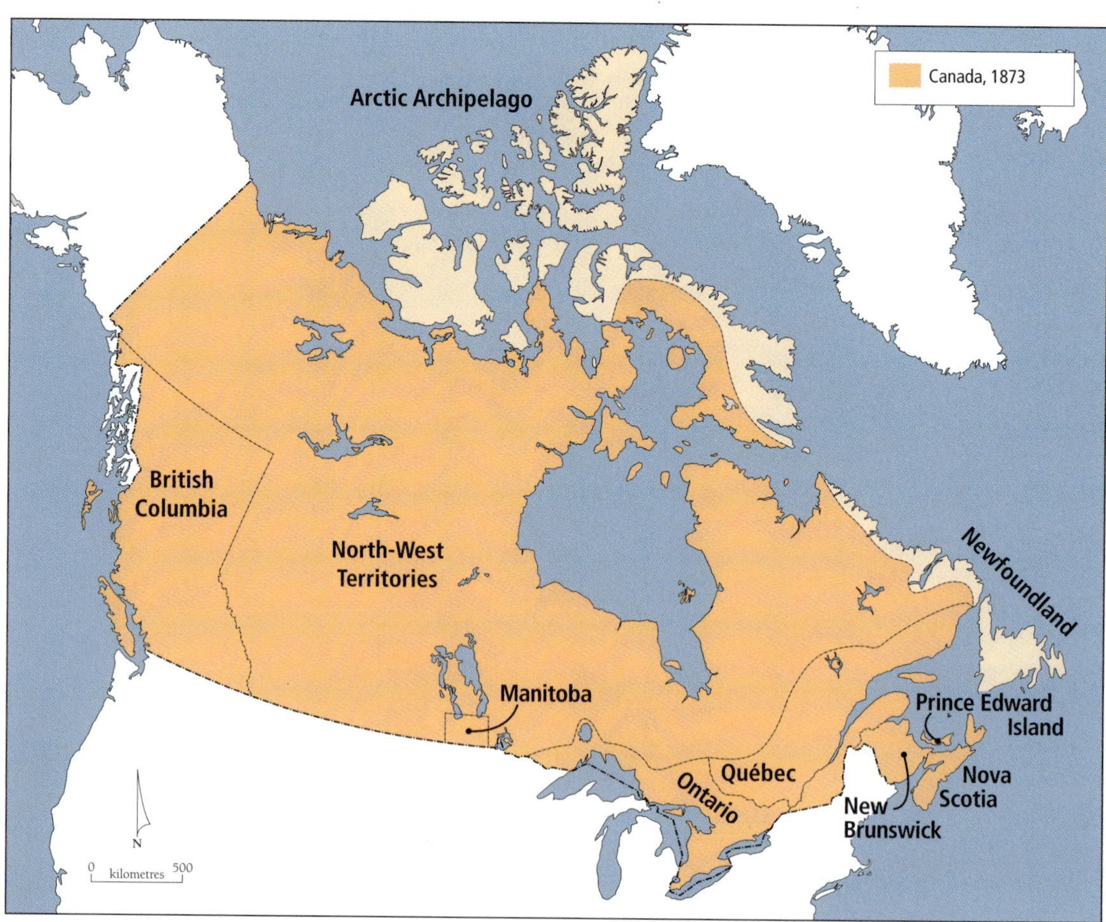

Figure 3.5 Canada, 1873. Within seven years of Confederation, Canada had obtained the Hudson's Bay Company lands (including the Red River Colony) and two British colonies (British Columbia and Prince Edward Island) had joined the new Dominion. For the North-Western Territory and Rupert's Land, the Crown paid the HBC £300,000, granted the Company one-twentieth of the lands in the Canadian prairies, and allowed it to keep its 120 trading posts and adjoining land. In 1870, these lands were renamed the North-West Territories. (Further resources: Student website, National Atlas section, Map 13. Website instructions are found on p. xx.)

between Ottawa and each British colony took place and treaties were signed with Indian tribes, but no such negotiations took place with the Red River Métis. The result was the Red River Rebellion of 1869–70, the formation of the Métis Provisional Government, and then negotiations with Ottawa that led to the entry of Manitoba into Confederation.

By 1880, Canada stretched from the Atlantic to the Pacific and north to the Arctic. While still part of the British Empire, Canada had begun the slow journey to independence and nationhood. In 1905 the provinces of Alberta and Saskatchewan were created, and much later, in 1949, Newfoundland joined Canada, completing the union of British North America into a single political entity. (The territorial evolution of Canada is illustrated in Figures 3.4 to 3.8 and Table 3.4.)

Within a decade after Confederation, Canada's territorial extent had burgeoned, making it the second largest country in the world, which posed a serious problem for the nation's leaders, who had to somehow transform this vast territory into a nation. Sir John A. Macdonald, Canada's first Prime Minister, resolved this in part by authorizing the construction of a transcontinental railway, the Canadian Pacific Railway.

National Boundaries

Well before Confederation in 1867, wars and treaties between Britain and the United States shaped many of Canada's boundaries. The southern boundary of New Brunswick, Québec, and Ontario was settled in 1783 when Britain and the United States signed the Treaty of Paris. Under this treaty, the United States gained control of the Indian lands of the Ohio Basin. Earlier, Britain had formally recognized the rights of Indians to these lands in the Royal Proclamation of 1763, which provided the constitutional framework for negotiating treaties with Aboriginal peoples. This recognition was the basis of Aboriginal rights in Canada (see 'Aboriginal Rights' in this chapter). Based on the fur trade route to the western interior, the boundary of 1783 passed through the Great Lakes to the Lake of the Woods.

In 1818, Canada's southern boundary was adjusted; it was set at 49° N from the Lake of the Woods to the Rocky Mountains. As for the northwestern boundary, there was some question as to where British territory ended and Russian territory began. British and Russians had come into contact through the fur trade. In the eighteenth century, Russian fur traders established trading posts along the Alaskan coast. Indians travelled along the Yukon River through the interior of Alaska to trade at these Russian forts. By the early nineteenth century, the Hudson's Bay Company had reached the tributaries of the Yukon River. In 1825, Britain and Russia set the northern boundary at 141° W (the Treaty of St Petersburg). Russia also agreed to relinquish to Britain its claims to the coastal regions south of 54°40' N to 42° N. In

Table 3.4 Timeline: Territorial Evolution of Canada

Date	Event
1867	Ontario, Québec, New Brunswick, and Nova Scotia unite to form the Dominion of Canada.
1870	The Hudson's Bay Company's lands are transferred by Britain to Canada. The Red River Colony enters Confederation as the province of Manitoba.
1871	British Columbia joins Canada.
1873	Prince Edward Island becomes the seventh province of Canada.
1880	Great Britain transfers its claim to the Arctic Archipelago to Canada.
1949	Newfoundland and Labrador joins Canada to become the tenth province.

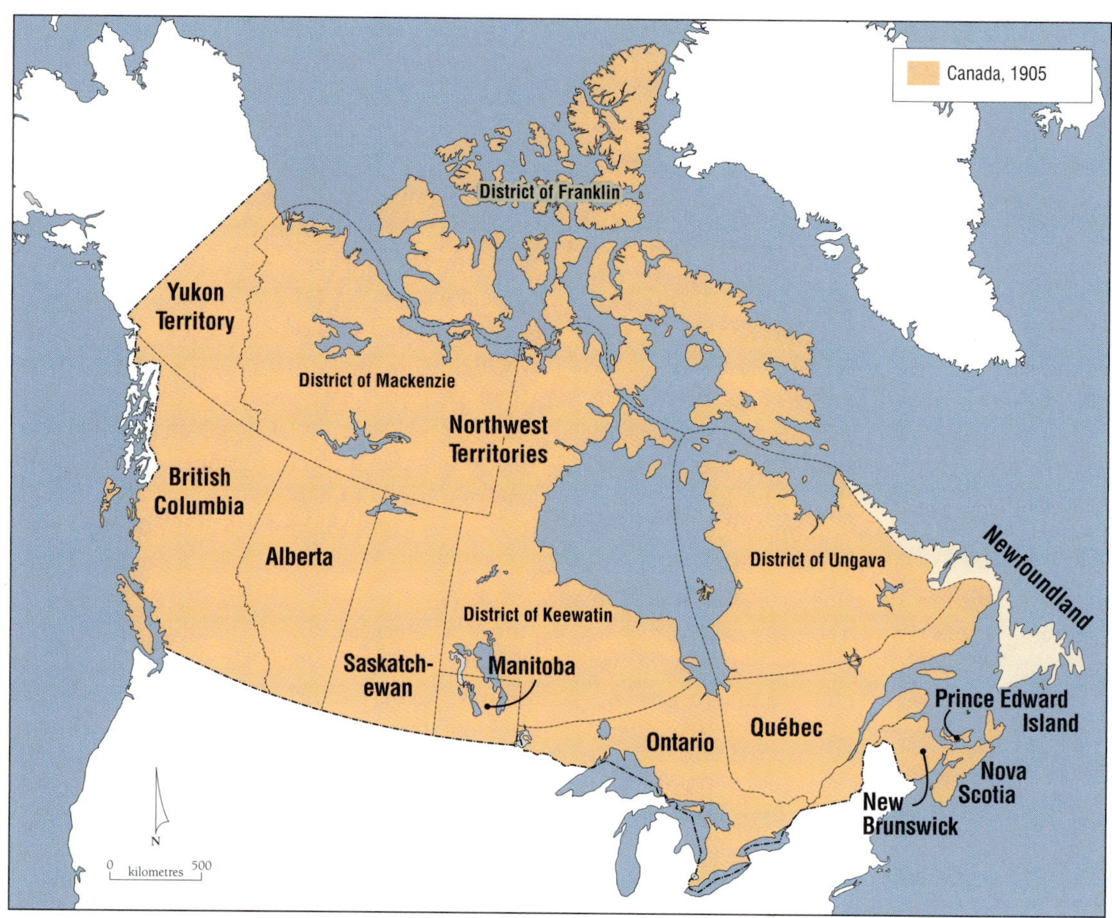

Figure 3.6 Canada, 1905. By 1905, two new provinces (Alberta and Saskatchewan) and one territory (Yukon) were created out of the North-West Territories. In 1905, what remained was renamed the Northwest Territories, and by this time the Arctic Archipelago (District of Franklin) had been incorporated into the NWT. Ontario, Québec, and Manitoba expanded their boundaries. (Further resources: Student website, National Atlas section, Map 13. Website instructions are found on p. xx.)

Atlantic Canada, the boundary between Maine and New Brunswick had not been precisely defined in 1783. Minor adjustments took place in 1842 when Britain and the United States concluded the Webster-Ashburton Treaty.

The last major territorial dispute between Britain and the United States took place over the Oregon Territory. The Oregon Territory stretched along the Pacific Coast from Alaska to Mexico—Mexico at that time extended northward to 42° N. In the early part of the nineteenth century, the Hudson's Bay Company established a fur-trading post at the mouth of the Columbia River, where it conducted trade with the local Indian tribes. In the 1830s, American settlers crossed the Rocky Mountains into the Columbia Valley, where they proceeded to cultivate the fertile soils of the Willamette Valley. Because both Britain and the United States had a foothold in the Oregon Territory, sovereignty was uncertain. Great Britain's claim was based on two factors: that (1) the Hudson's Bay Company had established the first settlement (Fort Vancouver) in the region, and (2) the Hudson's Bay Company exerted control over the land where the Indians trapped fur-bearing animals. The American claim was centred on the fact that the vast majority of settlers were Americans. In the final outcome, there was no doubt that occupancy was a more powerful claim to disputed lands than claims based on

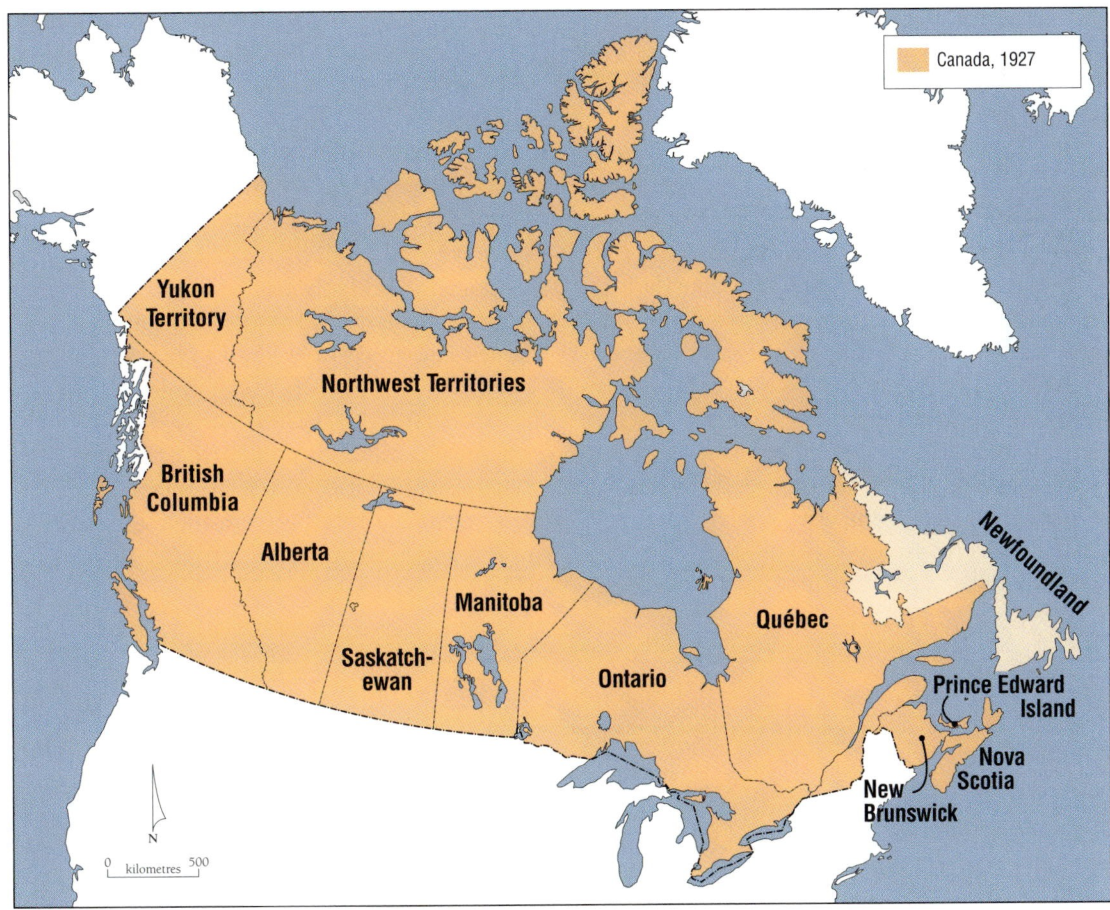

Figure 3.7 Canada, 1927. The boundary dispute between Newfoundland and Canada over the eastern boundary of Québec was resolved in Newfoundland's favour. Ontario, Québec, and Manitoba expanded again and attained their present boundaries. (Further resources: Student website, National Atlas section, Map 13. Website instructions are found on p. xx.)

exploration and the presence of a fur-trading economy. The boundary between British and American territory was set at 49° N with the exception of Vancouver Island, which extended south of this parallel.

Internal Boundaries

Since Confederation, the internal boundaries of Canada have changed (Figures 3.4–3.8). These changes have created new provinces (Alberta and Saskatchewan) and territories (Yukon and Nunavut). As well the boundaries of Manitoba, Ontario, and Québec were extended. In all cases, these political changes took land away from the Northwest Territories.

In 1870, the boundary of Manitoba formed a tiny rectangle comprising little more than the Red River Colony. The province's western boundary, while extended in 1881 and 1884, did not reach its present limit until 1912. By this time, Manitoba spread north to the boundary with the Northwest Territories (now Nunavut) and east to the Lake of the Woods.

Québec, too, received northern territories. In 1898, its boundary was extended northward to the Eastmain River and then eastward to Labrador. In 1912, Ottawa assigned Québec more territory that extended its lands to Hudson Strait. Canada also believed that the province of Québec should extend to the narrow coastal strip along the Labrador coast, while the colony of Newfoundland contended that Newfoundland owned all the land draining into the Atlantic Ocean. In 1927, this dispute between two British dominions (Canada and Newfoundland) went to London. The British government decided that the boundary between Québec and Labrador was not the narrow coastal strip proposed by Canada but rather the watershed of those rivers flowing into the Atlantic Ocean. While the Québec government has never formally accepted this ruling, Québec has conducted its affairs with Newfoundland as if Labrador were part of Newfoundland. For example, Hydro-Québec has made arrangements to purchase hydroelectric power produced in Labrador and the two provincial governments have discussed plans to produce more hydroelectric power in Labrador, sell it to Hydro-Québec, and then transmit the power through Québec to markets in New England.

In the years following Confederation, Ontario gained two large areas. In 1899, Ontario had its western boundary set at the Lake of the Woods (previously this area belonged to Manitoba), while its northern boundary was extended to the Albany River and James Bay. In 1912, Ontario obtained its vast northern lands, which stretch to Hudson Bay.

In 1905, Canada formed two new provinces, Alberta and Saskatchewan. The final adjustment to Canada's internal boundaries occurred in 1999 with the establishment of the government of the territory of Nunavut (Figure 3.8 and Table 3.5).

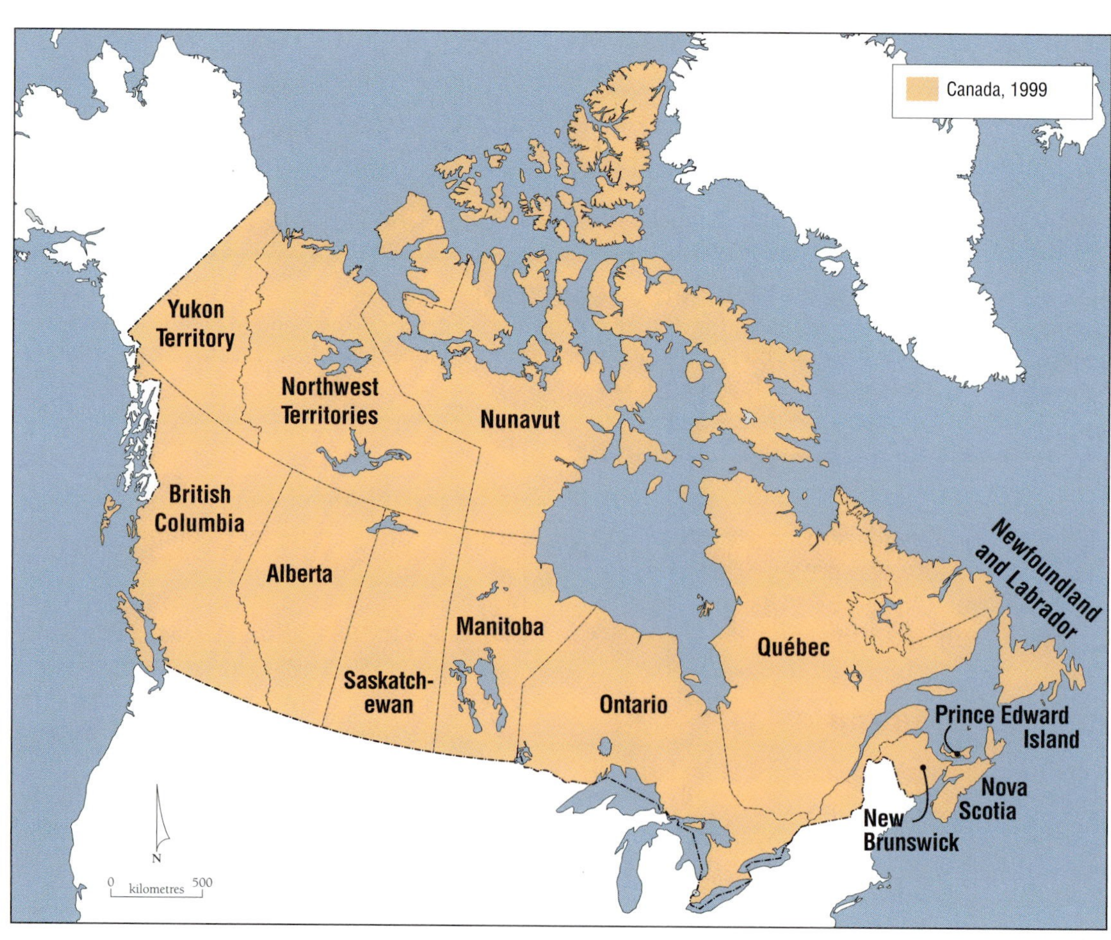

Figure 3.8 Canada, 1999. On 1 April 1999, Nunavut became a territory.

Table 3.5 Timeline: Evolution of Canada's Internal Boundaries

Date	Event
1881	Ottawa enlarges the boundaries of Manitoba.
1898	Ottawa approves extension of Québec's northern limit to the Eastmain River.
1899	Ottawa decides to set Ontario's western boundary at the Lake of the Woods and extend its northern boundary to the Albany River and James Bay.
1905	Ottawa announces the creation of two new provinces, Alberta and Saskatchewan.
1912	Ottawa redefines the boundaries of Manitoba, Ontario, and Québec, which extend their provincial boundaries to their present position.
1927	Great Britain sets the boundary between Québec and Labrador. Québec has never accepted this decision.
1999	A new territory, Nunavut, is hived off from the Northwest Territories in the eastern Arctic.

Faultlines

Canada's regional geography is partly defined by its faultlines, a notion introduced in Chapter 1. Tensions within Canadian society often begin as regional issues but soon turn into national issues. These disagreements are between one or more regions and Ottawa. The federal government, because it is charged with establishing national policies and programs, has the power to settle such issues. The challenge facing the federal government is to seek a balance between the regional factions, but such efforts rarely satisfy all parts of Canada. Nevertheless, the process of searching for solutions to these tensions informs Canadians of different views and, in doing so, lessens the tension—perhaps not initially but usually over time. Even so, Ottawa is most aware of the political power of Ontario and Québec and, whether real or not, federal policies seem to favour the two largest provinces, Ontario and Québec.

One recent example is the Kyoto agreement signed by Ottawa in December 2002. The Chrétien government saw the reduction of greenhouse gases as beneficial to all Canadians, but Alberta saw this agreement as placing it in a disadvantageous position because of its potentially negative impact on Alberta's heavy oil sands plants that discharge great amounts of carbon dioxide into the atmosphere. At the same time, Alberta interpreted the exemption of the Ontario automobile industry from reducing its greenhouse gas emissions as blatant favouritism. Ironically, Ottawa did little and Canada's greenhouse gas emissions actually increased. Early in 2007, the minority Conservative government woke up to the environmental 'crisis' and began to prepare legislation to regulate the emission of greenhouse gases.

Another example is the federal gun registration program that has created territorial/federal tensions. Ottawa sought to control firearms and, in that way, assist the police to combat crime. The six governments in Western Canada and the Territorial North oppose gun registration because the vast majority of their residents resent the registration system that included long guns (rifles and shotguns) and some refuse to participate. In 2006, the federal government proposed legislation, Bill C-21, to eliminate long guns from the registration list, but this legislation has failed to become law because three opposition parties support the registry.

The Centralist/ Decentralist Faultline

Geography poses powerful challenges to Canada's national unity. These challenges stem from Canada's vast space, economic competition between regions, regional trade patterns with the United States, and internal political differences between the provinces and Ottawa. In combination, these challenges to national unity are manifest as regional self-interest that often results in regional tensions. Sometimes these internal tensions are expressed by a group of individuals, such as fishers in Atlantic Canada, or through common interests among provinces, such as the oil-producing provinces. Whatever their form or geographic extent, these internal forces produce regional pressures that can be divisive.

Regional self-interest and resulting tensions (whether cultural, economic, or political) are a natural outcome of Canada's physical and human geography. For better or worse, regionalism is a fact of life in Canada and it may well be the most telling characteristic of Canada's national character. The regional challenge to Canada's east–west alignment, and therefore to Canada's national unity, may be summed up in four ways:

- Canadian regions are separated from each other by great distances, making trade and commerce between those regions more difficult (geographic forces).
- Regions compete with each other and provinces have trade barriers (forces derived from regional self-interest).
- Provinces compete over federal funding, since the division of political powers in the Canadian Constitution assigned costly services such as health care, education, and social services to provinces, and only the federal government has tax revenues large enough to pay for much of these services.
- Canadian regions are drawn into the economic orbit of the United States (continental economic forces).

The nature of Canada and the United States has always made trade a central issue between the two countries. For Canada, access to the American market is essential for its manufacturing and resource sectors. Canada's regions have their specific trade aspirations. As a consequence, the natural markets of North America have a north–south orientation. The Maritimes are geographically, historically, and economically tied to New England, while southern Ontario is linked to Michigan, Ohio, Pennsylvania, and New York. British Columbia is part of the Pacific Northwest and indeed has felt part of this international region referred to as Cascadia for many years.

Regional Tensions

The challenge of geography, therefore, has forced Ottawa to seek solutions to the problems bred by regional tensions, which are partly exacerbated by the regional implications of trade with the United States. Political solutions have attempted to overcome geography by creating transnational transportation systems, by fostering an industrial core, and by ameliorating regional disparities through equalization payments.

The first challenge facing the government of Sir John A. Macdonald was clear: a railway must span the country from the Atlantic Ocean to the Pacific Ocean. Without a more substantial presence, Macdonald's Conservative government feared that the US might annex the unoccupied parts of Western Canada (Vignette 3.3). From Ottawa's perspective, the railway would exert effective sovereignty over its western territories. Though the magnitude of this project was nearly beyond the capacity of the country, Ottawa proceeded. Its goals for this transcontinental railway were:

- to link the West with the rest of Canada;
- to settle the Canadian Prairies;
- to provide an export route for prairie grain;
- to create a market in the West for eastern industries.

At the same time as Ottawa secured its western borders, it launched a new economic initiative in 1879, the National Policy, which reinforced the centralist/decentralist faultline. The National Policy and other federal policies stimulated manufacturing in Ontario and Québec. Rightly or wrongly, Canadians living outside Central Canada believe that the concentration of political power in Ontario and Québec has had an unfair influence over national policies and therefore favoured economic development in Central Canada over that in the rest of the country. In the following pages, these two themes—economic and political power struggles—are explored from the perspective of Central Canada (the core) and the rest of Canada (the periphery). In the case of economic power, Central Canada refers to the industrial core in southern Ontario and southern Québec. In the case of political power, Central Canada refers to Ottawa and indirectly to the influence of Ontario and Québec in Ottawa. Together these two power struggles are the basis of the centralist/decentralist conflict or faultline.

Centralists advocate a strong central government, national policies that exert a political dominance over provinces, and a strong national economy (which means an expanding industrial core). Decentralists seek to strengthen the powers allocated to provinces. In particular, decentralists call for a devolution of federal powers to the provincial governments and the expansion and diversification of regional economies. Economic diversification is deemed necessary to generate more jobs and thereby encourage population growth. Eventually, it is thought that economic diversification will lead to larger populations in hinterland regions and thus more political representation, and therefore power, in Ottawa.

Economic Power Struggle

In 1879, following the creation of the National Policy, Canada's manufacturing industries were protected from foreign goods by high tariffs and bolstered by lower customs duties on raw materials. The goal was to strengthen Canada's economy by creating a national industrial base. In fact, this economic policy fostered the development of an economic core and a resource hinterland. For Central Canada, the benefits from the policy were considerable as it ensured the region's role as Canada's industrial heartland. It also resulted in the concentration of financial power in Montréal and Toronto. Hinterland regions were not so fortunate because they were compelled to sell their raw materials and foodstuffs on the world market and buy their manufactured goods in the domestic market. For the hinterland, this arrangement translated into 'selling their products cheap' and 'buying expensive goods from Central Canada'. Feelings ran high because people in the hinterland regarded the National Policy as an arrangement that benefited Central Canada at the expense of the rest of the country. They became suspicious of other national programs announced by Ottawa because they appeared to be designed to favour Central Canada and thereby to exploit other regions.

The National Policy accentuated the economic differences in various parts of the country. In doing so, it increased regional tensions, which resulted in the centralist/decentralist (or heartland/hinterland) divide. While each hinterland region faced a different set of economic relations with the industrial heartland of Canada, the basic economic relationship was the same—Central Canada produced the manufactured goods, while hinterlands produced raw materials and foodstuffs. But what prevented hinterlands from developing a manufacturing base under the National Policy?

Geography prevented manufacturers in the Maritimes from reaching markets in Central Canada and the growing market in the West. While the steel mill in Cape Breton did supply some of the steel rails for the construction of the Canadian Pacific Railway, most Maritime firms could not overcome the friction of distance to reach Central Canada and the West. Maritime industrialists also were no longer able to sell their products in the nearby market of New England because of high American tariffs. The combination of a small local market, great distance to the continental markets in Canada, and high American tariffs stalled economic development in the Maritimes. Before 1867, most Maritimers were uncertain about the benefit of Confederation; afterwards, many became convinced that union with Canada was a 'bad deal'.

At the same time, unrest in Western Canada was widespread in rural areas. Under the national economic policy, western farmers had to sell their grain on the world market where prices were low, but they had to purchase their farm machinery from manufacturers in Central Canada where prices were high. By the 1920s, this unrest took the form of political protest when the Progressive Party argued for free trade to allow farmers access to lower-priced American farm machinery. Geography compounded the problems facing grain farmers, who were located in the heart of North America. The cost of transporting their grain to foreign markets was very high, thereby making their returns even lower. How could farmers overcome the great distance to their grain markets, particularly those in Europe? They thought that the solution to this transportation problem lay in the construction of a railway to the nearest tidewater point—Hudson Bay! The Hudson Bay Railway, completed in 1929, never fulfilled the high expectations of western farmers because the savings derived from the shorter rail distance were offset by higher marine insurance costs due to the danger of icebergs. Farmers, however, believed the higher insurance charges were simply another trick by Central Canada to hold onto the grain trade at the expense of westerners.

British Columbia, though linked to the rest of Canada by the Canadian Pacific Railway, remained beyond the economic pull of Central Canada. British Columbia's economic order was driven by its geographic position on the west coast. The region's natural markets were overseas. At first, its raw materials, such as fish, timber, and minerals, were shipped to markets in the western United States, the British Isles, and the Far East. As well, because of British Columbia's geographic proximity to Western Canada, many of the Prairies' products were transported to the port of Vancouver for transshipment overseas. In 1914 the opening of the Panama Canal had a great impact on the forest industry in British Columbia because it provided a 'shorter' shipping route to markets in the British Isles, Western Europe, and the east coast of the United States. Like Canadians in Western Canada, British Columbians resented having to pay for higher-priced manufactured goods from Central Canada compared to those available just across the border.

The Territorial North suffered a different fate. Until World War II, Ottawa simply ignored it. Regarded by the federal government as a remote wilderness inhabited by Indians and Eskimos (later known as Inuit), the Territorial North remained a fur economy and did not participate in the events transforming Canada's emerging industrial society until well into the twentieth century. Ottawa's laissez-faire approach limited federal expenditures to Aboriginal peoples in the Territorial North. The few northern Royal North West Mounted Police posts (later the Royal Canadian Mounted Police) served as Ottawa's main expression of administration. Only when resource developments occurred, such as the Klondike gold rush, did Ottawa hurry to ensure a broader federal presence. After World War II, the federal government's laissez-faire policy was replaced by state involvement in northern development. This change was sparked by Prime Minister John Diefenbaker's 'northern vision', which translated into investments in northern infrastructure, especially highways leading to resources.

Political Power Struggle

While the core/periphery model emphasizes the core's economic dominance over the periphery, the core also exerts other forms of dominance, including political dominance. Within the context of Canada, Ottawa not only represents the core but, by favouring Central Canada (Ontario and Québec), extends the advantages of this political dominance to a particular area of Canada. This leads to a sense of alienation and frustration among the remaining regions. For example, in

Vignette 3.4 Federal/Provincial Powers

Under Canada's federal system, the powers of government are shared between the federal government and the 10 provincial governments. All provinces have the same powers. At the time of Confederation, the powers of the provinces were carefully enumerated (and thus limited), while those of the federal government were not limited. Provincial governments are responsible for education, health and welfare, highways, civil law (property and civil rights), local government, and natural resources, while Ottawa has a much wider mandate, including authority over defence and external affairs, criminal law, money and banking, trade, transportation, citizenship, and Indian affairs. The two levels of government were assigned joint jurisdiction over agriculture, immigration, and taxation. Territorial governments are assigned their powers from Ottawa, while municipal governments obtain theirs from the provincial governments. First Nations are seeking self-government. While First Nations' status is not yet defined by the federal government, Indian leaders often refer to a 'third level of government', meaning a political level somewhere between municipal and provincial.

the late 1970s, provincial–federal relations soured during debates over constitutional reform. The western provinces (Alberta, British Columbia, Manitoba, and Saskatchewan), supported by Nova Scotia and Newfoundland, had developed their own agenda for constitutional reform to obtain more powers for the provinces, including a reform of the Senate and the Supreme Court, to reflect regional interests. Québec Premier René Lévesque did not participate in these discussions, leaving it to the anglophone provinces to 'dismantle' Confederation. Ottawa's most consistent ally was Ontario, which confirmed hinterland charges that Ontario was the chief beneficiary of old federalism.

But does Ottawa really favour Central Canada or is it just a problem of balancing national against regional interests? The reality lies in the division of powers between the two levels of government (see Vignette 3.4). Can such a power imbalance exist in a federal state? How much power does a province need? How much power can the central government give up? These questions go to the heart of a federal state and the fragile relations between the central and regional governments. This tug of war also is played by the separatist movement in Québec, but for an end (independence) that is different from that sought by other provinces and premiers (more powers). In 1968, Jacques Parizeau likened the Canadian federation to a chicken being plucked by provincial governments led, of course, by Québec. When the last feather is plucked, the chicken will perish. Voilà, Québec has achieved its goal of independence (Nemni, 1994: 175). Québec is not the only province with a desire for more power, but such power in the rest of Canada focuses on financial power. In 2007, Premier Danny Williams of Newfoundland and Labrador 'declared war' on the federal government because, in Premier Williams's opinion, the 2007 version of equalization payments shortchanged his province. (For more on this dispute, see Vignette 9.5 and Figure 9.2.)

Just how are regions represented in Parliament? Canada's political geography is shaped by the election of representatives from various political parties to the House of Commons.[4] The electoral system is designed to reflect changes in Canada's population based on census figures each 10 years. In other words, the current apportionment of seats in the federal House of Commons is based on the

2001 census until 2011. In the 2006 federal election, the total number of seats was 308 (Table 3.6). The number elected and their electoral districts are determined by the size of Canada's population, while the assignment of electoral districts across the country is determined by Canada's population distribution. There are two exceptions. First, provinces must have at least the same number of seats in the House of Commons as in the Senate. Prince Edward Island has four seats in the Senate and therefore has four seats in the House of Commons, though its population only warrants one or possibly two seats. Second, each territory is allocated one seat irrespective of its population. Since Ontario and Québec have the largest populations, they also have the largest number of seats and, therefore, hold the political balance of power. The 2006 federal election results support this proposition, with the majority of seats in the House of Commons held by Ontario with 106 and Québec with 75—58.8 per cent of the total number of seats. This concentration of electoral power is further intensified by a concentration of political power within the governing party, which is naturally concerned with national issues and priorities, not regional ones. From the standpoint of those outside Central Canada, national policies appear to favour Ontario and Québec. The Senate is supposed to provide a regional balance in Canada's political system by giving a political voice to 'regions'. However, since its members are appointed by the Prime Minister largely on the basis of long-time service to the political party currently in power, senators are more concerned about party loyalty than regional interests. In short, the Senate fails to provide a regional counterweight to the House of Commons. Instead, this role falls to the provincial governments.

The struggle over political power takes place within the six geographic regions of Canada (Ontario, Québec, British Columbia, Western Canada, Atlantic Canada, and the Territorial North), which are far from equal. Ontario and Québec have most of Canada's population, its manufacturing industries, and national corporate headquarters. Of the six geographic regions, Ontario and Québec (Central Canada) are at the top of the economic and political hierarchy. In sharp contrast, the Territorial North is at the bottom. It has no industrial base; it has the smallest population of any geographic region; and it is far from the decision-making centres in Canada. Using population size and economic power (as measured by gross domestic product) as two indicators of the strength of regional power,

Table 3.6 Members of the House of Commons by Geographic Regions, 2006

Geographic Region	Members (number)	Members (%)	Population (%)	Difference	Seats per 1,000 population
Territorial North	3	0.9	0.3	0.6	1:33
Atlantic Canada	32	10.4	7.2	3.2	1:71
Western Canada	56	18.2	17.1	1.1	1:97
Québec	75	24.4	23.9	0.5	1:101
British Columbia	36	11.7	13.0	(1.3)	1:114
Ontario	106	34.4	38.5	(4.1)	1:115
Total	308	100.0	100.0		

Source: CBC.ca (2006).

the remaining regions fall between Ontario and the Territorial North (Table 1.2).

These regional variations in political and economic power make Canada a 'troublesome country to govern'. Each Prime Minister from Sir John A. Macdonald to Steven Harper has had to balance national economic interests against regional ones. Federal interventions in the marketplace are attempts to protect national interests, but they sometimes result in regional alienation.

Western Alienation

Western alienation represents a serious problem for Ottawa. It takes many forms but is based on western provinces' lack of power, whether real or perceived, to control their destiny. For Alberta, the fear is that Ottawa will 'steal' their oil wealth to benefit Ontario and Québec (Figure 3.9). The intensity of this regional alienation is strongest in Western Canada but particularly in Alberta and in British Columbia. Both regions complain about their lack of political power in Ottawa, which often translates into complaints that Ontario and Québec receive favoured treatment from the federal government. Since these two central provinces contain almost 63 per cent of Canada population and 59 per cent of the seats in the House of Commons, federal parties must have a strong foothold in one or both of these provinces to form the federal government. This demographic/electoral fact gives Ontario and Québec considerable political leverage in Ottawa. The National Energy Program was deeply resented by Alberta and while the program ended in 1984, the resentment and fear of a new version remains in the minds of Albertans.

National Energy Program

Resource development falls under provincial powers, yet, in 1980, the federal government imposed the National Energy Program on oil-producing provinces. Ottawa had three goals: energy security, greater Canadian ownership of the oil industry, and a greater amount of wealth produced by the oil industry for Ottawa. A fourth but unspoken goal was to keep oil prices low in Ontario and Québec. The National Energy Program lasted only four years (1980–4), but its chief legacy was the western oil-producing provinces' deep distrust of the federal government. Why was this so?

Without a manufacturing base, the Canadian hinterland has had to depend on its natural resources as an extremely important source of regional development and provincial tax revenue. Thanks to its geology, Alberta has abundant petroleum deposits. This fact, plus the sudden rise in oil prices in the 1970s, turned Alberta from a have-not province to a have province. Since natural resources fall under the jurisdiction of the provincial governments, the federal government has no authority to 'tax' natural resources, yet with its national energy policy, Ottawa was able to force oil-producing provinces to share their rapidly increasing oil revenue with the federal government. Ottawa acted in this way, knowing full well that it would lead to western

Figure 3.9 Hand in the pocket. Ontario calls for a 'sharing' of Alberta's oil wealth. (Michael de Adder, *The Halifax Daily News*)

alienation, because it believed the National Energy Program was in the national interest.

In the 1970s, the price of world oil rose rapidly because of the Organization of Petroleum Exporting Countries (OPEC) strategy: by curtailing the global supply of oil, the world price would rise. From 1972 to 1980, the world price of oil increased from US $2 a barrel to over US $20 a barrel. From 1973 to 1978, agreements were reached between the oil-producing provinces and the federal government to match the domestic price with the world price. Rising oil prices generated a huge profit for the oil companies and greatly increased oil revenues for Alberta and, to a much lesser degree, Saskatchewan and British Columbia. In 1979, however, Ottawa refused to match domestic oil prices with world prices, thereby creating a lower domestic price. For Central Canada, such a national policy gave its industrial firms a decided advantage compared to those across the US border. With world prices continuing to rise, subsidization of oil refineries in Atlantic Canada and Québec was necessary to meet the lower domestic price. The federal government needed a new energy strategy that addressed two key concerns. How could Ottawa pay for this price differential? Should not part of this 'windfall' that companies and provincial governments receive be redirected to the federal government?

Peter Smith, a professor of geography at the University of Alberta, denounced the National Energy Program as 'one more manifestation of the customary heartland outlook; what is good for Ontario and Québec has to be best for the rest' (Smith, 1982: 301–2).

Vignette 3.5 Equalization Payments

Regional equity is an important theme in Canada. In 1957, the federal and provincial governments recognized the importance of providing similar public services in each of the provinces. Equalization payments would provide the funds necessary to achieve this goal. As shown in Table 3.7, the distribution of these funds has favoured those provinces with weak economies. Equalization payments have ensured that per capita revenues of all provinces from shared taxes (personal income taxes, corporate income taxes, and succession duties) matched those of the wealthiest provinces. From 1993 to 2006, $125 billion or nearly $10 billion annually was transferred from Ottawa to the have-not provinces. About half of these payments had gone to Québec and 35 per cent to Atlantic Canada. In 2007–8, equalization payments amounted to nearly $12.8 billion. The territorial governments also received funds from the federal government, but these transfers fell under a separate program (Territorial Formula Financing).

In 2004, the federal government made a side agreement with Newfoundland and Labrador and Nova Scotia to allow both provinces to retain their equalization payments. Without this agreement, the two provinces would have seen their equalization payments decrease as their oil and gas revenues increased. Saskatchewan has requested a similar agreement (see Vignettes 8.3 and 9.5)

Equalization payments form a significant portion of the gross revenues for Québec and Atlantic Canada while Territorial Formula Financing provides almost the entire revenues for the three territories. While Québec receives the largest amount of equalization payments, these federal transfers amount to about 30 per cent of Québec's gross general revenues. For the four Atlantic provinces, however, these funds form about half of their gross general revenues. Fiscal dependency is even greater in the territories, with Territorial Formula Financing amounting to about 90 per cent of territorial revenues.

Equalization and Transfer Payments: A Fundamental Canadian Principle of Sharing?

The existence of a fiscal imbalance was the battle cry of the premiers following the 2006 election. From their perspective, equalization and transfer payments from the federal government needed to be increased. Surprisingly, the new federal government agreed, but the question was how? While all agreed that the federal government should provide more funds to provinces and territories to allow them to deliver education, health, and social programs to Canadians, the premiers could not agree on how to accomplish this task. For example, Ontario advocated an increase in transfer payments and no increase in equalization payments while many other provinces called for a substantial increase in equalization payments. Finally, Alberta and Saskatchewan strongly opposed including non-renewable resource revenues in the calculations of equalization payments.

This debate has its roots in Confederation when taxing and programs were divided between the two levels of government. By 2006, the original balance between taxing powers and program delivery had changed radically. First, the federal government, because of its greater taxing powers, generates much more revenue than do the provinces. Second, provinces are responsible for delivering what turned out to be very costly programs such as health care and post-secondary education. This disparity between the two levels of

Table 3.7 Total Equalization Payments ($ millions), 1993–4 to 2008–9

Year	PEI	NB	NL	NS	Man.	Que.	Sask.	BC	Total
1993–4	175	835	900	889	901	3,878	486	0	8,064
1994–5	192	927	958	1,065	1,085	3,965	413	0	8,605
1995–6	192	876	932	1,137	1,051	4,307	264	0	8,759
1996–7	208	1,019	1,030	1,182	1,126	4,169	224	0	8,958
1997–8	238	1,112	1,093	1,302	1,053	4,745	196	0	9,739
1998–9	238	1,112	1,068	1,221	1,092	4,394	477	0	9,602
1999–2000	255	1,183	1,169	1,290	1,219	5,280	379	125	10,900
2000–1	269	1,260	1,112	1,404	1,314	5,380	208	0	10,947
2001–2	256	1,202	1,055	1,315	1,362	4,679	200	240	10,309
2002–3	235	1,143	875	1,122	1,303	4,004	106	71	8,859
2003–4	232	1,142	766	1,130	1,336	3,764	0	320	8,690
2004–5	277	1,326	762	1,313	1,607	4,155	652	682	10,774
2005–6	277	1,348	861	1,344	1,601	4,798	82	590	10,901
2006–7	291	1,386	632	1,451	1,709	5,539	13	260	11,281
2007–8	294	1,308	477	1,477	1,826	7,160	226	0	12,768
2008–9 (projected)	310	1,294	197	1,492	2,003	7,622	0	0	12,918

Sources: Canada, Department of Finance (2006, 2007). Reproduced with permission of the Minister of Public Works and Government Services, 2007.

government in regard to revenues and expenditures has caused the so-called fiscal imbalance. Efforts to adjust this imbalance take place through equalization and transfer payments. Equalization payments are designed to ensure that the citizens in 'have-not' provinces enjoy a standard of public services comparable to those in 'have' provinces. This simple concept involves the collections of taxes by Ottawa and then the redistribution of some of those taxes to have-not provinces. While the equalization formula is complex, the federal government gives money to each province whose fiscal capacity is below a national standard. Fiscal capacity refers to a province's ability to raise revenue. At the moment, Alberta, Ontario, and Saskatchewan are the only provinces with this capacity—the remaining seven provinces receive various levels of equalization payments.

Transfer payments consists of two federal programs. They are the Canada Health Transfer (health care) and the Canada Social Transfer (post-secondary education, social assistance, and social services). Payments are based on a per capita basis. In 2005–6, $32 billion were sent to the provinces and territories under the Canada Health Transfer program and $15.5 billion under the Canada Social Transfer program.

Regional differences provide a major stumbling block to the premiers finding a solution. The objective of their June 2006 meeting was to agree on new funding arrangement for equalization and transfer payments that would resolve the fiscal imbalance. Yet, they could not find common ground on this matter because of provincial self-interest. As Premier Gordon Campbell of British Columbia observed: 'If you have a debate between principle and money, money is going to win out' (Martin, 2006: A7). Campbell went on to say that 'It's ironic, I think, that this whole argument about equalization, which should be something that unites the country, tends to be something that really exacerbates divisions' (Bryden, 2006). Without a common front, the federal government will have to make the final decision.

The Aboriginal/ Non-Aboriginal Faultline

In 1867, the British North America Act made Ottawa responsible for the Indian tribes. Later, Ottawa's responsibility was extended to all Aboriginal peoples, including the Inuit and Métis. For that reason, the Aboriginal/non-Aboriginal faultline is cast into an Ottawa versus Aboriginal peoples framework. In this model, Ottawa acts like a core, while Aboriginal Canadians living on the edge of Canadian society serve as the periphery. Originally, Ottawa's objective was the assimilation of Indian peoples into Canadian society. Instead, Aboriginal peoples were marginalized. Since the 1970s, Ottawa has adopted a more enlightened policy, stressing three elements: settling of outstanding land claims, recognizing Aboriginal right to self-government, and accepting that the concerns of each category of Aboriginal peoples (Indians, Métis, and Inuit) are different and that such concerns require specific solutions. Aboriginal peoples' struggle for power is rooted in two questions: Who are the Aboriginal peoples of Canada? What are Aboriginal rights?

Aboriginal Peoples

The Constitution Act, 1982 refers to Indians, Métis, and Inuit under the umbrella term **Aboriginal peoples**, that is, those now living in Canada who can trace their ancestry to the original inhabitants who were in North America before the time of contact with Europeans in the fifteenth century. Indians are further distinguished between status, non-status, and treaty Indians. People legally defined as **status (registered) Indians** are registered or entitled to be recorded as Indians, according to the Indian Act as amended in June 1985, and have certain rights acknowledged by the federal government, such as tax exemption for income generated on a reserve. In 1980, 317,000 Canadians were status (or registered) Indians. By 2005, the number of status Indians had grown to 748,000. **Non-status Indians** are people of Indian ancestry who are not registered as Indians and therefore have no rights under the Indian Act. **Treaty Indians** are status or registered Indians who are members of (or can prove descent from) a band that signed a treaty. They have a legal right to live on a reserve and participate in band affairs. Less than half live on reserves. The **Métis** are people of European and North American Indian ancestry. The **Inuit** are Aboriginal people located mainly in the Arctic. In 2001, 1,319,890 people reported Aboriginal ancestry but only 976,305 claimed to have an Aboriginal identity. This latter figure was composed of 608,850 Indians, 292,310 Métis, and 45,070 Inuit (Statistics Canada, 2003).

The Indians, Inuit, and Métis are a highly diversified population. One indication of their cultural diversity is linguistic classification. There were approximately 55 distinct Aboriginal languages (of 11 language families) spoken in Canada at the time of original contact. Five of the language families were spoken along the Pacific coast, while only two were spoken east of Manitoba and one (Inuktitut) in the Arctic (Figure 3.3). The largest language family is Algonkian. There are 15 distinct Algonkian-based languages, the most common of which are Cree and Ojibwa. Inuktitut, the Inuit language, has regional dialects and is spoken across the Canadian Arctic.

Another measure of Aboriginal diversity is self-identification. Many Indians prefer to identify themselves with the name of their tribal group, such as Cree or Iroquois, while others use the name of their First Nation (band) for a more precise identification. For example, the Cree occupy a vast territory that stretches from northern Québec to Alberta. There are many Cree tribes within that territory. A Cree living in northern Saskatchewan might identify himself or herself as a member of a Cree band, such as the Lac La Ronge band.

In 2004, the largest First Nation was southern Ontario's Six Nations of the Grand River (Iroquois) with a population of 22,086. The next five largest were the Mohawk of Akwesasne (10,052) at St Regis on the Ontario–Quebec border near Cornwall, the Blood (9,679) in southern Alberta, Kahnawake (Mohawk) (9,275) near Montréal, the Saddle Lake Cree reserve (8,253) outside Edmonton, and the Lac La Ronge Cree in northern Saskatchewan (7,840). Most First Nations people live on reserves (53 per cent) while the remainder live off-reserve. More and more First Nations people are moving to cities.

Aboriginal peoples are reclaiming their identity and place names. Some bands are relinquishing the names given to them by Europeans in favour of their original names, such as Anishinabe (for Ojibwa) and Gwich'in (for Kutchin). The landscape is also being reclaimed. For example, the Arctic community of Frobisher Bay, named after the English explorer, Martin Frobisher, is now Iqaluit, which means 'the place where the fish are', the capital city of Nunavut ('our land').

Aboriginal Rights

Aboriginal rights are group or collective rights that stem from Aboriginal peoples' occupation of the land before contact. Such rights apply most readily to status Indians and Inuit, while Métis are less well served. Aboriginal peoples' traditional attitudes and values towards land and wildlife are strikingly different from those held by most non-Aboriginal Canadians. For Canada's First Peoples, the land has not only economic value but also cultural, political, and spiritual value. Aboriginal attitudes and values are based on their former economic and social systems, that is, the largely subsistence system of hunting/fishing and gathering that was in place before contact with Europeans.

Land rights are the most fundamental Aboriginal rights. Indeed, it is from land rights that most other collective rights flow, such as self-determination and self-government. Aboriginal peoples first began to secure land rights or treaty rights before Confederation and this process continues today. Some Aboriginal peoples, including the Métis, are still negotiating with Ottawa.[5] Treaties set aside **reserve** land, held collectively by and for the benefit of the band. Until the late 1950s, all Indian reserves were under the absolute control and governance of the federal Department of Indian Affairs through local Indian agents, federal administrators who oversaw the perceived education, health, legal, and financial needs and business dealings of local band members.

The reasons for signing treaties varied depending on the historical context. Authorities acting on behalf of the Crown often signed treaties to secure Aboriginal peoples as allies during times of turmoil, such as the War of 1812, or simply to procure land for the growing numbers of settlers coming to Canada. During the expansion to the West, treaties were signed to provide a place for Indian tribes so that Indian wars common south of the border would not erupt in Canada. For Aboriginal peoples, treaties often promised reserves that would not be available to settlers and support during the transition from semi-nomadic hunting to sedentary farming. The numbered treaties for the Plains Indians therefore offered protection from the anticipated flood of settlers and some guarantee that the federal government would care for them now that their principal source of food, the buffalo, was gone. However, treaty assurances of federal assistance were often not met (see Carter, 2004; Brownlie, 2003).

The terms of each treaty varied, although they generally included cash gratuities and presents at the signing of the treaty, annual payments in perpetuity, the promise of educational and agricultural assistance, the right to hunt and fish on Crown land until such land was required for other purposes, as well as land reserves to be owned by the Crown in trust for the Indians. In Treaty No. 6 of 1876, for example, which covered much of central Saskatchewan and Alberta, the amount of land assigned to each tribe was determined by its population size, i.e., each family of five received one square mile. Reserves are collectively owned by First Nations bands, though legally they are owned by the Crown in trust for them.

Conflicting ideas as to the significance of treaties between the signing parties largely shaped Aboriginal and non-Aboriginal relations in Canada during the twentieth century. When treaties were signed, Crown authorities viewed them as vehicles for extinguishing Aboriginal rights and titles to land and thus for opening the land to agricultural settlement. Aboriginal peoples, however, viewed them as agreements between 'sovereign' powers to share land and resources. With such diverse perceptions of treaties, disagreements were inevitable. Added to the issue of perception, some Aboriginal groups never signed treaties, while other treaties were never fulfilled. The latter part of the twentieth century witnessed various movements to repair injustices related to Aboriginal land rights (for example, through land claims) and to recognize Aboriginal peoples' inherent right to self-determination (for example, through constitutional reform). Through these attempts and through the signing of modern treaties, Aboriginal peoples are seeking a new place in Canadian society (Vignette 3.6) in the hope of participating in the larger economy while still retaining their culture.

From Hunting Rights to Modern Treaties

History sometimes makes strange allies. Shortly after Pontiac, chief of the Odawa, led a successful uprising against the British in 1763, Britain decided to form an alliance with him and other Indian leaders.[6] For strategic reasons, George III issued the Royal Proclamation of 1763, which identified a part of British territory west of the Appalachian Mountains as Indian lands. At that time, Britain believed that it could claim 'uninhabited land', which the British defined as land without permanent occupation (that is, no cultivated land or permanent settlements). However, the British also believed that Indians had a limited ownership over the lands they inhabited, and that therefore such lands must be purchased from the owners. This somewhat ambiguous concept remains the basis of land claims by Canadian Aboriginal peoples (Figure 3.10).

The legal meaning of Aboriginal title to land has evolved over time. Until the 1970s,

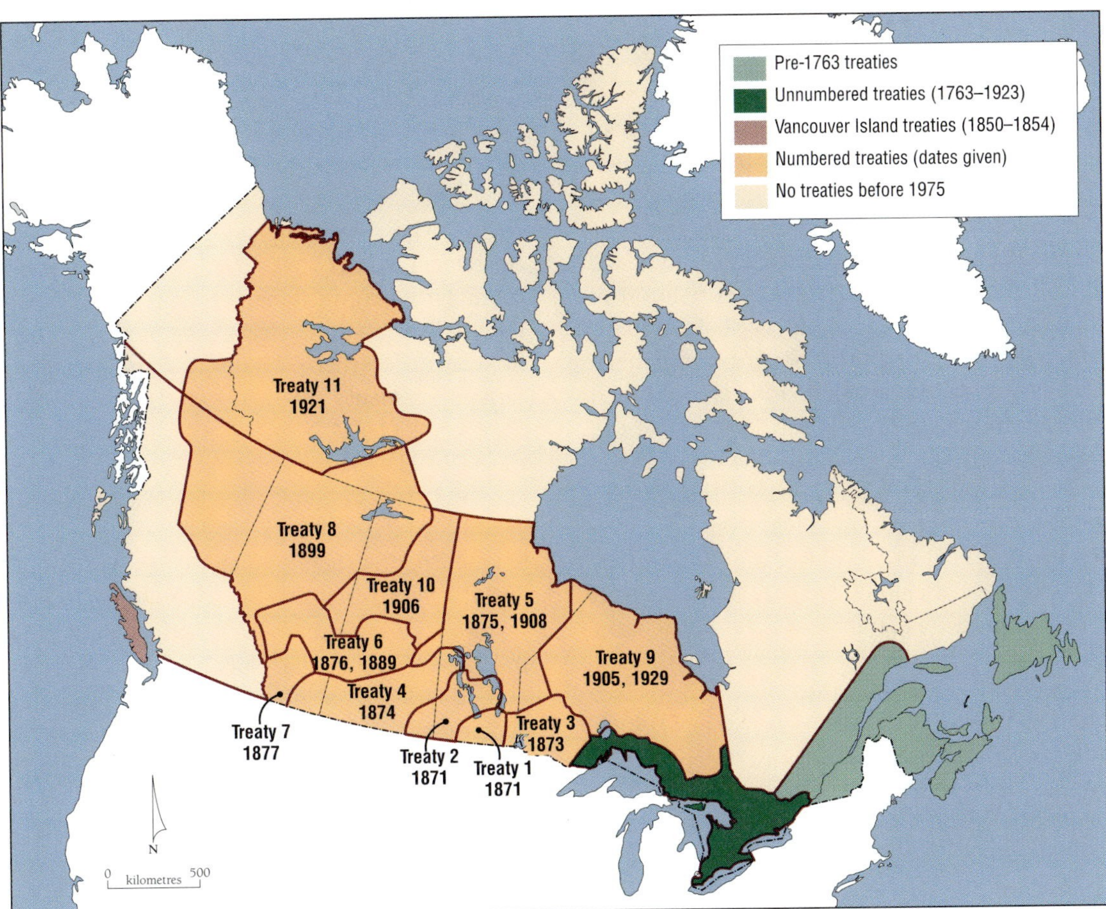

Figure 3.10 Historic treaties. The first treaties, made between the British government and Indian tribes, were 'friendship' agreements. In Upper Canada the Robinson treaties of 1850 set aside reserve lands in exchange for the title to the remaining lands. With the settlement of lands in the Canadian West, Indians became concerned about their future, so many of the 11 numbered treaties, which spanned a half-century from 1871 to 1921, included provisions for agricultural supplies. When the last numbered treaty was signed, many Aboriginal peoples in Atlantic Canada, Québec, and British Columbia were without treaties. (Further resources: Student website, National Atlas section, Map 14. Website instructions are found on p. xx.)

Ottawa recognized two forms of land rights. Reserve lands were one type of right or ownership, which the Canadian government held for Indian people. The second type was a usufructuary right to use Crown land for hunting and trapping. At that time, Crown lands (both provincial and federal) included most of Canada's unsettled areas. Indian, Inuit, and Métis families lived on Crown lands, continuing to hunt, trap, and fish. However, federal and provincial governments could sell such lands to individuals and corporations or grant them a lease to use the land for a specific purpose, such as mineral exploration or logging, without compensating the Aboriginal users of those lands. By the 1960s, many Aboriginal groups still did not have treaties with the Canadian government. Atlantic Canada, Québec, the Territorial North, and British Columbia contained huge areas where treaties had not yet been concluded. As a consequence, Aboriginal peoples had no control over developments on these lands.

A combination of events radically changed this situation. One factor was the emergence of Native leaders who understood the political and legal systems. They used the courts to force the federal and provincial governments to address the issue of Aboriginal rights and land claims. The first major event took place in 1969, when Ottawa proposed reforms to the Indian Act in its White Paper on Indian Policy. This galvanized treaty Indians into action. The White Paper proposed to treat all Canadians equally. For Indians, it meant the abolition of their treaty rights and the reserve land system. About the same time, the Nisga'a in northern British Columbia took their land claim, known as the *Calder* case, to court. In 1973, the Supreme Court of Canada narrowly ruled (by a vote of four to three) against the Nisga'a argument that the tribe still had a land claim to territory in northern British Columbia. However, in their ruling, six of the seven judges agreed that Aboriginal title to the land had existed in British Columbia at the time of Confederation. Furthermore, three judges said that Aboriginal title still existed in British Columbia because it had not been extinguished by the British Columbia government, while three other judges stated that Aboriginal title had been extinguished by the various laws passed by the British Columbia government since 1871. The seventh judge ruled against the Nisga'a claim on a legal technicality. The Supreme Court's narrow verdict and the legal opinion of three judges that Aboriginal title still existed changed the course of Aboriginal land claims in Canada. Now the federal government agreed that Aboriginal peoples who had not signed a treaty may very well have a legal claim to Crown lands.

In the early and mid-1970s the James Bay Project in northern Québec and the proposed Mackenzie Valley Pipeline Project in the Northwest Territories added fuel to the political fire over Aboriginal rights. The possible impact of these industrial projects on Aboriginal peoples was made clear through the Mackenzie Valley Pipeline Inquiry of 1974–6 (the Berger Inquiry) into possible environmental and socio-economic impacts and in the media. It was obvious that Aboriginal organizations were prepared to take action to defend their land claims. Their position in the 1970s was 'no development without land-claims settlements'. All these events changed both the public's views of Aboriginal rights and the government's position. At first grudgingly and then more willingly, governments, corporations, and Canadian society recognized the validity of Aboriginal land claims. The James Bay and Northern Québec Agreement in 1975 was the first modern land-claim agreement in Canada (Figure 3.11). Since then, eight comprehensive claims have been finalized in northern Canada: (1) the Inuvialuit Final Agreement (1984); (2) the Gwich'in Final Agreement (1992); (3) the Sahtu Final Agreement (1993); (4) the Nunavut Final Agreement (1993); (5) the Yukon Umbrella Final Agreement (1993); (6) the Nisga'a Final Agreement (2000); (7) the Dogrib (Tlicho) Final Agreement (2003); and (8) the Labrador Inuit Agreement (2005).

Many claims remain unsettled, especially in British Columbia. Until they are concluded, relations between those pursuing agreements and the federal government will remain strained. For example, virtually the entire province of British Columbia, except for Vancouver Island, is claimed by First Nations. In British Columbia, progress has been slow. Until 1992, the provincial government claimed that British occupancy had extinguished Aboriginal title. However, in 1992 the British Columbia government accepted the principle of Aboriginal land claims. The following year, Ottawa and Victoria agreed to a formula for paying outstanding land claims. The federal government would pay 90 per cent of the money needed to settle outstanding claims and the province would provide the land. In 2000, the Nisga'a Agreement was finalized, while other First Nations are negotiating with the British Columbia Treaty Commission. Progress has been slow. In March 2007, members of the Lheidli T'enneh First Nation voted to reject their Final Agreement. Two other First Nations—the Tsawwassen First Nation and the Maa-nulth First Nation—are close to final agreements, while another four First Nations—Sliammon, Yale, Yekooche, and In-SHUCK-ch—are currently negotiating towards final agreements.

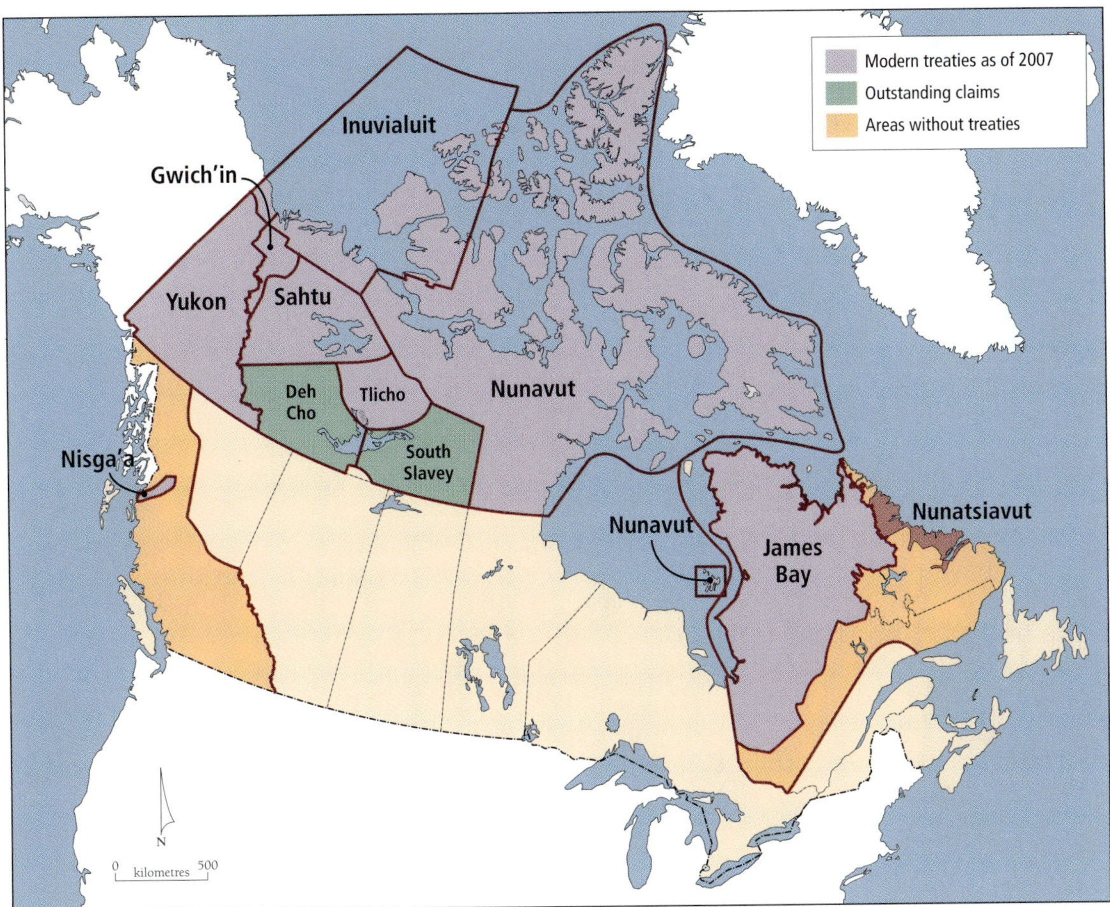

Figure 3.11 Modern treaties. The first modern treaty, the James Bay and Northern Québec Agreement, was signed in 1975. By 2007, the main areas without treaties were much of British Columbia, Labrador, and lands in Québec. (The original inhabitants of Newfoundland, the Beothuk, had perished from disease, encroachment, and slaughter by the early nineteenth century.)

Those Aboriginal groups who have concluded modern treaties are moving forward. They are able to focus on economic and cultural developments rather than expending their energies on land-claim negotiations. In 1993, the Nunavut agreement broke new ground by effectively establishing self-government over an entire territory. Since then, modern land-claim agreements, such as the Nisga'a Agreement, have included arrangements for self-government. As a result, a gap is emerging within the Aboriginal community between those who have a modern treaty and those who do not, as well as between those on reserve and those who live in urban areas among Canada's increasingly pluralistic majority society. Also, as with countries and with regions, some Aboriginal groups reside on lands that are rich in natural resources, resource developments, and development potential (e.g., oil and gas deposits, oil sands, pipelines, prime timber land) that provide a base for economic growth and considerable wealth, while many other groups live in areas with little resource potential where even subsistence from the land is marginal if not impossible.

Bridging the Aboriginal/ Non-Aboriginal Faultline

Aboriginal peoples are taking the control of their affairs away from Ottawa. Some Indian

Vignette 3.6 Modern Treaties

Modern treaties began in 1975 with the signing of the James Bay and Northern Québec Agreement (Figure 3.11). Since then, all modern treaties are either specific or comprehensive agreements between an Aboriginal group and the federal government. **Specific land-claim agreements** attempt to rectify shortcomings in the original treaty agreement with a band or seek to redress failure on the part of the federal government to meet the terms of the treaty. By 1990, over 500 specific claims had been filed with Ottawa. Most claims involved relatively small amounts of land. However, in 1992, a major agreement was signed between 26 First Nations in Saskatchewan and Ottawa. Called the Settlement of Treaty Land Entitlement in Saskatchewan, it involved a payment of $445 million over 12 years to allow these First Nations to purchase land that they should have received at the time of treaty or that they lost through Ottawa's mishandling of their affairs.

A comprehensive land-claim agreement is sought when a group of Aboriginal people, who have not yet signed a treaty, can demonstrate a claim to land through past occupancy. In 1984, the Inuvialuit of the Western Arctic became the first Aboriginal people to settle a comprehensive land claim with the federal government. In exchange for surrendering their Aboriginal claim to all this land, they received 91,000 km^2, $45 million (in 1977 dollars) in financial compensation, and guaranteed rights over resource management. Since 1984, comprehensive agreements have been signed between the government and the Gwich'in (1992), Sahtu (1993), Inuit of the Nunavut Settlement Area (1993), the Yukon First Nations (1993), Dogrib (2003), and Labrador Inuit (2005). A comprehensive agreement with the Nisga'a (2000) marked the first modern land-claim agreement in British Columbia.

and Inuit peoples have made substantial advances in economic development, while others have moved into the area of self-government. Unfortunately, some Aboriginal peoples, including the Métis, have not yet begun this process and remain on the margins of Canadian society. For most, fortunately, the process of change has started.

For historical reasons, while some Aboriginal peoples have made treaties, others have not. This difference is significant because a treaty, particularly a modern treaty (a comprehensive or specific land-claim agreement), provides the land, capital, and an administrative organization necessary to initiate this process of economic, social, and political change. In 1996, the report of the Royal Commission on Aboriginal Peoples identified two major goals: Aboriginal economic development and self-government. The gap between Aboriginal and non-Aboriginal societies will not be bridged until these goals are achieved. The principal factor is transferring power from Ottawa (the political power core) to the various Aboriginal communities (the politically weak periphery).

How well each Aboriginal community will manage its affairs is unknown, but breaking the dependency on the federal government will be an important step. Such a step will result in a new and more positive relationship between Aboriginal and non-Aboriginal Canadians. That a new relationship is unfolding is demonstrated by the new power to co-assess with the federal government industrial impacts on the land and people in the settlement areas of comprehensive land claims. The Labrador Inuit provide another example when they gained a share of the royalties from the Voisey's Bay nickel mine within their land-claim agreement.

The Immigration Faultline

The history of immigration to Canada is a complex and sometimes controversial topic. In the past, immigration was often an instrument of colonial power. After the British Conquest of New France, for example, the British government set the immigration policy and the French-speaking majority in Canada did not have a say in the shaping of this policy. This power lay exclusively with the colonial power. The British government's objective was to offset the large French-speaking population by encouraging large-scale immigration from the

Vignette 3.7 From a Colonial Straitjacket to Aboriginal Power

Until 1969, Canada's Aboriginal peoples were largely invisible to Canadians. They were outside the political process and thus denied access to the political decision-making process. In fact, status Indians did not receive the right to vote in federal elections until 1960. Most were geographically separated from urban Canadians, as many status Indians lived on reserves. Métis were in isolated communities, and Inuit still roamed the Arctic. Shunned by Canadian society, these marginalized peoples had been subjected to assimilation policies for nearly a hundred years. In 1969, Ottawa made one last attempt to assimilate the Indian people of Canada through its 'Statement of the Government of Canada on Indian Policy', more popularly known as the White Paper. The federal government proposed to eliminate the legal distinction between Indians and other Canadians by repealing the Indian Act and amending the British North America Act to remove those parts of the Act that called for separate treatment for Indians, and to abolish the Department of Indian Affairs. Ottawa believed that the separation of Indians from other Canadians was not only divisive but also made the Indians dependent on government and thereby held them back. The remedy was 'equality'. In the context of the 1960s, when oppressed people in other countries, including blacks in the United States and in South Africa, fiercely sought equality and the American Indian Movement railed against colonial suppression, Prime Minister Trudeau believed that the White Paper was the answer to Canada's Indian problem. And some Indian leaders supported this solution. As it turned out, many others did not. Reaction was swift. In the same year, Harold Cardinal published *The Unjust Society* and the following year, under his leadership, the Alberta chiefs published a formal rebuttal to the White Paper, commonly known as the 'Red Paper' and titled *Citizens Plus: A Presentation by the Indian Chiefs of Alberta to the Right Honourable P.E. Trudeau.*

Over the next decade, the debate over the place of Indians in Canadian society took several different directions. First, there was legal support for the Indian position, beginning with the *Calder* case in 1973 when the Supreme Court held that the Nisga'a had Aboriginal rights. Second, the election of the Parti Québécois in 1976 called for 'nation-to-nation' discussions between the province and the federal government. Aboriginal leaders seized the opportunity to present their demands in the same constitutional language. Third, recognition of the Aboriginal peoples and their rights in the 1982 Constitution Act dramatically enhanced their status and bargaining power. Fourth, the Constitution Act did not define Aboriginal Rights, leaving that task to negotiations or the courts. The courts have been active in defining Aboriginal rights. In 1997, the Supreme Court's landmark decision in the *Delgamuukw* case overturned the earlier decision denying that Indians in British Columbia had Aboriginal title. Furthermore, the Court ruled that Aboriginal title means that Indians have the right to the resources on those lands.

British Isles and outlawing immigration from France. In the case of the Acadians, the British, beginning in 1755, deported many of these people to England and other English colonies and many others fled to Québec or found their way to the Louisiana Territory. At the same time, the British sought to resettle the area with British subjects. Colonial-style immigration, therefore, not only generated tensions between the existing population and the newcomers, but it also imposed a way of life and a set of institutions on the existing population and often marginalized these people.

While the existing population historically viewed newcomers by asking how the new immigrants would benefit them and their society, the colonial power took the opposite position, i.e., how will the existing population benefit us? The economic, military, and social relationship between New France and the Huron Confederacy illustrates this point. In 1609, the Huron chiefs met with Samuel de Champlain to discuss both trade and a military alliance. The Huron had three objectives: (1) to gain access to European goods, including firearms, by supplying the French with beaver pelts; (2) to improve their material well-being with the trade goods and, in turn, trade these goods to more distant Indian tribes for profit; and (3) to strengthen their military position against their traditional enemies, the Iroquois, who were allied with the Dutch and later the English traders based in New York. The French had two objectives: (1) to secure a supply of furs; and (2) to convert the Huron to Christianity. At the height of the fur trade in the seventeenth century, New France greatly prospered and the Huron accounted for around half of the furs shipped to France (Dickason, 2002: 104). Trade was so important to the Huron tribes that when the French insisted that the Huron allow Jesuit missionaries to live among them as a condition for continued trade, the Huron reluctantly agreed. Unfortunately, the missionaries brought with them smallpox and other diseases that quickly swept through the Huron tribes, causing a sharp decline in their population. Another factor leading to the demise of Huronia was the French reluctance to trade firearms for furs. Since the English traders had no such hesitation, the Iroquois, though smaller in numbers, quickly became more powerful than the Huron and, with that military superiority, were able to harass the Huron fur brigades travelling to New France and later to attack and destroy Huron villages.

In this chapter, our attention is focused on the impact of immigration on Manitoba following the purchase of Hudson's Bay lands by Ottawa, and then on the subsequent settling of the Canadian Prairies by many people who were not of British ancestry. This historic period stretches from 1870 to 1914. During this time, although the face of colonialism had changed it had not softened, so immigrants had to conform to Canadian society. The experiences of the Métis of the Red River Colony and of the Doukhobors—very clearly not British immigrants—exemplify this demand to conform to the majority society.

Manitoba: The Métis against the Newcomers

With the transfer of the vast lands administered by the Hudson's Bay Company, Canada changed from a small territory to a truly continental country. While the boundary between Western Canada and the United States was determined by 1874, the survey of lands for agricultural settlement took place in the 1880s. The survey system, based on a township and range model, stamped a rectangular-shaped grid on the cultural landscape, thus determining the shape and placement of farms, roads, and towns (Vignette 3.2). As Moffat (2002: 204) points out, this survey system 'enabled the division of western lands among the HBC, the Canadian Pacific Railway (CPR) and homesteaders, and set aside two sections in each township for the future of local education.' However, Ottawa failed to recognize the landholdings of the Métis and relegated Indians to reserves. Ottawa did not inform the residents of the Red River Colony of its plans for the Red River Colony nor did the government recognize local landholdings. Events quickly spun out of control, resulting in the Red River Rebellion.

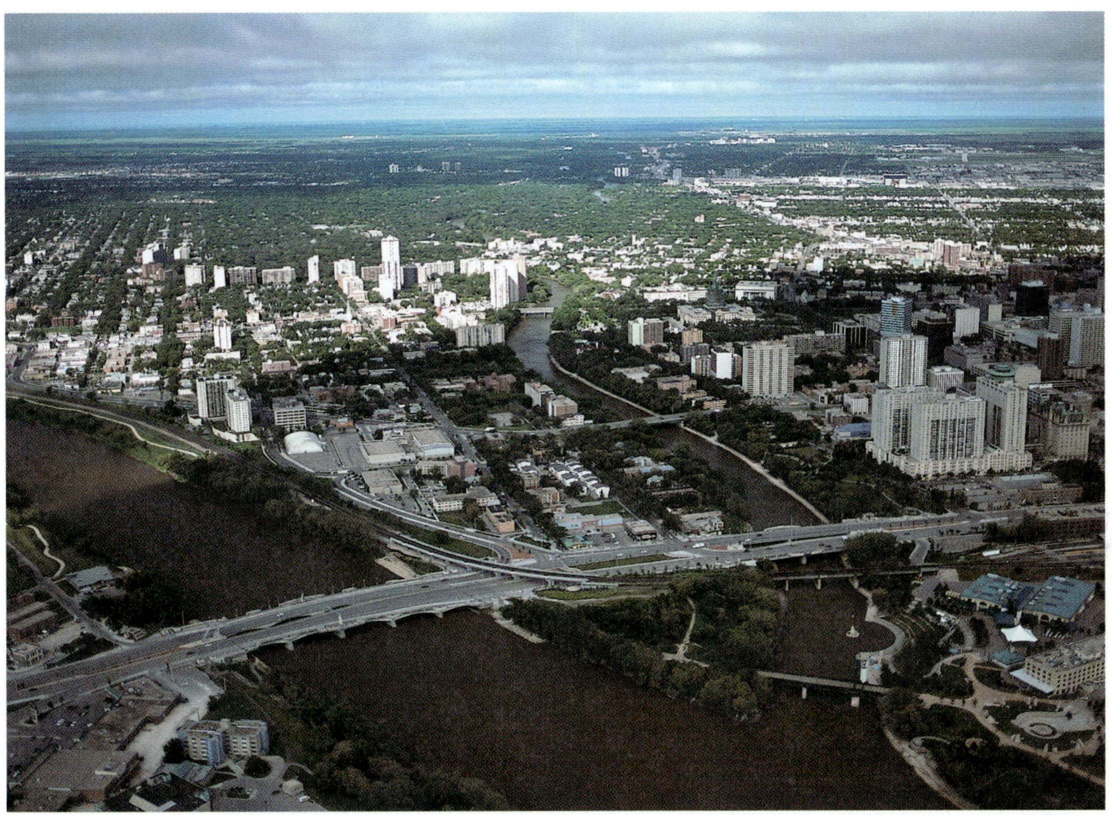

The confluence of the Red and Assiniboine rivers is known as the Forks. Today, the Forks lies in the heart of Winnipeg. In times past, the strategic location of the Forks provided Plains Indians with ready access by canoe to the lands south of the 49th Parallel and to the vast western interior. In 1738, the French explorer La Vérendrye established Fort Route at the Forks. With the founding of the Red River Colony in 1812, the Forks became the focal point of the Red River Colony. (Ron Garnett/AirScapes.ca)

The Red River Rebellion pitted the existing population of the Red River Colony against the newcomers whose agricultural system posed a deadly threat to the existing Métis hunting economy. Even before the arrival of settlers, surveyors sent by Ottawa ignored the long lot holdings of the Métis along the Red and Assiniboine rivers. In 1869, the Red River Colony was the only settled area of any size in the North-Western Territory, with a population of nearly 12,000 evenly divided between French- and English-speaking residents (Table 3.2). Most consisted of mixed-blood people, born of French and British fur traders and Indian mothers, who had settled in long lots along the banks of the two major rivers, and whose economy was based on the buffalo hunt and subsistence farming. By early 1869, news of the pending transfer of Hudson's Bay Company lands to Ottawa had reached the colony, and the arrival of land surveyors resulted in open hostility. When Canadian land surveyors began to survey lands occupied by the Métis, the Métis feared for their rights to those lands and even their place in the new society. Matters came to a boil when, in October 1869, Louis Riel put his foot on the surveyor's chain and told them to leave. Thus, the Red River Rebellion began.

Two months later, the Métis under Riel formed a provisional government and began to negotiate with Canada over the terms of entry into Confederation. During these negotiations, the concerns of the Métis were recognized. The Manitoba Act that led to the admission of Manitoba into Confederation as Canada's fifth province ensured the use of English and French languages within the government of the new province and established a dual system of Protestant and Roman Catholic schools.

Within a decade, the population balance was broken due to an influx of immigrants from Ontario, many of whom either belonged to or supported the views of the Orange Order, a Protestant fraternal organization with strong anti-Catholic beliefs. Some newcomers saw no place for the Métis and Indians in the emerging society, thus creating tensions between the existing population and the newcomers. From 1871 to 1881, Manitoba's population increased from 25,228 to 62,260, with most immigrants coming from Ontario, the British Isles, and the United States (Table 3.8). At this time, those of British ancestry formed 54 per cent of the population, other Europeans made up 17 per cent, Métis, 17 per cent, and Indians, 11 per cent (Census of Canada, 1882: Table III). The newly formed English-speaking majority focused their attention on the dual school system. By 1891, Manitoba's population exceeded 150,000 (Table 3.8). In an example of the tyranny of the majority, the English-speaking population argued that with so few French-speaking students, funding for the Catholic school was not warranted. In 1890, the government of Manitoba abolished public funding for Catholic schools. This decision took on national significance by becoming a critical issue between Québec and the rest of the country (see below).

What caused this influx? One reason was that Ontario no longer had a surplus of agricultural land and sons of farmers looked to the unsettled lands on the Great Plains of the United States and to Manitoba. A second reason was that the promise of a railway would make farming in Manitoba more viable. However, the completion of the Canadian Pacific Railway from Fort William on Lake Superior to Selkirk, Manitoba, was delayed and was only completed in 1882. Then, grain could be transported by rail and ship to eastern Canada and Great Britain rather than by the more circuitous steamship route to St Paul, Minnesota, and then by rail to New York. Wheat farming in Manitoba had become a profitable business because of advances in agricultural machinery, farming techniques, and rising prices for grain. Equally important, new strains of wheat, first Red Fife and then Marquis, both of which ripened more quickly than previous varieties of wheat, lessened the danger of crop loss due to frost. Marquis wheat, which matured seven days earlier than Red Fife, allowed wheat cultivation to take place in the parkland belt of Saskatchewan and Alberta where the frost-free period was shorter than in southern Manitoba.

Table 3.8 Population in Western Canada by Provinces, 1871–1911

Year	Manitoba*	Saskatchewan*	Alberta*
1871	25,228		
1881	62,260	21,652	9,875
1891	152,506	40,206	26,593
1901	255,211	91,279	73,022
1911	461,394	492,432	374,295

*The boundaries of Manitoba did not reach their present limits until 1881. Since Saskatchewan and Alberta became provinces in 1905, their populations for 1881, 1891, and 1901 were calculated from the censuses of Canada for 1881 and 1891.

Sources: Census of Canada 1880–81, vol. 1: 93–6; *Census of Canada 1890–91*, vol. 1: 112–13; Statistics Canada (2003).

The Doukhobors

The settling of Western Canada represents both a major migration effort and the formation of a new cultural landscape. By 1871, the newly founded country of Canada had three distinct rural landscapes—Ontario, with its small but contiguous farms concentrated in the fertile lowlands bordering Lake Erie and Lake Ontario and stretching north to the Canadian Shield; the Maritimes, with small pockets of farmland found in the river valleys of the hilly Appalachian Mountains, as well as the more extensive agricultural areas in Prince Edward Island and the Annapolis Valley; and Québec, with its long lots along the St Lawrence River. The emerging cultural landscape of Western Canada took two forms—the rectangular appearance of its rural landholdings and the symbols of ethnic diversity as expressed by farm buildings and churches.

Many immigrants came from non-English-speaking European countries, marking a break from the traditional sources of immigration, the British Isles and the United States. Others came from Ontario and, to a lesser degree, from other provinces. Within two decades of entering Confederation, Manitoba's population had increased by just over 600 per cent (Table 3.8). Most were of British stock, but substantial numbers of Mennonites and Icelanders had also come to Manitoba. At the same time, few settlers had reached Saskatchewan and Alberta, though the Métis had relocated in Saskatchewan, primarily around the settlement of Batoche on the South Saskatchewan River just north of Saskatoon. By 1895, Western Canada had a predominantly British population that had established political and social institutions.

In the next decade, the volume of immigrants from Central Europe, Scandinavia, and czarist Russia increased substantially. As peasants, they were prepared for the harsh physical conditions associated with breaking the virgin prairie land and were willing to deal with the psychological stress of living on isolated farmsteads in a foreign country where their native tongue was not accepted. As their numbers grew, the anglophone majority became concerned about these newcomers and their possible effect on the existing social structure. The demographic impact of the non-British migration to Western Canada is shown in the 1916 census (Table 3.9).

This wave of Central Europeans had tremendous implications for Western Canada. While most newcomers assimilated into the larger society, a few did not. Often these ethnic groups settled in one area where they were somewhat insulated from the larger society and where they attempted to maintain their traditional customs, language, and religion. The federal government, by providing land reserves for ethnic groups such as the Mennonites and Doukhobors, reinforced this tendency.

While they were successful farmers, the cultural differences between the more conservative Doukhobors and Canadian society were too great for the majority society to accept. While some Doukhobors were able to integrate into local society, the Community Doukhobors were not able to adapt. The Community Doukhobors remained faithful to their religious beliefs that emphasized communal living and minimized dealings with the state, including informing the state about births and deaths of their members.

These people were deeply religious Russian peasants who rejected both the practices and beliefs of the Russian Orthodox Church and the secular authority of the Czar. They were communalists and pacifists, and by refusing to serve in the army of czarist Russia they were considered by the state as outlaws and therefore needed to be punished. Consequently, they were persecuted by both the church and the state. Seeking to be left alone, the Doukhobors sought a place far from the forces of authority where they could practise their religion and communal lifestyle. In choosing to settle in Canada they were granted blocks of land and exemption from military service.

In 1899, the Doukhobors arrived in Canada and took possession of lands in Saskatchewan, and they soon built 57 villages. These lands were selected by representatives of the Doukhobors before the peasant settlers arrived in Canada. Through negotiations with the Canadian government, the Doukhobors

Table 3.9 Population of Western Canada, 1916

Ethnic Group	Western Canada	Percentage	Manitoba	Saskatchewan	Alberta
British	971,830	57.2	57.7	54.5	60.2
German	136,968	8.1	4.7	11.9	6.8
Austro-Hungarian	136,250	8.0	8.2	9.1	6.4
French	89,987	5.3	6.1	4.9	4.9
Russian	63,735	3.7	2.9	4.5	3.8
Norwegian	47,449	2.8	0.6	4.2	3.4
Indian (Aboriginal)	39,147	2.3	2.5	1.7	2.9
Ukrainian	36,103	2.3	4.1	0.7	1.8
Swedish	37,220	2.2	1.4	2.5	2.7
Polish	27,790	1.6	3.0	1.0	0.9
Jewish	23,381	1.4	3.0	0.6	0.6
Dutch	22,353	1.3	1.3	1.4	1.3
Icelandic	15,800	0.9	2.2	0.5	0.1
Danish	9,556	0.6	0.3	0.5	0.9
Belgian	9,084	0.5	0.8	0.4	0.4
Italian	5,348	0.3	0.3	1.0	0.9
Other	26,219	1.5	0.9	1.5	2.3
Total	1,698,220	100.0	100.0	100.0	100.0

Source: Census of the Prairie Provinces, 1916, Table 7. Data is adapted from the Statistics Canada website: http://www12.statcan.ca/english/census01/products/analytic/companion/age/provpymds.cfm.

had obtained four large blocks of land totalling 750,000 acres. The four colonies were located just west of Swan River, Manitoba (North Colony), and at Prince Albert (Saskatchewan Colony) and Yorkton, Saskatchewan (South Colony and Good Spirit Lake Annex). Ottawa had allowed them to receive blocks of land rather than individual homesteads, thus facilitating the building of villages and the establishment of a communal society. Farming was not only an economic activity, but it was also central to their religious beliefs, which emphasized the value of a simple, communal life. For example, Doukhobors shared in the returns from farming. In fact, no one person owned the land or the tools. In a land of individual landholdings and the pursuit of profit, the Doukhobors were seen as 'out of step' with the surrounding community.

As the land around their villages and allotments was settled by other newcomers, the Doukhobors came into closer contact with their neighbours, who often coveted the uncultivated areas found on the edges of the land reserves of the Doukhobors. Two factors were at play in their not cultivating all of their land. First, the grant of land reserves was quite generous and went far beyond their immediate needs. Second, their compact communal settlements and intensive land cultivation near their villages resulted in a particular land-use pattern that left lands far from the village underutilized. In sharp contrast, homesteaders were required to cultivate land on their quarter-sections and to erect farmhouses, thus creating the checkerboard pattern of rural settlement. Neighbours also were puzzled by and perplexed with their communal way of life that separated them from the rest of the population. As public resentment increased, the federal government took action. In 1905, Frank Oliver succeeded Clifford Sifton as Minister of the Interior. Oliver decided to enforce the Dominion Lands Act, so when the Doukhobors refused to swear an oath of allegiance to the Queen, Oliver had his excuse.

Failure to take such an oath had two implications. First, it suggested that these people were disloyal to the Queen. Second, it meant that the Doukhobors could not obtain title to their homestead lands. Under this pretext, Oliver used the Dominion Lands Act to cancel their right to land. A hard core of Doukhobors remained committed to the communal way of life, most of whom eventually moved to British Columbia; others abandoned the village life and took title to homesteads. The villages gradually lost members and lands. The South Colony just north of Yorkton, Saskatchewan, was the last holdout, but by 1918 it ceased to exist on Crown land. It persisted in a much reduced area on purchased land until 1938, as did other communal settlements established in the Kylemore and Kelvington areas.

One explanation for the ultimate failure of the Doukhobor experiment was that Canada's model of individual settlement was simply too rigid to accept a communal one. Yet with ownership of land, the Doukhobors could establish agricultural villages. Another explanation focused on their unwillingness to swear allegiance to the Queen. In reality, the Doukhobors were successful farmers and their villages were working. However, they could not fit into the existing culture, which required conformity to the laws—and the informal values and lifeways—of the country. Primarily for that reason, the Community Doukhobors were unable to find a place in Western Canada. They represent a classic example of a people being too different—too 'other'—from the majority to be allowed a comfortable space within the cultural landscape. Ironically, the village model of settlement was perhaps the most effective way of settling the Prairies in the late nineteenth and early twentieth centuries. Professor Carl Tracie (1996: xii) puts it this way:

> At the very time when the individual homesteader was struggling with the very real problems of isolation and loneliness, the Doukhobor settlements, whose compact form allayed these problems, were being dismantled by forces which could not accommodate the communal aspects of the group. Also, although the initial government concern was the survival of the Doukhobors, their very prosperity, based as it was on communal effort, may have worked against them since it illustrated the success of a system diametrically opposed to the individualistic system dictated by government policy and assumed by mainstream society.

The French/English Faultline

Although the ancestors of Aboriginal peoples were the first to occupy North America, two European powers—the French and the British—colonized Canada. Following the British military victories, the Treaty of Paris (1763) confirmed British hegemony over a French-Canadian majority. This historic fact underscores the dominant position of the British and a fundamental weakness in the vision of Canada as two founding nations. Relations between the French and English in North America began nearly 400 years ago. These two cultures have come to represent a major faultline in Canadian society. Since 1841, however, these two communities have had to work together, each dependent on the other. This interaction has done much to shape the cultural and political nature of Canada. Over the years, they have accomplished much together. Nevertheless, significant differences between the two communities exist and, from time to time, these differences flare into serious misunderstandings. Without a doubt, Canadian unity depends on the continuation of this relationship and the need for compromise. A brief examination of that relationship, as outlined in the following pages, will lead to a fuller understanding of the contemporary version of French/English differences, conflicts, and compromises.

The serious nature of this rift has profound geopolitical consequences for Canada. It is therefore crucial that we understand the origins and nature of this faultline. A well-known Canadian political columnist, Jeffrey Simpson, wrote:

> We can also hope that, in the 1980s, Canadians gained a deeper understanding of the faultlines running through their society, and that they will avoid measures that widen them, thereby concentrating on making new arrangements and reforming old ones, so that what the rest of the world rightly believes to be a successful experiment in managing diversity will endure and prosper. (Simpson, 1993: 368)

The beginning of formal French–English relations in Canada stretches back to the British conquest of the French on the Plains of Abraham in 1759, an event that remains a dark page in French-Canadian history. In 1760, the French Canadians watched the remnants of the French army and the French elite board ships to return to France. The French Canadians had no thought of leaving as they were born in the New World, but what would happen to them under British military rule? Would they, like their Acadian brethren, be deported to other British colonies? Britain did not need to take such drastic action as there was no French threat to Britain's North American possessions. In the Treaty of Paris, France ceded New France to Britain, which placed the French-Canadian majority under the British monarchy. While the English lived in cities in Québec and dominated the Québec economy, French Canadians lived mostly in rural areas where they successfully maintained their culture within a British North America, an achievement made possible because of two factors. First, the French Canadians were a large homogeneous population that occupied a contiguous geographic area. Second, Britain forged a close relationship with the former elites of New France (the Roman Catholic clergy and the landed gentry) to ensure the French Canadians' loyalty because Britain wanted to secure its northern colony against its restless American colonies to the south. This relationship between Britain and the French Canadians would be strengthened with the Québec Act of 1774.

The Québec Act, 1774

With the Québec Act of 1774, the unique nature and separateness of Québec were recognized, thus ensuring its place in British North America. This Act is sometimes described as the Magna Carta for French Canada.[7] Its main provisions ensured the continuation of the aristocratic seigneurial landholding system and guaranteed religious freedom for the colony's Roman Catholic majority and, by implication, their right to retain their native language.[8] This gave the most powerful people in New France a good reason to support the new rulers. The Roman Catholic Church was placed in a particularly strong position. Not only was the Church allowed to collect tithes and dues but its role as the protector of French culture went unchallenged. Therefore, the clergy played an extremely important role in directing and maintaining a rural French-Canadian society, a role further enhanced by the Church's control of the education system. The habitants (farmers) were at the bottom of French-Canadian society's hierarchy. They formed the vast majority of the population and continued to cultivate their land on seigneuries, paying their dues to their lord (seigneur) and faithfully obeying the local priest and bishop. The British granted another important concession, namely, that civil suits would be tried under French law. Criminal cases, however, fell under English law.

The seigneurial system formed the basis of rural life in New France and, later, in Québec. In 1774, there were about 200 seigneuries in the St Lawrence Lowlands. This type of land settlement left its mark on the landscape (the long, narrow landholdings and the vast estate of the seigneur) and on the mentality of rural French Canadians (close family ties, a strong sense of togetherness with neighbouring rural families, and staunch support for the Church). A habitant's landholding, though small, was the key to his family's prosperity, and by bequeathing the farm to his eldest son the habitant ensured the continuation of this rural way of life. In 1854, the habitant was allowed to purchase his small plot of land from his seigneur, but the last vestiges of this seigneurial system did not disappear until a century later. Even today, the landscape in the St Lawrence Lowlands shows many signs of this type of landholding.

While the heart of this new British territory was the settled land of the St Lawrence Lowlands, its full geographic extent was immense. Essentially, the Québec Act of 1774 recognized the geographic area of former French territories in North America. Québec's territory in 1774 was extended from the Labrador coast to the St Lawrence Lowlands and beyond to the sparsely settled Great Lakes Lowlands and the Indian lands of the Ohio Basin.

The Loyalists

The American War of Independence changed the political landscape of North America. Within the newly formed United States, a number of Americans, known as the **Loyalists**, remained loyal to Britain. Like the French-speaking people in North America, most of these Loyalists were born and raised in the New World. For them, North America was their homeland. During the revolution, they had sided with the British. They were hated by the American revolutionaries and lost their homes and property. As they were not welcome in the new republic, many Loyalists resettled in the remaining British colonies in North America, where Britain offered them land. The majority (about 40,000 Loyalists) settled in the Maritimes, particularly in Nova Scotia. About 5,000 relocated in the forested Appalachian Uplands of the Eastern Townships of Québec. A few thousand, including Indians who had supported Britain, took up land in the Great Lakes Lowlands along the northern shores of Lake Ontario. These Loyalists strengthened Britain's hold on its North American possessions, but those who settled in the major cities of Québec and in the Eastern Townships came from a different cultural world than the local francophone residents. Social and political tensions arose from time to time between the two cultural groups.

Within a few decades, the English-speaking settlers in the Great Lakes Lowlands grew in number. Soon they sought to control their own affairs so they could have a more British government with British civil law, British institutions, and an elected assembly. In the Constitutional Act of 1791, Québec was split into Upper and Lower Canada.

The Constitutional Act, 1791

In 1791, the British government passed the Constitutional Act (Canada Act) in an attempt to satisfy the political needs of its French- and English-speaking subjects. These were the main provisions of the Act: (1) the British colony of Québec was divided into the provinces of Upper and Lower Canada, with the Ottawa River as the dividing line, except for two seigneuries located just southwest of the Ottawa River; and (2) each province was governed by a British lieutenant-governor appointed by Britain. From time to time, the lieutenant-governor would consult with his executive council and acknowledge legislation passed by an elected legislative assembly.

In 1791, Lower Canada had a much larger population than Upper Canada. At that time, about 15,000 colonists lived in Upper Canada, most of whom were of Loyalist extraction, plus about 10,000 Indians, some of whom had fled northward after the American Revolution. Lower Canada's population consisted of about 140,000 French Canadians,

10,000 English Canadians, and perhaps as many as 5,000 Indians.

Following the Constitutional Act, Upper and Lower Canada each had an elected assembly, but the real power remained in the hands of the British-appointed lieutenant-governors. In Lower Canada the lieutenant-governor had the support of the Roman Catholic Church, the seigneurs, and the Château Clique. The **Château Clique**, a group consisting mostly of anglophone merchants, controlled most business enterprises and, as they were favoured by the lieutenant-governor, wielded much political power. In Upper Canada the **Family Compact**—a small group of officials who dominated senior bureaucratic positions, the executive and legislative councils, and the judiciary—held similar positions in commercial and political circles. While these two elite groups promoted their own political and financial well-being, the rest of the population grew more and more dissatisfied with blatant political abuses, which included patronage and unpopular policies that favoured these two groups. Attempts to obtain political reforms leading to a more democratic political system failed. Under these circumstances, social unrest was widespread.

In 1837 rebellions broke out. In Lower Canada Louis-Joseph Papineau led the rebels, while William Lyon Mackenzie headed the rebels in Upper Canada. Both uprisings were ruthlessly suppressed by British troops. The goal of both insurrections was to take control by wresting power from the colonial governments in Toronto and Québec and putting government in the hands of the popularly elected assemblies. In Lower Canada the rebellion was also an expression of Anglo-French animosity. While both uprisings were unsuccessful, the British government nevertheless sent Lord Durham to Canada as Governor-General to investigate the rebels' grievances. He recommended a form of responsible government and the union of the two Canadas. Once the two colonies were unified, the next step, according to Durham, would be the assimilation of the French Canadians into British culture. When Durham left in 1838, a second rebellion broke out in Lower Canada, but it was as unsuccessful as the first.

The Act of Union, 1841

In response to Durham's report, in 1841 the two largest colonies in British North America, Upper and Lower Canada, were united into the Province of Canada. This Act of Union gave substance to the geographic and political realities of British North America. The geographic reality was that a large French-speaking population existed in Lower Canada, while an English-speaking population was concentrated in Upper Canada. The political reality was twofold. Both groups had to work together to accomplish their political goals and neither group could achieve all its goals without some form of compromise. When the two cultures were forced to work together in a single legislative assembly, a new beginning to the French/English faultline surfaced.

The new governor, Sir Charles Bagot, was appointed by and reported to the Colonial Office in London. The governor had the authority to appoint members to a Legislative Council and an Executive Council. The only representative body was a Legislative Assembly. Even though Lower Canada (after union known as Canada East) was somewhat larger in economic strength and population size (670,000) than Upper Canada (Canada West, with a population of 480,000), each elected 42 members to the Legislative Assembly. The vast majority of inhabitants in Canada East and Canada West lived in the rural countryside. For example, in 1841, the principal towns of Montréal and Toronto had populations of about 40,000 and 15,000 respectively.

Demographic Shifts

By the time of the Constitutional Act in 1791, the balance of French- and English-speaking inhabitants of British North America had begun to tilt more and more in favour of the English. This demographic shift began after the American Revolution when thousands of Loyalists from the former American colonies relocated in British North America. In 1791, the European population of British North America was about 225,000 (mostly French Canadians). Some 162,000 (72 per cent) lived along the St Lawrence River in what is now the province of Québec. Perhaps as many as 50,000 (22 per cent) lived in Atlantic Canada. The remaining 15,000 (6 per cent) were scattered along the north shores of Lake Erie and Lake Ontario in what is now part of the province of Ontario. By this time, the Aboriginal population of Upper Canada, Lower Canada, and Atlantic Canada had declined substantially to about 25,000.

Within 50 years, not only had this geographic pattern changed but the balance of demographic power had shifted. While their numbers increased due to high fertility rates, French Canadians were no longer the majority in British North America because of the flood of British immigrants. In 1841, British North America had about 1.5 million inhabitants, 45 per cent of them located in Lower Canada, about 33 per cent in Upper Canada, and 12 per cent in the Atlantic colonies. Over the next 30 years, the balance continued to swing in favour of English-speaking regions, thanks to massive immigration from the British Isles. By 1871, Québec's population was only 34 per cent of Canada's population (Table 3.10).

As the country expanded its boundaries and more land was settled, Canada's English-speaking population grew, while Québec's French-speaking population diminished in proportion. Manitoba joined Confederation in 1870 with a population of almost 12,000, which was comprised largely of French- and English-speaking Métis. In 1871, British Columbia, with an estimated population of 28,000 British subjects, became a member of the Dominion of Canada. Beyond these provinces, Indian and Inuit peoples inhabited the land. The total Aboriginal population of all territory that would eventually become Canada was about 100,000 at the time of Confederation. In the subsequent decades, these new lands would be settled by Canadians, Europeans, and Americans. With few exceptions, English became the adopted language of these settlers. For a while, Manitoba was an exception, but when the English-speaking majority gained control of the government and the public institutions, the Métis found it difficult to maintain their culture and language.

Strained Relations

During these formative years, several events seriously strained relations between the Dominion's two founding peoples. Two cultures, French and English, were in firm opposition to each other. In the settling of the West, these two cultures clashed over language and

Table 3.10 Population by Colony or Province, 1851–71 (percentages)

Colony/Province	1851	1861	1871
Ontario	41.1	45.2	46.5
Québec	38.5	36.0	34.2
Nova Scotia	12.0	10.7	11.1
New Brunswick	8.4	8.1	8.2
Manitoba			< 0.1
British Columbia			< 0.8
Total per cent	100.0	100.0	100.0
Total population	2,312,919	3,090,561	3,525,761

Source: Wayne W. McVey and W.E. Kalbach, *Canadian Population* (Toronto: Nelson Canada, 1995), 38. Reprinted with permission of Nelson, a division of Thomson Learning: www.thomsonrights.com. Fax 800-730-2215.

religious rights. Four events illustrate the intensity of this power struggle:

- the Red River Rebellion, 1869–70;
- the North-West Rebellion, 1885;
- the Québec Jesuits' Estates Act, 1888;
- the Manitoba Schools Question, 1890.

The Red River Rebellion, 1869–70

In 1868, the British government passed the Rupert's Land Act, which would transfer the Hudson's Bay Company lands to the Crown, and in November of the following year the HBC signed the deed of transfer. At first, nothing changed for the Aboriginal peoples inhabiting the vast grasslands west of the Great Lakes, but Ottawa's plan was to settle these lands. With the arrival of surveyors and then settlers, the world of the hunters and trappers soon disappeared. The first to sense this threat were the French-speaking Métis who lived near the Red River. They became alarmed by the arrival of the land survey teams from Ottawa. The Métis of Assiniboia feared that they might lose their culture, religion, and freedom to hunt buffalo on the open Prairies, so they reacted swiftly. The Métis rebellion, led by Louis Riel, soon became a national issue, reopening differences between English, Protestant Ontario and French, Roman Catholic Québec.[9] Québec considered Riel a French-Canadian hero who was defending the Métis, a people of mixed blood who spoke French and followed the Catholic religion. Ontario, on the other hand, considered Riel a traitor and a murderer. For Canada, the larger issue was the place of French Canadians in the West. A compromise was achieved in the Manitoba Act of 1870. Accordingly, the District of Assiniboia became the province of Manitoba. Under this Act, land was set aside for the Métis, and the elected legislative assembly of Manitoba provided a balance between the two ethnic groups with 12 English and 12 French electoral districts. Equally important, Manitoba had two official languages (French and English) and two religious school systems (Catholic and Protestant) financed by public funds.

The North-West Rebellion, 1885

During the 1870s, many Ontarians settled in Manitoba while some Métis sought a new home on the open prairie. Seeking to remain hunters, one group settled along the South Saskatchewan River where they established a Métis colony around Batoche, about 60 km northeast of present-day Saskatoon. Batoche became the new centre of the French-speaking Métis in Western Canada. As settlers spread into Saskatchewan, the Métis again feared for their future. In 1884, a party of Métis went to Montana where Louis Riel, their old leader, was living and pleaded with him to return to Batoche and lead them again.

Convinced of his destiny, Riel accepted this challenge. Late in 1884, Riel sent a petition to Ottawa with various demands for all the inhabitants of the North-West—Indians, Métis, and whites. After Ottawa ignored his petition, Riel established a provisional government and began to organize the Métis into armed bands. Ottawa responded by sending a militia to suppress the rebellion. The militia travelled from Ontario to Saskatchewan in eight days, thanks to the new railway. Within a relatively short period, after Métis successes in battles at Duck Lake and Fish Creek, the Canadian militia captured Batoche, took Riel prisoner, and defeated the Indian bands that had taken part in the rebellion and were led by Plains Cree Chiefs Poundmaker and Big Bear. For Québec, the defeat of the Métis and the subsequent hanging of their leader, Louis Riel, not only represented a defeat for a French presence in the West but also widened the gulf between French and English Canadians.

The Québec Jesuits' Estates Act, 1888

In the nineteenth century religious and linguistic intolerance was widespread. For example, Protestant extremists in Ontario were ready to pounce on any perceived injustice to their cause. The Jesuits' efforts to obtain financial compensation for lands that the British took from them in 1763 and later transferred to Lower Canada proved to be such a case.

The Jesuit estates, which were granted under the French regime and used for schools and missions, were appropriated by Britain after the Conquest and given to Lower Canada in 1831. In 1838, Catholic bishops petitioned unsuccessfully for the return of the Jesuit estates. After Confederation, the ownership of the estates passed to the Québec government, with which the Jesuits began negotiating in 1871 for financial compensation. However, the archbishop of Québec argued that the money should be divided among Catholic schools rather than given in its entirety to the Jesuits, who wanted to establish a university in Montréal that would compete with Québec's Université Laval. Québec Premier Honoré Mercier asked Pope Leo XIII to act as arbiter in the dispute among the Roman Catholic hierarchy. In 1888, Québec's Legislative Assembly passed the Jesuits' Estates Act, which determined the division of the financial compensation: $160,000 went to the Jesuits, $140,000 went to the Université Laval, and $100,000 went to selected Catholic dioceses.

Ontario's Orange Order, a Protestant fraternal society, opposed the settlement, arguing that the arbiter, Pope Leo XIII, was an intruder into Canadian affairs and that public funds should not be used to support Catholic schools. In March 1889, the House of Commons debated the motion to disallow the Québec Jesuits' Estates Act and eventually voted against this motion. Similar anglophone, Protestant, anti-Catholic sentiment surfaced in Manitoba regarding the Manitoba Schools Question.

The Manitoba Schools Question, 1890

The British North America Act of 1867 established English and French as legislative and judicial languages in federal and Québec institutions. The remaining three provinces (New Brunswick, Nova Scotia, and Ontario) had only English as the official language. The question of French language and religious rights in acquired western territories first arose in Manitoba.

The French/English issue became the focal point for the entry of the Red River Colony (now Manitoba) into Confederation. Local inhabitants—mostly French-speaking, Roman Catholic Métis and the less numerous English-speaking Métis—were determined to have some influence over the terms that would include their community as part of Canada. One of their concerns was language rights, an issue that was ultimately resolved when a list of rights drafted by the provisional government became the basis of federal legislation. When the Red River Colony entered Confederation in 1870 as the province of Manitoba, it did so with the assurance that English- and French-language rights, as well as the right to be educated in Protestant or Roman Catholic schools, were protected by provincial legislation.

During the 1870s and 1880s, a large number of Anglo-Protestant settlers, mainly from Ontario, moved to Manitoba, causing the proportion of Anglo-Protestants in the population to increase and the proportion of French and Roman Catholic inhabitants to decrease. This demographic change created a stronger Anglo-Protestant culture in Manitoba. In 1890, the provincial government ended public funding of Catholic schools. From Québec's perspective, this legislation shook the very foundations of Confederation. Sir Wilfrid Laurier became Prime Minister in 1896 and, in the following year, Laurier negotiated a compromise agreement with the government of Manitoba. The compromise allowed for the teaching of Catholic religion in a public school when there were sufficient Catholic students. Similarly, if there were sufficient French-speaking students, classes could be taught in French.

One Country, Two Visions

The greatest challenge to Canadian unity comes from the cultural divide that separates French- and English-speaking Canadians and their respective visions of the country. In the early years of Confederation, events such as the Red River Rebellion, the North-West Rebellion, the Jesuits' Estates Act, and the Manitoba

Schools Question widened the French/English faultline. For French Canadians these events demonstrated the 'power' of the English-speaking majority and their unwillingness to accept a vision of Canada as a partnership between the two founding peoples. The root of each vision lies in the history of Canada.

One vision of Canada is based on the principle of two founding peoples. This vision originated in French-Canadian historical experiences and compromises that were necessary for the sharing of political power between the two partners. This vision began with the Conquest of New France, but its true foundation lies in the formation of the Province of Canada in 1841. From 1841 onward, the experience of working together resulted in a Canadian version of cultural dualism.

Henri Bourassa, an outstanding French-Canadian thinker (and Canadian nationalist) in the early twentieth century, was a strong advocate of cultural dualism. He wrote, 'My native land is all of Canada, a federation of separate races and autonomous provinces. The nation I wish to see grow up is the Canadian nation, made up of French Canadians and English Canadians' (quoted in Bumsted, 2007: 307). Bourassa argued that a 'double contract' existed within Confederation. Even today, Bourassa's 'double contract' is an essential element in the two founding peoples concept. He based the notion of a double contract on a liberal interpretation of section 93 of the BNA Act, which guarantees denominational schools. Bourassa expanded the interpretation of the religious rights to include cultural rights for French- and English-speaking Canadians. In more practical terms, Bourassa regarded Confederation as a moral contract that guaranteed French/English duality, the preservation of French-speaking Québec, and the protection of the language and religious rights of French-speaking Canadians in other provinces.

From a geopolitical perspective, Canada is a bicultural country. In one part the majority of Canadians speak English, and in another part French is the majority language. Thus, French culture predominates in Québec and has a strong position in New Brunswick. In addition to provincial control over culture, two other geopolitical factors ensure the dynamism of French in those provinces. One factor is the large size of Québec's population—the vitality of Québécois culture is one indication of its success. The second factor is the geographic concentration of French-speaking Canadians in Québec and adjacent parts of Ontario and New Brunswick. In New Brunswick the French-speaking residents, known as Acadians, constitute over one-third of the population. Federal bilingualism policies instituted in the late 1960s and 1970s also helped rejuvenate francophone minorities. Before these policies, assimilation into the much larger English-speaking culture had seriously weakened the position of francophones in all the provinces, except Québec, and in the two territories. While the attraction of joining the dominant anglophone culture remains, financial support from Ottawa for French educational and cultural facilities in the English-speaking provinces has ensured a place for bilingualism in all provinces and territories.

The Royal Commission on Bilingualism and Biculturalism was an attempt to bridge the gap between English and French Canadians. This Commission, set up in 1963, examined the issue of cultural dualism, that is, an equal partnership between the two cultural groups. But by the 1960s, Canada's demographics revealed a third ethnic force and the concept of duality no longer reflected reality. English-speaking Canada had changed. English-speaking Canada had evolved from a predominantly British population to a more diverse one with several large minority groups who also spoke other languages besides English, especially German and Ukrainian. Ottawa, in searching for a compromise, established two policies, bilingualism (1969) and multiculturalism (1971).

In the second vision, Canada consists of 10 equal provinces—yet this, too, is complex. On the one hand, this vision represents the simple notion based on provincial powers granted under the British North America Act, which ensured that Canada consists of a union of equal provinces, all of which have the same powers of government. Nonetheless, by assigning provinces, including Québec, powers over education, language, and other cultural matters within their provincial jurisdictions, Québec's French culture was secure from political tampering by the anglophone majority in the rest of Canada. Confederation then provided a form of collective rights for French culture within Québec. Under Canada's federal system, the powers of government are shared between the federal government and 10 provincial governments. But are all provinces really equal? As noted earlier, population size, geographic extent, and financial strength vary considerably, which is reflected in the need for equalization payments (Vignette 3.5).

The vision of 10 equal provinces may reflect an English-Canadian nationalism. For some time, English-speaking Canadians have been searching for their cultural identity. Before World War I, Canadians saw themselves as part of the British Empire. By the end of World War II, this perspective began to change. The maple leaf flag, adopted by Parliament in 1964, and 'O Canada', the new national anthem approved by Parliament in 1967 and officially adopted in 1980, were signs of this cultural change. While the Québécois culture was flourishing, thanks in part to generous provincial funding for the arts, English-speaking Canadians continued to lean heavily on American culture. Some looked with envy at the cultural accomplishments of the Québécois and wondered aloud if similar achievements in English-speaking Canada were possible. The answer was yes, providing the provincial governments offered similar financial support for the arts.

Compromise

Given the incompatibility of the two visions—two founding peoples versus 10 equal provinces—and the historical development of the country, Canadian politicians have had the unenviable task of trying to accommodate demands from different groups—especially French Canadians, new immigrants, and Aboriginal peoples—and from different regions without offending other groups or regions. As in the past, politicians have continued to struggle with this Canadian dilemma, but in reality there is no perfect solution, only compromise. With this object in mind the federal government has made many efforts in search of the elusive middle ground.[10] It seems the search for an acceptable compromise between the two opposing visions of Canada will never end, and perhaps that is a good thing because the process is more important than the end result. To understand the current struggle for compromise, it is important to understand the political, economic, and cultural developments that have taken place in Québec over the past five decades. These, and their effects on the English/French faultline, are outlined in the following pages.

Resurgence of Québec Nationalism

After World War II, Québec broke with its past. A rise of Québec nationalism had begun much earlier but gained political momentum during the **Quiet Revolution** of the early 1960s. This development was the result of four major events. The most important was the resurgence of ethnic nationalism, that is, a pride in being a Québécois. The second was Québec's joining the urban/industrial world of North America and the subsequent expansion in the size of its industrial labour force and business class. The third was the removal of the old elite. This reform movement was profoundly anticlerical in its opposition to the entrenched role of the Church in Québec society, particularly the Church's control over education. In many ways, this reform was based on the aspirations of the working and middle classes in the new Québec economy. The fourth was the state's aggressive role in the province's affairs.

With the election of Jean Lesage's Liberal government in 1960, the province moved forcefully in a new direction. It created a more powerful civil service that allowed francophones access to middle and senior positions that were often denied them in the private sector of the Québec economy, which was controlled by English-speaking Quebecers and American companies. It nationalized the province's electric system, thereby creating the industrial giant known as Hydro-Québec, now a powerful symbol of Québec's revitalized economy and society. In turn, Hydro-Québec built a number of huge energy projects that demonstrated the province's industrial strength. By 1968, this Crown corporation had constructed one of the largest dams in the world on the Manicouagan River. Called Manic 5, this dam demonstrated Hydro-Québec's engineering and construction capabilities. To Quebecers, Hydro-Québec was a symbol of Québec's economic liberation from the years of suffocation associated with Maurice Duplessis and his Union Nationale government, which had been closely tied to big businesses owned by English-speaking Canadians and Americans. Clearly, Lesage's political goal of becoming '*maîtres chez nous*' (masters in our own house) had materialized with the success of Hydro-Québec, thus sparking a growth in Québec nationalism. Québec's desire for more autonomy in its own affairs intensified with increased confidence. In short, a new society had arisen in Québec, a society that wanted to chart its destiny. Charles Taylor (1993: 4) summed up this new feeling as 'a French Canada which, after a couple of centuries of enforced incubation [under London and then Ottawa], was ready to take control once more of its history.' The political question Taylor raised is a simple one: Would this 'control' take place within the framework of Canada's political system or outside it?

Separatism

Separatism grew out of the Quiet Revolution. It is a form of ethnic nationalism that is popular with francophones but unpopular with anglophones and allophones (those whose first language is neither French nor English). In 1967, French President Charles de Gaulle visited Québec and ignited the forces of French-Canadian nationalism with his now famous words, '*Vive le Québec. Vive le Québec libre.*' From that moment on, separatism gained support and took on a political form. By the time of the first referendum on independence in 1980, the separatists formed a substantial minority within Québec's population, with perhaps as many as 20 per cent dedicated separatists and another 20 per cent strongly dissatisfied with their place within Canada.

Separatism has had two distinct branches, represented by the Front de libération du Québec (FLQ) and the Parti Québécois (PQ). The FLQ was a small fringe group within the separatist movement. It sought political change through revolutionary means, including bombing, kidnapping, and murder. However, the vast majority of separatists sought change through democratic means. The PQ, first elected to government in November 1976, was committed to a democratic solution by means of a referendum followed by negotiations with the rest of Canada. Referendums, the process of referring a political question to the electorate for a direct decision by general vote, are notoriously tricky political instruments, but Québec Premier René Lévesque, who had been the architect of Hydro-Québec as a minister in the Lesage government and in the late 1960s left the Liberals to form the PQ, offered Quebecers what he thought was a clear choice—the unpopular status quo or a bold new beginning under sovereignty-association. **Sovereignty-association** meant political separation but a new economic association with the rest of Canada. Lévesque recognized that the integrated nature of Canada and its east–west economic axis made continued economic ties with Canada essential for the survival of Québec.

Prior to the May 1980 referendum, the PQ vigorously tackled challenging economic problems and critical cultural issues, all of which had three purposes:

- to accelerate the modernization processes that began with the Quiet Revolution;
- to promote the Québécois culture;
- to demonstrate that a PQ government could run the affairs of an independent state.

The provincial government was involved in the marketplace, often through Crown corporations and government assistance for francophone business operations. The provincial government also promoted Québécois culture in a variety of ways. Under the Liberal government of Robert Bourassa, the French language had been declared the sole official language in 1974, but the PQ government went much further with Bill 101 in 1977.[11] This bill made it necessary for most Quebec children, regardless of background or preference, to be educated in French-language schools, and was a key measure in ensuring the supremacy of the French language in the province. Among its many goals, Bill 101 was designed to ensure that the children of new immigrants went to French schools and thus to guarantee that the French-speaking population of Québec would continue to grow.

Québec voters rejected the sovereignty-association referendum, with almost 60 per cent voting to remain in Canada, which suggests that just over half of the francophone voters stood with the 'Non' side, along with almost all the English-speaking residents. The rest of Canada responded with a collective sigh of relief, but separatism was far from dead. Several political events renewed separatist sentiment. One was the Constitution Act of 1982, which patriated the Constitution and gave Canadians the Charter of Rights and Freedoms. The Trudeau government accomplished this political feat at the cost of poisoning relations with Québec City by including the Charter of Rights and Freedoms in the Constitution. The Charter curtailed the power of the Québec government, and the Constitution was patriated without the approval of the Québec government. There were also the failed attempts of Brian Mulroney's Conservative government to achieve provincial unanimity for constitutional reform. The first attempt was the Meech Lake Accord, a package of constitutional revisions incorporating Québec's 'minimum' demands for political reform. This Accord was agreed to in principle by Ottawa and the 10 provinces in 1987, but two provinces, Manitoba and Newfoundland, failed to pass it within the required three years.

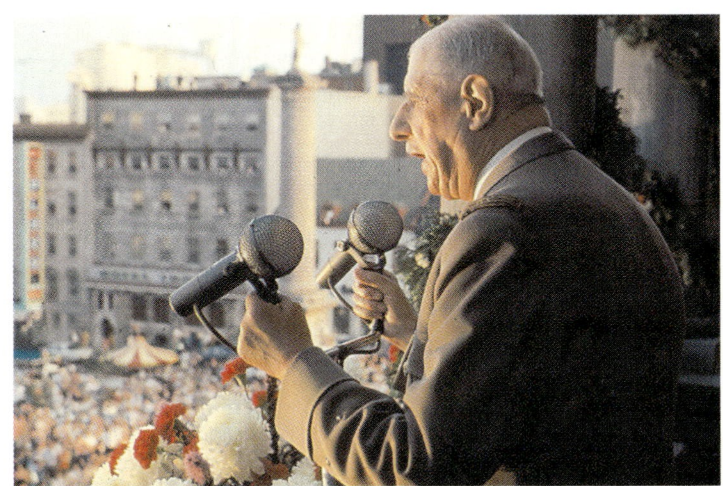

French President Charles de Gaulle during his incendiary 'Vive le Québec libre' speech in Montreal, 24 July 1967. (City of Montréal. Records Management and Archives)

Quebecers felt humiliated and rejected by the rest of Canada. In 1991, the Mulroney government attempted a second round of constitutional negotiations culminating in the Charlottetown Accord, which would give Québec distinct society status, the provinces more power and input on the selection of judges, and Aboriginal peoples the entrenched right to self-government, as well as provide for reform of the Supreme Court and the Senate to make these institutions more representative of provincial interests and power. In 1992, the Charlottetown Accord was roundly rejected in a national referendum and only narrowly approved in four provinces (but not Québec).

These last two political misadventures, as well as lingering bitterness from the patriation process, revived the spirits of the separatists, led by Jacques Parizeau and Lucien Bouchard. Parizeau's party, the PQ, returned to power in 1994, promising a referendum on sovereignty. While the referendum question referred to a new partnership with the rest of Canada, Parizeau believed that such an arrangement was impossible and saw only one solution—an independent Québec. In the previous year, the Bloc Québécois (a new federal party representing Québec interests), led by Lucien Bouchard (a former cabinet minister in the Mulroney government), formed the official opposition in the

House of Commons. The Québec public expressed their dissatisfaction with Ottawa by rejecting traditional political parties. This left Québec federalists in a vulnerable position where they grew steadily weaker and more disorganized. The federal Liberals could offer little help. In fact, Jean Chrétien, himself a Quebecer, was extremely disliked by many in Québec for a variety of reasons, including his role in the patriation of the Constitution in 1982. He had become a 'tête de turc' (a scapegoat for federal policies), a symbol of those Quebecers who, as federal ministers, put Québec 'in its place'. (For further discussion of the French/English faultline in Québec, see Chapter 6.)

The results of the 1995 referendum vote in Québec were extremely close. 'No—by a Whisker' screamed the headline of the *Globe and Mail* on the morning after the referendum of 30 October 1995. Québec came within 40,000 votes of approving the separatist dream of becoming an independent state (Vignette 3.8).

Moving Forward

The 1995 referendum was a low point in French/English relations, and its after-effects were many and varied. English Canada, dazed by the outcome, attempted to respond. Ottawa reacted almost immediately after the October referendum by passing a unilateral declaration that recognized Québec as a distinct society. The leader of the separatist forces, Jacques Parizeau, had shocked all Quebecers on the night of the referendum when he blamed the 'Oui' side's loss on 'money and the ethnic vote'. In an electrifying moment, the dark side of ethnic nationalism had been revealed. Other centrifugal forces were released, too, including the partitionists, who argued that 'if Canada is divisible, so is Québec.' The Cree in northern Québec threatened secession, and partitionists pressured dozens of municipalities around Montréal and Hull (since renamed Gatineau) to declare their allegiance to Canada.

By 1996, the federal government had decided to take a hard line with Québec, which included having the Supreme Court of Canada determine the conditions of separation. In the same year, the premiers attempted to address the unity issue. In a much more conciliatory manner, they announced the Calgary Declaration: 'the unique character of

Vignette 3.8 No—by a Whisker

The Results of the 30 October 1995 Referendum

The Question: 'Do you agree that Québec should become sovereign, after having made a formal offer to Canada for a new Economic and Political Partnership within the scope of the Bill respecting the future of Québec and of the agreement signed on June 12, 1995?'

The Answer (at 10:30 p.m. Eastern Time, 21,907 of 22,427 polls):

	Number	Per cent
No	2,294,162	49.5
Yes	2,254,496	48.7
Rejected	83,340	1.8
Total	4,631,998	100.0

Source: Globe and Mail, 31 Oct. 1995, A1. Reprinted with permission of *The Globe and Mail.*

Québec society with its French-speaking majority, its culture and its tradition of civil law is fundamental to the well-being of Canada.' In the typical fashion of Canadian provincial leaders, the premiers added to their Declaration that 'any power conferred to one province in the future must be available to all.' This Declaration was the third attempt at reconciliation with Québec since the patriation of the Constitution in 1982.[12] The next step was for each provincial government to pass the appropriate legislation, giving this declaration legal status. By July 1998, all provinces (except Québec) and territories had passed this resolution in their legislatures.

Changes are taking place in Québec, too. Separatism, while not gone, has lost its spark for the time being, so much so that the Parti Québécois was in disarray following the provincial election in the spring of 2007, having fallen to third, behind the governing Liberals and the Action Démocratique du Québec. Like other Canadians, Quebecers have grown weary of referendums, political bickering, and the resulting unsettling effect on the national and Québec economies. After a decade of weak economic performance, Québec and the rest of Canada are mainly concerned with economic matters, particularly high levels of unemployment and insufficient funds for education and health. Then, too, the threat of the English is a thing of the past. Quebecers, confident in their language and culture, are more comfortable and secure than ever before. Equally important, Ottawa is more comfortable with the idea of Québécois forming a 'distinct cultural group' or nation within Canada. In November 2006, the House of Commons overwhelmingly passed a motion by Prime Minister Harper that recognized Québécois as a nation within Canada.

Towards the Future

History and geography explain the basis of the French/English relationship. This dynamic relationship goes to the heart of the nation. Geography compels Canadians to recognize that Québec represents a distinct region of Canada in which a different language and culture dominate. History teaches Canadians that compromise leads to national unity, while conflicts drive a wedge between the two founding peoples of Canada.

Canada's history reflects over 200 years of French/English relations. Through conflicts and compromises, these innumerable interactions, both large and small, have shaped the essential components of the national character, namely, the capacity and willingness to find solutions to complex questions. Both Québec and the rest of Canada are different places and those changes have shaped the French/English faultline and the attempts at compromise in this relationship.

On the surface, reconciliation seems an impossible task, but political realities demand some form of compromise or at least a willingness to search for a solution. Perhaps Paul Villeneuve (1993: 104) was correct when he observed that 'for Canada, survival lies in the travelling toward an identity.' For most Canadians, this implies recognition of the deep divide between French and English Canada but also an acceptance of the importance of each vision. Canadians also have learned that the political gains achieved by having the dominant English-Canadian society impose its will over the minority French-Canadian society are short-term and will eventually weaken national unity. Perhaps the House of Commons recognition of Québécois as a nation within Canada in 2006 represents an important step forward.

Notes

1. Although national, provincial, territorial, and municipal governments exist in Canada, only the federal and provincial governments have powers that no other level of government can usurp.
2. Under the terms of the British North America Act, the Dominion of Canada was composed of four provinces (Ontario, Québec, New Brunswick, and Nova Scotia). Modelled after the British parliamentary and monarchical system of government, the newly formed country had a Parliament made up of three elements: the head of government (a governor-general who represented the monarch), an upper house (the Senate), and a lower house (the House of Commons). This Act was modified several times to accommodate Canada's evolving political needs and its gradual movement to independent nationhood. The patriation of Canada's Constitution in 1982 removed the last vestige of Canada's political dependence on the United Kingdom, although Canada still recognizes the British monarch as its symbolic head.

   The British North America Act was based on the highly centralized government of the United Kingdom in the 1860s. However, this Act assigned specific powers to the provinces in order to satisfy Québec's demand for control over its culture. The Canadian political system that emerged, therefore, allowed for regionalized politics. For example, political parties in the House of Commons sometimes serve regional interests. In the 1920s, the Progressive Party represented the concerns of farmers in Western Canada, while the pro-independence Bloc Québécois, which was formed in 1990, not only serves the interests of Québec but is also active in the separatist movement. Furthermore, while the House of Commons is based on the principle of representation by population, Senate membership is based on the principle of equal regional representation. However, because senators are appointed by the Prime Minister and not elected by the people in the different regions of the country, the Senate fails to provide a regional counterweight to the House of Commons.
3. A group of Irish Americans, known as Fenians, was struggling for Irish independence. They believed that attacking British possessions in North America would advance the cause of a free Ireland. Between 1866 and 1870, the Fenians launched several raids across the border into Canada. The United States did not encourage these raids and eventually forced the Fenians to disband. By the end of the American Civil War, Anglo-American relations again were strained because of Britain's tacit support for the Confederacy in the American Civil War. For that reason, the United States withdrew from the Reciprocity Treaty in 1866. This treaty, a free trade agreement between British North America and the United States, began in 1854; the subsequent years were prosperous ones for British North America. Its loss forced the Province of Canada to seek an alternative economic union with the other British colonies in North America.
4. Representation in the House of Commons is readjusted after each decennial (10-year) census in accordance with the Constitution Act, 1867 (formerly the BNA Act) and the Electoral Boundaries Readjustment Act (1985, as amended). On 13 June 1992, following the release of the population figures from the 1991 census, the Chief Electoral Officer of Canada published in the *Canada Gazette* the results of the calculations required by the Constitution Act, which meant an increase in the number of seats in the House of Commons from 295 to 301. The next readjustment took place in 2004 when seven new ridings were created for a total of 308 members of the House of Commons. The formula for determining the number of seats for each province and territory is available at:<http://www.elections.ca/scripts/fedrep/federal_e/repform_e.htm>.
5. The Manitoba Act of 1870 recognized the legal status of farms and other lands occupied by the Métis as 'fee simple' private property. As well, the Act provided that 1.4 million acres (566,580 ha) be reserved for the children of Métis. The land was allocated to these Métis in 240-acre (97-ha) parcels, plus 160 acres (65 ha) in 'scrip' for each adult head of a family. These lands were distributed after 1875, but much of the 'scrip' land was sold and then occupied by non-Métis. For more on this subject, see Tough (1996: ch. 6).
6. Pontiac, the Odawa chief in the Ohio Valley, led a successful uprising against the British in 1763.

By capturing the forts in the Ohio Territory, he exposed Britain's precarious hold on this region, which the British had just obtained from the French. However, Pontiac and his followers could not hold these forts against the British because they could no longer obtain ammunition and muskets from the French. Pontiac concluded that his best move would be to make peace with Britain. The British came to the same conclusion, though for other reasons. Without the help of Pontiac and the other chiefs in this region, Britain would lose control over these lands. Britain therefore had to form an alliance with them. With that objective in mind, George III announced an important concession to these Indians in the Royal Proclamation of 1763, namely, that the King recognized them as valued allies and that the land they used to hunt and trap was 'Indian land' within the British Empire.

7. In 1215, King John of England was forced to sign the Magna Carta. In this charter, he promised to stop interfering with the Church and the law and to consult regularly with the country's leaders before collecting new taxes.
8. While the French language was not recognized by the Québec Act, 1774, the Governor made use of the French language in conducting his business with local officials. For example, the judges appointed by the Governor had to know both languages in order to facilitate the business of the court. In short, while English was the official language of British North America, the British colony of Québec functioned in both the French and English languages.
9. Louis Riel was the Métis political and spiritual leader in the late nineteenth century. This controversial figure is considered both a Father of Confederation and a traitor to the country. Riel, who was born in the Red River Colony in 1844, studied for the priesthood at the Collège de Montréal. The founder of Manitoba and the central figure in both the Red River Rebellion (1869–70) and the North-West Rebellion (1885), he was captured shortly after the Battle of Batoche, where the Métis forces were defeated. After a trial in Regina, the jury found Riel guilty of treason but recommended clemency. Appeals were made to Manitoba's Court of Queen's Bench and to the Judicial Committee of the Privy Council. Both appeals were dismissed. A final appeal went to the federal cabinet, but the government of John A. Macdonald wanted Riel executed. Riel was hanged in Regina on 16 November 1885. His body was interred in the cemetery at the Cathedral of St Boniface in Manitoba.

   Riel's execution has had a lasting effect on Canada. In Québec, French Canadians felt betrayed by the Conservative government and federalism. Riel's execution was proof for Canada's French-speaking population that they could not count on the federal government to look after French-Canadian interests. It was also a blow against a francophone presence in the West. In Ontario the hanging of Louis Riel satisfied the anti-Catholic and anti-French majority. For the Orange Order (the Protestant fraternal society that blamed Riel for the death of one of their members, Thomas Scott, who was executed by a Métis firing squad during the Red River Rebellion), Riel's execution was long overdue. In the West, Riel's hanging resulted in the marginalization of both the Métis and Indian tribes, especially those who participated in the uprising.
10. A recent example is the political fallout from the Québec referendum. Prime Minister Chrétien sought to fulfill his verbal promises made in the closing days before the vote on the 1995 referendum. In a House of Commons resolution, the federal government proposed three concessions to Québec: (1) a veto over constitutional changes; (2) recognition of Québec's distinct society status; and (3) devolution of federal powers to Québec. In the case of the veto, Ottawa was prepared to 'lend' its constitutional veto to Québec, Ontario, Atlantic Canada, and the four western provinces. Not only was the federal government committing itself to seeking permission from these four regions before putting its stamp of approval on any constitutional change, it was also recognizing that Canada consisted of four major regions. The premiers of Alberta and British Columbia reacted negatively to that concept of regionalism. British Columbians in particular saw this arrangement as another example of Ottawa's failure to recognize the west coast as a 'distinct and powerful' part of Canada. The federal government retreated from this issue and quickly amended its resolution to extend the veto to British Columbia. In December 1995, this resolution passed in both the House of Commons and the Senate. It then became the law of the land that Canada consists of five major regions!
11. In 1974, the Québec Liberal government passed Bill 22 (Loi sur la langue officielle), which made French the language of government and the

workplace. English was no longer an official language in Québec. In 1977, the Parti Québécois government introduced a much stronger language measure in the form of Bill 101 (Charte de la langue française). This legislation eliminated English as one of the official languages of Québec and required the children of all newcomers to Québec to be educated in French. Four years later, Bill 178 required all commercial signs to use only French. The French language has made modest gains outside of Québec. In 1969, New Brunswick passed an Official Languages Act, which gave equal status, rights, and privileges to English and French, and the federal Parliament passed the Official Languages Act, which declared the equal status of English and French in Parliament and in the Canadian public service.

12. Separatists argue that Québec does not have enough powers, that is, Québec is subordinate to Ottawa. For separatists, the solution lies in independence, whether achieved through Parizeau's 'chicken-plucking' strategy or Bouchard's 'winning' referendum strategy. A federalist counter-argument is that, as a member of a federation, Québec has many 'exclusive' powers, such as power over language and education. However, circumstances may force Ottawa to make a decision that adversely affects some provinces while favouring others. In 1982, the patriation of the British North America Act, renamed the Constitution Act, 1867, was such a decision. The Constitution Act, 1982, which was entrenched at the same time, added to the British North America Act in several ways, but without a doubt the most important addition has been the Charter of Rights and Freedoms. These rights and freedoms strengthen the rights of individuals and weaken collective rights. Prime Minister Trudeau, who conceived of society as an agglomeration of individuals (not collectivities), whose rights accrued to them as individuals, saw the Charter as protecting individuals from governments that try to suppress individual rights. Then, too, there is the Supreme Court of Canada's changed role, which has become more proactive with the adjudication of Charter cases.

Challenge Questions

1. Why did the doctrine 'terra nullius' allow Europeans to consider North America 'unoccupied' and therefore open to European settlement?
2. Why did Britain encourage its North American colonies to unite?
3. What were Prime Minister John A. Macdonald's four goals for the Canadian Pacific Railway?
4. Why was the Doukhobor settlement in the Prairies a failure?
5. What are the centrifugal forces that are trying to pull Canada apart?
6. What is the difference between comprehensive and specific land-claim agreements?
7. Why are there two visions of Canada—as two founding peoples or as 10 equal provinces?
8. Henri Bourassa was a strong advocate of cultural dualism. Would Bourassa approve of the House of Commons recognition of the Québécois as a nation within Canada?

Key Terms

Aboriginal peoples
All Canadians whose ancestors lived in Canada before the arrival of Europeans; includes status and non-status Indians, Métis, and Inuit.

Château Clique
The political elite of Lower Canada, composed of an alliance of officials and merchants who had considerable political influence with the British-appointed governor; similar to the Family Compact in Upper Canada.

comprehensive land-claim agreement
Agreement based on territory claimed by Aboriginal peoples that was never ceded or surrendered by treaty. Such agreements extinguish the Aboriginal land claim to vast areas in exchange for a relatively small amount of land, capital, and the organizational structure to manage their lands and capital.

culture areas
Regions within which the population has a common set of attitudes, economic and social practices, and values.

Family Compact
A group of officials who dominated senior bureaucratic positions, the executive and legislative councils, and the judiciary in Upper Canada.

Inuit
People who are descended from the Thule. The Thule migrated into Canada's Arctic from Alaska about 1,000 years ago. The Inuit do not fall under the Indian Act, but are identified as an Aboriginal people under the Constitution Act, 1982.

Loyalists
Colonists who supported the British during the American Revolution. About 40,000 American colonists who were loyal to Britain resettled in Canada, especially in Nova Scotia and Québec.

Métis
People who have a mixed biological and cultural heritage, usually either French-Indian or English/Scottish-Indian. This 'mixing' between Indians and Europeans took place during the fur trade and continues today. Originally the term was more narrowly applied to French-Indian people who settled in the Red River area and who developed a distinct hunting economy and society based on the French language and the Roman Catholic religion.

non-status Indians
Those of Amerindian ancestry who are not registered as Indians under the Indian Act.

Quiet Revolution
A period in Québec during the Liberal government of Jean Lesage (1960 to 1966) that was characterized by social, economic, and educational reforms and by the rebirth of pride and self-confidence among the French-speaking members of Québec society, which led to a resurgence of francophone ethnic nationalism. During this time, the secular nationalist movement gained strength.

reserve
Under the Indian Act, reserves are defined as lands 'held by her Majesty for the use and benefit of the bands for which they were set apart; and subject to this Act and to the terms of any treaty or surrender'.

sovereignty-association
A concept designed by the Parti Québécois under the Lévesque government and employed in the 1980 referendum. This concept was based on the vision of Canada as consisting of two 'equal' peoples. Sovereignty-association called for Québec sovereignty but with a partnership with Canada based on an economic association.

specific land-claim agreements
Agreements designed to rectify shortcomings in the original treaty agreement with a band or that seek to redress failure on the part of the federal government to meet the terms of the treaty. Many of these have involved the unilateral alienation by the government of reserve land.

status Indians
Aboriginal peoples who are registered as Indians under the Indian Act.

terra nullius
During the eighteenth century, Europeans believed that this doctrine gave legal right to claim ownership of the land occupied by Indians and Inuit because the land was not cultivated nor were there permanent settlements.

treaty Indians
Aboriginal peoples who are descendants of Indians who signed a numbered treaty and who benefit from the rights described in the treaty. All treaty Indians are status Indians, but not all status Indians are treaty Indians.

Bibliography

Anderson, Robert B., and Robert M. Bone. 1995. 'First Nations Economic Development: A Contingency Perspective', *Canadian Geographer* 39, 2: 120–30.

Asch, Michael, ed. 1997. *Aboriginal and Treaty Rights in Canada: Essays on Law, Equality, and Respect for Difference*. Vancouver: University of British Columbia Press.

Bonnichsen, Robson, and Karen L. Turnmire, eds. 1999. *Ice Age Peoples of North America: Environments, Origins, and Adaptations of the First Americans*. Corvallis: Oregon State University Press.

Bryden, Joan. 2006: 'Provinces Remain Divided on Equalization', *Globe and Mail*, 8 June. At: <www.theglobeandmail.com/servlet/story/RTGAM.20060608.wequall0608/BNStory/National/home>.

Brownlie, Robin Jarvis. 2003. *A Fatherly Eye: Indian Agents, Government Power, and Aboriginal Resistance in Ontario, 1918–1939*. Toronto: Oxford University Press.

Bumsted, J.M. 2007. *A History of the Canadian Peoples*, 3rd edn. Toronto: Oxford University Press.

Canada. 1882. *Census of Canada 1880–1*, vol. 1. Ottawa: MacLean, Rogers & Company.

Canada. 2003. Elections Canada. At: <www.elections.ca/scripts/fedrep/federal_e/repform_e.htm>.

Canada, Department of Finance. 2006, 2007. Equalization Program. At: <www.fin.gc.ca/FEDPROV/eqpe.html>.

Carter, Sarah. 2004. '"We Must Farm To Enable Us To Live": The Plains Cree and Agriculture to 1900', in R. Bruce Morrison and C. Roderick Wilson, eds, *Native Peoples: The Canadian Experience*. Toronto: Oxford University Press.

CBC. 2006. Canada Votes. At: <www.cbc.ca/canadavotes>.

Coates, Ken. 1992. *Aboriginal Land Claims in Canada: A Regional Perspective*. Toronto: Copp Clark Pitman.

Cook, R. 1969. *Provincial Autonomy, Minority Rights and the Concept Theory, 1867–1921*. Studies of the Royal Commission on Bilingualism and Bilculturalism, no. 4. Ottawa: Queen's Printer.

———. 1993. *The Voyages of Jacques Cartier*. Toronto: University of Toronto Press.

Dickason, Olive Patricia. 2002. *Canada's First Nations: A History of Founding Peoples from Earliest Times*, 3rd edn. Toronto: Oxford University Press.

Fife, Robert, and Joël Bellevance. 2000. 'Cabinet Vows to Win over the West', *National Post*, 1 Dec., A6.

Fremlin, Gerald, ed. 1974. *The National Atlas of Canada*. Ottawa: Macmillan.

Garreau, Joel. 1981. *The Nine Nations of North America*. Boston: Houghton Mifflin.

Globe and Mail. 1995. 'No—by a Whisker', 31 Oct., A1.

Harris, R. Cole, and John Warkentin. 1991. *Canada before Confederation: A Study in Historical Geography*. Ottawa: Carleton University Press.

Kerr, Donald, and Deryck W. Holdsworth, eds. 1990. *Historical Atlas of Canada, Volume III: Addressing the Twentieth Century 1891–1961*. Toronto: University of Toronto Press.

Lower, J. Arthur. 1983. *Western Canada: An Outline History*. Vancouver: Douglas & McIntyre.

McVey, Wayne W., and W.E. Kalbach. 1995. *Canadian Population*. Toronto: Nelson Canada.

Marsh, James H., ed. 1988. *The Canadian Encyclopedia*, 2nd edn. Edmonton: Hurtig.

Martin, Don. 2006. 'Premiers Still a Divided Front on Equalization', *National Post*, 9 June 2006: A7.

Mitchell, Robert D., and Paul A. Groves. 1987. *North America: The Historical Geography of a Changing Continent*. Totowa, NJ: Rowman & Littlefield.

Moffat, Ben. 2002. 'Geographic Antecedents of Discontent: Power and Western Canadian Regions 1870 to 1935', *Prairie Perspectives* 5: 202–28.

Morton, D. 1983. *A Short History of Canada*. Edmonton: Hurtig.

Nemni, Max. 1994. 'The Case against Quebec Nationalism', *American Review of Canadian Studies* 24, 2: 171–96.

O'Handley, Kathryn, ed. 1994. *Canadian Parliamentary Guide*. Toronto: Globe and Mail Publishing.

Pielou, E.C. 1991. *After the Ice Age: The Return of Life to Glaciated North America*. Chicago: University of Chicago Press.

Riendeau, Roger. 2000. *A Brief History of Canada*. Markham, Ont.: Fitzhenry & Whiteside.

Saul, John Ralston. 1997. *Reflections of a Siamese Twin: Canada at the End of the Twentieth Century*. Toronto: Viking.

Simpson, Jeffrey. 1993. *Faultlines: Struggling for a Canadian Vision*. Toronto: HarperCollins.

Smith, P.J. 1982. 'Alberta Since 1945: The Maturing Settlement System', in L.D. McCann, ed., *Heartland and Hinterland: A Regional Geography of Canada*. Scarborough, Ont.: Prentice-Hall.

Statistics Canada. 1998. '1996 Census: Ethnic Origin, Visible Minorities', *The Daily*, 17 Feb. At: <www.statcan.ca/Daily/English/>.

———.2003. *Historical Statistics of Canada*. Ottawa: Statistics Canada Catalogue no. 11–516–XIE. At: <www.statcan.ca/english/freepub/11-516-XIE/sectiona/sectiona.htm>.

Taylor, Charles. 1993. *Reconciling the Solitudes*. Montréal and Kingston: McGill-Queen's University Press.

Tough, Frank. 1996. *As Their Natural Resources Fail: Native Peoples and the Economic History of Northern Manitoba, 1870–1930*. Vancouver: University of British Columbia Press.

Tracie, Carl J. 1996. *Toil and Peaceful Life: Doukhobor Village Settlement in Saskatchewan, 1899–1918*. Regina: Canadian Plains Research Centre, University of Regina.

Villeneuve, Paul. 1993. 'Allocution présidentielle: L'invention de l'avenir au nord de l'amérique', *Le géographe canadien* 37, 2: 98–104.

Warkentin, John, ed. 1968. *Canada: A Geographical Interpretation*. Toronto: Methuen.

Williams, Glyndwr. 1983. 'The Hudson's Bay Company and the Fur Trade, 1670–1870', *The Beaver* (Autumn): 4–81.

Wright, James V. 1995, 1999. *A History of the Native People of Canada*, vols 1 and 2. Ottawa: Canadian Museum of Civilization.

Wynn, G. 1990. *People, Places, Patterns, Processes: Geographic Perspectives on the Canadian Past*. Toronto: Copp Clark Pitman.

Further Reading

The historical geography of Canada recalls past events. Maps play a large role in this rediscovery of Canada's past. In 1970, several geographers and historians explored the idea of preparing a major Canadian historical atlas focused on social and economic themes. Four editors (R. Cole Harris, Volume I; R. Louis Gentilcore, Volume II; and Donald Kerr and Deryck W. Holdsworth, Volume III) brought this enterprise to a successful conclusion in 1993. These three volumes weave together the various historic strands that comprise Canada's historical geography and represent a rich legacy for Canadian scholars and students. For those studying *The Regional Geography of Canada*, these three volumes parallel the discussion found in this chapter and provide a wonderful source of essay topics.

Volume 1: Harris, R. Cole, ed. 1987. *Historical Atlas of Canada, Volume I: From the Beginning to 1800*. Toronto: University of Toronto Press.

Professor Harris and his collaborators have produced a detailed account of the arrival of Canada's First Peoples and then the French and British settlers. Each map and descriptive explanation provides an in-depth record of a particular historic event taking place in one region. Students will discover many interesting maps, including The Last Ice Sheet (Plate 1); Population and Subsistence (Plate 18); The Newfoundland Fishery, 18th Century (Plate 25); Maritime Canada, Late 18th Century (Plate 32); and The Seigneuries (Plate 51).

Volume 2: Gentilcore, R. Louis, ed. 1993. *Historical Atlas of Canada, Volume II: The Land Transformed 1800–1891*. Toronto: University of Toronto Press.

Under Professor Gentilcore's direction, a series of maps with text capture the essential historic events in nineteenth-century Canada. This volume examines agricultural settlement to mid-century as the building of a nation. Canada's economy remained heavily dependent on extraction and exportation of natural resources to Great Britain and the United States, though the development of a manufacturing base in Central Canada was supported, especially after 1879, by the high tariffs of the National Policy. Examples of the wide range of topics found in Volume II include Timber Production and Trade to 1850 (Plate 11); Unrest in the Canadas (Plate 23); and An Emerging Urban System, 1845, 1885 (Plate 45).

Volume 3: Kerr, Donald, and Deryck W. Holdsworth, eds. 1990. *Historical Atlas of Canada, Volume III: Addressing the Twentieth Century 1891–1961*. Toronto: University of Toronto Press.

The main historic themes examined by Professors Kerr and Holdsworth are the transformation of Canada from an agrarian society to an urban/industrial society and the crises of the Great Depression, World War II, and the immediate post-World War II period. Canada was changing into an industrial country within a North American context. American investment in Canada was growing and American branch plants dominated our manufacturing sector. During this time, Canada's population grew from 5 million to 18 million and this population was composed primarily of Canadians with a British, French, or European background. The number of Indians, Métis, and Inuit was relatively small, under half a million. Topics covered include The Changing Structure of Manufacturing (Plate 7); The Grain Handling System (Plate 19); and Population Changes (Plate 59).

The *Historical Atlas of Canada* is an excellent reference for scholars and students. Four more recent major events, however, are not covered—the rise of Aboriginal political power, the threat of separation of Québec from the rest of Canada, the influx of non-European immigrants to Canada, and the signing of the Canada–US Free Trade Agreement. These issues are discussed in Chapter 4.

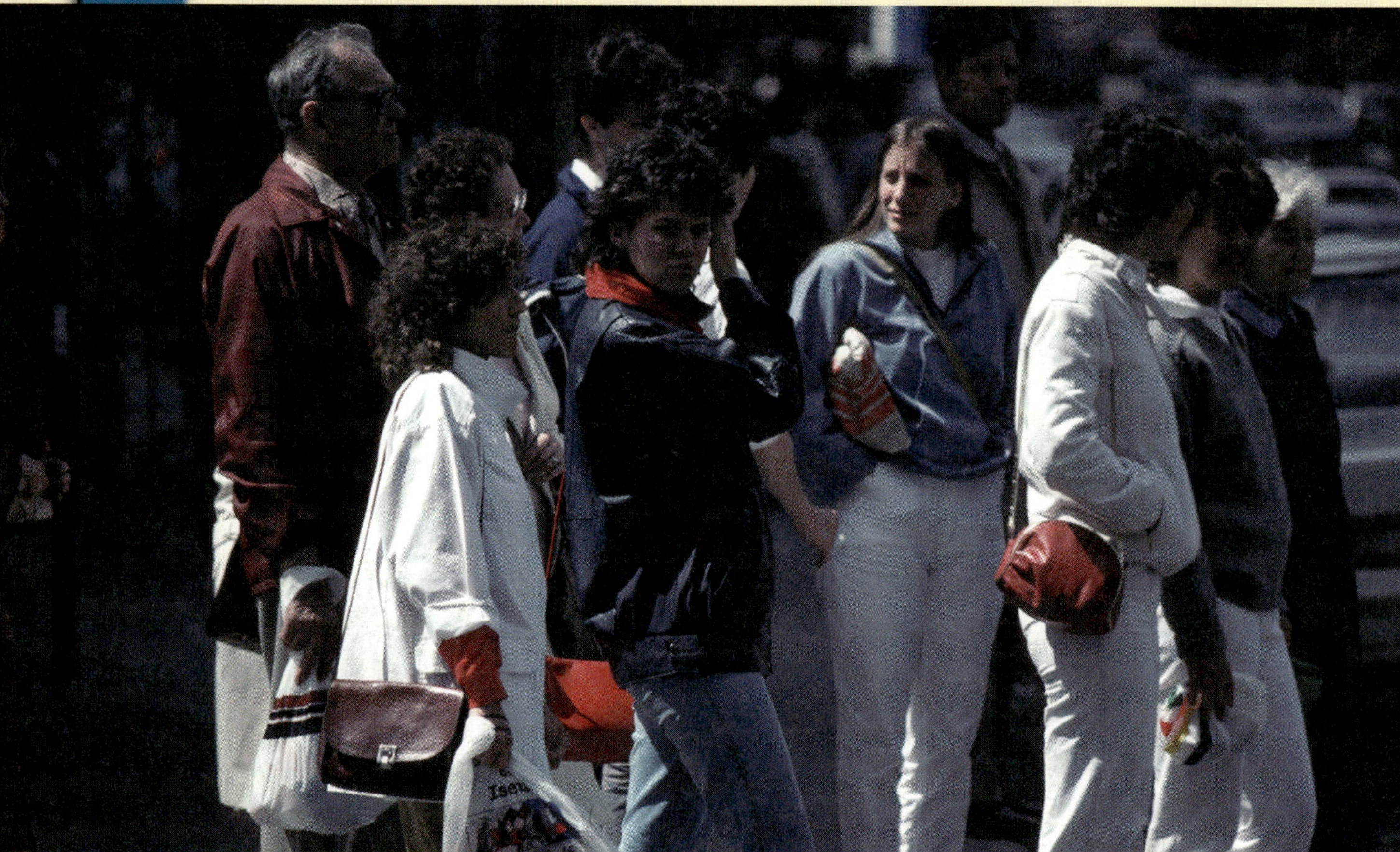

Montréal, Québec (Jean Bruneau/Valan Photos)

Overview and Objectives

Canada's human geography is changing in many ways. First, Canada's population continues to increase. Second, Canada's population increase is now driven by immigration rather than by natural increase. Third, Canada has become a pluralistic society composed of more than 200 ethnic groups from around the world. Fourth, trade liberalization has propelled Canada's economy into the North American and global economies while rising prices for energy and other natural resources have created an economic boom in Alberta. Fifth, Canada now has a north–south trade axis as well as an east–west one. These issues, as well as the demographic implications for Canada's four faultlines, are examined in this chapter, where we will:

- Discuss recent changes in Canada's population.
- Examine the factors causing Canada's population to increase.
- Analyze the implications of immigration for Canadian society.
- Consider the new reality of Canada's North American and global economy.
- Explore the importance of the energy boom for Canada's economy.
- Study the emergence of a north–south trade axis.
- Analyze the impact of demography on Canada's four faultlines.

CHAPTER 4

Canada's Human Face

INTRODUCTION

Canada is home to just over 32 million people. Beyond population size, Canada is undergoing profound demographic, economic, and social changes. Demographic and social changes are largely the result of a declining rate of natural increase (except for Aboriginal peoples) and the impact of immigration from non-European countries. Economic expansion in Canada's resource sector is driven by the strong demand and record high prices for energy and mineral products, while Canada's manufacturing sector is facing difficult times. From 2002 to 2005, employment in manufacturing had fallen by 8 per cent, or 187,000 workers (Mathieu, 2006). The impact of these economic developments has varied across Canada, with Western Canada and British Columbia benefiting from the resource boom while Ontario and Québec find their manufacturing sector struggling because of fierce competition from foreign-based firms with extremely low labour costs.

In this chapter, Canada's population size, density, and distribution are examined first, followed by its natural growth, immigration, age/sex structure, and age dependency. Next, cultural characteristics of Canadians (as measured by ethnicity, multiculturalism, language, and religion) are investigated. The importance of Canada's Constitution, with its emphasis on bilingual, pluralistic, and civic nationalism, becomes readily apparent in this discussion. The four faultlines are affected by demographic changes and the consequences of these impacts are examined. The chapter concludes by examining Canada's economic structure, that is, the primary, secondary, and tertiary sectors of the economy.

Canada's Population

Demography, the statistical study of human populations, including their size, age and gender structure, distribution, density, growth, and related socio-economic characteristics, is central to an understanding of regional geography. Today, Canada's society is not only larger, with an older age structure, but Canadian society is moving in new geographic, social, and religious directions. For the first time, Canada's population exceeds 32 million. Since 2001, most growth has taken place in three geographic regions—Western Canada, Ontario, and British Columbia. Equally important, the country's ethnic composition and cultural diversity now reflect fresh elements in our society. Canada has a substantial number of adherents to Islam, Sikhism, and Hinduism, its larger cities are home to many so-called visible minorities, and Chinese forms the third most commonly spoken language in Canada. Clearly, immigration plays a major role in these demographic and social changes. Other forces of demographic and social change include the rapid increase in Canada's Aboriginal population, the place of Québec within Confederation, and Canada's role within the North American community. These factors have sparked an economic and social revolution that has yet to run its course. The transformation of the Canadian transportation system, especially its major trucking and railway routes, from an east–west orientation to a north–south one marks a visible sign of the shift from a national economy to a continental one.[1]

Population Size

Since Confederation, Canada's population has increased from 3.4 million to 32.6 million, a nearly tenfold increase. Three primary factors accounting for this growth over the last 140 years are natural increase, population gained from territorial expansion, and immigration. For over a hundred years, natural increase was the primary factor accounting for Canada's population growth. In the latter part of the twentieth century, birth rates dropped significantly. Immigration, which has always played a role, became the dominant factor in population growth. Of course, Canada gained popu-

Founded in 1642, Montréal, Québec, is one of Canada's oldest cities. Currently Canada's second most populated city, until the 1970s Montréal was the centre of commerce and trade in the country. (Francis Lépine/ Valan Photos)

Toronto, Ontario, with a population approaching 5 million, is Canada's most populous city and serves as the economic engine for southern Ontario and the financial capital for Canada. Toronto has also become the nation's most culturally diverse city due to the immigration of new Canadians. (Search4Stock Inc.)

lation as its territory expanded. The last territorial expansion took place in 1949 when Newfoundland, with a population of 360,000 at that time, joined Confederation. With its current population, Canada can be considered a medium-sized country on the world stage—only one-quarter of the nations of the world have larger populations than Canada.

Vancouver, British Columbia, is Canada's third most populated city and serves as Canada's portal to Asian markets. In the foreground, the Cambie Street Bridge spans False Creek and leads to BC Place; to the west of the bridge is the Granville Island Public Market. (Al Harvey/The Slide Farm)

Population Density

As the second largest country in the world by geographic area, Canada's population density is one of the lowest in the world. **Population density** is determined by dividing the total number of people by the total land area. Canada has a population density of nearly 4 people per square kilometre, which means the country has an extremely low population density (but not the lowest in the world). Australia, another country with much of its dry lands unsuitable for settlement, has only 2.6 people per square kilometre. Mongolia, another large country with little land suitable for settlement, has an even lower density of 1.7. All other countries of the world have higher population densities. The United States, for example, has 30 people per square kilometre, while Bangladesh is one of the most densely populated countries with a density of 1,000 people per square kilometre.

The explanation for these variations is simple. Land varies greatly in its capacity to support human settlement. Most land in Canada lies beyond the northern limits of agriculture, and land in the Territorial North especially has a low capacity to support human life. The region's population density is only 0.03 people per square kilometre (Table 4.1). On the other hand, Ontario has a much larger population living in a much smaller geographic area. Consequently, Ontario's population density is the highest among the six geographic regions.

Population density figures are more meaningful if they are expressed as the amount of arable land per person. This measure is called physiological density. By this measure, Canada's physiological density is similar to that found in the United States.

Population Distribution

Population distribution is the dispersal of people within a geographic area. Canada's population is extremely unevenly distributed across the country (Figure 4.1 and Table 4.1). In fact, few nations have so much of their population concentrated in such a relatively small area of the country, while the rest of the country is almost vacant. So pronounced is this uneven population distribution that an American geography text described Canada's population as if it were 'drawn by a magnet toward the giant neighbor on the south, for they [Canada's inhabitants] are strikingly concentrated along the United States border' (Trewartha et al., 1967: 542). Canadian scholars often view this same distribution as consisting

Table 4.1 Population Size, Percentage, and Density, Canada and Regions, 2006

Geographic Region	Population	Population (%)	Land Area (000 km^2)	Population Density (per km^2)
Territorial North	101,310	0.3	3,778	0.03
Atlantic Canada	2,284,779	7.2	502	4.7
British Columbia	4,113,487	13.0	893	4.4
Western Canada	5,406,908	17.1	1,756	3.2
Québec	7,546,131	23.9	1,358	5.6
Ontario	12,160,282	38.5	917	13.4
Canada	31,612,897	100.0	9,203	3.5

Source: Statistics Canada (2007a). Adapted from Statistics Canada website URL: http://www40.statcan.ca/l01/cst01/demo02a.htm?sdi=population%20change%20provinces.

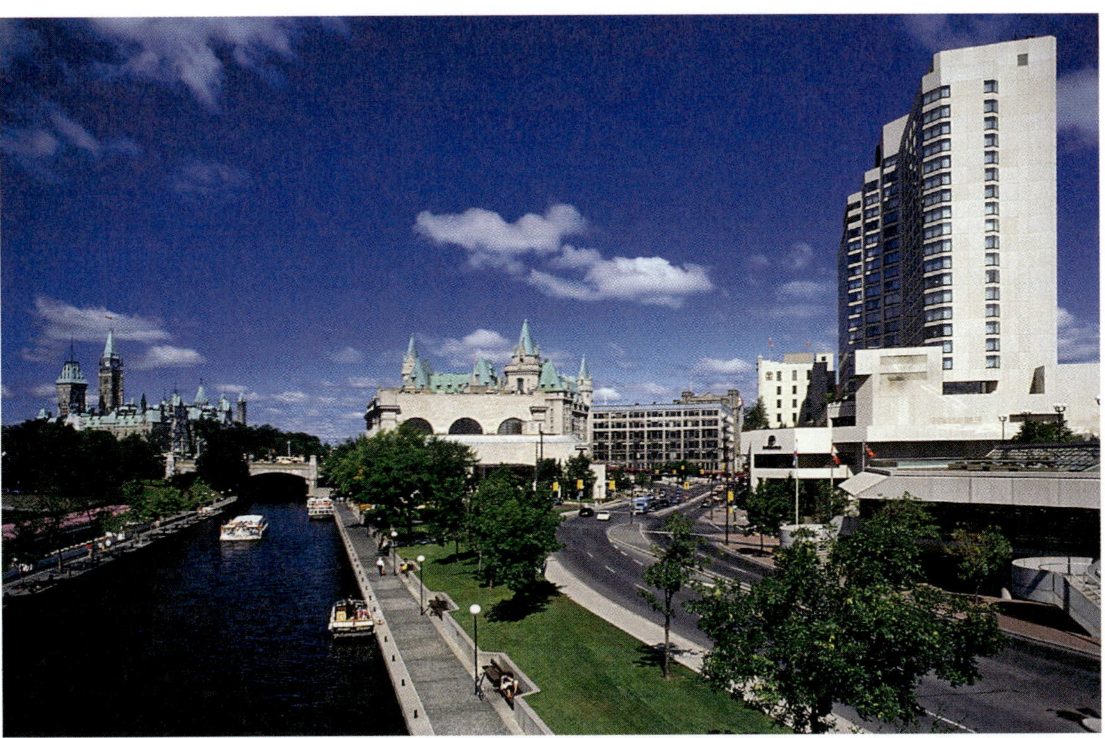

In the heart of Ottawa, Ontario, the Rideau Canal is both an urban recreational zone and a waterway for pleasure craft travelling from Kingston on Lake Ontario. Completed in 1832, the canal was originally designed as an alternative military route between Montréal and Kingston. Parliament Hill and the Château Laurier are two landmarks that now sit on either side of the canal. (Ivy Images)

of a national population core surrounded by a sparsely populated hinterland. The population core is sometimes described as Canada's national **ecumene** (inhabited area).

As Table 4.1 illustrates, Ontario has the highest density at 13.4 persons per km^2 followed by Québec at 5.6 persons per km^2. These two geographic regions combine to form a demographic core pattern by accounting for 62 per cent of Canada's population. Combined or individually, these geographic regions exert consider political force within the Canadian federation and, as a result, they are a source of much regional alienation. The emergence and stability of this demographic core are not surprising, given the historical position of Ontario and Québec in Canada, their geographic advantages, and their economic strength as measured by gross domestic product (GDP).

Population zones provide a more exact geographic picture of Canada's population distribution. As shown in Figure 4.1 and Table 4.2, four population zones vary in population size from very large (60 per cent of Canada's population) to very small (less than 1 per cent of Canada's population). Similarly, the four zones vary in population density from very dense to extremely sparse. The overall spatial pattern reinforces the image of a highly concentrated population core surrounded by more thinly populated zones.

Canada's core population zone lies in the Great Lakes–St Lawrence Lowlands where the bulk of Canada's population is found. As the most naturally favoured physiographic region, the Great Lakes–St. Lawrence Lowlands contains 19 million people (60 per cent of the national population) and almost three-quarters of Canada's major cities. This population core

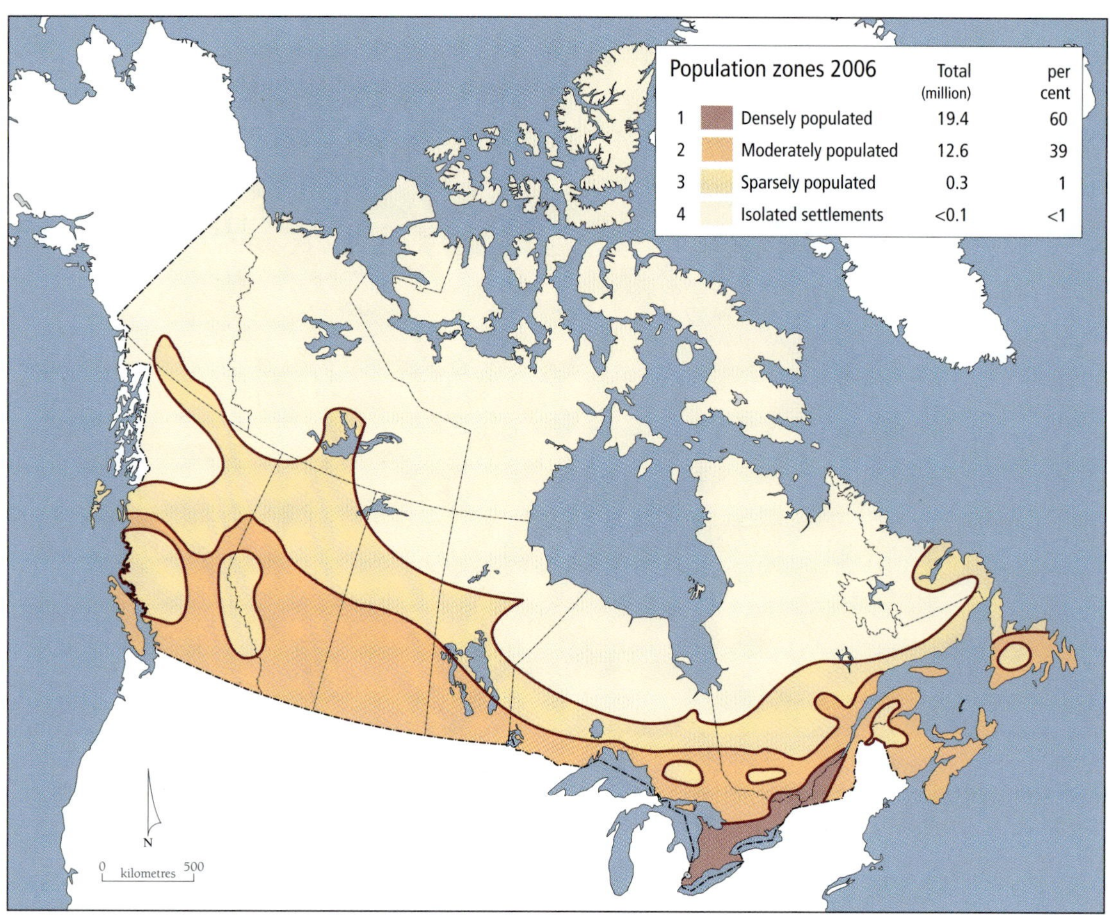

Figure 4.1 Canada's population zones. Canada's population is heavily concentrated in southern Ontario and southern Québec, where a favourable physical geography and an advantageous geographic location have resulted in a dense population. A secondary belt of population spans a southern strip of Canada. Together the densely and moderately populated zones account for 99 per cent of Canada's population. (Further resources: Student website, National Atlas section, Map 16. Website instructions are found on p. xx.)

includes Toronto, Montréal, Ottawa–Gatineau, Québec City, Hamilton, Oshawa, London, and Windsor, to name only some of the largest cities in the region. As Canada's most densely populated area, its economy is based on manufacturing and its agriculture lands contain the most fertile farmlands in Canada.

The secondary core zone extends in a narrow band across southern Canada. In general, its northern boundary corresponds with the polar edge of arable land. As the second most favoured zone, it occupies the more southerly portions of the Appalachian Uplands, the Canadian Shield, the Interior Plains, and the Cordillera. About 12 million Canadians (over one-third of the country's total population) live in this moderately populated zone. Canada's remaining major cities are located within this zone, including Vancouver, Edmonton, Calgary, Winnipeg, and Halifax. Within the secondary zone, some areas, such as southern Alberta and British Columbia, are growing quickly while other areas are subject to population losses, such as Newfoundland and Labrador. As a result, the population of the secondary zone is increasing slowly and unevenly.

The third zone, characterized by sparse population, is associated with a narrow band of the boreal forest that stretches across mid-Canada. Three physiographic regions are found

Table 4.2 Population Zones, 2006

Zone	Population (millions)	Percentage of Canada's population	Major City	Population of Major City
1. Core zone: densely populated	19.0	60	Toronto	5,113,149
2. Secondary zone: moderately populated	12.3	39	Vancouver	2,116,581
3. Sparsely populated zone	0.3	1	Fort McMurray*	51,496
4. Almost uninhabited zone with isolated centres	<0.1	<1	Labrador City	7,240

*Wood Buffalo Regional Municipality
Source: Statistics Canada (2007a, 2007b). Adapted from Statistics Canada website, Searched 18 February 2007, http://www40.statcan.ca/l01/cst01/demo33c.htm.

here—the Canadian Shield, the Interior Plains, and the Cordillera. As the third most populous zone, less than 1 per cent of all Canadians (about 300,000) live here. None of Canada's major cities are in this zone. The populations of the largest cities range between 10,000 and 40,000. Whitehorse and Yellowknife, as the capital cities of Yukon and the Northwest Territories, are administrative centres and **regional service centres**, since they also provide most of

Located on the west coast of Vancouver Island, Port Alberni remains an important resource town with a population of 25,297 (2006). The basis of its economy is the forest industry. In recent years, Port Alberni, like other forest-based resource towns, has fallen on hard times due to declining demand for forest products and to the automation of its production system. As a result, Port Alberni's population dropped by nearly 6 per cent from 1996 to 2001. Similar population declines in other resource towns reflect the troubled forest economy, resulting in a serious and perhaps ongoing population retreat from British Columbia's hinterland. (J.A. Wilkinson/Valan Photos)

Figure 4.2 Capital cities. With five exceptions, the capital cities of the 10 provinces and three territories are the largest urban centres in each political jurisdiction. The largest city in New Brunswick is Saint John; in Québec, it is Montréal; in Saskatchewan, Saskatoon; in Alberta, Calgary; and in British Columbia, Vancouver.

the service functions for their areas (Figure 4.2). These two cities, with populations of 20,461 and 18,700 respectively in 2006, also anchor the poleward edge of zone 3. **Resource towns** contain a single industry, such as a mine. Fort McMurray, Alberta, is an outstanding example of a booming resource town. As the hub of northern Alberta's oil sands extraction and exploration, Fort McMurray (which is within the Wood Buffalo Regional Municipality) is the largest city in the tertiary zone with a population exceeding 51,000.

Most of Canada's expansive territory is found in the last, almost uninhabited lands of the North. Here, because of the extremely cold climate, the presence of permafrost, and the polar ice pack, the land is inhospitable for settlement. This zone includes fewer than 100,000 inhabitants, and all live in small, isolated settlements. With a small population and a large geographic area, this population zone has the lowest population density. From a resource perspective, it is the least productive area in the country. Most commercial activities centre on exploitation of non-renewable resources, including mineral bodies and petroleum deposits. Unlike in the other zones in Canada, Aboriginal peoples are the majority in this zone. In spite of a high rate of natural increase among the Aboriginal population, the quaternary zone is affected by a net out-migration. Urban centres are small, most with populations under 5,000. In 2006, the iron-mining town of Labrador City had the largest population in the fourth zone with 7,200 inhabitants followed by the rapidly growing Iqaluit, the capital city of Nunavut, with 6,184. Unlike the resource town of Fort McMurray, Labrador City's population is decreasing. From 2001 to 2006, Labrador City saw its population decline by 6.5 per cent while Iqaluit, with its increasing public sector, had a remarkable jump in population of 18.1 per cent.

Urban Population

Canada is an urban country with 80 per cent of its population living in cities and towns. Statistics Canada (2007e) defines an urban area as having a population of at least 1,000 and no fewer than 400 persons per square kilometre. Canada's urban population has grown remarkably. For over a century, a shift from rural to urban places has been at work. From 1901 to 2006, the country's urban population has increased steadily from 37 per cent to 80 per cent (Figure 4.3).

What is the attraction of cities? First and of greatest importance, most business and employment opportunities are found in cities, especially large cities. As well, Canadians prefer to live in an urban setting where amenities are readily available. Cities are also important for other reasons. Major urban centres are at the cutting edge of technological innovation and capital accumulation. In the new world of the knowledge economy, manufacturing does not determine a city's prosperity; rather, the creativity of its business and university communities is the determining factor.

Despite the steady and in some instances remarkable growth of Canadian cities, all is not well in metropolitan Canada. Major urban centres are confronted with serious environ-

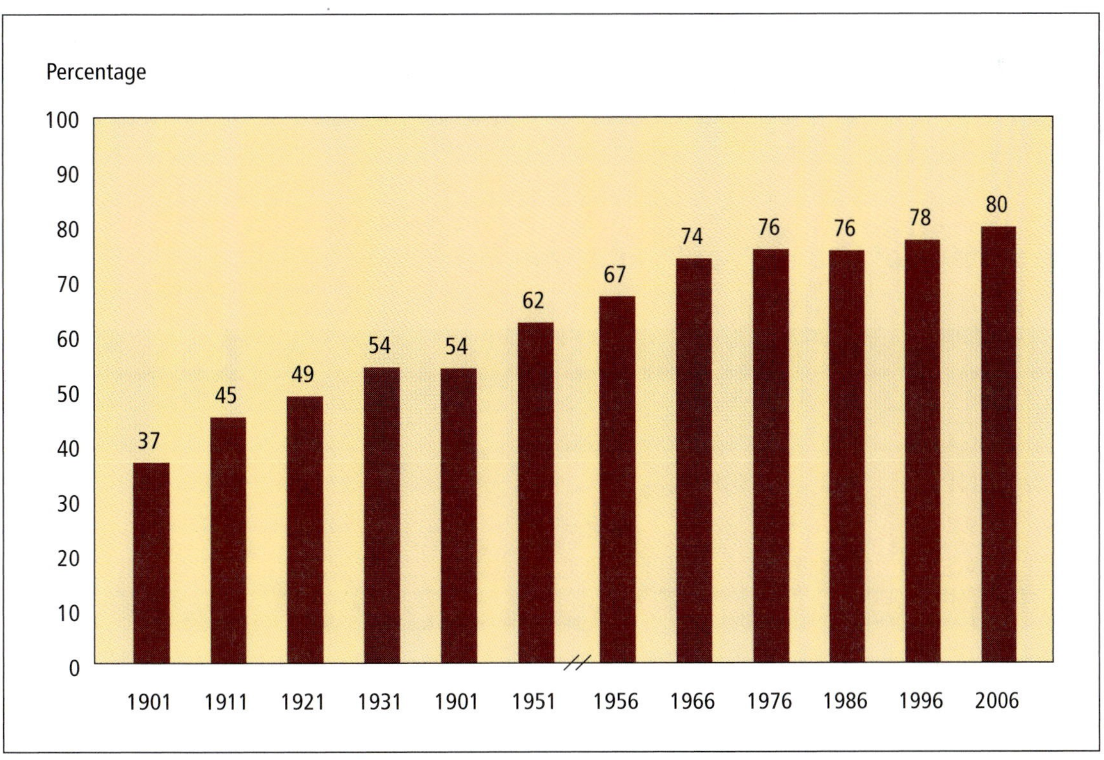

Figure 4.3 Percentage of Canadian population in urban regions, 1901–2006
Source: Adapted from Statistics Canada (2007c).

mental and social problems. In the twenty-first century, cities face the daunting task of finding solutions to these problems that are associated with the concentration of people in a relatively small area. Air pollution, traffic congestion, the homeless, water contamination, and waste disposal are among the leading challenges. Already these environmental and social problems have caused the quality of urban life in Canada's major cities to deteriorate.

Variation in Urban Population by Geographic Region

Urbanization is associated with economic development. For that reason, the rate of urbanization has varied across Canada. Not surprisingly, Ontario and British Columbia have the highest percentages of their populations classified as urban, while Atlantic Canada and the Territorial North have the lowest. For years, however, Ontario led all other geographic regions (Table 4.3).

Gradually, the gap is closing. By 2001, as shown in Table 4.3, British Columbia had the same percentage of urban population as Ontario. The 2006 census figures, when available, are expected to show that the urban percentage gap between geographic regions continues to decline, especially in Western Canada. In 1901, the rural nature of Western Canada was much stronger than in other geographic regions south of the sixtieth parallel. In 1901, Western Canada had less than 20 per cent of its population living in urban places. By 2001, the figure had jumped to almost 76 per cent. The shift from rural to urban communities, although most obvious in Western Canada, is a national phenomenon and is associated with the declining numbers involved in agriculture.

Census Metropolitan Areas

The emergence of extremely large cities across Canada is the latest outcome of urbanization. These cities serve as the economic and cultural anchors of their hinterlands. Statistics Canada defines **census metropolitan area** (CMA) as a very large urban area (known as the urban core) together with adjacent urban and rural areas that have a high degree of social and economic integration with the urban core. The urban core population of a CMA must be at least 100,000 based on the previous census.

Table 4.3 Percentage of Urban Population by Region, 1901–2001

Region	1901	1921	1941	1961	1981	1991	1996	2001
Ontario	40.3	58.8	67.5	77.3	81.7	81.8	83.3	84.7
British Columbia	46.4	50.9	64.0	72.6	78.0	80.4	82.1	84.7
Québec	36.1	51.8	61.2	74.3	77.6	77.6	78.0	80.4
Western Canada	19.3	28.7	32.4	57.6	71.4	74.4	74.4	75.7
Atlantic Canada*	24.5	38.8	44.1	50.1	54.9	54.1	52.8	53.9
Canada	34.9	47.4	55.7	70.2	76.2	77.2	77.9	79.7

*Newfoundland was not included in Atlantic Canada's figures until 1961.

Note: While comparable statistics are not available for the Territorial North, two observations are possible. (1) Prior to the 1950s, few people in this region lived in settlements. (2) By 2006, Statistics Canada (2007f) classified nine centres in the Territorial North as urban areas: Whitehorse in Yukon; Hay River, Inuvik, and Yellowknife in the Northwest Territories; and Iqaluit, Pangnirtung, Rankin Inlet, Cambridge Bay, and Arviat in Nunavut. Their total population in 2006 was 55,137, resulting in 54 per cent of the Territorial North's population defined as urban.

Sources: McVey and Kalbach (1995: 149); Statistics Canada (1997c, 2002a, 2007f).

Calgary, with over a million residents, is the headquarters for many energy companies. Several office towers are occupied by oil and gas companies. Petro-Canada built the reddish-coloured skyscraper that towers over downtown Calgary. (Photo by David Wise)

Edmonton's downtown core is bounded on two sides by the North Saskatchewan River. This aerial view illustrates the flat prairie landscape surrounding Alberta's provinicial capital, which is known as 'the gateway' to Canada's North. (Photo by Neil Koven Photography)

The proportion of Canada's population residing in census metropolitan areas has increased from 30.3 per cent in 1931 to 68 per cent, or 21.5 million Canadians, in 2006. In 1931, there were 10 CMAs: Halifax, Hamilton, Montréal, Ottawa, Québec, Saint John, Toronto, Vancouver, Windsor, and Winnipeg. By 2006, Canada had 33 census metropolitan areas (Table 4.4). Of the 21.5 million people residing in these CMAs, 14.1 million lived in one of the six metropolitan areas with a population of more than 1 million: Toronto, Montréal, Vancouver, Ottawa–Gatineau, and, for the first time topping the million mark, Calgary and Edmonton.

Since 1951, the populations of the five largest CMAs have increased remarkably. Over this 55-year period, Toronto has gained the greatest number of people and has consistently ranked in the top five cities by growth rate. By 1971, Toronto surpassed Montréal as Canada's largest metropolitan centre, and Toronto, with a higher growth rate, continues to outdistance Montréal in population size. From 2001 to 2006, for instance, Toronto's growth rate was 9.2 per cent while that of Montréal was 5.3 per cent. Yet, the greatest rate of increase among the census metropolitan areas—a phenomenal 19.2 per cent—took place in Barrie, Ontario. The five leading census metropolitan areas by growth rates for the last five years were Barrie, Calgary, Oshawa, Edmonton, and Kelowna. Only two CMAs suffered a decline—Saguenay (–2.1 per cent) and Saint John (–0.2 per cent).

Population Change

Population change has three components: births, deaths, and migration. **Population increase** is the sum of natural increase and net migration over a given period. The term **population growth** is used when this increase is expressed as a rate, that is, as a percentage change over time. **Natural increase** is the difference between the **crude birth rate** (CBR) and the **crude death rate** (CDR). CBR is the number of live births per 1,000 people in a given year. CDR is the number of deaths per 1,000 people in a given year. **Net migration** is the difference between in- and out-migration. Canada has a large influx of migrants, but a number of Canadians also leave the country, often for the United States. While reasons for out-migration vary, many Canadians move to the United States because of job opportunities and/or because of a more temperate climate. This trend may slow due to the concern of Americans for internal security following the attacks on New York and Washington in 2001, the American declaration of war on terrorism, and the invasion of Iraq.

Since 1851, Canada has enjoyed continuous population growth (Table 4.5). At first, high rates of natural increase and high levels of immigration propelled this growth. The highest rate occurred n 1911, when it reached 3 per cent. As more Canadians moved to cities, parents opted for smaller families, causing the rate to decline. During the 1950s and 1960s, however, the baby boom pushed the average annual rate of population increase to nearly 3 per cent (Vignette 4.1). After that short burst, the rate fell. From 2001 to 2006, the annual average rate was 1.1 per cent. Most population increase is now related to immigration rather than natural increase.

Natural Increase

Natural increase is determined by the number of births minus the number of deaths. Until 1986, most of Canada's population growth was due to natural increase. At that time, the number of immigrants began to exceed the number due to natural increase (Statistics Canada, 2007g). Until 1921, the majority of Canadians lived in rural settings where large families were the rule (Figure 4.2). As most available farmland became occupied, children of farm parents had no choice but to seek their fortunes elsewhere. Many went to the cities in search of work. Birth rates were extremely high, while death rates were low, allowing for a high rate of natural increase. In the 1870s, natural increase exceeded 2 per cent per year. At that rate, Canada's population would double every

Table 4.4 Population of Census Metropolitan Areas, 2006

Order	Census Metropolitan Areas	Province	Population	% Change, 2001–6
1	Toronto	Ontario	5,113,149	9.2
2	Montréal	Québec	3,635,571	5.3
3	Vancouver	British Columbia	2,116,581	6.5
4	Ottawa–Gatineau	Ontario–Québec	1,130,761	5.9
5	Calgary	Alberta	1,079,310	13.4
6	Edmonton	Alberta	1,034,945	10.4
7	Québec	Québec	715,515	4.2
8	Winnipeg	Manitoba	694,668	2.7
9	Hamilton	Ontario	692,911	4.6
10	London	Ontario	457,720	5.1
11	Kitchener	Ontario	451,235	8.9
12	St Catharines–Niagara	Ontario	390,317	3.5
13	Halifax	Nova Scotia	372,858	3.8
14	Oshawa	Ontario	330,594	11.6
15	Victoria	British Columbia	330,088	5.8
16	Windsor	Ontario	323,342	5.0
17	Saskatoon	Saskatchewan	233,923	3.5
18	Regina	Saskatchewan	194,971	1.1
19	Sherbrooke	Québec	186,952	6.3
20	St John's	Newfoundland and Labrador	181,113	4.7
21	Barrie	Ontario	177,061	19.2
22	Kelowna	British Columbia	162,276	9.8
23	Abbotsford	British Columbia	159,020	7.9
24	Greater Sudbury	Ontario	158,258	1.7
25	Kingston	Ontario	152,358	3.8
26	Saguenay	Québec	151,643	–2.1
27	Trois-Rivières	Québec	141,529	2.9
28	Guelph	Ontario	127,009	8.2
29	Moncton	New Brunswick	126,424	6.5
30	Brantford	Ontario	124,607	5.5
31	Thunder Bay	Ontario	122,907	0.8
32	Saint John	New Brunswick	122,389	–0.2
33	Peterborough	Ontario	116,570	5.1

Source: Statistics Canada (2007d). Adapted from Statistics Canada publications *A Profile of the Canadian Population: Where We Live, 2001 Census*, Catalogue 96F0030XIE2001001, http://www.statcan.ca/bsolc/english/bsolc?catno=96F0030XIE2001001 and *Portrait of the Canadian Population in 2006: Subprovincial population dynamics* http://www12.statcan.ca/english/census06/analysis/popdwell/Subprov3.cfm.

Vignette 4.1 The Baby Boom Effect

During the Great Depression and World War II, economic and social conditions did not favour large families. With the return of Canada's servicemen and women, a stable political world, and improving economic conditions, couples' attitude towards family formation took a positive turn. In 1946, the birth rate suddenly increased and it continued at this level until 1966. Demographers refer to this aberration as the baby boom. By the late 1960s, however, fertility rates decreased sharply. The end of the baby boom is sometimes described as the baby bust.

The baby boom lasted for 20 years. During that time of high fertility, there were almost 10 million births, which created a bulge in the age structure of Canadian society that continues to have both economic and social implications. Thus, while this demographic phenomenon was short-lived, it has left its mark on Canadian society (Foot, 1996). As consumers of goods and services, baby boomers have had a decided impact on the economy as they move through their life cycle. Companies have geared their products to meet the strong demand for goods and services created by baby boomers. In the early 1950s, the emphasis was on baby products and larger houses. In the 1960s, a similar age-related pressure was exerted on school facilities, creating a demand for more schools and teachers. As the baby boomers approach old age, the demand for health-care services is expected to rise. Companies are already targeting their advertisements at the growing number of retirees from the baby boom generation. Governments, on the other hand, are concerned about rising health-care costs associated with the expected increase in senior citizens.

30 to 35 years. Like many social phenomena, such rapid growth failed to sustain itself because circumstances changed dramatically. As Canada became an industrial country, fertility rates declined. Except for a short time after World War II, the crude birth rates declined steadily from about 45 births per 1,000 people per year to approximately 13 births per 1,000 people. Over the same period, the crude death rate dropped from about 20 deaths per 1,000 people per year to about seven. While there is no single explanation for these changes, public health measures such as water purification, improved nutrition, medical advances, and the development of public health-care systems across the country account for most of the sharp reduction in the **mortality rate**. Yet, the recent rebound in mortality rates to 7.3 per cent in 2004–5 signals the impact of Canada's aging population. The decline in the **fertility rate** is more complex. A series of social and economic changes caused parents to opt for smaller families. Among these changes, three stand out: the shift of people from rural areas to towns and cities; the sharp increase in the number of women in the labour force; and the widespread acceptance of family planning. Family planning was greatly helped by the birth control pill. Introduced in 1960, the birth control pill has played an important role in reducing the birth rate and was a factor in ending the baby boom.

By the twenty-first century, Canada's crude birth and death rates were extremely low. These vital rates are similar to those found in other industrial countries. Consequently, Canada has a low rate of natural increase. By 2004–5, the natural rate was 0.3 per cent, and every indication points to a further decline in the coming years (Table 4.6). These trends in birth and death rates, which are consistent with the experience of other industrialized countries, are accounted for in the demographic transition theory.

Table 4.5 Population Increase, 1851–2006

Year	Population (000s)	Percentage Change	Average Annual Rate
1851	2,436.3	—	—
1861	3,229.6	32.6	2.9
1871	3,689.3	14.2	1.3
1881	4,324.8	17.2	1.6
1891	4,833.2	11.8	1.1
1901	5,371.3	11.1	1.1
1911	7,206.6	34.2	3.0
1921	8,787.9	21.9	2.0
1931	10,376.8	18.1	1.7
1941	11,506.7	10.9	1.0
1951	14,009.4	21.8	1.7
1956	16,080.8	14.8	2.8
1961	18,238.2	13.4	2.5
1966	20,014.9	9.7	1.9
1971	21,568.3	7.8	1.5
1976	22,992.6	6.6	1.3
1981	24,343.2	5.9	1.2
1986	25,309.3	4.0	0.8
1991	27,296.9	7.9	1.6
1996	28,846.8	5.7	1.1
2001	30,007.1	4.0	0.8
2006	31,612.9	5.4	1.1

Sources: McVey and Kalbach (1995: 42); Statistics Canada (1997c, 2002a, 2007a).

The Demographic Transition Theory

The **demographic transition theory** is the most widely accepted theory that describes population change in industrial societies. This theory is based on the assumption that changes in birth and death rates occur as a society moves from a pre-industrial to an industrial economy. These demographic changes occur in four phases, each of which has a distinct set of vital rates that coincide with the phases in the process of industrialization (Table 4.7).

Canada did not experience a late **pre-industrial phase**, which was associated with a feudal society. A cursory examination of Canada's birth and death rates over the last 150 years reveals strong similarities to the early industrial, late industrial, and post-industrial phases. In the nineteenth century high birth rates and lower death rates were common. These vital rates are similar to those in the second phase of the demographic transition theory. Then Canada entered the late **industrial phase**, characterized by low death rates and a declining birth rate. By the 1980s, birth rates

Table 4.6 Canada's Rate of Natural Increase, 1851–2006

Year	Crude Birth Rate	Crude Death Rate	Natural Increase (%)	Natural Increase (000s)
1851	45	20	2.5	61
1861	45	20	2.5	81
1871	42	20	2.2	81
1881	40	19	2.1	90
1891	38	18	2.0	97
1901	35	16	1.9	97
1911	32	14	1.8	129
1921	29.3	11.6	1.8	160
1931	23.2	10.2	1.3	138
1941	22.4	10.1	1.2	145
1951	27.2	9.0	1.6	255
1961	26.1	7.7	1.8	335
1971	16.8	7.3	1.0	205
1981	15.2	7.0	0.8	200
1991	14.3	7.0	0.7	207
1995–6	12.6	7.1	0.6	163
2001–2	10.5	7.1	0.3	108
2004–5	10.5	7.3	0.3	103
2005–6	10.6	7.3	0.3	108

Note: Data on births and deaths are now recorded from 1 July to 30 June.
Sources: Adapted from Statistics Canada (1997d, 2003d, 2006c, 2007h, 2007i) and McVey and Kalbach (1995: 268, 270).

had declined significantly, resulting in little natural increase. Such rates characterize the **post-industrial phase** of the demographic transition theory. Based on a more precise measure of fertility, demographers argue that Canada's rate of natural increase has fallen below its replacement level (Vignette 4.2).

Age Structure

Age structure is one of the most basic characteristics of a population. As is the case with other developed countries, Canada's population is growing older. The demographic implications for Canada are most significant. These implications include a smaller proportion of children (under 15 years of age) and a smaller proportion of the population in the workforce (ages 15 to 64), plus a large percentage over 64 years of age. The economic implications are also significant because a smaller percentage of workers means a growing tax burden on these people and ever-increasing medical costs and demands by Canada's senior citizens.

Aging of developing nations is associated with the process of industrialization, urbanization, and increasing longevity. The shift from a young society to an aging one was swift and Statistics Canada predicts this aging trend will

Table 4.7 Phases in the Demographic Transition Theory

Phase	Birth and Death Rates	Rate of Natural Increase
Late pre-industrial	High birth and death rates	Little or no natural increase but possible fluctuations because of variations in the death rate
Early industrial	Falling death rates	Extremely high rates of natural increase
Late industrial	Falling birth rates	High but declining rates of natural increase
Post-industrial	Low birth and death rates	Little or no natural increase. Stable population

continue with the median age reaching almost 50 by 2056 (Table 4.8).

Age Dependency

Age dependency ratio is the ratio of persons in the 'dependent' age groups (under 15 and over 64 years) to those in the 'economically productive' age group (between 15 and 64 years). The assumption is that productive members of society are those between the ages of 15 and 64, while unproductive members are either too young (under 15) or too old (over 64) to make an economic contribution. The purpose of the age dependency ratio is to compare the number of dependants with the number of economically productive members of society, thus giving a rough measure of the economic burden on those in the economically productive age group.

In Canada, the age dependency ratio declined substantially from 1961 to 2001. In 1961, the ratio recorded 70 dependants per 100 people of working age. By 2006, it had declined to about 44 per 100. Why is the age dependency ratio declining? The answer lies in the baby boom sector of the population, which still remains in the economically productive age group (Vignette 4.1). However, by 2010, the leading edge of this bulge in the age structure of Canadian society will be joining Canada's senior citizens, causing the age dependency ratio to rise. From 2010 onward, the age dependency ratio will increase each

Vignette 4.2 The Concept of Replacement Fertility, Total Fertility, and Canada's Sex Ratio

The concept of replacement fertility refers to the level of fertility at which women have enough daughters to replace themselves. If women have an average of 2.1 births in their lifetime, then each woman, on average, will have given birth to a daughter and a son. The number 2.1 was determined to represent the minimum level of replacement fertility because, on average, slightly more boys than girls are born. In 1961, the Canadian total fertility rate was 3.8 births per woman of child-bearing age (15–49). Since then, the fertility rate has declined. The estimated 2007 total fertility rate was 1.6 births (CIA, 2007).

Nature has ensured that slightly more male than female babies are born. However, male mortality rates are higher than those for females. The 2001 **sex ratio**, defined as the number of males per 100 females in Canada's population, was 96.1, meaning that there are slightly more females than males.

Table 4.8 Canada's Aging Population as Defined by Median Age

1946	1966	1986	2006	2011	2056
27.7	25.4	31.4	38.8	40.11	46.9

Source: Statistics Canada (2006g). Adapted from Statistics Canada publication *The Daily*, Catalogue 11-001, Searched 7 Feb. 2007, http://www.statcan.ca/Daily/English/061026/d061026b.htm.

year, leaving fewer taxpayers to cover social costs, such as education and health care.

Immigration

Canada is a country of immigrants. Figure 4.4 illustrates the variation in annual immigration from 1901 to 2001. Ottawa encourages immigration for three reasons: (1) newcomers keep Canada's population increasing; (2) newcomers add valuable members to Canada's workforce; (3) Canada takes in refugees who are fleeing oppressive socio-political conditions in their homelands. Since 1992, some 200,000 immigrants have come to Canada each year (Table 4.9). A key social issue is not the number of immigrants but how well they enrich and add to the cohesiveness of Canadian society.

The proportion of Canada's population who were born outside the country has reached nearly 20 per cent, the highest level in 70 years. From 1901 to 2004–5, the annual number of people settling in Canada has var-

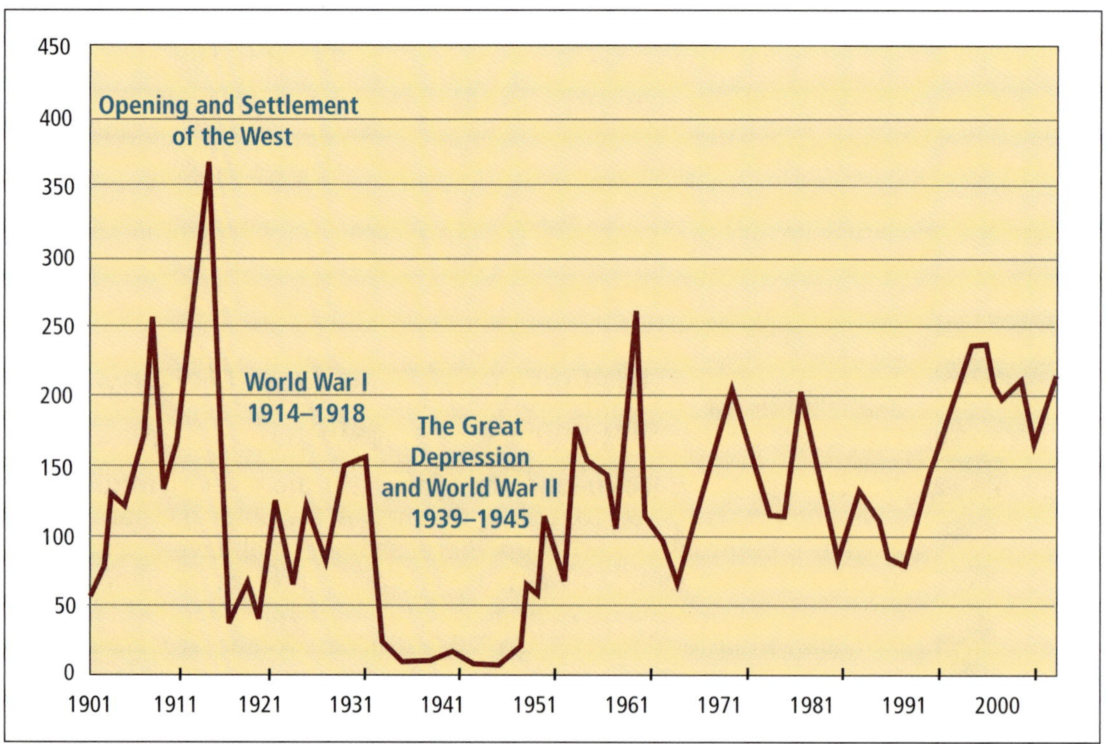

Figure 4.4 Annual number of immigrants admitted to Canada, 1901–2001
Source: Adapted from Statistic Canada (2003c: 2). Adapted from Statistic Canada *Canada's Ethnocultural Portrait: The Changing Mosaic, 2001 Census*, Catalogue 96F0030XIE2001008, Released 21 January 2003, Searched 28 January 2003, http://www12.statcan.ca/english/census01/products/analytic/companion/etoimm/canada.cfm.

ied: the greatest annual numbers of newcomers arrived in 1912 and 1913 (settling of the West) and the lowest numbers came in the years of the Great Depression and World War II. By 2004–5, the number had reached 245,000 (Statistics Canada, 2006b).

The changing proportion of immigrants born in Europe and Asia is shown in Figure 4.5. Table 4.9 provides a broader picture of the country of origin of newcomers. Most immigrants in recent years were born in Asia and the Middle East (58 per cent) with smaller numbers coming from Europe (20 per cent), the Caribbean and Central and South America (11 per cent), Africa (8 per cent), and the United States (3 per cent).

Why Immigration?

Canada needs immigrants. Since the 1970s, Canada's population increase has been driven by immigration. In the future, immigration is expected to play an even more significant role. Today, immigration represents close to 70 per cent of the population growth. Since natural increase is expected to continue to decline, Statistics Canada projects that the proportion of Canada's population growth attributed to natural increase will drop to zero by 2025 (Figure 4.6). At that time, only immigration is predicted to keep Canada's population increasing. The assumption underlying this projection is that the birth rate will continue to decline. While such a decline over the next 20 years seems reasonable, Statistics Canada did not forecast the baby boom (see Vignette 4.1). In other words, forecasting population trends is subject to error caused by unforeseen changes in human behaviour.

The demographic implications of immigration for Canada are significant, as the data in Table 4.10 suggest.

- Immigration ensures that Canada's population keeps increasing.
- Immigrants will form a larger and larger proportion of the national population, thus further increasing the cultural diversity of Canada's society.
- Canada's increasing diversity will call into question the validity of the vision of two founding peoples.
- Concentration of new Canadians in Canada's largest cities has changed the demographics of these cities, making them different from the rest of Canada.

Table 4.9 Newcomers by Place of Origin

	Period of immigration									
	Before 1961		1961–70		1971–80		1981–90		1991–2001	
	Number	%	Number	%	Number	%	Number	%	Number	%
Total Newcomers	894,465	100.0	745,565	100.0	936,275	100.0	1,041,495	100.0	1,830,680	100.0
United States	34,805	3.9	46,880	6.3	62,835	6.7	41,965	4.0	51,440	2.8
Europe	809,330	90.5	515,675	69.2	338,520	36.2	266,185	25.6	357,845	19.5
Asia	28,850	3.2	90,420	12.1	311,960	33.3	491,720	47.2	1,066,230	58.2
Africa	4,635	0.5	23,830	3.2	54,655	5.8	59,715	5.7	139,770	7.6
Caribbean, Central and South America	12,895	1.4	59,895	8.0	154,395	16.5	171,495	16.5	200,010	10.9
Oceania	3,950	0.4	8,865	1.2	13,910	1.5	10,415	1.0	15,385	0.8

Source: Statistics Canada (2003c). Adapted from Statistics Canada publication Canada's *Ethnocultural Portrait: The Changing Mosaic, 2001 Census*, Catalogue 96F0030XIE2001008, Released 21 January 2003, Searched 28 January 2003, http://www12.statcan.ca/english/census01/products/analytic/companion/etoimm/canada.cfm.

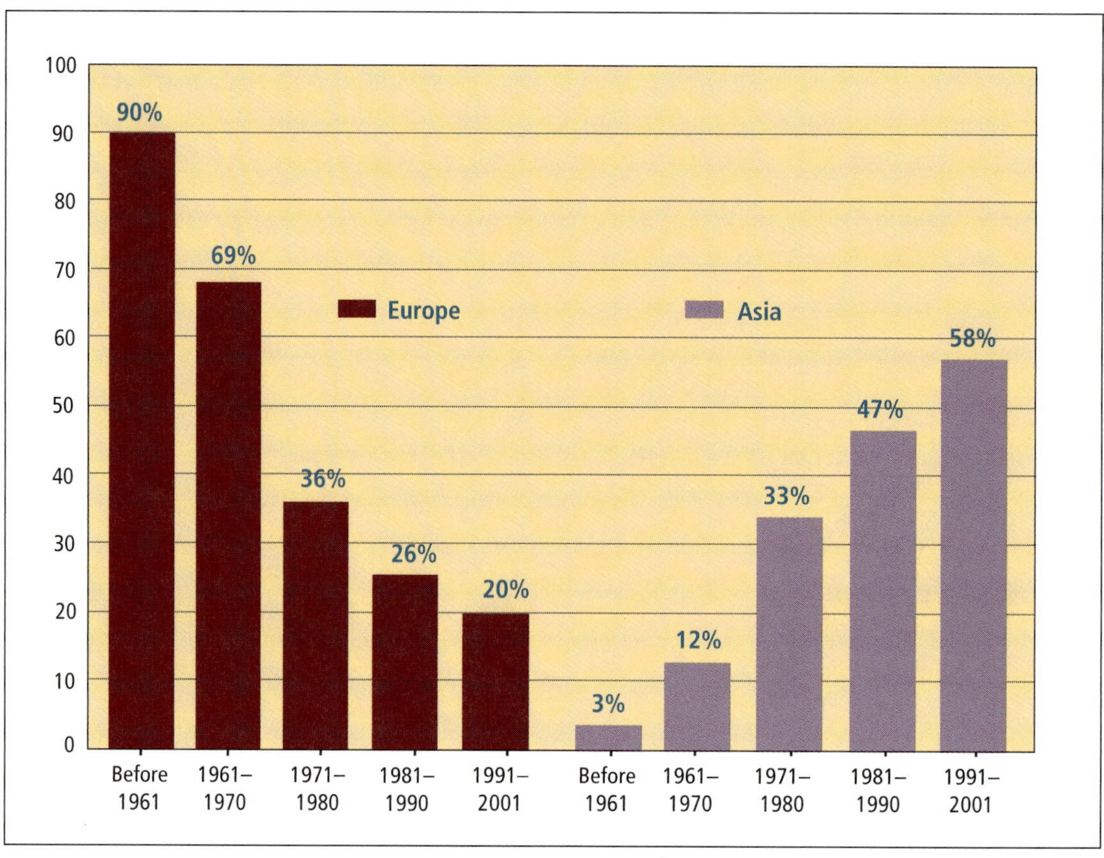

Figure 4.5 Proportion of immigrants born in Europe and Asia by period of immigration
Source: Statistic Canada (2003c: 3). Adapted from Statistic Canada publication *Canada's Ethnocultural Portrait: The Changing Mosaic, 2001 Census*, Catalogue 96F0030XIE2001008, Released 21 January 2003, Searched 28 January 2003, http://www12.statcan.ca/english/census01/products/analytic/companion/etoimm/canada.cfm.

Ethnicity

Canadian society is composed of many ethnic groups.[2] In fact, more than 200 different ethnic origins were reported in the 2001 census question on ethnic ancestry (Statistics Canada, 2003c). In the same census report, the 10 largest ethnic origins based on responses were Canadian (39.4 per cent), English (20.2 per cent), French (15.8 per cent), Scottish (14.0 per cent), Irish (12.9 per cent), German (9.3 per cent), Italian (4.3 per cent), Chinese (3.7 per cent), Ukrainian (3.6 per cent), and North American Indian (3.4 per cent). An **ethnic group** is made up of members of a population who share a culture that is distinct from that of other groups. Each group has a common identity, shared values, and cultural/linguistic/religious bonds and symbols. **Culture** is the learned collective behaviour of a group of people.

In spite of these varied origins, time erodes the ethnic connection with the country of origin. For Canadians born and raised in this country, the connection with their original (foreign) ethnic origins is tenuous at best (Beaujot, 1991: 297). Place, as cultural geographers insist, plays a critical role in the development of a regional/national identity. This phenomenon is well known. Consider, for example, the attachment of the French born in New France who had no interest in returning to France in 1763 because their lives were centred on the New World. They were no longer French but had become French Canadian. Not surprisingly, then, the ethnic selection of 'Canadian' by nearly 40 per cent of Canadians

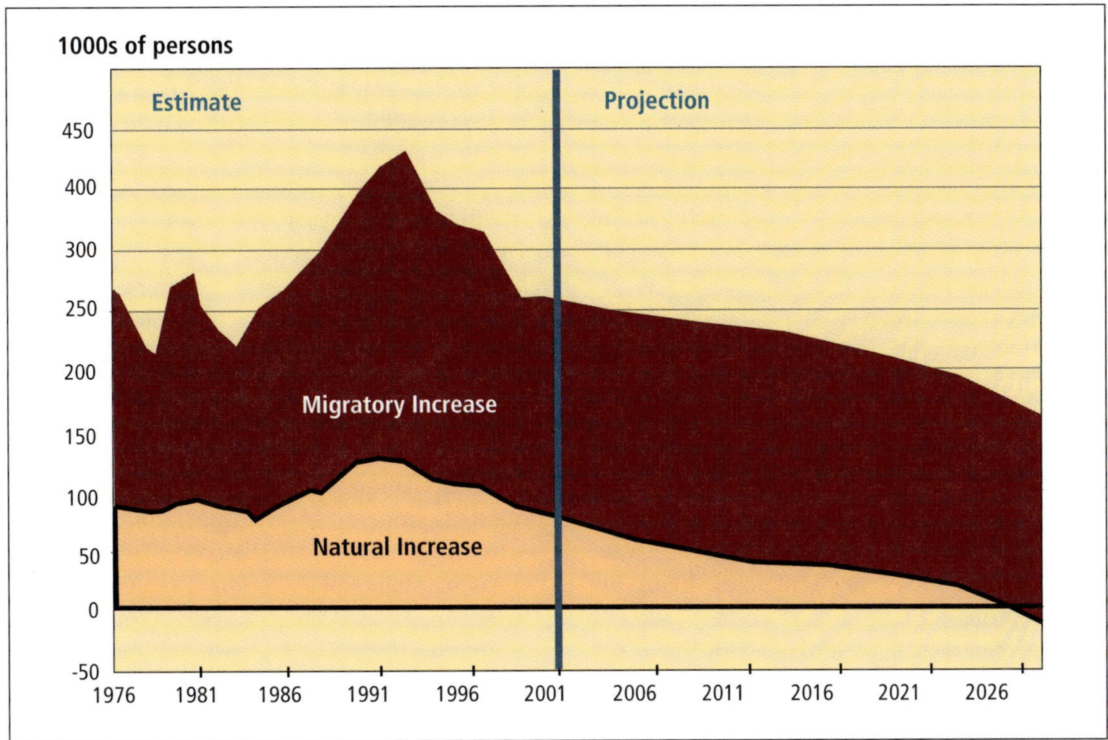

Figure 4.6 Immigration: An increasingly important component of population growth in Canada
Note: The projection is based on medium assumptions: fertility rates at 1.5; immigration at 225,000 new immigrants per year; a gradual increase in life expectancy from 77 to 80 years for men and 83 to 85 for women. The projection suggests that natural increase will be negative by 2026, with deaths exceeding births.
Source: Statistics Canada (2003k). Adapted from Statistics Canada, *Learning resources: Civics and Society – Declining birth rate and the increasing impact of immigration*, Searched 28 June 2003, http://www.statcan.ca/english/kits/issues/charts/chart3.htm.

Table 4.10 Demographic Impact of Immigration

Year	Births	Deaths	Natural Increase	Immigration	Difference
1992–3	403,107	196,967	206,140	266,890	60,750
1993–4	389,286	207,528	181,758	227,860	46,102
1994–5	382,870	212,830	170,040	216,988	46,948
1995–6	372,453	209,746	162,707	224,881	62,174
1996–7	357,313	217,220	140,093	193,452	53,359
1997–8	345,423	217,688	127,435	194,459	67,024
1998–9	338,963	222,538	116,425	205,469	89,044
1999–2000	333,954	229,138	104,816	200,000	95,184
2000–1	327,187	231,232	95,955	255,999	160,044
2004–5	337,856	234,645	103,211	244,579	141,368
2005–6	343,517	234,914	108,603	254,359	145,756

Sources: Adapted from Statistics Canada (2003d, 2006b).

is attributed to the geographic notion of place overriding the concept of ethnicity (Table 4.11). To put it differently, the resettlement of people in a new place causes their ethnicity to fade with time, especially if marriages take place with members from other groups. Loyalty to the 'old country' tends to fade with the second generation of immigrant families. This process of putting down cultural roots is a normal process as long as the newcomers are accepted into the mainstream society. Of course, there are cases of newcomers finding it difficult to put down roots in their new country. As we saw in Chapter 3, some 100 years ago the Doukhobors found that their collective way of farm life, as well as their strict adherence to their faith and Russian language, was not accepted by other Canadians living in Western Canada.

The emergence of a third cultural force within Canada broke the dominance of the British and French ethnic groups with the flood of non-British and non-French migrants settling the Prairies, and some of these people eventually moved to the towns and cities of industrial Canada. After World War II, immigrants from all over the world came to Canada in ever-increasing numbers. Canadian society now consisted of a number of ethnic groups whose members shared a sense of identity based on country of origin, language, religion, tradition, and other common experiences. Overarching these shared ethnic experiences, however, was a sense of belonging to Canada, and since 1971 Ottawa has promoted multiculturalism, the idea that a cultural identity and a national one are not mutually exclusive. The results of this policy are reflected in Table 4.11.

Estrie, formerly called the Eastern Townships, lies in the Appalachian physiographic region. Dairy farms are found in the rolling countryside, which is surrounded by wooded uplands. Hay is the principal crop and it is used as feed for dairy cows. (Clara Parsons/Valan Photos)

In Québec, a strong sense of cultural identity exists among the people of French origin, but it is tempered by centuries of life in North America. The term 'Québécois', if it refers to the descendants of the original French settlers, more accurately reflects their ethnic background and continues to drive that province's society and politics. History and geography are the causes of this sense of ethnic nationalism. Nationalism is nurtured by Québec's place, not in Canada but in North America. As a French enclave in the English-speaking North American continent, Québec naturally fears for its cultural existence. Its close proximity to the United States underscores its precarious position in North America much more than its 'sheltered' position within Canada. The United States exerts an enormous cultural impact on the world through the mass media and now, increasingly, through the Internet. Within North America, American culture poses the major threat to French- and English-Canadian cultures. This fear of assimilation is the basis of ethnic nationalism in Québec, a nationalism generally expressed by the Québécois in one of two ways: (1) Québec society is distinct within Canada; (2) only through independence will Québec be fully protected and emancipated. Not surprisingly, concern over change to the existing way of life by accommodating new cultural ways exists across Canada but was most openly expressed by Hérouxville's municipal council when they announced their Code of Standards for newcomers (see Vignette 1.1).

Dual loyalties exist in all geographic regions, but they are strongest among Québé-

Table 4.11 Ethnic Origins of Canadians

2001			1996		
	Number	%		Number	%
Total population	**29,639,030**	**100.0**	**Total population**	**28,528,125**	**100.0**
Canadian	11,682,680	39.4	Canadian	8,806,275	30.9
English	5,978,875	20.2	English	6,832,095	23.9
French	4,668,410	15.8	French	5,597,845	19.6
Scottish	4,157,215	14.0	Scottish	4,260,840	14.9
Irish	3,822,660	12.9	Irish	3,767,610	13.2
German	2,742,765	9.3	German	2,757,140	9.7
Italian	1,270,369	4.3	Italian	1,207,475	4.2
Chinese	1,094,700	3.7	Ukrainian	1,026,475	3.6
Ukrainian	1,071,055	3.6	Chinese	921,585	3.2
North American Indian	1,000,890	3.4	Dutch (Netherlands)	916,215	3.2

Note: Table shows total responses. Because some respondents reported more than one ethnic origin, the sum is greater than the total population or 100 per cent.
Source: Statistics Canada (2003c). Adapted from Statistics Canada publication *Canada's Ethnocultural Portrait: The Changing Mosaic, 2001 Census*, Catalogue 96F0030XIE2001008, Released January 21, 2003, Searched 28 January 2003, http://www12.statcan.ca/english/census01/products/analytic/companion/etoimm/canada.cfm.

cois when they look to their provincial government to protect their interests rather than to the federal government. Dual loyalties also exist among recent immigrants whose ties to their home country remain strong and who retain a passport from their former country. Dual passports have the advantage of making travel for pleasure or business more convenient, but dual citizenship is not legally recognized in all countries, which can lead to serious difficulties. Then, too, there is the question of loyalty. As the former Minister of Citizenship and Immigration, Judy Sgro, stated, 'We need to be loyal to one country as far as your citizenship. Your heart can be where you were born, but I think the commitment to Canada has to be strong and I think dual citizenship weakens that' (Woods, 2006).

Language

Language is a key component of ethnicity. Indeed, language represents the most durable link to the past and the tool for maintaining a culture. Canada's two official languages are English and French. Approximately 85 per cent of Canadians use one of the two official languages. The remainder, who speak other languages, eventually choose English and/or French. New Canadians who speak neither official language hold the key to the future balance between English and French. Where they settle and what official language they learn impacts on the French and English societies. From a political perspective, issues surrounding the two official languages are a delicate matter. In fact, this topic is the basis of a population faultline discussed later in this chapter.

A number of French organizations across Canada aim to protect and promote their language and culture. Such organizations have received federal funding to finance their cultural and educational activities. This public support, plus the presence of French-language radio and television programs, has resulted in a rejuvenation of these provincial organizations.

Many languages, other than French or English, are spoken across the country, especially in larger cities where new immigrants tend to settle. Without the official-language status enjoyed by French and English, however, other languages are often difficult to maintain. The survival of these other languages in Canada is bolstered somewhat by continued immigration and the efforts of non-official language groups to maintain their languages. However, Canadian-born members of ethnic groups—the second and subsequent generations of immigrants—usually lose the language of their parents and grandparents. The same loss of language has affected Aboriginal Canadians. Newhouse (2000: 402–3) sees the emergence of English as the lingua franca among Aboriginal peoples. By 2001, only three of the 50 languages spoken by Canada's Aboriginal peoples had a secure future. They are Cree, with 80,000 speakers (down 7,500 from 1996), Inuktitut, with 29,700 (up nearly 2,000), and Ojibwa, with 23,500 (down 2,400).

The Basilica of Notre Dame was opened in 1829 in what is now 'Old Montréal'. At that time, the Basilica was the largest religious edifice in North America. The interior of the Basilica features finely sculpted polychrome wood decorated in gold leaf and a star-spangled vault. While now surrounded by tall buildings, the physical presence of the Basilica of Notre Dame is reminder of the Roman Catholic Church's powerful role in the history of French-Canadian society. (Kennon Cooke/ Valan Photos)

Religion

Religion is another key element of culture. The changing nature of Canada's religious composition is indicative of the changing cultural mix of people. The recent shift in Canada's religious makeup is linked with the large number of immigrants arriving from non-Christian countries. Over the past 30 years, the number of Canadians who subscribe to religions such as Islam, Hinduism, Sikhism, and Buddhism has increased substantially.

Culture is not only a durable link to the past but also provides the institutional organization to preserve ethnicity. Religious organizations provide an institutional structure that consolidates people of similar beliefs. One example is the Roman Catholic Church. In the early history of Canada, the Roman Catholic Church helped sustain Catholicism, the French language, and the French-Canadian way of life. While its role has diminished in recent decades, the Church was the dominant cultural force among francophones from the conquest of New France to World War II. Not only did it give spiritual direction to French Canadians within Québec but it also organized its parishes to sponsor group immigration to more isolated areas of Québec (such as the Clay Belt in northwestern Québec), to other provinces, and to New England. In these group migrations, priests provided the leadership for the move and for organizing the new settlement, including its educational, social, and religious structures.

Religious freedom is the right of all Canadians. This right was inscribed in the British North America Act and is contained in the Constitution Act of 1982. In fact, many immigrants have come to Canada seeking religious freedom. The Hutterite Brethren, for example, are a small Christian group who live in agricultural colonies. Like the Amish and Mennonites, they place pacifism at the core of their faith. In 1899, Hutterites fled Europe to the United States. The Hutterites immigrated to Canada in 1918 because the US government refused to exempt them from military service. Initially, Hutterite colonies were formed in Manitoba and Alberta. Later, colonies were established in Saskatchewan. Hutterites number about 30,000 and the majority live and work in one of the 300 agricultural colonies where they remain committed to their religion, a communal lifestyle, and pacifism (Statistics Canada, 2003j).

Canada is thought of as a Christian country. This image was certainly true in 1867, but today Canada is a much more religiously diverse nation. It is true that most Canadians are Christians. For example, in the 1960s nearly 90 per cent of Canadians declared themselves to be Christian (though some may not have been active church members). By 2001, this figure had dropped to 77 per cent (Statistics Canada, 2003j). Roman Catholics remain the largest group of Christians, comprising in 2001 44 per cent of all Canadians, followed by Protestants at 29 per cent. Canadians with no religious affiliation make up 17 per cent of the population. At the same time, the numbers of Muslims, Hindus, Buddhists, and Sikhs had more than doubled from 1991 to 2001, making up 5 per cent of Canada's population (ibid.).

Multiculturalism

Multiculturalism, the cornerstone of Canada's new and distinctive demographic character, is a product of immigration and, strangely enough, biculturalism. In an ideal pluralistic society, tolerance and respect for different cultural and ethnic groups are practised by all citizens and supported by law. In many ways, multiculturalism is the opposite of **ethnocentricity**. But tolerance and respect of others are not an automatic outcome of life in pluralistic societies. Canadian tolerance and respect for cultural and ethnic diversity, such as it is, were learned the hard way, going back centuries to often bitter (and sometimes violent) conflicts between French- and English-speaking Canadians. For Canada to survive as a nation, the two antagonists had no choice but to become partners. One product of this so-called partnership was biculturalism. The other path to nation-building—a classic European-style nation-state founded on a single common ethnicity and language—was not possible in the northern half of North America. Reaching an accord (not a solution) between the two founding peoples was not a simple task and disputes continue to emerge, as discussed in Chapter 3. Since dominance is not possible, then compromise becomes a political necessity and eventually the search for compromise becomes engrained as a national trait. In his book, *Reflections of a Siamese Twin: Canada at the End of the Twentieth Century*, John Ralston Saul described Canada as a 'soft' country to capture this flexible trait (see 'Faultlines within Canada' in Chapter 1 for more on this subject).

Multiculturalism is a form of cultural pluralism; in its Canadian form multiculturalism simply reflects the demographic reality created by liberal immigration policies. Canada has a pluralistic society. Cultural pluralism means that Canadians are not of any one cultural background, race, or heritage. But Canadian identity, flexible as it is, includes core values that are defined by Canada's history and geography. Three examples are: (1) government is based on British parliamentary institutions; (2) two official languages ensure a place for French as well as English, which also means that other languages have no standing in the political and public affairs of Canadian society; (3) Aboriginal peoples, but especially First Nations people, have special rights, most of which flow out of treaties, as we have seen in Chapter 3.

Book IV of the report of the Royal Commission on Bilingualism and Biculturalism, *Cultural Contributions of the Other Ethnic Groups* (1970), recommended that Ottawa recognize the multiplicity of Canada's population. The following year, in 1971, the federal Liberal government made multiculturalism official policy and in 1972 the cabinet position of Minister of State for Multiculturalism was created. Government funding to ethnic organizations and for publishing and other educational programs soon followed. In 1988, the federal government passed the Canadian Multiculturalism Act, which is designed to encourage greater human understanding and stronger bonds among Canadians of different cultural backgrounds and ethnic origins, and today official multiculturalism is under the purview of the Department of Canadian Heritage.

Multiculturalism is a distinctively Canadian approach to social equality in nation-building, encouraging respect for cultural diversity. Ideally, multiculturalism is a shared vision of people of diverse cultural backgrounds seeking to live together in harmony. In reality, tensions exist within Canadian society, and four of these are articulated by the faultlines we have discussed. Multiculturalism should not lead to barriers between citizens and thereby threaten the cohesion of Canadian society. Most Canadians, but particularly new Canadians, recognize that multiculturalism binds Canada's diverse populations together. Canadians as a whole see no conflict between citizens having three identities—a cultural (ethnic/religious/language) identity, a regional identity (province/city/First Nation), and a national one. The national identity is based on tradition and law, specifically the Constitution, which includes Canada's Charter of Rights.

Yet, as Andrew Coyne has written (2006: A19):

> It's [multiculturalism] a slippery word, with multiple meanings. Is it, as it is sometimes used, merely a synonym for the observed fact of ethnic and cultural diversity? Is it the ideology that all cultural norms are of equal moral value, the dreaded cultural relativism? Or is it the policy of official multiculturalism, complete with grants for folk-dancing and heritage language training?

So far, Canada has remained a coherent society, with different ethnic, religious, and social groups sheltered under this broad concept of multiculturalism. But are there dangers? With Muslim extremists blowing up a railway station in Spain and bombing the underground transit system in London, England, European Union countries are having second thoughts (Vignette 4.3). Holland and Germany have apparently turned their backs on multiculturalism and are hardening their immigration policies. Even some prominent Canadians are questioning the concept because of their fear of ethnic ghettos and parallel societies. In 2005, Bernard Ostry, who was responsible for drafting and implementing the federal multicultural policies in the 1970s, expressed concerns about multiculturalism:

> The multiculturalism policy of the 1970s is under increasing criticism. This policy, developed from grassroots, was intended to assure new citizens full participation in the nation's life. Some now claim that it has had the opposite effect, that it has encouraged minorities to retreat to their own corners. (Ostry, 2005: A25)

It certainly is arguable whether minorities in Canada are retreating into ethnic enclaves (Walks and Bourne, 2006). More likely, economic and social factors are at play, which means that these minorities are just enjoying the experience of living near others of the same socio-cultural background. Over time, their children and grandchildren will find a place in Canada. But the question remains, 'might multiculturalism increase ethnic group identification at the expense of Canadian social cohesion?' Without doubt, tensions have arisen from time to time in Canada as people of different cultures, languages, and racial origins have expressed themselves when they encounter what they perceive as barriers within Canadian society, but peaceful discord is not in itself a failure of multiculturalism but part of a process of social interplay necessary to expose and resolve differences. Canadian history is on the side of both multiculturalism and immigration because immigrants—particularly their children—have found a place within Canadian society. While this has not always been easy, the opportunity to find a place remains an essential element of Canadian society.

In sum, the issue confronting the security of Canada is not multiculturalism, but rather the potential for violent behaviour on the part of a handful of extremists or from those who have brought too much unsorted cultural baggage from their countries of origin. The 1985 bombing of an Air India flight from Montreal to London en route to Bombay (Mumbai), which killed 329 passengers and crew, most of whom were Canadians, and gang violence in the recent past involving disaffected Vietnamese and Jamaican youth are cases in point. Then, too, ongoing disputes with First Nations over treaty lands keep resurfacing, with the confrontation at Caledonia, Ontario, that flared up 2006 and remained unresolved by mid-2007, and the 2007 blocking of the main rail line between Toronto and Montreal by the Mohawks of Tyendinaga in eastern Ontario only the most recent. There is no single answer to why citizens attack fellow citizens, but, rightly or wrongly, their motivation is based on deeply held feelings of injustice stemming from Canada's colonial past or from Old World disputes. As Prime Minister Stephen Harper announced at the World Urban Forum held in Vancouver in June 2006:

> The recent arrests of 17 people in southern Ontario and the Toronto area on charges related to terrorism have reminded us the potential for hate-filled violence in Canada is very real. It has led to some commentary, to the effect, that Canada's open and culturally diverse society makes us more vulnerable to terrorist activity. I believe the exact opposite is true. Canada's diversity, properly nurtured, is our great strength. (Bohn, 2006: A6)

Population Trends and Demographic Faultlines

The uneven geographies of population and social change are well known to Canadian geographers. Bourne and Rose (2001) have discussed such changes (see 'Further Readings' in Chapter 1). But do population trends provide the basis for the exploration of the four demographic tensions within Canada? In our discussion, these four faultlines focus on the implications of immigration for Canada

Vignette 4.3 European Experience with Multiculturalism and Islamic Terrorism

The presence of religious extremists troubles both Muslim immigrants to European countries and their host citizens. In July 2005, London bombings by young British Muslims came as a shock. How could these young men, born and raised in Britain, carry out suicide attacks on their fellow citizens? In their minds, is this part of the struggle between Islam and the West? Equally troubling is the widespread alienation that British Muslims feel, as revealed by the United Kingdom government's estimate that 26 per cent of Britain's 1.6 million Muslims felt no loyalty to Britain (Phillips, 2006).

and its cities, Aboriginal population growth, the changing French/English balance, and shifts in regional population. As Bourne and Rose point out, these population trends began some time ago and are already having an impact on Canada and its regions. Atlantic Canada, but especially Newfoundland and Labrador, faces a massive out-migration with considerable implications for the regional economy. Former federal cabinet minister John Crosbie put it this way:

> Christ, everywhere you turn around there's somebody leaving for Alberta. That they've got somewhere to go to better themselves and get work is not looked upon negatively. But we're exporting the ones that have the most initiative and are the most adventuresome and this weakens us. (Martin, 2005: A18)

The rapidly expanding Aboriginal population is clearly outpacing the growth in the rest of Canada's population. The myth of the vanishing Indian has been replaced with the re-emerging Indian. While the population of reserves continues to increase, many have moved to cities. Over the past 20 years, cities in Western Canada have experience substantial increases in the number of Canadians of Indian, Métis, and even Inuit ancestry.

The opposite situation faces the French/English balance. While the percentage of French-speakers in Québec remains above 80 per cent, in the rest of the country it has declined from 29 per cent of the national population in 1951 to 23 per cent in 2001, and the French-Canadian population is growing very slowly.

Immigration is, without a doubt, a most complex and somewhat sensitive subject. Canada has always been a land of immigrants. The annual inflow of immigrants now outnumbers those born in Canada. With many immigrants coming from different cultural and religious countries, the settlement of most immigrants into Canada's three major cities has made an enormous impact on the very nature of these cities, the city landscapes, and the emergence of ethnic enclaves. Sociologists such as Herberg (1989) and Li (2003) see immigration as a tension between maintaining the sense of a common identity and heritage and the need to incorporate some elements of the cultures of the immigrants into mainstream Canadian culture. Geographers such as Hiebert (2000; Hiebert and Ley, 2003) and Ley (1999; Ley and Hiebert, 2001) examine one aspect of immigrant culture, namely the effect of transplanting foreign landscapes into Canadian cities. Immigration, therefore, is one of the most significant elements of change in Canadian society. The Canada described by Kenneth Hare in 1968, which was on the cusp of major social, economic, and political change, is gone (see 'Further Reading' in Chapter 1). No one would recognize Toronto 'the Good' of the 1950s in the cosmopolitan Toronto of today. Toronto and other cities have undergone dramatic transformations of their cityscapes and their ethnic/racial makeup.

Newcomers and Old-Timers

Immigration is not a new phenomenon. Three major movements can be described. First, Canadians who claim Aboriginal ancestry have deep roots that go back thousands of years. Second, the original European founders of modern Canada, the British and French settlers, formed the basis of the two Canadian solitudes and were joined by other European newcomers in the late nineteenth and early twentieth centuries. Third, and most recently, waves of immigrants have come from numerous non-European countries, especially in South Asia, the Far East, and the Caribbean.

Everyone is a newcomer—at first. With the exception of Aboriginal peoples, children of newcomers become old-timers (Vignette 4.4 discusses why most Aboriginal children do not become 'old-timers'). This pattern of inclusiveness—some would say integration—forms the basis of the evolving Canadian society. Inclusiveness still leaves geographic space for cultural differences, as expressed by the province of Québec, by the territory of Nunavut, and by cultural enclaves in Canadian cities. Aboriginal peoples are a special case. As Peter Li (2003: 9) pointed out, 'Over time, earlier immigrants and their descendants become old-timers of the land, and their charter status places them in a privileged position vis-à-vis the newcomers who must accept the conditions of entry and rules of accommodation laid down by those who came before them.'

Vignettes 3.7 and 4.4 explain why the 'newcomer to old-timer' model is not applicable to Aboriginal peoples, who for the most part were excluded from participating in mainstream Canadian society until recently. For example, only after 1959 were status Indians allowed to vote in federal elections without losing their Aboriginal status, and it was not until 1958 that an Indian band—the Tyendinaga Reserve near Belleville, Ontario—was allowed *any* control over band funds (Dickason, 2002: 376). The Inuit and Métis, while not excluded legally, had little opportunity and encouragement to participate in political affairs, and, like the status Indians, sought some form of autonomous existence within Canada. The place of Aboriginal peoples within Canadian society has evolved over time. (For a more complete discussion, refer to the subsection entitled 'Aboriginal Peoples' in Chapter 3.)

Vignette 4.4 Invisible Canadians

Peter Li's description of the transition from newcomer to old-timer status, cited above, refers to immigrants who have chosen to join Canadian society and who are welcomed to join Canadian society. For the most part, the newcomers accepted integration into Canada's economy and society. Aboriginal peoples, unlike immigrants, did not choose to join Canadian society; rather, they were casualties of colonialism who had lost their economies and governments, and, in large measure, their own cultures. In today's world, urban Aboriginal peoples may fit the Li model but those living on reserves or in Native communities are more inclined to seek some form of autonomous government that protects their culture. Reserves and band governments represent one version while Nunavut and its territorial government provide another example of self-government.

Geography, too, has played a role in making Aboriginal peoples 'invisible' on the human landscape of Canada. For instance, Aboriginal peoples, by living on reserves and in other isolated locations, were geographically isolated from the rest of Canada. Until Aboriginal peoples relocated to settlements, they were unable to access social programs and services. Oppressed by the Indian Act, the first version of which was cobbled together out of earlier statutes from the Canadas in 1876, status Indians had to endure assimilation programs like residential schools, witnessed the destruction of their cultures through the outlawing of cultural practices such as potlatch and the sun dance, and, by means of the pass law, endured the indignity and economic consequences of confinement to their reserves. After World War II, change began—*very slowly*. In 1973, the *Calder* case led to the federal government recognizing that Aboriginal peoples had a legal claim to Crown lands that their ancestors had occupied. Since 1970, Ottawa has settled nearly 400 specific land claims, but the pace of their resolution is very slow—on average 10 claims are settled each year (St Germain and Sibbeston, 2006: v). With 855 specific land claims still outstanding, this unfinished business could take another 85 years to complete (ibid.). No wonder Indian leaders are frustrated.

Since newcomers often come from a different cultural world, they may bring with them different languages, customs, foods, and religious beliefs from those found in mainstream Canada. While these cultural imports often enrich Canadian society, a few have caused a cultural clash between some members of the two groups. On top of these cultural differences, since many recent immigrants come from non-European countries, racism can add another element to these cultural tensions. Visible minorities make up a major and growing share of the populations of Canada's major cities. Bourne and Rose (2001: 115–16) argue that 'the challenge of overcoming exclusionary practices, including those based on racism in all its forms, and in particular the impact of these practices on labour market opportunities for youth, becomes all the more urgent.'

Immigration brings benefits, including maintaining Canada's population, supplying much needed skilled labour to Canada's workforce, and bringing investment capital. As well, the newcomers add to the cultural richness of Canadian society (Vignette 4.5). Yet, another side to immigration exists in the minds of the old-timers. The two primary concerns are that the newcomers will somehow lessen their economic position in Canadian society, and that, coming from a different cultural, linguistic, and racial background, they will remake Canadian society in their image. This latter fear is reinforced by the concentration of immigrants in Canada's three major cities, where they may create cultural enclaves that prevent their integration into Canadian society.

Since 11 September 2001, new concerns have emerged. First, as part of a campaign of violence against the West, a few Muslims who are **jihadists** may slip through Canada's immigration screening system. Some Muslims, perhaps with a certain degree of justification, believe that the West (and the United States in particular) is waging war against them, not just in the deserts and mountains of Iraq and Afghanistan but through the extent to which Western cultural, political, and economic values are exported to and imposed on people throughout the world. For this reason, some followers of fundamentalist Islam have sought to fight this expansionist American hegemony by all available means, including sporadic and persistent acts of terrorism against their own people and against targets in the West or targets that represent what they revile of Western and American influence. Second, homegrown Muslim terrorism may be a new element in Canadian society, as occurred in England with the suicide bombing in 2006 of the London underground by British-born Islamic terrorists. But homegrown terrorists are an aberration and hopefully a short-term one.

Vignette 4.5 Russell Peters, a Premier South Asian-Canadian Comic

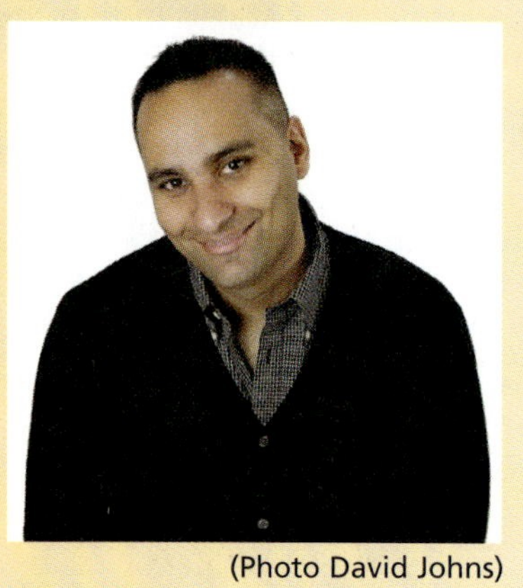
(Photo David Johns)

Russell Peters is a Canadian comedian who explores attitudes towards race (his and others) in a way that is challenging and cheeky. Born and raised in Ontario by parents who moved to Canada from India, Peters combines his experiences of growing up 'as the only brown child on the block' with insightful observations on race relations today. Now a North American star who has toured globally, Peters delights in exposing the humorous side of clashing cultures in Canada and the United States.

The history of immigration has shown two remarkable developments over time. First, Canadian society absorbs some of the cultural imports and multiculturalism fosters a social expression of the cultural contribution made by newcomers. Second, recent immigrants, especially their children, have shown a remarkable capacity not only to integrate into Canadian society, but also to reshape it. The cohesiveness of Canadian society depends on such integration and on an acceptance by all Canadians of cultural adaptation within the society to Old World cultures and religions. Such adaptation, of course, includes the ability to laugh at oneself and to appreciate the humour of other people and within other cultures. Russell Peters, a standup comedian who pokes fun at racism in Canada, exemplifies this positive side to immigration (Vignette 4.5).

Regional Patterns of Immigration

Immigrants are not evenly dispersed across Canada. Over half of all immigrants reside in Ontario; most others live in British Columbia, Québec, and Alberta. The reason is largely economic and partly social. Immigrants within these four provinces have been drawn to the major cities where fellow immigrants help them to find housing and jobs. Meniy Zewde, for example, a parking lot attendant from Addis Ababa, was attracted to social factors found in Toronto but not in Saskatoon (Jiminex and Lunman, 2004):

> I went to Saskatoon when I first came here 12 years ago but I was so lonely and it was hard to find work. Here [Toronto], there is a large community of Ethiopians. I go to the Evangelical Church of Ethiopia and we have lots of Ethiopian restaurants.

The geographic result is that nearly 95 per cent of new Canadians reside in Canada's census metropolitan areas. The regional pattern of immigration therefore takes on an urban/rural dichotomy. Ottawa has expressed some concern about this dichotomy. In October 2002, Denis Coderre, at that time the federal Immigration Minister, proposed that prospective immigrants sign a social contract promising to

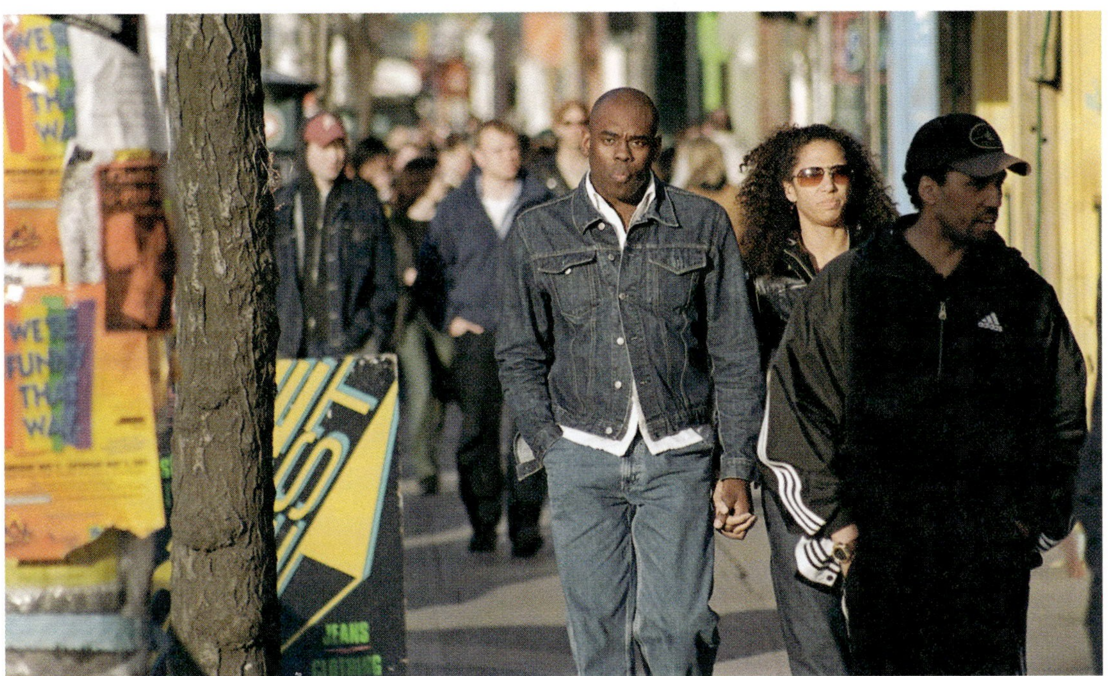

Canada's major cities attract immigrants from throughout the world. (Copyright Andrew Rubtsov)

live for three to five years in cities, towns, and rural communities outside of Canada's three major cities in return for Canadian citizenship. Such immigrants would be offered jobs and the right to bring their families to Canada. The underlying assumption is that the newcomers would fit into these communities and decide to stay after their contractual period expires (Fife, 2002). Coderre's proposal never took effect.

Canada's three major cities attract most immigrants. Nearly three-quarters of immigrants to Canada during the 1990s settled in Toronto, Montréal, and Vancouver (Figure 4.7). Large ethnic concentrations exist within these cities, forming neighbourhood communities with names such as Little Italy and Chinatown. Toronto attracts the largest share of new immigrants. Over 2 million immigrants live in the census metropolitan area of Toronto, making up 42 per cent of its population. The trend towards settling in Toronto is increasing: the 1991 census reported that 40 per cent of immigrants who arrived in the 1980s came to Toronto, but in the next decade this had risen to 43 per cent. A similar trend is found in Vancouver, but this is not so true for Montréal (Figure 4.7). Close to 21 per cent of immigrants had settled in other census metropolitan areas in 2001. Nearly 4 per cent of new immigrants came to Ottawa–Gatineau and Calgary, while Edmonton received 2.5 per cent and Hamilton nearly 2 per cent. A more complete analysis of immigration to Canada's major cities is found in Chapters 5–7.

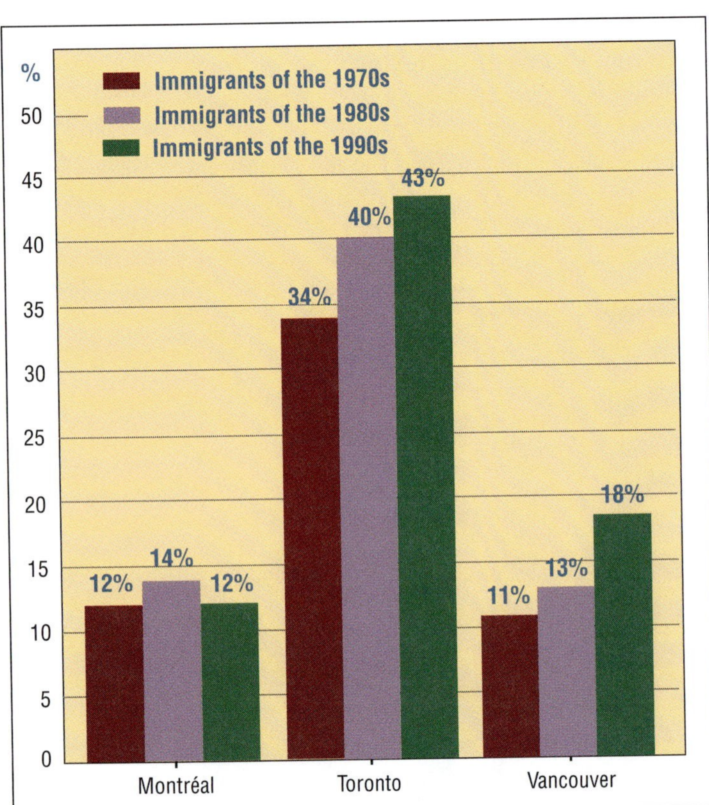

Figure 4.7 Share of immigrants in Montréal, Toronto, and Vancouver, 1981, 1991, and 2001

Source: Statistics Canada (2003c: 6). Adapted from Statistic Canada publication *Canada's Ethnocultural Portrait: The Changing Mosaic, 2001 Census*, Catalogue 96F0030XIE2001008, Released 21 January 2003, Searched 28 January 2003, http://www12.statcan.ca/english/census01/products/analytic/companion/etoimm/canada.cfm.

Aboriginal Population

The population figures for Aboriginal peoples (Indians, Inuit, and Métis) have gone through great changes since European contact. Prior to contact, Aboriginals within the territory that became Canada were quite numerous, especially among the sedentary peoples of the Northwest Coast and in parts of what would become southern Ontario, where rich resources (salmon on the Northwest Coast) and agriculture supported sizable communities. However, less than 100 years ago, the Aboriginal population had declined to about 100,000 because of the effects of European diseases and the loss of vast hunting grounds that had previously been their means of survival. But during the past five decades, Aboriginal peoples have experienced a population explosion. The total Aboriginal population in Canada now exceeds 1 million, many of whom now live in cities. For example, 43 per cent of registered Indians lived off-reserve in 2004. Professor John Richards of Simon Fraser University argues that the most important social issue in the country is the growing number of urban Aboriginal peoples—an argument that has little appeal to Aboriginal leaders, whose attention remains focused on better conditions for people living on reserves and in land-claim regions (Janigan, 2004:14).

The general population trends from pre-contact to the present illustrate three major demographic phases (Table 4.12). When Jacques Cartier sailed into Baie de Chaleur in 1541, the Indian and Inuit population in Canada may have been as high as 500,000. The exact figure will never be known. What we have are only estimates. By reconstructing the land's capacity to support wildlife and therefore also hunting societies, anthropologist James Mooney (1928: 7) estimated that about 220,000 Indians and Inuit lived in Canada at the time of contact. More recently, scholars have revised this figure upward. Dickason (2002: 45) and Denevan (1992: 370) estimate that the number of Aboriginal peoples living in Canada was closer to half a million. Whatever the exact figure, initial contact with Europeans resulted in a rapid depopulation of Aboriginal peoples. Factors include loss of hunting grounds and therefore food shortages, increased warfare, the spread of new diseases from Europe among the Indian tribes, and, in some instances, the intentional slaughter of Native people by the European newcomers. Communicable diseases, such as smallpox, caused great suffering and many deaths among Indian tribes. Epidemics sometimes quickly reduced the size of tribes by half. The result was that by the time of Confederation, Aboriginal peoples numbered about 100,000.

A demographic rebound of the Aboriginal population began in the 1940s and has just shown signs of slowing down (Figure 4.8). In 2001, there were approximately 1.3 million Canadians of Aboriginal descent (by Aboriginal identity, however, the figure was just under one million). The outcome of this increase has not only meant greater numbers but also a greater proportion of Aboriginal peoples within Canadian society. In 1951, Aboriginal peoples comprised less than 2 per cent of Canada's population. By 2001, this proportion had reached 4.3 per cent. Many factors contributed to this population increase; especially important was their relocation to

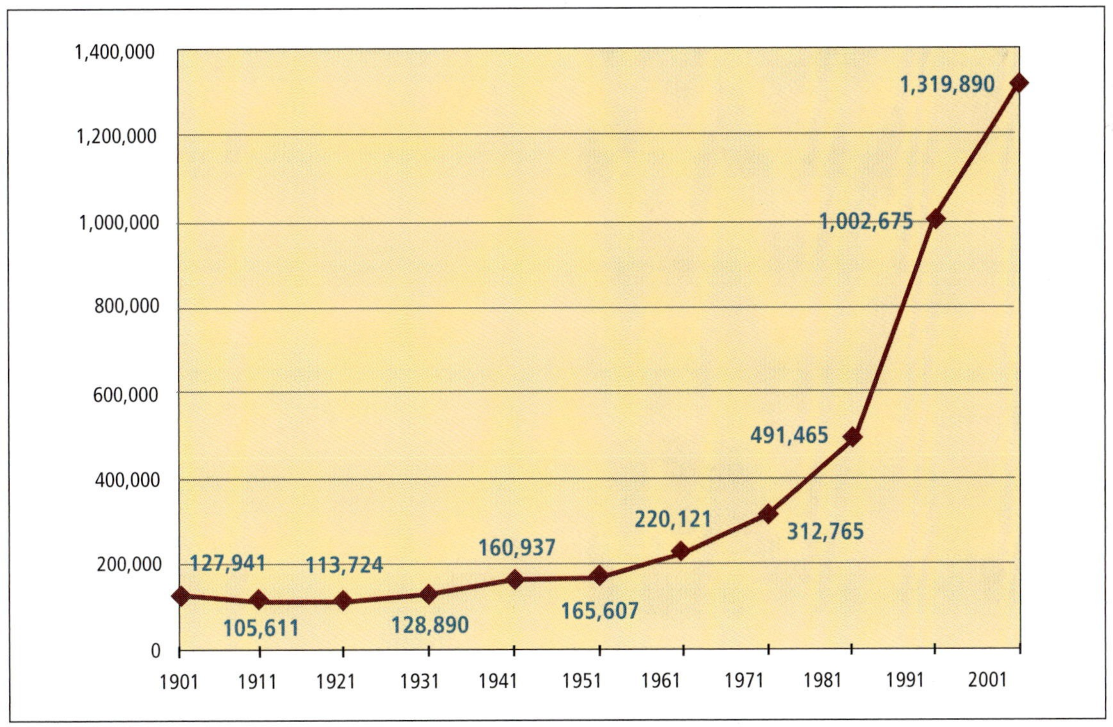

Figure 4.8 Aboriginal population by ancestry, 1901–2001

Source: Statistics Canada (2003h). Adapted from Statistics Canada publication *Aboriginal peoples of Canada: A demographic profile, 2001 Census*, Analysis series 96F0030XIE2001007, Released 21 January 2003, Searched 28 January 2003: http://www12.statcan.ca/english/census01/products/analytic/companion/abor/contents.cfm.

settlements where food supplies and medical care were available.

Since World War II the rate of natural increase of Aboriginal peoples has greatly exceeded the national figure. For much of that time, the rate of natural increase was about 3 per cent thanks to high birth rates and low death rates. The death rate for Aboriginal peoples dropped swiftly with the introduction of modern medicine. While communicable diseases, such as tuberculosis, continue to plague Native peoples, their mortality rates have declined below the national average. However, this low rate is somewhat misleading because the Aboriginal population is so much younger than the national average. Access to medical services and food supplies provides a partial explanation for the drop in the mortality rates.

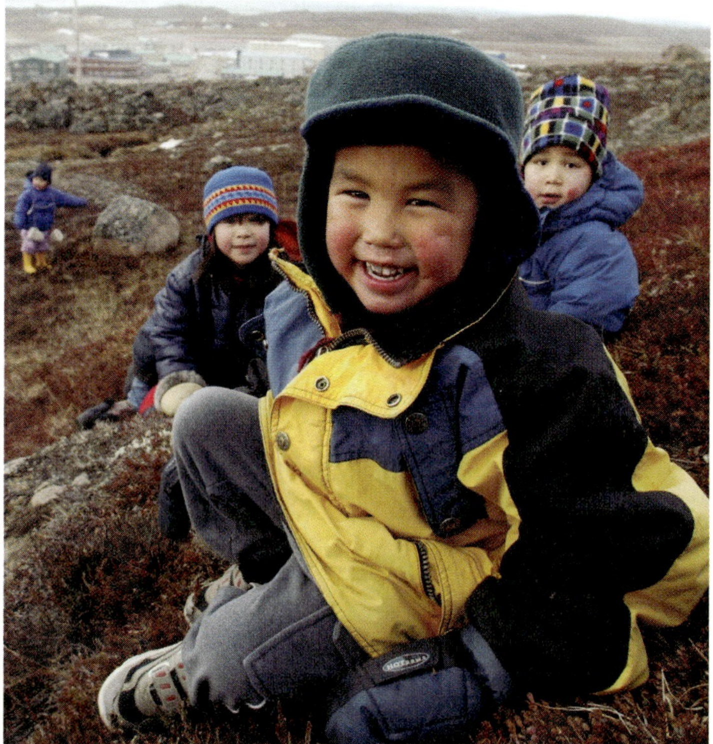

Simone Arnatsiaq plays with friends from a daycare on the tundra-covered hills looking over Iqaluit, the capital of Nunavut. One of the challenges facing Nunavut is its rapidly growing population. Nunavut has an extremely young population with nearly 40 per cent of its people under the age of fifteen. Comparatively, in the rest of Canada, 20 per cent of the population is under the age of fifteen. (CP/Paul Chiasson)

As the Aboriginal death rate declined, the birth rate remained high. The fertility rate for Aboriginal Canadians remains more than double the national rate. While the national fertility rate is below replacement level, the rate for Aboriginal peoples is much higher. In the past decade, the national rate was about 13 births per 1,000 people while the Aboriginal rate was close to 30 births per 1,000 people. This fertility difference has led to other demographic differences, including a more youthful Aboriginal demographic structure. While only 21 per cent of the Canadian population is under the age of 15, the figure for Aboriginal peoples is 38 per cent—almost double. While their rate of natural increase is high, Native Canadians face very high infant mortality and suicide rates, which indicate a social malaise that has affected the general health of that population. Nevertheless, based on the current demographic trends in Aboriginal society, the Aboriginal population is likely to continue increasing, though at a slower rate. As Aboriginal parents opt for smaller family units, the gap between the birth and death rates will diminish, as will the high rate of natural increase.

The political implications of this high rate of natural increase are profound. Already the rapid population increase has dispelled the myth of the 'vanishing Indian', instead assuring the continued existence of Aboriginal peoples. Aboriginal political power is, in part, linked to population size, so with the percentage of Aboriginal peoples in Canada increasing (perhaps exceeding 5 per cent by 2006), the federal, provincial, and territorial governments will have to heed the political voice of Native leaders more carefully. A rapid population increase can also have negative impacts, such as placing greater demands on scarce resources. For example, Aboriginal peoples are already facing increased housing demands. Another implication of population increase has a regional component. Since the distribution of Aboriginal peoples across Canada is uneven, the greatest increases in their numbers will take place in Western Canada, followed by Ontario and British Columbia (Figure 4.9). (Although the proportion of Aboriginal people to the total

Table 4.12 Major Phases for the Aboriginal Population in Canada

Phase	Characteristics
Pre-contact (1000–1500)	The Aboriginal population in Canada was at least 200,000 and possibly as large as 500,000. This population may have varied in size due to the carrying capacity of the land, which, in a hunting society, is controlled by the availability of game (food). For instance, natural conditions, especially weather, could affect the size and migration routes of animal populations.
Early Contact (1500–1940)	Aboriginal peoples who came into contact with Europeans were exposed to new diseases, and these new diseases often spread across the land prior to the arrival of Europeans in a particular place. Population losses were heavy. Loss of hunting lands also added to their demise. By the end of the nineteenth century, the Aboriginal population was just over 100,000.
Late Contact (1940–present)	The Aboriginal population now exceeds 1 million with a growing number living in cities. High fertility and low mortality account for this remarkable population rebound. The net result was a population explosion. While the Aboriginal population is likely to increase at a rate well above the national average in the coming decades, its natural rate of increase is expected to diminish due to a declining fertility rate.

Vignette 4.6 Aboriginal Ancestry and Identity

The 2001 census produced two population figures for Aboriginal peoples. One figure derived from the census question on ethnic origin. For 2001, the Aboriginal population based on this census question was 1,319,890. The second figure of 976,301 came from the census question based on Aboriginal identity. The identity question consisted of three main categories: North American Indian, Métis, and Inuit. A fourth category included other Aboriginal identities, such as 'registered Indian'.

	2001	1996	Percentage growth 1996–2001
Total: Aboriginal ancestry[1]	1,319,890	1,101,960	19.8
Total: Aboriginal identity	976,305	799,010	22.2
North American Indian[2]	608,850	529,040	15.1
Métis[2]	292,310	204,115	43.2
Inuit[2]	45,070	40,220	12.1
Multiple and other Aboriginal responses[3]	30,080	25,640	17.3

(1) Also known as Aboriginal origin.
(2) Includes persons who reported a North American Indian, Métis, or Inuit identity only.
(3) Includes persons who reported more than one Aboriginal identity group (North American Indian, Métis, or Inuit) and those who reported being a registered Indian and/or band member without reporting an Aboriginal identity.

Source: Statistics Canada (2003g). Adapted from Statistics Canada publication *The Daily*, Catalogue 11-001, Released 12 March 2003, http://www.statcan.ca/Daily/English/030311/d030311a.htm.

population of Ontario is low, as shown in Figure 4.9, Ontario's much larger population means that a significant number of Aboriginal people—just under 190,000 as of 2001—reside in that province.) The five provincial governments in these regions will face considerable pressure from those status Indians seeking a resolution to their land claims and from the non-status Indians and Métis who seek to better their way of life.

French/English Balance

French- and English-speaking Canadians represent the traditional duality of Canadian society. This dual relationship has deep historic roots. With the establishment of the Province of Canada in 1841, the relationship flowered into a partnership with English-speaking Canada West and French-speaking Canada East sharing political power. In 1867, the British North America Act declared that both French and English were official languages of the new Dominion of Canada. The assignment of language rights to provinces ensured that the Québec government could protect and promote the French language.

Language is the essential element of French heritage and ensures the preservation of French culture. It is from this perspective that francophones interpret the social and political significance of the French/English balance. If the census of Canada reports that the proportion of French-speaking Canadians is declining, then francophones become concerned. If a Royal Commission (such as that on Bilingualism and Biculturalism, 1963–71) provides evidence that the French language is in 'deadly danger', then francophones become alarmed and want their governments to take action.

Linguistic rights for both French and English as the official languages give them a special place in Canadian society. One implication is that new Canadians whose language is neither French nor English gravitate to one linguistic group or the other. Within Canada, English is

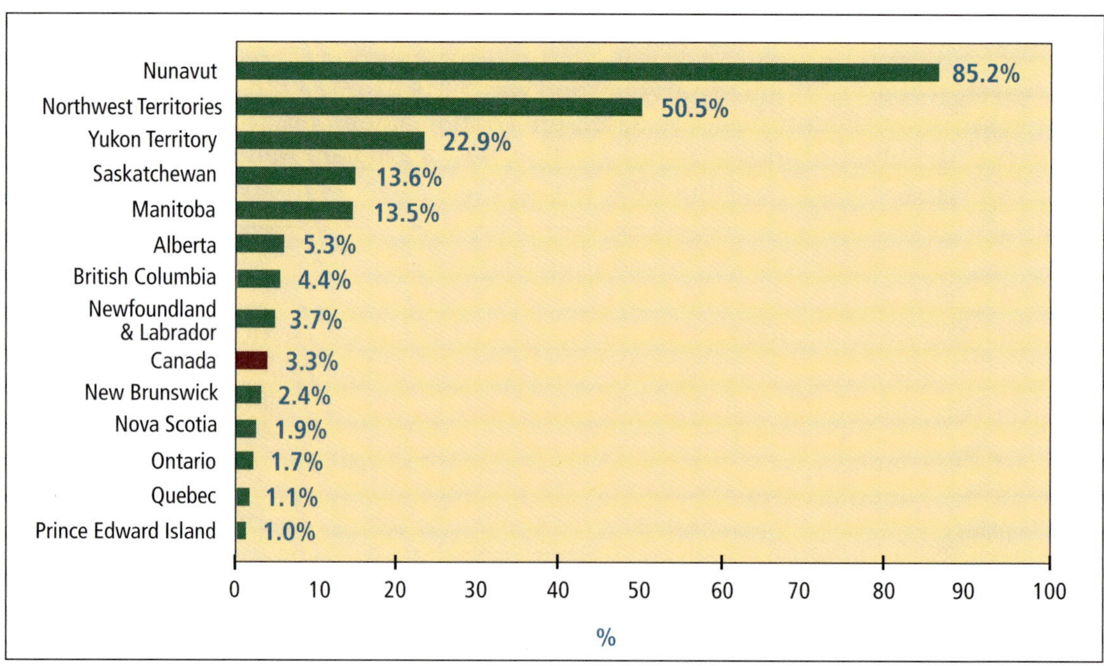

Figure 4.9 Percentage of Aboriginal population by Aboriginal identity, provinces and territories, 2001

Source: Statistics Canada (2003h). Adapted from Statistics Canada publication *Aboriginal peoples of Canada: A demographic profile, 2001 Census*, Analysis series 96F0030XIE2001007, Released 21 January 2003, Searched 28 January 2003, http://www12.statcan.ca/english/census01/products/analytic/companion/abor/contents.cfm.

the dominant language. English is also the business language of North America and, to a large degree, the world. Immigrants therefore have a compelling reason to learn English. Also, most immigrants settle in English-speaking areas and so are naturally drawn into the anglophone linguistic group. Outside of Québec and New Brunswick, the French language is losing ground. Anne Gilbert (Gilbert, 2001: 173) points to the decision of the government of Ontario to reject the concept of recognizing both English and French as official languages as a missed opportunity to showcase Canada's serious commitment to its dual languages.

The most recent Statistics Canada figures (2001) support this argument. The proportion of Canadians outside of Québec speaking English at home was 85.6 per cent, the proportion speaking a non-official language was 11.7 per cent, and those speaking French was 2.7 per cent. However, a number of Canadians speak both English and French. The number of bilingual Canadians is increasing. From 1951 to 1996, the percentage of Canadians who could speak both official languages increased from 12 per cent to 17 per cent (Statistics Canada, 1997b: 11).

French/English dualism is a fundamental aspect of the geopolitical nature of Canada. As Jacques Bernier (1991: 79) of Université Laval stated: 'Canada's duality is intrinsic, and as long as it is not clearly recognized and dealt with, the issue of Canadian unity will remain.' This duality is a political concept embedded in the historical relationship between the two cultures. The main indicator of the stability of this French/English dualism is language; in other words, the stability of this concept depends on a relatively constant number of Canadians speaking each language. But how should we measure duality? Should mother tongue or household language hold the key? Perhaps the population size of the minority language group should be the determining factor? Or does the number of bilingual Canadians hold sway?

The French/English balance can be measured by a number of factors relating to language use. In the 2001 census, Canadians were asked three questions about language—their mother tongue, home language, and bilingual skills. Mother tongue is the first language learned that is still understood, while home language is the language most often spoken in the home. In that year, 17.5 million Canadians declared that their mother tongue was English, while 20 million indicated that the language spoken at home was English (Statistics Canada, 2002d). Furthermore, Canadians who were able to speak English totalled 22.5 million, or 83 per cent. Since most immigrants today learn English but have a different mother tongue, the number of Canadians who can speak English is much higher than indicated by mother tongue or home language. In comparison, 6.8 million Canadians stated that their mother tongue was French, but only 6.5 million spoke French at home. Many Canadians can speak both official languages and 8.5 million or 32 per cent of Canada's population can speak French.

French-speaking Canadians are distributed in three major geographic areas. The principal one is the province of Québec, where 83.1 per cent of the population speaks French in the home (Statistics Canada, 2002d). A second francophone area is in New Brunswick, where Acadians and other French-speakers form 30.3 per cent of the population. A minor cluster exists in the third area, Ontario, where 2.7 per cent speak French at home. Most of these live in northeastern Ontario. The number of French-speakers is increasing in Canada but most of this increase has occurred in Québec (Table 4.13). Québec is the only geographic region where the number of people speaking French at home is greater than those declaring French as their mother tongue. This difference suggests that some immigrants living in Québec are adopting the French language and assimilating into the francophone culture.

The number of French-speaking Canadians is increasing at a rate slower than the national rate of population growth. Since 1951, the number of Canadians whose mother tongue is French has increased from 4 million to 6.8 million. Within Québec, the increase is even greater, jumping from 3.3 million in

Table 4.13 Population with French Mother Tongue, 1951–2001

Year	Canada (000s)	Québec (000s)	Rest of Canada (000s)	Canada (%)	Québec (%)	Rest of Canada (%)
1951	4,069	3,347	722	29.0	82.5	7.3
1961	5,123	4,270	854	28.1	81.2	6.6
1971	5,794	4,867	926	26.9	80.7	6.0
1981	6,178	5,254	924	25.7	82.5	5.2
1991	6,562	5,586	976	24.3	82.0	4.8
1996	6,637	5,747	970	23.5	81.5	4.5
2001	6,782	5,802	980	22.9	81.4	4.4

Sources: Harrison and Marmen (1994: Table 2.1); Statistics Canada (1997b, 2002d).

1951 to 5.8 million in 2001. Even in the predominantly English-speaking areas of Canada, the number of francophones has increased from 722,000 to 980,000. Over the same time span, however, the proportion of Canadians indicating that French is their mother tongue has dropped from 29 per cent to below 23 per cent (Statistics Canada, 2002d). Finally, the number and percentage of bilingual Canadians have increased sharply. In 1981, 3.7 million Canadians spoke both languages, while nearly 5 million were in this category by 2001.

The French/English duality remains strong at a national level, but the increasing numbers of Canadians whose mother tongue is neither French nor English reflects the arrival of immigrants who speak neither official language (Table 4.14). Within Québec, the low fertility level among francophones means that the adoption of French by newcomers is extremely critical for the long-term maintenance of the French language.

Between 1996 and 2001, the English-speaking population in Canada grew by about 2.6 per cent compared with only 1.1 per cent for the French-speaking population. As a result, the proportion of Canadians who were French-speaking slipped to 22.9 per cent by 2001. Still, the number of French-speaking Canadians within Québec is rising. Two reasons account for this increase in Québec: the natural increase of the Québécois and the increasing numbers of allophones who become French-speaking. Until the Québec government passed a series of language laws restricting the use of English, most allophones were choosing to learn English. The reason was simple—it gave them greater economic mobility because they recognized that English is spoken in other parts of Canada and the United States.

Among the Québécois, language laws are very popular because they ensure the place of French in Québec, particularly in Montréal. While such legislation runs against the current of bilingualism in the rest of Canada, Québec governments (both Liberal and Parti Québécois) support the principle of a unilingual province in order to ensure the survival of the French language in Canada and North America. While language laws are not popular with all new immigrants, they are required by Québec law to send their children to French schools.

Québec language laws, particularly the law requiring all signs to be in French, have annoyed many English-speaking citizens of the province. In 1988, the Supreme Court of Canada ruled that Québec's French-only sign regulation violated the Charter of Rights. Francophones, ever concerned about the preservation of their culture, reacted strongly, taking to the streets, declaring 'Ne touchez pas à la loi

Table 4.14 Population by Mother Tongue, by Percentage, 1951–2001

Year	English	French	Other
1951	59.1	29.0	11.8
1961	58.5	28.1	13.5
1971	60.2	26.9	13.0
1981	61.4	25.7	13.0
1991	60.4	24.3	15.3
1996	59.9	23.5	16.6
2001	59.1	22.9	18.0

Sources: Harrison and Marmen (1994: Table 2.1); Statistics Canada (1997b, 2002d).

101.' The Québec Premier, Robert Bourassa, finding himself on the horns of a dilemma, chose to circumvent the Supreme Court's ruling by invoking the 'notwithstanding' clause (section 33) in the Charter. Later, the provincial government amended its sign legislation so that signs could be in both languages as long as the French is larger than the English. Nevertheless, ever since language legislation favouring the French language was first passed in 1974, and especially after the PQ's introduction of Bill 101 in 1977, English-speaking Quebecers have felt less comfortable living in that province. In fact, these language laws, combined with two referendums, have caused many anglophone Quebecers to migrate to other provinces. In 1971 there were 789,000 anglophones in Québec, but 30 years later the number had dropped below 600,000 (Statistics Canada, 2002d).

Core/Periphery Populations

Canada's population is unevenly distributed across the country's regions. The population remains concentrated in Central Canada, though population shifts have occurred in Western Canada and British Columbia. Atlantic Canada, on the other hand, has lost ground. Therefore, the hinterland has increased its share of the national population, but that increase has been geographically uneven (Table 4.15). Ontario has continued to grow rapidly while Québec's rate of population increase has slowed. The reasons for this distribution in population and population growth can be found in the country's historical migration and settlement patterns, which can be characterized as population shift from the east to west and from rural to urban. More recent figures from Statistics Canada support these population trends, with the greatest gains occurring in Western Canada, Ontario, Québec, and the Territorial North and the only loss taking place in Atlantic Canada (Table 4.16).

All indications suggest that current regional trends will continue. Most population gains are anticipated to take place in Western Canada, Ontario, and British Columbia, though the rate of increase may remain highest in the Territorial North. Certainly, if Atlantic Canada continues to experience high levels of out-migration, its share of the national population will continue to decline. Over the last five years, the rate of increase of Ontario and Western Canada was equal, at 6.6 per cent, with British Columbia following at 5.3 per cent. In 2006, these three geographic regions contained nearly 69 per cent of Canada's population. How much of Canada's population will be concentrated in Ontario, Western Canada, and British Columbia by the time of the 2011 census will be determined by job opportunities. In 2001, British Columbia

Table 4.15 Regional Population of Canada by Percentage, 1901–2006

Region	1901	1921	1941	1961	1981	1991	1996	2001	2006
Ontario	40.6	33.4	32.9	34.2	35.4	37.0	37.3	38.0	38.5
Québec	30.7	26.9	29.0	28.8	26.4	25.3	24.8	24.0	23.9
British Columbia	3.3	6.0	7.1	8.9	11.3	12.0	12.9	13.1	13.0
Western Canada	7.8	22.2	21.0	17.5	17.4	16.9	16.6	16.9	17.1
Atlantic Canada	16.7	11.4	9.8	10.4	9.2	8.5	8.1	7.7	7.2
Territorial North*	0.9	0.1	0.2	0.2	0.3	0.3	0.3	0.3	0.3
Canada (millions)	5.4	8.8	11.5	18.2	24.3	27.3	28.9	30.0	31.6

*In 1901, the Yukon population included many associated with the Klondike gold rush. By 1911, most had left. This accounts for the abrupt change in population for the North after 1901. Also, even though the present provinces of Alberta, Saskatchewan, and much of Manitoba still belonged to the Northwest Territories in 1901, their populations were assigned to the Prairie region.
Sources: McVey and Kalbach (1995: 46); Statistics Canada (1997c, 2002a, 2007a).

and Western Canada combined held 29.9 per cent of the national population, by 2006 the figure had grown to 30.1 per cent. This small but steady population shift to the West has not translated into more political power in Ottawa, at least not on paper, so Ontario and Québec maintain the demographic power to dominate the political landscape of the early twenty-first century. With this in mind, the discussion of the six geographic regions begins with the two largest regions by population, namely Ontario and Québec, followed by British Columbia, Western Canada, Atlantic Canada, and the Territorial North.

Canada's Economic Face

The Canadian economy is growing at a pace well ahead of other G-8 countries, but how is this growth affecting its labour force, employment levels, and the unemployment rate? Over the 10-year period ending in 2005, Canada's

Table 4.16 Regional Population Shifts, 2001 to 2006

Geographic Region	2001	2006	Change	% Change
Atlantic Canada	2,285,729	2,284,779	–950	–0.1
Québec	7,237,479	7,546,131	308,652	4.3
Ontario	11,410,046	12,160,282	750,236	6.6
Western Canada	5,073,323	5,406,908	333,585	6.6
British Columbia	3,907,738	4,113,487	205,749	5.3
Territorial North	92,779	101,310	8,531	9.2
Canada	30,007,094	31,612,897	1,605,803	5.4

Source: Statistics Canada (2007a). Adapted from Statistics Canada website URL: http://www40.statcan.ca/l01/cst01/demo02a.htm?sdi=population%20change%20provinces.

labour force increased from 15.4 million to 17.3 million, or by over 12 per cent (Table 4.17). During this period, the unemployment rate dropped from 8.3 per cent to 6.8 per cent. With economic growth across Canada uneven, regional differences exist in employment and unemployment rates. Most economic growth took place in Alberta, Ontario, and British Columbia. Large numbers of migrants from other provinces to those three provinces suggest that many are attracted to job opportunities. The unemployment rates by province reflect a regional pattern with the highest rates in Atlantic Canada and the lowest rates in Western Canada (Table 4.18). Then, too, the number of women entering the labour force has been a major source of increased labour supply for the Canadian economy. In 2005, over 8 million working women made up nearly half of the labour force (Table 4.19). Finally, Canada's open economy has exposed its manufacturing firms to foreign competition, compelling them to undergo restructuring (Vignette 4.7).

Much of Canada's recent economic success is related to the rising demand for resources, especially energy resources. However, the rising Canadian dollar has made exports from Canada's manufacturing industries more expensive in foreign markets but it has increased the value of energy and mineral exports. The impact on Canada's economy has varied regionally: Western Canada and British Columbia have benefited from rising prices for energy and mineral resources while the manufacturing industries of Ontario and Québec have found it more difficult to compete with foreign firms in the domestic market and to export their products to the United States.

Economic Structure

Like other modern industrial countries, Canada's economic structure has evolved from an agrarian to an industrial economy (see Vignette 4.8). By structure, we mean the relative share of primary, secondary, tertiary, and quaternary economic activities in the economy (Tables 4.20 and 4.21). As Canada developed an industrial base in the late nineteenth century, many of its workers were involved in secondary activities. However, this workforce was concentrated in major cities, especially those in the manufacturing belt of Central Canada. The rest of the country was involved in **primary activities** such as agriculture, forestry, and mining. This created a geographic division in Canada's economic structure, which took on the appearance of an industrial core and a resource hinterland. For this reason, the economic structures of Canada's six regions are examined in Chapters 5–10 to illustrate their positions within Canada's economic structure. For example, core regions such as Ontario have the highest percentage of workers in the secondary sector (manufacturing) while

Table 4.17 Labour Force Characteristics, 1998–2005 (000s)

	1998	1999	2000	2001	2005
Population 15 years and over	23,671.1	23,969.0	24,284.9	24,617.8	25,805.5
Labour force	15,417.7	15,721.2	15,999.2	16,246.3	17,342.6
Employed	14,140.4	14,531.2	14,909.7	15,076.8	16,169.7
Unemployed	1,277.3	1,190.1	1,089.6	1,169.6	1,172.8
Participation rate (%)	65.1	65.6	65.9	66.0	67.0
Unemployment rate (%)	8.3	7.6	6.8	7.2	6.8

Sources: Statistics Canada (1998a, 2003a, 2006a).

Table 4.18 Annual Unemployment Rates by Province, 1986–2006

Province	1986–9	1996	2000	2002	2006
Newfoundland and Labrador	18.4	25.1	16.7	16.9	14.8
Prince Edward Island	15.3	13.8	12.0	12.1	11.0
New Brunswick	12.6	15.5	10.0	10.4	8.8
Nova Scotia	12.1	13.3	9.1	9.7	7.9
Québec	11.2	11.8	8.4	8.6	8.0
British Columbia	10.1	9.6	7.2	8.5	4.8
Ontario	7.9	9.1	5.7	7.1	6.3
Manitoba	8.3	7.9	4.9	5.2	4.3
Alberta	8.6	7.2	5.9	5.3	3.4
Saskatchewan	7.4	7.2	5.2	5.7	4.7
Canada	9.5	10.1	6.8	7.7	6.1

Sources: Scotiabank (1998: 12); Statistics Canada (2003e, 2007k).

peripheral regions such as the Territorial North lead in the primary sector.

Major shifts in the percentage of workers in the primary, secondary, tertiary, and quaternary sectors are shown in Table 4.21. These shifts mark three fundamental transitions in the nature of the Canadian economy—from agrarian or pre-industrial to early industrial, from early industrial to late industrial, and from late industrial to post-industrial. The post-industrial economy is marked by the growth of high-technology industries such as telecommunications.

The country's post-industrial economy is characterized by an expanding tertiary/quaternary workforce that accounts for three-quarters of the workers, a slowly declining secondary or manufacturing sector that employs 20 per cent of the labour force, and a shrinking primary or resource sector that now accounts for only 4 per cent of the labour force (Table 4.21). As discussed in Vignette 4.7, the **primary sector**

Table 4.19 Men and Women in the Labour Force, 2005

	Male	Female
Population 15 years and older	12,692,600	13,112,900
Employed	8,594,700	8,098,800
Participation rate (%)	72.8	61.8
Unemployment rate (%)	7.0	6.5
Employment rate (%)	67.7	57.8

Source: Statistics Canada (2006a). Adapted from Statistics Canada website http://www40.statcan.ca/l01/cst01/gblec02a.htm?sdi=exports%20country, Searched 18 June 2006.

Vignette 4.7 Restructuring Continues

Since the Canada–US Free Trade Agreement in 1989, companies have had to find ways to remain competitive in the North American marketplace. In addition, trade barriers with other countries have been greatly reduced, and the Canadian dollar has appreciated. Consequently, Canadian manufacturing firms, but especially those employing unskilled or semi-skilled labour, were hard hit by foreign competition. Firms in Canada reacted as follows:

- American firms closed their smaller Canadian plants and supplied their Canadian markets from their larger American plants.
- Canadian plants looked to offshore locations to reduce their labour costs.

This relocation trend continues: Goodyear Tire & Rubber Co. closed its Québec tire factory in 2006, and in 2007 North America's largest T-shirt maker, Gildan Active Wear Inc., shut two factories in Québec, stating that factory costs in Central America are 25 per cent lower than in Québec (*National Post*, 2006, 2007). Relocation is now affecting skilled workers and office workers because Canadian-based firms can recruit such employees at much lower wages in developing world countries. In February 2007, Nortel announced that it was downsizing its Canadian operations and transferring jobs offshore to places such as Mexico and Turkey where skilled labour is available at much lower costs (*Globe and Mail*, 2007).

Vignette 4.8 Stages of Development

The term 'stages of development' refers to the evolution in the industrial structure of nations and the diffusion of industry from Western Europe to other countries. Technological advancements and the liberalization of trade have led to major changes in the economic structure of industrial states. Canada, for instance, has lost most of its labour-intensive manufacturing to countries with lower wages, but Canada has expanded its industries requiring highly skilled workers.

Modern industrial nations often have gone through three stages of development: pre-industrial, industrial, and post-industrial. In the pre-industrial stage, primary activities, led by agriculture, dominate the economy and occupy most of the labour force. The industrial stage occurs when manufacturing activities surpass those of the primary sector of the economy. In this stage, most of the workers are involved in secondary activities (such as processing raw materials and manufacturing goods) and tertiary activities (such as providing a variety of services to the public). The last stage is the post-industrial stage, in which the economy is dominated by tertiary and quaternary activities. Table 4.20 provides a fuller description of the types of economic structures in each stage.

Table 4.20 Types of Economic Structure

Economic Sectors	Characteristics
Primary	Economic activities that are concerned with natural resources of any kind, such as fishing, forestry, mining, trapping, and agriculture.
Secondary	Economic activities that process, transform, fabricate, or assemble the raw materials derived from primary activities, or that are high-tech knowledge-based activities that reassemble, refinish, or package manufactured goods. Examples include automobile manufacturing, meat-packing, pulp and papermaking, and software manufacturing.
Tertiary	Economic activities that involve the sale and exchange of goods and services, including professional services, retail sales, and education.
Quaternary	Economic activities that deal with the handling and processing of knowledge and information that lead to decision-making by companies and governments. Such activities are often found in the research and development units of companies engaged in primary, secondary, and tertiary activities.

in Canada includes resource activities, the **secondary sector** focuses on manufacturing activities, and the **tertiary/quaternary sector** provides a wide variety of services.

Primary Sector

The primary sector forms a tiny but important part of Canada's economy. In the nineteenth century, the primary sector dominated Canada's economy and accounted for approximately half of the labour force (Table 4.21). As workers' wages rose, companies looked to ways to increase output per worker. This was achieved by substituting machinery for labour. Two developments then took place—output increased and the primary labour force decreased. Yet, the resource economy continues to play an important role on the world stage by providing close to 30 per cent of the value of Canadian exports. Resource industries are found in all six geographic regions. However, their importance to the regional economies varies. The primary sector forms a larger part of the economies of peripheral regions, with the Territorial North most dependent on its resource industry. Within the primary sector, the prices for energy resources have risen sharply. Since 2000, the

Table 4.21 Sectoral Changes in Canada's Labour Force, 1881–2006

Development Stage	Year	Primary	Secondary	Tertiary*
Agricultural	1881	51	29	19
Early industrial	1901	44	28	28
Late industrial	1961	14	32	54
Post-industrial	2006	4	20	76

*Because data for the quaternary sector are not available, percentages for the tertiary sector conventionally include both tertiary and quaternary jobs.

Sources: Adapted from McVey and Kalbach (1995: Table 10.3); Statistics Canada (2006d).

price of oil, for example, increased from around $20 US a barrel to $80 US a barrel in September 2007. Oil-producing provinces such as Alberta, Saskatchewan, and Newfoundland and Labrador have greatly benefited from this energy boom.

Secondary Sector

With the establishment of the Canada–US Free Trade Agreement (1989), NAFTA (1994), and the World Trade Organization (1995), the Canadian manufacturing industry underwent severe restructuring. Some jobs were lost, first to the American South, then to Mexico and to low-wage countries in Asia, as Canadian firms could not compete with the foreign firms. Canadian manufacturers with labour-intensive operations, such as those producing textiles, shoes, and garments, saw their sales drop as retail stores purchased lower-priced imported goods. The problem was simple: labour costs in China and many other foreign countries were only a fraction of Canadian wages. For the companies, the solution was simple: shift operations to countries with low wages. Other jobs simply disappeared because the rationale for American branch plants in Canada ceased when Canadian tariffs on imported goods dropped. As a result, many American branch plants in Canada ceased to operate because it was more efficient to consolidate their operations within the United States. Ontario and Québec continue to see their manufacturing industries under great pressure.

In the 1990s, two positive developments occurred in the secondary sector. First, automobile assembly production and automobile parts operations expanded, generating many new jobs directly and indirectly. The now defunct Auto Pact was largely responsible for this expansion in Canada's manufacturing industry (though the General Motors of Canada plant in London, Ontario, that produces lightweight combat vehicles depends on US military contracts). Second, new manufacturing activity emerged around the application of new technology to Canadian firms. As Barnes et al. (2000: 11) explain:

> The impact of computer and information technology is visible at every turn in Canada's urban centers, and even the countryside bristles with receiver and transmitter towers. Behind these visible markers of the knowledge economy are a series of industries that produce hardware and software, and firms that incorporate such technology into their products and production processes.

New technology takes the form of high technology (high tech), which is well established in four major areas: biotechnology, information technology, materials technology, and transportation technology. Biotechnology involves harnessing the power of DNA, the double-helix-shaped chemical strand at the centre of each living cell. Genetically modified canola seeds, for example, are widely used by farmers in Western Canada because of higher yields and resistance to disease. Information technology includes information-based, computer-driven, and communications-related activities. The Ottawa area, the Silicon Valley of the North, specializes in telecommunications but also has many companies that produce other forms of hardware, such as computer chips, and a variety of software. The software (program) makes use of the hardware (computer) to generate a product, such as text. Materials technology refers to new metal alloys, plastic-coated metals, and laminated glass. Many of these products are produced in Ontario and Québec for various manufacturing companies ranging from automobile to aircraft companies. Transportation technology has created high-speed trains, super tankers, and modern passenger aircraft. Québec-based Bombardier produces both high-speed trains and modern passenger aircraft.

All these high-tech companies have three important qualities: (1) they are value-added industries; (2) they employ highly skilled and highly paid workers; and (3) they depend on and frequently conduct or support pure research (sometimes in conjunction with research scientists at universities or government research centres) that may result in the development of new commercial products.

Most high-tech firms are located in the major cities of Canada's industrial heartland. The three leading centres are Toronto, Montréal, and Ottawa. In the case of Ottawa, high-tech firms such as Nortel are located on its outskirts, at Kanata. High-tech firms employ large numbers of computer programmers, and the geographic location of these programmers revealed that, in 1997, nearly half were working in Ontario. Québec had 27 per cent, followed by Western Canada at 13 per cent, British Columbia at 9 per cent, and Atlantic Canada at 3 per cent (Gower, 1998). By 2005, Ontario continued to account for half of the high-tech industry while Québec had slipped to 20 per cent. On the other hand, Western Canada, British Columbia, and Atlantic Canada posted gains with 15, 10, and 5 per cent respectively (Statistics, Canada, 2007j). While many high-tech firms were adversely affected by the economic downturn that took place from 2000 to the middle of 2003, sales increased from 2003 to 2005 (ibid.).

Still, high-tech companies are seeking to reduce their costs of production by locating some of their operations offshore and then selling the lower-cost product in Canada.

Tertiary Sector

The tertiary sector is expanding rapidly and taking on new functions. In fact, the service industry fuelled job growth in Canada for much of the twentieth century (Table 4.21). Today, three-quarters of Canadian workers are engaged in tertiary functions. Only the United States exceeds this figure and is slightly more 'tertiarized' than Canada.

As society has become more urban and its labour force more specialized, the tertiary sector has grown and diversified. The service industry includes accommodations, communications, education, finance, health, insurance, personal services, public administration, and transportation. As these service functions have grown, some scholars have split this section into two parts, adding the quaternary sector to include decision-making occupations such as research and senior management jobs. Since research and senior management positions are part of most companies, the task of recording such positions as belonging to the 'quaternary sector' is difficult. Other approaches divide the tertiary sector into public and private employment or by low- and high-skilled jobs. Low-skilled employees are found in both the public and private sectors. In the private sector, low-paid workers are common in retail stores, fast-food outlets, and gasoline service stations. Often these are part-time workers with no pension plan or job security. Clearly, they represent the low end of the service industry. Furthermore, such jobs are not restricted to those with limited education. Sessional lecturers at Canadian universities also fall into the low end of the service industry. The high end of the service industry in terms of pay and job security falls to those with skills valued by society. Professional and senior public officials earn high wages and look forward to a secure future. Such jobs provide stability for the individual, his or her family, and the larger community. Professionals include accountants, dentists, doctors, engineers, and lawyers, who occupy the top pay category, followed by professors, teachers, and nurses.

Summary

Since World War II, the human face of Canada has changed. Much change took place after the 1960s when a new economic and social climate arose. Trade liberalization triggered a new economic arrangement within North America. The separatist movement in Québec, the influx of Asian, African, and Caribbean migrants, and the search by Aboriginals for a place in (or apart from) Canadian society have changed the social and political dynamics of Canada. Like other developed countries, Canada's economy has transformed from an industrial base (manufacturing, resource extraction) to a post-industrial stage where most of the jobs are to be found in the growing service sector. But this transformation has taken place within a regional context whereby some regions have enjoyed greater economic development and experienced faster-growing populations than less-favoured regions. This transformation has also been affected by Canada's geographic position within North America. Having the United States, the most powerful nation in the world, as a neighbour and trading partner has resulted in close economic ties and a strong dependency on the American economy.

Major economic changes in Canada stem from the country's economic integration with the United States and, to a lesser degree, from increased involvement in the global economy. Economic restructuring, sometimes painful, took place following the free trade agreement with the US in 1989. It continues under NAFTA and the World Trade Organization, but at a slower pace. Economic restructuring has also affected the Canadian labour force. Three notable labour changes have taken place: more women are employed; more workers are engaged in tertiary occupations; and more people are now unemployed.

Equally important changes stem from Canadian social and political forces, such as the Constitution with its emphasis on bilingual, pluralistic, and civic responsibility. Also, demographic changes have resulted in the reduction of Canada's rate of natural increase. Since 1967 the Canadian government has extended immigration opportunities to people from all over the world and increased the annual number of immigrants coming to Canada. In turn, new Canadians, particularly those from non-European and non-Christian countries, have altered the ethnic and religious makeup of Canada. In response, Ottawa initiated a policy of multiculturalism to create harmonious relations among Canada's various ethnic groups and to promote social equality. Such a policy runs counter to the concept of duality between the two founding peoples, the French and the English, and makes francophones, especially in Québec, feel their cultural status and political power are threatened.

Changes to Canada's economy and society have affected Canada's four critical tensions or faultlines: (1) Aboriginal/non-Aboriginal, (2) core/periphery, (3) French/English, and (4) newcomers/old-timers. Key demographic changes are occurring along these divides, reflecting internal forces and revealing population and political shifts. The first faultline is characterized by a much higher rate of natural increase among the Aboriginal population. The second faultline involves changes in the sizes of Canada's regional populations. Since Confederation, population growth has occurred in all regions but at varying rates. The Ontario, Québec, and Atlantic Canada shares of the national population have declined, while those of British Columbia, Western Canada, and the Territorial North have increased. The third faultline has experienced a faster rate of population increase among English-speaking Canadians than French-speaking Canadians; the explanation lies not in fertility rates but in immigration. The fourth faultline—between new Canadians and those born in Canada—is reflected in the federal policy of multiculturalism and is played out daily in Canada's major urban centres, where the vast majority of new Canadians settle.

The following chapters, 5 through 10, focus on the regional nature of Canada, with each chapter devoted to exploring one of Canada's six geographic regions. Friedmann's version of the core/periphery model provides a guide for the ordering of these six regions (Table 1.3). The regional discussion therefore begins with Ontario and Québec, which represent Canada's economic core. These two chapters are followed by chapters on British Columbia, Western Canada, Atlantic Canada, and the Territorial North. While many of the same topics and issues appear in each of the six geographic regions, each is examined from the perspective of the region under discussion. Each regional chapter will illustrate the physical diversity within the region and show how this diversity affects human activities and settlement. Each chapter also has a section on the region's historical development to enrich our sense of the present by providing a link with the past. One recurring theme is the relationship among the six regions. This relationship is cast in the core/periphery model, but it is changing due to continental economic integration (NAFTA) and global trade (the WTO).

Notes

1. The American railway industry has been going through a major restructuring process since the 1970s. Canada joined this process as the flow of goods between Canada and the United States increased in the 1990s. This pressure came from the increased trade stemming from the Canada–US Free Trade Agreement in 1989 and later with the North American Free Trade Agreement in 1994. While the vast majority of the trade is between Canada and the US, trade with Mexico is anticipated to increase in the twenty-first century. After CN purchased Illinois Central it then merged with the Burlington Northern Santa Fe Corporation in 1999. The new CN offers its customers shorter transit times and access to key ports in North America. Canadian exports to the US make up most of CN's traffic. These exports include petroleum and chemicals, forest products, automobiles and automobile parts, grain, coal, minerals, and fertilizer.

   Canadian National Railway had its origins in the amalgamation of five financially failing railways: from 1917 to 1923, the Grand Trunk, the Grand Trunk Pacific, the Intercolonial, the Canadian Northern, and the National Transcontinental were combined to form the publicly operated Canadian National Railway. In 1993, the Canadian government privatized this Crown corporation.
2. Race, unlike ethnicity, is based on physical characteristics. Racial types are frequently assigned a set of social characteristics, which is known as 'stereotyping'. Sociologists define 'race' as the socially constructed classification of persons into categories on the basis of real or imagined physical characteristics such as skin colour. Others consider 'race' a means of creating major divisions of humankind on the basis of distinct physical characteristics.

Challenge Questions

1. Is there an end in sight to the difficulties facing Canada's manufacturing sector?
2. From 2001 to 2006, which two geographic regions have increased their share of Canada's population and which three have seen their share of Canada's population decrease?
3. The former Minister of Citizenship and Immigration, Judy Sgro, stated that 'We need to be loyal to one country as far as your citizenship. Your heart can be where you were born, but I think the commitment to Canada has to be strong and I think dual citizenship weakens that.' Should Canada abolish the option to have dual citizenship for the reasons she suggested?
4. Of the four sectors of the economy, which ones have experienced an increase in labour force percentage, which ones have seen a decline, and which one has not been identified and therefore not been measured separately by Statistics Canada?
5. In recent years, the natural rate of population increase for Aboriginal peoples has been significantly above the national average. Does the demographic transitional theory offer anything to suggest that the rate may slow down?
6. Why does Peter Li's model of newcomers eventually becoming old-timers not apply to Aboriginal peoples?
7. The higher Canadian dollar is good news for the energy industry and bad news for the manufacturing industry. Why?
8. Why does Ottawa encourage immigration? Explain your agreement or disagreement with this policy.

Key terms

age dependency ratio
The ratio of the economically dependent sector of the population to the productive sector, arbitrarily defined as the ratio of the elderly (those 65 years and over) plus the young (those under 15 years) to the population of working age (those 15 to 64 years).

census metropolitan area
An urban area with a population of at least 100,000, together with adjacent smaller urban centres and even rural areas that have a high degree of economic and social integration with the larger urban area.

crude birth rate
The number of births per 1,000 people in a given year.

crude death rate
The number of deaths per 1,000 people in a given year.

culture
The sum of attitudes, habits, knowledge, and values shared by members of a society and passed on to their children.

demographic transition theory
The historical shift of birth and death rates from high to low levels in a population. The decline in mortality precedes the decline in fertility, resulting in a rapid population growth during the transition period.

demography
The scientific study of human populations, including their size, composition, distribution, density, growth, and related socio-economic characteristics.

ecumene
The portion of the land that is settled.

ethnic group
People who have shared awareness of a common identity and who identify themselves with a particular culture.

ethnocentricity
The viewpoint that one's ethnic group is the centre of everything, against which all other groups are judged. Ethnocentricity assumes that one's own group is superior.

fertility rate
The number of births per 1,000 people in a given year; also called crude birth rate (not to be confused with the 'general fertility rate', which is the number of live births per 1,000 women who are of child-bearing age—15 to 44 years—in a given year).

industrial stage
Each major change in the evolution of the capitalist economic system is called a stage or phase. The industrial stage marks the shift from a predominantly agricultural economy to an industrial one.

jihadists
Within Islam, 'jihad' connotes a wide range of meanings, from an inward spiritual struggle to a political struggle to further the Islamic cause. 'Jihad by the sword' refers to a holy war, and those Muslims who pursue acts of terror and guerrilla warfare for a particular political-religious cause are sometimes referred to as jihadists.

mortality rate
The number of deaths per 1,000 people in a given year; also called crude death rate.

natural increase (decrease)
The surplus (or deficit) of births over deaths in a population in a given time period.

net migration
The net effect of immigration and emigration on a country's population in a given period.

old-age dependency ratio
Similar to the age dependency ratio except the old-age dependency ratio focuses only on those over 64.

population density
The total number of people in a geographic area divided by the land area; population per unit of land area.

population distribution
The dispersal of a population within a geographic area.

population growth
The rate at which a population is increasing or decreasing in a given period due to natural increase and net migration; often expressed as a percentage of the original or base population.

population increase
The total population increase resulting from the interaction of births, deaths, and migration in a population in a given period of time.

population strength
Equates population size with economic and political power.

post-industrial stage
The post-industrial phase or stage in capitalism marks the shift from an industrial economy based on manufacturing to an economy in which service industries, particularly high-technology industries, become the dominant economic activities.

pre-industrial stage
Each major change in the evolution of the capitalist economic system is called a stage. The pre-industrial phase or stage identifies an economic system that predates capitalism.

primary sector
Economic sector involving the direct extraction of natural resources such as agriculture, fishing, logging, mining, and trapping.

regional service centres
Urban places where economic functions are provided to residents living within the surrounding area.

resource towns
Urban places where a single economic activity focused on resource extraction (e.g., mining, logging, oil drilling) dominates the local economy; single-industry towns.

secondary sector
Sector of the economy involved in processing and transforming raw materials into finished goods; the manufacturing sector of an economy.

sex ratio
The ratio of males to females in a given population; usually expressed as the number of males for every 100 females.

tertiary/quaternary sector
Economic sector engaged in services such as retailing, wholesaling, education, and financial and professional services; the quaternary sector, for which statistical data are not at present compiled, involves the collection, processing, and manipulation of information.

urban population areas
Communities with economic and social functions that differentiate them from rural places; the common practice of defining urban population is by a specified size that assumes the presence of urban economic and social functions. In 1996, Statistics Canada considered all places with a combination of a population of 1,000 or more and a population density of at least 400 per km^2 to be urban areas. People living in urban areas make up the urban population. People living outside of urban areas are considered rural residents and, by definition, constitute the rural population.

Bibliography

Badets, Jane, and Tina W.L. Chui. 1994. *Canada's Changing Immigrant Population*. Catalogue no. 96–311E. Ottawa: Statistics Canada.

Barnes, Trevor J., John N.H. Britton, William J. Coffey, David W. Edginton, and Glen Norcliffe. 2000. 'Canadian Economic Geography at the Millennium', *Canadian Geographer* 44, 1: 4–24.

Bataïni, Sophie-Hélène, and William J. Coffey. 1998. 'The Location of High Knowledge Content Activities in the Canadian Urban System, 1971–1991', *Cahiers de Géographie du Québec* 42, 115: 7–34.

Beaujot, Roderic. 1991. *Population Change in Canada: The Challenges of Policy Adaptation*. Toronto: McClelland & Stewart.

Bernier, Jacques. 1991. 'Social Cohesion and Conflicts in Quebec', in Guy M. Robinson, ed., *A Social Geography of Canada*. Toronto: Dundurn Press.

Bissoondath, Neil. 1994. *Selling Illusions: The Cult of Multiculturalism*. Toronto: Penguin Books.

Bohn, Glenn. 2006. 'Harper: Diversity Will Protect Us', *National Post*, 20 June, A6.

Bourne, Larry S., and David F. Ley, eds. 1993. *The Changing Social Geography of Canadian Cities*. Montréal and Kingston: McGill-Queen's University Press.

——— and Damaris Rose. 2001. 'The Changing Face of Canada: The Uneven Geographies of Population and Social Change', *Canadian Geographer* 45, 1: 105–19.

Britton, John N.H., ed. 1996. *Canada and the Global Economy: The Geography of Structural and Technological Change*. Montréal and Kingston: McGill-Queen's University Press.

Burgess, Bill. 2000. 'Foreign Direct Investment: Facts and Perceptions about Canada', *Canadian Geographer* 44, 2: 98–113.

Buzzelli, M. 2000. 'Toronto's Postwar Little Italy: Landscape Change and Ethnic Relations', *Canadian Geographer* 44, 3: 298–301.

Cairns, Alan C. 2000. *Citizens Plus: Aboriginal Peoples and the Canadian State*. Vancouver: University of British Columbia Press.

———, John C. Courtney, Peter Mackinnon, Hans J. Michelmann, and David E. Smith, eds. 1999. *Citizenship, Diversity and Pluralism: Canadian and Comparative Perspectives*. Montréal and Kingston: McGill-Queen's University Press.

Cameron, Duncan, and Mel Watkins, eds. 1993. *Canada under Free Trade*. Toronto: James Lorimer.

Cardinal, Harold. 1969. *The Unjust Society*. Edmonton: Hurtig.

Cartwright, Don. 1996. 'The Expansion of French Language Rights in Ontario, 1968–1993: The Uses of Territoriality in a Policy of Gradualism', *Canadian Geographer*, 40, 3: 238–57.

Ceh, S.L. Brian. 1997. 'The Recent Evolution of Canadian Inventive Enterprises', *Professional Geographer* 49, 1: 64–76.

Chartrand, Paul L.A.H., ed. 2003. *Who Are Canada's Aboriginal Peoples?* Saskatoon: Purich.

Chase, Steven, and Brent Jang. 1999. 'CP Chops 1,900 Rail Jobs, Warns of More Carnage', *Globe and Mail*, 22 July, A1.

CIA. 2007. *The World Fact Book: Canada*. Washington, DC. At: <www.cia.gov/library/publications/the-world-factbook/print/ca.html>.

Coffey, W.J. 1994. *The Evolution of Canada's Metropolitan Economies*. Montréal: Institute for Research on Public Policy.

——— and Richard G. Shearmur. 1997. 'The Growth and Location of High Order Services in the Canadian Urban System, 1971–1991', *Professional Geographer* 49, 4: 404–18.

——— and ———. 1998. 'Factors and Correlates of Employment Growth in the Canadian Urban System, 1971–1991', *Growth and Change* 29: 44–66.

Cohen, A. 1990. *A Deal Undone: The Making and Breaking of the Meech Lake Accord*. Vancouver: Douglas & McIntyre.

Denevan, William M. 1992. 'The Pristine Myth: The Landscape of the Americas in 1492', *Annals, Association of American Geographers* 82, 3: 369–85.

Dickason, Olive Patricia. 2002. *Canada's First Nations: A History of Founding Peoples from Earliest Times*, 3rd edn. Toronto: Oxford University Press.

Dominion Bureau of Statistics. 1953. *Population: General Characteristics*. Vol. 1, Ninth Census of Canada. Ottawa: Queen's Printer.

Fife, Robert. 2002. 'Migrants Must Spread Out: Ottawa', *National Post*, 22 June, A1, A9.

Flanagan, Tom. 2000. *First Nations? Second Thoughts*. Montréal and Kingston: McGill-Queen's University Press.

Fleras, Augie, and Jean Leonard Elliott. 1999. *Unequal Relations: An Introduction to Race, Ethnic, and Aboriginal Dynamics in Canada*, 3rd edn. Scarborough, Ont.: Prentice-Hall.

Fong, Eric, and Milena Gulia. 1997. 'The Effects of Group Characteristics and City Contexts on Neighbourhood Qualities among Racial and Ethnic Groups', *Canadian Studies in Population* 24, 1: 45–66.

——— and ———. 2000. 'Neighbourhood Changes within the Canadian Ethnic Mosaic, 1986–1991', *Population Research and Policy Review* 19: 155–77.

——— and Kumiko Shibuya. 2000. 'The Spatial Separation of the Poor in Canadian Cities', *Demography* 37, 4: 449–59.

——— and R. Wikes. 2003. 'Racial and Ethnic Residential Patterns in Canada', *Sociological Forum* 18, 4: 577–602.

Foot, David. 1996. *Boom, Bust and Echo: How to Profit from the Coming Demographic Shift*. Toronto: Macfarlane, Walter & Ross.

Francis, Diane. 2002. 'Tellier Keeps His Eye on Horizon', *National Post*, 24 Aug., FP1, FP4.

Frideres, James S., and René R. Gadacz. 2001. *Aboriginal Peoples in Canada: Contemporary Conflicts*, 6th edn. Toronto: Prentice-Hall.

Gartley, John. 1994. *Focus on Canada: Earnings of Canadians*. Catalogue no. 96–317E. Ottawa: Statistics Canada.

Gilbert, Anne. 1999. *Espaces franco-ontariens, essai*. Ottawa: Le Nordir.

———. 2001. 'Le français au Canada, entre droits et géographie', *Canadian Geographer* 45, 1: 175–9.

——— and Joan Marshall. 1995. 'Local Changes in Linguistic Balance in the Bilingual Zone: Francophones de l'Ontario et Anglophones du Québec', *Canadian Geographer* 39, 3: 194–218.

Globe and Mail. 2007. 'Nortel Cleans House Again, Cuts 2,900 Jobs', 7 Feb. At: <www.theglobeandmail.com/servlet/story/RTGAM.20070207.wnortel0207/BNStory/Business/home>.

Gower, Dave. 1998. 'The Labour Market for Computer Programmers', *The Daily*, Statistics Canada, 10 June. At: <www.statcan.ca /Daily/English/980610/d 980610.htm>.

Halli, S.S., and L. Driedger, eds. 1999. *Immigrant Canada: Demographic, Economic and Social Challenges*. Toronto: University of Toronto Press.

Harrison, Brian, and Louise Marmen. 1994. *Focus on Canada: Languages in Canada*. Catalogue no. 96–313E. Ottawa: Minister of Industry, Science and Technology.

Harvey, David. 1989. *The Urban Experience*. Baltimore: Johns Hopkins University Press.

Haydamack, Brent. 1998. 'Regional Trajectories of Technological Change in Canadian Manufacturing', *Canadian Geographer* 42, 1: 2–13.

Healy, Theresa. 1993. 'Selected Plant Closures and Production Relocations: January 1989–June 1992, Ontario', in Cameron and Watkins (1993: Appendix I).

Herberg, Edward N. 1989. *Ethnic Groups in Canada: Adaptations and Transitions*. Toronto: Nelson.

Hébert, Raymond M. 1998. 'Identity, Cultural Production and the Vitality of Francophone Communities Outside Québec', in Leen d'Haenens, ed., *Images of Canadianness: Visions on Canada's Politics, Culture, Economics*. Ottawa: University of Ottawa Press.

Hewitt, Kenneth. 2000. 'Safe Place or Catastrophic Society? Perspectives on Hazards and Disasters in Canada', *Canadian Geographer* 44, 4: 325–41.

Hiebert, Daniel. 1994. 'Canadian Immigration: Policy, Politics, Geography', *Canadian Geographer* 38, 3: 254–8.

———. 2000. 'Immigration and the Changing Canadian City', *Canadian Geographer* 44, 1: 25–43.

——— and David Ley. 2003. 'Assimilation, Cultural Pluralism and Social Exclusion among Ethnocultural Groups in Vancouver', *Urban Geography* 24, 1: 16–44.

Holmes, John. 1997. 'In Search of Competitive Efficiency: Labour Process Flexibility in Canadian Newsprint Mills', *Canadian Geographer* 41, 1: 7–25.

Hyndman, Jennifer, and Margaret Walton-Roberts. 2000. 'Interrogating Borders: A Transnational Approach to Refugee Research in Vancouver', *Canadian Geographer* 44, 3: 244–58.

Indian Chiefs of Alberta. 1970. *A Presentation by the Indian Chiefs of Alberta to Right Honourable P.E. Trudeau*. Edmonton: Indian Association of Alberta.

Ip, Greg. 1996. 'Jobs Cut Despite Hefty Profits', *Globe and Mail*, 6 Feb., A4.

Jackson, Andrew. 1993. 'Manufacturing', in Cameron and Watkins (1993: ch. 5).

Janigan, Mary. 2004. 'Can It Be Solved', *Maclean's*, 19 Jan., 14.

Jiminez, Marina, and Kim Lunman. 2004. 'Canada's Biggest Cities See Influx of New Immigrants', *Globe and Mail*, 19 Aug.

Kalbach, Madeline A., and Warren E. Kalbach. 1995. 'Ethnic Diversity and Persistence as Factors in Socioeconomic Inequality: A Challenge for the Twenty-first Century', *Toward XXI Century: Emerging Socio-Demographic Trends and Policy Issues in Canada*, 147–60. Proceedings of the 1995 Symposium Organized by the Federation of Canadian Demographers.

Kobayashi, Audrey. 1993. 'Multiculturalism: Representing a Canadian Institution', in J. Duncan and David Ley, eds, *Place/Culture/Representations*. New York: Routledge, 205–31.

———. 1994. *Women, Work, and Place*. Montréal and Kingston: McGill-Queen's University Press.

——— and B. Ray. 2000. 'Civil Risk and Landscapes of Marginality in Canada: A Pluralist Approach to Social Justice', *Canadian Geographer* 44, 4: 401–17.

Ley, David. 1996. *The New Middle Class and the Remaking of the Central City*. Toronto: Oxford University Press.

———. 1999. 'Myths and Meanings of Immigration and the Metropolis', *Canadian Geographer* 43, 1: 2–18.

_____ and Daniel Hiebert. 2001. 'Immigration Policy as Population Policy', *Canadian Geographer* 45, 1: 120–5.

Li, Peter S. 1988. *Ethnic Inequality in a Class Society*. Toronto: Wall and Thompson.

———, ed. 1990. *Race and Ethnic Relations in Canada*. Toronto: Oxford University Press.

———. 1996. *The Making of Post-War Canada*. Toronto: Oxford University Press.

———. 1998. *The Chinese in Canada*. Toronto: Oxford University Press.

Little, Bruce. 1997. 'Are We Better Off under the Liberals?', *Globe and Mail*, 26 Apr., B1, B5.

Long, David, and Olive Patricia Dickason, eds. 2000. *Visions of the Heart: Canadian Aboriginal Issues*. Toronto: Harcourt Canada.

McArthur, Keith. 1999. 'Bata Pioneers Feel the Pain of Plant Closing', *Globe and Mail*, 20 Oct., B1.

McCallum, J. 1995. 'National Borders Matter: Canada–U.S. Regional Trade Patterns', *American Economic Review* 85, 3: 615–23.

Mackenzie, Suzanne, and Glen Norcliffe. 1997. 'Restructuring in the Canadian Newsprint Industry', *Canadian Geographer* 41, 1: 2–6.

McKenna, Barrie. 1998. 'More Firms Flock to Mexico', *Globe and Mail*, 8 July, A1.

MacLachlan, Ian, and Ryo Sawada. 1997. 'Measures of Income Inequality and Social Polarization in Canadian Metropolitan Areas', *Canadian Geographer* 41, 4: 377–97.

McRoberts, Kenneth. 1997. *Misconceiving Canada: The Struggle for National Unity*. Toronto: Oxford University Press.

McVey, Wayne W., and W.E. Kalbach. 1995. *Canadian Population*. Toronto: Nelson Canada.

Martin, Don. 2005. 'The Paycheque Exiles', *National Post*, 6 Sept., A18.

Mathieu, Emily. 2006. 'Manufacturing Remains a Black Mark on Jobs Data', *National Post*, 21 June, WK3.

Monture-Angus, Patricia. 1999. 'Considering Colonialism and Oppression: Aboriginal Women, Justice and the "Theory" of Decolonization', *Native Studies Review* 12, 1: 63–94.

———. 2000. 'Lessons in Decolonization: Aboriginal Overrepresentation in Canadian Criminal Justice', in Long and Dickason (2000: 361–86).

Mooney, James. 1928. *The Aboriginal Population of America North of Mexico*. Smithsonian Miscellaneous Collections. Washington: Smithsonian Institution.

Nash, Alan. 1994. 'Some Recent Developments in Canadian Immigration Policy', *Canadian Geographer* 38, 3: 258–61.

National Post. 1999a. 'Shirt Maker Closes after 44 Years', 1 Feb., C3.

———. 1999b. 'Paragon to Stop Making Diapers at Canadian Plant', 1 May, D2.

———. 2006. 'Cutback Costs Cut into T-shirt Maker's Q4', 8 Dec, FP2

———. 2007. 'Goodyear Slams on Brakes in Quebec', 5 Jan., FP1.

Newhouse, David R. 2000. 'From the Tribal to the Modern: The Development of Modern Aboriginal Societies', in Ron F. Laliberte et al., eds, *Expressions in Canadian Native Studies*. Saskatoon: University of Saskatchewan Extension Press, 395–409.

Nicol, Heather, and Greg Halseth, eds. 2000. *(Re)Development at the Urban Edges*. Publication Series No. 53, Department of Geography. Waterloo, Ont.: University of Waterloo.

Norcliffe, Glen. 1996. 'Foreign Trade in Goods and Services', in Britton (1996: ch. 2).

Norris, Mary Jane. 2000. 'Aboriginal Peoples in Canada: Demographic and Linguistic Perspectives', in Long and Dickason (2000: 167–236).

Olson, R., and Audrey Kobayashi. 1993. 'The Emerging Ethnocultural Mosiac', in Bourne and Ley (1993: 138–52).

Ostry, Bernard. 2005. 'Canada's Ethnic Makeup Is Branching Out in New Directions', *Globe and Mail*, 15 Nov., A25.

Parker, Paul. 1997. 'Canada–Japan Coal Trade: An Alternative Form of the Staple Production Model', *Canadian Geographer* 41, 3: 248–66.

Peters, Evelyn J. 2000a. 'Aboriginal People and Canadian Geography: A Review of the Recent Literature', *Canadian Geographer* 44, 1: 44–55.

———. 2000b. 'Aboriginal People in Urban Areas', in Long and Dickason (2000: 237–70).

Phillips, Melanie. 2006. 'The Country That Hates Itself', *National Post*, 16 June, A13.

Pooler, James A. 2000. *Hierarchical Organization in Society: A Canadian Perspective*. Burlington, Ont.: Ashgate.

Preston, Valerie, and Lucia Lo. 2000. 'Canadian Urban Landscape Examples—21: "Asian Theme" Malls in Suburban Toronto: Land Use Conflict in Richmond Hill', *Canadian Geographer* 44, 2: 182–90.

Randall, Stephen J., and Herman W. Konrad, eds. 1996. *NAFTA in Transition*. Calgary: University of Calgary Press.

Rashid, Abdul. 1994. *Focus on Canada: Family Income in Canada*. Catalogue no. 96–318E. Ottawa: Statistics Canada.

Reid, Scott. 1992. *Canada Remapped: How the Partition of Quebec Will Reshape the Nation*. Vancouver: Arsenal Pulp Press.

———. 1993. *Lament for a Notion: The Life and Death of Canada's Bilingual Dream*. Vancouver: Arsenal Pulp Press.

Reitz, Jeffrey, and Raymond Breton. 1995. *The Illusion of Differences*. Ottawa: C.D. Howe Institute.

Robinson, Allan. 1999. 'Falconbridge to Cut 140 Jobs Despite Better Profit Outlook', *Globe and Mail*, 20 Oct., B3.

Royal Commission on Aboriginal Peoples. 1996. *Report of the Royal Commission on Aboriginal Peoples*, 5 vols. Ottawa: Minister of Supply and Services.

Royal Commission on Bilingualism and Biculturalism. 1970. Report. *Book IV: Cultural Contributions of the Other Ethnic Groups*. Ottawa: Queen's Printer.

Scotiabank. 1998. *Global Economic Outlook*. Toronto: Bank of Nova Scotia.

Sharpe, Bob. 2000. 'Geographies of Criminal Victimization in Canada', *Canadian Geographer* 44, 4: 418–28.

Slocombe, D. Scott. 2000. 'Resources, People and Places: Resource and Environmental Geography in Canada 1996–2000', *Canadian Geographer* 44, 1: 56–66.

Statistics Canada. 1991. *Canada Year Book 1992*. Ottawa: Minister of Industry.

———. 1992a. *Immigration and Citizenship*. Catalogue no. 93–316. Ottawa: Minister of Supply and Services.

———. 1992b. *Age, Sex and Marital Status*. Catalogue no. 93–310. Ottawa: Minister of Industry.

———. 1992c. *Census Divisions and Subdivisions*. Catalogue no. 93–303. Ottawa: Minister of Supply and Services.

———. 1992d. *Census Metropolitan Areas and Census Agglomerations*. Catalogue no. 93–303. Ottawa: Minister of Supply and Services.

———. 1993a. *Ethnic Origin*. Catalogue no. 93–315. Ottawa: Minister of Supply and Services.

———. 1993b. *Mobility and Migration*. Catalogue no. 93–322. Ottawa: Minister of Supply and Services.

———. 1996a. *Canadian Economic Observer: Historical Statistical Supplement 1995/96*. Catalogue no. 11–210. Ottawa: Minister of Supply and Services.

———. 1996b. *Canada Year Book 1997*. Ottawa: Minister of Industry.

———. 1997a. '1996 Census: Immigration and Citizenship', *The Daily*, 4 Nov. At: <www.statcan.ca/Daily/English/>.

———. 1997b. '1996 Census: Mother Tongue, Home Language and Knowledge of Languages', *The Daily*, 2 Dec. At: <www.statcan.ca/Daily/English/>.

———. 1997c. *A National Overview: Population and Dwelling Counts*. Catalogue no. 93–357–XPB. Ottawa: Minister of Industry.

———. 1997d. *Mortality—Summary List of Causes, 1995*. Catalogue no. 84–209–XPB. Ottawa: Minister of Industry.

———. 1997e. Selected Income Statistics for Individuals, Families and Households, 1991 and 1996 Censuses. Ottawa. At: <www.statcan.ca/english/Pgdb/People/Families/famil61c.htm>.

———. 1998a. Labour Force, Employment and Unemployment. Ottawa. At: <www.statcan.ca/english/Pgdb/Economy/Economic/econ10.htm>.

———. 1998b. Recent Immigrants by Country of Last Residence. Ottawa. At: <www.statcan.ca/english/Pgdb/People/Population/demo08.htm>.

———. 1998c. Single and Multiple Ethnic Origin Responses, 1996 Census. Ottawa. At: <www.statcan.ca/english/Pgdb/People/Population/demo28a.htm>.

———. 1998d. '1996 Census: Education, Mobility and Migration', *The Daily*, 14 Apr. At: <www.statcan.ca/Daily/English/>.

———. 1998e. '1996 Census: Ethnic Origin, Visible Minorities', *The Daily*, 17 Feb. At: <http://www.statcan.ca/Daily/English/>.

———. 1998f. 1981–1996 Census: Labour Force Activity, 1998. Ottawa. At: <www.statcan.ca/english/census96/mar17/labour/table6 t6p59a.htm>.

———. 2000. *Canadian Economic Observer: Historical Statistical Supplement 1999/00*. Catalogue no. 11–210–XPB. Ottawa: Minister of Industry.

———. 2002a. Census of Canada 2001—Census Geography. Highlights and Analysis: Canada's 2001 Population. Ottawa. At: <www12.statcan.ca/English/census01/>.

———. 2002b. *Census of Canada 2001: Age and Sex for Canada, Provinces and Territories 2001*. Ottawa. At: <www12.statcan.ca/English/census01/>.

———. 2002c. Census of Canada 2001: Population by Sex for Canada, Provinces and Territories, 2001. Ottawa. At: <www12.statcan.ca/English/census01/>.

———. 2002d. 'Census of Population: Language, Mobility and Migration', *The Daily*, 10 Dec. At: <www.statcan.ca/Daily/English/02120/d021210a.htm>.

———. 2003a. 'Canadian International Merchandise Trade', *The Daily*, 20 Feb. At: <www.statcan.ca/Daily/English/030220/d030220a.htm>.

———. 2003b. 'Financial Statistics for Enterprises', *The Daily*, 27 Feb. At: <www.statcan.ca:80/Daily/English/030227/d030227b.htm>.

———. 2003c. *Census of Canada 2001—Canada's Ethnocultural Portrait: The Changing Mosaic*. Analytical Series 96F0030XIE2001008. Ottawa, 21 Jan. At: <www12.statcan.ca/english/census01/products/analytic/companion/etoimm/canada.cfm>.

———. 2003d. Components of Population Growth, 28 Feb. Ottawa. At: <www.statcan.ca/English/Pgdb/demo33a.htm>.

———. 2003e. Labour Force, Employed and Unemployed, Numbers and Rates. Ottawa. At: <www.statcan.ca/english/Pgdb/labor07a.htm>.

———. 2003f. 'Overcoming Distance, Overcoming Borders: Comparing North American Regional Trade', *The Daily*, 16 Apr. At: <www.statcan.ca/Daily/English/030416/d030446h.htm>.

———. 2003g. 'Census of Population: Earnings, Levels of Schooling, Field of School Attendance', *The Daily*, 12 Mar. At: <www.statcan.ca/Daily/English/030311/d030311a.htm>.

———. 2003h. *Census of Canada 2001—Aboriginal Peoples of Canada: A Demographic Profile*. Analysis series 96F0030XIE2001007. Ottawa, 21 Jan. At: <www12.statcan.ca/english/census01/products/analytic/companion/abor/contents.cfm>.

———. 2003i. 'Labour Force Survey', *The Daily*, 12 Apr. At: <www.statcan.ca/Daily/English/030404/d030404a.htm>.

———. 2003j. *Census of Canada 2001—Religions in Canada*. Analysis series 96F0030XIE2001015. Ottawa. At: <www12.statcan.ca/english/census01/products/analytic/companion/rel/contents.cfm>.

———. 2003k. Learning Resources: Civics and Society—Declining Birth Rate and the Increasing Impact of Immigration. Ottawa. At: <www.statcan.ca/english/kits/issues/charts/chart3.htm>.

———. 2003l. *Profile of the Canadian Population by Mobility Status: Canada, a Nation on the Move*. Ottawa. At: <www12.statcan.ca/english/census01/products/analytic/companion/mob/contents.cfm>.

———. 2004a. 'Exports of Goods on a Balance-of-Payments Basis', *Canadian Statistics, International Trade*, 22 Jan. At: <www.statcan.ca/english/Pgdb/gblec04.htm>.

———. 2004b. 'Study: Immigrants in Canada's Urban Centres', *The Daily*, 18 Aug. At: <www.statcan.ca/Daily/English/040818/d0818b.htm>.

———. 2006a. Imports, Exports and Trade Balance of Goods on a Balance-of-Payments Basis, by Country or Country Grouping. Ottawa. At: <www40.statcan.ca/l01/cst01/gblec02a.htm?sdi=exports%20country>.

———. 2006b. Components of Population Growth, by Province and Territory, July 1, 2005 to June 30, 2006. Ottawa. At: <www40.statcan.ca/l01/cst01/demo33c.htm>.

———. 2006c. *Annual Demographic Statistics 2005*. Catalogue no. 91–213–X1B. Otawa. At: <www.statcan.ca/english/freepub/91-213-XIB/0000591-213-XIB.pdf>.

———. 2006d. 'Labour Force Survey', *The Daily*, 10 Feb. At: <www.statcan.ca/Daily/English/060210/d060210a.htm>.

———. 2006e. Exports of Goods on a Balance-of-Payments Basis, by Product. Ottawa. At: <www40.statcan.ca/l01/cst01/gblec04.htm>.

———. 2006f. Labour Force, Employed and Unemployed, Numbers and Rates, by Province. Ottawa. At: <www40.statcan.ca/l01/cst01/labor07c.htm>.

———. 2006g. 'Canada's Population by Age and Sex', *The Daily*, 26 Oct. At: <www.statcan.ca/Daily/English/061026/d061026b.htm>.

———. 2007a. Population by Year, by Province and Territory. At: <www40.statcan.ca/l01/cst01/demo02a.htm?sdi=population%20change%20provinces>.

———. 2007b. Community Profiles. At: <www12.statcan.ca/english/census06/data/profiles/community/Search/SearchForm_Results.cfm?Lang=E>.

———. 2007c. Portrait of the Canadian Population in 2006: Subprovincial Population Dynamics: Canada's Population Becoming More Urban. At: <www12.statcan.ca/english/census06/analysis/popdwell/Subprov1.cfm>.

———. 2007d. Portrait of the Canadian Population in 2006: Subprovincial Population Dynamics: Vast Majority of Canada's Population Growth Is Concentrated in Large Metropolitan Areas. At: <www12.statcan.ca/english/census06/analysis/popdwell/Subprov3.cfm>.

———. 2007e. 2006 Census Dictionary. At: <www12.statcan.ca/english/census06/reference/dictionary/index.cfm>.

———. 2007f. Dwelling Counts, for Canada, Provinces and Territories, and Census Subdivisions (Municipalities), 2006 and 2001 Censuses—100% Data. At: <www12.statcan.ca/english/census06/data/popdwell/Filter.cfm?T=308&S=1&O=A>.

———. 2007g. Population and Growth Components (1851–2001 Censuses). At: <www40.statcan.ca/l01/cst01/demo03.htm>.

———. 2007h. Births and Birth Rate, by Province and Territory. At: <www40.statcan.ca/l01/cst01/demo04b.htm>.

———. 2007i. Deaths and Death Rate, by Province and Territory. At: <www40.statcan.ca/l01/cst01/demo07b.htm>.

———. 2007j. 'Annual Survey of Software Development and Computer Services', *The Daily*, 8 Feb. At: <www.statcan.ca/Daily/English/070208/d070208c.htm>.

———. 2007k. Labour Force, Employed and Unemployed, Numbers and Rates. At: <www40.statcan.ca/l01/cst01/labor07c.htm>.

St Germain, the Honourable Gerry, and the Honourable Nick Sibbeston. 2006. *Negotiations or Confrontation: It's Canada's Choice*. Senate Report of the Standing Committee on Aboriginal Peoples. At: <www.parl.gc.ca/39/1/parlbus/commbus/senate/com-e/abor-e/rep-e/rep05dec06-e.pdf>.

Steinhart, David. 1999. 'Levi to Shut Plants in Cornwall, U.S.: Brantford Loses Jobs Too', *National Post*, 23 Feb., C1.

Thériault, J. Yvon, ed. 1999. *Francophonies minoritaires au Canada: état des lieux*. Moncton: Éditions d'Acadie.

Thorpe, Jacqueline. 2003. 'Red Tide Stops at 49th Parallel', *National Post*, 27 Feb., FP1, FP4.

Trewartha, G.T., A.H. Robinson, and E.H. Hammond. 1967. *Elements of Geography*, 5th edn. New York: McGraw-Hill.

United Nations. 1994. *Human Development Report 1994*. New York: Oxford University Press.

Walks, R. Alan, and Larry S. Bourne. 2006. 'Ghettos in Canada's Cities? Racial Segregation, Ethnic Enclaves and Poverty Concentration in Canadian Urban Areas', *Canadian Geographer* 50, 3: 273–97.

Woods, Alan. 2006. 'Dual Citizenship Faces Review', *National Post*, 21 Sept. At: <www.canada.com/nationalpost/news/story.html?id=fb2d75ab-8880-4945-8537-1508186a4964&k=61921>.

Young, R.A. 1989. 'Political Scientists, Economists, and the Canada–US Free Trade Agreement', *Canadian Public Policy* 15, 1: 49–56.

Further Reading

Bunting, Trudi, and Pierre Filion, eds. 2006. *Canadian Cities in Transition: Local Through Global Perspectives*, 3rd edn. Toronto: Oxford University Press.

Canada is a highly urban society. Its cities are complex, dynamic, and open to change. In *Canadian Cities in Transition*, Professors Bunting and Filion have brought together Canada's leading urban geographers to discuss why and how Canadian cities work or don't work. The underlying theme is change and the range of topics is broad.

Cities, as a reflection of society, are constantly changing in appearance, function, and problems. Cities are also a reflection of national and global forces, such as economic restructuring, telecommunications advances, and international migration. The book has eight main sections: from mercantilism to globalism, the structure of urban development, zonal geography of Canadian cities, city functions (employment, housing, and retailing), governance and planning, the urban natural environment, pressing issues, and future transitions. Students might be attracted to the comparison of Canadian and American cities found in Chapter 2. Kim England and John Mercer challenge the idea of a 'North American city', arguing that Canadian cities are unique. Another interesting—but distressing—chapter deals with a social problem found in large cities—urban homelessness. Since urban poverty is not discussed in *The Regional Geography of Canada*, reading Chapter 25 will expose the student to this persistent problem. R. Alan Walks examines what has become a permanent feature of the urban landscape in Canada and explains how destitute urban dwellers often are casualties of our market economy. He considers the reasons for this urban phenomenon and explores several solutions, including increased federal spending on new rental housing.

The section titled 'The Urban Natural Environment: The Grassroots of Sustainability' is new to this third edition of *Canadian Cities in Transition*. As the title implies, urban geographers embrace the environmental imperative to 'think globally, act locally'. Of special interest here is Stephen D. Murphy's 'Why Micro-Scale Urban Ecology Matters' (Chapter 22).

Parliament Hill as seen from Gatineau, Québec (Ronald F. Boisvert)

Overview and Objectives

Among Canada's six geographic regions, Ontario ranks first in economic output and population size. In 2006, Ontario accounted for 39 per cent of Canada's industrial output and population (Figure 1.1). As the economic powerhouse of Canada, Ontario produces most of Canada's manufacturing goods. As well, the largest population cluster in Canada extends from the US border at Niagara Falls northward to Hamilton, Toronto, and then on to Oshawa. Toronto, the largest city in Canada, serves as the nation's financial headquarters and is becoming an international site for opera and live theatre. In total, these cultural, economic, and demographic factors define Ontario as Canada's 'core' region. Yet, Ontario consists of two distinct economies—a robust economic core and growing population in the south and a struggling resource extraction economy and declining population in the north. Much of Ontario's economic growth can be attributed to its automobile industry and its integration into the North American market, so that today the region's primary trade axis has been reoriented from east–west to north–south. This chapter will:

- Describe Ontario's physical geography and historical background.
- Present the basic characteristics of Ontario's population and resources.
- Discuss the economic and environmental challenges facing Ontario.
- Identify Ontario's dichotomy—an industrial core in the south and a resource hinterland in the north.
- Examine the importance of the manufacturing sector, especially the automobile industry, to Ontario's economy.
- Analyze the significance to Ontario of several important trade agreements.
- Focus on Ontario's changing role within Canada and North America.

CHAPTER 5

Ontario

INTRODUCTION

Indications of Ontario's prominent position within Canada are its dominant position in manufacturing, its leadership role in financial matters, and its emergence as the cultural centre for English-speaking Canada. The auto industry in southern Ontario is the major secondary industry in Canada, while Toronto serves as the headquarters of most corporations. In 2005, automakers and auto parts manufacturers accounted for nearly 20 per cent of Canada's exports, and in the same year 37 per cent of Canadian head offices were located in Toronto (Statistics Canada, 2006b; Brown and Beckstead, 2006).

Ontario is a geographic paradox, however, because southern Ontario and northern Ontario are so dissimilar, with very different physical conditions that have formed the basis of two distinct economies. The upbeat description of Ontario as the 'Province of Opportunity' applies to only one part, southern Ontario. Southern Ontario's economy has generated jobs and business opportunities. Since the Canada–US Free Trade Agreement (FTA) in 1989, Ontario's manufacturing industries have been drawn more and more into the North American economy. As well, its positive economic environment has attracted people from across Canada and around the world to its cities. The manufacture of automobiles has been a significant part of southern Ontario's economy and constitutes the Key Topic in this chapter. While southern Ontario is the industrial and population heartland of the province, northern Ontario is an old resource hinterland, where economic and population growth is stalled. Because so many young people are moving to more prosperous areas in Canada, the population of northern Ontario is losing a valuable segment of its labour force, leaving its cities and towns with disproportionately large numbers of older and very young people. In 2006, northern Ontario had a population of just under 800,000, while southern Ontario had a population of nearly 12 million.

Ontario within Canada

Ontario, by contributing 39 per cent of Canada's gross domestic product (GDP), remains the foremost economic region in the nation (Figure 5.1). Ontario's strength comes from its central location within Canada where most of the nation's population is concentrated. With this numerical strength dating to Confederation, plus ample support from Ottawa, Ontario established itself as the leading manufacturing area in Canada. Its products supplied the nation, thus reinforcing an east–west transportation axis. With the FTA, more and more goods have gone to the United States. In this way, a north–south trade axis has emerged to supplement the traditional east–west axis. In fact, according to Thomas Courchene, one of Canada's leading economists, Ontario has already evolved into one of North America's key economic regions and may become the new 'heartland' of North America (Courchene, 1998).

Four natural resources—agriculture, forests, minerals, and water—spurred the province's past economic development, and the processing of these products created a strong industrial base. The Great Lakes provide low-cost water transportation and, at Niagara Falls, low-cost hydroelectric power. But today Ontario faces two challenges. One challenge is energy. Not only does Ontario require more energy, but the cost of energy has risen rapidly. A second challenge confronts the manufacturing sector, which is coping with fierce competition from countries with much lower labour costs and the increasing value of the Canadian dollar. The high dollar value, which by September 2007 had climbed to 98 cents on the US dollar, the highest level in more than 30 years, makes manufacturing exports to the chief market, the United States, more expensive.

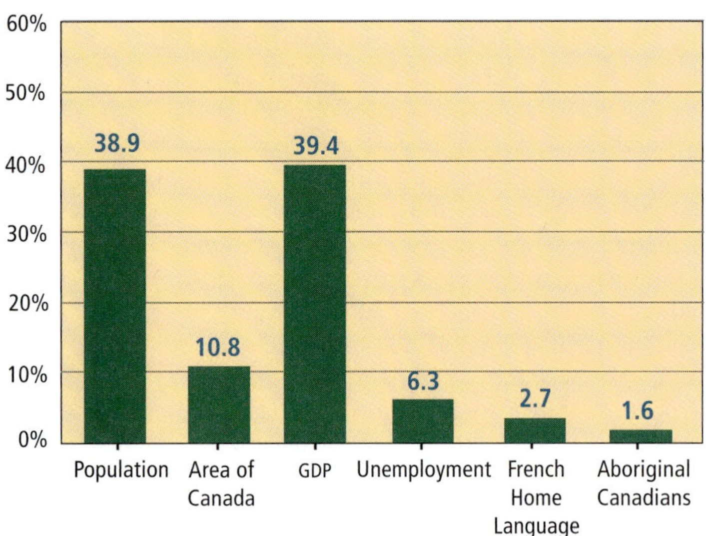

Figure 5.1 Ontario, 2006. Ontario's economic and political strength within Canada is revealed by its share of the nation's GDP and population. With 12.2 million people, Ontario comprises the largest single market in Canada and holds the demographic key to political power in Ottawa, and its economy, by accounting for more than 39 per cent of national GDP, exhibits a high level of productivity. Franco-Ontarians and Aboriginal peoples form small minorities. Ontario's unemployment rate of 6.3 per cent in 2006 is just slightly above the national average of 6.1 per cent but well above the much lower rates found in the western provinces, with Alberta's rate at 3.4 per cent (Table 4.18).
Sources: See Tables 1.1, 1.7, and 4.18.

Over the past five years, shifts have occurred in Ontario's manufacturing sector. Manufacturers in Ontario have benefited from the resource boom in Western Canada and British Columbia, which has created a demand for certain manufactured goods. From 2001 to 2006, the greatest gains in Ontario's manufacturing sector occurred in machinery needed for construction and mining projects in the West. Demand for other products, such as office furniture, was also strong. During the same period, other sectors of the manufacturing industry did not fare so well—automobile manufacturing has not grown while the demand for forest products has declined and textile production has continued to drop.

Ontario's position within Canada goes beyond its economic role. Ontario remains the political linchpin in Confederation and has become the cultural sparkplug for English-speaking Canada. Its cultural and political role is attributed largely to Ontario's large and affluent population. With 39 per cent of Canada's population, Ontario sends more representatives to the House of Commons than any other province and it has the capacity to support the arts, including the Canadian Opera Company, and professional sports teams not found elsewhere in Canada—the Toronto Blue Jays (baseball), the Toronto Raptors (basketball), and Toronto FC (soccer).

Ontario's political power in Ottawa often draws the federal government to seek solutions, whether to ease access across the US border or to provide financial support for Ontario's automobile industry. Since Ontario's economic future lies in expanding trade with the United States, Ontario is interested in having Ottawa pursue trade policies with the United States that support its industrial base,

The Four Seasons Centre in Toronto is the first building of its kind in Canada built specifically for opera and ballet. In August 2006, the Centre opened with a performance of Wagner's monumental Ring cycle (*Der Ring des Nibelungen*) by the Canadian Opera Company, which attracted opera and Wagner enthusiasts from around the world. *Source:* Website for the Canadian Opera Company: <www.coc.ca/house/house.html>. Diamond and Schmitt Architects (Photo by Tim Griffith)

facilitate ease of border crossings for its goods, and promote the well-being of its manufacturing industry, especially the automotive sector. Both the federal and provincial governments have responded. Their investments demonstrate a strong commitment to manufacturing, especially the automobile industry. For example, the federal and provincial governments have made substantial contributions to the Ontario Automotive Investment Strategy, thus ensuring a partnership with automobile companies to strengthen the industry's competitiveness in the global marketplace.

Canadians living in less prosperous and less powerful regions often see Ontario in a different light, perhaps precisely because of Ontario's economic success, dominant cultural role, and political power. Westerners and Maritimers have accused Ontario of using its political power to maintain its economic mastery over the rest of the country. Québecers have pointed to the striking difference between the two provinces' unemployment rates as proof that Ottawa favours Ontario. Ontarians reject these criticisms, claiming that Ontario has done its share to shoulder the 'burden of Confederation'. For example, over the years Ontario has been the major contributor to equalization payments allocated to have-not provinces. In recent years, Ontario has complained about contributing more tax dollars to Ottawa than it receives in federal transfer payments. Indeed, on the issue of fiscal imbalance, the Ontario government strongly opposed the Atlantic Accord, a side deal on equalization arranged by the Martin Liberal government in 2005 with Nova Scotia and Newfoundland and Labrador whereby the oil and gas revenues of these two provinces are not taken into account in determining equalization payments (for more discussion on this subject, see Vignette 9.5 in Chapter 9).

Ontario's Physical Geography

Ontario is larger than most countries (Figure 5.2), stretching out over 1 million km^2.

Extending over Ontario are three of Canada's physiographic regions (Great Lakes–St Lawrence Lowlands, Canadian Shield, and Hudson Bay Lowland), and three of the country's climatic zones (Arctic, Subarctic, and Great Lakes–St Lawrence) (Figures 2.1 and 2.6). Manitoba lies to its west, Hudson and James bays to its north, while Québec is on its eastern boundary, marked in part by the Ottawa River. This central location within Canada and its close proximity to the industrial heartland of the United States have facilitated Ontario's economic development.

Southern Ontario is located in the southernmost part of Canada. Windsor, for example, is at latitude 42° N and Toronto is close to 44° N. Southern Ontario lies in the Great Lakes–St Lawrence climatic zone, which has a moderate continental climate. This climate is noted for long, hot, and humid summers followed by short, cold winters. Annual precipitation is about 1,000 mm. The greatest amounts of precipitation occur in the lee of the Great Lakes, where winter snowfall is particularly heavy (see Vignette 5.1). The Great Lakes modify temperatures and funnel winter storms into this

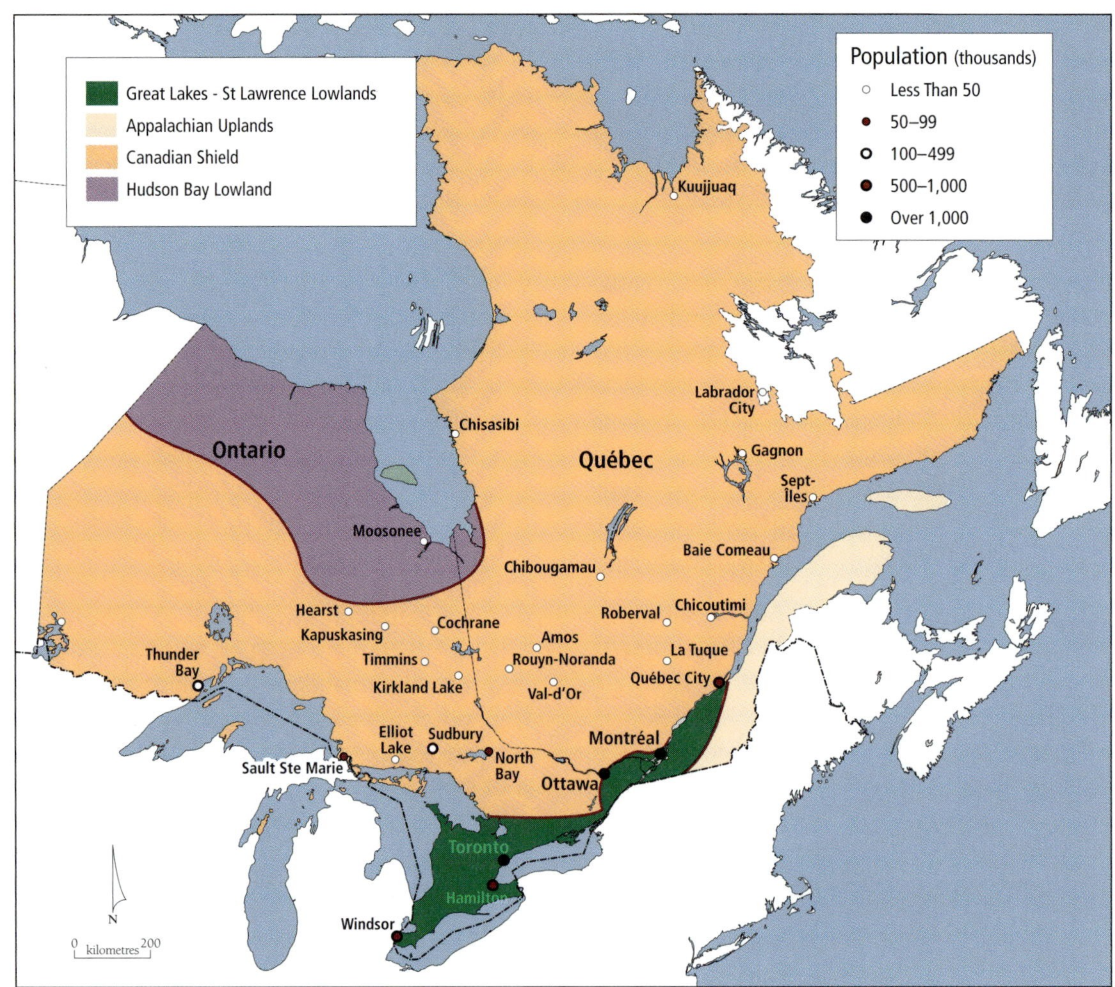

Figure 5.2 Central Canada. Ontario and Québec have the two largest provincial economies and populations in Canada. Toronto and Montréal, the two largest cities in Canada, are located in Canada's manufacturing belt, which stretches from Windsor to Québec City. Canada's most productive agricultural lands also are located in this manufacturing belt. (Further resources: Student website, National Atlas section, Map 17. Website instructions are found on p. xx.)

region. During the winter, this region experiences a great variety of weather conditions.

Southern Ontario is the most favoured physical area in Canada. Not surprisingly, the vast majority of the province's population, nearly 12 million or 93 per cent, live in southern Ontario. The area is underlain by slightly tilted sedimentary rocks, which are covered by a thin deposit of glacial till. Except for the Niagara Escarpment, there is little relief topography. Formed since deglaciation, the Niagara Escarpment is an erosional remnant of the more resistant sedimentary rocks that extends from the Bruce Peninsula to Niagara Falls. This escarpment provides the most spectacular scenery in southern Ontario and a nature trail known as the Bruce Trail extends some 725 kilometres along its limestone escarpment. Prior to settlement, a mixed forest vegetation flourished in this temperate continental climate (Figure 2.7). With a long growing season, ample precipitation, and fertile soils, the southern Ontario lowland has the most productive agricultural lands in Canada. The Canadian Shield, however, marks an abrupt end to agricultural land. At this point, northern Ontario begins.

Northern Ontario has fewer natural assets than southern Ontario. With a much weaker economy, only 7 per cent of Ontarians live in this part of the province. Northern Ontario lies in two physiographic regions: the Canadian Shield and the Hudson Bay Lowland. Located in much higher latitudes, its winters are longer and colder than those experienced in southern Ontario. For the most part, northern Ontario lies north of 46° N and extends almost as far north as 57° N. Even along its southern edge, short summers make crop agriculture vulnerable to frost damage. In addition to a difficult climate for agriculture, the rocky Canadian Shield has very little agricultural land. Its rugged, hilly topography is dotted with lakes. Climate, soils, and physiography combine to limit agriculture in northern Ontario. Unlike Québec, the Ontario section of the Canadian Shield has relatively low elevations, thus limiting the opportunities for hydroelectric power developments. However, this subregion does have vast forests, superb scenery (which supports tourism), and extensive mineral wealth. As a resource hinterland, northern Ontario's economy is almost entirely dependent on these three resources.

Vignette 5.1 Ontario's Snowbelts

Ontario's snowbelts are legendary. On the upland slopes facing Lakes Huron and Superior and Georgian Bay, huge snowfalls in the 300 to 400 cm range fall each winter from November to late March. The uplands on the northeastern shores of Lake Superior receive the greatest total snowfall amounts of any area in Ontario (exceeding 400 cm annually). Much of the snowfall in snowbelt areas can be attributed to cold northwesterly to westerly winds blowing off the lakes and ascending the highlands. As the arctic air travels across the relatively warmer Great Lakes, it is warmed and moistened. Snow clouds form over the lakes and, once onshore, intensify as the air is then forced to ascend the hills to the lee of the lakes, triggering heavy snowfalls. Areas on the downslope side of the higher ground to the lee of the lakes receive less than half the annual snow totals of the upslope snowbelt areas. For example, Toronto, Hamilton, and other places to the lee of the Niagara Escarpment are snow-shadow regions with winter amounts of 100 to 140 cm. In the snowbelt regions, snowfall accounts for about 32 per cent of the year's total precipitation; but in the snow-sparse area of southwestern Ontario around Windsor and Chatham, the snow contribution is only about 13 per cent.

Because Ontario has the largest population of all the provinces, it has the greatest number of seats in the House of Commons—and therefore is seen as holding the balance of power in Canada. After the 2006 general election, Ontario held 34 per cent of the seats in the House of Commons (106 out of 308 seats). (© Perry Mastrovito/Corbis)

The Niagara Escarpment, stretching over 700 km from Queenston on the Niagara River to Tobermory on the Bruce Peninsula, is the most significant landform in southern Ontario. The Escarpment reaches over 350 metres above the surrounding land at various points. (Victor Last, Geographical Visual Aids)

Environmental Challenges

Like other industrial regions, Ontario is faced with two major environmental challenges—air pollution and water pollution. Without more robust action by governments, these two problems will only get worse, causing serious harm to the environment and to the health of its citizens. Both types of pollution represent the hidden costs of our industrial world. Cities are confronted with another environmental challenge: how to dispose of their garbage. Toronto's solution has been to export its garbage to a dump site in Michigan (Vignette 5.2)

The densely populated area of the Golden Horseshoe—which includes Toronto and Hamilton—has significant health problems caused by smog. Smog becomes a serious health hazard in the summer months when high temperatures and air inversions combine to keep the smog hanging over the city as a brown haze. The word 'smog' is actually a combination of 'smoke' and 'fog'. Smog is the most visible form of air pollution. It is a brownish-yellow hazy cloud caused when heat and sunlight react with various pollutants emitted by industry, vehicle exhaust, pesticides, and oil-based home products. A study by the Ontario Medical Association estimated that illnesses caused by smog cost Ontario more than $1 billion a year in hospital admissions, emergency room visits, and absenteeism. Even more startling, the OMA study (2005) concluded that about 5,800 people die prematurely in Ontario each year because of smog-related respiratory problems.

In 2003, the Ontario government recognized the need to deal with the smog problem. Most air pollution comes from automobile exhaust and coal-burning thermoelectric plants. In terms of air pollution, Ontario Power Generation ranks first (Table 5.1) because of its coal-burning generating plants near Toronto. The provincial government proposed to close its five coal-burning thermoelectric generating stations. The coal plants contribute 17 per cent of Ontario's electrical consumption while other energy sources do

Vignette 5.2 Toronto's Garbage: An Unwanted Export to Michigan

As North American cities grow, the problem of disposing of their urban waste becomes more complicated and expensive. After Toronto filled its local dump sites in 2000, the city arranged to ship its waste across the border to a site in Michigan. Toronto's waste disposal contract with Michigan is in effect until 2008; however, many Americans are seeking options to end the contract sooner. The process of shipping Toronto's waste to another country has resulted in many problems and complaints from Michigan residents who resent the dumping of foreign garbage in the United States.

not contribute to smog (nuclear power provides 49 per cent of Ontario's electricity, hydro 25 per cent, natural gas 7 per cent, and other, including wind and solar, 2 per cent). In September 2006, Ontario Power Generation submitted a request to Ottawa for permission to build a new nuclear reactor at Darlington.

In September 2006, the federal government announced its Green Plan 2, which emphasizes improving air quality and reducing smog. Vehicle emissions are one of the principal sources of smog. For that reason, the federal government has targeted automobile manufacturers, who must, by 2010, greatly reduce gasoline emissions to the 'California' level. California's cities are greatly affected by smog, causing the state to impose the strictest regulations on vehicle emissions in North America.

The Lakeview generating station near Toronto has closed while the plants at Atikokan and Thunder Bay were to cease operations in 2007.

With demand for electricity increasing each year, Ontario realized that closure of all its coal plants would result in energy shortages and the possibility of widespread brownouts. For that reason, the closure of the remaining

In 2007, the Ontario government planned to replace its Thunder Bay coal-burning generating station (left) with a gas-fired unit and to close its Atikokan plant (right). The Thunder Bay station has burned low-sulphur lignite coal from Western Canada and low-sulphur sub-bituminous coal from the United States. (Courtesy of Ontario Power Generation)

Table 5.1 Greenhouse Gas Emissions, 2004

Rank	Company	Million Tonnes CO_2	Province
1	Ontario Power Generation	24.9	Ontario
2	Transalta Utilities Corporation	22.7	Alberta
3	Saskatchewan Power Corporation	13.7	Saskatchewan
4	Alberta Power Ltd	12.0	Alberta
5	Nova Scotia Power Inc.	10.6	Nova Scotia
6	Syncrude Canada	10.4	Alberta
7	Suncore Energy Inc.	8.6	Alberta
8	EPCOR Generation Inc.	6.9	Alberta
9	Petro-Canada	5.7	Alberta
10	Dofasco Inc.	4.9	Ontario
Total top 10 companies		120.2	
Total Canadian industrial emissions		278.9	
Total, all emissions		758.0	

Source: Environment Canada (2006). Reprinted with permission of the Minister of Public Works and Government Services Canada, 2007.

two coal-fired stations, the Nanticoke plant on the north shore of Lake Erie and the Lambton plant near of Sarnia, will be delayed until clean-air energy alternatives are in place by 2010. Ironically, these two stations—located in the heart of southern Ontario—produce most of the smog-causing pollutants that affect the health of residents of southern Ontario. Yet, the province does not want to cease electric production from its remaining coal plants until clean-air energy alternatives are in place—and their construction will take time. Here are some examples:

- Two natural-gas generating stations that will replace the electrical output from the Lambton plant are not expected to be operative until 2009.
- More hydroelectricity is coming by 2010 with the anticipated completion of the Niagara Tunnel Project.
- In late April 2007 the Ontario government gave approval to the plan of a California-based company to construct a 365-hectare, 40-megawatt solar farm near Sarnia that will be the largest solar source of electricity in North America. Yet, on a sunny day this project, once in operation, will provide enough power for only 10,000–15,000 homes (Hamilton, 2007).
- Even more time is required for increasing the capacity of existing nuclear generating plants.

Water pollution constitutes a serious problem across Ontario. Industrial effluents, farm chemicals, and livestock waste are the main culprits. At the local level, in May 2000, the contamination of drinking water with E. coli at Walkerton provided a deadly example with seven deaths and many illnesses. At the regional level, the integrity of the Great Lakes has been compromised by human activities. Over 35 million people, many of whom live in large urban cities such as Chicago, Detroit, Toronto, and Montréal, depend on the Great Lakes/St Lawrence for water supply and sewage disposal. The area surrounding Lake Michigan, Lake Erie, and Lake Ontario contains many industries and factories that, in the past, dumped their toxic wastes into the Great Lakes and buried their chemical wastes along the shore. Later, these chemical wastes began to leak into the lakes. Measures were taken to regulate these polluters, but pollution from the runoff from agricultural lands, the waste from cities, and toxic discharges from industry continue to affect the Great Lakes. The Great Lakes are also an integral part of the St. Lawrence Seaway. Ocean-going ships have carried unwanted passengers such as zebra mussels and sea lampreys, which have had a negative impact on domestic fisheries.

Canada and the United States have recognized the need to regulate the discharge of toxic wastes into the Great Lakes. The Great Lakes Water Quality Agreement, first signed in 1972 and renewed in 1978, expresses the commitment of Canada and the United States to restore and maintain the chemical, physical, and biological integrity of the Great Lakes Basin (see Figure 5.3). By the beginning of the 1990s, cleaner water and increasing fish populations indicated that these joint efforts were making solid progress. However, between 1998 and 2002, industrial discharges of pollutants increased by 21 per cent. Growing levels of phosphorus, primarily from waste water containing detergents and from agricultural runoff containing high levels of chemical fertilizers, are contributing to the creation of a 'green slime', which is a dead zone where only toxic organisms can survive. In addition, the careless introduction of exotic species is resulting in sea lampreys, Asian carp, and zebra mussels squeezing out native species and thereby radically changing Great Lakes ecosystems.

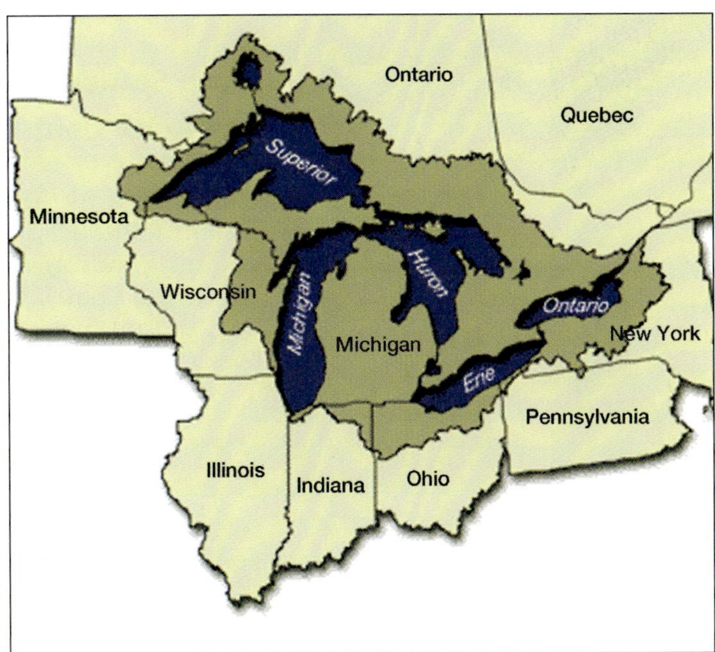

Figure 5.3 The Great Lakes. The darker area represents the drainage basin of the Great Lakes.
Source: Great Lakes Information Network, at: <www.great-lakes.net/lakes/>, Great Lakes Commission, Ann Arbor, Michigan. © GLIN

Ontario's Historical Geography

Since its beginning as a British colony, Ontario has been the counterweight to Québec. When the American Revolution began in 1775, Ontario—except for the French settlement around Detroit—was a densely forested wilderness inhabited by a few fur traders and Indians (Vignette 5.3).

After losing the American colonies, loyal British subjects returned to England or relocated in one of Britain's colonies. In 1782–3, Loyalists moved northward to Nova Scotia and New Brunswick, while others resettled in the Eastern Townships of Québec. A smaller number, perhaps as many as 10,000 Loyalists and members of the Six Nations who fought alongside British troops, settled on land along the St Lawrence River upstream from French-speaking habitants, colonizing the wilderness area that later became known as Upper Canada. In 1784, Britain awarded its Indian military allies, the Six Nations, with the 'Haldimand Tract' (named after the commander of the British forces). The Haldimand Tract consisted of a strip of land on each side of the Grand River totalling 385,000 hectares (Darling, 2007). These Loyalists were followed by American, British, and European newcomers, who flooded into Upper Canada over the next half-century in search of land. In less than 100 years, this natural forest landscape was transformed into a British agricultural colony.

Vignette 5.3 French Settlements at Detroit and Windsor in the Eighteenth Century

In 1701, the French established a fort (Fort Ponchartrain) at what is now downtown Detroit. The purpose of Fort Ponchartrain was to control the Ohio Territory. The land across the Detroit River (now Windsor) was first settled in 1748. An agricultural settlement named Petite Côte (Little Coast), this is the oldest continually inhabited settlement in Ontario. By 1751, Detroit had a French population of 600 with approximately 150 living at Petite Côte. In 1760, British troops took control of this fort and the surrounding settled area.

Sources: Detroit (2007); Wikipedia (2007).

Until the War of 1812, many settlers had come from the United States. They, like most other colonists, sought land in the fertile lowlands of the Great Lakes. When hostilities between Britain and the United States escalated, the Americans launched an attack on British North America. The War of 1812 effectively ended the influx of American settlers into Upper Canada. After the war ended with the signing in December 1814 of a treaty of stalemate at Ghent, an increasing number of settlers came from the British Isles, especially Ireland and Scotland. By 1851, Canada West (as Upper Canada had become with the Act of Union of 1841) had reached a population of 952,004 with 86 per cent living in rural settings (Statistics Canada, 2007b). By Confederation, virtually all the arable land in the Great Lakes Lowlands was cleared of forest by settlers who created an agricultural landscape. With a severe land shortage, some turned further north to try their luck in the few pockets of arable land within the Canadian Shield, but few were successful. Others migrated to towns and cities where they became absorbed into the urban workforce. Many more took their chances by participating in the last great American land rush that took place west of the Mississippi River. By 1867, the demographic balance of power clearly remained with the rural community—urban areas, such as villages and towns, contained less than 20 per cent of the population of Ontario.

When Canada West joined Confederation in 1867, it was renamed Ontario (Figure 3.4). At that time, the geographic extent of Ontario was about 100,000 km^2—just a fraction of its present size—but as Canada grew in size, acquiring more territory from Great Britain, both Ontario and Québec benefited by obtaining some of these new lands (Figures 3.5–3.7). In both cases, however, these new lands had little immediate value for economic development and settlement because they were carved out of two physiographic regions (the Canadian Shield and the Hudson Bay Lowland) that were far from markets and had little or no agricultural potential. Since Confederation, the borders of Ontario have been extended three times. These boundary changes have greatly increased the geographic size of the province, but not its agricultural lands. The first expansion occurred in 1874 when Ontario's boundaries were pushed northward to about 51° N and westward towards the Lake of the Woods. The second expansion took place in 1889. Until then, both Manitoba and Ontario claimed the land around the Lake of the Woods. The Ontario government won this bitterly contested dispute and was able to enlarge the province once more. At the same time, Ontario's northwest boundary was adjusted to the Albany River, which flows into James Bay. As a result of this adjustment, Ontario gained access to James Bay. In 1912, the final boundary modification occurred when the District of Keewatin south of 60° N was assigned to Ontario and Manitoba. As a result, Ontario extended its political boundary to the northwest, stretching from Manitoba to Hudson Bay at the latitude of 56°51' N. Through these boundary adjustments, Ontario reached its present geographic extent of 1 million km^2.

The Birth of an Industrial Core

The economic essence of Ontario is its industrial base. Manufacturing in southern Ontario (then Upper Canada) began in the nineteenth century and received a boost in 1854, when Britain and the United States signed the Reciprocity Treaty, which allowed trade between the British North American colonies and the United States. By 1867, Ontario had the largest population of Canada's four founding provinces and a fledgling industrial base. Small manufacturing outfits existed in small villages and often consisted of less than five employees (i.e., a village blacksmith or miller). Larger manufacturing activities took place in towns and cities, where sawmill, gristmill, and distillery operations were commonplace. At that time, most manufacturing activities were dependent on water power, so most were located near a stream or river.

The Americans allowed the Reciprocity Treaty to lapse in 1866, cutting off the access of Upper Canada to the American market. Confederation and the 1879 National Policy of Prime Minister Macdonald, however, enabled Ontario to secure the Canadian markets. Under the National Policy high tariffs were imposed on imported manufactured goods, which allowed manufacturing in southern Ontario to flourish. (For more about the Reciprocity Treaty and the National Policy, see Chapters 1 and 3.) The consequences of Confederation and protective tariffs for Ontario were threefold: (1) the creation of a Canadian market for

In 1794, York became the capital of Upper Canada. Despite its political status, this frontier village remained on the western edge of British settlement that stretched westward from Lower Canada along the north shore of Lake Ontario. By 1812, York had only 700 residents. This painting (dated 1804) illustrates a group of houses strung along the shore of Toronto Bay. Beyond this narrow strip of cleared land lies the original forest of southern Ontario. (*Library and Archives Canada C34334*)

View of King Street [Toronto], Looking East, 1835 by Thomas Young. By the time of this painting, Toronto had a population of nearly 10,000. In the previous year, York had been renamed Toronto. (*Library and Archives Canada 1669*)

Ontario products; (2) an increase in the size of the more successful manufacturing companies; and (3) the growth of the industrial workforce in Ontario. In the rest of the country, however, prices for manufactured goods (made in Ontario) were generally higher than in the adjacent areas of the United States. The regional price variations reflected two factors: Canadian transportation costs and Ontario's more limited economies of scale in comparison to those of US manufacturers.

The National Policy of high tariffs therefore provided the impetus that established a national industrial core in southern Ontario while the rest of the country—except for Québec—was relegated to a domestic market for these manufactured goods. This economic arrangement made necessary the east–west transportation axis and, in doing so, favoured Ontario and Québec, thus laying the groundwork for western alienation and Maritime dissatisfaction with Confederation. Not until the Free Trade Agreement between Canada and the United States was signed in 1989 did this national core/periphery relationship undergo significant change and a north–south transportation axis emerge as a prominent aspect of Canada's manufacturing trade. (See Chapter 1 for more on the FTA.)

Ontario Today

While Alberta's rate of economic growth is now the highest in Canada, Ontario remains Canada's economic engine. A measure of Ontario's industrial success is its enormous energy demand. Ontario's powerful economy is fuelled by electricity generated within the province by hydroelectric installations at Niagara Falls and along the Ottawa River, by four thermal coal plants using coal from Pennsylvania and West Virginia, and by nuclear generators in plants at Pickering and Darlington along Lake Ontario east of Toronto and at Tiverton on the Bruce Peninsula. Still, with an expanding industrial base and an ever-increasing population, Ontario requires more energy. To meet that demand, electricity is imported from Québec and natural gas from Alberta. With the record high prices of coal, natural gas, and uranium expected to continue, Ontario is facing a double whammy—energy shortage and higher energy costs.

Ontario is not a homogeneous region. The province contains three physiographic regions: the Great Lakes–St Lawrence Lowlands, the Canadian Shield, and the Hudson Bay Lowland. Each physical environment has a distinct and different economy and settlement pattern. Northern Ontario, for example, has the characteristics of a resource hinterland, while southern Ontario is the epitome of an industrial core. As discussed earlier in this chapter, geography goes a long way towards explaining this spatial variation. With this dichotomy in mind, the following economic and population statistics for Ontario apply best to southern Ontario:

- Ontario produces 39 per cent of Canada's GDP.
- Its annual output exceeds $300 billion.
- Its average personal income is well above the national average.
- With 39 per cent of Canada's population, it has the largest population of the six geographic regions.

Ontario's economic success has been facilitated by transportation systems that promoted trade for the province with other provinces and other countries. The continental rail and highway networks are essential to trade between provinces and with the United States, and the St Lawrence Seaway and the Welland Canal have facilitated trade with the US and other foreign countries (Vignette 5.4).

Ontario's Spatial Divide: Southern and Northern Ontario

Ontario's economic and physical structures take on a distinctive geographic pattern, thus creating a spatial divide. On the one hand, primary activities are found mainly in northern Ontario's Canadian Shield; on the other hand, secondary and tertiary activities are concentrated in the Great Lakes region of southern Ontario.

Northern Ontario lies in two physiographic regions that offer little for agriculture and settlement. The resource industries—but especially forestry and mining—are the staple economic activities of northern Ontario. Initial but not final processing of these resources takes place within this resource hinterland. Examples include transforming timber into lumber and paper products and the smelting of nickel and other ores. In all cases, this form of manufacturing creates a more valuable product that is less expensive to ship to distant markets and provides jobs in single-industry towns. All resource economies in Canada depend on global prices and access to the US market. Unfortunately, the forest industry in northern Ontario, like other forest regions in Canada, has suffered from the US duty on softwood lumber and then from the collapse of the US housing market. For these reasons, single-industry communities in northern Ontario are hurting from the depressed forest industry.

Southern Ontario, occupying the southwestern half of the Great Lakes–St Lawrence Lowlands, is situated in a most favourable natural environment. Southern Ontario is home to Canada's most productive agricultural lands and to Canada's largest concentration of manufacturing and service industries. The area lies in the most southerly latitudes in Canada, enjoys a moderate continental climate, and has easy access to US markets. Manufacturing firms take advantage of close proximity to the large US market just across the border. Among its many assets, two stand out. Farmers benefit from the long growing season, hot, wet summer weather, and rich soils, making southern Ontario the centre of agriculture in Canada by value of production. Local urban markets encourage farmers to engage in intensive market gardening. Wine production in the Niagara area is another element of the varied agriculture found in southern Ontario. Finally, its economic position within Canada is unmatched. A large internal market, proximity to the manufacturing belt of the United States, and its firmly established role as the financial and service capital of Canada provide southern Ontario with the right conditions for its large and varied secondary and tertiary activities.

The role of proximity cannot be underestimated. For example, trucks can deliver Ontario goods to firms in Michigan, Ohio,

Vignette 5.4 The Welland Canal

The Welland Canal connects Lake Ontario and Lake Erie, thus allowing ocean-going ships to enter into the heart of North America. To avoid the Niagara River and its famous falls, the first canal builders faced the daunting task of constructing a canal across the Niagara Peninsula, a distance of some 44 km. In doing so, a series of lift locks were needed to overcome a difference in elevation of nearly 100 metres between Lake Ontario and Lake Erie. The first canal, opened in 1829, was dug by hand. A series of locks made from hand-hewn timbers connected a series of creeks and lakes. As the size of ships increased, the original canal proved inadequate and a new canal was built in 1845. Within 40 years, larger ships required a third renovation, which was opened in 1887. The present canal was completed in 1932. In 1973, a new channel was constructed to bypass the city of Welland. The Welland Canal has been part of the St Lawrence Seaway since 1959 and is operated by the St Lawrence Seaway Management Corporation.

The Welland Canal is a strategic link between Lake Ontario and Lake Erie that provides a water route around Niagara Falls. With ever increasing traffic, the lock system was divided into two at several places, as shown in this aerial photograph. In 2006, approximately 40 million tonnes of goods passed through these locks. The three leading products were grain, iron ore, and coal. (Al Harvey/The Slide Farm)

Pennsylvania, and New York in a matter of hours, allowing a just-in-time production system to flourish on both sides of the border. Backups at the major border crossings—especially between Windsor and Detroit—caused by heightened security concerns in the US have become a major problem for the 'just-in-time' delivery of goods. With much of Canada's $1.7 billion worth of daily exports to the US coming from Ontario, hampering the flow of people and goods across the border to satisfy Washington's concerns about security and terrorism will have a negative impact on Ontario—the only question is 'how much'. The answer to that question is still unknown, largely because of the difficulty of collecting information on the cost. Estimates, such as the 2004 report to the Ontario Chamber of Commerce, suggested a figure of $5.25 billion for Ontario. More recently, Len Crispino, chief executive of the Ontario Chamber of Commerce, stated: 'That border now has become more of a choke point, rather than a conduit for trade' (French, 2007). He estimated that Canada loses as much as $8 billion Canadian every year to border delays.

Trade

Ontario is well positioned to engage in trade, both domestically and internationally. From northern Ontario, forest products and minerals are shipped to both national and international markets. In southern Ontario, manufactured goods ranging from aircraft to steel products are produced for both these markets. Restructuring of Ontario's manufacturing industry took place after 1989 when the FTA was signed. Canadian firms, especially those with high labour costs in their manufacturing process, were faced with a highly competitive market that sometimes resulted in plant closure and/or relocation to the United States or Mexico.

The FTA and then NAFTA, along with Ontario's geographic location within North America, have allowed the region's business firms to penetrate the huge US market and that has led to greater integration into the North American economy. For example, Ontario's exports to the United States in the 1980s were of about the same value as those to the rest of Canada; but by 1998, Ontario's exports to the US soared to two and a half times the value of exports to the rest of Canada (Courchene, 1998: 276–7). Driven by automobile shipments to the US market, this trade gap continues to widen in the first years of the twenty-first century as the sheer size of the US market has tilted the province's trade in a new direction.

The international trade dimension of the Ontario economy is evident in the 2005 export figures. In that year, the leading Canadian exports by value were automobiles and their parts, and the value of motor vehicle exports reached $88 billion (Statistics Canada, 2006b). Since over 90 per cent of these automotive exports came from assembly and parts plants in Ontario, the importance of this industrial sector to Canada and Ontario cannot be overestimated. Furthermore, a growing automobile industry has a ripple effect on the economy of southern Ontario by creating additional demand for steel, rubber, plastics, aluminum, and glass products. Of course, a slowdown in automobile production would have a negative impact on southern Ontario's economy.

Southern Ontario is, in reality, a northern extension of the American manufacturing belt. With US automobile plants just across the border in Detroit, automobile manufacturing spread across the Detroit River into Windsor. Initially this northward diffusion of the automobile industry took the form of branch plants owned by the Big Three (General Motors, Ford, and Chrysler). Until the 1965 Auto Pact, Ottawa placed high tariffs on imported automobiles and their parts. In doing so, the federal government encouraged American and other foreign firms to open branch plants that produced similar goods, but because of the size of the Canadian market these were smaller manufacturing plants than those in the United States and various European countries. The net effect was higher prices for Canadian consumers due to higher production costs, the high cost of shipping these goods long distances to regional markets in Canada, and the provincial/federal taxes on automobiles that were higher than those in the United States. Ontario, however, benefited from the expansion of its industrial base.

The Canada–US Auto Pact was designed to integrate Canada's automobile industry into the North American market. At that time, over 90 per cent of Canadian automobile production was controlled by the Canadian branch plants of the Big Three automobile companies. These companies wanted to rationalize their production and marketing of automobiles within North America and thus be better prepared for competition from foreign producers. While the Auto Pact ended in 2001, this Canada–US agreement not only altered the nature and purpose of automobile manufacturing in Ontario, but it also marked a shift in federal policy towards more liberal trade arrangements and towards economic continentalism in North America (discussed in Chapter 1). One important feature of the Auto Pact is that Canadian production could not fall below the 1964 automobile output. Today, Ontario accounts for 14 per cent of automobile production in North America compared to roughly half that figure in 1964.

Most of Ontario's exports to the United States cross the Detroit River on the Ambassador Bridge and through the Detroit–Windsor Tunnel. In fact, automobile trade between Windsor and Detroit plays a critical factor in this international trade and accounts for 30 per cent of Canada's trade with the United States.

The New World Economic Order

During the 1980s and 1990s, two significant changes in the rules of trade propelled Canada into a new world economic order. Ottawa elected to abandon its closed economy based on high tariffs. Simply put, Canada's prosperity depended on assured access to the US market. The consequences for Ontario were severe, forcing a sometimes painful restructuring of its manufacturing sector. Restructuring meant that companies had to become more efficient because of fierce competition from foreign countries, many of which had extremely low costs of production. Canadian-based firms had two solutions: (1) reducing the size of their labour force but maintaining their production level and (2) moving part or all of their operations offshore where lower labour and other factory costs exist. As part of this restructuring, many American branch plants consolidated their operations in the United States where these companies could take advantage of a longer production run and therefore reduce production costs through economies of scale.

The first change, dating to the immediate post-war period, was the liberalization of world trade through the creation of the loosely organized General Agreement on Tariffs and Trade (GATT), to which Canada was one of the original signatory nations. Under the terms of the GATT, countries agreed to reduce barriers to international trade. In 1940, for example, the average tariff on manufactured goods had been about 40 per cent. Since 1947, however, a series of tariff reductions were achieved through agreements made by members of GATT, which grew over the years to include most trading nations of the world. By the eighth agreement in 1990, the average tariff was 5 per cent (Dicken, 1992: 153). In January 1995 a more institutionalized and formal international trade regime, the World Trade Organization (WTO), was created out of the final agreement of the Uruguay Round of the GATT, which concluded in December 1993. Today the main impediments to trade are not tariffs but non-tariff barriers such as import quotas ('voluntary' export restraints) and health regulations.

The second major impact on Canadian trade was the Canada–US Free Trade Agreement. The FTA committed Canada and the United States to an integrated North American economy. In 1994, the FTA was superseded by a similar agreement, the North American Free Trade Agreement (NAFTA), which brought Mexico into what effectively has become a continental trade bloc. Since tariffs between Canada and the United States are decreasing at a faster rate under NAFTA than under the WTO, NAFTA gives Canadian manufacturing firms an opportunity to adjust to the new North American market before being fully exposed to world trade. By creating larger operations, Canadian firms hope to achieve economies of scale that will decrease their per unit costs of manufacturing and therefore allow them to become more competitive. NAFTA also exposed Canadian manufacturing firms to competition from Mexican-based firms whose labour costs are much lower and whose environmental regulations are less stringent (see Vignette 4.6).

One result of the FTA and NAFTA is that trade between Canada and the United States is increasing, thereby making Canada more dependent on the United States. From this perspective, Canada has hitched its wagon to one horse, which is especially true for southern Ontario (Norcliffe, 1996: 32). The consequences of this trade dependency are threefold: (1) booms and slumps in the American economy affect Canada more directly than they did before FTA/NAFTA; (2) Canada does not have unlimited access to the US market because Washington can still limit Canadian products from entering its market; and (3) Canada's long-term economic fortunes are even more closely tied to the American economy than was the case before these trade agreements. Canada, but particularly Ontario, has benefited from closer trade ties with the United States. During the late 1990s, for instance, the booming American economy stimulated Ontario's exports of high value, ranging from automobiles to high-tech products. Yet, in 2000, when the American economy entered a downturn, Canada's exports to the United States remained at a high level and Ontario's economy continued to grow well into 2005. By 2006, however, a slowdown in manufactured exports, especially automobiles, appeared and this trend may continue for the next few years. Manufactured exports dropped for two reasons. First, the United States market witnessed a rise in foreign imports from China and other countries with very low labour costs, which made it difficult for Ontario firms to compete in the American market place. Second, the rising value of the Canadian dollar vis-à-vis the American dollar made Ontario products more expensive in the United States.

These trade agreements, while generally considered successful, have raised several controversial issues within the country and led to bitter disputes with the country's main trading partner, the US. For Ontario, plant closures have been a sore point while trade disputes have affected grain and lumber exporting regions of Canada. Since 1989, several bitter disputes between Canadian grain farmers and lumber companies and their American counterparts have taken place. Each time, the United States government has pressured Ottawa into accepting a 'voluntary quota' on Canadian exports to the United States. The 2006 softwood lumber agreement provides another example of such an agreement.

Ontario's Industrial Structure

Geographers regard an economy as having a particular industrial structure based on its economic activities (see Tables 4.20 and 4.21). These activities traditionally were divided into three categories: primary, secondary, and tertiary. Each geographic region has a distinct industrial structure that provides a measure of its place within Canada. The allocation of workers to these sectors constitutes a measure of the industrial structure of Canada's regions. As expected, Ontario leads all regions in terms of secondary employment and follows the other five regions in the percentage of its employment in the primary (extractive) sector. The Territorial North, on the other hand, has the smallest proportion of its labour force in secondary activities and the largest percentage in its primary sector (Table 1.5).

Employment in Ontario's primary sector is small (1.5 per cent of the total workforce) and shrinking. In 1995, the primary sector accounted for 3 per cent of the total workforce. Of the major resource activities, agriculture and forestry are under enormous pressure due to low global prices and barriers imposed by the United States. As a result, the family farm, the original backbone of Canadian agriculture, is in dire straits. From 1999 to 2005, for instance, the number of agricultural workers in Ontario dropped from 114,000 to 93,000. The forest industry, which relies heavily on exports to the American market, has faced severe trade restrictions imposed by Washington. Ontario lumber companies have had to restructure their operations to seek greater efficiencies, but restructuring has also resulted in downsizing of their workforce and the closure of pulp mills and sawmills. Companies engaged in fishing and mining have also had to reduce the size of their labour force to remain competitive in the North American and world economy. As a result, the primary sector in Ontario has suffered the greatest workforce decline.

Ontario contains the bulk of Canada's secondary industry. This sector accounted for 23.6 per cent of the total workforce in the region in 1995 and 2005. Over this 10-year period, some annual fluctuations took place. In 2002, for instance, 25.1 per cent of the total workforce was in the secondary sector. The secondary sector consists of heavy industry (iron and steel production as well as the processing of raw materials), light industry (automobile, aircraft, and textile manufacturing), and high-technology industry (fibre optics, software, and Internet applications). The interconnection between these industrial activities is strong. For instance, the automobile industry requires products from heavy industry (steel and nickel) and specialized devices from high-technology firms (digital and computer equipment). The importance of the automobile industry to Ontario is further explored below.

The tertiary sector[1] has become the dominant form of economic activity in all modern industrial states. In 2005, the tertiary sector accounted for 74.4 per cent of Ontario's total workforce. Most of Canada's tertiary industry is located in southern Ontario. The service industry has expanded in many directions to meet the needs of its customers. The complexity of the tertiary sector is revealed in Table 5.2, with tertiary categories ranging from trade to health care. Employment data for 2005 illustrate both the dominance of the tertiary sector in Ontario's economy and the importance of the secondary sector.

Agriculture is an important part of the economy of southern Ontario. Many farmers are engaged in dairy operations while others focus on fruit and vegetable production. As shown in the photograph, many dairy farmers plant hay, which is used for winter feed. (Barrett & MacKay Photography, Inc.)

Agriculture and Climate

Southern Ontario's climate is dominated by its long, warm summer that extends from May to September. Tropical air masses that originate in the Gulf of Mexico extend over this area, resulting in hot and humid weather. Winter, on the other hand, takes hold for three to four months from mid-November to March when occasional invasions of Arctic air masses bring cold weather. Such weather is associated with clear, sunny days. During the early spring and late fall, unsettled, cloudy weather dominates, with minimum temperatures often falling below the freezing point. Proximity to the Great Lakes affects the local weather by funnelling air masses into this region and by increasing local precipitation due to air masses absorbing moisture from the surface of the Great Lakes. During the winter, high winds funnelled along the Great Lakes storm track often accompany cold spells, driving the wind chill to a dangerous level. Fortunately, most storms are short-lived and followed by relatively mild weather. Annual precipitation ranges around 100 cm. Warm, moist air masses from the Gulf of Mexico provide most of the moisture for the region, which falls as convectional and frontal precipitation.

Because of its mild climate and fertile soils, southern Ontario accounts for much of Canada's agricultural output by value. Southern Ontario has over half of the highest-quality agricultural land (known as Class 1) in Canada. With a combination of fertile soils and a warm summer, farmers in southern Ontario account for more than $7 billion of agricultural products each year. In 2006, Ontario had just over 57,000 farms, a 4.2 per cent decrease over the past five years compared to the national decline of 7.1 per cent (Statistics Canada, 2007c). This decline in the number of farms is related to increase in farm size from 92 hectares per farm in 2001 to 94 hectares per farm in

Table 5.2 Employment by Industrial Sector in Ontario, 2005

Industrial Sector	Workers (000s)	Workers (%)
Primary	**128**	**2.0**
Agriculture	93	1.5
Secondary	**1,509**	**23.6**
Manufacturing	1,064	16.6
Tertiary	**4,761**	**74.4**
Trade	995	15.6
Transport	289	4.5
Finance	452	7.1
Professional services	443	6.9
Management	286	4.5
Education	428	6.7
Health care	626	9.8
Culture	301	4.7
Accommodation	364	5.7
Other services	257	4.0
Public administration	322	5.0
Total	6,398	100.0

Source: Statistics Canada (2006c). Adapted from Statistics Canada website, http://www40.statcan.ca/l01/cst01/labor21c.htm.

2006, which is necessary for farmers to achieve economies of scale (ibid.).

Located south of 44° N, southern Ontario has abundant moisture and a long, warm-to-hot summer permitting the production of highly specialized crops such as grain corn, soybeans, and sugar beets as well as a wide variety of fruits, grapes, and vegetables that cannot be grown in other parts of Canada. In the Niagara Fruit Belt, a unique set of growing conditions has provided the basis for grape production, resulting in high-quality wine production (Vignette 5.5). Equally important, the large urban population provides a stable local market for its dairy, livestock, soft-fruit, and vegetable products. Although the amount of cropland is relatively small, farming operations in southern Ontario are much more intense than in Western Canada, where extensive grain crops dominate farming operations. The difference between the two agricultural areas is revealed in the size of farms and the types of crops. In 2006, the average size of farms in Ontario, 94 ha, compared to an average of 587 ha for farms in Saskatchewan (ibid.). Grain farming and cattle ranching dominate in Western Canada, while there is a much more diversified agricultural land use in southern Ontario, where corn, barley, and winter wheat are the important crops along with highly specialized ones such as tobacco and vegetables. Corn and grain often serve as fodder for the hog, dairy, and beef livestock farms that operate in Ontario.

Agricultural land use varies within southern Ontario due to subtle but important physical differences. For example, soils are less fertile and the growing season is shorter east of Toronto than southwest of Toronto. There are three highly specialized agriculture zones west

Vignette 5.5 The Niagara Fruit Belt

The Niagara Fruit Belt lies on the narrow Ontario Plain that extends from Hamilton to Niagara-on-the-Lake. The Ontario Plain has fertile, sandy soils suitable for most types of agriculture. Lake Ontario marks its northern limit while the Niagara Escarpment forms its southern edge. This small agricultural zone contains the best grape and soft-fruit growing lands in Canada, and it now is home to vineyards that account for most of Canada's quality wines, including the unique ice wine.

While the Niagara Fruit Belt occupies a northerly location for grape vines and soft-fruit trees, local factors have more than offset the threat of frost at 43° N latitude. These factors are:

- Air drainage from the Niagara Escarpment to Lake Ontario reduces the danger of both spring and fall frosts.
- The waters of Lake Ontario are warm in autumn and proximity to the Niagara Fruit Belt helps to moderate advancing cold air masses.
- In the early spring, the cool waters of Lake Ontario keep temperatures low, thereby delaying the opening of the fruit blossoms until late spring when the risk of frost is much lower.

The Niagara Fruit Belt extends about 65 km between Hamilton and Niagara-on-the-Lake. The Niagara Fruit Belt is one of the major fruit and grape producing areas in Canada. Most vineyards are located on the slopes of, or below, the Niagara Escarpment. For many years, hardy vines that produced low-quality grapes resulted in a poor quality (but inexpensive) wine. With the Free Trade Agreement, Canada had to remove its tariffs, making it difficult to compete with foreign wines. Since that time, farmers in the Niagara Fruit Belt have been successful in growing the finest varieties of grapes and have been able to make some of the finest wines in the world. (Barrett & MacKay Photography, Inc.)

of Toronto: the Essex–Kent vegetable area, the Norfolk Tobacco Belt, and the Niagara Fruit Belt. All lie in southwestern Ontario where the growing season is relatively long. These zones, located south of 43° N, are the southernmost lands in Canada.

Challenges to Ontario's Agriculture

Ontario farmers have had to adapt to the new economy. For instance, prior to 1995, feed grain for Ontario's livestock industry was imported from Western Canada. However, with the elimination of the Crow Benefit (a transportation subsidy), western feed grain is now too expensive for Ontario farmers, resulting in the expansion of meat-packing plants in Western Canada and the shipping of the finished product to other parts of Canada, the United States, and other foreign markets.

Marketing boards are a key element in providing controlled markets and prices for Ontario's dairy and poultry farmers. While government-regulated marketing boards are safe for now, the United States—a strong advocate for market-determined farm supply and demand—wants them abolished. The US argues that marketing boards are violating free trade rules under NAFTA and the WTO. Still, these boards are popular with farmers because they ensure stable prices and markets for producers. Without such protection, Ontario dairy farmers may have difficulty competing with American farm products. Yet, the federal US government and some states provide agricultural subsidies. For example, milk production is subsidized and trade barriers keep foreign milk from competing in the US domestic market (Sumner and Balagtas, 2002).

Agriculture products exported to the United States may be subjected to inspections and thus cause border delays. The US Department of Agriculture plans to impose a fee of $5.25 per crossing for trucks hauling agriculture products to defray the cost of inspecting agriculture products for pests, animal diseases, and bioterrorism (Macleod, 2006). This plan, if imposed, would make the cost of Canadian products higher in the US, which may cause lost of market share.

Urban encroachment poses another challenge to agriculture. Since the economic value of urban/industrial classified land is higher than land designated for agricultural uses, urban/industrial encroachment continues to spread into the rural landscape. Although Ontario has over half of the best agriculture land in Canada, the spread of towns and cities over the last decade alone has destroyed 8 per cent of this irreplaceable resource.

Manufacturing

Southern Ontario accounts for almost half of all manufacturing jobs in Canada. Several factors account for this concentration but the paramount one is the availability of a skilled and hard-working labour force, which is so essential to producing quality products. Since the FTA, the face of manufacturing has changed, and so has southern Ontario's mix of manufacturing activities. Textile firms and other manufacturers that required semi-skilled labour working at or near the minimum wage have largely disappeared. New to the scene are high-technology companies that depend on highly skilled workers who can command high wages. Ontario is at the leading front for Canada's high-technology industry (Vignette 5.6).

Many urban centres, both large and small, house important sectors of southern Ontario's manufacturing industry. With the exception of Windsor and Oshawa, most automobile manufacturing takes place in smaller centres, namely St Thomas, Ingersoll, Cambridge, Brampton, Alliston, and Oakville. All these communities have two features: excellent access to major highways and a relatively short distance to their parts suppliers in Michigan, Ohio, Pennsylvania, New York, Québec, and southern Ontario itself. Heavy industry is concentrated in Hamilton where two steel companies, Dofasco and Stelco, are based.

Historically, four factors led to the successful development of manufacturing in southern Ontario. Using the automobile industry as an example, it becomes apparent how each factor contributed to the growth of manufacturing in Ontario. The first factor is Ontario's geographic advantage—close proximity to America's manufacturing belt, which led American industries to locate branch plants in Canada early in the twentieth century. This spillover effect first took place in the Windsor–Detroit area in 1904, when the Ford Motor Company established an automobile assembly plant in Windsor. The second factor was trade restriction on foreign manufactured goods (imposed by the National Policy in 1879). Automobile parts from Ford's plant in Detroit had to be ferried across the Detroit River and assembled in an old carriage factory in Windsor. Access to the markets in the British Empire provided a third advantage. American branch plants in Canada could take advantage of lower tariffs for Canadian-made products in the British Empire. As a result, Canadian-built Ford cars were sold not only in Canada but also in various parts of the British Empire. General Motors followed this same strategy by opening a branch plant in Oshawa. The size of its domestic market provided southern Ontario with its fourth location advantage for the automobile industry. Most cars were sold in southern Ontario, thus minimizing transportation costs.

Canada's financial industry is concentrated along Bay Street where many corporate headquarters, legal firms, insurance companies, and stockbrokers have their offices in the large office towers built by Canada's leading banks. (Wikipedia, at: <en.wikipedia.org/wiki/Bay_Street>.

With the liberalization of trade, manufacturing in Canada changed. The first step was the Auto Pact. By the early 1960s, the Big Three automobile companies proposed a common market between Canada and the US for the production and marketing of their products. In 1965, Ottawa and Washington passed the appropriate legislation for the Auto Pact. The second step took place in 1989 when the Canada–US Free Trade Agreement brought other components of the economy under a single North American market, thus removing sometimes high tariffs between the two countries. Since American firms had already benefited from economies of scale, their costs of production were considerably lower than those of similar Canadian firms. As well, the American firms were well established in the marketplace and dislodging them was a daunting task. Canadian-based manufacturing firms were forced to play catch-up and many did not survive. American firms often closed their smaller and therefore less efficient branch plants and served their Canadian customers from their American factories. Some Canadian firms merged in order to attain a size deemed necessary to compete in the North American and global markets. The purpose was twofold: to compete in the world marketplace and to avoid a takeover by a large foreign company. Mergers, however, often resulted in the closure of offices to avoid duplication and, thus, in the reduction of the number of employees.

Like many towns in southern Ontario, Cambridge had to adjust to this new economic environment. Created in 1973 by the amalgamation of Galt, Preston, and Hespeler, Cambridge is an industrial centre just west of Toronto. Since 1989, the economy of Cambridge has experienced strong growth. While some firms have closed their operations, others have established themselves in Cambridge. With over 400 manufacturing firms, Cambridge has a solid industrial core. These firms range from textile companies to technology-based service industries. In 1985, Cambridge became an important automobile

Vignette 5.6 The Wave of the Future: Clusters and High Technology

High-technology companies are found in many places across Canada but particularly in Canada's major urban centres. In fact, Britton (1996b: 266) states that 'Canadian urbanization is the key to understanding the location of technology-intensive activities.' Marc Busch takes this idea further. His answer is 'clusters' (Atkin, 2000: C1). A cluster is a place where institutions, companies, and individuals have a commitment and enthusiasm for innovative technological research along with the capital necessary to develop and market the product. High-tech clusters are often anchored around a university or a public research agency like the National Research Council. Most high-technology companies are found in Toronto, Montréal, and Ottawa rather than in Saskatoon, Halifax, or Victoria. The latter do have high-tech operations, but they are on the edge of the high-tech world and depend on a niche operation. For Saskatoon, it is agricultural biotechnology.

assembly centre for Toyota Canada. Since then, Toyota has expanded its automobile production. Employment at its assembly plant jumped from 2,000 workers in 2002 to 4,340 workers in 2005, making Toyota the city's largest employer. The presence of an automobile assembly plant has had several benefits for Cambridge: a stable employment base, high-paying jobs, and spinoff effects due to purchases by the company and its employees.

For Cambridge, the FTA had its downside. Fierce competition resulted in plant closures. Savage Shoes, Inglis, Franklin Manufacturing, InterRoyal Corp., and Dobbie Industries Ltd have shut their doors, laying off hundreds of workers. Savage Shoes simply could not compete with lower-cost shoes from developing countries, where labour costs are a fraction of those in Canada. After the FTA, the American owners of the Inglis household appliance plant, the InterRoyal furniture factory, and the Dobbie wool-spinning mill decided to close their branch plants and produce their entire output from their much larger plants in the United States. One American branch plant, Rockwell Automation, has managed to survive, but with a smaller labour force. Before the FTA, this firm produced electromechanical devices used in automated factories in Canada. To survive in the North American market, the plant had to find a single product for the global market. Now it is the only Rockwell plant in the world that produces medium-voltage products and electrical generators, 70 per cent of which are exported to the United States and other foreign countries. Niche production for the North American market allowed this plant to achieve economies of scale, and now its products can compete for business in all parts of the world.

Research In Motion, a leader in the wireless communications industry, has its corporate headquarters in Waterloo, Ontario. The company is the creator of the BlackBerry wireless solution, which consists of software, service, and hardware. BlackBerry smartphones support wireless e-mail, corporate data applications, mobile telephone, text messaging, and more. Launched in 1999, the BlackBerry solution is now used by over eight million people. (Photo courtesy of Research In Motion.)

Restructuring of the Manufacturing Sector

Within Canada, the FTA committed the country's manufacturing sector to become more competitive. Restructuring was the order of the day. Firms unable to adjust to the highly competitive marketplace in North America would be forced to close. For most firms, adjusting to greater competition meant lowering the costs of production by creating larger manufacturing plants and by increasing the productivity of workers. As part of this economic restructuring, American branch plants lost their raison d'être (MacLachlan, 1996: 195–214; Norcliffe, 1994: 2–17). A few adjusted their operations by specializing in one aspect of the manufacturing process to serve the North American market. In this way, a few branch plants were able to achieve economies of scale, which reduced their costs of production per unit of output. Most companies simply ceased production, thus allowing the parent plants in the United States to supply the Canadian market.

A new geography of manufacturing is emerging in Canada but especially in Ontario. The FTA exposed Ontario's manufacturing sector to fierce foreign competition, causing restructuring of Canadian firms into a more 'lean and mean' profile. Writing about the organizational restructuring resulting from the FTA and NAFTA, MacLachlan expressed concern that 'southern Ontario stands the greatest risk of employment losses because of its large secondary manufacturing sector and continued dependence on foreign-owned subsidiaries for employment in key manufacturing industries' (MacLachlan, 1996: 195).

The dual forces of restructuring and relocation have two devastating effects. First, these forces have resulted in plant closures, especially in the textile and shoe manufacturing sector, and they have kept wage increases low, such as in the non-unionized automobile parts firms, and even have caused wage rollbacks. Plant closures have sent shivers across Ontario's manufacturing belt because of the 'trickle-down effect'. For example, cuts in motor vehicle output at General Motors, Ford, and DaimlerChrysler assembly plants translate into job losses in automobile parts firms. In short, the increased competition brought on by NAFTA and the WTO will continue to challenge and sometimes threaten the province's manufacturers—a sector of the economy that Ontario heavily relies on. Consequently, the level of uncertainty for manufacturing is high. Statistics support this fear. Statistics Canada (2006a) reported a loss of nearly 100,000 manufacturing jobs in Canada from August 2005 to August 2006. This represents a decline of nearly 5 per cent in the total number of workers in the manufacturing industry (from 2.2 million to 2.1 million). Equally disturbing, manufacturing jobs, most of which are in Ontario, declined by 10.5 per cent from 2002 to 2006 (Lovely, 2006). Perrin Beatty (2006), president of Canadian Manufacturers and Exporters, sees Ontario manufacturers under siege partly because of a rapidly appreciating dollar, but also because of skyrocketing energy and commodity prices, intense competition from around the world, and increasing skilled labour shortages. According to Beatty, the future is not bright because manufacturers have few choices to survive—find cost-cutting solutions, turn to outsourcing, or relocate in countries with lower labour costs.

The squeeze on Ontario's manufacturing industry is hopefully a short-term phenomenon. The 10-year trend from 1995 to 2005 is encouraging. During that time, as noted earlier, the percentage of workers in Ontario's manufacturing sector, while fluctuating annually, remained at 23.6 per cent. Over the same period, the percentage of workers in the primary sector dropped from 3 per cent to 2 per cent while the tertiary sector rose from 73.4 per cent to 74.4 per cent.

The Dutch Disease Arrives in Ontario

Why would Alberta's rapidly expanding economy have a negative impact on Ontario's export-based industry? The explanation lies in a new economic theory known as the **Dutch Disease**. This theory attempts to explain why a rapidly expanding resource sector can push the country's exchange rate to higher levels, thus making the price of goods in foreign countries higher. Canada provides an example of the Dutch Disease. In 2002, Canada's dollar was valued at 64 cents to the US dollar. By September 2007, the Canadian dollar had reached 98 cents US, thus making Canadian exports more expensive in the United States. The regional implications are clear. While Americans have no choice but to import Alberta oil and pay the higher price, Ontario manufacturers have to compete in the US market against local and foreign manufacturers.

Key Topic: The Automobile Industry

The automobile industry remains the key manufacturing activity in Ontario. Within North America, by 2004 Ontario had become the leading automobile producer among states and provinces, surpassing Michigan.

Why is automobile production so important? Peter Dicken (1992: 268) put it best:

> The significance of the industry lay not only in its sheer scale but also in its immense spin-off effects through its linkage with numerous other industries. The motor vehicle industry came to be regarded as a vital ingredient in national economic development strategies.

The rapid expansion of the auto industry in southern Ontario dates back to the 1965 Auto Pact. In that year, Canada produced nearly 900,000 vehicles. By 2000, the figure had reached 2.5 million. Over the past five years, annual production has varied between 2.5 and 2.6 million cars and trucks. Most of these are exported. At this level of production, Canada accounts for 14 per cent of North America's vehicle output. Ontario's automotive manufacturing industry remains the anchor of manufacturing in the province. About 150,000 people are employed by assembly and parts firms. These well-paid workers produce high-quality motor vehicles that account for 12 per cent of the country's gross domestic product.

The Auto Pact

The Canada–US Automotive Products Agreement of 1965, more commonly called the Auto Pact, is an example of a successful single-industry production-sharing agreement between Canada and the United States. Before the Auto Pact, the auto industry in Canada had small, high-cost plants, each a smaller version of the much larger plants in the United States. Even though Canadian wages for auto workers were lower than those for American workers, Canadian automobile prices were considerably higher than for the same models in the United States. From Canada's perspective, the Auto Pact served three objectives:

- It secured guarantees that Canadian automobile plants would not close.
- It brought to Canadian plants the advantage of economies of scale by allowing them to specialize in a few types of automobiles that would supply the North American market.
- It reduced the price of cars for Canadian customers.

The agreement called for Canada to eliminate the 15 per cent tariff on imported automobiles and US parts for Canadian manufacturers, and for the US to eliminate its corresponding tariff. The agreement included special protection for Canadian plants. Canada was guaranteed a minimum level of automobile production based on production levels for each type of automobile manufactured in Canada in 1964, that is, the ratio of vehicle production to sales in Canada for each class of vehicle had to be at least 75 per cent or the percentage attained in the 1964 model year, whichever was the highest. Furthermore, value added in vehicles produced in Canada had to be no less than the dollar amount achieved in 1964.

Under the terms of the Auto Pact, qualified motor vehicle manufacturers (Ford, General Motors, and DaimlerChrysler) were able to import both vehicles and automotive parts duty-free into Canada. Other motor vehicle manufacturers, however, had to pay a

6.1 per cent import duty. Japan and the European Union appealed to the WTO, arguing that the Auto Pact discriminated against imported vehicles. In July 2001, as a consequence of a WTO ruling in favour of Japan and the EU that it did indeed amount to an unfair trade practice, the Auto Pact, which helped build Canada's automobile assembly and parts industry, ceased to exist. Its demise helped Toyota and Honda expand their market share in North America.

The Growth of the Auto Industry

The automobile industry drives the Ontario economy. With 150,000 workers employed by the automobile assembly and parts firms, nearly one in seven Canadian manufacturing jobs depends directly on this industry. Wages in the assembly plants are relatively high for semi-skilled workers, but somewhat lower for workers in parts firms. Nevertheless, these incomes enable workers to make substantial purchases, thereby stimulating southern Ontario's retail sector. In addition, the automobile industry accounts for nearly one-quarter of Canada's merchandise exports. (Most of Canada's remaining exports are primary products, such as energy, grain, and forest products.)

From 1995 to 2002, exports of automotive products (passenger autos and chassis, trucks and other motor vehicles, and motor vehicle parts) increased in value from $62.9 billion to $97 billion (Statistics Canada, 2003c), but by 2005 the value of these exports dropped to $88 billion (Statistics Canada, 2006b). During this period, Ontario's strong hold on automobile production was largely due to:

- Canadian assembly plants had a higher productivity than comparable US plants. In 2001, Canadian workers had a 7 per cent advantage over American workers in the time required to produce an automobile (Brant, 2002: FP5).
- The low Canadian dollar allowed Canadian exports of cars and trucks to cost less than similar vehicles produced in the United States.
- Health-care costs are covered by Ontario while these costs are part of the wage package in the United States.

Both the federal and Ontario governments supported the automobile industry by supporting research into advanced technologies and by providing training in automotive manufacturing technologies at Ontario universities. For example, McMaster University and the University of Waterloo have established an Automotive Manufacturing Innovation initiative with 35 industry partners and support from the Ontario government. Another example is the launching of Auto21 at the University of Windsor with funding coming from the federal government.

After 2002, however, two problems surfaced. First, North America developed a mismatch between production and sales, forcing companies to close less productive plants. The North American automobile industry had the capacity to assemble 23 million units a year, but sales in 2002 fell below 18 million units and sales predictions of 17 million units for 2003 pointed to layoffs and plant closures (Keenan, 2003: B4). New, more efficient and flexible assembly plants were constructed and, by 2005, these additional plants added another 3 million units of capacity. Second, with the exchange rate on the Canadian dollar rising from 64 cents US in late 2002 to 96 cents in September 2007, the price of Canadian exports to the United States rose significantly. If the Canadian dollar retains this position or continues to rise, automobile exports to the United States may be affected, as will investments in new assembly plants and auto part firms.

The changing nature of the Canadian automotive industry has seen a shift from the domination of the Big Three to the rising power of Japanese-based firms. A May 2007 report by Statistics Canada confirmed this shift:

> The downsizing of the traditional Big 3 firms has been widely-publicized, reflecting the traumatic impact on their employees and their suppliers. Less appreciated, however, is how rapidly overseas firms have ramped up production in Canada to offset many of the losses.

Japanese-based manufacturers have filled much of the gap left by the traditional North American automakers in production in Canada. In 2006, Japanese automakers produced just over 900,000 automobiles in Canada, double their level in 1998. As a result, their share of the domestic market in production jumped from 16 per cent to more than one-third (36 per cent) during this period. (Statistics Canada, 2007f)

Automobile Parts Firms

The automobile industry consists of two separate operations: the assembly of automobiles and trucks and the production of their parts. In addition to these two operations are manufacturing firms that supply semi-processed materials. In southern Ontario, fabricating firms produce steel, rubber, plastics, aluminum, and glass parts for automobile assembly and parts plants in Canada and the United States. Then there are the service firms, ranging from the advertisers and designers to the sales and service staff. In short, the auto industry is a final-product type of manufacturing, and as such, its added value reaches a maximum.

The automobile parts industry employs some 100,000 workers in Canada. By being highly efficient and strategically located, automobile parts firms can operate on a just-in-time principle—auto components are produced in small batches and quickly delivered as needed to their customers. In turn, this allows the assembly plants to achieve considerable savings by reducing their inventories, warehousing space, and labour costs.

In the late 1980s, the Big Three automobile manufacturers decided to subcontract their parts business. This practice of subcontracting parts is called outsourcing. Outsourcing had two advantages for automobile companies. First, outsourcing allowed automobile manufacturers to concentrate on assembling automobiles, thereby reducing their costs and improving the quality of their product. The second advantage was that automobile parts companies were not unionized and therefore had lower wages. This wage differential—and the savings it provides the auto manufacturers—is the main reason why General Motors, Ford, and DaimlerChrysler continue to divert work to parts firms. In Ontario, the Canadian-based Magna International has grown into the third largest auto parts company in North America.

The 'Canadian advantage' encouraged the expansion of Canadian-based auto parts firms. This advantage was based on Canadian workers' higher productivity as compared to their American counterparts, Canada's health-care program, and a weak Canadian dollar. By 1996, the average North American vehicle contained about $1,500 of Canadian-made parts, up 27 per cent since the late 1980s (Jestin, 1996: 10). Employment in auto parts production reached a peak in 1999. However, since 2001, employment has slipped back to mid-1990s levels. Two post-2001 events altered the 'Canadian advantage'. First, the strengthening of the Canadian dollar caused the exchange advantage to largely disappear. Second, access to the US market became more complicated because of security concerns related to the 11 September 2001 attacks on New York and Washington. Delays at the border increased costs for truckers and compromised the 'just-in-time' system for the parts industry. The border issue is a particularly critical matter for the parts industry because many items cross the border multiple times before they are transformed into parts for the assembly plants. Some appreciation of trans-border movement of parts is revealed in the fact that new vehicles are composed of over 8,000 individual parts.

Automobile Assembly Plants

Automobile assembly plants are concentrated in southern Ontario where transportation links to the major markets of Canada and the United States are readily available and driving distances are short (Figure 5.4). All Canadian-based automobile companies sell most of their cars and trucks in the United States. In 2005, approximately 85 per cent of all vehicles produced in Ontario were sold in the United States. Three companies, General Motors, Ford, and DaimlerChrysler, accounted for 60 per cent of the North American vehicle production in 2005.

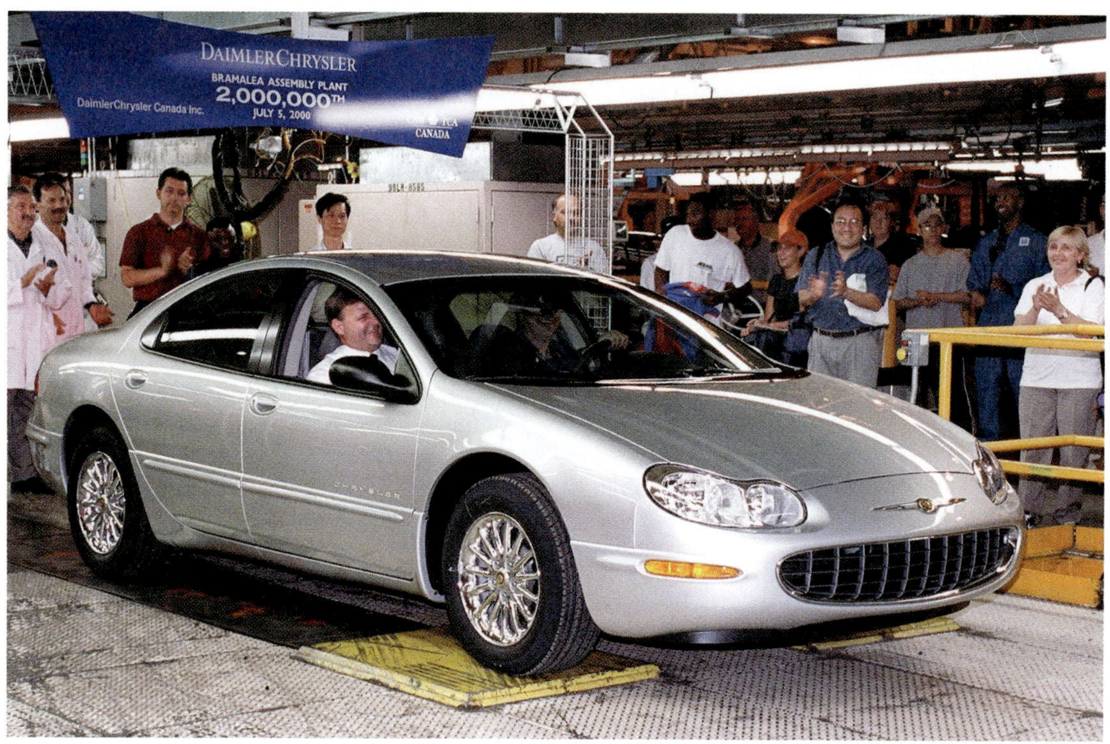

A new car roles off the assembly line at the Chrysler plant in Brampton, Ontario. (CP/Tannis Toohey)

Ten years earlier, the Big Three dominated the automobile market by controlling 90 per cent of car and truck sales. Competition from Japanese and Korean automobile companies with assembly plants located in North America, including southern Ontario, is fierce. These Asian-based firms have captured 40 per cent of Canada's market and an equivalent part of the American market.

While Japanese automobile factories are located in the United States, the Canadian advantages mentioned earlier—a highly motivated workforce, the lower value of the Canadian dollar, and the existence of a public medical system—have attracted substantial Japanese investment in Ontario. In American-based assembly plants, the cost of providing medical insurance is often built into the wage agreement.

With the closing of the General Motors plant at Sainte-Thérèse near Montréal in 2002, all auto assembly plants are located in southern Ontario close to both the Canadian and American markets (Table 5.3). In the long run, this closure will affect auto-parts firms based in the Montréal area because these suppliers tend to locate their factories near assembly plants to reduce transportation costs. For example, in 1996 Magna International decided to build a truck frame plant in St Thomas, Ontario, from which truck frames could be delivered to any of three General Motors truck assembly plants within one day. One of these assembly plants is located at Oshawa, Ontario, while the other two are located in Pontiac, Michigan, and Fort Wayne, Indiana.

Troubled Times: Lose Some, Win Some

A fierce fight for the highly prized North American market is underway. The closing of the General Motors plant at Sainte-Thérèse near Montréal in 2002 signalled the start of restructuring by the Big Three. By 2006, the Big Three's hold on this market had slipped while Honda and Toyota, the two leading Japanese automobile manufacturing compa-

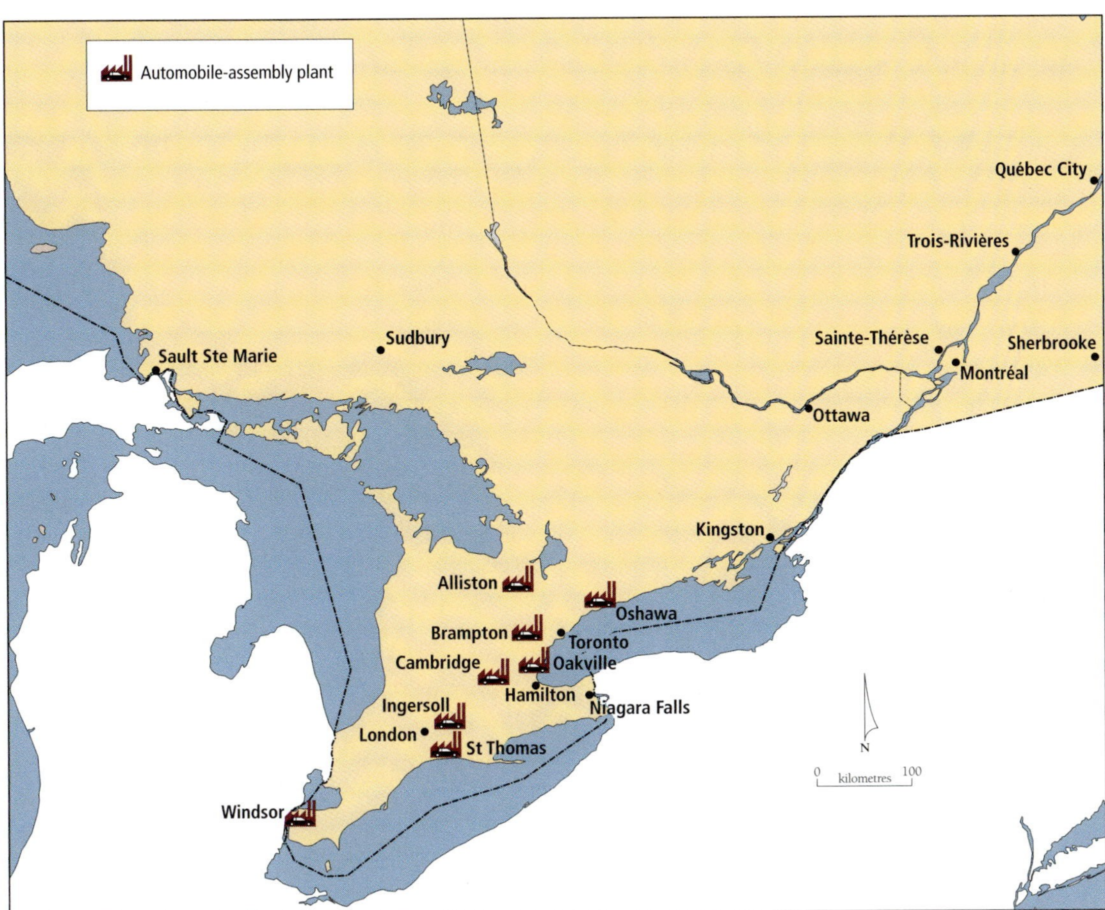

Figure 5.4 Automobile assembly centres in Ontario. Canada's auto assembly plants are located in southern Ontario. The Sainte-Thérèse plant north of Montréal closed in 2002. (Further resources: Student website, National Atlas section, Maps, 18, 19, and 20. Website instructions are found on p. xx.)

nies, had grabbed more of the market. For instance, the top five passenger cars sold in 2004 consisted of four Japanese cars and one General Motors vehicle. The Big Three dominate in the sale of light trucks, minivans, and SUVs. For the Big Three, the recent drop in their sales is due to (1) a shift from the larger vehicles to smaller, more fuel-efficient cars because of higher gasoline prices; (2) customer satisfaction with the performance and quality of Japanese vehicles, which results in both new sales and return sales; and (3) the new demographics, i.e., the baby boomers are becoming empty-nesters and no longer require large vehicles such as minivans and SUVs.

The Big Three automobile companies are in trouble. Their new strategy is to reduce production and cut costs by closing older, less efficient assembly plants and replacing them with more efficient and flexible factories. Yet, in 2003, DaimlerChrysler, troubled by overproduction, abandoned its plans to build a state-of-the-art assembly plant at Windsor. At the same time, the company announced that it would close its Dodge Ram van assembly plant in Windsor, thus eliminating 1,000 jobs (Brent, 2003: FP5). Since 2003, the bad news has continued. Ford announced in 2006 that it was closing its Essex engine plant in Windsor while General Motors plans to shut its powertrain plant in St Catherines (Van Praet, 2006). In February 2007, DaimlerChrysler declared that the company would eliminate as many as 2,000 jobs at its plants in Brampton (Keenan, 2007). In spite of these cutbacks, most plant closures are occurring in the United States, and

Table 5.3 Automobile Assembly Plants in Southern Ontario, 2005

Location	Company	Employment	Product
Oshawa	General Motors Canada	9,643	Chevrolet Silverado, GMC Sierra, Impala, Pontiac Grand Prix, Buick Allure/LaCrosse
Windsor	DaimierChrysler Canada	5,522	Dodge Grand Caravan, Town and Country, Pacifica
Cambridge	Toyota	4,340	Matrix, Lexus
Alliston	Honda	4,300	Civic Sedan, Acura MDX, Pilot, Ridgeline
Brampton	DaimierChrysler Canada	4,200	Chrysler 300, Dodge Magnum, Dodge Charger
Oakville	Ford Canada	3,557	Freestar, Mercury Monterey
Ingersoll	GM/Suzuki Canada (CAMI)	2,775	Chevrolet Equinox, Pontiac Torrent
St Thomas	Ford Canada	2,578	Crown Victoria, Mercury Grand Marquis, Mercury Marander

Source: Robertson (2005). Reprinted with permission from *The Globe and Mail*.

Ontario is making some gains. For instance, Ford will transfer production of its Lincoln Town Car from St Thomas to Oakville in 2008 and begin assembling a new vehicle, the Ford Edge. Another example is that General Motors, in a surprise move, plans to produce the Camaro, a 'muscle' car, at Oshawa. The reductions by the Big Three have parts plants feeling the pinch and layoffs have been announced for firms in Stratford, St Mary's, Newmarket, and Toronto (Van Praet, 2006).

Asian automakers are capturing more and more of the North American market. While Honda and Toyota are scouting for sites to build manufacturing facilities in North America, GM and Ford are closing plants and slashing jobs after losing a steady stream of customers to foreign-based rivals. In Ontario, Honda and Toyota are expanding their production capacity, thus creating more jobs and more demand for parts. Toyota is building an assembly and parts plants at Woodstock with production of the RAV4 sports utility beginning in 2008. Honda's new engine plant near Alliston will also be in production by 2008. Why are Honda and Toyota expanding their operations in Ontario when most of their sales are in the United States?

The primary reason is that there is a strong demand for their automobiles. Expansion in Ontario is encouraged by support from both the federal and provincial governments plus the strong work ethic of small-town workers. For these reasons, Japanese manufacturers see Ontario as one of its preferred production sites in North America.

Northern Ontario

As an old resource hinterland, northern Ontario is troubled by a sluggish economy, a declining population base, and high unemployment rates. Even northern Ontario's transshipment role between Ontario and Western Canada has weakened as more products, such as grain and potash, are shipped to Vancouver for sale in Asian countries. With a stagnant economy, northern Ontario's population has three demographic characteristics that are strikingly different from southern Ontario:

- an aging population;
- a net out-migration, especially of younger members of its population;
- few immigrants.

The physical base of northern Ontario is considerable but its population consists of fewer than 750,000 people. In percentage terms, then, northern Ontario comprises 87 per cent of the geographic area of Ontario but has less than 7 per cent of Ontario's population. With virtually no agricultural base, most people live in towns and cities that lie along the two transportation routes that link Montréal and Toronto with Winnipeg. Both of these trans-Canada routes cross the rugged Canadian Shield. The southern route includes the Canadian Pacific Railway and the Trans-Canada Highway. The Trans-Canada Highway connects North Bay with Sudbury, Sault Ste Marie, Nipigon, Thunder Bay, Dryden, and Kenora. The northern route includes the Canadian National Railway and a major highway. This northern highway connects North Bay, Timmins, Kirkland Lake, Cochrane, Kapuskasing, and Nipigon to Thunder Bay.

Mining, forestry, and tourism are the major economic activities, although many people are employed in the public sector. Like other resource hinterlands, the development of northern Ontario's economy was linked to external markets. Resource exploitation in northern Ontario began after the construction of railways in the late nineteenth century. Soon thereafter, resource development took place along the lines of the Canadian Pacific Railway (CPR) and the Canadian National Railways (CN). The same pattern of resource development took place along the Temiskaming and Northern Ontario Railway (also known as the Ontario Northland Railway).[2] The railway extends northward from North Bay to New Liskeard, Kirkland Lake, and Cochrane, which is on the CN main line. Northern Ontario exports minerals and forest products, including gold, nickel, newsprint, and lumber. These resource products contribute less than 10 per cent of the value of Ontario's exports to foreign countries.

Unlike the northern hinterland of Québec, the natural conditions in northern Ontario are not conducive to major hydroelectric developments. While a number of rivers flow across northern Ontario to James Bay and Hudson Bay, the gentle slope of the land, especially in the Hudson Bay Lowland, does not make the construction of hydroelectric dams feasible. A few established sites for hydroelectric production, such as along the Ottawa River and at Abitibi Canyon about 100 km north of Cochrane, are important, but they cannot meet the energy demands in Ontario. The much higher elevations in the Canadian Shield area of Québec provide the necessary natural drop to drive the turbines that produce hydroelectric power. Since Québec has a surplus of electrical power, some of it is transmitted to southern Ontario.

Northern Ontario has a strikingly different settlement pattern from that of southern Ontario (Vignette 5.7). Long distances separate the four major cities. For example, the distance from North Bay to Sudbury is over 100 km; from Sudbury to Sault Ste Marie is about 300 km; and from Sault Ste Marie to Thunder Bay is about 500 km. The explanation for this oasis-like settlement pattern is due to northern Ontario's physical geography. The rocky terrain of the Canadian Shield discourages continuous settlement; a series of isolated settlements are located at mining sites, pulp plants, and key transportation junctions. Similar to other old hinterlands, the urban centres of northern Ontario are declining and the populations of single-industry towns like Cobalt, Kirkland Lake, and Porcupine rose and fell as the cycle of resource exploitation ran its course.

Mining Industry

Northern Ontario's mining industry is centred in the Canadian Shield, which provided ideal geological conditions for the formation of hard-rock minerals such as gold, nickel, and copper. In 2005, Ontario was the leading Canadian producer of gold and nickel. The annual value of metallic minerals was nearly $5 billion (Natural Resources Canada, 2006b). Most production comes from Red Lake (gold), Hemlo (gold), Wawa (gold), Manitouwage (gold), Marathon (gold), Thunder Bay (gold, copper, zinc), and Sudbury (nickel, copper). In a few years, Ontario's first diamond mine

Mining forms a major part of northern Ontario's economy. The Williams gold mine near Marathon is the largest gold-producing mine in Canada. In 2002, Canada accounted for 5.8 per cent of world gold production, which amounted to 6 million ounces. (Barrett & MacKay Photography, Inc.)

(the Victor Mine), located in the Hudson Bay Lowland some 500 kilometres north of Timmins, may open. For the past several years, prices for gold, nickel, and copper have been very high, allowing mines to operate at full production. For those reasons, mining remains the most robust primary activity in northern Ontario.

The first important mineral discovery took place in 1883 when the copper-nickel ores of the Sudbury area were discovered during the building of the Canadian Pacific Railway. In 1903, a rich silver deposit near Cobalt was detected during the construction of the Temiskaming and Northern Ontario Railway. Rich gold deposits were uncovered near Timmins in 1909 and at Kirkland Lake in 1911. While new mineral finds kept the mining economy of Ontario going, the overall pattern was one of boom and bust. During the 1960s, a large lead-zinc deposit was discovered in Timmins. A decade later, rich gold deposits were found at Marathon (the Hemlo gold mine) and at Pickle Lake. Promising deposits, including diamonds at three locations (Victor, Marathon, and Wawa), are currently being evaluated for commercial production. Overall, however, the labour force is declining. The main reason is increased workforce productivity as a result of greater mechanization of mining processes.

Minerals are a non-renewable resource that is depleted over time. Therefore, mining communities can have a short lifespan. A mine closure can occur without warning, causing the sudden demise of a single-industry centre. Over the years, a number of ore bodies have been exhausted or production has ceased because new, lower-cost mines have been discovered. These two circumstances have forced some mines to close, resulting in a great outflow of miners and their families. For instance, Elliot Lake's uranium mine was closed, not because its ore was exhausted but because open-pit uranium mines in northern Saskatchewan could produce uranium oxide at a much lower price. Ontario Hydro, the principal buyer of Elliot Lake uranium, decided to purchase its supplies from the uranium mines in northern Saskatchewan. Other mines that have closed include an iron mine at Steep Rock and the silver mines at Cobalt. While former mining towns may remain, they undergo a drastic downsizing. With a stock of low-priced houses and an attractive wilderness setting, Elliot Lake made a modest recovery as a retirement centre. However, its population has declined steadily and is now only about 12,000 (see Table 5.7).

Forest Industry

The forest industry of northern Ontario produces approximately $15 billion of products each year, with 60 per cent exported to markets in the United States. Ontario is one of the leading exporting provinces of softwood lumber. Access to the American market is critical. Access is controlled by agreements between Canada and the United States. The most recent agreement, the 2006 Softwood Lumber Agreement (SLA), limits Canadian softwood exports. Under this agreement, Canadian lumber firms are allocated 34 per cent of softwood lumber sales in the US market, which is roughly Canada's recent share. (For a fuller discussion of the SLA, see Vignette 7.6, 'The 2006 Softwood Lumber Agreement'). Each province would be allocated a share of those exports based on 2004–5 exports. Ontario was allocated around 9 per cent of the US softwood lumber exports (CTV, 2006).

Across Ontario, the northern coniferous forest is classified into two regions: the boreal barrens and the boreal forest regions. The boreal barrens region is found in the Hudson Bay Lowland. This forest region, which represents a transition between the boreal forest and

Vignette 5.7 Population Distribution in Northwestern Ontario

The large, sprawling Thunder Bay census division of northwestern Ontario stretches from Lake Superior to Hudson Bay and James Bay. This division has less than 150,000 people. As is typical of northern Ontario, most people live in a few large urban centres while the remaining territory is virtually empty. Northwestern Ontario's population geography has three distinct zones:

- Zone 1: Most people (67 per cent) reside in the City of Thunder Bay. In 2006, this major CMA population cluster had 122,907 people.
- Zone 2: Most of the remainder (27 per cent) live south of the CN rail line where a number of small towns, villages, and Native reserves are located.
- Zone 3: North of the CN rail line lies the 'empty' population zone, which is beyond the reach of the national highway system and has only 2 per cent of the region's population.

Beyond these towns and cities lies a sparsely populated area where fewer than 10,000 people live. Cree and Ojibwa Indians live on isolated reserves or in small Native settlements in this vast region of scrub forest and muskeg. Only a few gold mines, trapping cabins, and fishing lodges dot the landscape. For the Indian residents, fishing and hunting are an important source of food, while trapping and guiding provide cash income. Since this cash income usually does not meet their basic needs, transfer payments supplement their income. The development of the Victor diamond deposit near the Attawaspiskat First Nation reserve provides a potential economic boost for the band because de Beers has agreed to hire local workers for both its construction and production operations. This open pit mine is scheduled to open in 2008.

Northern Ontarians are dissatisfied with their lot. Such grumbling is common in resource hinterlands. Claiming to be isolated and ignored by the provincial government, some of its citizens are calling for a new province called Mantario (Matteo, 2006a).

the tundra, has no commercial value. Scattered patches of spruce and tamarack are surrounded by poorly drained land known as muskeg. The commercial forest industry is located in the boreal forest region that extends from Québec to Manitoba. The geographic limits of the boreal forest region closely parallel those of the Canadian Shield. The main species are black and white spruce, tamarack (larch), balsam fir, Jack pine, white birch, and poplar. Some 50 communities depend on the forest industry based on the boreal forest region. By volume of wood cut, Ontario ranks just behind British Columbia and Québec. Ontario's mills produce pulp and paper, lumber, fence posts, and plywood. The pulp and paper industry in northern Ontario accounts for about 25 per cent of the national production, and Ontario is a leading exporter of newsprint and pulpwood to the United States. In fact, most pulp and paper firms operating in northern Ontario are American-owned companies.

Road access is an extremely important factor in the forest industry. Trucks must bring the pulpwood to the mill and then the final product must be shipped to market. Pulp and paper mills are located in communities that have access to transportation networks—Thunder Bay, Sault Ste Marie, Sturgeon Falls, Kenora, Fort Frances, Dryden, Marathon, Iroquois Falls, Kapuskasing, and Terrace Bay.

Technology in the forest industry has advanced over the years and greatly increased the productivity of workers in the mills and in the woods. When loggers first arrived in northern Ontario, they had only axes and saws. A logger with such equipment might cut five to 10 trees a day. By the middle of the twentieth century, the technology had advanced. Chainsaws then enabled a logger to cut 50 to 70 trees a day. Mechanical harvesters, which can harvest hundreds of trees per day, have now replaced the chainsaw.

Transportation of timber to mills has also changed. In the past, logging often took place in the winter months. Horses pulled the logs to frozen rivers where they were stored until the river melted in the spring. Tractors later took over the task of hauling the logs to the river. In the spring, the logs were floated to sawmills, which were often located at the mouths of the rivers. Today, most timber is used in the pulp and paper industry. These huge plants require a steady supply of pulpwood. This demand and the technological changes that have taken place have altered the nature of logging in four ways: (1) logging has become a year-round affair; (2) most logs are hauled to the mills by trucks; (3) trees are harvested before they reach maturity; and (4) mechanical tree harvesters are employed and usually clear-cut a wooded area.

Today, the forestry industry faces several challenges. The first challenge is to maintain a balance between logging and the regeneration of the forest. Since over 90 per cent of Ontario's forest lands are owned by the provincial government, private companies must obtain forest leases. The Ontario government, like most other provincial governments, has insisted that these private companies take on the responsibility for restoring trees to logged areas through forest management agreements. Efforts to hand-plant seedlings are helping to speed up the process of reforestation, but how successful these efforts have been can only be known 20 years from now. The practice of granting forest companies long-term timber leases is based on the assumption that it is in the companies' self-interest to manage the forest well. It remains to be seen whether this assumption and the corresponding leasing policy will ensure the protection and regeneration of Ontario's forests.

A second challenge is the changing nature of the boreal forest from a predominantly coniferous forest to a broadleaf one. Since the 1950s, the volume of timber harvesting has more than tripled. This level of logging has put so much pressure on the boreal forest that original species are disappearing. As a result, the boreal forest is shifting from coniferous to broadleaf species. Spruce, pine, and fir have been replaced by poplar and birch, which are both pioneer species, well adapted to regenerate in clear-cut portions of the boreal forest. The long-term consequences of this species shift are unknown, but such a shift clearly illustrates how forest companies have replaced forest fires as the main agent of change.

A third challenge facing the forest industry is the age of its pulp and paper plants and a drop in demand for these products. Many mills in Ontario were built before World War II and continue to use old technology, which results in much higher discharges of toxic wastes into the environment. The consequences of such practices can be life-threatening. The worst cases of massive toxic discharges into streams and rivers occurred in the 1960s. At that time, a chemical plant that prepared the bleaching solution for the pulp plant at Dryden was releasing effluents containing mercury into the English-Wabigoon river system.[3] Since then, forestry mills have updated their operations to ensure a cleaner and safer natural environment.

The fourth challenge is related to the US market for softwood lumber. The forest industry is in trouble because of the low prices for softwood lumber. In 2006, the sharp downturn in the US housing market produced a double whammy: fewer exports to the US and a tax on Canadian softwood lumber exports. The tax, which comes into effect when the price drops below US $355 per thousand board feet, is part of the new softwood lumber agreement. As US demand for lumber declines, so does the price of lumber. From May 2006 to May 2007, the price of softwood lumber had dropped from US $367 to US $224, thus triggering the maximum export tax of 15 per cent, which is collected by Ottawa (Anderson, 2006; Export Development Canada, 2007).

These challenges have placed the forest economy of northern Ontario in a tailspin. In one year, from 1 April 2005 to 31 March 2006, nine mills closed (Natural Resources Canada, 2006). Kenora and Thunder Bay suffered two losses each—Kenora lost a sawmill and a newsprint plant while a fine paper mill and sawmill disappeared from Thunder Bay's industrial landscape. Only one new investment—a mere $14 million for a new boiler and kiln at Chapleau in northeastern Ontario—took place during this time frame. While mill closures have occurred in all regions of the country, the majority of closures took place in Ontario and Québec. Conversely, forest companies have invested most heavily in Western Canada and British Columbia. The problem for northern Ontario has been many old plants with higher production costs than elsewhere, thus making them candidates for closure.

Ontario's Urban Geography

Ontario is the most highly urbanized province in Canada. Almost 9.5 million people—nearly 85 per cent of the province's total population—live in towns and cities. As well, many of Canada's largest cities are located in Ontario. For example, 10 of the 25 largest cities in Canada are in Ontario. With the exceptions of Sudbury and Thunder Bay, these census metropolitan areas (CMAs) are located in southern Ontario.

Since cities are the engines of the Canadian economy, Ontario's urban geography provides an enormous economic advantage. Cities are where employers set up for business and thus are where most employment is located. The pattern of urban growth from 2001 to 2006 varied considerably for these CMAs (Table 5.4). The fastest-growing cities in Ontario were Oshawa, Toronto, Guelph, and Ottawa–Gatineau. Sudbury and Thunder Bay,

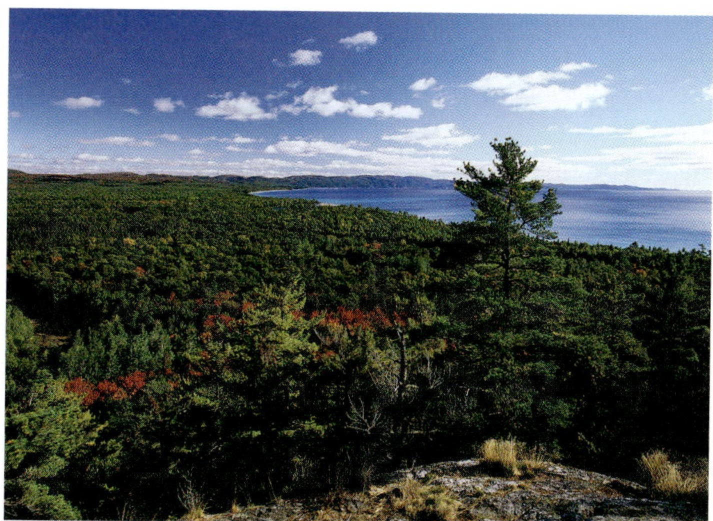

The boreal forest stretches across most of northern Ontario, providing the basis for a forest industry. The boreal forest, in combination with its lakes and rugged landscape, produces spectacular scenery that attracts many tourists. (© Derek Trask/Corbis)

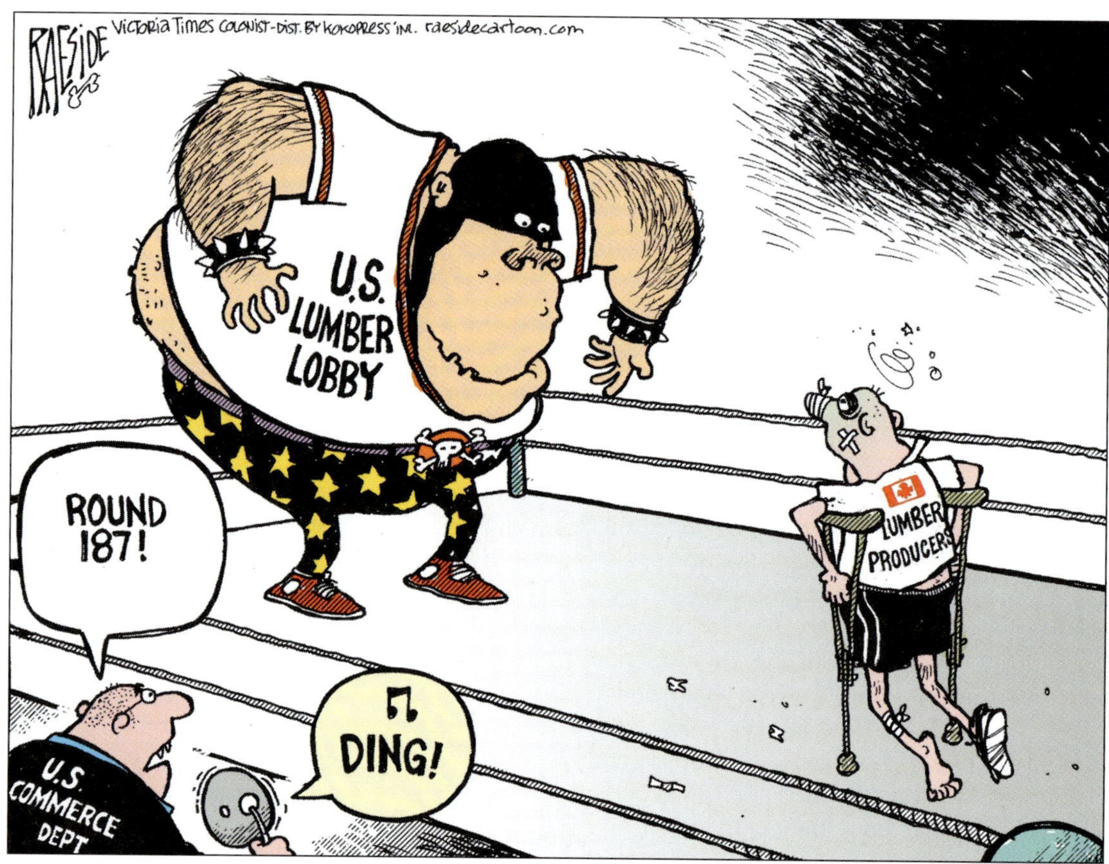

Figure 5.5 US lumber lobby wins again. The objective of the US lumber lobby is to curtail imports of Canadian softwood. The Canadian forest industry, over the years, has suffered from duties imposed by the US government. Just as the latest softwood lumber agreement was reached in 2006, US prices for softwood lumber dropped sharply. The combination of the US duties followed by falling prices has badly hurt the forest economy in northern Ontario, where plant closures and layoffs are far too common. (© Adrian Raeside)

both located in the resource hinterland of the Canadian Shield, showed practically no growth from 2001 to 2006, well below the growth rate of all the CMAs in southern Ontario. In fact, in the previous five-year period, 1996–2001, these two cities suffered a population decline due to the stagnant economy found in northern Ontario.

The eight largest cities in Ontario are located in southern Ontario (Figure 5.6), and of these the four largest (Toronto, Ottawa, Hamilton, and London) form the core of the three urban clusters in southern Ontario, namely, the Golden Horseshoe, southwestern Ontario, and the Ottawa Valley. Each of southern Ontario's three major urban clusters has a population of at least 1 million. As mentioned earlier, northern Ontario has fewer and smaller cities, each separated by long distances. It has two major cities (Sudbury and Thunder Bay) and no single urban cluster. Sudbury, the largest centre, had only 158,258 residents in 2006, while Thunder Bay's population was 122,907.

The Golden Horseshoe

The Golden Horseshoe obtained its name because of its horseshoe-like shape around the western end of Lake Ontario and its outstanding economic performance over the years. As a result, this tiny area of the Great Lakes Lowlands forms the most densely populated area of Canada. The Golden Horseshoe

Table 5.4 Census Metropolitan Areas in Ontario, 2001–6

Census Metropolitan Area	Population 2001 (000s)	Population 2006 (000s)	Change (%)
Peterborough	110.9	116.6	5.1
Thunder Bay	122.0	122.9	0.8
Brantford	118.1	124.6	5.5
Guelph	117.3	127.0	8.2
Kingston	146.8	152.4	3.8
Sudbury	155.6	158.3	1.7
Oshawa	296.3	330.7	11.6
Windsor	307.9	323.3	5.0
St Catharines–Niagara*	377.9	390.3	3.5
Kitchener	414.3	451.2	5.2
London	432.5	457.7	5.1
Hamilton	662.4	692.9	4.6
Ottawa–Gatineau*	1,067.8	1,130.8	5.9
Toronto	4,682.9	5,113.1	9.2
Total	9,043.7	9,691.8	7.2

*Statistics Canada has combined St Catharines and Niagara Falls as a single census metropolitan area although these cities still exist as separate political jurisdictions, as do, of course, Ottawa, Ontario, and Gatineau, Québec.

Source: Statistics Canada (2007d). Adapted from Statistics Canada *Population and Dwelling Counts (On-line) – Highlight Tables, 2001 Census*, Catalogue 93F0051XIE, http://www.statcan.ca/bsolc/english/bsolc?catno =93F0051XIE and *Population and Dwelling Counts, 2006 Census*, Catalogue 97-550-XWE2006002 http://www.statcan.ca/bsolc/english/bsolc?catno=97-550-XWE2006002.

extends from the US border at Niagara Falls northward to Hamilton, Toronto, and then on to Oshawa. Nearly 7 million Canadians live, work, and play in Canada's largest population cluster, and many visitors come either as tourists or on business trips. Accounting for nearly one-quarter of Canada's population, the Golden Horseshoe contains numerous towns and cities, including Toronto, Hamilton, Oshawa, St Catharines, Niagara Falls, Burlington, Oakville, Pickering, Ajax, and Whitby. Toronto, the largest city in Canada, is its urban anchor, while Hamilton, with its steel plants, is the focus of heavy industry (Vignette 5.8), and Oshawa is Canada's leading automobile-manufacturing city.

Toronto

As the largest city in Canada, Toronto dominates the urban landscape of southern Ontario and plays a key role in Canada's urban hierarchy. Toronto is the financial capital of Canada. The city houses the main offices of national and international banks and investment firms, as well as the Toronto Stock Exchange. Over the years, Toronto's population growth has outpaced that of most other cities, and by 2006 the population of the Greater Toronto Area had reached 5 million. Over the last decade, the main factor driving Toronto's growth has been the arrival of immigrants from foreign countries. Between 1996 and 2001, nearly half a million new Canadians settled in Toronto. At the same time, people and businesses spilled over Toronto's boundaries, causing a geographic expansion of this urban growth. The spread of Toronto's population and businesses into neighbouring areas is partly due to lower land values and rents. Land values and rents are highest in downtown Toronto. Cities surrounding Toronto have much lower land values and can therefore attract businesses by offering lower office rents.

As the province's major cultural and entertainment centre, Toronto is a hub for the entertainment industry, which ranges from live theatre to professional sports, and plays an essential role in shaping Canadian culture. Tourists flock to Toronto to enjoy these world-class cultural and entertainment activities. Just how dependent the cultural and service industries are on tourists was revealed by the outbreak of SARS in April 2003, which led to 23 deaths and, in part as a result of a travel advisory issued by the World Health Organization, was estimated to have cost the city $1 billion in that month alone (Brean, 2003).

Like other major cities in North America, Toronto is trying to cope with a rapidly expanding population, much of which has moved into

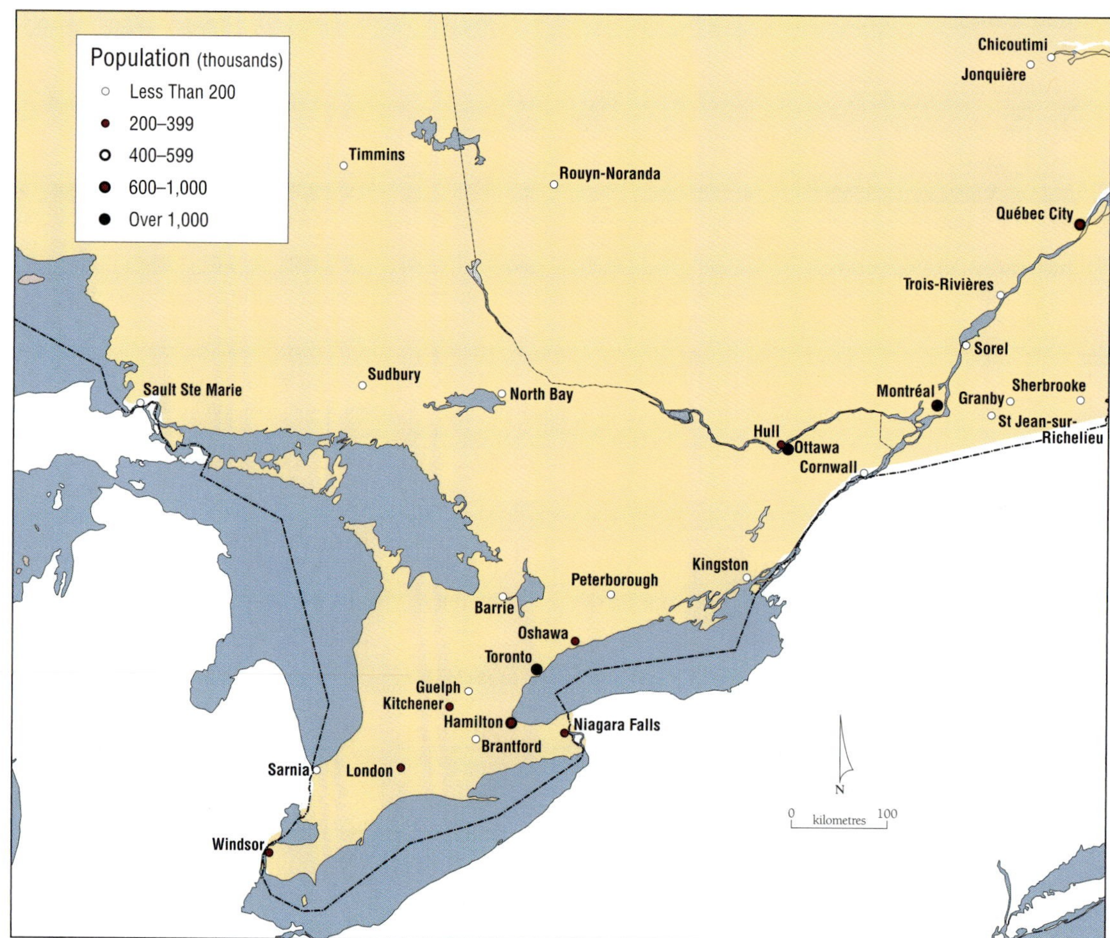

Figure 5.6 Major urban centres in Central Canada. Most large urban centres in Central Canada are located in the Great Lakes–St Lawrence Lowlands, especially in southern Ontario.

Vignette 5.8 Hamilton: Steel City

Hamilton, the third largest city in Ontario, is situated at the west end of Lake Ontario on Burlington Bay. Hamilton is Canada's largest steel producer and ranks high in industrial production.

Originally, however, Hamilton was a textile town. Streams flowing down the Niagara Escarpment provided power for the mills, and the city's location at the head of navigation on Lake Ontario gave it access to other towns and cities.

Steel production flourished with the railway boom on the Canadian Prairies in the early 1900s. Hamilton's steel industry languished during the Great Depression but during both World War I and World War II, military demand for steel created boom conditions.

After World War II, Hamilton's steel industry shifted its production to the rapidly expanding consumer market, especially for appliances and automobiles. Two of Canada's largest steel firms (Stelco and Dofasco) are located in Hamilton. Stelco employs some 5,000 workers while Dofasco has 7,400 employees. In 2003, Stelco fell on hard times, and in May 2004 the company filed for bankruptcy protection. By 2006, Stelco was back in business after restructuring and recapitalizing.

adjoining urban areas. One effort to deal with the administration of this urban area was to create a super-city known as the Greater Toronto Area (GTA). In 1998, the City of Toronto and the surrounding regional municipalities of Halton, Peel, York, and Durham officially formed a single city government with the hope that solutions to problems arising from rapid population growth and geographic expansion of the urban population could be found.[4] While many opposed this approach to mega-urban government, the objective is to reduce administrative overlap and thereby create a more efficient urban government. Whether or not this single administration approach is succeeding in serving such a large population remains to be seen. However, many problems remain, including traffic congestion. With so many people living outside of downtown Toronto but working in the city centre, massive numbers of them commute daily, creating traffic jams during the morning and afternoon rush hours. Traffic congestion leads to time lost travelling to and from work. A radical solution would be charging a toll on automobile traffic into and out of Toronto. London, England, imposed such a tax and it resulted in a reduction in traffic flow and a decrease in travel time.[5]

Interior of the Toronto Eaton Centre (Dennis Flood, www.dennisflood.com)

Since the 1960s, immigrants have been coming to Canada from non-European countries in large numbers. In 2001, 16 per cent of immigrants came from China, 11 per cent from India, and 6 per cent from Pakistan. These photos depict two thriving communities in Toronto: Chinatown and Little India, the Southeast Asian community in Toronto's east end. (©Megan Mueller)

As the most popular destination for immigrants to Canada, Toronto has experienced remarkable ethnic diversification over the past 30 years, to the point where visible minorities are hardly in the minority. Visible minorities comprised 37 per cent of Toronto's population in 2001 compared to 26 per cent in 1991. By 2001, Toronto had over 2 million immigrants, who accounted for 44 per cent of the city's population. In recent years, most have come from China and India. The origin of immigrants by place of birth reveals that, in 2001, nearly 18 per cent of the 792,030 immigrants came from China and Hong Kong, with another 10 per cent from India and 7 per cent from the Philippines (Table 5.5).

Toronto is known as a city of neighbourhoods, partly because immigrant groups have clustered in certain neighbourhoods. Little Italy and Little Portugal are older, well-established ethnic neighbourhoods. More recent Asian, African, and Caribbean neighbourhoods are emerging. Recent immigration into Toronto has had an impact on its cityscape, including 'foreign' architecture and commercial activities designed to meet the demands of these new Canadians. Asian theme malls represent a dramatic change to Toronto's space. Asian theme malls consist of a large number of small retail outlets, many of which are either Chinese restaurants or specialty stores. Unlike other suburban malls, no anchor store exists and, rather than a single developer who leases space, each retail unit is owned by the operator of that retail space. The failed development of an Asian theme mall in Richmond Hill marks the resistance by local residents to a changing urban space with a new ethnic identity (Preston and Lo, 2000: 182–90). Toronto Island also represents a distinct neighbourhood born out of confrontation with city planners who had decided to transform Toronto Island into public open space. While originally a sandbar attached to the mainland, this district of Toronto now consists of a series of islands, most of which are dedicated to parkland. However, residential communities, both of which resemble bohemian enclaves, exist on two islands (Ward's Island and Algonquin Island).

Ottawa Valley

Ottawa–Gatineau is not only a major population cluster in Canada, but it is also an important area where both official languages are used. As the nation's capital, federal government operations are found on both sides of the Ottawa River. In 2001, Ottawa had a population of just over 800,000 while Hull (now Gatineau) in Québec had nearly 260,000 citizens. Together, Ottawa–Gatineau represents the fourth largest metropolitan area in Canada, and in 2006 its population totalled 1,130,761. On the Ontario side, Greater Ottawa is the third largest urban cluster in Ontario. Much of its growth has come from in-migration from other parts of Canada and from foreign countries. Many are attracted by the employment opportunities offered by the federal government and the business community. In 2001, 18 per cent of its population was foreign-born compared to 15 per cent in 1991. Visible minorities accounted for 14 per cent of population, with the three leading groups being blacks, Chinese, and Arabs/West Asians (Table 5.6).

The federal government, located in the national capital, is the major employer, followed by the high-technology sector. The federal government requires a wide variety of goods and services in its daily operations. This demand provides an opportunity for many small and medium-sized firms in the Ottawa Valley. By locating its departments and agencies in both Ottawa and Gatineau, the federal government has ensured that Ottawa's economic orbit extends to a number of small towns on both the Ontario and Québec sides of the Ottawa River. Most people live on the Ontario side, especially in Ottawa. Other urban centres on the Ontario side include Nepean, Gloucester, Kanata, and the municipalities of Rockcliffe and Vanier. Around Ottawa, in its periphery, are a number of smaller centres, including Pembroke,

Table 5.5 Top 10 Countries by Birth of Immigrants to Toronto, 1991–2001

Country of Birth	Immigrated 1991–2001	%
China, People's Republic of	85,345	10.8
India	81,845	10.3
Philippines	54,885	6.9
Hong Kong, Special Administrative Region	54,805	6.9
Sri Lanka	50,425	6.4
Pakistan	39,265	5.0
Jamaica	25,355	3.2
Iran	23,840	3.0
Poland	21,555	2.7
Guyana	20,800	2.6
Total of all immigrants to Toronto, 1991–2001	792,030	100.0

Source: Statistics Canada (2003f). Adapted from Statistic Canada publication *Canada's Ethnocultural Portrait: The Changing Mosaic*, 2001 Census, Catalogue 96F0030XIE2001008, Released 21 January 2003, Searched 28 January 2003, http://www12.statcan.ca/english/census01/products/analytic/companion/etoimm/canada.cfm.

Cornwall, Brockville, and Kingston. However, with the exception of Pembroke, these cities are more closely tied to economic and social activities in Toronto and Montréal.

As the capital of Canada, Ottawa is the focus of national and international affairs and has subsequently developed into a national administrative centre with a large federal civil service. In its early days, the forest industry played a strong economic role, but it was eventually overshadowed by the activities of the federal government and then the high-tech industry. Still, the geographic location made this an ideal place for a pulp and paper mill: pulpwood was easily harvested from the interior forest lands and then transported along the Ottawa and Gatineau rivers to the Eddy Pulp and Paper Mill in Hull, Québec. An added advantage of this site is the availability of low-cost electrical power from the hydroelectric installation on the Gatineau River near the Chaudière Falls. Today, the E.B. Eddy plant produces fine and coated paper.

Described in the press as the 'Silicon Valley of the North', Ottawa has become an industrial leader in high technology. Ottawa firms specialize in telecommunications, computers, software and computer services, and electronics (Britton, 1996b: 266). Northern Telecommunications (Nortel), the leading high-tech company, specializes in fibre optics. In the 1990s, the demand for high-tech products seemed limitless and high-tech companies expanded rapidly. For example, in 1997, Nortel announced plans for major expansion in the Ottawa area. Over a four-year period, Nortel planned to invest $250 million and hire nearly 5,000 new employees (McCarthy, 1997: B4). Such a commitment demonstrates the attractiveness of Ottawa as a location for the high-tech industry, but a sharp downturn in demand for its products beginning in 2000 not only put Nortel's expansion plans on hold but put the company close to bankruptcy. The dot-com bubble burst and Ottawa's high-tech business was on the downslide. The leading company, Nortel, had recovered by 2003, but with a much smaller operation and a greatly reduced labour force. Other leading figures in Ottawa's high-tech world failed and some have been purchased by larger companies outside of Canada. As James Bagnall (2003: FP5) reported, 'With software maker Corel now in the hands of a California-based venture capital firm and fibre-optics star JDS Uniphase set to shift its headquarters next month to San Jose, Ottawa is swiftly reverting to its former role as a technology R&D centre, serving the whims of multinationals based elsewhere.' However, such views proved too pessimistic: after several years of sluggish sales and painful cost-cutting, Canada's high-tech industry bounced back. In Ottawa, the recovery has been rapid. According to Mark Evans (2006), the number of high-tech companies in the Ottawa area has increased by 800 to a record 1,800. In 2005, 2,000 new employees were added to these firms. In 2006, Dell announced an expansion of its customer call centre that will double the number of its employees (Avery, 2006).

Southwestern Ontario

Southwestern Ontario represents the third major urban cluster in southern Ontario. About 1 million people live in this area, which extends from Lake Erie to Lake Huron. Although this region does not have a major metropolitan city, London, with a population of 432,000 in 2001, is the largest city in southwestern Ontario. London is close to Toronto, Buffalo, and Detroit and the main urban centres within southwestern Ontario, Cambridge, Kitchener, London, and Waterloo, are near London. Windsor and Sarnia are on the edge of this urban cluster.

London is the unofficial capital of southwestern Ontario. It provides administrative, commercial, and cultural services for the larger region. London is also the headquarters of several insurance companies, including London Life Insurance Company. Among its

Table 5.6 Top 10 Ethnic Origins in Ottawa–Hull*, 2001

Ethnic Origin	Total—Single and Multiple Responses**	Single Responses	Multiple Responses**
Canadian	463,280	231,095	232,180
French	272,085	58,455	213,625
English	200,900	39,645	161,255
Irish	183,130	25,010	158,120
Scottish	152,215	18,705	133,510
German	63,295	10,370	52,925
Italian	37,440	16,950	20,485
Chinese	31,595	26,895	4,705
North American Indian	25,780	3,830	21,950
Polish	23,695	7,560	16,135
Total population	1,050,755	596,345	454,415

*Now known as Ottawa–Gatineau.

**Respondents who reported multiple ethnic origins are counted more than once as they are included in the multiple responses for each origin they reported. For example, a respondent who reported 'English and Scottish' would be included in the multiple responses for English and for Scottish.

Source: Statistics Canada (2003f). Adapted from Statistic Canada publication *Canada's Ethnocultural Portrait: The Changing Mosaic*, 2001 Census, Catalogue 96F0030XIE2001008, Released 21 January 2003, Searched 28 January 2003, http://www12.statcan.ca/english/census01/products/analytic/companion/etoimm/canada.cfm.

economic activities, London is noted for its manufacturing, including the production of armoured personnel carriers and diesel locomotives by General Motors. London, therefore, has a sound and growing industrial foundation based on insurance, manufacturing, and high-tech industries. In 2003, Magna announced plans to build a new auto-parts plant that will supply car seats for the Equinox, a new sports utility vehicle produced at Suzuki/GM's Cami assembly plant in Ingersoll. These manufacturing firms pay relatively high wages to their employees. Consumer spending by these employees supports a strong retail sector.

Automobile assembly plants are located in Cambridge, Ingersoll, St Thomas, Alliston, and Windsor (Table 5.3). A new Toyota assembly plant is planned for Woodstock. Auto-parts plants play an important role in the economy of southwestern Ontario while a number of high-tech firms, particularly in the Cambridge, Kitchener, and Waterloo areas, add to the region's economic diversity. These three cities form what is known as Canada's 'Technology Triangle' where innovative technology is often developed in research institutes or universities. Such technology is then used to create commercial products. Leading technology companies include Research In Motion Ltd, Open Text Corp., Dalsa Inc., Automated Tooling Systems, and Com Dev Ltd. The following description by Bathelt and Hecht (1990: 225) captured the essence of this technological zone in the 1990s and remains accurate today.

> All [high-tech firms] are surprisingly strongly tied, in terms of input and output linkages, to economic activities within a 100-km radius, which includes the Toronto area. Skilled labour and local residence/education are the strongest location factors. The relatively favourable industrial climate in the Cambridge Technological Triangle region should enable key technology industries to flourish in the area.

Cities of Northern Ontario

In sharp contrast to cities in southern Ontario, those urban centres in northern Ontario find their economies stalled and their populations declining. Like most downward transitional regions, the resource base is losing its economic strength for three reasons: (1) as the best mineral and timber resources have been exploited, resource production is now more costly; (2) depressed prices for lumber have resulted in mill closures; and (3) as companies introduce more and more technology, fewer workers are required.

Timmins, located in the mineral-rich Canadian Shield, is in the centre of a gold belt, while Sudbury is in a world-famous nickel belt, where the mining and smelting of nickel and copper ores are core functions. Both cities began as single-industry towns, but over time they have become important regional centres with many service industries and more diversified economies, factors that have supported community stability. Even with these efforts, however, the two communities have been unable to maintain their populations over the past 10 years. For example, from 1996 to 2006, the populations of Sudbury and Timmins declined by 4.3 per cent and 3.0 per cent respectively (Statistics Canada, 2002, 2007d). For the past five years, Sudbury has enjoyed a slight revival in its population growth while Timmins continued to decline (Table 5.7).

Sault Ste Marie is a steel town located on the Great Lakes between Lake Superior and Lake Huron. Like most other centres in northern Ontario, its population trend over the last 10 years has been downward. For instance, Sault Ste Marie's population declined from 83,619 in 1996 to 80,098 in 2006 (Statistics Canada, 2002, 2007d). Along with Sudbury, North Bay, and Thunder Bay, Sault Ste Marie saw its population increase slightly from 2001 to 2006 (Table 5.7). The town's major firm is the Algoma steel mill, which has struggled to succeed in a highly competitive marketplace. Algoma is located at some distance from its major industrial customers in southern

Ontario and the United States. With high transportation costs, the Algoma mill is at a geographic disadvantage. However, Algoma has made use of local railway and seaway connections to secure raw materials, such as iron ore, and to market its steel products. In 2001, Algoma Steel fell on hard times and filed for protection under the Companies' Creditors Arrangement Act. Algoma Steel was able to recover and is now a profitable firm. The measures were hard on workers, however, including a 9 per cent reduction in pay and a reduction of the workforce by 18 per cent (Bertin and Keenan, 2003: B1). The smaller but more efficient Algoma operation focuses on direct strip production, which has positioned Algoma as one of the leaders in the North American hot rolled sheet market.

Thunder Bay, at the western end of Lake Superior, is a major transshipment point and a key element in the east–west transportation system. Bulky products—particularly grain, iron, and coal—are shipped to Thunder Bay for loading onto lake vessels. In recent years, however, this function has diminished. In the past, for example, iron from mines at Atikokan passed through Thunder Bay on its way to steel mills located along the shores of the Great Lakes. This mining operation is now defunct. Grain and potash from the Canadian Prairies are the major products handled at Thunder Bay. The Canadian Wheat Board sends prairie grain by rail to Thunder Bay and then by grain carrier to eastern ports for export to foreign countries. Until 1996, Ottawa heavily subsidized grain shipments under the Crow Benefit. With the elimination of that subsidy, the advantage of sending grain by rail to Thunder Bay has diminished, thus weakening its grain transportation role. (For more on this subject, see Chapter 8.)

Table 5.7 Urban Centres in Northern Ontario, 2001–6

Centre	Population 2001	Population 2006	Change (%)
Sudbury	155,601	158,258	1.7
Thunder Bay	121,986	122,907	0.8
Sault Ste Marie	78,908	80,098	1.5
North Bay	62,303	63,424	1.8
Timmins	43,686	42,997	–1.6
Kenora	15,838	15,177	–4.2
Temiskaming Shores*	12,904	12,927	–0.2
Elliot Lake	11,956	11,549	–3.4
Kapuskasing	9,238	8,509	–7.9
Kirkland Lake	8,616	8,248	–4.3
Total	521,036	524,094	0.6

*Temiskaming Shores is the newly restructured area comprising the three former municipalities of New Liskeard, Haileybury, and Dymond Township.

Source: Statistics Canada (2007d). Adapted from Statistics Canada publication *2001 Census: Population and Dwelling Counts, for Census Metropolitan Areas and Census Agglomerations, 2001 and 1996 Censuses*, Catalogue 93F0050XCB2001013, http://www.statcan.ca/english/IPS/Data/93F0050XCB2001013.htm and Population and Dwelling Counts, 2006 Census, Catalogue 97-550-XWE2006002 http://www.statcan.ca/bsolc/english/bsolc?catno=97-550-XWE2006002.

Since 1996, Thunder Bay's population has declined. Its population dropped from 126,643 in 1996 to 122,907 in 2006 (Statistics Canada 2002, 2007d), although, as shown in Table 5.7, the population of Thunder Bay increased slightly (by 0.8 per cent) from 2001 to 2006. Overall, thanks to marginal population increases in the larger centres—Sudbury, Thunder Bay, Sault Ste Marie, and North Bay—the cities in northern Ontario just held their own over the most recent census period (2001–6), but the smaller centres—Timmins, Kenora, Temiskaming Shores, Elliot Lake, Kapuskasing, and Kirkland Lake—all lost population.

Cities and towns in northern Ontario are struggling to diversify their economies. One option being pursued is tourism. Many small private firms—motels, summer camps, and fishing and hunting lodges—have sprung up (especially in smaller communities) to meet the demand from the growing number of tourists coming to experience the wilderness of northern Ontario. The urgency to diversify is driven by the continuing reduction in the size of the primary labour force. For example, the workforce at the nickel mines in Sudbury has fallen by half over the last 30 years. Sudbury has reacted by aggressively seeking to expand its service industry. The city has met with some success, partly because of its strategic location at the junction of three highways—the Trans-Canada Highway, Highway 69 (which leads to Toronto), and Highway 144 (which connects Sudbury with Timmins)—and partly because of its efforts to have federal and provincial agencies (the federal National Tax Data Centre and provincial Department of Mines) and institutions (Laurentian University and the Science North Museum) locate in Sudbury.

Many First Nations are found in northern Ontario. Some are situated in remote locations, such as the Kashechewan First Nation reserve. The residents of Kashechewan Reserve, located near James Bay some 450 km north of Timmins, were evacuated three times in 2005–6 because of spring floods and polluted drinking water. The federal government was considering relocating the reserve to higher ground, but another option was to move the reserve to the Timmins area where access to drinking water, health, housing, schools, and employment would be greatly improved. Such an integrationist approach has some merit, but social engineering strategies also involve certain dangers. Reaction from Mayor Demeules of Smooth Stone, who was not consulted about the proposed relocation, was not enthusiastic about the proposed relocation to his community. With the closing in July 2006 of a timber mill, 230 employees lost their jobs. As the mayor stated: 'There's no work for my own people. You don't take a community in crisis and dump them on another community in crisis. This has to be a win-win situation. Our community has to be respected' (Campion-Smith, 2006). However, culture trumped economics. On 30 July 2007 Kashechewan Chief Solomon signed an agreement with Ottawa to redevlop the community at its present location.

Ontario's Future

Ontario's future will be dictated by the performance of its labour force and its industrial economy. While recent rates of economic and population increases are now greater in Alberta and British Columbia, Ontario's economy and population size continue to rank first among Canada's geographic regions. For that reason, Ontario will remain Canada's core industrial region for the foreseeable future. Even the 2004–6 **net interprovincial migration** figures, which strongly favour Western Canada and British Columbia over Ontario, do not signal an end to Ontario's dominant position within Canada, but such figures, if sustained, could mark the beginning of a new regional balance of power (Table 5.8).

Ontario's position within the domestic and North American economies depends on two key factors:

- its continued role as Canada's leading manufacturing region, which is indicated by its position as a core within Canada's economy;

- its capacity to compete as a manufacturer within the North American market, which will be measured by the performance of its automobile industry.

By succeeding in both arenas, Ontario will secure its place within the North American economy by developing a stronger north–south economic axis while maintaining its traditional centrality as the core in the east–west axis.

Ontario, strong as it is, faces economic challenges. Dangers lurk in the global economy, especially in the forestry and manufacturing industries. A declining market for softwood lumber and paper products spells trouble for northern Ontario. Both sawmills and pulp and paper plants are likely to face a difficult future with more shutdowns possible. Manufacturers also are looking at a gloomy future. Fierce competition from foreign manufacturing countries with low labour costs will continue to force Ontario-based firms to make hard decisions, such as to reduce their labour costs, to find niche markets, to relocate to low-wage countries, or to cease operations. Within North America's automobile manufacturing sector, the Big Three are reducing their production to better match their market share. Ontario has felt a few of these cuts, and the province also has made some gains, thanks to its highly productive labour force, federal and provincial support, and Canada's public health-care program. Major gains came from Toyota and Honda. Toyota will begin production of motor vehicles at Woodstock, the midpoint between Cambridge and London, while Honda will have a new engine plant near Alliston.

Ontario faces three other challenges:

- an energy crisis and the prospect of much higher electrical rates;
- a serious health problem caused by industrial smog that threatens to engulf Toronto and other cities in southern Ontario.
- Settling of outstanding Indian land claims, especially the Ipperwash and Caledonia disputes.

The first two challenges are interrelated because Ontario depends heavily on electricity produced by coal-burning thermal plants. The government of Ontario has recognized these challenges, but the province is reluctant to close its main coal-burning thermal plants because they account for a substantial portion of Ontario's power requirements and because Ontario's energy needs are increasing. Without closure, smog will continue to be a problem, causing respiratory health problems for resi-

Table 5.8 Net Interprovincial Migration, July 2004–June 2006

Geographic Regions	2004–5 Interprovincial Migration	2005–6 Interprovincial Migration	Two-Year Total
Western Canada	8,200	39,397	47,597
British Columbia	7,456	3,779	11,235
Territorial North	−729	−1,417	−2,146
Québec	−2,332	−8,155	−10,487
Atlantic Canada	−4,220	−8,281	−12,501
Ontario	−8,375	−21,391	−29,766

Note: The figure for Western Canada masks the importance of Alberta. In 2005–6, Alberta received 57,105 net migrants while Manitoba and Saskatchewan lost 8,635 and 9,073 respectively.
Source: Statistics Canada (2006d, 2007e). Adapted from Statistics Canada website http://www40.statcan.ca/l01/cst01/demo33a.htm.

dents of southern Ontario. The energy/health solution lies in a combination of importing electricity and closing its coal-burning plants.

The third challenge—land claims—is in many ways more complex. In Ontario, Indians were granted land hundreds of years ago, beginning with the Haldimand Grant of 1784 in southwestern Ontario and followed in 1850 by the two Robinson treaties that covered large areas west of Lake Huron and north of Lake Superior. When disputes arise, however, reaching an agreement is challenging because the 'facts' are often lost in time (see Vignette 5.9). The slow pace of resolution is frustrating and two land disputes (Ipperwash and Caledonia) have resulted in violent protests by Indians from the Stony Point and Six Nations reserves. In the case of the Ipperwash dispute, the facts are relatively clear, i.e., Stony Point Reserve was taken from the Stony Point Indians in 1942 to serve as a military training camp, which was named Camp Ipperwash. After the war, the reserve was supposed to be returned but the Department of National Defence decided to keep the camp to train cadets. Despite many promises from Ottawa to return the land, no action took place. By 1993, the Stony Point Indians were utterly frustrated with the failure of Ottawa to respond and the Indians occupied Camp Ipperwash. In September 1995, Indian protestors moved into Ipperwash Provincial Park, where a confrontation with the Ontario Provincial Police took place, ending in the shooting of an unarmed Native protestor, Dudley George, by an Ontario police sniper. In 2003, the Ontario government asked Justice Sidney Linden to conduct a public inquiry into the circumstances surrounding the 1995 death of Dudley George, including the role of the then Premier, Mike Harris. In May 2007, Justice Linden issued the Ipperwash Inquiry Report, in which he concluded that Mike Harris, the former Premier, did not order the Ontario police into the Ipperwash Provincial Park to remove the Indian protestors. At the same time, Justice Linden called for the immediate return of Camp Ipperwash to the Kettle and Stony Point First Nation. Jim Prentice, the federal Minister for Indian Affairs and Northern Development, responded by stating that, 'We'll do something immediately' (*National Post*, 2007).

Figure 5.7 The Ipperwash conflict. The former Ontario Premier, Mike Harris, reportedly wanted a quick resolution to the occupation of the Ipperwash Provincial Park. The province purchased the land that forms the provincial park in 1932 from private interests who had bought the land in 1928 from the Stony Point band, although at that time no Indian band had control over its own finances or resources. Former Ontario Attorney General Charles Harnick claimed to have heard Premier Harris say he wanted the 'f—king Indians out of the park', a comment that Justice Sidney Linden, in his report, was inclined to believe was made despite Harris's denial (Gary Clement, *National Post*, 2007: A10. © Gary Clement).

Unlike the Ipperwash dispute, which is relatively cut and dried in regard to who has the right to the land, the basis of the Caledonia protest and Aboriginal occupation of a housing development under construction, which began in early 2006, dates to the original Haldimand Grant of 1784. This Aboriginal claim hinges on Crown misuse of lands granted to the Indians, yet also in play are nineteenth-century disagreements among the Iroquois of the Six Nations and over two centuries of misunderstandings, misdealings, and arbitrary policy changes on the part of officials and governments. As Vignette 5.9 suggests, these issues can be intractable. As of mid-2007 the occupation continued and no negotiated settlement between the Six Nations and the provincial and federal governments was in sight.

Vignette 5.9 Historical Timeline of the Caledonia Dispute

By no means are all specific land claims by Native groups simple and straightforward. The following outline of the historical evolution of the Six Nations claim to a 40-hectare parcel owned by a land developer at Caledonia, Ontario, near Hamilton, demonstrates just how complex these claims can become and suggests why, in many instances, settlements have been achieved at such a slow rate.

1784
For their loyalty to the British Crown during the American Revolution, the Six Nations (Iroquois Confederacy) are allowed to 'take possession of and settle' a strip of land nearly 20 kilometres wide along the Grand River, from its source to Lake Erie, totalling about 385,000 hectares. This is the Haldimand Grant, named after the commander of the British forces.

1792
Lt.-Gov. John Graves Simcoe reduces land grant to the Six Nations to 111,000 hectares.

1796
Six Nations grants its chief, Joseph Brant, the power of attorney to sell off some of the land and invest the proceeds. The Crown opposes the sales but eventually concedes.

1835
The Crown approaches Six Nations about developing Plank Road (now Highway 6) and the surrounding area. Six Nations agree to lease half a mile of land on each side for road, but do not surrender the land. Lt.-Gov. John Colborne agrees to the lease but his successor, Sir Francis Bond Head, does not. After 1845, despite the protests of Six Nations, Plank Road and surrounding lands would be sold to third parties.

1840
The government recommends that a reserve of 8,000 hectares be established on the south side of the Grand River and the rest sold or leased.

18 Jan. 1841
Six Nations council agrees to surrender for sale all lands outside those set aside for a reserve, on the agreement the government would sell the land and invest the money for them. A faction of Six Nations petitions against the surrender, saying the chiefs were deceived and intimidated. Six Nations would challenge that claim in a 1995 law-suit and it is part of the basis for the current protest.

June 1843
A petition to the Crown said Six Nations needed a 22,000-hectare reserve and wanted to keep and lease a tier of lots on each side of Plank Road and several other tracts of land in the Haldimand area.

18 Dec. 1844
A document signed by 47 Six Nations chiefs appears to authorize sale of land to build Plank Road.

15 May 1848
The land where the current development, Douglas Creek Estates, now sits is sold to George Marlot Ryckman for 57 pounds and 10 shillings and a Crown deed is issued to him.

continued

Vignette 5.9 continued

1850
The Crown passes a proclamation setting out extent of reserve lands—about 19,000 hectares agreed to by the Six Nations chiefs.

1924
Under the Indian Act, the Canadian government establishes an elected government on the reserve.

1992
Henco Industries Ltd purchases a company that owned 40 hectares of what it would later call the Douglas Creek Estates lands.

1995
The Six Nations sue the federal and provincial governments over the land. The developer calls it 'an accounting claim' for 'all assets which were not received but ought to have been received, managed or held by the Crown for the benefit of the Six Nations'. Henco Industries also claims that the original Haldimand Grant was merely a licence to occupy the lands, with legal title remaining with the Crown. The Six Nations dispute that claim.

July 2005
The subdivision plan for Douglas Creek Estates is registered with title to the property guaranteed by the province of Ontario.

28 Feb. 2006
A group of Six Nations members takes over the housing project, erecting tents, a teepee, and a wooden building.

Spring–Fall 2006
Native occupation continues, with frequent blockade of Highway 6, disruption of electricity to three communities, and through court challenges and non-Native counter-protests and periodic confrontations.

June 2006
Ontario government buys out Henco interest in disputed property for $12.3 million and promises to compensate for loss of future profits.

15 Oct. 2006
Several hundred demonstrators gather in Caledonia to protest occupation and what they call police inaction. Dozens of police officers form lines between Native and non-Native protestors. Three people arrested.

Sources: Adapted from CBC News (2006a, 2006b).

Notes

1. As a general rule, wages in the primary and secondary sectors are often higher than the national average, while wages in certain tertiary jobs are below the national average. In fact, many jobs in the service sector are associated with low hourly wages and part-time employment. These are often disparagingly referred to as 'hamburger-flipping' jobs or McJobs.
2. At the beginning of the twentieth century, the Ontario government sought to develop the northeastern section of the province, especially agricultural lands in the Clay Belt, forest stands, and mineral deposits in the Canadian Shield. Between 1903 and 1909, the government-financed Temiskaming and Northern Ontario Railway was built from the Canadian

Pacific Railway line at North Bay northward to the small town of Cochrane on the National Transcontinental Railway (now the CN). Over the next decade, branch lines were extended to Cobalt, Timmins, and Iroquois Falls. Overall, the Ontario government was pleased with the railway's impact on resource development. Plans were made to build the railway farther north—into the Hudson Bay Lowland. By 1932, the Temiskaming and Northern Ontario Railway stretched from Cochrane to Moosonee at the southern tip of James Bay, but further developments did not occur in the resource-scarce Hudson Bay Lowland.

3. In 1970, mercury was discovered in the fish near the Grassy Narrows Reserve, which is about 500 km downstream from the pulp mill. Levels of methyl mercury in the aquatic food chain were 10 to 50 times higher than those in the surrounding waterways (Shkilnyk, 1985: 189). These levels were similar to those found in the fish of Minamata Bay, Japan. Over 100 residents of this Japanese village died from mercury poisoning in the 1960s, and over 1,000 people suffered irreversible neurological damage. Because they depend on fish and game, the Ojibwa at the Grassy Narrows Reserve ate fish on a daily basis and many complained of mercury-related illnesses. While, unlike the Minamata incident, no one died, the economic and social impact on the Ojibwa was nevertheless an industrial tragedy of immense proportions.
4. Toronto has had various forms of regional government. About 40 years ago, the Ontario government established a form of regional government by combining the City of Toronto with the surrounding centres of Etobicoke, North York, Scarborough, York, and East York into the municipality of Metropolitan Toronto (Metro Toronto). These six jurisdictions became one large urban area for regional planning purposes, but each retained its city government. As Metro Toronto continued to grow, it spread into adjacent jurisdictions. In 1988, the Ontario government created the Greater Toronto Area (GTA). Often called the Metro Toronto Region, it consists of Metro Toronto and the regional municipalities of Halton, Peel, York, and Durham. In 1997, the provincial government passed a bill to change the municipal government structure to ensure more efficient, cost-effective services. The legislation came into effect in 1998, amalgamating six former municipalities (Etobicoke, York, Toronto, East York, North York, and Scarborough) and the regional municipalities into the new City of Toronto.
5. Toronto may solve its traffic problem by imposing a tax on automobiles entering the city's downtown. The first world-class city to experiment with a road toll was London, England. In 2002, London began charging drivers of automobiles £5 a day (about $12) to enter and leave the centre of London between 7 a.m. and 6:30 p.m. on weekdays, and this has resulted in a sharp decline in the volume of traffic, a great improvement in urban mobility, and a modest reduction in air pollution.

Challenge Questions

1. Ontario has two distinct economies—a robust economic core in the south and a struggling resource extraction economy in the north. What are the physical factors underlying these two strikingly different economies?
2. What physical factors make the Niagara Peninsula so successful in grape production?
3. In 2003, the provincial government proposed to close its five coal-burning thermoelectric generating stations and thus reduce smog in southern Ontario. What prevented Ontario from acting on its proposal?
4. What three demographic factors are causing northern Ontario's population to decline?
5. The forest economy of northern Ontario is in a tailspin. What challenges does the industry face?
6. Japanese automobile companies are expanding production in Ontario. Assuming you are the owner of one of these companies, why are you making the decision to locate new plants in Ontario rather than in the US, where most of your North American car sales take place?
7. Do you think that the recent increase in net interprovincial migration to Western Canada and British Columbia and the net loss in Ontario, as shown in Table 5.8, is a flash in the pan or a long-term trend?
8. Why was the Stony Point Reserve 'borrowed' in 1942 for military purposes still not returned by June 2007?
9. In 'Further Readings', does Professor Courchene argue that Ontario's economic future lies with North America or Canada? Do you agree?

Key Terms

Dutch Disease
A concept describing the apparent relationship within a country between its expanding energy resource sector and a subsequent decline in the manufacturing sector. In 1977, the term first appeared in *The Economist* to describe the phenomenon of a declining manufacturing sector in the Netherlands, which at the same time was enjoying increased revenues from the export of its natural gas (en.wikipedia.org).

Free Trade Agreement
Trade agreement between Canada and the United States enacted in 1989.

just-in-time principle
System of manufacturing in which parts are delivered from suppliers at the time required by the manufacturer.

net interprovincial migration
Annual estimates of net migration by provinces and territories determined by the number of people arriving and leaving each province and territory as permanent residents; based on administrative records, namely Child Tax Benefit data and income tax records

North American Free Trade Agreement
Trade agreement between Canada, the United States, and Mexico enacted in 1994.

outsourcing
Arrangement by a manufacturing firm to obtain its parts from other firms.

restructuring
Economic adjustments made necessary by fierce competition; companies driven to reduce costs often resort to reducing the number of workers at their plants.

value added
The difference between a firm's sales revenue and the cost of its materials.

Bibliography

Amrhein, Carl, and Harold Reynolds. 1997. 'Using the Getis Statistic to Explore Aggregation Effects in Metropolitan Toronto Census Data', *Canadian Geographer* 41, 2: 137–48.

Anderson, Fiona. 2006. 'Dropping Lumber Demand Threatens Canadian Exporters', *StarPhoenix* (Saskatoon), 23 Sept., D7.

Atkin, David. 2000. 'In a Hot-Wired World, Everything's Personal', *National Post*, 15 Nov., C1, C6–7.

Avery, Simon. 2006. 'In Ottawa, It's a Low-Budget Tech Rebound', *Globe and Mail*, 8 Nov. At: <www.theglobeandmail.com/servlet/story/RTGAM.20061108.wottawaa1108/BNStory/Technology/home>.

Bagnall, James. 2003. 'Ottawa Slipping Off the Tech Map', *National Post*, 22 Aug., FP5.

Barnes, Trevor J., John N.H. Britton, William J. Coffey, David W. Edgington, Meric S. Gertler, and Glen Norcliffe. 2000. 'Canadian Economic Geography at the Millennium', *Canadian Geographer* 44, 1: 4–24.

Bathelt, Harald, and Alfred Hecht. 1990. 'Key Technology Industries in the Waterloo Region: Canada's Technology Triangle (CTT)', *Canadian Geographer* 34, 3: 225–34.

Beatty, Perrin. 2006. 'Manufacturing in Crisis', *National Post*, 22 June, FP19.

Bertin, Oliver. 1998. 'Agreement Freezes Boeing Wages', *Globe and Mail*, 11 July, B2.

——— and Greg Keenan. 2003. 'Algoma to Axe 600, Cut Costs in Latest Bid to Stay Profitable', *Globe and Mail*, 15 May, B1.

——— and Jeff Sallot. 1998. 'Boeing Plant Set to Close', *Globe and Mail*, 4 June, B1.

Blackbourn, Anthony, and Robert G. Putnam. 1984. *The Industrial Geography of Canada*. London: Croom Helm.

Bluestone, B., and B. Harrison. 1982. *The De-industrialization of America*. New York: Basic Books.

Brean, Joseph. 2003. 'Outbreak Predicted to Cost City $1-billion', *National Post*, 3 May, A6.

Brent, Paul. 2002. 'Car Output Still Higher in Canada', *National Post*, 15 Aug., FP5.

———. 2003. 'Daimler Calls Off Windsor Plant Plans', *National Post*, 23 May, FP5.

Britton, John N.H., ed. 1996a. *Canada and the Global Economy: The Geography of Structural and Technological Change*. Montréal and Kingston: McGill-Queen's University Press.

———. 1996b. 'High Tech Canada', in Britton (1996a: ch. 14).

Brown, M., and D. Beckstead. 2006. 'Head Office Employment in Canada, 1999 to 2005', *Canadian Economic Observer*. Ottawa: Statistics

Canada Catalogue no. 11–010. At: <www.statcan.ca/english/ads/11-010-XPB/pdf/jul06.pdf>.

Campbell, Murray. 1996. 'Galt 25 Years Later', *Globe and Mail*, 13 Apr., D1–2.

Campion-Smith, Bruce. 2006. 'Mayor Fumes at Kashechewan Move', *Toronto Star*, 9 Nov. At: <www.thestar.com/NASApp/cs/ContentServer?pagename=thestar/Layout/Article_Type1&c=Article&cid=1163026213121&call_pageid=968332188774&col=968350116467>.

Canada. 1994. *1993 Canadian Minerals Yearbook: Review and Outlook*. Ottawa: Ministry of Natural Resources.

Canada. 2006. 'Net Merchantable Volume of Roundwood Harvested by Category and Province/Territory, 1940–2005', National Forestry Database Program, Table 5.1. At: <www.nfdp.ccfm.org/compendium/products/tables_index_e.php>.

Canadian Press. 2003. 'Imports Threaten Camco Inc. Plant', *National Post*, 5 Feb., FP7.

CBC News. 2000. 'Inside Walkerton: A Water Tragedy', CBC Backgrounder, 26 May. At: <www.cbc.ca/news/indepth/walkerton/>.

———. 2003. 'Camco Closing Hamilton Plant; 800 Jobs Cut', 4 Dec. At: <www.cbc.ca/stories/2003/10/17/camco_031017>.

———. 2006a. 'Caledonia Land Claim: Historical Timeline', 1 Nov. At: <www.cbc.ca/news/background/caledonia-landclaim/historical-timeline.html>.

———. 2006b. 'Caledonia Land Claim: Timeline', 1 Nov. At: <www.cbc.ca/news/background/caledonia-landclaim/index.html>.

Ceh, S.L. Brian. 1996. 'Temporal and Spatial Variability of Canadian Inventive Companies and Their Inventions: An Issue of Ownership', *Canadian Geographer* 40, 4: 319–37.

Crone, Greg. 1999. '560 Jobs Lost as Libbey Closes Ontario Plant', *National Post*, 5 Jan., C1.

CTV News. 2006. 'Canada, U.S. Agree on Softwood Framework', 26 Apr. At: <www.ctv.ca/servlet/ArticleNews/story/CTVNews/20060426/softwood_deal_060426/20060426?hub=Canada>.

Dabrowski, Wojtek. 2003. 'Ottawa Says No to Bailout of Navistar Plant', *National Post*, 17 May, FP3.

Darling, Graham. 2007. 'Land Claims and the Six Nations in Caledonia Ontario', Centre for Constitutional Studies. At: <www.law.ualberta.ca/centres/ccs/Current-Constitutional-Issues/Land-Claims-and-the-Six-Nations-in-Caledonia-Ontario.php>.

Dear, M.J., J.J. Drake, and L.G. Reeds, eds. 1987. *Steel City: Hamilton and Region*. Toronto: University of Toronto Press.

Detroit. 2007. History Detroit: 1701 to 2001. At: <www.historydetroit.com/places/fort_ponchartrain.asp>.

Dicken, Peter. 1992. *Global Shift: The Internationalization of Economic Activity*, 2nd edn. New York: Guilford Press.

Donald, Betsy. 2002. 'The Permeable City: Toronto's Spatial Shift at the Turn of the Millennium', *Professional Geographer* 54, 2: 190–203.

Economist, The. 1977. 'The Dutch Disease', 26 Nov., 82–3.

———. 1996. *Country Profiles: Canada*. London: The Economist Intelligence Unit.

Environment Canada. 2006. Information on Greenhouse Gas Sources and Sinks. Ottawa. At: <www.ec.gc.ca/pdb/ghg/onlineData/dataAndReports_e.cfm>.

Erwin, Steve. 2001. 'Labour Gears Up for Post-Auto Pact Era', *National Post*, 2 Jan., D3.

Evans, Mark. 2006. 'The "Quiet Boom": High-Tech Turnaround Drives Ottawa, Waterloo Growth', *National Post*, 31 Jan. At: <www.canada.com/nationalpost/financialpost/story.html?id=2a0b745b-eded-4269-af0f-f454d25a8ca6>.

Export Development Canada. 2007. Commodity Price Update, 28 May. At: <www.edc.ca/english/docs/commpric_e.pdf>.

Francis, R. Douglas, Richard Jones, and Donald B. Smith. 1996. *Destinies: Canadian History Since Confederation*, 3rd edn. Toronto: Harcourt Brace.

French, Cameron. 2007. 'U.S. Security-related Delays at Border Cost Billions, Canadian Producers Say', *International Herald Tribune*. At: <www.iht.com/articles/2007/03/05/business/canada.php>.

Gad, Gunter. 1991. 'Toronto's Financial District', *Canadian Geographer* 35, 2: 203–7.

Gertler, Meric S. 1998. 'Negotiated Path or "Business as Usual"? Ontario's Transition to a Continental Production Regime', paper presented at the annual meeting of the Association of American Geographers, Boston, 17 Mar.

Gilbert, Anne, and Joan Marshall. 1995. 'Local Changes in Linguistic Balance in the Bilingual Zone: Francophones de l'Ontario et Anglophones du Québec', *Canadian Geographer* 39, 3: 194–218.

Hamilton, Tyler. 2007. 'Ontario Goes Solar', *Toronto Star*, 26 Apr. At: <www.thestar.com.article/207415>.

Harris, Richard. 1999. *Unplanned Suburbs: Toronto's American Tragedy, 1900 to 1950*. Baltimore: Johns Hopkins University Press.

Herod, Andrew. 2000. 'Implications of Just-in-Time Production for Union Strategy: Lessons from the 1998 General Motors-United Auto Workers Dispute', *Annals, Association of American Geographers* 90, 3: 521–47.

Hill, Bert. 1999. 'Nortel to Spend $400 M to Stay on Fibre-optic Cusp', *National Post*, 3 Nov., C3.

Holmes, John. 1983. 'Industrial Reorganization, Capital Restructuring and Locational Change: An Analysis of the Canadian Automobile Industry in the 1960s', *Economic Geography* 59: 251–71.

———. 1996. 'Restructuring in a Continental Production System', in Britton (1996a: ch. 13).

Ibbitson, John. 2001. *Loyal No More: Ontario's Struggle for a Separate Destiny*. Toronto: HarperCollins.

Jack, Ian. 2000. 'EU, Japan Gang Up on Canada', *National Post*, 12 Aug., D1, D6.

Jestin, Warren. 1996. *Global Economic Outlook*. Toronto: Scotiabank.

Keenan, Greg. 2003. 'Layoffs Mount among Auto Makers', *Globe and Mail*, 26 Mar., B4.

———. 2006. 'Oshawa Wins Hot Competition for Camaro Plant', *Globe and Mail*, 19 Aug. At: <www.theglobeandmail.com/servlet/story/LAC.20060819.RGM19/TPStory/Business>.

———. 2007. 'Chrysler to Axe 20% of Canadian Work Force', *Globe and Mail*, 8 Feb. At: <www.theglobeandmail.com/servlet/story/RTGAM.20070208.wchrysler08/BNStory/Business/home>.

Konadu-Agvemang, Kwadwo. 1999. 'Characteristics and Migration Experience of Africans in Canada with specific reference to Ghanaians in Greater Toronto', *Canadian Geographer* 43, 3: 400–14.

Kreutzwiser, Reid, and Rob de Loë. 2004. 'Water Security: From Exports to Contamination of Local Water Supplies', in Bruce Mitchell, ed., *Resource and Environmental Management in Canada*, 3rd edn. Toronto: Oxford University Press, ch. 6.

Krueger, Ralph. 2000. 'Trials and Tribulations of the Canadian Fruit-Growing Industry', *Canadian Geographer* 44, 4: 342–54.

Little, Bruce, and Greg Keenan. 1996. 'Ontario's Economic Future Is the Sum of Its Auto Parts', *Globe and Mail*, 2 Mar., A1, A5.

Lorch, Brian J. 1999. 'New Format Development and Its Impact on the Commercial Structure of Thunder Bay', in K. Jones, ed., *Case Studies of the Impact of International Retailing in Canada*. Toronto: Ryerson Polytechnic University, ch. 5.

Lovely, Warren. 2006. 'Canadian Job Market Dries Up', *CIBC World Markets*, 8 Sept., Toronto. At: <research.cibcwm.com/res/Eco/Ecoresearch.html>.

McCarthy, Shawn. 1997. 'Ottawa Trades Bureaucrats for Bytes', *Globe and Mail*, 30 Aug., B1, B4.

MacDonald, N.B. 1980. *The Future of the Canadian Automotive Industry in the Context of the North American Industry*. Working Paper no. 2, Science Council of Canada. Ottawa: Minister of Supply and Services.

Macleod, Steven. 2006. 'US Agriculture Plan Comes at a Price for All Cross-Border Traffic', *Daily News: Canadian Transportation & Logistics*, 14 Sept. At: <www.ctl.ca/issues/ISArticle.asp?id=60259&issue=09142006>.

McKnight, Tom L. 1992. *Regional Geography of the United States and Canada*. Englewood Cliffs, NJ: Prentice-Hall.

MacLachlan, Ian. 1996. 'Organizational Restructuring of U.S.-Based Manufacturing Subsidiaries and Plant Closure', in Britton (1996a: ch. 11).

Marsh, James H., ed. 1988. *The Canadian Encyclopedia*, 2nd edn. Edmonton: Hurtig.

Matteo, Livio Di. 2006. 'Breakaway Country', *National Post*, 6 Sept., FP19.

———, J.C. Herbert Emery, and Ryan English. 2006. 'Is It Better to Live in a Basement, an Attic or to Get Your Own Place? Analyzing the Costs and Benefits of Institutional Change for Northwestern Ontario', *Canadian Public Policy* 32, 2: 173–96.

Matthew, Malcolm R. 1993. 'The Suburbanization of Toronto Offices', *Canadian Geographer* 37, 4: 293–306.

National Post. 2007. 'Ipperwash Report', 1 June, A14.

Natural Resources Canada. 2006a. *Mill Closures and Mill Investments in the Canadian Forest Sector: The State of Canada's Forests, 2005–2006*. Ottawa. At: <www.nrcan.gc.ca/cfs-scf/national/what-quoi/sof/sof06/mergers_e.html>.

———. 2006b. 'Mineral Production of Canada, by Province and Territory'. Ottawa. At: <mmsd1.mms.nrcan.gc.ca/mmsd/production/production_e.asp>.

Norcliffe, Glen. 1994. 'Regional Labour Market Adjustments in a Period of Structural Transformation: An Assessment of the Canadian Case', *Canadian Geographer* 38, 1: 2–17.

———. 1996. 'Mapping Deindustrialization: Brian Kipping's Landscape of Toronto', *Canadian Geographer* 40, 3: 266–73.

Ontario Chamber of Commerce. 2004. Cost of Border Delays to Ontario. Prepared by OCC Borders and Trade Development Committee. At: <http://occ.on.ca/Policy/Reports/121>.

Ontario Medical Association. 2001. *The Illness Costs of Air Pollution in Ontario: A Summary of Findings*. Toronto. At: <www.oma.org/phealth/icap.htm>.

Ontario Ministry of Energy. 2007. McGuinty Government Coal Replacement Strategy. At: <www.energy.gov.on.ca/index.cfm?fuseaction=english.news&back=yes&news_id=100&backgrounder_id=75.>

Paterson, J.H. 1994. *North America: A Geography of the United States and Canada*. New York: Oxford University Press.

Polèse, Mario, and Richard Stren, eds. 2002. *The Social Sustainability of Cities and the Management of Change*. Toronto: University of Toronto Press.

Preston, Valerie, and Lucia Lo. 2000. 'Canadian Urban Landscape Examples—21: "Asian Theme" Malls in Suburban Toronto: Land Use Conflicts in Richmond Hill', Canadian Geographer 44, 2: 182–90.

Putnam, D.F., ed. 1952. *Canadian Regions: A Geography of Canada*. Toronto: Dent & Sons.

Relph, Edward. 1991. 'Suburban Downtowns of the Greater Toronto Area', *Canadian Geographer* 35, 4: 421–5.

———. 2002. *The Toronto Guide: The City, the Fringe, the Region: An Illustrated Interpretation of Toronto's Landscapes*. Revised and updated for the Canadian Association of Geographers' annual meeting, Toronto, May. Toronto: Centre for Urban and Community Studies at the University of Toronto.

Robertson, Grant. 2005. 'Cars, Trucks, and Jobs', *Globe and Mail*, 15 Sept., B18.

Rumney, Thomas. 2001. *The Geography of Canada Bibliography Series: Vol. 4, Ontario*. Plattsburgh, NY: Plattsburgh State University, Center for the Study of Canada.

Shaw, Anthony B. 1999. 'The Emerging Cool Climate Wine Regions of Eastern Canada', *Journal of Wine Research* 10, 2: 79–94.

Shkilnyk, A.M. 1985. *A Poison Stronger Than Love: The Destruction of an Ojibwa Community*. New Haven: Yale University Press.

Shrubsole, Dan, and Milford Green. 1997. 'The Actual and Perceived Effects of Floodplain Land-Use Regulations on Residential Property Values in London, Ontario', *Canadian Geographer* 41, 2: 166–77.

Smith, Jamie, Beth Lavender, Heather Auld, David Broadhurst, and Tim Bullock. 1998. 'Adapting to Climate Variability and Change in Ontario', in *The Canada Country Study: Climate Impacts and Adaptation*, vol. 4. Ottawa: Environment Canada.

Stanford, Quentin H., ed. 1998. *Canadian Oxford World Atlas*, 4th edn. Toronto: Oxford University Press.

Statistics Canada. 1982. *Census Metropolitan Areas and Census Agglomerations*. Catalogue no. 95–903. Ottawa: Minister of Supply and Services.

———. 1992. *Census Metropolitan Areas and City Agglomerations*. Catalogue no. 93–303. Ottawa: Minister of Supply and Services.

———. 1996. *Labour Force Annual Averages 1995*. Catalogue no. 71–220–XPB. Ottawa: Statistics Canada.

———. 1997a. *Historical Overview of Canadian Agriculture*. Catalogue no. 93–358–XPB. Ottawa: Industry Canada.

———. 1997b. *A National Overview: Population and Dwelling Counts*. Catalogue no. 93–357–XPB. Ottawa: Industry Canada.

———. 1997c. *Canada Year Book*. Catalogue no. 11–402–XPE/1997. Ottawa: Minister of Industry.

———. 1997e. 1996 Census: Nation Tables—Population by Mother Tongue, Showing Age Groups, for Canada, Provinces and Territories, 1996 Census—20% Sample Data, 2 Dec. At: <www.statcan.ca/english/_census96/>.

———. 1997f. '1996 Census: Mother Tongue, Home Language and Knowledge of Languages', *The Daily*, 2 Dec. At: <www.statcan.ca/Daily/English/>.

———. 1998a. *Canadian Economic Observer*. Catalogue no. 11–010–XPB. Ottawa: Statistics Canada.

———. 1998b. '1996 Census: Ethnic Origin, Visible Minorities', *The Daily*, 17 Feb. At: <www.statcan.ca/Daily/English/>.

———. 1998c. '1996 Census: Aboriginal Data', *The Daily*, 13 Jan. At: <www.statcan.ca/Daily/English/>.

———. 2002. 2001 Census: Population and Dwelling Counts, for Census Metropolitan Areas and Census Agglomerations, 2001 and 1996 Censuses, 16 July. At: <www.statcan.ca/english/IPS/Data/93F0050XCB2001013.htm>.

———. 2003a. 'Top Three Fruit and Field-Grown Vegetable Crops by Area, Province', *Canadian Statistics, Census of Agriculture*. At: <www.statcan.ca/english/Pgdb/econ101a.htm>.

———. 2003b. 'Total Area of Farms, Land Tenure and Land in Crops, Provinces', *Canadian Statistics, Census of Agriculture*. At: <www.statcan.ca/english/Pgdb/econ124i.htm>.

———. 2003c. 'Exports of Goods on a Balance-of-Payments Basis', *Canadian Statistics, International Trade*. At: <www.statcan.ca/english/Pgdb/gblec04.htm>.

———. 2003d. Canadian Statistics: Population of Census Metropolitan Areas. At: <www.statcan.ca/english/Pgdb/demo05.htm>.

———. 2003e. Census of Canada 2001—Labour, Employment and Unemployment, by Industry, by Province. At: <www.statca.ca/english/Pgdb/labor21b.htm>.

———. 2003f. *Census of Canada 2001—Canada's Ethnocultural Portrait: The Changing Mosaic*. Analytical series 96F0030XIE2001008, 21 Jan. At: <www12.statcan.ca/english/census01/products/analytic/companion/etoimm/subprovs.cfm>.

———. 2003g. Canadian Statistics: Distribution of Employed People, by Industry, by Province. At: <www.statcan.ca/english/Pgdb/labor21b.htm>.

———. 2003h. Labour Force, Employed and Unemployed, Numbers and Rates. At: <www40.statcan.ca/l01/cst01/labor07c.htm>.

———. 2006a. 'Labour Force Survey', *The Daily*, 8 Sept. At: <www.statcan.ca/Daily/English/060908/d060908a.htm>.

———. 2006b. 'Exports of Goods on a Balance-of-Payments Basis', *Canadian Statistics, International Trade*. At: <www40.statcan.ca/101/cst01/gblec04.htm>.

———. 2006c. Canadian Statistics: Distribution of Employed People, by Industry, by Province. At: <www40.statcan.ca/101/cst01/labor21c.htm>.

———. 2006d. Components of Population Growth, by Province and Territory. At: <www40.statcan.ca/l01/cst01/demo33a.htm>.

———. 2007a. Labour Force, Employed and Unemployed, Numbers and Rates. At: <www40.statcan.ca/l01/cst01/labor07c.htm>.

———. 2007b. Population Urban and Rural, by Province and Territory: 1851 to 2001. At: <www40.statcan.ca/l01/cst01/demo62g.htm>.

———. 2007c. 'Census of Agriculture Counts 57,211 Farms in Ontario', *2006 Census of Agriculture*. At: <www.statcan.ca/english/agcensus2006/media_release/on.htm>.

———. 2007d. Population and Dwelling Counts, for Census Metropolitan Areas and Census Agglomerations, 2006 and 2001 censuses—100% data. At: <www12.statcan.ca/english/census06/data/popdwell/Table.cfm?T=201&S=3&O=D&RPP=150>.

———. 2007e. Components of Population Growth, by Province and Territory. At: <www40.statcan.ca/l01/cst01/demo33a.htm?sdi=migration>.

———. 2007f. 'Study: Canada's Changing Auto Industry', *The Daily*, 17 May. At: <www.statcan.ca/Daily/English/070517/d070517c.htm>.

Stewart, Walter. 2003. 'A Late Great Lake?', *Canadian Geographic* 123, 5: 36–46.

Steinhart, David. 1999. 'Ste-Thérèse Truck Plant Reopens under Paccar Label', *National Post*, 3 Aug., C7.

Sumner, Daniel A., and Joseph V. Balagtas. 2002. 'United States' Agriculture Systems: An Overview of U.S. Dairy Policy', in H. Rogenski, J. Fuquay, and P. Fox, eds, *Encyclopedia of Dairy Science*. At: <aic.ucdavis.edu/research1/DairyEncyclopedia_policy.pdf>.

Van Alphen, Tony, and Rick Westhead. 2004. 'Stelco Job Cuts Loom in Fight to Stay Alive', *Toronto Star*, 30 Jan., A1, A18.

Wang, Shuguang. 1999. 'Chinese Commercial Activities in Toronto CMA: New Developments, Patterns and Impacts', *Canadian Geographer* 43, 1: 19–35.

Weintraub, Sidney, and Christopher Sands. 1998. *The North American Auto Industry under NAFTA*. Washington: Center for Strategic and International Studies Press.

Wikipedia. 2007. 'Windsor'. At: <en.wikipedia.org/wiki/Windsor%2C_Ontario#History>.

Yeates, Maurice. 1975. *Main Street: Windsor to Quebec City*. Toronto: Macmillan.

Further Reading

Courchene, Thomas J., with Colin R. Telmer. 1998. *From Heartland to North American Region State: The Social, Fiscal and Federal Evolution of Ontario*. Monograph Series on Public Policy, Centre for Public Management. Toronto: Faculty of Management, University of Toronto.

From Heartland to North American Region State received the Donner Prize for the best book on public policy in 1998. While the book was written 10 years ago, the issue of Ontario's place within North America remains a critical issue, and Ontario's automobile industry is an integral part of the North American manufacturing system.

The premise of this book by two economists is simple: Ontario has become a regional economic force within the North American economy. From a neo-liberal economic perspective, Ontario's interests now lie with the continental economy, not the national one. Put differently, Ontario benefited from Canada's national economy with its high tariffs but now the province's self-interest lies in the North American economy. How did this happen?

For over 100 years, Ontario stood for a strong central government and the national economy. Both elements worked in Ontario's favour, placing the province at the apex of Canada's core/periphery economy. In fact, these elements were the 'power pillars' that reinforced Ontario's dominant place within the Canadian economy. The province could count on Ottawa to further its interests. For example, the federal government negotiated the Auto Pact with the United States and this agreement has benefited Ontario enormously. Similarly, the National Energy Program worked in Ontario's favour at the expense of three periphery provinces—Alberta, Saskatchewan, and British Columbia. From a spatial perspective, Ontario was the industrial core (along with Québec), while the rest of the country formed the resource hinterland. Federal high-tariff policies encouraged manufacturing in Ontario and discouraged such developments in the rest of the country.

In 1989, the Free Trade Agreement pushed Ontario into the North American economy. Courchene and Telmer argue that Ontario's interests have changed. Ontario lost its economic stranglehold on the rest of the country and now its economic self-interest lies in finding a place in the larger North American economy. To achieve such a goal, Ontario would benefit from a weak central government. Trade figures support Courchene and Telmer's view that Ontario is now more closely allied with the United States. For example, in 1981, Ontario's exports to other provinces were approximately equal to exports to the rest of the world (which effectively means the United States). In 1994, exports to the rest of the world were 2.1 times greater than exports to other provinces. From a spatial perspective, Canada is now witnessing a series of regional economic alliances with other countries. All provinces are importing more manufactured goods from foreign countries, thereby breaking their dependency on Ontario's manufactured goods. At the same time, all provinces, but particularly Ontario, see their trade with the United States growing by leaps and bounds. Economic data clearly show that Ontario's trade has increased greatly with the United States, but much of this increase is due to greater sale of automobiles to American customers. The political reason for such sales is not due to the Free Trade Agreement but to the Auto Pact. (Of course, since this book was written the Auto Pact has been cancelled as a result of a WTO ruling.) Nevertheless, Courchene and Telmer demonstrate that a north–south trade axis links Ontario with the United States. Extending their economic logic across Canada, geography (distance) dictates that a series of regional trade arrangements with the United States and other foreign countries must emerge.

Québec City, Québec (Ron Watts/Firstlight.ca)

Overview and Objectives

Within Canada, Québec ranks second in economic output and population size, accounting for almost 20 per cent of the country's GDP and 24 per cent of its population in 2006 (Figure 6.2). Québec has both a northern resource hinterland and a southern industrial core. Manufacturing and high-technology industries make Québec an important part of Canada's industrial heartland, while forestry and hydroelectric production are found in Québec's hinterlands. Québec, like Ontario, is seeking a place in the North American economy. As home to most Canadians of French origin, Québec represents a distinct cultural and linguistic region of Canada. Hydro-Québec continues to play a prominent role in the province's economic and social transformation. The James Bay Hydroelectric Project places Québec in the forefront of hydroelectric production and, at the same time, led to the first modern land-claim agreement (1975), which was followed 26 years later by the Paix des Braves. The James Bay and Northern Québec Agreement also laid the groundwork for the political evolution of the Inuit of Québec, who are close to achieving a new form of public self-government within a province. This chapter will:

- Describe Québec's physical geography and historical roots.
- Present the basic elements of Québec's population and economy in the context of the region's physiography.
- Examine Québec's cultural and economic development within the context of the core/periphery model.
- Discuss Québec's position within Canada, its francophone culture, and its separatist movement.
- Examine the effects of the James Bay and Northern Québec Agreement and the Paix des Braves on the Cree.
- Focus on Hydro-Québec's strategy of developing low-cost electrical power in northern Québec to stimulate industrial activities in southern Québec.
- Consider Québec's changing role within Canada and North America.

CHAPTER 6

Québec

INTRODUCTION

By virtue of its geography and history, Québec occupies a special place in Canada. To begin with, the St Lawrence River opened North America to French settlement, exploration, and the fur trade. This great river, which figures so prominently in Québec's geography, continues to play an essential role in Québec's life. The Port of Montréal owes much of its past economic prosperity to its role as a major transshipment point for goods entering North America (Vignette 6.1).

Québec is a large geographic region with four of Canada's physiographic regions found in its territory (Figure 2.2). Each physiographic region has a different settlement and land-use pattern. The St Lawrence Lowlands region is the agricultural, industrial, and population core of Québec. The Canadian Shield extends from the St Lawrence Lowlands to Hudson Strait. It, along with the Hudson Bay Lowland found at the southern tip of James Bay, represents a resource frontier. The Appalachian Uplands region, which borders on New Brunswick and three American states—Vermont, New Hampshire, and Maine—lies south of the St Lawrence Lowlands and extends eastward to the Gaspé Peninsula. It is divided into two areas: an agricultural/industrial area in Estrie (the Eastern Townships) and a downward transitional resource hinterland in Gaspésie.

Québec's Cultural Place

History has made this corner of North America the homeland of the French-speaking people. The French language and culture have generated a strong sense of belonging among its francophone citizens, forming the basis of ethnic pride, loyalty, and nationalism, which from time to time fuels the desire for an independent political state. Such feelings are particularly strong among the 'old stock'—the descendants of some 10,000 settlers who migrated from France in the seventeenth and eighteenth centuries. While their first loyalty is often to Québec, most Québécois also have a strong attachment to Canada. This sense of dual loyalty is unique in Canada and it accounts for Québec's interest in the concept of a political partnership within Confederation (for a fuller discussion of this concept, see the section, 'One Country, Two Visions', in Chapter 3). Federal politicians have avoided referring to Québec as a **nation** because they feared the acceptance of the word 'nation' would encourage the separatist movement as well as alienate the other provincial governments, which traditionally have subscribed to the concept of 10 equal provinces. In November 2006, however, Parliament passed a motion by Prime Minister Harper that recognized the Québécois as a nation within Canada.

Vignette 6.1 The St Lawrence River

The St Lawrence River provides a natural waterway into the interior of North America and, consequently, played a key role in the history of New France. Today, the St. Lawrence is an essential part of Québec's transportation system. Cities along its shores, particularly Québec City and Montréal, have benefited greatly from the waterway's role as a major trading route. Montréal's favourable location on the St Lawrence gives it an economic advantage and fuels the city's growth.

While waters from the Great Lakes empty into the St Lawrence River, ships could not reach the ports along the Great Lakes. As part of the St Lawrence Seaway, a system of canals, locks, and dredged channels were built to allow the passage of ocean-going ships into the heart of North America. For ocean-going ships, two major obstacles had to be overcome: the Lachine Rapids near Montréal, and Niagara Falls, which prevented ships from passing from Lake Ontario to Lake Erie. The building of the Lachine Canal (1825) and the Welland Canal (1829) allowed ships to reach the Great Lakes. Over time, these canals and locks were improved to accommodate the increasing size of ships and barges.

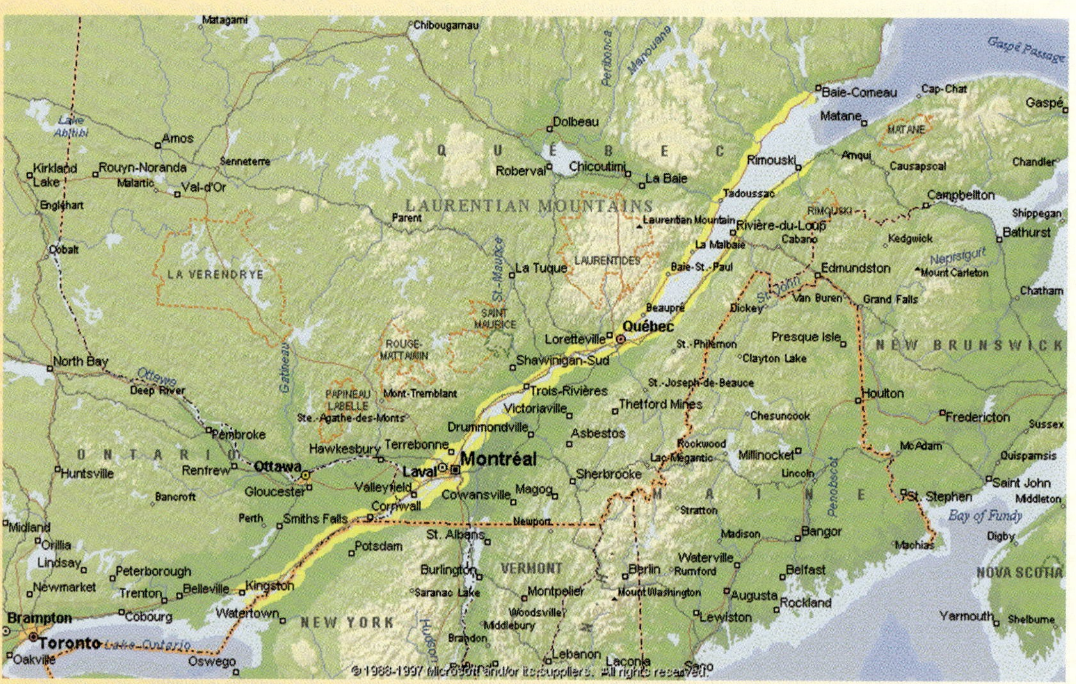

Figure 6.1 The St Lawrence River and Lake Ontario
Source: <www.aquatic.uoguelph.ca/rivers/stlawmap.htm>.

After World War II, there were vast improvements in ocean transportation. This was the era of the 'supertanker'. Even the average size of freighters had increased substantially. These larger ships required much greater depth of water than their predecessors to avoid running aground. Consequently, more and more ocean-going ships were unable to sail past Montréal and enter the Great Lakes. The solution to this transportation bottleneck was the St Lawrence Seaway, which, when it was inaugurated in 1959, allowed ocean vessels to travel from the Atlantic Ocean to the Great Lakes. While Montréal lost its natural advantage as a transshipment point, the city remains an important port, especially for container traffic.

Québec's culture is largely derived from the historical experience of **francophones** living in North America for over 400 years and of their being Canadians for 140 years. Events of the distant past, including massive immigration of British subjects to Québec in the early nineteenth century, the suppression of the rebellions in Lower Canada of 1837–8, and the 1885 execution of Louis Riel, and more recent ones, such as conscription crises during the two world wars, the War Measures Act of 1970, and the **patriation** of the Canadian Constitution in 1982 over the opposition of the Québec government, fuelled the desire for a separate state. Québec's role as the geographic heart of French language, customs, and heritage in North America is widely recognized in Canada and the US. Many French-speaking Canadians from the rest of Canada and Franco-Americans visit Québec to renew their sense of ethnicity. In sum, Québécois history and geography give Québec its raison d'être and its vision of Canada as a nation of two founding peoples. Canada's cultural duality has led to tensions between French- and English-speaking Canadians that have often strained the bonds of Confederation. (See Chapters 3 and 4 for a broader discussion of the French/English faultline and the two competing visions of Canada.)

The vast majority of people living in Québec speak French. In 2006, Québec's population was 7.6 million, a sizable increase from 7.2 million in 1996. Over 83 per cent of Quebecers declared French as their mother tongue (Figure 6.2). These Quebecers are known as the **Québécois**, although in recent years the term is also applied to all Québec residents. The remaining Quebecers include Aboriginal peoples, **anglophones** (those whose mother tongue is English), and **allophones** (those immigrants whose mother tongue is neither English nor French). Anglophones are concentrated in Montréal, Estrie, and Outaouais (the Ottawa Valley). Allophones are concentrated mainly in Montréal. In northern Québec, Cree and Inuit form the majority. From time to time, social tensions surface between French- and English-speaking **Quebecers**, and between Aboriginals and the Québec government.

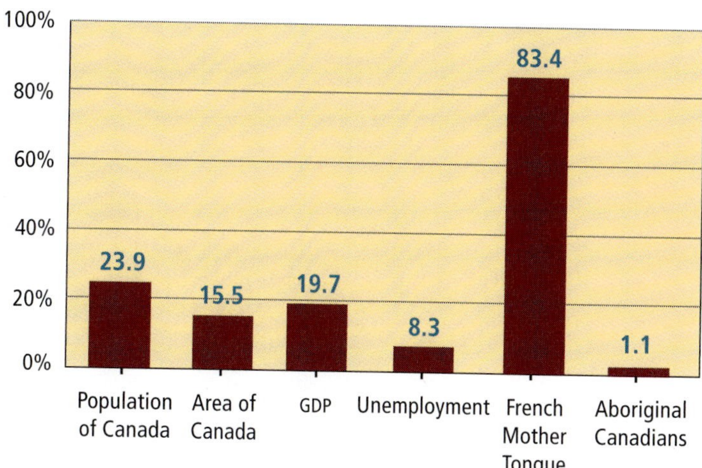

Figure 6.2 Québec, 2006. By population size and gross domestic product, Québec is the second-ranking geographic region in Canada. *Sources:* Tables 1.1, 1.7, and 4.18.

Québec's Place within Canada

Québec's demographic and economic position within Canada is considerable. Ranking second among the 10 provinces, Québec accounted for 19.7 per cent of all the goods and services produced in this country and 23.9 per cent of Canada's population in 2006 (Figure 6.2). Yet, Québec has been losing ground to the rest of Canada. In 2001, Québec produced 21.1 per cent of Canada's GDP, and its share of Canada's population dropped slightly from 24 per cent in 2001. Most economic and population gains are taking place in British Columbia and Western Canada. For example, the combined 2006 populations of British Columbia and Alberta (7,686,215) now exceed those of La Belle Province (7,546,100).

Concern over sluggish economic growth and low birth rates is not new to Québec. In 1995, the separatist leader, Lucien Bouchard, called on Québécoise women to have more children (Vignette 6.2). Eleven years later, Bouchard, who had left politics for the business world, was worried about what he saw as an economic malaise in Québec. He led a group of business leaders to prepare a manifesto, *Manifest—pour un Québec lucide* [For a clear-eyed vision of Quebec]. The manifesto

Figure 6.3 Less sovereignty, more association. In the weeks before the 1995 referendum in Québec, the low birth rate among the francophone population became an issue, causing the separatist leader, Lucien Bouchard, to call on Québécoise women to have more children. Cartoonist Brian Gable of the *Globe and Mail*, in a play on words, made use of the separatist call for 'sovereignty-association' with Canada. Since then, Québec has introduced new family support programs and the birth rate has increased. (*Globe and Mail*, 17 Apr. 1997, A16. Reprinted with permission from *The Globe and Mail*.)

sees the province slipping behind other provinces and states, charting a different path for the province's economic future and calling for greater involvement of the private sector. Quebecers, these business leaders believe, must work harder because other provinces and states have higher levels of worker productivity (Bouchard et al., 2006).

Why has Québec's position within Canada weakened since Confederation? To begin with, Québec's geopolitical position within Canada has changed from one of four provinces in 1867 to one of 10 provinces and three territories in 2008. Second, Québec's demographic position has shrunk from 32 per cent of Canada's population in 1871 to 23.9 per cent in 2006. Third, Québec's economy was hurt, at least initially, by the rise of separatism in the 1970s, which culminated in the referendum of 1980. The main damage was due to the perceived unattractive investment climate, which meant little capital flowed into the province, and a number of corporate headquarters moved from Montréal to Toronto.

Since 1989 and the Canada–US Free Trade Agreement, Québec's economy has operated within the American and global economic system. Consequently, the forestry and manufacturing industries are facing strong competition and even trade restrictions from other countries. Three examples illustrate this point. In 2001, Washington introduced restrictions on softwood lumber from Canada entering the United States. These restrictions were designed to ensure that American lumber firms would retain their traditional market share, but their impact on the forest industry in Québec has been very damaging, causing closure of sawmills and pulp mills and the layoff of thousands of workers. Manufacturing companies also have felt the sting of a sluggish world market. Bombardier, one of the world's leading aircraft manufacturers, has faced a slumping market for its business jets and stiff competition from Brazil. In 2002 and 2006, the company had to reduce its size and, in doing so, had to lay off hundreds of employees in its Montréal plant. As Pierre Beaudoin, president and chief operating officer of Bombardier Aerospace, stated:

> The restructuring of the airline industry continues, with relatively few orders for regional jets in the 70- to 90-seat jet category being awarded in recent years. This situation should improve, as attested by the numerous sales campaigns we are actively pursuing. However, we must be prudent and manage proactively our CRJ700/900 jets production schedule in the short term to ensure we achieve our goal of increased profitability and our success in the long term. This means making difficult but necessary decisions. (Parkinson, 2006)

The closure of Noranda's brand new magnesium plant at Danville in 2003 follows a similar pattern—its inability to compete in the American market with low-cost imports from China.

Both within and outside Québec, that province's economic position has often been compared to that of Ontario, and, inevitably, Québec is found wanting. In recent years the Québec economy has expanded at a slower rate than Ontario's economy, which accounts for out-migration and relatively high unemployment rates. For example, in 2005, Québec had an unemployment rate of 8.3 per cent while Ontario's rate was 6.6 per cent (Table 4.18).

However, not all the economic news is negative. The liberalization of trade has had many positive impacts on Québec's economy. First, the worldwide trade system has forced existing manufacturing firms to become more efficient and more conscious of quality control. In turn, these firms, including the apparel industry, have been able to maintain market share in Canada and gain a foothold in the

The Port of Montréal became a year-round port in 1964. Since then, Montréal has obtained more and more of the North Atlantic container traffic. In 2003, Montréal handled 38 per cent of container traffic on the Atlantic coast compared to 35 per cent for New York City. (Search4Stock, Inc.)

American market. Second, the chief advantage for aerospace, metal refining, and pharmaceutical firms to locate in Québec—especially in the Montréal manufacturing core—is that these companies can easily serve both North American and global customers. Bombardier provides such an example in its manufacturing of trains. In 2006, Bombardier won $2.5 billion worth of train contracts in Canada, England, and South Africa.

The challenge for Québec is to continue on the path to create a more robust economy within its industrial core and under an atmosphere of harmony between its francophone majority and its minority groups. In the first years of the twenty-first century, Québec has moved forward on a number of fronts. The pace of economic activities is growing—Montréal is Canada's east coast hub for container traffic between Europe and North America (Padova, 2006). Petroleum refining is now the second largest manufacturing enterprise in Québec. The refining business is based on Middle East oil, which is shipped by tankers to Montréal and Québec City. Cultural development is moving forward largely because of Montréal's expanding role as a centre of the arts and theatre and Québec City's Winter Carnival, making it an international tourist destination. Since the 1995 referendum, the rapprochement between the French/English/allophone communities has turned Québec, but particularly Montréal, into a centre of cosmopolitan lifestyle unique in North America. Relations with Aboriginal peoples, especially with the Mohawks and Cree, have been tense. For the Cree, relations are moving towards a closer partnership with the provincial government as a result of hydroelectric development in the James Bay region, which offers benefits both for the province and for the Cree, for whom subarctic Québec is their traditional homeland.

Québec's Physical Geography

Québec, the largest province in Canada, has a wide range of natural conditions. Its climate varies from the mild continental climate in the St Lawrence Valley to the cold arctic climate found in Nunavik (Inuit lands lying north of the fifty-fifth parallel). For the most part, the weather across Québec falls within a predictable range, but, like other regions of Canada, Québec has had its share of extreme weather events. While Québec commonly experiences heavy rainfall and freezing rain, two recent storms stand out. Extremely heavy rains in July 1996 in the Saguenay region caused extensive flooding and resulted in loss of life, extensive damage, and the evacuation of 15,000 people (Vignette 6.2). Two years later, freezing rain struck southern Québec and eastern Ontario, causing great damage and disrupting the electrical power. The ice storm began on 5 January 1998 and continued for several more days. Over 4 million people were without electricity, many for several weeks (Vignette 6.3).

Four of Canada's physiographic regions extend over the province's territory—the Hudson Bay Lowland, the Canadian Shield, the Appalachian Uplands, and the Great Lakes–St Lawrence Lowlands—and each has a different resource base and settlement pattern (Figure 5.2).

The heartland of Québec lies in the St Lawrence Lowlands. Formed from the Champlain Sea some 10,000 years ago, this physiographic region provides the best agricultural land in Québec. Settlers from France began farming along the edge of the St Lawrence River some 400 years ago. New France was established within this physiographic region and, by means of the St Lawrence, spread into the interior of North America. It remains the cultural core of Québec. The St Lawrence Lowlands offers several natural advantages that give the region its central role in modern Québec. Arable land is one such advantage. By far the best agricultural land is in a small area between Montréal and Québec City. As well, most industrial plants are located in this same area.

The Appalachian Uplands physiographic region is a northern and considerably less prominent extension of the Appalachian Mountains. The Appalachian Mountains origi-

Vignette 6.2 **The Great Flood of 1996**

In July 1996, a savage rainstorm took place some 250 km north of Québec City in the Saguenay region. Lasting over a four-day period from 17–21 July 1996, a record 277 mm of rain fell. Flooding was widespread. The tributary streams and rivers of the Saguenay River overflowed their banks and eroded the surrounding land and highway system. Over a hundred houses, farmsteads, and buildings were destroyed, forcing the evacuation of some 15,000 people. Seven were killed in the flood, including two children killed in their sleep when a mudslide buried their house. Dams and roads were washed out. Electricity and telephone services were disrupted. L'Anse Saint Jean, La Baie, Jonquière, Chicoutimi, and Hébertville were hardest hit. The loss of property from this storm was over $600 million.

nate in Georgia and divide New England from the Ohio Valley. While this geological feature reaches into Atlantic Canada, its topography is much modified and consists more of rugged hills and rolling plains than a mountain chain. Most arable land in this region is in Estrie, where the Loyalists settled in the late eighteenth century. By the end of the nineteenth century, the ancestors of the Loyalists had largely been replaced by French-speaking farmers. On the Gaspé Peninsula, small communities dot the coastline, reflecting the inhabitants' orientation to the sea and the fishing industry. Because the local economy is so weak, many people are unemployed. Others combine a resource activity, such as farming, fishing, or logging, with part-time employment in the villages and towns. Because of the area's spectacular scenery, tourism has become an important source of income during the summer on the Gaspé Peninsula. Mining and forestry are other economic activities in this physiographic

Vignette 6.3 **The Great Ice Storm of 1998**

On 5 January 1998, the worst ice storm in Canada's history struck eastern Ontario, southern Québec, and the Maritime provinces. It also affected the neighbouring New England states. While freezing rain occurs regularly in Canadian winters, this ice storm was unusually long. On the first day of the storm, about 20 mm of freezing rain fell on the Montréal area. Similar amounts fell in Kingston, Ottawa, Sherbrooke, and Trois-Rivières. These five cities and their surrounding rural communities felt the full fury of nature. But, unlike the one-day ice storm of 1961 that affected the Montréal area, the storm of 1998 continued throughout the next five days. The weight of the ice snapped hydroelectric poles and caused high-voltage transmission towers to collapse. The electrical system in eastern Ontario and southern Québec failed. More than 4 million people were without electrical power for at least 36 hours. About 500,000 people in Estrie were without power for several weeks. The physical damage was enormous, perhaps reaching $500 million. Economic losses to farmers and business also totalled millions of dollars. Hydro-Québec faced considerable reconstruction of its damaged transmission system. The storm impacted negatively both human activity and the natural landscape (many trees were lost or cut down as a result).

region. With the exception of the Lake Champlain gap in the Appalachian Uplands, easy road access to the populous parts of New England is blocked by these rugged uplands. The Lake Champlain gap has therefore become a very important north–south transportation link between Montréal and points south in the US, especially New York City.

As the largest physiographic region in Québec, the Canadian Shield occupies over three-quarters of the province's territory (Figure 5.2). The Canadian Shield is noted for its forest products and hydroelectric production—it has most of the hydroelectric sites in Canada because of a combination of heavy precipitation, large rivers, and significant changes in elevation. Near Montréal and Québec City, the Canadian Shield is a recreational playground where the rolling, rugged, forested upland with its numerous lakes has become a popular site for urbanites and tourists. Further north, single-industry towns based on mining and forestry dot the landscape. Beyond the commercial forest zone lie the lands of the Cree and Inuit, where the Cree must coexist with the massive La Grande hydroelectric project. The huge hydroelectric reservoirs initially resulted in a serious environmental problem for the Cree—the microbial breakdown of large amounts of submerged vegetation creates high levels of methylmercury contaminants. These contaminants enter the aquatic food chain and eventually find their way into fish. The Cree, who traditionally have consumed large quantities of fish, were confronted with the risk of mercury poisoning, but, as the microbial breakdown eased, the mercury problem has gradually diminished. In arctic Québec, a number of Inuit settlements are found along the coasts of Hudson Bay and Hudson Strait. The Cree and Inuit communities are unable to provide sufficient employment opportunities, and consequently, these Native communities have high unemployment rates. The Québec government supports a hunting-and-trapping program, thus encouraging many Cree families to stay on the land. This popular program has important cultural and social values.

The Hudson Bay Lowland extends from Ontario along the southeast edge of James Bay near Rupert River. By far, this lowland is the smallest physiographic region in Québec. Few people live here.

Environmental Challenges

Québec, as an old industrial region, is confronted with a series of environmental problems, most of which stem from the discharge of agricultural and industrial wastes into the atmosphere and water bodies or from the construction of huge hydroelectric dams. Even its so-called pristine north is dotted with abandoned mining camps and exploration sites where toxic chemicals lie unattended (Duhaime et al., 2005). Given this dismal record, Québec, like other regions in Canada, faces a massive cleanup job. Yet, rather than shirking the challenge, Québec has wholeheartedly embraced the environmental challenge and the Kyoto Accord but, like other provinces, expects Ottawa to provide much of the cash. With the passage of the 2006 federal budget, federal cash is to come from the $1.5 billion Canada Eco-Trust and Clear Air Fund. Québec will receive about $350 million (CTV News, 2007).

When Canada ratified the Kyoto Protocol in 2002, Ottawa under the Liberal government of Jean Chrétien agreed to reduce its annual emissions of greenhouse gases between 2008 and 2012 by a level 6 per cent below its actual emissions in 1990. However, since 2002, emissions have not decreased but increased. In 2006, the Conservative government of Stephen Harper stated that Canada cannot meet its Kyoto targets based on 1990 emissions. Yet, Québec is unfazed by Ottawa's position and remains a steadfast supporter of Kyoto. Unlike other provinces and territories, Québec has embraced the concept of 'users pay'. Consequently, as of 1 October 2007, Québec motorists will pay an extra 0.8 cent a litre increase for gasoline. The revenue from this tax is anticipated to generate $200 million

Hydro-Québec continues to construct dams and hydroelectric power stations. Along the Eastmain River, trees have been removed beyond the new and higher water level in the reservoir completed in 2006. (Hydro-Québec)

per year and these funds will be used to reduce greenhouse gas emissions and to encourage public transportation (Dougherty, 2007). While polls consistently reveal that support for the Kyoto Protocol is strongest in Québec, higher gasoline prices at the pump may not be as popular. For instance, a recent Canada-wide poll reveals that 62 per cent of Canadians polled were 'not really' or 'not at all' willing to pay more for gasoline as a way to cut emissions, while 35 per cent were somewhat or very willing to pay more for gas (Woods, 2007). Still, with Québec leading the way on the carbon tax, other provinces and territories may follow suit.

In the most populated and industrial area near the St Lawrence River, both humans and other forms of biological life are threatened by polluted waters. Once a pristine river, the St Lawrence is now badly contaminated. The reason is simple. Over the years, great quantities of industrial toxic wastes have been deposited directly into the river. The result today is a river with high levels of toxic chemicals—polychlorinated biphenyls (PCBs) and dichlorodiphenyltrichloroethane (DDT), as well as heavy metals such as lead, mercury, and cadmium. More specifically, the waste contaminants from the aluminum and magnesium smelters—known as polycyclic aromatic hydrocarbons (PAHs)—have had a serious impact on the ecological system of the St Lawrence. Take, for example, the impact on the beluga whales. At one time, the population of the St Lawrence beluga whales totalled almost 10,000. Today, the population is under 1,000. Even more disturbing is the link to the health of local populations who obtain their drinking water from the St Lawrence River. According to a scientific paper by Martineau and others (2002), the human population living along the estuary of the St Lawrence River has rates of cancer higher than those found in people in the rest of Québec and Canada, and some of these cancers have been epidemiologically related to PAHs.

Environmental challenges extend into Québec's so-called pristine north. While few people live in this vast area, mineral exploration companies have failed to clean up their sites. Inuit and Innu hunters in the northeast have reported that mineral exploration companies have littered the landscape with abandoned machinery, chemicals, and oil barrels (Duhaime et al., 2005: 262).

Québec's Historical Geography

Relative to other parts of North America, the history of Québec is long, rich, and complicated by the period of British rule and then the search for a place within Confederation (see Tables 6.1–6.3). Beneath the surface of this search lies the deeply fractured French/English faultline. Québec's historical geography can be divided into three periods: New France, British occupation, and Confederation.

New France, 1608–1760

The introduction of the French to North America began in 1534 when Jacques Cartier sailed into the Bay of Chaleur and set foot on the shores of the Gaspé Peninsula. Yet, the first permanent French settlement in Québec was established in 1608, when Samuel de Champlain founded a fur-trading post at the site of Québec City, thus establishing the French colony in North America. Although France eventually lost its North American colony, it left a cultural legacy in the form of the French language and the Catholic religion, and a French stamp on the landscape with its unique settlement pattern and arrangement of farms into long lots. Village after village was built around a Catholic church, and ownership of the surrounding farmland, arranged in long lots based on the French seigneurial system.

During the seventeenth and eighteenth centuries, France had control over vast areas of North America. Its core, however, was the St Lawrence Valley, from which New France developed a vast fur-trading empire. The wealth from the fur trade was enormous and was the reason for France's interest in the New World. Almost every male French settler wanted to participate in the fur trade, which left only a few to clear the forest and till the land. For example, several canoes full of beaver pelts could make a man extremely wealthy compared with the meagre returns obtained from the back-breaking toil of clearing land and breaking the soil. Frenchmen who were coureurs de bois (trappers) often lived with the Indians. A few were extremely successful and returned to France to enjoy their good fortune. Others remained in the fur trade or settled in New France.

Geography played a part in New France's success both as a fur-trading empire and as an agricultural colony. The St Lawrence River provided a route to the interior, which gave the French explorers and fur traders an advantage over their English rivals, who had to contend with crossing the Appalachian Mountains. With further exploration of the interior of North America, the French were able to expand their territory and secure more and better fur-trading routes. Thus, by portaging from the Great Lakes to the Ohio River and then to the Mississippi, the famous French explorer, René-Robert Cavelier, Sieur de La Salle, reached the mouth of the Mississippi River at New Orleans in 1682. Early in the eighteenth century, French fur traders, led by Pierre Gaultier de Varennes, Sieur de La Vérendrye, established a series of fur-trading posts in Manitoba. These trading posts, especially Fort Bourbon on the Saskatchewan River at Cedar Lake (just east of the present-day community of The Pas), made it convenient for Indians to trade their furs with French traders rather than travel farther to the British fur-trading posts along the Hudson Bay coast.

New France also established a successful agricultural society. Once the land was cleared, the fertile soils in the St Lawrence Lowlands provided a solid basis for essentially feudal agricultural settlement. Farming took hold in New France, particularly following the efforts of Jean Talon, the greatest administrator of New France. By the early eighteenth century, the French had turned to farming, leaving the lure of the wilderness to a relatively small number of more daring souls. By then, New

France had existed for over 100 years, and many of its people had been born and raised in the New World. They were no longer 'French' but *Canadiens*.

By the middle of the eighteenth century, almost all the lands in the St Lawrence Lowlands were under cultivation. Farmlands stretched in a continuous belt from Québec City to Montréal. Like the feudal agricultural system in France, peasant farmers (**habitants**) worked the land, paying their lords (**seigneurs**) both in kind and in labour. Through the efforts of the habitants and their seigneurs, New France became a successful agricultural colony, but one with a system of landownership and rural life that was quite different from that of the British colonies along the Atlantic seaboard (a difference that would eventually cause Britain to split Québec into Upper and Lower Canada in 1791 [see Chapter 3]).

The Seigneurial System

When the first Intendant of New France, Jean Talon, arrived in New France in 1665, he encountered a population of only 3,000 inhabitants, most of whom were men engaged in the fur trade. Talon had been instructed by Louis XIV to create a feudal agricultural society resembling that of rural France in the seventeenth century. Talon undertook three measures to achieve this goal. First, he recruited peasants from France. Second, he sent for young women—orphaned girls and daughters of poor families in France—to provide wives for the men of the colony. Third, he imposed the French feudal system of landownership, known as the seigneurial system. In the seigneurial system, huge tracts of land were granted to those favoured by the king, namely, the nobility, religious institutions of the Roman Catholic Church, military officers, and high-ranking government officials. The seigneur was obliged to swear allegiance to the king and to have his tenants cultivate the lands on his estate. The tenants owed certain obligations to their seigneur: paying yearly dues (*cens et rentes*) to their seigneur, working the seigneur's land, especially in regard to road maintenance (*corvée*), and paying rent for using the seigneur's grinding mill and bake ovens (*droit de banalité*).

In the days of New France, the seigneurial system offered several advantages: (1) it encouraged the establishment and functioning of an agricultural society; (2) it provided access to the seigneur's mill where the habitants could grind their grain; and (3) the mill served as a centre of defence for the habitants when under attack.

By 1760, there were approximately 200 seigneuries. Seigneuries, which were usually 1 by 3 leagues (5 by 15 km) in size, were generally divided into river lots (rangs). These long, rectangular lots were well adapted to the St Lawrence Valley for several reasons, the most important of which was that each habitant had access to a river, either a tributary of the St Lawrence or the river itself. At that time, most people and goods were transported along the river system in New France. For that reason, river access was vital for each habitant family.

After the British Conquest of New France, the seigneurial system was retained by the British more for political reasons—to gain the support of the principal source of power (the seigneurs and the Roman Catholic Church)—than for economic reasons. By the early nineteenth century, however, agriculture in Lower Canada had become a commercial venture, making the seigneurial system an anachronism. The seigneurial system was abolished in 1854 by the legislature of the Province of Canada.

British Colony, 1760–1867

Following the defeat of the French in 1760, the British ruled Québec for over 100 years. The British governor was installed at Québec City along with a regiment of British troops, while the fur trade continued to flourish and the agricultural economy went unchanged. Most French Canadians were peasant farmers. After the Conquest, their life on the land remained much the same. Their social and economic lives revolved around the parish church and a landholding system centred on the seigneuries. Life in the towns and cities, however, changed radically due to a massive influx of British immi-

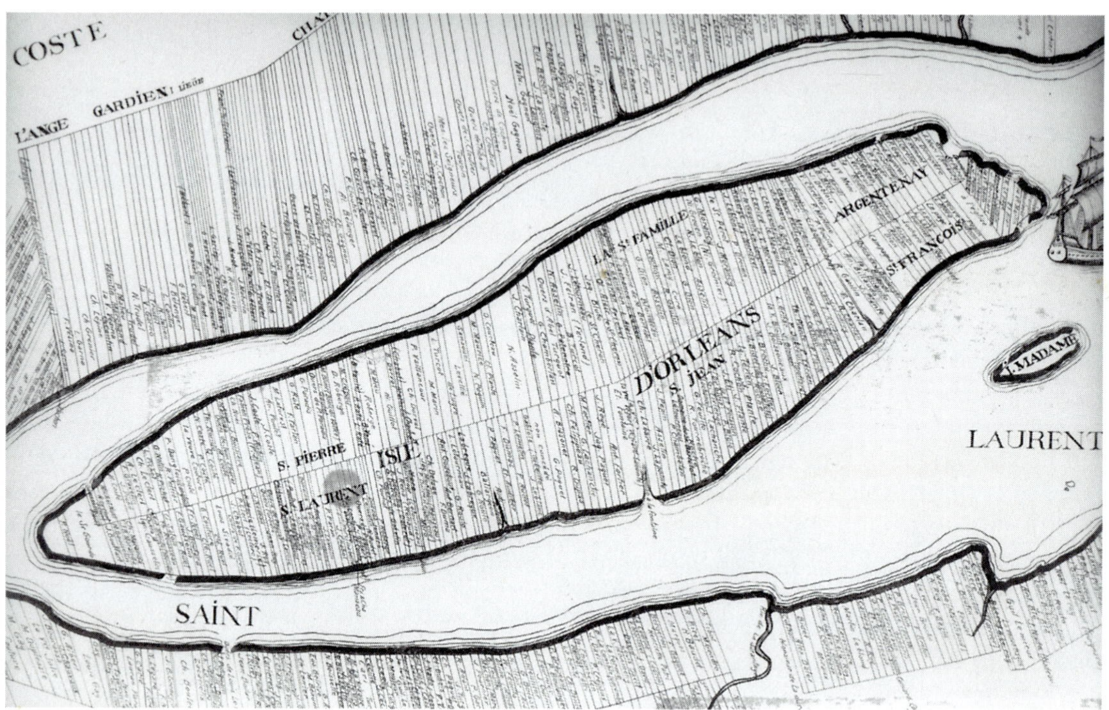

Historic map of Île d'Orléans, which is situated in the St Lawrence River near Québec City. Île d'Orléans was one of the first areas settled by the French and its *habitants* produced food for themselves and their seigneur. Île d'Orléans clearly illustrates the *rang* land tenure system of New France, which took the form of long, narrow lots. The purpose of the *rang* was to provide each *habitant* access to the river transportation system. (*Library and Archives Canada NM0048248*)

grants, the powerful political position of English Canadians, and their control of the commercial and industrial sectors of urban places. By 1851, French Canadians accounted for only about half of the population of Montréal. From 1851 to 1951, enormous demographic changes took place. The saving grace for the Canadiens was their high fertility rate, which was known as the 'revenge of the cradles' (*revanche des berceaux*). Even though this rate began to slow, for 100 years Québec's birth rate remained higher than the birth rate in the rest of Canada. Thus, Québec's slower decline in fertility helped to compensate for the province's out-migration and for the growing English-speaking population.

Land hunger forced many **French Canadians** to migrate. By the middle of the nineteenth century, many had left the St Lawrence Lowlands due to a land shortage. Birth rates were so high in this rural society that there was not enough land left for the children of farm families. French Canadians migrated in three directions: to the Appalachian Uplands, where they either purchased farms from English-speaking farmers or found jobs in textile mills; to the Canadian Shield, where they tried to exist on extremely marginal agricultural land; or to New England's industrial towns, where most were employed in textile factories. By the early twentieth century large numbers of French Canadians, perhaps as many as one million, had left Québec for the United States, while only a small number settled in the Canadian West that was calling out for homesteaders. Through all of these economic and political changes, the vast majority of French Canadians maintained their language and Catholic religion. They turned to the Roman Catholic Church for both spiritual and political leadership. The Church, in turn, encouraged the people to stay on the land, far from the secularizing influences in towns and cities, where the English Protestants lived.

Table 6.1 Timeline: Historical Milestones in New France

Year	Geographic Significance
1534	Jacques Cartier sailed into the Bay of Chaleur and claimed the land for France. The following year, Cartier discovered the mouth of the St Lawrence River, which provided access to the interior of North America.
1608	Samuel de Champlain, described as the 'Father of New France', founded a fur trading post near the site of Québec City. Champlain was instrumental in the development of the fur economy, which provided the initial economic basis for New France.
1642	Paul de Comedey de Maisonneuve established Ville-Marie on Île de Montréal, which is strategically situated at the confluence of the Ottawa and St Lawrence rivers. Later, Ville-Marie was renamed Montréal.
1759	The struggle between France and England over North America saw the British defeat the French army on the Plains of Abraham. The final battle between the French and English forces ended with the capture of Montréal by the British. In 1763, the formal surrender of New France to England took place with the Treaty of Paris.

By the 1830s, political unrest was growing in both Upper and Lower Canada. Each colonial government was headed by a governor who was appointed in England and had absolute power. The governor administered the colony along with leading members of the community. This cozy arrangement not only concentrated power in the hands of a few but also led to blatant abuse by powerful elites. In Upper Canada, the political elite was known as the Family Compact, while in Lower Canada it was the Château Clique. In 1837, rebellions broke out in each colony (Vignette 6.4). Both rebellions, which called for political reforms and the curbing of the political power of the elites, were crushed by the British army.

Britain, however, was determined to remedy the political situation in its two colonies. In an attempt to identify the main sources of discontent in Upper and Lower Canada, the British government sent Lord Durham, a politician, diplomat, and colonial administrator, to North America. Durham recognized that the political solution lay in an elected government, where power was dispersed among elected representatives rather than concentrated among an appointed elite. Durham also observed the French/English faultline, which he described as 'two nations warring in the bosom of a single state'. In Durham's report he recommended responsible (elected) government and the union of English-speaking peo-

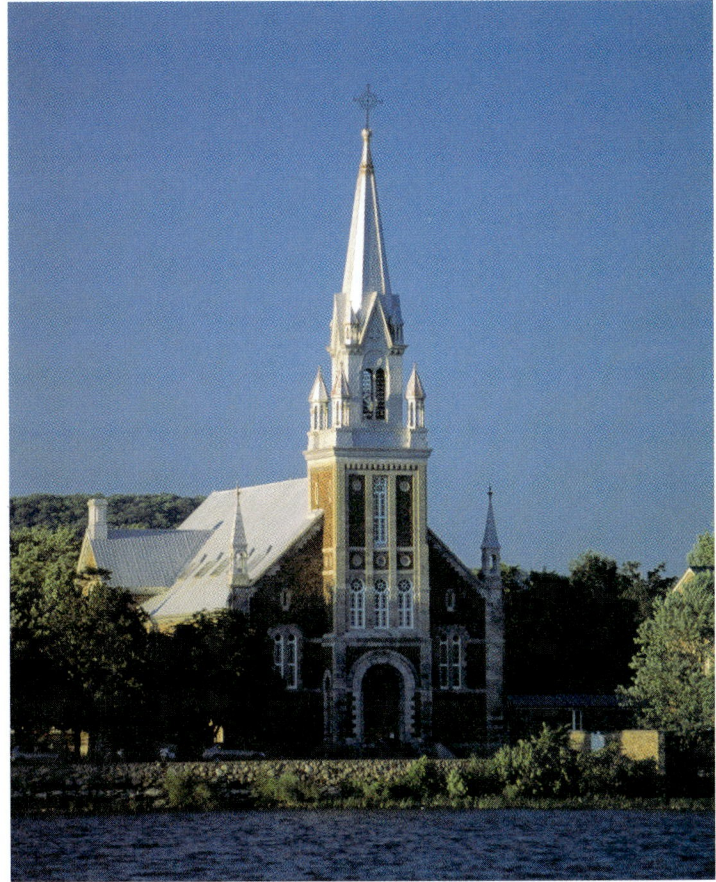

Roman Catholic churches are found in most communities across Québec, signifying the dominant role that the Roman Catholic Church once played in the social life and political affairs of the province. By the time of the Quiet Revolution, however, the Church's influence in Québec affairs had greatly diminished. (Search4Stock Inc.)

Vignette 6.4 The Rebellions of 1837–8

In Lower Canada, Louis-Joseph Papineau, a lawyer, seigneur, and politician, was the leader of the French-speaking majority in the Assembly of Lower Canada. In 1834, Papineau issued a list of grievances known as 'The Ninety-Two Resolutions'. At this time, the economy was depressed and tensions between the French-Canadian majority and the British minority were growing. Papineau sought to shift political power from the British authorities to the elected Assembly of Lower Canada. He planned to use his majority in the Assembly to pass legislation, including tax bills. The British government rejected 'The Ninety-Two Resolutions'—it was just a matter of time before an armed uprising broke out. When it did, the British reacted with force. Even with the strong support of rural areas, Papineau and his Patriotes were soundly defeated. Nearly 300 rebels were killed in six battles. Papineau fled to the United States. A rebellion based on the same popular objections to elite rule also took place in Upper Canada at the same time and it, too, was crushed. Following a second uprising in Lower Canada in November 1838, the British captured hundreds of rebels and ultimately 12 men were sentenced and executed and another 58 were exiled to Australia (Bumsted, 2007: 169). The British government sought to remedy the unrest in both of its colonies. It began this process with a fact-finding mission headed by Lord Durham. The result and Britain's solution was the Act of Union in 1841.

ple in Upper Canada with the French-speaking settlers of Lower Canada.[1] Lord Durham believed that assimilation of the French was desirable and possible, claiming that the French Canadians were 'a people with no literature and no history' (Mills, 1988: 637). He recommended that English be the sole language of the new Province of Canada, and that a massive immigration of English-speaking settlers be launched in order to create an English majority in Lower Canada. In response to Durham's recommendations, the Act of Union was passed by the British Parliament in 1841, uniting the two colonies into the Province of Canada and thus creating a single elected assembly.

Lower Canada was now known as Canada East, and in spite of the political changes and the flood of immigrants from the British Isles, the French Canadians were not assimilated. Under the new form of British administration, the task of maintaining their French culture and language was not easy, but it was achieved because of several factors, the most important being their strong desire to remain Catholic and French-speaking. A second factor was the institutional support provided by the Roman Catholic Church. By providing spiritual guidance and schooling in French, the clergy played an essential role in cultural preservation. Geography and demography were other factors essential to the survival of French Canada in the nineteenth century. The overwhelming number and concentration of French-speaking people in Canada East provided a critical mass necessary for cultural survival. A high birth rate and high rate of natural increase for the *Canadiens* ensured an expanding population of French Catholics. Other factors were the rural nature of the French-speaking population, which isolated them from English-speaking residents of the major cities, and the emergence of a French-Canadian intellectual elite whose writings preserved the history and literature of French Canada. One of the most popular novels in early Québec was *Jean Rivard* (1874), published not long after Confederation. Written by Antoine Gérin-Lajoie, *Jean Rivard* is the story of a young French Canadian in Canada East who is advised by his *curé* on the

Table 6.2 Timeline: Historical Milestones in the British Colony of Lower Canada

Year	Geographic Significance
1763	The Treaty of Paris awards New France to Great Britain.
1774	The British Parliament passed the Québec Act, which recognized that Québec, as a British colony, had special rights, including use of the French language, the Catholic religion, and French civil law.
1791	The British Parliament approved the Constitutional Act that created two colonies in British North America called Upper and Lower Canada
1841	Based on the Durham report, the British Parliament passed the Act of Union that reunited Upper and Lower Canada into a single colony and made English the official language of the newly formed province of Canada.

advantages of becoming a farmer rather than a lawyer. The novel promotes the virtue of living in a harsh and remote land, which is superior (in God's eyes) to living in an urban centre with its many worldly temptations.

Confederation, 1867–Present

Confederation, achieved in 1867, sought to unite two cultures—English and French—within a British parliamentary system. For Québec, Confederation provided a political framework offering three benefits: an economic union with Ontario, Nova Scotia, and New Brunswick; a political environment where Roman Catholicism and, to a lesser degree, the French language were guaranteed protection by Ottawa; and provincial control over education and language. George-Étienne Cartier, one of the Fathers of Confederation and a French-Canadian leader, viewed these provincial powers as a way for Québec to shape its own destiny within Confederation. Cartier may have identified a fourth benefit—since Québec and Ontario often had mutual economic interests, they could, by working together, influence federal policies and thereby shape the future of Canada.

Confederation also led to the expansion of the geographic size of Québec (Figures 3.4 and 3.6). Since Confederation, Québec's geographic size has increased greatly. It is now 1.5 million km^2. As Canada acquired more territory from the British government, Ottawa assigned to Québec parts of Rupert's Land lying north of the St Lawrence drainage basin. In 1898, the Québec government received the first block of Rupert's Land. The second was obtained in 1912.[2] Some of this land, however, was claimed by another British colony, Newfoundland. In its argument, Newfoundland demanded all of Québec's territory that drained into the Atlantic Ocean. Though the Imperial Privy Council of Britain awarded this land, known as Labrador, to Newfoundland in 1927 (Figure 3.7), to this day the Québec government does not recognize the decision.

Prior to World War II, Québec continued to project an image of a rural, inward-looking, Church-dominated society. In 1960, the Quiet Revolution unleashed the force of change that drove Québec into a modern industrial state. As with other social revolutions, the origin of changes began long before 1960. Yet, the election of the Liberal government of Jean Lesage marked a dramatic transformation in government, French-Canadian society, and the place of French within Québec. His government initiated major political innovations that accelerated the process of social and economic change. In effect, the provincial government replaced the Catholic Church as the leader and protector of French culture and language in Québec. The Quiet Revolution instilled a sense of pride and accomplishment among Quebecers. (See Chapter 3 for more on the Quiet Revolution.)

The main reforms of the Lesage government were hinged on state intervention in the Québec economy through Crown corporations, and on the expansion of a French-speaking provincial civil service. The government's principal achievements were:

- the nationalization of private electrical companies under Hydro-Québec;
- the modernization and secularization of the education system, making it accessible to all;
- the investment of Québec Pension Plan funds in Québec firms, thereby stimulating the francophone business sector;
- the establishment of Maisons du Québec (quasi-embassies) in Paris, London, and New York, thus signalling to Ottawa that the Québec government wanted to represent Québec interests to the rest of the world.

With these accomplishments behind them, Quebecers felt confident about their future. Lesage's slogan, '*Maîtres chez nous*' (Masters in our own house), became a reality. For federalists in Québec, these achievements proved that a strong Québec could function within Canada, but for separatists they were not enough. The rise of separatism in Québec signalled that some Québécois felt only an independent Québec could adequately represent French-Canadian interests. For them the slogan became '*Le Québec aux Québécois*' (Québec for the Québécois). After two referendums, the election of a Liberal government in 2003, and the slide to third place by the separatist Parti Québécois in the 2007 provincial election, it would appear that Quebecers have, for the time being, turned away from pursuing political separation and are focusing more on economic and social concerns. Then, too, Québec nationalism, which is at the core of the separatist movement, has shifted somewhat from the goals and values of the 'old stock' francophones and has become more inclusive, i.e., embracing all French-speaking Quebecers, including immigrants (allophones) and even bilingual anglophones.

Québec Today

Québec is a modern industrial society effectively operating within the North American and global economies. Immigration has made its society more diversified and has created a substantial allophone community, especially in Montréal. Montréal, an extremely cosmopoli-

Table 6.3 Timeline: Historical Milestones in Confederation

Year	Geographic Significance
1867	The Dominion of Canada was formed, the new state consisting of Québec, Ontario, New Brunswick, and Nova Scotia.
1898	Ottawa extended Québec's northern boundary to the Eastman River, thus expanding Québec's territory well beyond its core area of the St Lawrence Lowlands into the Cree lands of James Bay in the Canadian Shield.
1912	Ottawa added the Territory of Ungava to Québec, thus extending Québec to Inuit lands of Nunavik. With the addition of these two northern lands in 1898 and 1912, Québec's territory more than doubled.
1927	In settling a dispute between Canada and Newfoundland, Britain rejected Canada's claim that the boundary should be placed just inland from the shore. Instead, Britain declared that the boundary between its two colonies followed the Hudson Bay and Atlantic Ocean watersheds. Québec does not recognize this boundary.

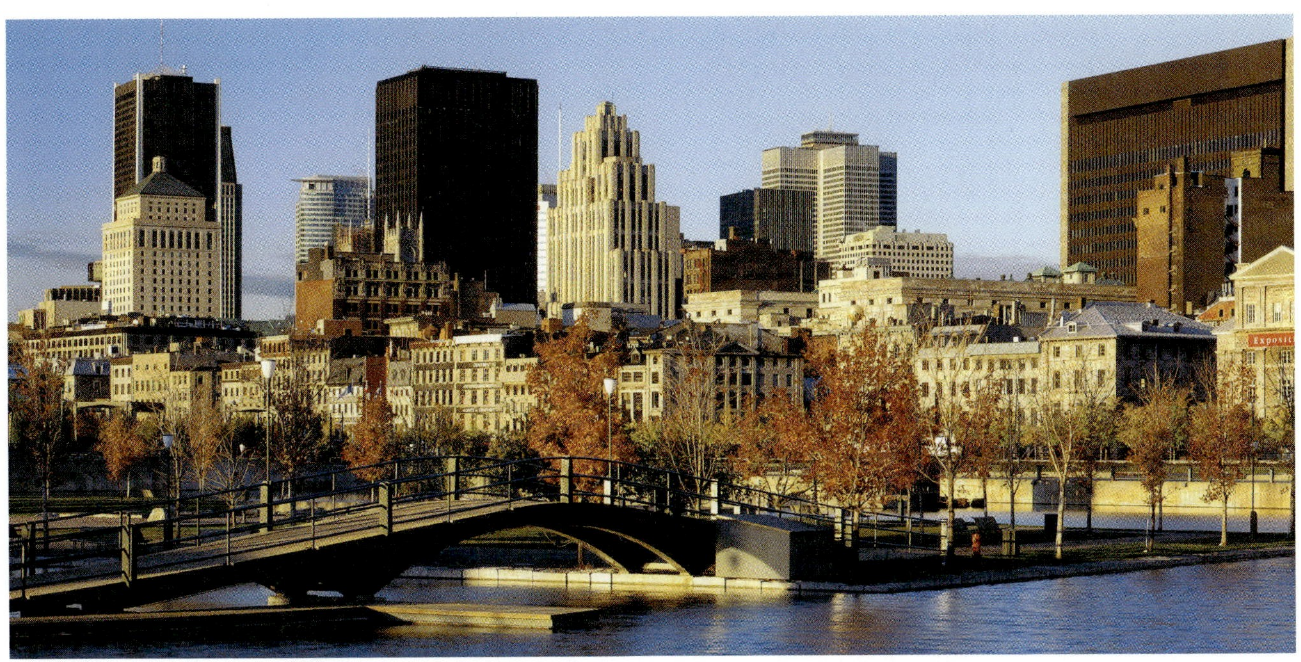

Montréal skyline from the Old Port, with the St Lawrence River in the foreground. (Stéphan Poulin).

tan city, is the engine of economic and cultural growth within Québec.

Québec has a varied resource base, including a large number of untapped hydroelectric sites, a solid manufacturing industry sparked by growth in the new economy (pharmaceuticals, aviation, and the multimedia), and a growing tourist industry. However, Québec faces two challenges. First, Québec's share of Canada's economy and population is slipping. If the current trends continue, both Western Canada and British Columbia will surpass Québec before the midpoint of the twenty-first century. The political implications are considerable. Even though Québec is guaranteed 75 seats in the House of Commons, Québec's political clout will diminish as the total number of seats increases to accommodate population increases in other geographic regions.[3] Second, the creation of a North American market has challenged and will continue to challenge the province, forcing industries to adapt to a continental market.

As a modern industrial society, Québec exhibits the customary economic and social characteristics that other such societies do. For instance, most Quebecers, like their counterparts in the rest of Canada, live in urban settings. Québec also has an extremely low rate of natural increase. Attracting immigrants to the province ensures a positive rate of population growth, but some worry that immigrants could change the linguistic balance between French- and English-speaking Quebecers. In fact, before the language laws were introduced, almost all non-French-speaking immigrants chose to speak English and send their children to English schools. The language laws now require immigrants to send their children to French schools, thereby ensuring the growth of the French-speaking population.[4]

Today, language continues to shape Québec's economy and politics, as well as the province's relationship with Ottawa and the other provinces. French is the issue around which most other concerns are framed—in other words, most political, social, and economic issues in Québec are seen from the perspective of language and culture. The French/English faultline is therefore a crucial component of Québec and must be well understood if it is to be adequately addressed.

A Decisive Moment: The 1995 Referendum

The 1995 referendum was a political watershed for Canada. The question was: 'Do you agree that Quebec should become sovereign, after having made a formal offer to Canada for a new Economic and Political Partnership within the scope of the Bill respecting the future of Québec and of the agreement signed on June 12, 1995?' While one can debate the meaning of the referendum question, all would agree that Quebecers came very close to separating from Canada—the 'Non' side won by a very slim majority. If there had been a 'Oui' majority, Premier Jacques Parizeau's strategy would have been to immediately declare Québec an independent nation.

The deep language divide within Québec society is shown clearly in the results of the referendum. In 1995, the political map of Québec showed strong support for the 'Oui' side in rural Québec where the francophone population dominates. In fact, support for independence was strongest among Québec francophones, with estimates placing it as high as 60 per cent, and weakest among anglophones and allophones. Only four areas of Québec—Montréal, the Ottawa Valley, Estrie, and northern Québec—supported the federalist side, but these areas combined to have the majority of the voters. In the first three areas, there is a high proportion of anglophones and allophones, and in northern Québec, Cree and Inuit peoples make up the vast majority. At that time, the Cree had been particularly outspoken on the issue of sovereignty, arguing that they have the right to determine their political future (just like Québec), whether it means remaining part of Canada or joining an independent Québec.

Figure 6.4 Referendum fatigue. Following the 1995 referendum, Quebecers had grown weary of the political wrangling over Québec's place in Confederation. Economic issues had become more important to the average citizen. However, although the issue of Québec's separation from Canada is on the political back burner for the time being, this does not mean that the sovereignty movement is dead. (*Globe and Mail*, 31 Oct. 1995, A24. Reprinted with permission from *The Globe and Mail*.)

After the Storm

Separation is a complex and extremely personal political issue. After the 1995 referendum, Québec had had enough—at least for a while. The federal government, however, was still edging away from the brink. To strengthen its hand, Ottawa went to the Supreme Court of Canada for a reference on independence, and the Court ruled, on 20 August 1998, that under both Canadian and international law, Québec (or any other province) cannot declare its independence unilaterally, thus bringing to an end the threat of separation by decree. Nevertheless, the Supreme Court accepted the concept that separation was possible through negotiation after a 'clear majority' voted for separation.

After 1995, Quebecers had grown weary of the political wrangling over Québec's place in Confederation (Figure 6.4). For the average Quebecer, economic, social, and health issues have become more important. Still, recent federal and provincial elections reveal the ambivalence of Québec politics. In 2003, the Liberal Party won the provincial election, displacing the ruling Parti Québécois. Yet, in the federal election of 2006, the Bloc Québécois received the greatest share of the popular vote in Québec (42 per cent) and won 51 seats (68 per cent of the seats in Québec. So one separatist party wins while another loses! Perhaps the famous French phrase sums it up: *Plus ça change, plus c'est la meme*. On top of these political events, Prime Minister Harper recognized Québécois as a nation within Canada (Vignette 6.5). And then, as if this were not confusing enough, in the 2007 provincial election the Liberal Party was returned to power, but this time with a minority government as the Action Démocratique du Québec, a right-leaning nationalist/populist party, slipped into second place and the PQ was relegated to third in the provincial pecking order.

Québec's Economy

Québec's economy is the second largest of the regions in Canada. Both Ontario and Québec are heavily dependent on the manufacturing sector. The rising exchange rate for the Canadian dollar and fierce competition from low-wage countries have placed great pressure on the manufacturing sector. From 2003 to 2005, manufacturing declined by 68,000 in

Vignette 6.5 The 'N' Word

For years, federal politicians have avoided calling Québec a nation. With a language and territory different from the rest of Canada, Québec meets the basic requirements of 'cultural nation' within Canada. In November 2006, the House of Commons broke rank with the past by overwhelmingly passing a motion by Prime Minister Harper that recognized the Québécois as a nation within Canada.

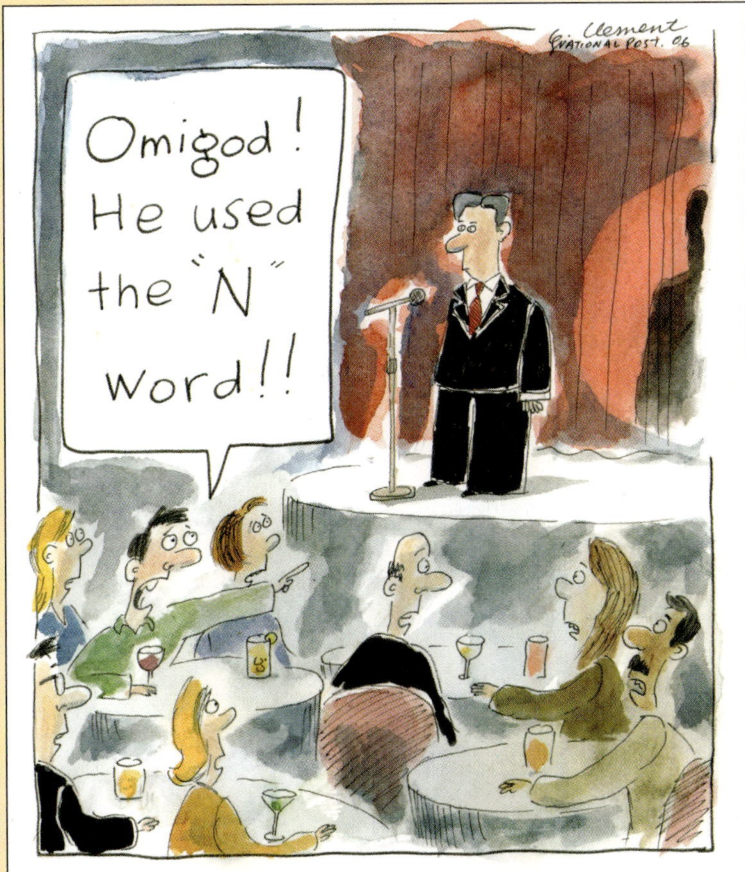

(© Gary Clement)

Quebec and 61,000 in Ontario, accounting for just under 90 per cent of the net loss in manufacturing jobs in Canada (Ferrao, 2006).

In its report on the sectoral outlook for Québec, Service Canada (2006) confirmed that Québec's manufacturing sector was in trouble:

> The manufacturing sector has faced multiple challenges since the beginning of the decade. They are reflected in its poor results in terms of output and especially employment. Between 2003 and 2005, employment across the sector declined at an average annual rate of 1.7 per cent. Never has there been such a prolonged decline, other than during periods of recession. Over the past three years, output, measured in real terms, has risen by a meagre 0.5 per cent. However—and this is an encouraging fact—manufacturing output has begun to recover in the last two years, albeit modestly, after declining for three consecutive years between 2001 and 2003.

Fortunately, other sectors of Québec's economy have done much better, and Québec continues to have a strong and diversified economy that is heavily dependent on foreign trade. Québec accounts for much of Canada's exports: 25 per cent of information technologies, 55 per cent of aerospace production, 30 per cent of pharmaceuticals, 40 per cent of biotechnology, and 45 per cent of high-tech exports. Access to the American market is crucial. Improved access after 1989 has had a positive impact on Québec's economy. Whether primary products, such as hydroelectricity, semi-processed materials, such as magnesium, or manufactured products, such as aircraft, most production is exported to the US, with a lesser amount to the rest of Canada and world markets. Dairy products represent Québec's principal interprovincial export.

Prior to the Free Trade Agreement in 1989, approximately half of Québec's trade went to other parts of the country and half was in exports. By 2002, nearly two-thirds of trade was outside the country, especially to the United States. This trend has continued. The intervention of both the provincial and federal governments to support the revitalization of Québec's manufacturing sector and to help it compete in the North American and world markets has paid dividends. This revitalization strategy promoted key industrial sectors (including the aerospace, pharmaceutical, and metal-processing industries), and encouraged industrial development in southern Québec by offering firms low-cost electrical power. Yet, trade is a two-edged sword and governments can no longer shelter their economies from foreign producers. Imports from developing countries with low wages have undercut Canadian producers and sometimes pushed Canadian firms into bankruptcy. As Canada lowered its tariffs on imports from developing countries, Québec's apparel and textile firms faced stiff competition—many firms either closed or relocated offshore to take advantage of lower labour costs. Global trade has meant that low-priced magnesium imports from China have hurt the magnesium refining industry in Québec.

The spatial aspects of Québec's economy are similar to Ontario, and, like Ontario, Québec can be divided into two economic areas: an agricultural and manufacturing core in southern Québec and a resource-based periphery in the north. Québec's core is associated with the fertile Great Lakes–St Lawrence Lowlands and the hinterland with the Canadian Shield (Figure 2.1). Consequently, Québec is similar to Ontario in its array of resources and access to the US and global markets. Yet, there are differences. The climatic conditions in Ontario's part of the Great Lakes–St Lawrence Lowlands provide for a longer growing season, thus giving Ontario an advantage. On the other hand, Québec's part of the Canadian Shield contains much better physical conditions for the production of hydroelectricity. Then, too, the close proximity of the Canadian Shield to Montréal and Québec City provides excellent recreation sites—lakes for summer activities and ski hills for winter, thus making the area one of North America's most popular recreational tourist areas.

The Canadian manufacturing belt extends from Windsor to Québec City, and the St Lawrence Seaway connects these two industrial cores (Vignette 6.1). Proximity to Canada's greatest trading partner, the United States, is extremely important for manufacturing. Windsor and other centres in southern Ontario have had the advantage of proximity to the American automobile-manufacturing centre of Detroit. Montréal, on the other hand, is farther from major American manufacturing cities. Nonetheless, many firms in Canada's manufacturing belt supply products to other manufacturing companies in this belt, indicating a high degree of economic integration. An example of such integration is the auto-parts industry. Over 20 firms in the Montréal area produce automobile parts and then ship them to assembly plants in Ontario and the United States. General Motors of Canada did have an assembly plant in Sainte-Thérèse that made the Pontiac Firebird and Chevrolet Camaro sports cars, but this assembly plant was closed in 2002.

Over the years, Québec-based firms have specialized in certain industrial sectors, including apparel and textiles, high-tech industries, metal refining, printing, and transport equipment. The apparel and textile industries formed the traditional heart of manufacturing in Québec, but since the FTA these industries have faced stiff competition from abroad because of the reduction in tariffs on apparel and textiles, which has allowed imports from low-wage countries to grab a major share of the North American market. In 2002, the final blow came with China's entry into the World Trade Organization (WTO). As a result, quotas on textiles and garments disappeared. With Canadian quotas on Chinese textiles removed in 2003, imports into Canada soared. In 1995, for example, China accounted for 6 per cent of textile imports, but, by 2005, Chinese imports had reached 46 per cent (Wyman, 2006). The resultant decline in such manufacturing in Québec is a common phenomenon found in the rest of Canada and in other countries. By 2007, China was clearly the world's biggest and least expensive producer of quality garments, thus driving Québec and other Canadian garment firms from the marketplace.

High-tech industries, which require a highly skilled labour force, have fared much better. In fact, many economic gains in Québec have come from high-tech firms, which account for 45 per cent of Canada's high-tech exports. Because Montréal has a critical mass of high-tech companies, several universities, and strong provincial support, it is the most important centre for the new economy in Canada. Leading components are found in aerospace, biotechnology, fibre optics, and computers (both hardware and software). Québec provides nearly half of Canada's information technologies, aerospace, and pharmaceutical exports as well as 40 per cent of its biotechnology exports. Its aerospace firms, led by Bombardier and CAE Inc., have been particularly successful in recent years. For example, Bombardier, while struggling in the airplane sector, obtained three large orders for trains in 2006 (Vignette 6.6). CAE is a global leader in advanced simulation and control equipment for aircraft and ships. The company employs about 4,000 highly trained workers in the Montréal area and another 2,000 in other parts of the world. Approximately one-

Vignette 6.6 Bombardier Rebounds

Over the last five years, times were tough for Bombardier, especially its airplane sector. In 2006, Bombardier Inc., the world's largest train maker, did very well. The company won contracts worth US $1.65 billion for a rapid rail transit system in South Africa. The order came after Bombardier had won two other large train orders in 2006: a Cdn $469 million order from London in the UK and a $674 million order for new subway cars from the Toronto Transit Commission.

Source: Grant (2006).

third are engaged in research and engineering, designing, and testing new products, such as a robot that strips paint from aircraft bodies. Its main product, however, is a flight simulator manufactured at its plant in Saint-Laurent near Montréal. Unlike most manufacturing firms, most of CAE's sales do not go to the United States but to customers in Europe, the Middle East, and Asia.

Industrial Structure

As a core region of Canada, Québec has the second largest number of workers—23 per cent of the country's workers. Its industrial structure has many similarities with that of Canada's other core region, Ontario (Table 6.4). First, both regions saw their workforces increase in size. From 1995 to 2005, Québec's workforce increased from 3.2 million to 3.7 million. Second, the division of Québec's and Ontario's industrial labour forces into the three principal sectors (primary, secondary, and tertiary) indicates that less than 3 per cent of workers in each province are engaged in the primary sector, just under one-quarter of workers are in the secondary or processing industries, and approximately three-quarters are in tertiary industries. Third, the tertiary sectors of Québec and Ontario increased their share of the total labour force from 1995 to 2005 at the expense of their primary and secondary sectors.

But what aspect of the Québec and Ontario industrial structures distinguishes them from those of the other geographic regions? The answer lies in their secondary sectors. As core regions, both Québec and Ontario have strong secondary industrial sectors dominated by manufacturing. For instance, in 2005, Québec's secondary sector accounted for 22 per cent of its labour force compared to 24 per cent in Ontario (Table 6.5). In comparison, the four other geographic regions, the so-called hinterland, had much smaller secondary sectors. In 2005, Western Canada had 17 per cent of its labour force in the secondary sector, Atlantic Canada 16 per cent, British Columbia 18 per cent, and the Territorial North approximately 2 per cent. At the same time, the tertiary sectors of Québec and Ontario were at or below 75 per cent while the tertiary sectors of the other geographic regions, with the exception of Western Canada, exceed 75 per cent.

Southern Québec

Southern Québec is the economic, social, and political core of Québec, while northern Québec is a sparsely populated resource hinterland. The relationship between the two regions is a provincial version of the core/periphery model and overlaying that model

Table 6.4 Employment by Industrial Sector in Québec, 2005

Industrial Sector	Québec Workers (000s)	Québec Workers (%)	Ontario Workers (%)	Difference (%)
Primary	99.2	2.7	2.0	0.7
Agriculture	60.8	1.6	1.5	0.1
Secondary	826.7	22.2	23.6	−1.4
Manufacturing	615.7	16.6	16.6	0.0
Tertiary	2,791.4	75.1	74.4	0.7
Total	3,717.3	100.0	100.0	

Source: Statistics Canada (2006c). Adapted from Statistics Canada website Searched 12 September 2006, http://www40.statcan.ca/101/cst01/labor21c.htm.

Table 6.5 Comparison of Ontario and Québec Industrial Structures, 1995–2005

Sector	Ontario 1995 (%)	Ontario 2005 (%)	Percentage Difference	Québec 1995 (%)	Québec 2005 (%)	Percentage Difference
Primary	3.0	2.0	(1.0)	3.5	2.7	–0.8
Secondary	23.6	23.6	(0.0)	23.0	22.2	–0.8
Tertiary	73.4	74.4	1.0	73.5	75.1	1.6
Total	100.0	100.0		100.0	100.0	
Workers (000s)	5,231	6,398	22	3,204	3,717	16

Sources: Statistics Canada (1996b, 2006c).

are the francophone presence in southern Québec and an Aboriginal majority in northern Québec.

The more physically favoured lands are found in southern Québec, especially in the valley of the St Lawrence River. Southern Québec contains two physiographic regions: the Appalachian Uplands and the St Lawrence Lowlands. Bordered on the north by the Canadian Shield and on the south by the United States, southern Québec is only a small part of the territory of Québec, but it contains over 90 per cent of Québec's population and agricultural lands, and is the industrial heartland of the province. The human and physical characteristics of each physiographic region in southern Québec are presented below.

Appalachian Uplands

The Appalachian Uplands region lies west of the Atlantic Ocean and north of the United States. Its northern edge faces the St Lawrence Lowlands. Because of this geography, the Appalachian Uplands region was settled at different times and in different ways. First settlements took place along the Gaspé coast. In the sixteenth century the waters off the Gaspé Peninsula attracted fishers from Spain, Portugal, England, and France. Even today, fishing in the Atlantic Ocean provides a way of life, though many supplement their income with farming and wage employment in the small coastal settlements. The main settlements are scattered along the Gaspé coast and the south shore of the lower St Lawrence River, leaving the rugged interior with few settlements. The largest urban centres are along the south shore. Rimouski, with a population of nearly 50,000 in 2006, is by far the largest city in the region. Rivière-du-Loup and Matane are medium-sized towns with populations roughly half the size of Rimouski. Towns are much smaller along the Gaspé coast, as the area has a very limited resource base. Percé, now an important tourist town, is the largest of the small centres along the Gaspé coast with a population of about 4,000. This French-speaking area has place names like New Richmond, New Carlisle, and Chandler, the legacy of early English settlers, some of whom had relocated along the Gaspé coast after the American Revolution. Over the last half-century, the lack of jobs caused many to migrate out of this area, which seriously reduced the size of English-speaking communities. In 2006, English-speaking residents constituted less than 5 per cent of the region's population.

Estrie (the Eastern Townships) is a pocket of communities in the rolling land of the Appalachian Uplands located east of Montréal. Estrie was settled after the American Revolution by British Loyalists. The British organized the surveyed land into rectangular townships rather than the French long lots found in the St Lawrence Valley. Much of the land was ill-suited for cultivation. Within several generations, the more marginal lands were abandoned when English-speaking owners left to look for jobs in Montréal and Boston or to try homesteading on the American frontier. Because of a land shortage in the St Lawrence Valley, French Canadians began to move into the Eastern Townships. French-Canadian migrants, often sons of farmers living in the St Lawrence Lowlands, either bought or took over abandoned farms from the original English-speaking settlers.

The physical geography of Estrie is much more conducive to economic development and agricultural settlement than are the Gaspé coast and south shore of the St Lawrence. From the 1870s to the 1970s, mining at Thetford Mines and Asbestos was a key sector of the regional economy. Since the 1970s, asbestos mining has fallen on hard times because this mineral, formerly used as insulation in the construction trade, has proven hazardous to human life. Still, mining of this deposit continues. Agriculture and forestry have also proven to be durable in this area. Dairy farming is pursued in the broad valleys, and logging in the forested uplands. Overall, Estrie is the most prosperous area of the Appalachian Uplands. Sherbrooke, which has grown over the years and is the largest urban centre in the Appalachian Uplands, exemplifies the relative well-being of the region. With a population of 187,000 in 2006, Sherbrooke has become an important regional centre (Table 6.6). From 2001 to 2006, its population increased by 6.3 per cent, making it the second fastest-growing city in Québec after Gatineau (6.8 per cent). Montréal was third at 5.3 per cent. This demographic increase over the years is reflected in its economic growth. Proximity to Montréal has often worked in Sherbrooke's favour, allowing it to engage in the textile industry in the nineteenth century and now in the high-technology industries.

St Lawrence Lowlands

Most of Québec's agricultural and industrial production is in the St Lawrence Lowlands. In fact, this area functions as the province's core. It contains Québec's largest market and is close to transportation networks and the St Lawrence River, which facilitate trade with foreign countries.

The warm to hot summer climate coupled with abundant rainfall makes the St Lawrence

Table 6.6 Major Cities and Census Metropolitan Areas in Québec, 2001–6

Census Metropolitan Area	Population 2001 (000s)	Population 2006 (000s)	Change (%)
Montréal	3,451.0	3,635.6	5.3
Québec City	686.6	715.5	4.2
Gatineau	226.7	242.1	6.8
Sherbrooke	176.0	187.0	6.3
Saguenay	154.9	151.6	–2.1
Trois-Rivières	137.5	141.5	2.9
Total	4,832.7	5,073.3	5.0

Source: Statistics Canada (2007). Adapted from Statistics Canada publications *Population and Dwelling Counts, 2001 Census*, Catalogue 93F0050XCB2001013, http://www.statcan.ca/bsolc/english/bsolc?catno=93F0050X2001013 and *Population and Dwelling Counts, 2006 Census*, Catalogue 97-550-XWE2006002 http://www.statcan.ca/bsolc/english/bsolc?catno=97-550-XWE2006002.

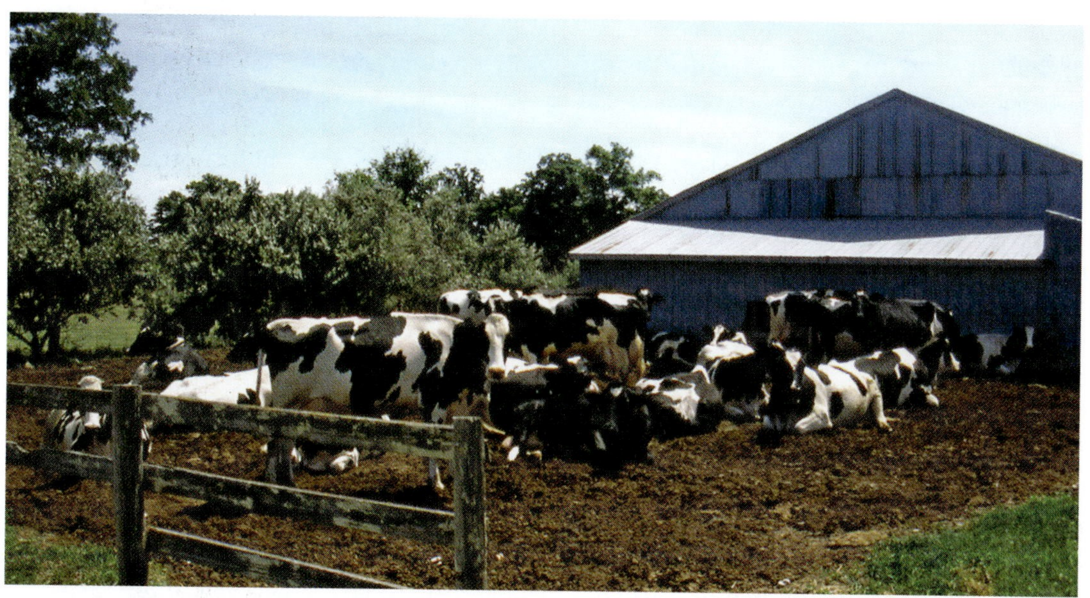

In 2004, approximately 40 per cent of Canada's milk cows were located in Québec. While most of Québec's industrial milk and cream is consumed within the province, some is exported to other provinces, especially Ontario. (Ivy Images)

Lowlands the most favoured region in Québec for agricultural activities. Livestock farms specializing in cattle, hogs, or sheep are common, while some farmers concentrate on dairy, poultry, and egg production. Livestock farmers grow forage crops for winter feed. During the summer, the cattle graze on pastures. Specialized crops, particularly vegetables and fruit, are also popular and are sold mostly in the major urban centres of the province.

Though farmers in this region engage in a variety of agricultural activities, dairy and vegetable farming predominate. With 37 per cent of Canada's one million milk cows, Québec leads in the production of dairy products. While each province has its own milk marketing board, the National Milk Marketing Plan calculates the allocation of milk quota for each province, with Québec receiving nearly 40 per cent of the Canadian market. Moreover, the processing of agricultural products provides an added benefit for the Québec economy. For instance, large butter and cheese firms rely on milk products from the dairy farms. Often the processed food products are designed for the provincial market. Because of a ready market for unpasteurized cheese products in Québec, a few cheese firms began producing *fromage au lait cru*, cheese made with unpasteurized milk. These popular cheese products compete favourably with European imports, including popular brands from France.

Dairy farmers have fared relatively well under Québec's fluid milk marketing board and a national marketing system that allocates producers a share of the Canadian market for milk. In 2002, cash sales from milk and cream amounted to $1.5 billion in Québec. Under NAFTA and WTO rules, however, Canada is under pressure to dismantle its marketing boards. Without a regulated dairy industry, dairy farmers in Québec would face stiff competition from American dairy imports. In 2002, the World Trade Organization ruled that Canadian dairy exports are illegally subsidized because the national marketing system's price for Canadian milk production is above market price. While 95 per cent of Canada's total milk production is consumed domestically, the value of Canadian dairy exports amounts to about $250 million. According to the WTO ruling, Canada will have to reduce its exports by around half of the current figure of $250 million.

Manufacturing is concentrated in the Montréal area but extends eastward to Estrie and northeastward to Québec City. As part of the Canadian manufacturing axis, this industrial zone serves as the engine driving both the provincial and the national economies. The manufacturing sector is very sensitive to global trade. While information technology, aerospace, pharmaceutical, and biotechnology firms have benefited from the liberalization of world trade and NAFTA, they are now tied to world demand. In order to stay ahead of foreign competitors, these firms have spent heavily in research, making Montréal the leading research centre in Canada (Canada's Innovation Leaders, 2006).

Since 1989, these firms have expanded, although in the early years of the twenty-first century some businesses, such as the aerospace industry, are contracting due to a reduction in worldwide demand. These high-technology industries employ highly skilled and well-paid workers. Most of Québec's manufacturing sector is in Montréal, where high-tech companies are now the leading edge in manufacturing. These companies were not only responsible for Montréal's impressive economic recovery in the late 1990s but are also transforming the city into one of the leading high-tech centres in North America.[5] This shift from traditional to high-tech manufacturing has two consequences for Montréal's labour force. First, the demand for highly skilled workers is increasing, creating labour shortages and a search for skilled immigrants. Second, the demand for semi-skilled workers is decreasing, thereby contributing to the relatively high unemployment rate in the region and province.

The traditional sector, composed of textile, knitting, leather, and clothing firms, was once the mainstay of manufacturing, but this sector has been declining for some time and its future is uncertain. As a labour-intensive industry where education and a good command of English or French is not necessary, many low-income, immigrant families have members, especially females, working in this industry. Now struggling to compete with foreign imports, the traditional manufacturing firms have sought to lower their costs of production by reducing the size of their labour force, introducing more efficient machines, and seeking niche markets in North America. While Montréal remains the focus of Canada's clothing and textile industry, its grip on market share is slipping for two reasons. First, established firms are either shifting their production from Canada to countries with low wages, or these firms are subcontracting various elements of production in these low-wage countries. Second, in accordance with the WTO rulings, Ottawa has continued to reduce tariffs on textiles, clothing, and related products from developing countries. While these measures may stimulate economic development in these foreign countries, including China, the negative impact on Québec firms is contributing to the collapse of the textile and garment industry in Montréal and other Québec centres.

With the automobile and aerospace industries facing a slump in demand, both industries have had to reduce production capacity. In 2002, Québec lost its only automobile-assembly plant when GM closed its plant at Sainte-Thérèse, laying off more than 1,000 workers. In an agreement with the provincial government, GM has committed to increase its purchases of auto parts from Québec firms to lessen the economic blow caused by the assembly-plant closure and to invest in developing lightweight auto parts using aluminum and manganese materials. In 2002, GM purchased $1.6 billion of auto parts from Québec firms and that figure was expected to climb significantly (MacAfee, 2003: FP5).

Québec's aerospace industry has become a major player on the world scene. Assisted by generous funding from Ottawa and the Québec government, Bombardier has become Canada's leading aerospace firm. The firm accounts for around one-quarter of business jet sales in the world and employs nearly 20,000 workers. Bombardier produces passenger jets (the Challenger and the Regional Jet) and smaller business jets (Learjets). The Regional Jet, a stretched version of the Challenger, is the leading aircraft for short-haul markets in North America and Europe. The success of this firm is due largely to its ability to sell its product in the global market. Bombardier and other aerospace firms face a difficult future in the short run. While jet aircraft sales exploded in the late 1990s and peaked in 2001, sales have declined sharply. Bombardier, for instance, saw its sales plummet from 162 passenger jets in 2001 to 77 in 2002 (Silcoff, 2003: FP5), primarily the result of a near-mortal blow with the crashing of passenger jets into the World Trade Center in New York and the Pentagon outside Washington by terrorists. With the United States declaring war on terrorists, the volume of business and tourist travel dropped dramatically and public confidence in air travel has been slow to return. Like other major airline companies, for example, Air Canada faced a difficult future with fewer air travellers and stiff competition from smaller airlines like West Jet. In May 2003 and again in May 2004, Air Canada just escaped from bankruptcy, but after reducing the size of its staff and by renegotiating wage rates with its union members, Air Canada is again a strong airline company.

By the end of 2005, manufacturing in Montréal and the rest of the St Lawrence Lowlands had just endured a challenging period of economic adjustment. In the greater Montréal region, both traditional and high-technology firms reduced their operations and some plants closed. During this painful period of adjustment, economic growth in the eastern segment of Canada's manufacturing core slowed. Fortunately, the tertiary sector expanded, thus allowing the total labour force to expand.

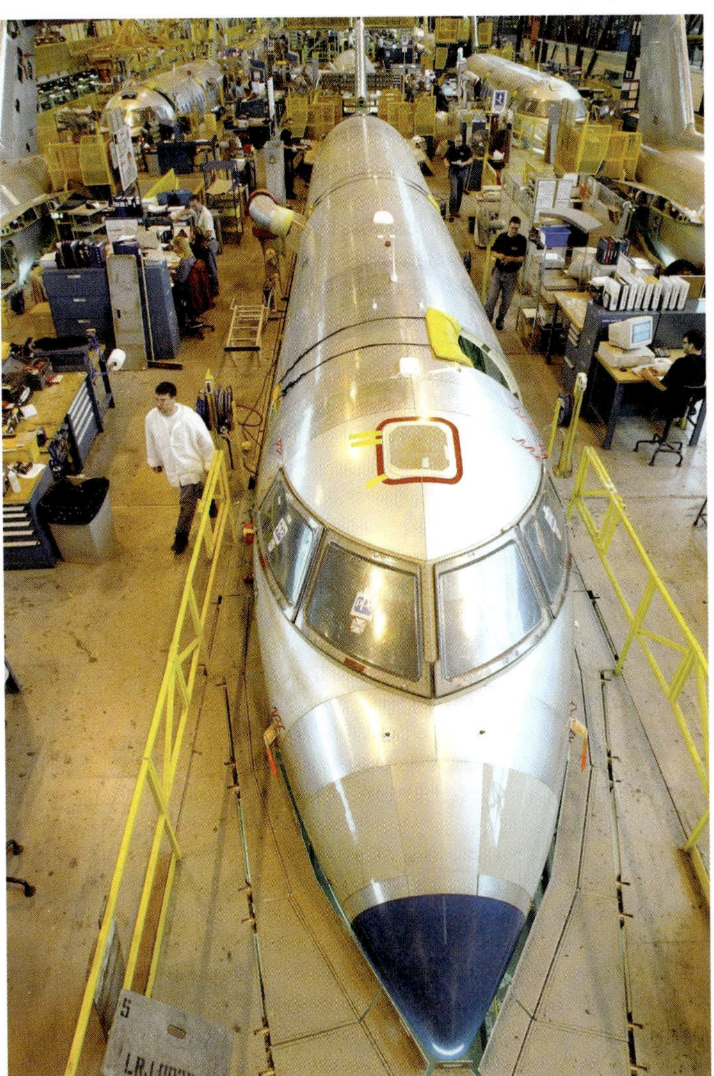

In Bombardier's Montréal factory, a new jet aircraft will soon be completed. Passenger jets produced by Bombardier are the Challenger and Regional aircraft. (CP/Ryan Remiorz)

Key Topic: Hydro-Québec

Since the Quiet Revolution, successive Québec governments have played an active role in shaping the province's industrial economy. This approach, initiated by the Lesage government, has pursued two goals—one economic, the other political. Its economic objective has been to stimulate economic growth through state intervention in the marketplace. Its political goal is to increase Québec's ownership and control of its economy within the francophone business community. The most direct way to achieve these two goals was to create public (Crown) corporations. The best example of the success of this industrial development strategy is Hydro-Québec. Now the prize Crown corporation in Québec, Hydro-Québec has played a fundamental role in achieving these two goals and, in so doing, has helped transform Québec's economy and build a strong French-speaking business community. Today Hydro-Québec is Canada's largest electric utility and its production of electric energy ranks first among Canada's provinces (Table 6.7). This Crown corporation generates an annual profit of $3 billion and pays a dividend of around $2 billion to the province. The success of Hydro-Québec is based on the great natural advantage for producing hydroelectric power in Québec's Canadian Shield (Vignette 6.7).

Created in 1944, Hydro-Québec was a minor force in the Québec economy until the Lesage government came to power in 1960. At that time, the Québec government announced its intention to purchase the private electricity companies in Québec and place them under the umbrella of Hydro-Québec. The Crown corporation soon extended its activities to cover the whole province by purchasing the shares of nearly all remaining privately owned electrical utilities and taking over their debts. With a virtual monopoly to generate and distribute electricity in the province, Hydro-Québec undertook the task of expanding Québec's hydroelectric capacity. In the 1960s, Hydro-Québec built dams on the Manicouagan River, creating the huge Manic-Outardes hydroelectric complex. The design and construction of this massive project were carried out by Québec firms and workers, thereby creating highly specialized engineering firms capable of designing and building huge hydroelectric projects.

Hydro-Québec harnessed the vast water resources of the Canadian Shield and transported this electrical energy from the dams on the Manicouagan River to Québec City and Montréal. In order to transmit the complex's annual production of about 30 billion kWh over a distance of nearly 700 km, Hydro-Québec became the first utility in the world to transmit electricity at very high voltages (735 kv), which reduces the amount of electrical energy lost in the transmission process. This breakthrough technology later enabled Hydro-Québec to transmit electrical power from Churchill Falls[6] and James Bay to markets in southern Québec and New England.

Following the success of the Manicouagan River complex, Hydro-Québec embarked on an even larger project—the James Bay Project. As it turned out, the James Bay Project, which includes La Grande, Great Whale, and Nottaway river basins, was the crowning success of Hydro-Québec, but also its bête noire. This was the largest hydroelectric project in the world and would transmit power over long distances never attempted before. Québec could now sell massive amounts of power to the United States and, in so doing, recoup much of the construction costs. The natural environment, however, was dramatically altered by river diversions and by the creation of huge reservoirs. The project negatively affected the fish stocks and the prime wildlife

Table 6.7 Quick Facts About Canada's Hydropower

- Canada is the world's biggest producer of hydropower, generating 353 TWh/year and producing nearly 13 per cent of the world's output.
- There are 475 hydropower plants in Canada; 233 of these plants have a generating capacity of more than 10 MW and represent 99 per cent of total capacity.
- Québec, British Columbia, Ontario, Newfoundland and Labrador, and Manitoba are the highest producers of hydropower in Canada. Together, these provinces produce 97 per cent of the country's total hydropower generation.
- With the exception of Prince Edward Island, every province in Canada has some hydropower capacity.
- Two thirds of the electricity produced in Canada comes from hydropower, making it the most important electricity source in the country.
- Hydropower accounts for 97 per cent of Canada's renewable electricity generation.

Source: Canadian Hydropower Association, www.canhydropower.org.

Vignette 6.7 Québec's Major Waterpower Resources

Vignette 6.7

Hydroelectric developments depend on three factors: precipitation, topography, and access to market. Two of the factors work well for Québec's Canadian Shield—annual precipitation exceeding 800 mm and favourable topography such as high elevations, lakes that store water, and steep drops to sea level. Access to markets has been solved by the innovation of high-voltage transmission lines. Two of Canada's principal hydroelectric generating stations form part of Hydro-Québec's electrical system.

- LG-2 on La Grande Rivière, Québec (5,328 mw)
- Churchill Falls on the Churchill River, Labrador (5,225 mw)
- Gordon M. Shrum on the Peace River, BC (2,416 mw)
- LG-4 on La Grande Rivière, Québec (1,650 mw)
- Kettle Rapids on the Nelson River, Manitoba (1,255 mw).

The main advantages of hydroelectric developments are: the long life of the facilities, low operating costs, and zero air pollution. However, there are drawbacks: the initial high capital investment, the long construction period, and the socio-environmental costs. When river valleys used for hunting and fishing by Aboriginal peoples are flooded, the loss of wildlife habitat and fishing grounds has serious impacts on their economy and culture. Such losses represent a very high but hidden social cost.

habitat along the river valleys and lakes. The hunting and trapping lifestyle of the local Cree was disrupted by this construction project and the resulting alterations to the natural landscape. Although Hydro-Québec made efforts to minimize the environmental damage and compensated the Cree, Inuit, and Naskapi through the James Bay and Northern Québec Agreement, the Cree strongly opposed further hydroelectric developments in northern Québec. In 2001, however, the Cree and the government of Québec signed an agreement to permit further hydroelectric development in the James Bay region (see the section on 'Hydro Power' later in this chapter for further discussion on these topics).

Industrial Strategy and Regional Linkages

With hydroelectric power developments in Québec's Canadian Shield, an industrial strategy and regional linkages were born. The vast electrical power generated by the first phase of the James Bay Project provided the provincial government with an opportunity to attract energy-hungry industries into southern Québec by offering them special, low electricity rates. Such an industrial strategy takes on the spatial form of the core/periphery model where the hinterland supplies the energy for industrial users in the core. An early version of this strategy took place in 1957 when Reynolds Aluminum built a smelter at Baie-Comeau because Hydro-Québec provided it with low-cost power. In return, the company added to the value of production in the province and, more importantly, it provided jobs in an economically depressed area. Today, Reynolds Aluminum, known locally as Société canadienne des métaux Reynolds, employs approximately 2,500 workers.

This industrial strategy took hold in the 1960s when the provincial government decided to take a more active role in Québec's economy and when Hydro-Québec had a large surplus of power. The surplus came from Québec obtaining the rights to the electrical

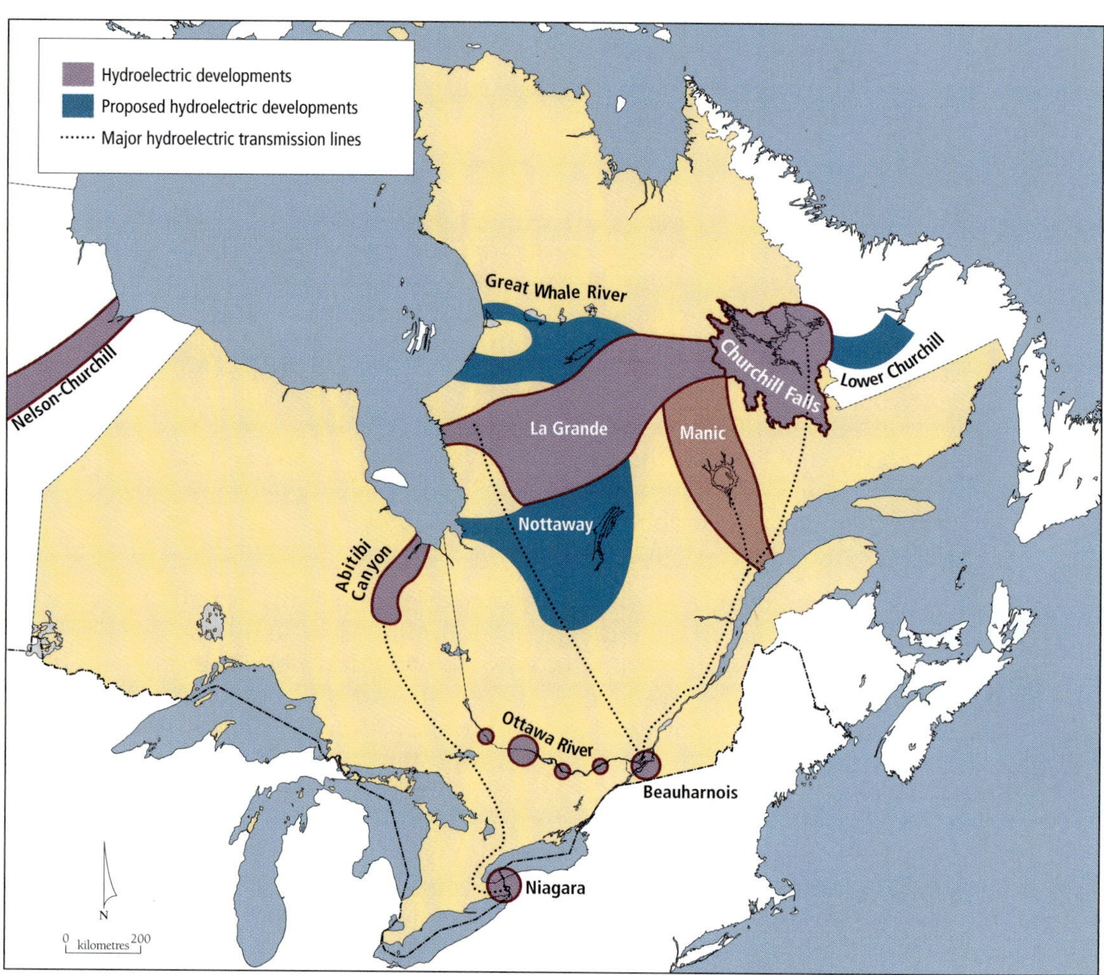

Figure 6.5 Hydroelectric power in Central Canada. Québec's dominant role in the production of hydroelectric power is due to two natural factors: (1) the heavy annual precipitation in northern Québec, and (2) the high elevation of the Canadian Shield in Québec. While plans for the Great Whale River were cancelled, work on the Eastmain River, in the northern section of Nottaway, nears completion. Waters of the Eastmain Basin will be diverted into La Grande Basin. (Further resources: Student website, National Atlas section, Map 17. Website instructions are found on p. xx.)

power produced at Churchill Falls in Labrador and from Hydro-Québec's recently completed Manic-Outardes hydroelectric complex on the Manicouagan River, and it was used to lure industrial companies into southern Québec. With the completion of the first—La Grande—phase of the James Bay Project in 1985, the provincial government took a more aggressive stance by offering risk-sharing contracts and electric-power rebates to industrial firms requiring large amounts of power in their processing operations, such as metallurgical companies. As a result, industrial firms either expanded their operations or new firms located along the St Lawrence River.

Hydro-Québec is able to provide industrial firms with low-cost energy for four reasons:

- Northern Québec can produce vast quantities of low-cost electrical power.
- Hydro-Québec has a long-term contract to buy power from Churchill Falls in Labrador at 1969 prices.
- Hydro-Québec has control over its price structure and can set extremely low power rates for its industrial cus-

tomers. In fact, these rates were so low that American firms complained that they represented a subsidy and therefore gave Canadian-based firms an unfair advantage in the US market.

- With prices for natural gas more than tripling since 2000, Hydro-Québec's competitive position in New England has improved because natural gas thermoelectric plants can no longer produce power at a lower per unit cost than imported electricity.

As for future natural gas prices, Michael Zenker, an industry analyst, states: 'The rising demand for gas, coupled with flat production, has tripled prices in the last four years' (Energyshop.com, 2007). Zenker also notes that if a substantial amount of offshore liquefied natural gas arrives on the east coast where LNG facilities are planned or under construction, shortages in the New England market may be reduced, thus reducing the rate of anticipated price increases in the next decade.

Québec's industrial strategy attracted metallurgical firms. One international firm, Norsk Hydro, took up the government's offer.[7] Norsk Hydro is involved in a large magnesium-smelting operation near Trois-Rivières and such smelting operations require huge amounts of electrical power. In 1988, Norsk Hydro signed a 25-year contract with Hydro-Québec to purchase electrical power at very low rates (Vignette 6.8). By the mid-1990s, the demand for magnesium die castings for automobile engines was growing at an annual rate of 15 per cent and the price of magnesium rose to over $1.30 a pound, causing Norsk Hydro to expand its smelter on the south shore of the St Lawrence River. By 2001, automobile produc-

Vignette 6.8 Norsk Hydro Canada Inc.

In 1988 Norsk Hydro signed a 25-year contract with Hydro-Québec for electrical power at special rates. Norsk Hydro buys hydro power at rates that fluctuate with the world price of magnesium. For the first three years, the company's magnesium smelter at Bécancour was eligible for electrical power rebates of up to 50 per cent of consumption. The purpose was to help defray the start-up costs of the smelter during its first three years of operation.

The Norwegian state-owned company spent $600 million to build its magnesium smelter at Bécancour. Production began in 1990. By 1991, Norsk Hydro had captured 24 per cent of the American market. Its magnesium metal products are used by automobile parts firms. An American firm, Magnesium Corp. of America, complained to the US Commerce Department that Norsk Hydro used unfair trade practices by selling its products on the US market below the cost of production. This allegation was based on the fact that Norsk Hydro obtained its electrical power at rates below the cost of producing that power. In 1992, the US Commerce Department placed a 32.7 per cent anti-dumping duty on Norsk Hydro's magnesium exports to the United States. This duty was later withdrawn after Hydro-Québec agreed to remove its electric rebates, but a countervailing duty of 4.48 per cent remains. In spite of the duty, Norsk Hydro made plans in 1997 to double its total capacity at its Bécancour smelter from 43,000 tonnes of magnesium products to 86,000 tonnes by 2000. At that time, employment at the smelter was anticipated to increase from 360 workers to 430. Since 2001, competition from China has forced Norsk Hydro to curtail its expansion plans.

Source: Yakabuski (1997: B6). Reprinted with permission from *The Globe and Mail*.

tion began to slow and magnesium prices, which had reached $1.50 a pound, began to fall. At the same time, magnesium imports from China began to undercut North American producers. As China grabbed a larger and larger share of the North American market, the magnesium refining industry in Québec was threatened. In 2003, Norsk Hydro's plant at Bécancour and the Noranda state-of-the-art magnesium plant at Danville lost their profit margin because the North American price for magnesium dropped below $1 a pound. For the Danville plant, the lower price was a disaster. In the late 1990s, Noranda assumed that world prices for magnesium would remain around $1.50 a pound. With the Québec government investing 20 per cent into the construction of Noranda's $1.3 billion magnesium plant at Danville, neither company nor government officials foresaw that the price of magnesium would fall so precipitously. In the spring of 2003, Noranda announced that it would temporarily close its plant and not reopen it until the price for magnesium reaches $1.30 a pound. Noranda may have a long wait because China can supply the North American magnesium market at prices well below $1 a pound.

Northern Québec

Northern Québec lies beyond the ecumene of the St Lawrence Valley. Here are the traditional homelands of the Cree, Naskapi (Innu), and Inuit. Its resource-based economy responds to foreign demand for such products. As a resource hinterland lying in the Canadian Shield, the main economic activities are associated with three natural resources: forests, minerals, and water. Today, such economic activities require relatively little labour. Consequently, the region is sparsely populated with a few large mining towns such as Chibougamai. Further north in the James Bay region, most people are Aboriginal Quebecers. While tourism flourishes in the Laurentides, which draw visitors from the nearby major population centres of Montréal and Québec City, most of northern Québec is too remote for tourists. Attempts at agricultural settlement have had marginal success in the Lac Saint-Jean area, but the short growing period and thin soils associated with the Canadian Shield prohibit commercial agriculture (Vignette 6.9). Northern Québec's geography is best suited for producing hydroelectric power, as well as for mining and forestry. Power is transmitted to industries and residences of the St Lawrence Lowlands and beyond to Ontario and New England. Low-cost electricity makes the smelting of bauxite (alumina ore) a major processing industry in the Saguenay region and along the north shore of the St Lawrence. Alcan, the world's second-largest aluminum producer, continues to expand its production capacity in Québec. Low-cost electrical power and an ocean shipping route to the Atlantic Ocean provide key location factors.

In 2002, industrial expansion at Alma demonstrated Alcan's commitment to the region, where low-cost electrical energy keeps production costs low and where ocean shipping to import the raw material, bauxite, and to export the finished product minimizes transportation costs. Alma is now one of Alcan's principal aluminum smelters, with a capacity 1.9 million tonnes a year. Located in the Lac Saint-Jean area, access to the Saguenay River provides an ocean link to its sources of bauxite and to world markets for its aluminum products. In 2005, Alcan and four partners announced plans to expand their Alouette aluminum plant at Sept Îles, making it the largest smelter in the world. Electric power is transmitted from Churchill Falls, and the port at Sept Îles, a natural deep-water harbour on the north shore of the St Lawrence, facilitates shipment.

Agriculture

The settlement of the Clay Belt demonstrates the difficulty of farming in the northern area of Québec. The Clay Belt occupies an enormous area of northwestern Québec and northeastern Ontario. The Canadian Shield acquired this relatively thick layer of sand, silt, and clay sediments near the end of the last ice age about

Agricultural Settlement of the Clay Belt in Québec

By the late nineteenth century, the growing rural population in the St Lawrence Valley began to move into the Clay Belt of northern Québec in search of arable land. French-Canadian migrants, often sons of farmers in the St Lawrence Lowlands, tried to clear small pockets of land in the Canadian Shield. By the 1880s, settlers had reached the Little Clay Belt around Lake Temiskaming. Agricultural settlement followed the construction of the Grand Trunk Railway (now Canadian National) from Québec City to the Little Clay Belt and then across the Canadian Shield to Winnipeg.

The Québec government and the Roman Catholic Church encouraged settlers to take up 'free' lands in both the Little Clay Belt around Lake Temiskaming and the Great Clay Belt around Lake Abitibi. Part of their concern was that thousands of French Canadians were migrating to New England and would be assimilated into American society. The Québec government and the Roman Catholic Church interpreted this migration as a 'loss' to French Canada. For that reason, they made special efforts to attract French Canadians to the last remaining agricultural lands in Québec—the so-called Clay Belt—where the French language and heritage could be maintained.

The Clay Belt is located within the Canadian Shield of northern Ontario and Québec. French-Canadian farmers began to settle these lands in the late nineteenth century. Life was hard and many farms were abandoned by the end of the twentieth century. (Victor Last, Geographical Visual Aids)

10,000 years ago. The rivers flowing to Hudson and James bays were blocked by the remnants of the Laurentide ice sheet. Gradually, a huge glacial lake called Lake Barlow-Ojibway was formed and lasted for several thousand years over an area of about 200,000 km^2. Lake sediments, especially minute particles of clay, sand, and silt, were deposited. When the glacial lake drained, much of the land was covered by peat, making it unsuitable for agriculture. In fact, less than 5 per cent of the Clay Belt contains arable land, only a portion of which has been cultivated. As a result, farmland in the Clay Belt is in scattered pockets, separated by large areas of forested country.

Efforts to settle land in the Clay Belt began in the late nineteenth century (Vignette 6.9), but serious land settlement took place during the Great Depression of the 1930s. At this time, both the Québec and Ontario governments encouraged a 'back to the land' movement in the Clay Belt. In Québec, the Catholic Church took a lead role by organizing settlements in Catholic parishes. Farms were small in size and had little arable land. As they were far from urban markets, these settlers were destined for a hard life with few prospects for developing commercial farming operations.

In Québec, the resettlement program had two goals: to combat unemployment in the cities of southern Québec and to ensure that the settlers followed a French-Catholic lifestyle. The provincial government and the Catholic Church gave little attention to the long-term prospect of commercial farming in the Clay Belt; rather, the emphasis was on defusing the unemployment problem in Québec and 'saving souls'. At its peak, agricultural colonization resulted in fewer than 15,000 farms. Most of these farms were small and achieved little commercial production.

Since World War II, agriculture in the Clay Belt has undergone three changes: (1) the emergence of a local market for dairy products in the nearby urban centres; (2) the establishment of a dairy marketing board, thereby providing price stability for farmers; and (3) the consolidation of most farms into fewer but larger farms. The trend to farm consolidation is revealed by the tenfold reduction in the number of farms in the Clay Belt from 15,000 in 1936 to 1,500 in 2001. Farm consolidation was driven by economic pressures. For instance, a commercial dairy farm needs sufficient land for summer grazing and for the production of fodder crops for winter feed. As well, a commercial family farm must have a minimum of 100 milk cows and modern milking machinery in order to achieve viability according to economies of scale. Though the changes undergone in the Clay Belt have improved the prospects for commercial farmers, with individual farmers benefiting from larger dairy farms and the regular payments from the provincial fluid milk marketing board, the potential for commercial agriculture in northern Québec remains extremely limited.

Forest Industry

Québec has 22 per cent of Canada's productive forest lands. In terms of productive forest, Québec ranks first among Canada's geographic regions, but it ranks second behind British Columbia in terms of total volume of wood cut. In 2004, for instance, Québec accounted for 23 per cent of Canada's harvested logs while British Columbia reached 43 per cent (Canada, 2006). However, Québec leads British Columbia in terms of pulp and paper production and the output of newsprint. Québec's advantage is its close proximity to major US cities, especially New York, where the demand for paper and newsprint is extremely high.

Québec has attempted to increase the value of its forest products through more local processing. Over the last 15 years, the manufacture of wood into higher-value products has increased. In fact, wood processing and paper manufacturing represent the most important manufacturing sector in the province by value of production. While the value of forest products has increased, the introduction of advanced woodcutting machinery and computerized mills has reduced demand for forest workers. This inverse relationship between technology and labour is particularly troublesome for workers in northern Québec, where few employment alternatives exist.

In total, Québec accounts for nearly 760,000 km^2 of forest lands. These lands are found in each of the four physiographic regions of Québec, but the vast majority of commercial forest is in the Canadian Shield south of 53° N. The boreal forest is concentrated in the southern half of the Canadian Shield, while mixed and hardwood forests are found in the St Lawrence Lowlands and the Appalachian Uplands. The Hudson Bay Lowland has few commercial stands. While timber is cut in the northern hinterland, most logs are processed at mills located at the mouths of tributaries flowing into the St Lawrence River. For instance, logs are floated down the St Maurice River to the pulp and paper mill at Trois-Rivières. The main exceptions are the Lac Saint-Jean and Saguenay regions, where there are many sawmills and eight large pulp and paper mills. Québec has a total of 64 pulp and paper mills.

While the forest industry contributes significantly to the province's economy in both value of production and employment, it has fallen on hard times. Falling prices for paper and high duties on softwood exports to the US have resulted in a contracting industry. For example, from 2001 to 2005, the number of sawmills declined from 291 to 262. More bad news arrived in 2006 when Domtar closed three sawmills in the Abitibi area until softwood lumber prices increase. A measure of the decline is revealed in the value of forest production, which was $18 billion in 2001 but had dropped to $13.5 billion by 2003 (Québec, 2003b, 2006a). With over 80 per cent of Québec's forest products exported to the United States, dependence on that single market is crucial. With the United States imposing a 27 per cent duty on Canadian softwood lumber from 2001 to 2006, the Québec lumber industry suffered. In late 2006, the US duty was removed, but the US housing industry had collapsed, causing the price for softwood lumber to drop precipitously.

Mining Industry

Mining has always been important in northern Québec. In 2005, the value of its metallic mineral production was $2.1 billion, placing Québec behind Ontario and British Columbia. Québec accounted for 16 per cent of Canada's $13.3 billion of metallic mineral production. Several northern Québec communities are single-industry towns and rely on mining for their existence. Mine (and town) closures are not unusual events. The cyclical nature of the mining industry, due to its dependence on world markets, poses a problem for resource communities. Low demand means layoffs and even the closure of mines and ore-processing mills. For example, the iron mine at Gagnon was closed in 1985, spelling the end to this single-industry town. This boom-bust cycle driven by fluctuations in world prices is particularly hurtful to the narrowly based economies of resource hinterlands. The world demand for mineral products reached a peak in the 1960s, declined sharply in the 1980s, recovered somewhat by the late 1990s, and then began to slide downward in 2002. Since then, demand (and prices) for minerals has climbed significantly. Such price fluctuations have had a profound impact on Québec's iron-mining industry. Québec iron mines account for much of the iron ore produced in North America. Québec's low-grade iron ore is converted into high-grade iron pellets and these pellets are the main raw material for North American and European blast furnaces and electrical mini-mills. The iron pellets also allow the mining companies to reduce their transportation costs from the mine site to the nearest ocean shipping point (Port Cartier or Sept-Îles). In 1982, the slowdown in the world economy caused the Iron Ore Company of Canada to close its pellet plant at Sept-Îles, thus concentrating its pellet-making operations in Labrador City. Sept-Îles remains the key transshipment point for iron ore (from rail to ship).

Commercial mineral deposits in the hard rock of northern Québec often consist of gold, iron, and copper. There are two main mining areas within this portion of the Canadian Shield. In the west, gold and copper are mined at Noranda and Val-d'Or, the centres for much of this production. In the northeast, iron is mined. Deposits of iron were first reported in 1895 by A.P. Low of the Geological Survey of

Canada, the first geologist to investigate the region's mineral potential. At that time, however, these deposits had no commercial value because more accessible mines could supply the needs of the iron and steel companies.

All that changed in the 1940s. American steel producers could no longer count on domestic supplies of iron ore, so they sought more reliable sources, including those in northern Québec and Labrador. Through a process of market integration initiated by US steel companies, Québec's resource hinterland became dependent on a particular group of steel companies in the industrial heartland of the United States for its economic well-being. Two mining companies, Québec Cartier Mining Company and the Iron Ore Company of Canada, developed iron mines in isolated areas of northern Québec and Labrador. By 1947, plans were laid for an open-pit mine in northern Québec near the border with Labrador. The Iron Ore Company built a town (Schefferville) for miners and their families; transmitted power from Churchill Falls to operate the mine and the town; and built a railway (the Québec North Shore and Labrador Railway) to deliver the iron ore to the port at Sept-Îles, from where the ore was eventually transported to supply US steel mills in Ohio and Pennsylvania.

The demand for iron ore rose in the 1960s, resulting in the establishment of three more mining towns—Wabush and Labrador City in Labrador and Fermont, Québec. At the same time, Québec Cartier Mining Company built a similar iron-mining operation by constructing the Cartier Railway from Port Cartier on the St Lawrence River to the resource town of Gagnon. But by the 1980s world steel production had surpassed the demand, causing a severe slump in the demand for iron ore. To add to this economic problem, US steel plants were now less efficient than the new steel mills in Brazil, Canada, Korea, and Japan, and lower-cost iron mines had opened in Australia and Brazil. As lower-priced steel from these countries undercut the price of US steel, American steel companies had to reduce their output, close plants, and sell their shares in the two mining companies. The repercussions for the mine workers in Québec and Labrador were severe. Production from these northern mines fell by half in the early 1980s and hundreds of workers were laid off. In 1983 and 1985, the mines at Schefferville and Gagnon were closed. During the 1990s, both companies restructured their operations to reduce costs. As demand from American iron and steel plants increased in the 1990s, these two companies were ready to supply this North American market. By 2002, the Iron Ore Company of Canada operated a pellet plant at Labrador City and the mines at Fermont, Wabush, and Labrador City, while the Quebec Cartier Mining Company had one mine at Mont-Wright and a pellet plant at Port Cartier.

Hydro Power: The James Bay Project

The massive James Bay Project calls for the production of hydroelectricity from all the rivers that flow into James Bay from Québec territory. This hydroelectric project was announced in 1971 by Premier Robert Bourassa. The James Bay Project is divided into three separate river basins (La Grande, Great Whale, and the Nottaway-Broadback-Rupert basins). The project involves about 20 rivers and affects an area one-fifth the size of Québec. Construction of the first phase of the James Bay Project, La Grande, began in 1972 and was completed in 1985 at a cost of $15 billion (Bone, 2003: 128). The La Grande project involved diverting waters from three other rivers (Eastmain, Opinaca, and Caniapiscau) into La Grande Rivière. Electrical energy generated from the three power stations in La Grande Basin is more than 10,000 MW each year.

The first phase of the James Bay Project, however, raised considerable controversy. It evoked an unprecedented response from Aboriginal peoples and environmental organizations. For them, the James Bay Project unleashed social and environmental problems that remain unresolved. For example, the project resulted in an unexpected high level of

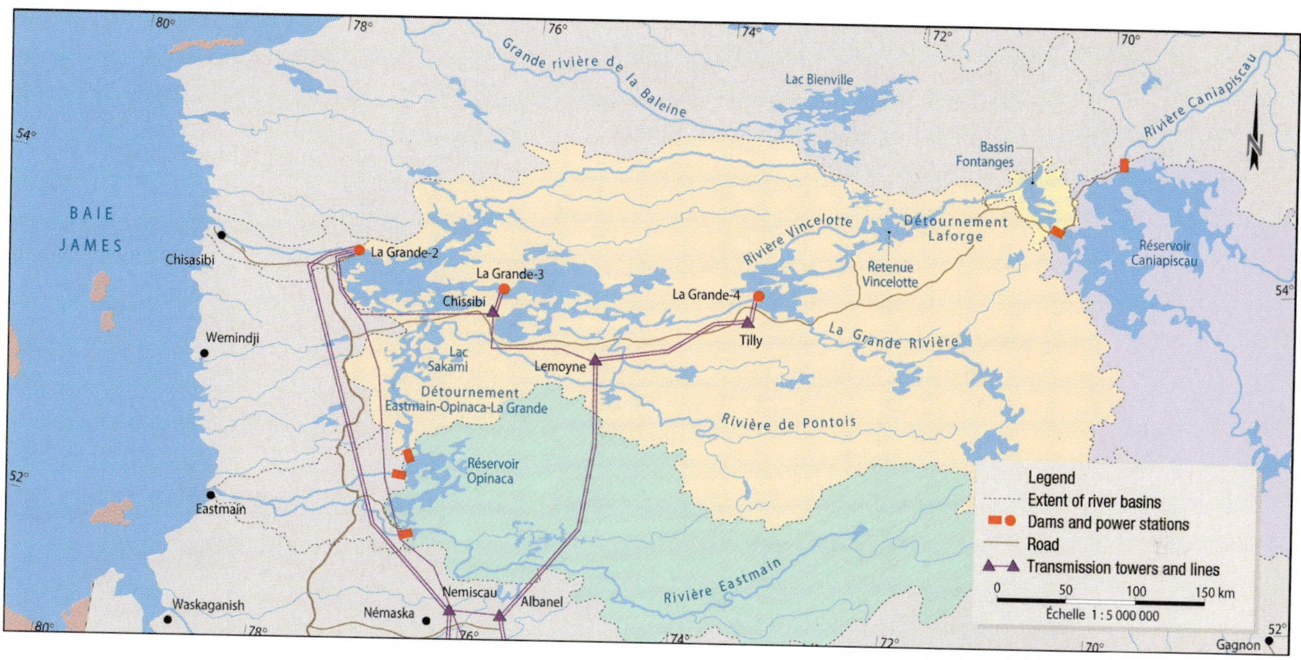

Figure 6.6 Phase 1 of the James Bay Project: La Grande Basin and Eastmain-1. La Grande's flooded lands (blue-coloured reservoirs) lie east of the LG-2 dam. Downstream is the newly constructed community of Chissibi. In 2007, waters from the Eastmain River were diverted into La Grande Basin.
Source: Hydro-Québec (2005).

mercury in the reservoirs, which has had serious implications for the Cree, who consume fish on a regular basis. Hydro-Québec maintains that the environmental impacts have been mitigated to an acceptable level through modifications to the design of the project. Remaining environmental impacts, such as high mercury content in the waters of the reservoirs, the Crown corporation contends, will diminish over time.

In 1985, the second phase of the James Bay Project was announced. The Great Whale River project, located just north of the La Grande Basin, was to consist of three powerhouses, four reservoirs, and the diversion of two rivers. In addition, another generating station (LG-1) was to be located at the mouth of the La Grande Rivière. Opposition from the Cree and the Sierra Club, who mounted a joint public relations campaign, had an effect on public opinion in New England and New York—the principal markets for Québec electric power—and when a new natural gas pipeline from Alberta to New England came on stream to provide a low-cost alternative the government of Québec announced in 1994 that the project would not proceed until the demand (price) for electricity in New England improved. From the very beginning, the Cree opposed the James Bay Project because of its effect on their hunting grounds. The Cree, joined by the Inuit of arctic Québec, forced a land-claim settlement known as the James Bay and Northern Québec Agreement.

The James Bay and Northern Québec Agreement

In 1971, 6,000 Cree in northern Québec lived as eight bands scattered across 375,000 km^2 of rivers and forest. They were under the administration of the federal Department of Indian Affairs and Northern Development. The James Bay Project threatened to flood their lands. This threat united the eight bands. When construction began in 1972, the Cree asked the Inuit to join them in taking legal action to halt the construction until the Cree and Inuit land claims were addressed. This action forced the Québec government and the Aboriginal claimants to the bargaining table. The result was the **James Bay and Northern Québec Agreement** (JBNQA). Under this agreement, both the federal and Québec governments became responsible for providing the 'treaty' benefits. As the first modern land-claim agreement in Canada, this 1975 agreement provided land, cash, and the power to administer cultural matters (education, health, and social services) to Aboriginal peoples. In exchange, the Cree and Inuit surrendered their Aboriginal claims to northern Québec and agreed to allow construction of the La Grande project to proceed.

In combination, these events—the negotiations, the agreement, and the construction project—have forever altered the lives of the Cree and Inuit. Both groups, now living in settlements, are more involved in the modern industrial society than ever before. Many are employed in businesses run by Cree and Inuit organizations, while others work in construction activities in the growing Cree and Inuit settlements and for Hydro-Québec. Still others are

The James Bay Project generates huge amounts of electrical energy. At the same time, the project has altered the natural environment by diverting rivers, changing their seasonal flow, and creating enormous reservoirs in Québec's Canadian Shield. (Ivy Images)

involved in the administration of their cultural affairs through the Cree Regional Authority and the Kativik Regional Government. In comparison with other Aboriginal peoples in Québec, the economic situation of the Cree and Inuit is much improved and certainly much better than that of those Québec Indians who do not have the benefit of such an agreement (Simard et al., 1996). Still, the Cree felt that both Ottawa and Québec had failed to honour their responsibilities under the JBNQA. When Québec announced plans for the second phase of the James Bay Project in 1985, relations between the Cree and the Québec government were so confrontational that the Cree actively opposed the Great Whale River project and took the Québec government to court over a number of issues related to the earlier agreement. Without a doubt, tensions between the Cree and the Québec government reached a low point in the latter part of the twentieth century, marking a particularly strained Québec version of the Aboriginal/non-Aboriginal faultline.

Turnaround

In 2001, the Cree reached an agreement with the Québec government for the economic development of the resources of northern Québec. This agreement, known as the Paix des Braves, opens the door to a major diversion of the Rupert and Eastmain rivers into La Grande Basin, thus adding more water for its hydroelectric plants. The acceptance of such an agreement is an astonishing reversal for the Québec Cree, who had bitterly opposed the project and who had mounted numerous national and international protests against further hydroelectric developments in their traditional lands. Some argue that the Cree now recognize that their participation in northern economic development is their only option. But feelings run high because efforts to protect the land for a hunting and trapping lifestyle have faltered. Paul Dixon, the Cree trapper representative, put it this way: 'They [Québec] promised the traditional way of life would continue undisturbed. Today, the whole territory has been slated for development' (Roslin, 2001: FP7).

Some, especially younger Cree, have chosen an urban lifestyle. For them, living on the land is no longer a viable option. Some say that the Cree leaders had to make a deal. Faced with a rapidly growing population, high unemployment rates, a critical shortage of public housing, and a desperate need for sewer and water systems, the Cree leaders had to seek an agreement with the Québec provincial government. Québec wanted to develop the northern resources and the Cree needed revenue to operate their communities and to find work for their people. Under the terms of the agreement, the Cree receive $3.6 billion over 50 years (roughly $70 million a year), but these funds release Québec from its obligations for economic and community development associated with the James Bay and Northern Québec Agreement. A newly created Cree Economic Development Agency will administer these funds with a mandate to foster growth of Cree businesses. The agreement also required the Cree to drop their lawsuits against the Québec government for failure to meet its obligations under the James Bay and Northern Québec Agreement. Whether or not the Cree will benefit from this model of economic development remains to be seen. What is clear, however, is that similar agreements are taking place across the country and all are designed to allow Aboriginal people to participate in economic development taking place in their traditional lands.

The Latest Addition to the James Bay Project

By 2007, the latest addition to the huge James Bay hydroelectric project saw the completion of the $2 billion Eastmain-1 project. Located south of the original development on La Grande River, the three generating units of Eastmain-1 have a total installed capacity of 480 megawatts. Construction began in 2002 following the 2001 Paix des Braves agreement with the Québec Cree. The main components of the Eastmain-1 project are the powerhouse, the main dam across the Eastmain River, the spillway, and the dikes for reservoir closure (see photographs). A 315-kV transmission line

Panoramic view of the Eastmain spillway and dam. (Hydro-Québec, 2006)

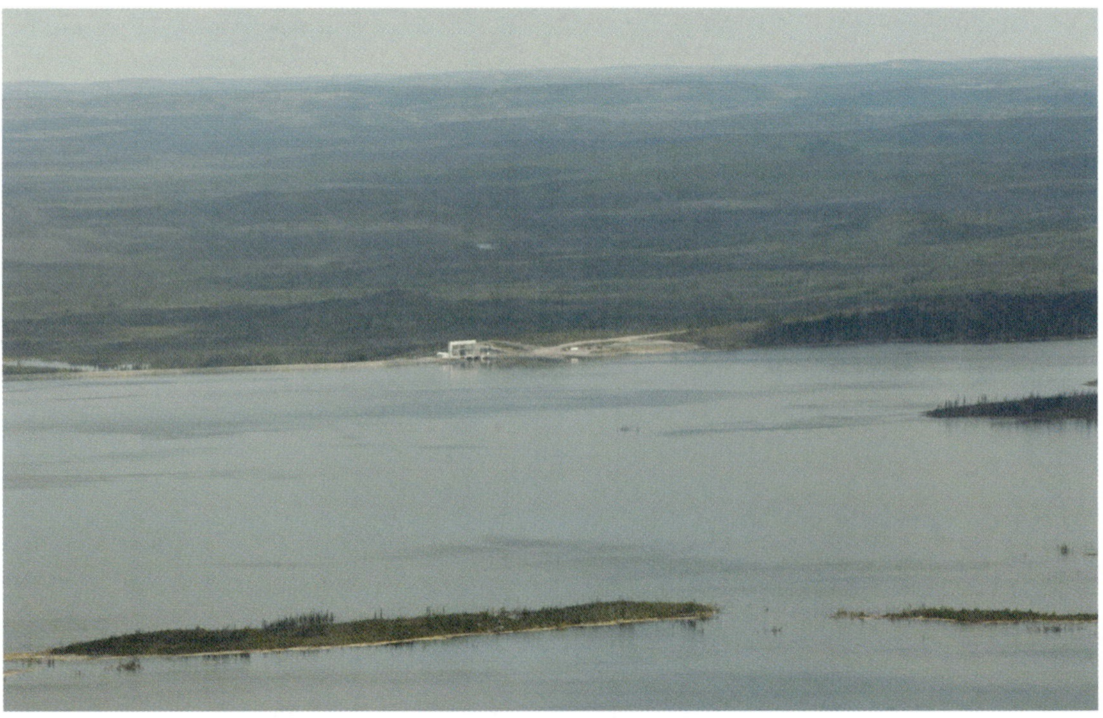

Impoundment of Eastmain-1 reservoir was completed in May 2006. (Hydro-Québec, 2006)

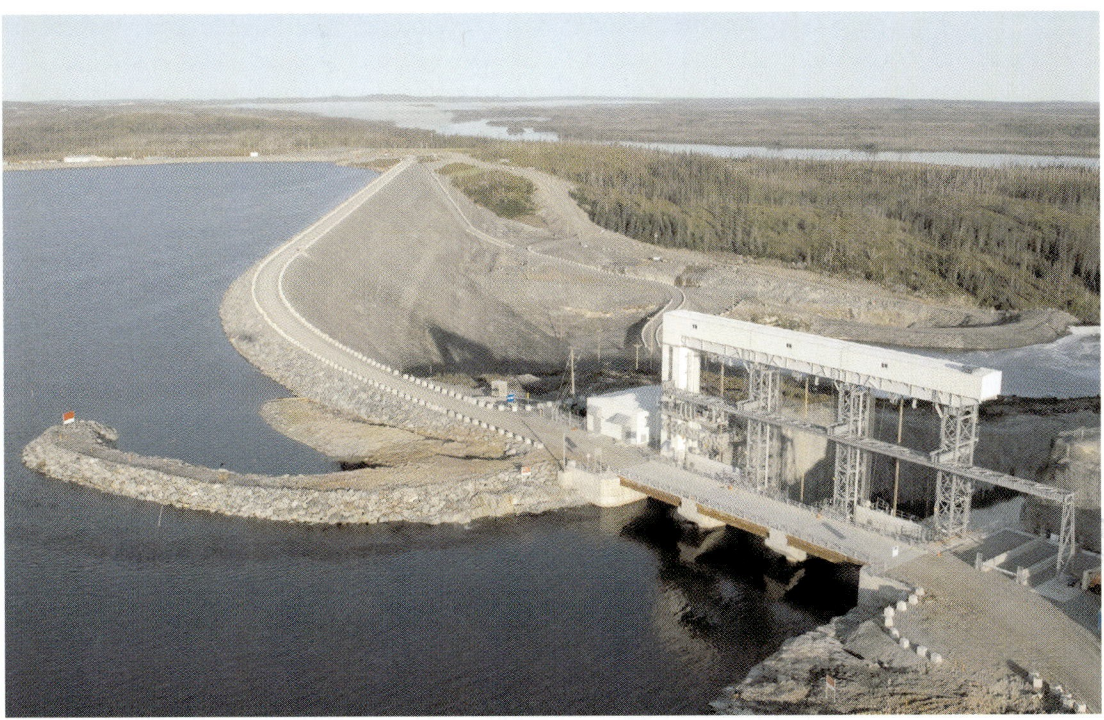

By August 2005, the land along the edge of the future reservoir of the Eastmain River had been cleared of trees. The removal of trees will help to prevent the decomposition of vegetative material and mercury pollution once the reservoir is full. (Hydro-Québec, 2006)

links the powerhouse to Nemiscau substation. With the completion of the dam in 2006, water from the Eastmain River formed a reservoir 35-km long with a total area of 603 km^2. After passing through the Eastmain powerhouse, water from the reservoir flows into Opinaca Reservoir and then into Robert Bourassa Reservoir. The same water is utilized at three generating stations—Eastmain-1, Robert Bourassa or La Grande-2-A, and La Grande-1—before flowing into James Bay.

Tourism

In Canada and in the rest of the world, tourism has become an extremely important economic sector. Québec is no exception. In fact, Québec has combined its natural beauty, historic past, and francophone culture to draw more and more tourists each year for over a decade, and tourist dollars spent in the province also have increased. In 2004, 7.2 million tourists visited different areas in Québec. Approximately half originated in the province. Of those coming from outside of the province, nearly 40 per cent came from the United States, 33 per cent from other provinces, and 26 per cent from other foreign countries (Québec, 2006b). Most tourists in Québec come from other regions of Canada and from the United States, France, the United Kingdom, Germany, and Japan. Since Québec is ideally located to cater to tourists from Ontario, New England, and Western Europe, its future as a world-class tourist destination seems secure.

The centres of Montréal and Québec City are major attractions for tourists seeking an urban vacation with a francophone atmosphere, while the Laurentides, a low range of mountains bounded by the Saguenay, St Lawrence, and Ottawa rivers, attract visitors looking for a summer or winter playground in the forests and lakes of the Canadian Shield. In addition to the natural beauty, the Laurentides have several advantages as an all-season recreational area. The Laurentides are but a short distance from Montréal and relatively close to New England and to major American cities such as Boston and New York. The climate in this area is ideal for the tourist business—the summers are hot, and heavy snowfall provides ideal conditions for winter sports. For example, Mount Tremblant Resort is a world-class tourist destination for both winter and summer activities. Tourism is the major business in the Laurentides, and the towns and villages cater to the tourists who flock there to enjoy water sports in the summer, skiing and snowboarding in the winter, and the numerous restaurants and shops all year long.

While Québec has the setting for summer and winter recreation activities, the tourist industry is vulnerable to external events. With the terrorist attacks on New York and Washington and the resulting war on terrorism announced by Washington, the tourist industry lost some American tourists. The steadily rising Canadian dollar, which had climbed to a 30-year high by June 2007, sent another shiver through the tourist industry.

Québec's Urban Geography

Over 80 per cent of Québec's population lives in urban centres. Two of the largest metropolitan cities in Canada, Montréal and Québec City, are located in the province. Other major population centres are Gatineau, Sherbrooke, Saguenay, and Trois-Rivières. In total, these six urban clusters have a population of 5.1 million (Table 6.6).

As in Ontario and most other regions of Canada, migration has played a major role in Québec's urbanization. Push and pull factors have attracted rural Quebecers to cities. The key push factors in rural Québec have been limited job opportunities, a shrinking labour force in the primary sector, and an increasing number of young people entering the workforce.

From 2001 to 2006, the population of the major cities, with the exception of Saguenay, grew, led by Gatineau (formerly Hull), Sherbrooke, and Montréal (Table 6.6). The population growth in Gatineau, across the Ottawa River from the nation's capital, was largely due to the employment and business opportunities generated by the federal government. Saguenay (formerly Chicoutimi–Jonquière) saw its population drop by 2.1 per cent. Located in the resource hinterland of the Lac Saint-Jean region, this city has been losing population for several decades. The downward trend is related to its two main economic activities, the aluminum plants and the forest mills. As in other resource hinterlands, employment prospects in these two industries have been declining, especially in the forest industry.

Montréal

Montréal, the metropolis of the province, is the industrial, commercial, and cultural focus of Québec. Montréal is the largest city in the province and the second largest city in Canada. Reflecting its bilingual nature, Montréal has four universities—Université de Montréal, McGill University, Concordia University, and Université du Québec à Montréal. Beyond its role within Québec, Montréal is recognized as one of the world's great cities and the second largest French-speaking city in the world. In 2006, Montréal became the first North American city to join the recent UNESCO City of Design network, which acknowledges the city's leading role in literature, music, food, cinema, folk art, and digital arts. Climate too affects Montréal. Unlike Toronto, Montréal experiences a longer winter where snow-covered streets are a fact of life.

For a century and a half, Montréal was the largest and most prosperous city in the Dominion. Within Canada today, this historic city has slipped behind Toronto in size and importance. Within Québec, however, Montréal remains the dominant city. In 2006, Montréal had a population of 3.6 million, making it the largest population concentration in Québec.

Like other great cities, Montréal is surrounded by smaller cities and towns. Such centres are often called satellite towns because they fall within the trade orbit of the larger city. Trois-Rivières, a major urban centre located on the St Lawrence over 100 km northeast of Montréal, is a satellite city. Smaller centres within 50 km of Montréal, such as Sainte-Hyacinthe, St-Jérôme, and Salaberry-de-Valleyfield, are even more closely tied to that metropolis. Laval and Longueuil serve as bedroom communities for Montréal. In most cases, urbanites have fled to the suburbs, attracted by lower-cost housing and the amenities of suburban life. These suburbanites often work in Montréal and live within commuting distance. Each weekday morning and late afternoon, commuters generate huge volumes of slow traffic and crowded buses in and around Montréal.

Given its strategic location and economic size, Montréal serves as the transportation hub of Québec, making it the regional core of the province and the rest of Québec the periphery. At the national scale, Montréal is part of the national core because it is part of Canada's manufacturing belt.

Following the Free Trade Agreement, Montréal's manufacturing sector had to respond to strong foreign competition. Labour-intensive manufacturing firms experienced great difficulty in competing with foreign firms that had substantially lower labour costs. Montréal's manufacturing firms made two major changes: labour-intensive plants substituted machinery for workers to increase productivity, and high-technology firms expanded. By the late 1990s, Montréal's economy had become much more specialized in aerospace, computers, fibre optics, multilingual software, telecommunications, and other areas of industrial research and development. The provincial government has taken a leading role by providing subsidies for high-tech firms that relocate to Montréal and other Québec cities. In 2006, Montréal and Québec City were ranked first and fifth among Canada's top research communities (Canada's Innovation Leaders, 2006).

Historical Context

Geography and history lie behind Montréal's success over the years. Montréal's strategic location on the St Lawrence River was a key factor in its initial growth as a transportation route and as a source of electrical energy, though today most power comes from tributaries flowing into the St Lawrence. In the 1820s, the Lachine Canal near Montréal provided the energy to drive the grist mills. Early in this century, power was transmitted to Montréal first from the hydroelectric installation on the St Maurice River at Shawinigan and then from the Beauharnois power site on the St Lawrence River.

Outdoor staircases in wintertime in downtown Montréal. (Stéphan Poulin).

Montréal benefited from the fur trade. Wealth accumulated by partners in the North West Company, which was headquartered in

Montréal, provided much of the capital necessary for new industrial ventures in manufacturing, resource development, and transportation. At the time of Confederation, Montréal was the largest and most prosperous city in Canada. With the introduction of high protective tariffs in 1879, Montréal's industrial base increased, thus propelling the city's economic and demographic growth. As Montréal's transportation and manufacturing industries expanded, its role as a financial and cultural centre solidified. By the end of the nineteenth century, Montréal was Canada's transportation hub and financial centre. For example, the construction of railway lines, particularly the Grand Trunk Railway, confirmed Montréal's role as a transportation centre and facilitated the development of railway repair depots and the manufacture of engines and railcars. This dominant role continued for 100 years after Confederation. In 1967, Montréal was the premier city in Canada and the undisputed headquarters of Canada's textile and financial enterprises. In the years that followed, Montréal's demographic and economic growth continued, but at a slower pace than that of Toronto.

Montréal and Toronto

Beginning in the 1970s, Toronto quickly replaced Montréal as the premier city in Canada. Montréal was no longer the largest city and the financial capital of the country. The principal reason for this shift in metropolitan power was the strong economic growth in Toronto and in southern Ontario powered by the Auto Pact and the resulting expansion of the automobile industry. A secondary factor was the economic and demographic fallout from the political unrest in Québec and the very real possibility of Québec separating from Canada. The unsettled political environment leading up to the 1980 Québec referendum worried the anglophone community. Many anglophones and some corporations moved to Toronto. While the francophone business community and the provincial government kept the Montréal economy growing, by 1981 Toronto had a population of 3.0 million compared to Montréal's 2.8 million (Table 6.8). Three other economic factors explain Montréal's slow growth and Toronto's much faster growth:

1. Montréal's economy was much more dependent on labour-intensive manufacturing (which was in decline) such as that in the textile industry.
2. Montréal's industrialization had begun much earlier than Toronto's and, for that reason, its manufacturing firms tended to be older and less efficient than those in Toronto.
3. Toronto has a national hinterland while Montréal has a provincial one and, for that reason, Toronto has a much larger 'external' demand for its wholesale services industry.

For those reasons, Toronto's growth rate exceeded that of Montréal. Even in the 10-year period from 1951 to 1961, when both cities grew at phenomenal rates—Montréal at 4.3 per cent per year and Toronto at 5.1 per cent—Toronto's rate was higher. From 1971 to 1981, the rate of population increase slowed in both cities, with Montréal's close to zero. During that time, Montréal's annual growth rate was only 0.3 per cent, while Toronto's was 1.9 per cent. This gap continued in the early 1990s with the annual rates for Toronto and Montréal at 1.9 per cent and 1.3 per cent, respectively. By 2005, Toronto's population had reached 5.3 million while Montréal's was 3.6 million (Statistics Canada, 2006).

Québec City

Québec City has the next largest urban concentration in the province, totalling nearly 700,000 people. Close to the Laurentides, Québec City has a magnificent physical setting on high banks just above the St Lawrence River. It is the only walled city in North America and features buildings over 300 years old. In 1985, Québec City was selected as a World Heritage Site by UNESCO.

The economic base of Québec City revolves around three functions. First, it is a

Table 6.8 Population Change: Montréal and Toronto, 1951–2006 (000s)

Year	Montréal	Toronto	Difference
1951	1,539	1,262	277
1961	2,216	1,919	297
1971	2,743	2,628	115
1981	2,828	2,999	–171
1991	3,127	3,893	–766
2001	3,426	4,683	–1,257
2006	3,636	5,113	–1,477

Sources: Statistics Canada (2002, 2007). Adapted from Statistics Canada publication *Population and Dwelling Counts, 2001 Census*, Catalogue 93F0050XCB2001013, Released 16 July 2002, http://www.statcan.ca/bsolc/english/bsolc?catno=93F0050X2001013.

government and university town. As the seat of government for the province, Québec City employs a large number of civil servants. Second, it has become a world-class tourist centre. The Old World charm of Québec City draws tourists from around the world, while special events such as its Winter Carnival are very popular. Third, Québec City is only minutes away from excellent seasonal recreation areas—from skiing in the winter to water sports in the summer. The economic base of Québec City remains heavily dependent on its political and cultural roles, but it also has a number of other economic functions: it is a port and rail centre, as well as a centre for resource processing, metal fabricating, and manufacturing. High-tech industries now play an important role in Québec City's economy, including the highly specialized photonics and optics sector.

Beyond the Urban Core

Québec's urban geography takes on a distinct regional character. Southern Québec has experienced modest growth in four of its five CMAs. Only Saguenay suffered a population loss from 2001 to 2006 (Table 6.6). Beyond the urban core, most cities and towns suffered population losses caused by out-migration (Table 6.9). This demographic pattern is associated with weak regional economies. From 2001 to 2006, population declines took place in the urban centres of northern Québec, the North Shore, and the Gaspé Peninsula. Two forces were at play.

First, the economies in these outlying areas of Québec are contracting. Within the

With the ancient wall enclosing Old Québec in the foreground, the Château Frontenac dominates the background. On 3 July 2008, Québec City will celebrate its 400th anniversary since its founding by French explorer Samuel de Champlain. (Luc-Antoine Couturier Photographe)

Held each year in Québec City, the Québec Carnival is the largest winter carnival in the world, ranking just behind the famous carnivals in Rio and New Orleans. The origins of the Québec Carnival go back to the beginning of New France when a tradition of winter revelry took hold to offer relief from the long, cold winters. The modern version of a winter carnival began in 1955. The Québec Carnival involves many activities ranging from night parades, snow sculptures, and the popular canoe race on the frozen St Lawrence River, as shown above. (CP/Clement Allard)

Table 6.9 Population of Cities of Northern Québec, the North Shore, and Gaspé Peninsula, 2001–6

City	2001	2006	% Change
Northern Québec			
Saguenay	154,938	151,643	−2.1
Rouyn–Noranda	39,621	39,924	0.8
Alma	32,930	32,603	−1.0
Val-d'Or	32,423	32,288	−0.4
North Shore			
Baie-Comeau	30,401	29,808	−2.0
Sept-Îles	27,623	27,827	−0.7
Gaspé Peninsula			
Gaspé	14,932	14,819	−0.8
Percé	3,614	3,419	−9.5

Source: Statistics Canada (2007). Adapted from Statistics Canada publication *Population and Dwelling Counts, 2006 Census*, Catalogue 97-550-XWE2006002 http://www.statcan.ca/bsolc/english/bsolc?catno=97-550-XWE2006002.

new North American marketplace, resource companies that export most of their products are under great pressure to reduce their costs, which translates into reducing the size of the labour force or, in the worst-case scenario, closing their operations. Urban centres are affected by the negative spinoff from industrial restructuring, which results in a smaller population and, therefore, fewer customers. This ripple effect may reduce the size of the market beyond the point of profitability and thus force some businesses to close.

Second, the demographic outcome of a declining regional economy where few job opportunities exist normally leads to out-migration, especially by the younger and more educated members of the local population. Such migration to other places causes a contraction of the local population and a loss of potential community leaders, and signals a general economic malaise that results in a relentless downward spiral of both the regional economies and urban populations.

The French/English Faultline in Québec

After the Quiet Revolution, the Québec government took legislative measures to ensure that the French language and the Québécois culture prospered within the province. From the Québec government's perspective, such action was necessary because Québec was surrounded by a sea of English-speaking North Americans. Accordingly, the Québec government passed a series of language laws that obliged businesses to use French and required allophone parents to send their children to French schools. For a time, these language laws were a major sore point between French- and English-speaking Quebecers. By the early years of the twenty-first century, however, language issues are less of a sore point within Québec because of a general acceptance of the primacy of the French language and the growing number of anglophones, allophones, and francophones who are bilingual. As a sign of the well-being of the French language in Québec, Lucien Bouchard and the other authors of the manifesto, *Pour un Québec lucide*, have called on the government to encourage the learning of English: 'The Government must also make far greater effort to ensure that all Quebeckers speak and write English, as well as a third language' (Bouchard et al., 2006: 8). Perhaps the former 'raw' language faultline within Québec no longer exists?

Québec's Future

Can Québec remain one of the two central pillars of Canada's industrial axis? As one of the original British colonies to form Canada in 1867, Québec's position within Confederation is weakening as it loses its share of Canada's population and economic output. The reason is simple: while Québec's economy and population have expanded, other regions of Canada—Ontario, British Columbia, and Western Canada—have expanded at a more rapid rate. For this reason alone, Québec's unique culture and language take on even more importance. As the heartland of francophones in Canada and North America, Québec has a special role to play. Cultural events like the St Jean Baptiste festival evoke a sense of ethnic nationalism and a love of the land and its people, which is popularly expressed as '*j'ai le goût du Québec*'. Within this cultural context, the French language serves as a linchpin. Yet, economic stability is necessary to maintain cultural well-being.

Except for Atlantic Canada, Québec's economy and population have not keep pace with the rest of Canada. Has Québec lost its steam? From 1995 to 2005, Ontario's labour force increased by 22 per cent while Québec's lagged behind at 16 per cent (Table 6.5). If this trend continues, then its place within Canada will be seriously eroded.

Quebecers can sense a problem. Support for the view that a second revolution may be in the offing is found in the 2006 manifesto, *Pour un Québec lucide*, in which former Premier Lucien Bouchard and eleven other Québecers declared that Québec faces three threats: (1) a rapid demographic decline due to a declining birth rate; (2) low-cost imports from Asian countries, led by China and India, which could displace Québec manufacturers; and (3) too much government regulation of the economy (ibid.). The results of the 2007 Québec election, which left the separatist Parti Québécois—the party Bouchard formerly led—in third place, also seem to suggest that the province may be heading in a new direction.

Québec possesses great vitality for fresh and innovative political, economic, and cultural undertakings. With a self-confidence formed from the enormous changes the province has undergone over the past half-century, Québec is looking at new, innovative solutions to old problems. One example is the proposed *public* self-government proposal of the Inuit of Nunavik. If successful, this concept of a semi-autonomous state within a province would mark a political breakthrough that would create a third level of government within a province, and such a political innovation could be applied to the northern areas of other Canadian provinces with Aboriginal majorities (Vignette 6.10). The question is, can Quebecers apply the same innovative approach to their economy?

Vignette 6.10 Mapping the Road to Nunavik

Nunavik extends over a vast arctic land in northern Québec. This land of permafrost is the homeland of some 11,000 Québec Inuit. Like other Aboriginal peoples, the Québec Inuit are seeking a form of regional autonomy that would respond to their needs, desires, and aspirations. To achieve that goal, three formidable challenges had to be resolved. First, how can a province contain an autonomous government? This begs the question of the division of powers between the province of Québec and the soon-to-be formed government of Nunavik. Second, how can Nunavik (a non-ethnic government) treat all its residents equally and still promote the Inuit culture? Lastly, how can 11,000 people living in 14 communities scattered over 660,000 km^2 govern themselves and also generate sufficient revenue to pay for their government?

The road to Nunavik is difficult but not impossible. Somehow the structure, operations, powers, and design of this new form of government within a Canadian province can be achieved. A 2007 agreement among the governments of Canada, Québec, and the Québec Inuit set the terms for an autonomous government within Québec. Québec will be the first province to share political power with a regional government and, by this example, will have created a unique model for other provinces with large Aboriginal populations to follow. Since 2001, negotiations continued between Ottawa, Québec City, and Makivik Corporation, which represents the Inuit. In 2003, a framework agreement was struck with the purpose of creating a new form of government in Nunavik. By 2007, an Agreement-in-Principle was signed by federal and Québec officials. Legislation is being drafted and will be introduced in the two parliaments. By 2008, the government of Nunavik will be born.

Sources: Report of the Nunavik Commission (2001); Koperqualuk (2006); personal communication with Donat Savoie, chief federal negotiator for Nunavik, 10 Feb. 2007.

Notes

1. By responsible government, Lord Durham meant a 'political system in which the Executive is directly and immediately responsible to the Legislature, in which the ministers are members of the Legislature, chosen from the party which includes the majority of the elected representatives of the people' (Lucas, 1912: I, 138).
2. In 1912, Québec gained northern territories inhabited by the Inuit and Cree. Ottawa ceded these lands to Québec with the understanding that the Québec government would be responsible for settling land claims with the Aboriginal peoples in these territories. The Cree in northern Québec, in response to the separatist claim to territorial independence, have declared that they have the right to secede from Québec. They argue that if Québec has the right to secede from Canada, then the Cree have the right to secede from Québec. From a geopolitical perspective, the partitioning of Canada or Québec makes sense only to those supporting ethnic nationalism.
3. For the 2006 federal election the House of Commons had 308 seats. Quebecers kept their 75 ridings because of a law that guarantees each province no fewer seats than it had in 1976. Reapportionment takes place every 10 years based on population figures from the census. The last reapportionment, based on the 2001 census, added eight new seats to the Commons for Ontario, BC, and Alberta.
4. Camille Laurin, the father of Bill 101, declared that French was the province's only official language. Bill 101 required the children of immigrants to go to French schools and made the presence of French compulsory in the workplace and on commercial signs.
5. BioChem Pharma Inc. is one of Canada's largest pharmaceutical research firms. Its plant is located in Laval, a community just a few kilometres north of Montréal. It was founded as a small research company, but with the discovery of a drug to treat the HIV/AIDS virus it soon became the crowning jewel in Québec's biotech industry. As a world leader in developing drugs for AIDS, cancer, and hepatitis, this Québec high-tech firm was recently sold for $5.9 billion in a share swap with Shire Pharmaceuticals Group PLC, based in the United Kingdom. BioChem will continue to function as a research pharmaceutical company at Laval but its products will be marketed through Shire. For Shire, this merger represents a major step towards creating the largest specialty pharmaceutical company in the world. The merger will broaden and diversify Shire's products and revenue base and strengthen the development of new drugs (Leger, 2000: C1, C6).
6. The case of Churchill Falls is an interesting one. The divide between the Atlantic Ocean and Hudson Bay marks the Labrador–Québec boundary. For historical reasons, the Québec government does not formally recognize this boundary, but it does treat the area as part of Newfoundland. Newfoundland owns the large Churchill Falls hydroelectric project, but because of the high cost of transmitting the electrical power from Churchill Falls across the Strait of Belle Isle to Newfoundland, virtually all this power is purchased by Hydro-Québec. Hydro-Québec then transmits it across Québec to markets in the Great Lakes–St Lawrence Lowlands and the United States.
7. The 13 companies with risk-sharing contracts are: Norsk Hydro Canada Inc., Aluminerie Alouette Inc., Quebec-Cartier Mining Co., Cafco Industries Ltd, Timminco Ltd, QIT-Fer et Titane Inc., PPB Canada Inc., Reynolds Metals Co., Argonal, Hydrogenal, SKW Canada Inc., ABI Inc., and Aluminerie Lauralco Inc.

Challenge Questions

1. What physiographic region is considered the homeland of the Québécois?
2. The Canadian Shield found in northern Québec has several natural advantages to produce hydroelectric power. What are these natural advantages and do they exist in Ontario's portion of the Canadian Shield?
3. What are the political implications for Québec if its share of Canada's population and GDP continues to decline?
4. Do you think that the manifesto, *Pour un Québec lucide*, offers Québec a viable new direction that will stem the decline of Québec's share of Canada's population and GDP?
5. Hydro-Québec exports much electricity to New England. If the price of natural gas increases, will this play to Hydro-Québec's advantage or disadvantage?
6. Why was the 'Paix des Braves' so fundamental to continued development of the James Bay Project and why did the Cree support this agreement?
7. What type of industrial firms have established factories in the St Lawrence Lowlands as a result of Québec's strategy to offer low electrical power rates?
8. Given the negative impact of imported goods on Québec's textile and garment firms, should Ottawa continue to support the WTO efforts to eliminate tariffs on textiles and garments imported from developing countries?
9. Why does the definition(s) of Québec as a 'nation' jangle the nerves of Canadians in other provinces?
10. If Nunavik gains its desired form of self-government, will this political development result in a new level of government within Québec and therefore within Canada's other provinces?

Key Terms

allophones
Quebec residents whose mother tongue is not French, English, or one of the Aboriginal languages.

anglophones
Those whose mother tongue is English.

francophones
Those whose mother tongue is French.

French Canadians
Canadians with roots to Quebec and who likely still speak French.

habitants
French peasants who settled the land in New France under a form of feudal agriculture known as the seigneurial system. After the British Conquest, the seigneurial system continued until the mid-nineteenth century, marking a significant difference between Upper and Lower Canada.

James Bay and Northern Québec Agreement
The 1971 announcement of the James Bay Project triggered a series of events that quickly led to a negotiated settlement and, in 1975, an agreement between the Cree and Inuit of northern Québec and the federal and Québec governments. In this modern treaty, Aboriginal title was surrendered by the Inuit and Cree of northern Québec in exchange for specific rights, including self-government and benefits (cash and financial support for the hunting economy).

nation
A territory that is politically independent; a group of people with similar cultural characteristics and a shared historical experience that is self-consciously aware of itself as a unique group.

patriation
The act of bringing legislation, especially a Constitution, under the authority of the autonomous country to which it applies.

Quebecers
Essentially an English translation of the French word 'Québécois'.

Québécois
A term that has evolved from referring to French-speaking residents of Québec to all residents of Québec.

seigneurs
Members of the French elite—high-ranking officials, military officers, the nascent aristocracy—who were awarded land in New France by the French king. A seigneur was an estate owner who had peasants (habitants) to work his land.

Bibliography

Beauregard, Ludger, ed. 1980. 'Numéro spécial: La problématique géopolitique du Québec', *Cahiers de géographie du Québec* 24, 61: 1–185.

Bone, Robert M. 2003. *The Geography of the Canadian North*, 2nd edn. Toronto: Oxford University Press.

Bothwell, Robert. 1995. *Canada and Quebec: One Country, Two Histories*. Vancouver: University of British Columbia Press.

Bouchard, Lucien, et al. 2006. *Manifest—Pour un Québec lucide*. At: <www.pourunquebeclucide.com/cgi-cs/cs.waframe.content?topic=28226&lang=1>.

Bradbury, John H. 1982. 'State Corporations and Resource-Based Development in Quebec, Canada, 1960–1980', *Economic Geography* 58, 1: 45–61.

Bumsted, J.M. 2007. *A History of the Canadian Peoples*, 3rd edn. Toronto: Oxford University Press.

Bussière, Luc. 1995. 'Une économie en perte de vitesse', *Provincial Outlook: Economic Forecast* 10, 1: 21–9.

Caldwell, Gary, and Eric Waddell, eds. 1982. *The English of Québec: From Majority to Minority Status*. Institute québécois de recherche sur la culture. Sainte-Foye: Les Presses de l'Université Laval.

Canada. 1993. *Canada Year Book 1994*. Ottawa: Minister of Industry, Science and Technology.

———. 2006. Statistics Canada: Canadian Statistics—Net Merchantable Volume of Roundwood Harvested. At: <www.statcan.ca/english/Pgdb/prim41.htm>.

———. 2006. Net Merchantable Volume of Roundwood Harvested by Category and Province/Territory, 1940-2005, National Forestry Database Program, Table 5.1. At: <www.nfdp.ccfm.org/compendium/products/tables_index_e.php>.

Canada Centre for Remote Sensing. 1998. Saguenay Flood 1996. At: <otter.ccrs.nrcan.gc.ca/ccrs/tekrd/rd/apps/hydro/saguenay/saquente.html>.

Canada's Innovation Leaders. 2006. 'Canada's Top 20 Research Communities 2006', *National Post*, Research infosource supplement, 3 Nov., 7.

Centre for Research and Information on Canada. 2003. Portraits of Canada 2002. At: <www.cric.ca>.

Chevalier, Jacques. 1993. 'Toronto–Ottawa–Montréal: Concentrations majeures Canadiennes de l'innovation par la recherche-développement', *Le géographe Canadien* 37, 3: 242–57.

Coffey, W.J. 1996. 'Make or Buy: Internalization and Externalization of Producer Service Inputs in the Montreal Metropolitan Area', *Canadian Journal of Regional Science* 19, 1: 25–48.

——— and Réjean Drolet. 1994. 'La décentralisation des services supérieurs dans la région métropolitaine de Montréal, 1981–1989', *Le géographe Canadien* 38, 3: 215–29.

——— and M. Polèse. 1989. 'The Role of Cultural Barriers in the Location of Producer Services: Some Reflections on the Toronto–Montreal Rivalry and the Limits to Urban Polarization', *Canadian Journal of Regional Science* 14: 433–46.

——— and ———. 1993. 'Le déclin de l'empire montréalais: regard sur l'économie d'une métropole en mutation', *Recherches sociographiques* 34: 417–37.

Conference Board of Canada. 2000. *Provincial Outlook* (Winter).

Cook, Ramsay. 1976. *Canada and the French-Canadian Question*. Toronto: Macmillan.

Courville, Serge. 2000. *Le Québec: geneses et mutations du territoire*. Sainte-Foye: Les Presses de l'Université Laval.

CTV News. 2007. 'Harper Goes to Quebec to Announce Green Fund', 12 Feb. CTVNews.ca. At: <www.ctv.ca/servlet/ArticleNews/story/CTVNews/20070212/harper_charest_070212/20070212?hub=CTVNewsAt11>.

De Benedetti, George J., and Maurice Beaudin. 1996. 'Linguistic Minority Communities' Contribution to Economic Well-being: Two Case Studies', *Canadian Journal of Regional Science* 19, 2: 175–92.

Doloreux, David. 1998. 'Politique technopolitaine et territorie: le cas de Laval', *Canadian Journal of Regional Science* 21, 3: 441–60.

———. 1999. 'La pépinière d'entreprises dans le contexte d'un parc scientifique: l'exemple du Centre Québécois d'Innovation en Biotechnologie à Laval, Québec (Canada)', *Canadian Geographer* 43, 4: 423–32.

Dougherty, Kevin. 2007. 'Quebec the First to Announce Carbon Tax', *National Post*, 7 June, A1, A10.

Duhaime, Gérard, Nick Bernard, and Robert Comtois. 2005. 'An Inventory of Abandoned Mining Exploration Sites in Nunavik, Canada', *Canadian Geographer* 49, 3: 260–71.

Energyshop.com. 2007. Natural Gas Prices—Historical and Forecast. Richmond Hill, Ont. At:

<www.energyshop.com/es/homes/gas/gaspriceforecast.cfm?>.

Ferrao, Vincent. 2006. 'Recent Changes in Employment by Industry', *Perspectives on Labour and Income* (Statistics Canada) 7, 1. At: <www.statcan.ca/english/freepub/75-001-XIE/10106/art-1.htm>.

Gibbens, Robert. 2000. 'Sept-Iles Pellet Plant Likely Set for Startup', *National Post*, 4 July, C4.

Grant, Tavia. 2006. 'Bombardier Wins $1.65-billion in Contracts', *Globe and Mail*, 28 Sept. At: <www.theglobeandmail.com/servlet/story/RTGAM.20060928.wbombardier_v2_0928BNStory/Business>.

Hamelin, Louis-Edmond. 1998. 'L'entièreté du Québec: le cas du Nord', *Cahiers de géographie du Québec* 42, 115: 95–110.

Hecky, Robert E. 1987. 'Methylmercury Contamination in Northern Canada', *Northern Perspectives* 15, 3: 8–9.

Higgins, Benjamin. 1986. *The Rise and Fall of Montreal?* Moncton: Instititut canadien pour le développement regional.

Hodgins, Bruce W., Richard P. Bowles, James L. Hanley, and George A. Rawlyk. 1974. *Canadiens, Canadians and Québécois*. Scarborough, Ont.: Prentice-Hall.

Hornig, James F., ed. 1999. *Social and Environmental Impacts of the James Bay Hydroelectric Project*. Montréal and Kingston: McGill-Queen's University Press.

Hydro-Québec. 2005. Water Resource System: La Grande Rivière. At: <translate.google.com/translate?hl=en&sl=fr&u=http://www.hydroquebec.com/&sa=X&oi=translate&resnum=1&ct=result&prev=/search%3Fq%3Dwww.hydroquebec.com%26hl%3Den%26client%3Dfirefox-a%26channel%3Ds%26rls%3Dorg.mozilla:en-US:official%26sa%3DG>.

———. 2006. Eastmain-1 Hydroelectric Development. At: <www.hydroquebec.com/eastmain1/en/batir/fiche_7.html>.

———. 2006. Eastmain-1 Hydroelectric Development. At: <www.hydroquebec.com/eastmain1/en/batir/etapes_photos.html?list=4>.

Joyal, André, and Laurent Deshaies. 2000. 'Réseaux d'information des PME en milieu non métropolitain', *Cahiers de géographie du Québec* 44, 122: 189–210.

Kaplan, David H. 1994. 'Two Nations in Search of a State: Canada's Ambivalent Spatial Identities', *Annals, Association of American Geographers* 84, 4: 585–606.

Koperqualuk, Lisa, ed. 2006. 'Nunavik: Makivik's Relation with the Nunavik Government', *Nunavik* 9: 21–3.

Lasserre, Jean-Claude. 1999. 'Pour comprendre la stagnation et les mutations des traffics sur le Saint-Laurent: une évaluation comparée des portes continentals nord-américaines', *Cahiers de géographie du Québec* 43, 118: 7–42.

Ledent, J., L. Bourne, and F. Dansereau. 1999. 'Comparative Development in Montreal and Toronto/Une Analyse Comparée du Développement à Montréal et Toronto', *Canadian Journal of Regional Science* 22 (special issue): 1–2.

Leger, Kathryn. 2000. 'Biochem Sold for $5.9-billion', *National Post*, 12 Dec., C1, C6.

Lo, Lucia, and Carlos Teixeria. 1998. 'If Québec Goes . . . The "Exodus" Impact?', *Professional Geographer* 50, 4: 481–98.

Louder, Dean, and Eric Waddell. 1992. *French America: Mobility, Identity and Minority Experience across the Continent*. Baton Rouge: Louisiana State University Press.

Lucas, C.P. 1912. *Lord Durham's Report of the Affairs of British North America*, vol. 1. Oxford: Clarendon Press.

MacAfee, Michelle. 2003. 'GM's $400M Quebec Pact to Spur 3,000 Jobs', *National Post*, 1 Mar., FP5.

McCutcheon, Sean. 1991. *Electric Rivers: The Story of the James Bay Project*. Montréal: Black Rose Books.

Manzagol, Claude, and Christopher R. Bryant. 1998. *Montréal 2001: visages et défis d'une métropole*. Montréal: Presses de l'Université de Montréal.

Marsh, James H., ed. 1988. *The Canadian Encyclopedia*, 2nd edn. Edmonton: Hurtig.

Martineau, Daniel, Karin Lemberger, André Dallaire, Philippe Labelle, Thomas P. Lipscomb, Pascal Michel, and Igor Mikaelian. 2002. 'Cancer in Wildlife, a Case Study: Beluga from the St. Lawrence Estuary, Québec, Canada', *Environmental Health Perspectives* 110, 3:285–92.

Mills, David. 1988. 'Durham Report', in Marsh (1988: 637–8).

National Post. 1999. 'Alcan to Invest Another $200M in Quebec Smelter', 4 Nov., C2.

Nemni, Max. 1994. 'The Case against Quebec Nationalism', *American Review of Canadian Studies* 24, 2: 171–96.

Noël, Michel. 1997. *The Native Peoples of Québec*. Ottawa: University of Ottawa Press.

Parkinson, David. 2006. 'Bombardier Axes 1,330 Jobs in Montreal and Belfast', *Globe and Mail*, 24 Oct. At: <www.theglobeandmail.com/servlet/story/RTGAM.20061024.wbombardier1024/BNStory/Business/home>.

Peluso, Tony, Laurie Baker, and Paul J. Thomassin.

1998. 'The Siting of Ethanol Plants in Quebec', *Canadian Journal of Regional Science* 21, 1: 73–88.

Padova, Allison. 2004. *Trends in Containerization at Canadian Ports*. Ottawa: Library of Parliament. At: <www.parl.gc.ca/information/library/PRBpubs/prb0575-e.htm>.

Phillips, David. 1993. *The Day Niagara Falls Ran Dry!* Toronto: Canadian Geographic and Key Porter Books.

Polése, Mario. 2000. 'Is Quebec Special in the Emerging North American Economy?', *Canadian Journal of Regional Studies* 23, 2: 187–212.

——— and Martin Roy. 1999. 'La dynamique spatiale des activités économiques au Québec: analyse pour la période 1971–1991 fondée sur un découpage centre-périphérie', *Cahiers de géographie du Québec* 43, 118: 43–71.

——— and Richard Stren, eds. 2000. *The Social Sustainability of Cities: Diversity and the Management of Change*. Toronto: University of Toronto Press.

Québec. 1998a. Pour donner au monde le goût du Québec: Résumé de la politique de développement touristique, Gouvernement du Québec, Ministère du Tourisme. At: <www.tourisme.gouv.qc.ca/francais/mto/publications/poldevtour.html>.

———. 1998b. Le tourisme au Québec: enjeux et orientations, Gouvernement du Québec, Ministère du Tourisme. At: <www.tourisme.gouv.qc.ca/francais/mto/enjeux.html>.

———. 2003a. Les Forêts, Le Québec forestier/Portrait statistique: Le secteur forestier dans l'économie, Gouvernement du Québec, Ministère des Ressources naturelles. At: <www.mrn.gouv.qc.ca/english/forest/investors/index.jsp>.

———. 2003b. The Québec Forest Industry, Québec Forest Industries Association. At: <www.aifq.qc.ca/english/stats/papers01.html>.

———. 2006a. The Québec Forest Industry, Québec Forest Industries Association. At: <www.cifq.qc.ca/html/english/sciage/statistiques.php>.

———. 2006b. Québec Statistiques Touristiques: Le tourisme au Québec en bref—2004. At: <www.google.ca/search?hl=en&client=firefox-a&rls=org.mozilla%3Aen-US%3Aofficial_s&q=le+tourisme+au+quebec+en+bref&btnG=Search&meta=>.

Québec City and area. 2006. Québec Heritage. At: <www.quebecheritage.com/en/>.

Report of the Nunavik Commission. 2001. *Let Us Share*. Québec City: Conseil du Trésor, Gouvernement du Québec.

Richardson, Boyce. 1975. *Strangers Devour the Land: The Cree Hunters of the James Bay Area versus Premier Bourassa and the James Bay Development Corporation*. Toronto: Macmillan.

Rioux, C., J.-C. Michaud, B. Urli, and L. Grossilin. 1998. 'Développement local et décisions collectives: le cas du Québec-côtier', *Canadian Journal of Regional Science* 21, 3: 365–74.

Rose, Damaris, and Marc Villemaire. 1997. 'Reshuffling Paperworkers: Technological Change and Experiences of Reorganization at a Quebec Newsprint Mill', *Canadian Geographer* 41, 1: 41–60.

Roslin, Alex. 2001. 'Cree Deal a Model or Betrayal?', *National Post*, 10 Nov., FP7.

Rumney, Thomas. 1996. *The Geography of Canada Bibliography Series: Vol. 1, Quebec*. Plattsburgh, NY: Plattsburgh State University, Center for the Study of Canada.

Saint-Laurent, Diane. 2000. 'Approches biogéographe de la nature en ville: parcs, espaces verts et friches', *Cahiers de géographie du Québec* 44, 122: 147–66.

Salée, Daniel. 2003. 'Quebec's Changing Political Culture and the Future of Federal–Provincial Relations in Canada', in Harvey Lazar and Hamish Telford, eds, *Canada: State of the Federation 2000/01: Canadian Political Culture(s) in Transition*. Montréal and Kingston: McGill-Queen's University Press.

Salisbury, Richard Frank. 1986. *A Homeland for the Cree: Regional Development in James Bay, 1971–1981*. Montréal and Kingston: McGill-Queen's University Press.

Scott, Richard T. 2001. 'Becoming a Mercury Dealer: Moral Implications and the Construction of Objective Knowledge for the James Bay Cree's Aboriginal Autonomy and Development in Northern Quebec and Labrador', in Colin H. Scott, ed., *Aboriginal Autonomy and Development in Northern Quebec and Labrador*. Vancouver: University of British Columbia Press.

Service Canada. 2006. Sectorial Outlook: Province of Québec. At: <www150.hrdc-drhc.gc.ca/imt/regional/english/sec_out/2006-2008/introduction/index.html>.

Silcoff, Sean. 2003. 'Landry Willing to Backstop Bombardier', *National Post*, 22 Jan., FP3.

Simard, Jean-Jacques, et al. 1996. *Tendances Nordiques: Les changements sociaux 1970–1990 chez les Cris et les Inuit du Québec: Une enquête statistique exploratoire*. Québec City: GETIC, Université Laval.

Simard, Majella. 1998. 'Les théories de développement régional et la contribution des ressources dans le démarrage des petites localités en voie de dépeuplement: le cas du Bas Saint-Laurent', *Canadian Journal of Regional Science* 21, 1: 127–50.

Simard, Martin. 2000. 'Développement local et identité communautaire: l'exemple du quartier Saint-Roch à Québec', *Cahiers de géographie du Québec* 44, 1–2: 167–88.

Simeon, Richard. 1988. 'Meech Lake and Shifting Conceptions of Canadian Federalism', *Canadian Public Policy* 14: S7–24.

Simpson, Jeffrey. 1993. *Faultlines: Struggling for a Canadian Vision*. Toronto: HarperCollins.

Stanford, Quentin H., ed. 1998. *Canadian Oxford World Atlas*, 4th edn. Toronto: Oxford University Press.

Statistics Canada. 1982. *Census Metropolitan Areas and Census Agglomerations with Components*. Catalogue no. 95–903. Ottawa: Minister of Industry.

———. 1992. *Census Divisions and Subdivisions*. Catalogue no. 93–304. Ottawa: Minister of Supply and Services.

———. 1996a. *Labour Force Annual Averages 1995*. Catalogue no. 71–220–XPB. Ottawa: Statistics Canada.

———. 1996b. Canadian Statistics: Distribution of Employed People, by Industry, by Province. At: <www.statcan.ca/english/Pgdb/labor21b.htm>.

———. 1997a. *A National Overview: Population and Dwelling Counts*. Catalogue no. 93–357–XPB. Ottawa: Industry Canada.

———. 1997b. 1996 Census: Nation Tables—Population by Mother Tongue, Showing Age Groups, for Canada, Provinces and Territories, 1996 Census—20% Sample Data, 2 Dec. 1997. At: <www.statcan.ca/english/census96/>.

———. 1997c. '1996 Census: Mother Tongue, Home Language and Knowledge of Languages', *The Daily*, 2 Dec. At: <www.statcan.ca/Daily/English/>.

———. 1998a. *Canadian Economic Observer*. Catalogue no. 11–010–XPB. Ottawa: Statistics Canada.

———. 1998b. '1996 Census: Aboriginal Data', *The Daily*, 13 Jan. At: <www. statcan.ca/Daily/English/>.

———. 1998c. '1996 Census: Ethnic Origin, Visible Minorities', *The Daily*, 17 Feb. At: <http://www.statcan.ca/Daily/English/>.

———. 2002. 2001 Census: Population and Dwelling Counts, for Census Metropolitan Areas and Census Agglomerations, 2001 and 1996 Censuses, 16 July. At: <www.statcan.ca/english/IPS/Data/93F0050XCB2001013.htm>.

———. 2003. Canadian Statistics: Distribution of Employed People, by Industry, by Province. At: <www.statcan.ca/english/Pgdb/labor21b.htm>.

———. 2006. Canadian Statistics: Population of Census Metropolitan Areas. At: <www40.statcan.ca/101/est01/demo05a.htm>.

———. 2007. Population and Dwelling Counts, for Census Metropolitan Areas and Census Agglomerations, 2006 and 2001 Censuses—100% data. At: <www12.statcan.ca/english/census06/data/popdwell/Table.cfm?T=201&S=3&O=D&RPP=150>.

Thibodeau, Jean-Claude, and Yvon Martineau. 1996. 'Essaimage technologique en région périphérique: étude de cas', *Revue Canadienne des sciences régionales* 19, 1: 49–64.

Thompson, Robert. 2000. 'IBM Canada Growing in Quebec', *National Post*, 30 Nov., C6.

Trent, John E., Robert Young, and Guy Lachapelle, eds. 1996. *Québec–Canada: What Is the Path Ahead/Nouveaux sentiers vers l'avenir*. Ottawa: University of Ottawa Press.

Turgeon, Pierre. 1992. *La Radissonie: le pays de la Baie James*. Montréal: Libre Expressions.

Vandermissen, Marie-Hélène, Paul Villeneuve, and Marius Thériault. 2001. 'L'évolution de la mobilité des femmes à Québec entre 1977 et 1996', *Cahiers de géographie du Québec* 45, 125: 211–43.

Villeneuve, Paul. 1992. 'Un Québec en révolution tranquille', in Roger Brunet, ed., *Géographie universelle*. Paris: Hachette and Reclus, 357–73.

———. 1997. 'Le Québec et l'intégration continentale: Un processus à plusieurs vitesses et à directions multiples', *Cahiers de géographie du Québec* 41, 114: 337–47.

——— and G. Côté. 1994. 'Conflits de localization et étalement urbain: y a-t-il un lien?', *Cahiers de géographie du Québec* 38, 105: 397–412.

Woods, Alan. 2007. 'Poll Finds Kyoto Backing', The Star.com, 1 June. At: <www.thestar.com/News/article/220293>.

Wyman, Diana. 2006. 'Trade Liberalization and the Canadian Clothing Market', *Canadian Economic Observer*. Catalogue no. 11–010. Ottawa. At: <www.statcan.ca/english/ads/11-010-XPB/pdf/dec06.pdf>.

Yakabuski, Konrad. 1997. 'Norsk to Expand Quebec Plant', *Globe and Mail*, 12 June, B6.

———. 1998. 'Alcan Plans $2.2-billion Smelter', *Globe and Mail*, 20 Feb., B1.

Further Reading

Bourassa, Robert. 1985. *Power from the North*. Scarborough, Ont.: Prentice-Hall.

Power from the North not only tells the story of the James Bay Project from the pro-development perspective of Premier Robert Bourassa but the book also provides contextual insights into Québec society (and for that matter Canadian society) and its view of Aboriginal peoples and their place (or lack of place) within Canada at that time. In the late 1960s, Bourassa grasped the concept of selling low-cost hydroelectricity from northern Québec to energy-deficient New England and of luring industry to Québec with the promise of low electrical prices. What Bourassa and his political counterparts in the rest of Canada did not understand was the threat of another claim to Crown lands. The courts began to recognize (but did not define) Aboriginal title to traditional hunting lands. Fifteen years earlier, a similar megaproject—the iron ore project—went forward in northern Québec without a peep from its Aboriginal peoples. Why? In the 1950s, the Cree, Inuit, and Naskapi (Innu) were not organized or sufficiently aware of their legal options. Living on the land, these Aboriginal peoples paid little attention to those in the south—as long as they stayed in the south.

However, times were changing. The North was gradually drawn into a resource hinterland for the provincial, national, and global economies. The James Bay Project was just one of several northern megaprojects. Young Cree leaders emerged and challenged the plan of the provincial government to harness the rivers of the north. Outside of Québec, other political developments were taking place. One was the 1973 *Calder* decision, in which the concept of Aboriginal title to traditional hunting lands gained support from the Supreme Court and opened the door to further discussion. Shortly thereafter, the federal government shifted from its original position of denial and agreed that Aboriginal title may indeed have a legal validity. Accordingly, the federal government announced a new approach to Aboriginal claims to Crown land, establishing specific and comprehensive land-claim negotiations. Nearly 10 years after the James Bay and Northern Québec Agreement, the first comprehensive land-claim agreement—the Inuvialuit Agreement—came into law in 1984.

But what exactly was the James Bay Hydroelectric Project? The entire project called for the development of the water resources of Québec's northern hinterland in order to provide low-cost power to industries willing to locate along the St Lawrence River and to generate sufficient revenues from long-term contracts with utility companies in New England to pay for the project. In addition, Hydro-Québec, while challenged to build such a massive project, would demonstrate once and for all the ability and capacity of Quebecers to undertake and complete such huge construction efforts. For Québec, so thought Bourassa, this was the project of the century. So far, only the first of three phases have been completed, though waters from the Rupert River were diverted in 2006 into the La Grande Basin. The first phase harnessed the waters of La Grande, Eastmain, and Caniapiscau basins, leaving uncompleted the second phase (the Great Whale Basin) and the third phase (the Nottaway, Broadback, and Rupert basins).

Five themes resonate in *Power from the North*. The first three themes reflected the pro-development perspective of Premier Robert Bourassa while the last two were 'unexpected' consequences that still resonate within Québec:

- a strong pro-development sentiment;
- a continental energy policy benefiting both Québec and New England;
- a provincial development strategy based on supplying low-cost electrical energy to industrial firms willing to relocate to southern Québec;
- a triggering of an angry Aboriginal response to this project that violated their traditional hunting grounds;
- a legal battle that resulted in the first modern land-claim agreement, the James Bay and Northern Québec Agreement.

Students should recognize that *Power from the North* describes a classic example of old-style development schemes. While huge quantities of electrical power were generated from this project and then shipped to southern markets, the James Bay Project was very controversial because of its impact on the traditional hunting economy of the Cree and on the environment (McCutcheon, 1991; Richardson, 1975; Salisbury, 1986; Turgeon, 1992). With the reconfiguration of the landscape by huge dams, river diversions, and reservoirs, the environment has changed dramatically and the Cree have been drawn into a new way of life where the traditional hunting economy is less prominent, especially among young Cree. But several decades ago, Cree and environmental groups banded together to fight this project tooth and nail to save the traditional hunting economy of the Cree and, by doing so, save the environment. In this sense, the James Bay Project took on a larger dimension, that is, a bitter choice between development and the well-being of the environment and Aboriginal peoples. In 1999, James Hornig edited a collection of papers on James Bay some 30 years after the project was announced. With the benefit of time, these articles present a more balanced and less inflammatory assessment of the social and environmental impacts of the James Bay Project.

Finally, in 2001, the Paix des Braves agreement marked a U-turn in relations between the Cree and the Québec government. No longer enemies but now partners, the two parties plan to develop northern resources together. The first joint endeavour was the $2 billion Eastmain-1 Project, which began in 2002 and was competed in 2007. In exchange for this agreement to develop the resources of northern Québec, provincial dollars are flowing to the Cree communities to help pay for new houses, roads, and other urban amenities, and Cree workers found jobs in Eastmain construction work and business contracts. While not all Cree are happy with this arrangement, the majority voted to support the Paix de Braves agreement.

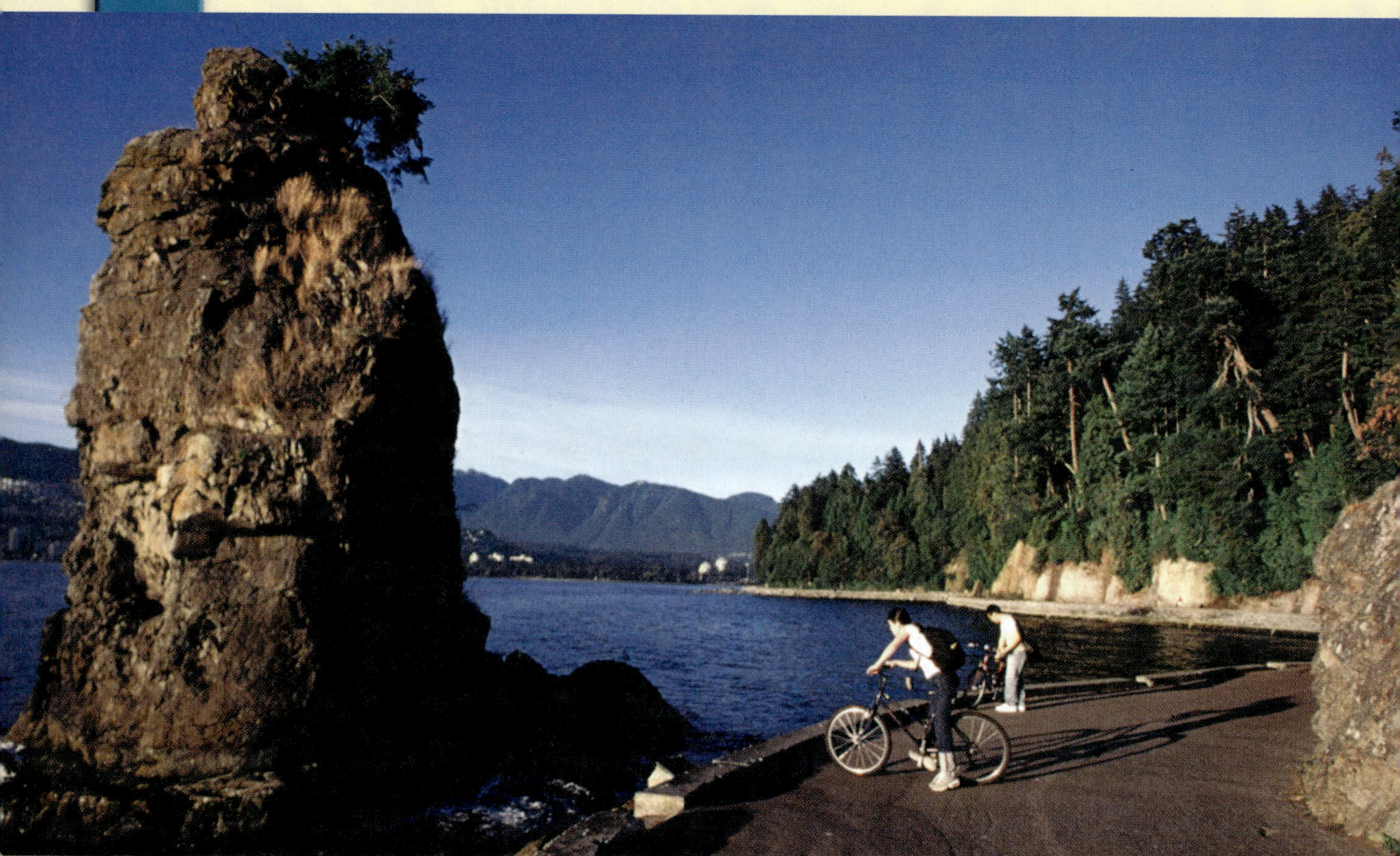

The Stanley Park Seawall (Al Harvey/The Slide Farm)

Overview and Objectives

British Columbia is an emerging economic power and political force within Canada. Its geographic location on the Pacific coast permits BC to benefit from trade in two ways—from the expanding economies of China and other Pacific Rim countries, and from the booming economy of Western Canada. The Port of Vancouver plays a key role in trade between North America and the Pacific Rim.

Historically, economic development began much later than in Québec and Ontario, but rapid growth is narrowing the gap, making BC a close rival of the two core industrial regions of Canada. At first, this Pacific coast province's economy depended on its natural resources and on gaining access to those resources. By the twenty-first century, BC's economy had become much more diversified and less reliant on natural resources for growth.

BC has two distinct subregions. For a long time, the heavily populated southwest corner formed BC's economic core, while the rest of the province remained a resource hinterland. Some areas within the periphery, however, are emerging as economic hubs. The Okanagan, with the cities of Kelowna, Penticton, and Vernon, forms one such hub. Across BC but particularly in the resource hinterland, First Nation land claims represent unfinished business. This chapter will:

- Describe British Columbia's physical geography and history.
- Present the basic elements of BC's economy and population.
- Examine these elements within BC's physical setting and the core/periphery model.
- Consider two faultlines—Aboriginal/non-Aboriginal and centralist/decentralist.
- Focus on the forest industry and its changing role in BC's economy.
- Consider the importance of international trade, the Asian-Pacific Gateway, and Prince Rupert's new role as a major port for container cargo.
- Ask whether BC remains an upward transitional region or has become an industrial core.

CHAPTER 7

British Columbia

INTRODUCTION

British Columbia has become a powerful force within Canada. Much of its economic and demographic strength remains concentrated in the southwest corner, especially in the Greater Vancouver region, and this important city, which reaps much of the benefits of tourism and trade in the region, continues to drive the BC economy. The interior of BC has been slow to develop, largely because of limited transportation links to Vancouver. Unlike Vancouver and Victoria, where most high-technology and service industries are located, the rest of British Columbia relies heavily on resource industries, especially the forest sector. Natural gas production now outranks forestry. The tourism industry, by capitalizing on BC's varied and attractive physical geography, has sparked new investment and generated more service jobs. The United States remains the principal trading partner with BC, though trade with China, Japan, and other Pacific Rim countries is turning BC into the crossroads for Asian trade—certainly that is the goal of the Asian-Pacific Gateway transportation projects.

What makes British Columbia so attractive for industry and people? While much can be attributed to its mild climate and rich resources, British Columbia's strategic position on the Pacific coast has made it a trade centre and a participant in the lucrative Alaska tourist cruise voyages. Much of its population increase is due to the arrival of immigrants from Asia who, in turn, have created business and established trade with Asian countries. Canadians from other provinces also have been attracted to the mild climate and economic opportunities. Population is a measure of political power, i.e., the larger the population, the greater the political clout in Ottawa. Yet, political friction between BC and the federal government prevents this simple relationship from taking form. The struggle for political power and respect is ongoing and underscores the centralist/decentralist faultline discussed in this chapter. From time to time, various signs of disenchantment with Ottawa emerge. To those living in British Columbia, the province fits comfortably into the Pacific Northwest. In a sense, the Rocky Mountains are more than a physical divide. One expression of this regionalism that transcends the forty-ninth parallel is the concept of Cascadia—Oregon, Washington, and British Columbia. In part, Cascadia is a reaction to the negative feelings towards Ottawa. Within BC there is another faultline—the Aboriginal/non-Aboriginal faultline—whereby Aboriginal peoples are demanding more power through land-claim agreements.

The distinguishing economic activity in BC is its forest industry, which is the Key Topic in this chapter. While the forest industry is no longer the principal economic force in the province, the industry continues to lead in

A new Pitt River Bridge between Port Coquitlam and Pitt Meadows, about 25 km east of Vancouver, to be completed in 2009, will improve access to Vancouver by means of the North Fraser Perimeter Road. This is one of the Asian-Pacific Gateway projects. (Copyright © Province of British Columbia. All rights reserved. Reprinted with permission of the Province of British Columbia.www.ipp.gov.bc.ca.)

value of exports. The forest industry is recovering from the US softwood lumber duties and now faces record low prices, repeated summer forest fires, and massive destruction of its mature lodgepole pine stands by the pine beetle. The long-term consequences of these developments—US duties, low prices, pine beetles, and forest fires—on the forest industry are difficult to gauge. Consolidation is one response. Smaller firms suffered the most from the US duties on softwood lumber, while larger forest products companies were able to increase their efficiency and continue to export to the United States.

British Columbia within Canada

British Columbia is an emerging giant within Canada's economic system. The region has undergone rapid economic growth, with trade playing a crucial role. Trade with Pacific Rim countries is expanding and, in 2006, BC signed the **British Columbia–Alberta Trade, Investment, and Labour Mobility Agreement** (TILMA) that will reduce barriers between the two provinces. Other provinces are seriously considering joining this innovative agreement. BC is ideally situated for trade with Pacific Rim countries. In fact, British Columbia's economy has benefited from the rapidly expanding Asian economies led by China and India. In recent years a large number of Chinese immigrants to BC (especially to Vancouver) brought with them skills, capital, and Asian business connections, thereby stimulating the region's economy. In the past, this economy was propelled by its vast array of natural wealth—fish, forests, and minerals. More recently, natural gas production has taken first place in the resource sector. Though exploitation of these resources remains very important, BC's economy is now powered by other sectors, especially high technology and the tourist industry. Construction of facilities for the 2010 Winter Olympics is adding to the booming BC economy.

BC's strong place in Canada is revealed by its share of Canada's GDP and population (Figure 7.1). Since 2001, British Columbia's rate of economic and population growth has outperformed the rates found in Canada's two core regions, Ontario and Québec. With ever-increasing immigration, especially from China and Hong Kong, BC's population has become more diverse. The City of Richmond, located just 25 minutes from both downtown Vancouver and the US border, is a popular destination for Chinese immigrants.

British Columbia is home to 198 First Nations, about one-third of all First Nations in Canada. The original inhabitants of BC represent the greatest diversity of Indian peoples in Canada. For example, seven of Canada's 11 unique language families are located exclusively in BC—more than 60 per cent of the country's Aboriginal languages. Yet, most Indian peoples are on the margins of BC's economy. Change may be in the air, however. The effort of the Campbell River First Nation to participate in the growing tourist business associated with Alaskan cruise ships is one example of change. The growing population of Indian peoples, with their ongoing land claims, makes them an important minority within British Columbia. In fact, those few bands with land-claim agreements are finding a place in BC's economy.

In 2001, Statistics Canada reported that 4.4 per cent of BC's population was reported as Aboriginal while French-speaking residents of BC formed only 0.4 per cent (Table 1.7). The 2006 census results are anticipated to show an even larger percentage gap between the Aboriginal and French-speaking residents of BC, chiefly the result of the rapidly increasing Aboriginal population.

As in Ontario and Québec, a core/periphery spatial arrangement exists. This geographic dichotomy is best expressed by British Columbia's population distribution: over 60 per cent of BC's residents are concentrated in the southwest corner of the province. The major urban centres, including Vancouver and Victoria, are within this core. Beyond this population core is more than 90 per cent of the province's territory. Much of this rugged and mountainous hinterland is sparsely populated, though there are growing population clusters in the Okanagan and Thompson valleys. As well, population centres are located along the north coast around Prince Rupert, in the central interior at Prince George, and in the Peace River country. A few interior towns, such as Trail, are located near the US border.

In British Columbia, natural resources remain important but they no longer dominate the province's economy. For instance, 50 years ago forestry alone accounted for 50 per cent of the provincial economy and for most of the employment. By the late 1990s, forestry comprised only 17 per cent of BC's economy and 14 per cent of employment (Barnes and Hayter, 1997: 5). According to 2005 statistics, only 9 per cent of BC workers have jobs in resource harvesting and extracting industries such as agriculture, fishing, forestry, and mining compared to 13 per cent in 1990 (BC Stats, 2006b). Yet, the processing of timber, ore, and fish still generates most of the manufacturing jobs in BC.

From an environmental perspective, BC faces a dilemma. Tourism has grown in importance and, with the 2010 Winter Olympic Games and the associated construction work, is reaching new levels. For instance, the expansion of the Sea-to-Sky Highway that winds along Howe Sound from North Vancouver to Squamish and then on to Whistler has met with opposition from environmentalists and

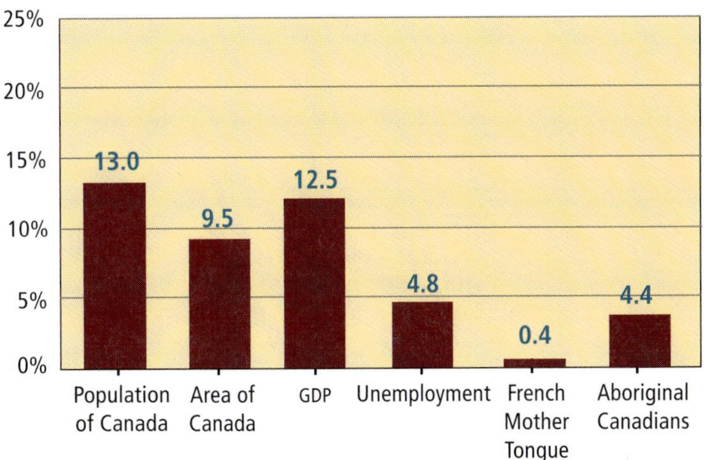

Figure 7.1 British Columbia, 2006. BC's strong place in Canada is revealed by its share of Canada's GDP and population. The Aboriginal population and French by mother tongue show the weak position of Aboriginal peoples and French-speaking Canadians in the province. The small percentage of Aboriginal peoples, however, does not reflect the extent of First Nations' potential ownership of land in BC through ongoing land-claim negotiations.
Sources: Tables 1.1, 1.7, and 4.18.

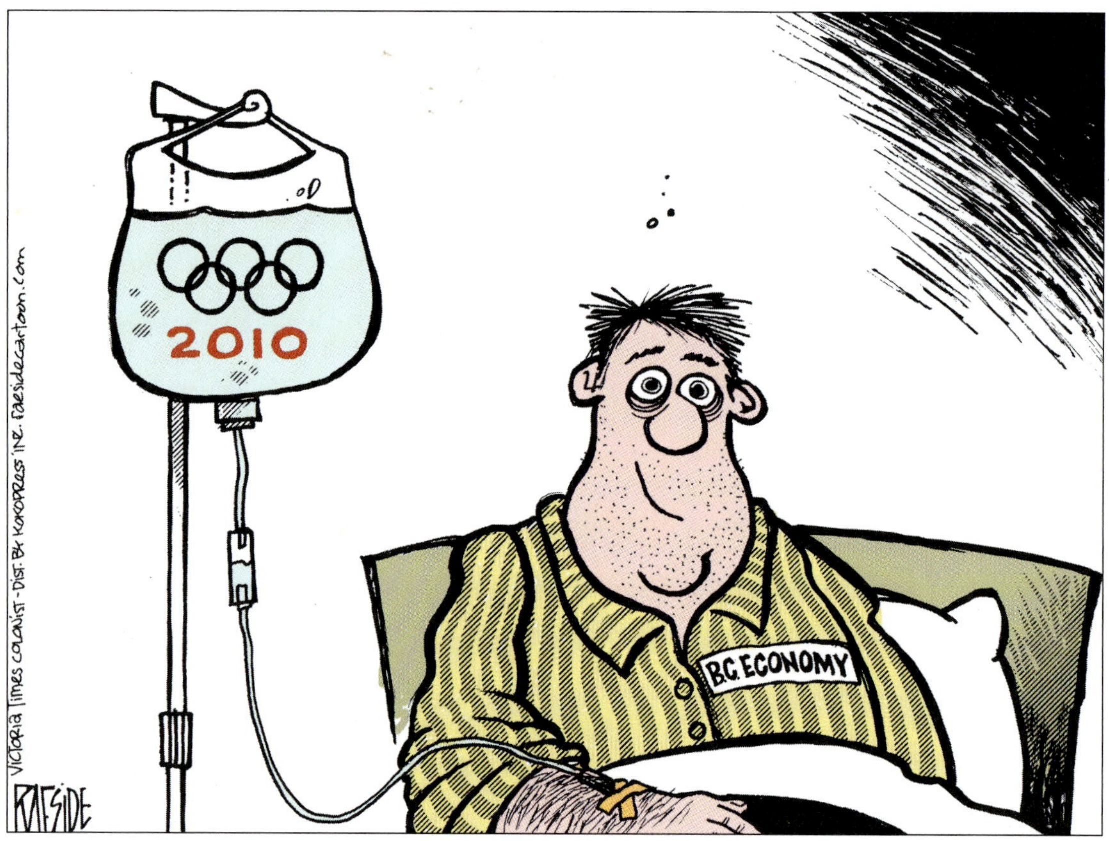

Figure 7.2 Vancouver wins Winter Olympics bid. British Columbia received a much needed boost when Vancouver was selected to host the 2010 Olympic and Paralympic Winter Games. In 2004, the cost was projected to be $1.3 billion, but by 2006 the Office of the Auditor General of British Columbia estimated the cost would reach $1.8 billion, and construction of facilities and upgrading of roads, including the highway to Whistler, could double the cost. On the positive side, the Games could generate up to $10 billion in direct economic activity. Vancouver will provide the facilities for hockey, figure skating, curling, and speed skating while alpine skiing and Nordic sports will take place in Whistler, 115 km north of Vancouver. Whistler boasts the largest ski area in North America. (© Adrian Raeside)

residents of West Vancouver. The basic question is how such development is affecting BC's pristine wilderness. The challenge remains how to balance the need for more tourist development and the desire to maintain a natural landscape and an urban environment free of air and water pollution.

British Columbia's Physical Geography

The spectacular physical geography of British Columbia is perhaps the region's greatest natural asset. The variety of its physiographic features is unprecedented. Then, too, British Columbia is famous for its mild west coast climate. The combination of two contrasting climates (west coast and interior climates) with mountainous terrain has resulted in a wide variety of natural environments or ecosystems. Three examples of natural diversity are rain forests along the coast, desert-like conditions in the Interior Plateau, and alpine tundra found at high elevations in many BC mountains.

This physical contrast between the wet BC coast and its dry interior is largely due to the effect of the Coast Mountains on precipitation. Easterly flowing air masses laden with moisture from the Pacific are forced to rise sharply over this high mountain chain, and consequently most moisture falls as orthographic precipitation on the western slopes while little precipitation reaches the eastern slopes (Vignette 7.1).

The climate of the west coast is unique in Canada. Winters are extremely mild and freezing temperatures are uncommon. Summer temperatures, while warm, are rarely as high as temperatures common in the more continental and dry climate of the Interior Plateau of British Columbia. Moderate temperatures, high rainfall, and mild but cloudy winters make the west coast of British Columbia an ideal place to live and a popular retirement centre for those Canadians wanting to escape long cold winters.

The Pacific Ocean has a powerful impact on BC's climate, resource base, and transportation system. Unlike in Atlantic Canada, the continental shelf in BC extends only a short distance from the coast. Within this narrow zone there are many islands, the largest being Vancouver Island followed by the Queen Charlotte Islands (Haida Gwaii). The riches of the sea include salmon, which return to the rivers, such as the Fraser and the Skeena, to complete their life cycle. Most of BC's natural wealth, however, is not in the sea but in the province's diversified physical geography, which provides valuable resources, particularly forests, minerals, and rivers.

BC's Physiographic Regions

Most of British Columbia lies in the physiographic region known as the Cordillera—a combination of mountains, plateaus, and valleys. British Columbia, unlike Atlantic Canada, has a very narrow and sometimes deep continental shelf. BC's continental shelf is, in fact, a submerged mountainous-like ocean bottom. A small portion of northeastern British Columbia, known as the Peace River country, extends into the Interior Plains (Figure 2.1). Here, the geological structure is part of the Western Sedimentary Basin.

The Cordillera is a complex physiographic region. Extending from southern British Columbia to Yukon, this region encompasses over 16 per cent of Canada's territory. As

Vignette 7.1 BC's Precipitation: Too Much or Too Little?

British Columbia receives the greatest amount of precipitation along its Pacific coast. Inland, the annual precipitation decreases sharply. In simple terms, there are two precipitation areas in British Columbia—one in the Pacific climatic zone, which receives from 800 to 2,000 millimetres of precipitation per year, the other in the Cordillera climatic zone, where less than 800 millimetres fall each year. These figures are average amounts of rain and snow. Fluctuations do occur. In August 2006, for instance, the west coast received very little precipitation. The small fishing/tourist town of Tofino, which is situated on the windward outer coast of Vancouver Island and is noted for its heavy rainfall, ran out of drinking water for the first time in its history. Two months later, in November, warm, moist subtropical air from Hawaii brought high temperatures and record amounts of moist air to the BC coast. Known as the Pineapple Express, heavy rainfall from this subtropical air mass flooded the low-lying Chilliwack area in the Fraser Valley, forcing an evacuation order for the residents of 200 houses. In early June 2007, the melting of record snowfall in the mountains and heavy rains caused the Skeena and Fraser rivers in the northern part of the province to overflow their banks in areas around Smithers, Terrace, and Prince George. As the heavy snowpack in the mountains melts and runs into the Fraser River, flooding in the densely populated low-lying areas, including the deltaic island of Richmond, poses a serious threat.

explained in Chapter 2, the Cordillera was formed by severe folding and faulting of sedimentary rocks. This coastal zone is subject to earthquakes because of tectonic movement.

The Cordillera has at least 10 mountain ranges, the most prominent of which are the Coast Mountains, which extend northward from Vancouver to the Alaskan panhandle, and the Rocky Mountains, which stretch from the Canada–US border almost to Yukon (Figure 7.3). Other north–south mountain ranges include the Insular Mountains that rise above the sea to form the Queen Charlottes and Vancouver Island. The zone between the Insular Mountains and the Coast Mountains forms the Inland Passage. Sheltered by Vancouver Island and the Queen Charlotte Islands, the Inland Passage is the only body of water along the shores of British Columbia, Washington, Oregon, and California protected from the direct impact of the Pacific Ocean. Because of this protection, cruise ships, ferries, and private vessels travel along the Inland Passage in the waters of Georgia and Hecate straits.

Two more mountain ranges, which lie in the southern interior of BC, are the Cascade Mountains, in south-central BC along the US border, and the Columbia Mountains. The Columbia Mountains consist of three parallel, north–south mountain ranges—the Purcell, Selkirk, and Monashee—and a fourth range, the Cariboo Mountains, forms the Columbia's northern extension. Further north are four mountain ranges—the Hazelton, Skeena, Ominica, and Cassiar mountains.

The Interior Plateau separates the Coast Mountains from the mountains of the interior (Figure 7.3). North of the Interior Plateau is the Stikine Plateau. The topography of these plateaus consists of gently undulating land with occasional deeply trenched river valleys. The Fraser Canyon, one of BC's best-known landforms, drops about 300 to 600 m below the Interior Plateau and extends from just south of Quesnel to Hope. Just south of Lytton, the canyon walls rise about 1,000 m above the river. This rocky gorge is called Hell's Gate. The Fraser Canyon is a major fault zone that separates the Cascades from the Coast Mountains.

Climatic Zones

British Columbia has two climatic zones, the Pacific and the Cordillera. Because of the extremely high elevations in the Coast Mountains, few moist Pacific air masses reach the Interior Plateau. The spatial variation in precipitation is remarkable. Heavy orographic precipitation occurs along the western slopes of the Insular and Coast mountains where 2,000 mm of precipitation fall annually (Vignette 7.1). In sharp contrast, the Interior Plateau receives less than 400 mm per year. In the Thompson Valley and Okanagan Valley of the Interior Plateau, hot, dry conditions result in an arid climate with sagebrush in the valleys and ponderosa pines on the valley slopes.

The Pacific coast of British Columbia has the most temperate climate in Canada, dominated by the constant flow of moist Pacific air masses across its terrain. The result is mild and often wet, cloudy weather. Because eastward-moving Pacific air masses must rise above the Insular Mountains and the Coast Mountains, orographic precipitation frequently occurs (Vignette 2.9). This, combined with frontal precipitation, gives the west coast a high level of annual precipitation. Known locally as 'liquid sunshine', the annual precipitation along the west coast of Vancouver Island can exceed 3,000 mm in some locations. Further east, the annual precipitation declines. At Vancouver, located near the Coast Mountains, the annual precipitation declines to about 1,000 mm. Victoria, however, lies in the partial rain shadow of the Insular Mountains of Vancouver Island and thus avoids the heavy orographic rainfall that affects Vancouver. Consequently, the capital city receives nearly 40 per cent less rainfall than Vancouver.

As a mountain climate, the Cordillera climatic zone consists of a number of microclimates. Changes in elevation (height above sea level), latitude (from 49° N to 60° N), variation in topography from mountain ranges to plateaus, and distance from the Pacific Ocean to the Rocky Mountains (600 km) control each microclimate, and in turn affect vegetation and soil conditions within these microclimates. In general, these areas become drier as distance from the Pacific Ocean increases and cooler as either elevation or latitude increases. The range of natural vegetation and soil types is enormous. In the Thompson Valley near Kamloops, an arid microclimate results in a desert-like grassland vegetation with chernozemic soils. In sharp contrast, huge trees in the coastal rain forest grow in podzolic soils. Different still are the higher elevations of the Rocky Mountains extending beyond the treeline, where grasses, mosses, and lichens grow on cryosolic soils. For much of northern British Columbia, the northern coniferous forest (evergreen, needle-leaf trees) is the most common natural vegetation.

Because the Cordillera stretches from southern British Columbia to Yukon, the climate turns into a Subarctic mountain climate north of 58° N. Higher elevations in these latitudes have Arctic climatic conditions. South of 58° N, however, summers are longer and warmer, partly because the eastward-moving Pacific air masses are more dominant and partly because of greater annual solar energy.

Natural Vegetation Zones

The natural vegetation of British Columbia is far from homogeneous (Figure 2.7). Along the west coast, the mild, wet climate encourages a rain forest. The Insular Mountains of Vancouver Island and the Coast Mountains along the mainland both rise abruptly from the Pacific Ocean and therefore receive much rainfall throughout the year. The land responds with lush evergreen and deciduous trees. Characteristic tree species are western hemlock, Douglas fir, western red cedar, and Sitka spruce. Under this vegetation cover, podzolic soils are common. These soils are strongly acidic and low in plant nutrients, making it necessary for farmers to use fertilizers. Fertilizers are used because nutrients are washed away by heavy rainfall, and the needles of the coniferous trees, when dropped, add to the ground cover and produce an acidic soil. This coastal region contrasts with the sagebrush and yellow grasses in the lower elevations of the southern Interior Plateau, where higher temperatures occur. Ponderosa pines are the predominant species in the southern Interior Plateau; at higher elevations and in more northerly areas, lodgepole pine forest replaces ponderosa pine along with other commercially valuable trees, including white and Engelmann spruce, Douglas fir, western hemlock, western red cedar, and two true firs—amabilis fir and subalpine fir. The series of forested mountain ranges, aligned north to south, include the Ominica Mountains located north of Prince George and the Columbia Mountains centred on Revelstoke. Both mountain ranges are situated west of the Rocky Mountain Trench while the Rocky Mountains lie to its east (see Figure 7.3).

Because of the rugged nature of the Cordillera, there is little arable land. Only about 2 per cent of the province's land is classified as arable. British Columbia's largest area of cropland lies outside of the Cordillera physiographic region in the Peace River country. Within the Cordillera, most arable land is in the Fraser Valley, while a smaller amount can be found in the interior, especially the Okanagan and Thompson valleys. This shortage of arable land poses a serious problem for British Columbia. With urban developments spreading onto agricultural land, British Columbia is losing some of its most productive farmland.

Environmental Challenges

British Columbia faces several environmental challenges. These challenges can be divided into those caused by human activities and those caused by nature. Human activities have subjected the seemingly limitless natural riches to mismanagement and wasteful practices that have led to resource loss, environmental degradation, and land-use conflicts. One challenge comes from clear-cutting of the forest. Clear-cutting, the harvesting method most widely used in BC whereby every tree within a large area is cut down, remains a controversial practice. In the past, huge areas were stripped of their trees. Now provincial regulation restricts such logging to 40–60 hectares. The alternate to clear-cutting is selective cut-

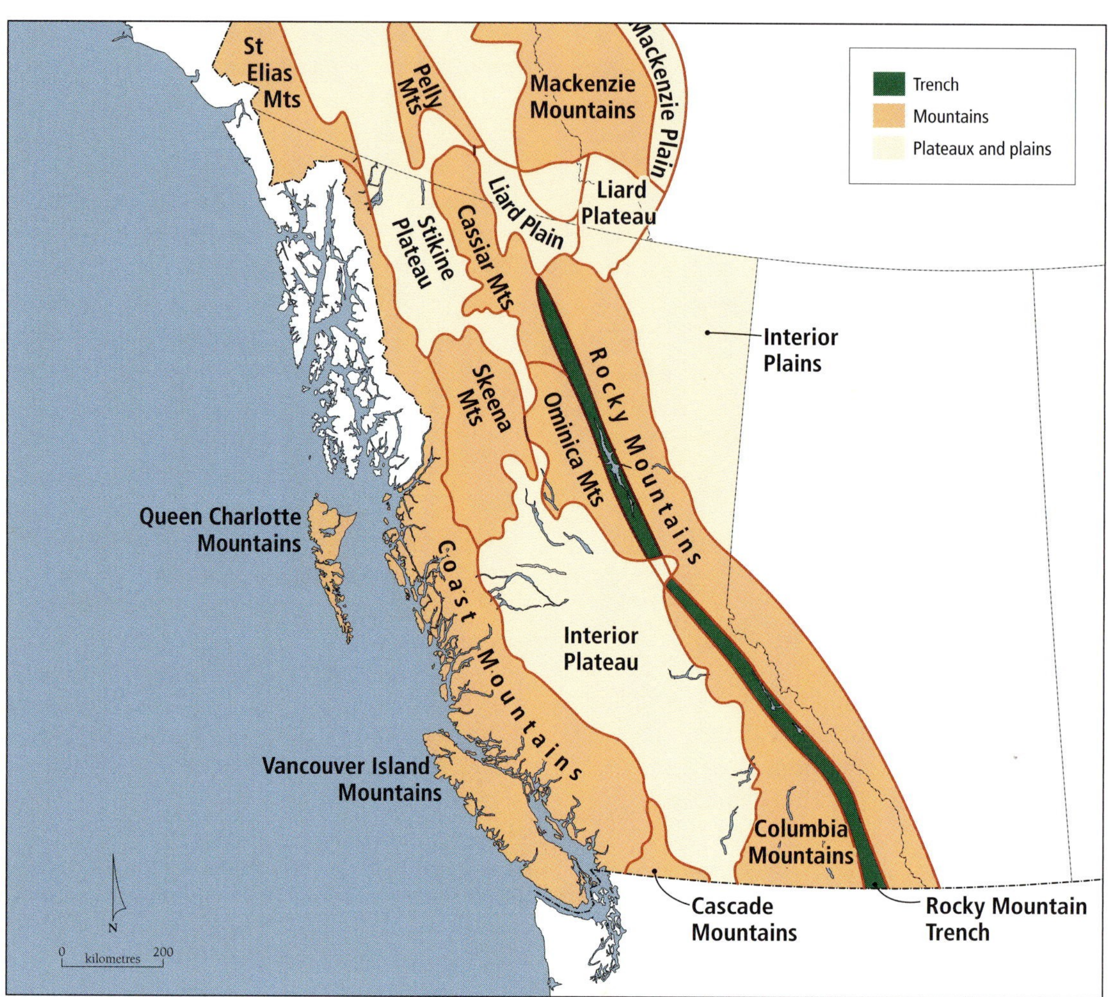

Figure 7.3 Physiography of the Cordillera. British Columbia's complex physiography is evident in the physiographic subregions of the Cordillera. The difficulty of constructing east–west transportation routes can be appreciated if one considers the number of north–south mountain ranges that must be traversed. For that reason, the importance of the Pacific Ocean for transportation prior to the completion of the CPR in 1885 becomes clear. As well, the grain of the land makes north–south land transportation construction into the Pacific Northwest of the United States relatively simple. Herein rest the natural factors behind the political concept of Cascadia and the feeling that BC is California North. (Further resources: Student website, National Atlas section, Map 25. Website instructions are found on p. xx.)

ting. However, logging companies claim that selective logging is too expensive and would not allow them to compete in world markets. In the past, clear-cutting extended right to the banks of streams and rivers, leaving surrounding land vulnerable to rapid soil erosion and stream sedimentation. Under such conditions, fish habitat is damaged and spawning grounds may be destroyed. The Nisga'a claim (Nisga'a, n.d.) that salmon spawning grounds along the Nass River have been damaged by such logging practices, thus significantly reducing the salmon run.

Changing weather patterns pose the chief natural challenge. Exceptionally dry summers over the last five years, for example, have resulted in vast forest fires. Forests in the interior become tinder dry, and the Okanagan Valley has three characteristics that can lead to fires spreading rapidly—strong winds, steep slopes, and a combination of dry underbrush, grass, and lodgepole pine trees. On 23 August

2003, such a fire reached Kelowna. As a forest fire raged out of control and swept into its suburbs, 10,000 residents had to abandon their homes and more than 250 houses were destroyed (Boei and O'Brian, 2003).

Another natural impact on the environment has been the destruction caused by the pine beetle on the lodgepole and ponderosa pine trees in the Interior Forest. Vast areas have been affected and there is no way to stop the spread of the pine beetle. The lifespan of an individual mountain pine beetle is about one year. Pine beetle larvae spend the winter under bark, feeding on the tree, which is often a mature lodgepole pine. The adult pine beetle emerges from an infested tree and seeks another host. Logging activities are accelerating to salvage these trees because the beetles do not affect the wood's strength, its gluing characteristics, or its ability to be finished. While the accelerated logging means more timber for the sawmills in the short run, the loss of such a huge section of the Interior Forest means that mature timber will be in short supply in the future. For instance, a lodgepole pine matures in approximately 80 years. Why is the pine beetle so active? In the past, cold winters kept this pest under control. In recent years, however, pine beetles have spread and multiplied as a result of milder winters. Temperatures of at least –38 degrees C for four days or longer are required before the beetles freeze to death. If cold winters return, then the pine beetle infestation will be controlled. On the other hand, if these mild winters are a feature of global warming, then there will be no stopping the pine beetle. Already, the pine beetle has spread across the BC border into Alberta.

While the BC government has allowed massive clear-cutting of beetle-infested stands, the problem for the lumber companies is that the wood is deteriorating very quickly. An alternative solution is to turn the beetle-killed trees into electricity. BC plans to convert the dead trees into electricity by using the wood to fuel local electric generators situated at sawmills and other strategic locations (VanderKlippe, 2007a). Like other bioenergy plans, the BC plan will require subsidies. In this case, the cost of electricity produced from hydroelectric stations is much less. For example, the cost at a sawmill where the wood is found would be around $82 a megawatt. When transportation of the raw material is factored in, the cost rises dramatically—to about $120 a megawatt (ibid.).

Social Issue: Unfinished Land Claims

Who owns British Columbia? Some 200,000 Aboriginal people live in British Columbia and few First Nations in the province have signed treaties. The pre-Confederation Douglas treaties on Vancouver Island and Treaty No. 8 of 1899

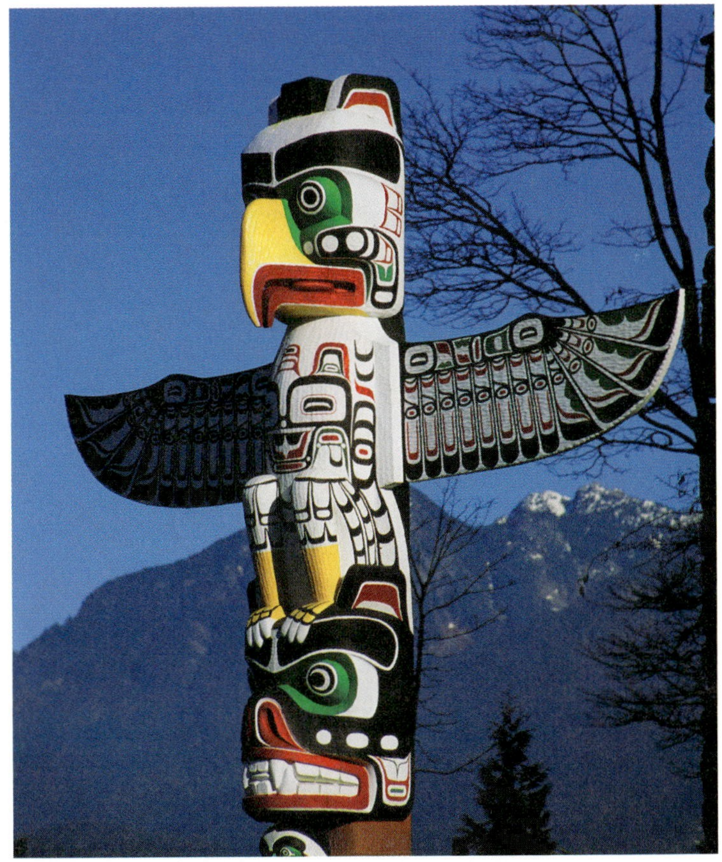

Totem poles not only represent Indian family kinships and stories, but support their land claims, which are based on their ancestors' occupancy of British Columbia. This well-known totem pole is located at Prospect Point in Stanley Park. In the background are the North Shore mountains. (Search4Stock Inc.)

extended into northeast British Columbia extinguished Aboriginal rights to the land in exchange for various benefits for the Indians. The 14 Douglas treaties focused on providing land reserves for Indians along with their right to hunt and fish on Crown land. Treaty No. 8, propelled by the Klondike gold rush of 1898, had two purposes: (1) to open access to miners and prospectors heading to Yukon across the Canadian Northwest, and (2) to allow the Indians to continue their traditional way of life and receive annual payments.

Until 1990, the government of British Columbia was dragging its feet on recognizing land claims. By 1993, British Columbia accepted the concept of Aboriginal title. The next step was to establish a system to settle the claims of BC's 197 Indian bands, which, in total, exceeded the size of BC. Claims to traditional lands are, through the process of negotiations, whittled down to around 10 per cent of the claimed territory. In exchange for land surrender, cash and other benefits and rights are acquired by the First Nation. The stakes are very high, and the complex negotiations mean that progress is slow. By November 1994, 42 First Nation groups representing about 70 per cent of the Aboriginal population had entered the negotiation process.

BC's Negotating Process for Land Claims

Ownership of Crown land is at the heart of the BC land claim issue. The federal and provincial governments are both involved in the negotiations at numerous 'tables' with BC First Nations. These negotiations are under the direction of the British Columbia Treaty Commission. In 1993, a six-stage negotiation process was established. The six stages are:

1. Statement of Intent
2. Table Readiness
3. Framework Agreement
4. Agreement-in-Principle
5. Final Agreement
6. Implementation of Treaty

The Nisga'a and the Lheidli T'enneh First Nation were among the first to file with the British Columbia Treaty Commission. In 1999, the first modern land-claim agreement in BC (the Nisga'a Agreement) was ratified by the Nisga'a, the provincial legislature, and the federal Parliament (Vignette 7.2). The political dimensions of the Nisga'a Agreement, which date back many decades and include the landmark 1973 *Calder* decision by the Supreme Court of Canada, have been so critical that Frank Cassidy (1992: 11) described it as 'a centrepiece in the historical development of the province of British Columbia'.

In 2006, a second Final Agreement was reached with the Lheidli T'enneh First Nation, whose land is near and within Prince George. Both agreements include access to resources, such as fish and forests, and allow for commercial sale. In the Lheidli T'enneh Statement of Intent, they claimed nearly 4.6 million hectares of land. This First Nation filed its Statement of Intent on 16 December 1993 and reached a Final Agreement on 26 July 2006 (Lheidli T'enneh, 2007a). The agreement would have awarded the band about $13 million and 4,000 hectares of land in addition to fishing and resource rights and a form of self-governance, but in late March 2007 the 234 voting members failed to give the agreement sufficient support (Hainsworth, 2007).

While negotiations are exceedingly slow, the process is extremely important. Once all the land claims are addressed, the tension underlying the Aboriginal/non-Aboriginal faultline in BC will be reduced, and First Nations will be able to focus their energies on improving the economic and social well-being of their members.

British Columbia's Historical Geography

Indians lived along the Pacific coast of British Columbia for over 10,000 years before European explorers reached the northern Pacific coast in the mid-eighteenth century. The Spanish had already sailed northward from Mexico to California, but the Russians were the first to reach Alaska and establish fur-trading posts along its coast. In 1778, Captain

Vignette 7.2 Facts about the Nisga'a Agreement

Location:
The Nass Valley extends from the Pacific Ocean across and beyond the Coast Mountains approximately 150 km inland. This valley is the homeland of the Nisga'a. Prince Rupert is the closest large city and it lies some 100 km southwest of the headwaters of the Nass Valley.

Population:
The total population of the Nisga'a is almost 10,000, with approximately 6,000 living in British Columbia and the rest in other parts of Canada and the United States. Some 2,400 Nisga'a reside in the Nass Valley in the villages of New Alyansh (Gitlakdamiks), Canyon City (Gitwinksihikw), and Kincolith (Gingolx).

Agreement:
The Nisga'a and the federal and BC governments approved and signed the agreement in 1999. In exchange for surrendering their traditional lands, the Nisga'a received title to almost 2,000 km^2 in the Nass Valley of British Columbia; access to forest and fishery resources; self-government powers, including a Native judicial system and policing; and nearly $500 million in cash, grants, and program funds from Victoria and Ottawa. British Columbia supplies the land and 30 per cent of the cash settlement with the federal government paying 70 per cent of the cash settlement. Under this agreement, the Nisga'a are no longer under the Indian Act and therefore cease to receive benefits as status Indians.

James Cook established Britain's interest in this region by sailing into Vancouver Island's Nootka Sound, where he and his sailors found the Nootka village of Yuquot. The Nootka, now known as Nuu-chah-nulth, fished for salmon and hunted the sea otter. Upon landing, Cook engaged in trade for sea otter pelts, which opened up a profitable trade with China, although Cook, among the greatest of nautical explorers, did not live to see this trade flourish—he was killed in a skirmish with natives in the Hawaiian Islands on the return voyage. After the Royal Navy published Cook's record of his voyage, British and American traders came to the Pacific Northwest to seek the highly valued sea otters. Russian fur traders, based in Alaska, also harvested sea otters. Spain, which considered the lands Spanish territory, was disturbed by these interlopers and sent a fleet northward from Mexico in 1789. At Nootka Sound on the west coast of Vancouver Island, the Spanish seized several ships and built a fort to defend their claim. In 1792, Captain George Vancouver of Britain's Royal Navy sailed around Vancouver Island. In the following year, Alexander Mackenzie of the

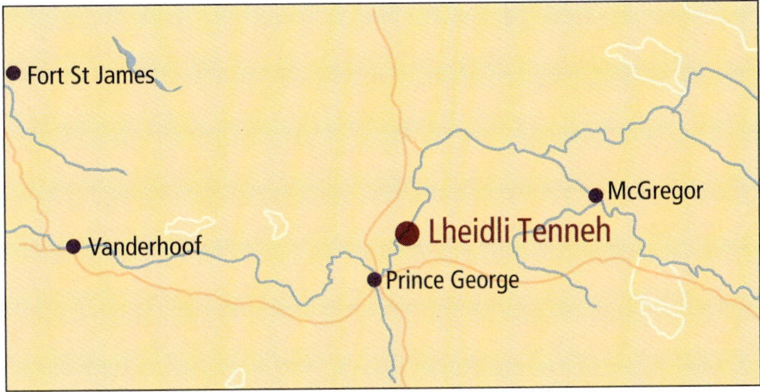

Figure 7.4 Lheidli T'enneh First Nation. The Lheidli T'enneh are Carrier Indians, most of whom live on a reserve near the confluence of the Fraser and Nechako rivers. According to their website, 'approximately 120 people [live] on reserve with a further 100 people living in Prince George.' Most of the remainder of this First Nation live in other BC communities, while some others live in other parts of Canada. (Lheidli T'enneh, 2007b; BC Treaty Commission, 2007)

Captain James Cook's ships were moored in Nootka Sound, as depicted in a watercolour by M.B. Messer. Four years earlier, the Spanish explorer Juan Hernandez sailed along the BC coast. However, there is the possibility that Francis Drake, while on a secret mission to find a western entrance to the Northwest Passage, reached these waters in 1579 (Hume, 2000: B1). (*Library and Archives Canada C11201*).

North West Company travelled overland from Fort Chipewyan to just south of Prince George and then to the Pacific coast near Bella Coola, which is just over 400 km north of Vancouver. Under the Nootka Convention (1794), the Spanish surrendered their claim to the Pacific coast north of 42° N, leaving the British and Russians in control.

In the early nineteenth century, the North West Company established a series of fur-trading posts along the Columbia River. From 1805 to 1808, Simon Fraser, a fur trader and explorer, explored the interior of British Columbia on behalf of the North West Company. He travelled by canoe from the Peace River to the mouth of the Fraser River. As elsewhere, the strategy of the North West Company was to develop a working relationship with local Indian tribes based on bartering manufactured goods for furs. After 1821, when the North West Company merged with its rival, the Hudson's Bay Company, the HBC took charge of the Oregon Territories, which extended from the mouth of the Columbia River to Russia's Alaska.

In 1843, American settlers began to arrive on the coast from the eastern part of the United States. In the same year, the HBC relocated its main trading post from Fort Vancouver at the mouth of the Columbia River to Fort Victoria at the southern tip of Vancouver Island. The increasing number of American settlers who came west along the Oregon Trail represented a challenge to the authority of the Hudson's Bay Company. A few years later, the United States claimed the Pacific coast northward to Alaska, where Russian fur-trading posts existed. In 1846, Britain and the United States agreed to place the boundary between the two nations at 49° N and then to follow the channel that separates Vancouver Island from the mainland of the United States. While the loss of the Oregon Territories was substantial, Britain was fortunate to hold onto the remaining lands administered by the Hudson's Bay Company. Britain recognized that its hold on these lands through the HBC was tenuous and could not withstand the political weight of the growing number of American settlers. Indeed, without the presence of the HBC and Britain's negotiating skills, Canada might well have lost its entire Pacific coastline.

The gold rush of 1858 brought about 25,000 prospectors from California to the Fraser River. Prospectors walked upstream along the sandbanks and sandbars of the Fraser and its tributaries, panning for placer gold (small particles of gold in sand and gravel deposits). The major finds were made in the BC interior, where the town of Barkerville was built near the town of Quesnel. By 1863, Barkerville had a population of about 10,000, making it the largest town in British Columbia. To ensure British sovereignty over territory north of 49° N, the British government established the mainland colony of British Columbia in 1858 under the authority of Sir James Douglas, who was also governor of Vancouver Island. In 1866, the two colonies were united.

Confederation

By the 1860s, the British government was actively encouraging its colonies in North America to unite into one country. Once the first four colonies were united in 1867, Ottawa adopted the British strategy to create a transcontinental nation. An important part of that strategy was to lure British Columbia into the 'national fabric'. The Canadian Pacific Railway was the first expression of this national policy.[1] Ottawa promised to build a railway to the Pacific Ocean within 10 years after British Columbia joined Confederation. In Fort Victoria, however, some wanted to join the United States. By the middle of the nineteenth century, British Columbia had developed significant commercial ties with Americans along the Pacific coast. San Francisco was the closest metropolis and, with its railway to New York, offered the simplest and quickest route to London. In 1859, Oregon became a state and Washington was soon to follow. Commercial links with the United States were growing stronger. But the majority of people in Fort Victoria wanted to remain British. In 1871, British Columbia chose to become a province of Canada (Figure 3.5). British Columbians, however, had to wait 14 years for the Canadian Pacific Railway to reach the Pacific coast at Port Moody on Burrard Inlet, near Vancouver. Later, the railway was extended 20 km westward to the small sawmill town of Vancouver, where there was a better harbour and terminal site for the railway (Figure 7.5).

When British Columbia joined Confederation in 1871, its official population was 36,247 (McVey and Kalbach, 1995: 35). This figure likely underestimated the number of Indians and prospectors who were living in the more remote areas of the province.[2] At the time of the first comprehensive census of British Columbia in 1881, there were approximately 4,200 Chinese, 19,000 white settlers (American, Canadian, and British), and about 30,000 Indians. Most British settlers lived around Fort Victoria. Beyond Fort Victoria, the vast majority of the inhabitants were Aboriginal peoples. In the previous decade, about 25,000 prospectors, mainly Americans, had been scattered in small camps along the Fraser River or at Barkerville. Most Americans had left at the end of the gold rush.

Post-Confederation Growth

At first, Confederation had little effect on British Columbia. The province was isolated from the rest of Canada, Canada's fledgling factories, and even the halls of power in Ottawa. Goods still had to come by ship from San Francisco or London. When the Canadian Pacific Railway was completed in 1885, British Columbia truly became part of the Dominion, and BC's role as a gateway to the world began.

The main line of the Canadian Pacific Railway and its many branch lines were responsible for the formation of many of the province's towns and cities and for providing access to its forest and mineral wealth. The Esquimalt and Nanaimo Railway, the Canadian National Railways, and British Columbia Rail (BC Rail) added to the rail network in British Columbia. In 1886, the Esquimalt and Nanaimo Railway was built, connecting the coalfields of Nanaimo with the capital city of Victoria, and stimulated logging and sawmilling along its route. With the closure of the coal mines in the early 1950s, the railway lost its main function and now plays a minor role in the transportation system of Vancouver Island. By 1914, the Grand Trunk Pacific Railway (later the Canadian National Railways) provided a trans-Canada rail route to Prince Rupert and an alternative rail service to Vancouver. The Pacific Great Eastern Railway

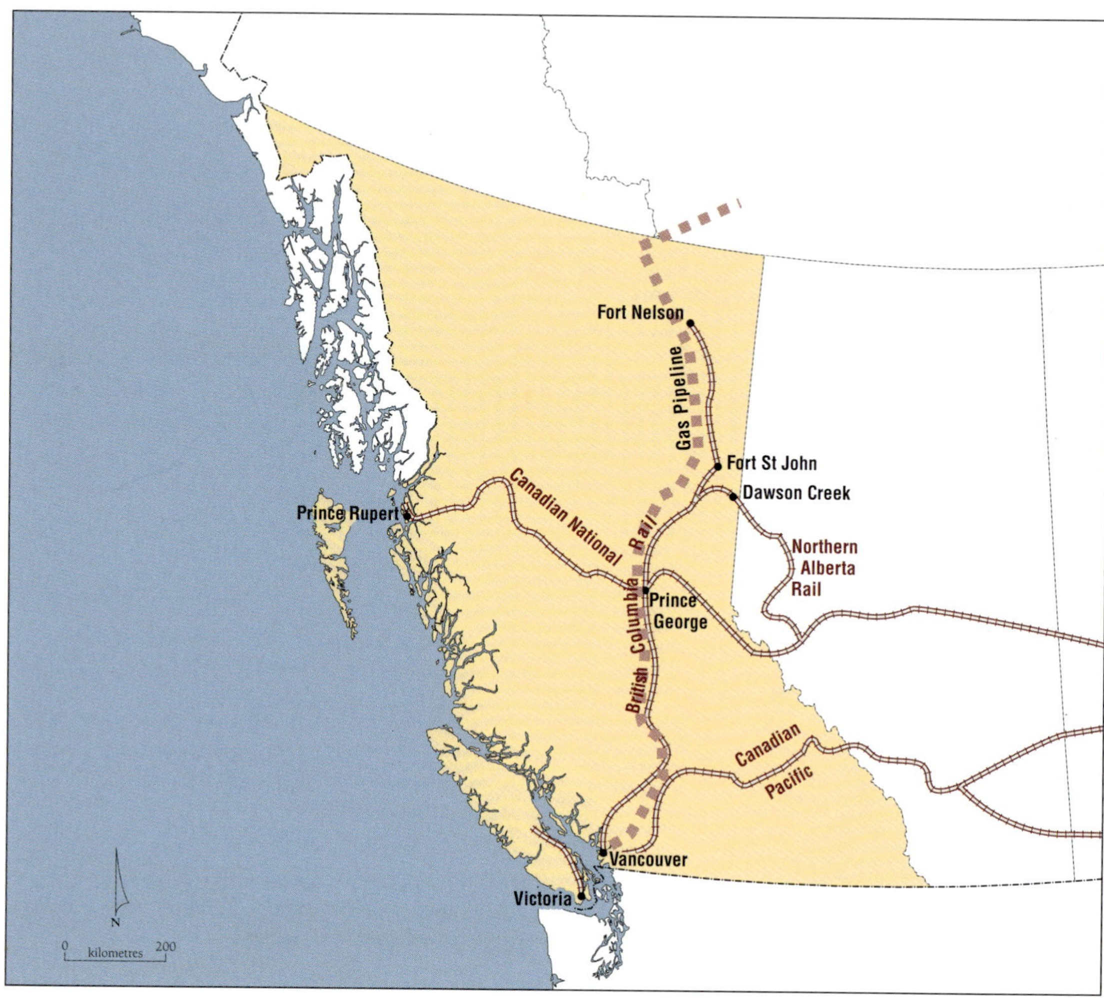

Figure 7.5 Railways in British Columbia. Railways have played a key role in BC's economic development. The first railway to cross the mountains of the Cordillera was the Canadian Pacific Railway in 1885. To open the northern interior of the province, Victoria built the Pacific Great Eastern Railway (now known as British Columbia Rail). After several extensions, British Columbia Rail joined North Vancouver to Fort Nelson in 1971. An eastern link joins Dawson Creek to the Northern Alberta Railway. BC Rail was owned and operated by the province from 1918 to 2004, when it was sold to Canadian National Railway. In 2007, Prince Rupert, the western terminus of the CN line, had the capacity to handle container traffic, thus relieving the congestion at the Port of Vancouver and providing another transportation route into the interior of North America.

(later BC Rail) was incorporated in 1912 but laid few rails until the early 1950s, when the provincial government made a commitment to complete the railway in order to facilitate resource development in the interior of British Columbia. By 1956, this rail line extended from North Vancouver to Prince George. Two years later, BC Rail extended its rail system to Dawson Creek and Fort St John in the Peace River country, and by 1971 it reached Fort Nelson in BC's forested northeast.

After the completion of the CPR, Vancouver grew quickly and soon became the major centre on the west coast. By 1901, Vancouver had a population of 27,000 compared to Victoria's 24,000. As the terminus of the transcontinental railway, Vancouver became the transshipment point for goods produced in the interior of BC and Western Canada. As coal, lumber, and grain were transported by rail from the interior of British Columbia and the Canadian Prairies, the Port of Vancouver spearheaded economic growth in the southwest part of the province. It was then possible to tap the vast natural resources of BC and ship them to world markets. By the twentieth century, Vancouver had become one of Canada's major ports. Located on Burrard Inlet, Vancouver has an excellent harbour. Unlike Montréal, it is an ice-free harbour, thanks to the warm Pacific Ocean. As Canada's major Pacific port, Vancouver became the natural transportation link to Pacific nations. With the opening of the Panama Canal in 1914, British Columbia's resources were more accessible to the markets of the United Kingdom and Western Europe.

Between the completion of the Canadian Pacific Railway in 1885 and the end of World War I in 1918, British Columbia underwent a demographic explosion. By 1921, BC had over half a million inhabitants. The province had been gradually transformed from a fragile political entity in 1871 into a self-confident political and social region within Canada. During this time, the populations of both European and Asian origin increased about tenfold. The European population had reached nearly half a million, while the Asian population was about 40,000. At the same time, a combination of disease and social dislocation caused the number of Aboriginal people to decline sharply, perhaps from 40,000 to 20,000. This magnitude of demographic decline was matched in other regions of Canada. While a number of factors were at work, smallpox, tuberculosis, and other communicable diseases caused the greatest losses. (See Chapter 4 for more details.)

While BC's economy and population continued to grow in the 1920s, the Great Depression of the 1930s caused the province's economy to stall and unemployment to rise sharply. Economic disaster struck British Columbia in 1929. Exports of Canadian products from BC, so necessary for its economic

Vancouver has a magnificent harbour that has facilitated trade with Pacific Rim countries. With Vancouver in the background, a cruise ship sails through Burrard Inlet and passes under Lions Gate Bridge. Stanley Park is located on the south side of Lions Gate Bridge while West Vancouver is on the north side. (Al Harvey/The Slide Farm)

well-being, slowed and prices dropped. In the Prairies, the collapse in agricultural prices was accompanied by prolonged drought, turning the land into a dust bowl. Many Prairie farmers abandoned their farms and fled to British Columbia, adding to the burden of unemployment in that province.

World War II and the Post-War Economic Boom

World War II called for full production in Canada, thereby pulling British Columbia's depressed resource economy out of the doldrums. Military production, including aircraft manufacturing, greatly expanded BC's industrial output. As well, resource industries based on forestry and mining (especially coal and copper) were producing at full capacity.

When the war ended in 1945, BC's resource boom continued. With world demand for forest and mineral products remaining high, the provincial government focused its efforts on developing the resources of its hinterland, the central interior of British Columbia. The first step was to create a transportation system from Vancouver to Prince George, the major city in the central interior. The highway system was improved and extended from Prince George to Dawson Creek in 1952. But the completion of the Pacific Great Eastern Railway to Prince George in 1956 and then to Dawson Creek in 1958 opened the country, allowed for exports to foreign markets, and triggered economic growth, especially in the forest industry. With rail access to Vancouver, forestry, as well as other resource industries, expanded rapidly, thereby leading to the integration of this hinterland into the BC and global industrial core.

Over the past two decades, BC's increasing economic strength, partly driven by the Asian economy, has outpaced that of all other regions in Canada. Trade is a dynamic force propelling British Columbia's economy, and as China and other Pacific Rim countries have led the world in economic growth, this growth has created new markets for Canadian products shipped through Vancouver. Trade opportunities are almost endless between British Columbia and the population of 2.5 billion people in the Pacific nations. Vancouver and, to a lesser degree, Prince Rupert and other Pacific ports serve as trade outlets for coal, lumber, potash, and grain from the interior of British Columbia and the Prairie provinces to reach world markets. In BC's export-oriented economy, it is usually much less expensive to ship the raw material than the finished product. This has led to trade focused on resources. High labour costs, a relatively small local market, and distance from world markets have inhibited the development of manufacturing in BC.

British Columbia Today

Over the years, the economy of British Columbia has diversified and become more service-oriented. Its dependence on the resource industry has diminished and the so-called new economy based on the tertiary sector has expanded, especially in the larger urban centres. Leading components of the new BC economy are tourism, filmmaking, and high technology. The construction industry, driven by a strong demand for housing and the Winter Olympics, adds another dimension to BC's booming economy. Innovative industries, such as fuel cell development by the federally funded National Research Institution and Ballard Power Systems, are at the cutting edge of new technology. International trade is also a key element in BC's economic structure. Products from both British Columbia and Western Canada flow through the Port of Vancouver to Asian markets. Job growth from 2001 to 2005 reflects the new BC economy, with the greatest increase in the construction industry and the least in manufacturing. Forestry was the only industry to suffer job loss (White et al., 2006).

British Columbia's physical geography provides a solid basis for a tourist industry. In fact, BC is recognized as a world-class tourist destination. The Inland Passage is a popular route for cruise ships heading for Alaska. The BC section of the Inland Passage refers to the protected waterway lying between the BC mainland and Vancouver Island and the Queen Charlotte Islands. The Winter Olympics in Vancouver and the ski resort of Whistler in 2010 is sure to spark tourism, with the economic fallout spilling over into other sectors of BC's economy. Filmmaking, too, takes advantage of BC's physical geography. Vancouver is known as 'Hollywood North', making the Vancouver area a major film production centre. In turn, these activities have created a new sector of employment and business opportunities. The high-technology industry has made the greatest impact on BC's economy. It has two components, manufacturing and service. Manufacturing involves those processes that require a substantial amount of technology. Three examples are telecommunications, pharmaceuticals, and scientific instruments. The service sector includes the application of high technology to customers through computer, engineering, and medical services.

BC's population increase has caused a shift of the population centre of Canada towards the west and to BC in particular. In turn, BC has gained more seats in the House of Commons. This has changed relationships between central and eastern Canada and British Columbia. In short, economic, demographic, and political power has shifted to the west.

Population and the Centralist/ Decentralist Faultline

One measure of BC's expanding power lies in its rapid population growth. In 1969, Premier W.A.C. Bennett boasted that 'With the population of British Columbia growing at twice the rate of the rest of Canada, the presence of British Columbia as an economic region of its own is more obvious as each day passes' (Francis et al., 1996: 428). In this case, the numbers support the politician's statement. In 1901, the population of this Pacific province amounted to only 3 per cent of the Canadian total. By 2006, British Columbia's 4.3 million residents constituted 13 per cent of Canada's population.

British Columbia is the only province that has made substantial population gains in each decade since Confederation. In comparison to Ontario, British Columbia has had a higher rate of population increase from 1951 to 2006. From 1991 to 1996, BC's annual rate of increase was 2.7 per cent, well above Ontario's rate of 1.3 per cent and almost two and a half times greater than the national average of 1.1 per cent. From 1996 to 2006, this trend continued, though in recent years Alberta's rate of increase has taken first place.

Most of BC's population increase comes from immigration and interprovincial migration. Since the 1970s, more Canadians have resettled in British Columbia than in any other province. Many immigrants also select British Columbia (after Ontario and Québec) as their new home. Many Chinese immigrants came to British Columbia before Britain turned its colony of Hong Kong over to China. This significant influx of immigrants has changed the cultural makeup of the province. From 1981 to 2001, the share of the population of visible minority groups (mainly Chinese and South Asians) in Vancouver's census metropolitan area jumped from 14 per cent to 37 per cent (Statistics Canada, 2003b).

Many immigrants have brought considerable wealth and economic connections to invest in British Columbia. Asian investment, particularly from Japan and more recently from China, has had an important impact on British Columbia's economy (Edginton, 1994: 32). Since the 1960s, Japanese multinational firms have invested heavily in mining, forestry, and tourism in BC. Japanese companies made a massive investment in the Northeast Coal Project in Tumbler Ridge. This project was designed to supply coal to the steel plants in Japan (Bone, 1992: 139–40). When the price of coal dropped, the arrangement ended, but the mines reopened in 2004 when the price of coal increased. As well, Japanese investors have made many real estate purchases in the Vancouver area rather than in larger centres like Toronto and Montréal because of the large influx of Japanese tourists to Vancouver (Edginton, 1996). In the early 1990s, Japanese investors (Nippon Cable) developed a ski resort just north of Kamloops, turning it into the second largest ski resort in British Columbia, after Whistler (Jim Miller, personal communication, 28 Apr. 1998).[3]

BC's growing population has not only affected its economic status and cultural diversity; it has also bolstered its political might. The political importance of British Columbia's rapid population increase lies in the increased number of seats held by BC members in the House of Commons. However, since the redistribution of the number of seats assigned to each province occurs after each 10-year census, a 10-year lag exists. For instance, the increase in BC's population from 2001 to 2011 will not result in more seats in the House of Commons until several years after 2011. In 1872, British Columbia had six representatives in the House of Commons. With BC's steady population increase, its number of members of the House of Commons increased to 36 members in the 2006 election. According to the 2006 population estimates by Statistics Canada, British Columbia formed 13.0 per cent of the national population. In addition, the so-called grandfather clauses inflate the number of seats in Atlantic Canada and Québec. For instance, Atlantic Canada had a total of 32 seats in the 2006 election while BC had 36. Based on population, each seat in Atlantic Canada represented an average of 73,300 residents while BC members, on average, represented 118,200 residents. This method of allocating the number of members to the House of Commons poses a serious irritant in BC's relations with Ottawa, thus reinforcing grievances that fuel the centralist/decentralist faultline.

Industrial Structure

Is British Columbia a core or a periphery? No doubt this Pacific province is an emerging powerhouse within Canada, but its economy still has a relatively small manufacturing base. Although its manufacturing remains small and relies on the processing of primary products, BC is no longer seen as a resource-based economy; at the same time, the province does not have a strong manufacturing base like Ontario and Québec. In 2005, the secondary sectors in Ontario and Québec comprised 23.6 and 22.2 per cent of employment, while BC's secondary sector accounted for only 17.7 per cent—and much of that figure consisted of the booming construction industry (Table 7.1). BC, like Western Canada, is somewhere between a resource-based economy and an industrial one. In Friedmann's classification, BC is an upward transitional region (Table 1.3). Yet, the shift towards a more balanced economy is revealed in a comparison of the 1991 and 2005 economies. In 1991, 17 per cent of the province's GDP came from the extraction and processing of natural resources while this figure had fallen to 14 per cent by 2005 (BC Stats, 2006b: 5). In 2005, BC's service sector accounted for 75 per cent of provincial GDP and 79 per cent of total employment (ibid., 6). On the other hand, BC's exports exhibit one of the classic characteristics of a resource hinterland: close to 90 per cent of its resource exports are either unprocessed or semi-processed goods. Coal, lumber, and natural gas are the three leading exports and they fall into that category. Prices for coal and natural gas were at all-time highs in 2007 while lumber prices were near the bottom of their price cycle.

Clearly, British Columbia is in a state of change, the leading edge of which is in the Vancouver/Victoria urban cluster, where high-technology and tourism services are concentrated. Beyond that cluster, the resource economy remains strong. Paradoxically, the industrial structure of this resource-rich province is more oriented to service activities than to primary employment. Over the years, the percentage of workers in the primary sector has declined for two reasons. First, the actual numbers have dropped. From 2003 to 2005, for example, the number of primary workers slipped from 79,000 to 77,000 (ibid., 4). Second, the total number of workers has increased: over the 2003–5 period, from 2,016,000 to 2,131,000. During this time span, the greatest rate of increase—40 per cent—was in the construction industry, driven

Table 7.1 Employment by Industrial Sector in British Columbia, 2005

Industrial Sector	BC Workers (000s)	BC Workers (%)	Ontario Workers (%)	Differences by percentage points
Primary	76.2	3.6	2.0	1.6
Secondary	376.5	17.7	23.6	–5.9
Manufacturing	198.2	9.3	16.6	–7.3
Tertiary	1,677.8	78.7	74.4	4.3
Total	2,130.5	100.0	100.0	

Source: Statistics Canada (2006c). Adapted from Statistics Canada website http://www40.statcan.ca/l01/cst01/labor21c.htm, Searched 12 September 2006.

by the Winter Olympics and an unprecedented housing boom. Resource extraction efforts, however, have not decreased, but firms have increased labour productivity by substituting machines for labour.

Manufacturing in British Columbia is based on the processing of forest and other primary resources. By value, just over half of all exports from British Columbia are forest products, while food items contribute another 11 per cent. Coal, grain, and lumber exports have transformed the Port of Vancouver into Canada's major shipping centre. Most resources, however, are exported in a raw or semi-processed form, and the key to expanding the manufacturing sector in British Columbia lies in greater processing of its resources, especially to the final product stage. Such diversification would greatly increase both the value of production and the size of the labour force. Yet, the simple fact is that the value of manufacturing has increased slowly. In 1999, the value of manufacturing was $37 billion; in 2005 it had reached $43 billion worth of goods.

In 2005, the tertiary sector accounted for 78.7 per cent of employment in BC (Table 7.1). Not only does service employment in BC exceed the Ontario average by 4.3 percentage points but most subsections within this sector outstrip their Ontario counterparts. Employment in transport, trade, finance, and service

Table 7.2 Comparison of Ontario and British Columbia Industrial Structures, 1995–2005

Sector	Ontario % 1995	Ontario % 2005	Percentage Difference	BC % 1995	BC % 2005	Percentage Difference
Primary	3.0	2.0	–1.0	4.7	3.6	–1.1
Secondary	23.6	23.6	0.0	18.1	17.7	–0.4
Tertiary	73.4	74.4	1.0	77.2	78.7	1.5
Total	100.0	100.0		100.0	100.0	
Workers (000s)	5,231	6,398	1,167 or 22%	1,762	2,131	369 or 21%

Sources: Statistics Canada (1996b, 2006c).

indicates the relative strength of the service industry. The service sector owes its strength to several factors, including British Columbia's growing domestic market, strong participation in world trade, a vibrant tourist industry, and its **producer services** firms.

The evidence is clear—British Columbia's economy is changing, but growth in the manufacturing sector remained elusive until recently. In 1982, a leading Canadian geographer, John Bradbury, saw BC as a resource hinterland that also had several characteristics common to a core region. But as Bradbury pointed out, BC lacked a traditional manufacturing sector centred on textiles that provided jobs for unskilled and semi-skilled workers. In the past two decades, however, BC has experienced growth in high technology. As the front edge of the structural shift taking place in the manufacturing industry across Canada, high-tech firms are playing a greater role in BC's economy and are providing jobs for highly skilled workers. High technology involves cutting-edge research in the manufacture of electronics, telecommunications equipment, and pharmaceuticals. For example, Ballard Power Systems Inc. is a world leader in the development of fuel cells that produce electrical power with virtually no pollution. BC's high-technology industry ranked third in 2005 behind forestry and construction, with nearly 65,000 employees.[4]

No doubt, manufacturing is a fundamental element in Friedmann's theoretical version of a core region (see Chapter 1, especially Vignette 1.4). But times are changing and the traditional manufacturing activities dependent on unskilled and semi-skilled workers are in trouble because such activities as textile manufacturing are shifting offshore to countries where wages are much lower. High-tech manufacturing employing highly skilled workers is the wave of the future for industrial countries like Canada. Most old-style manufacturing in Canada is located in Ontario and Québec. As discussed in Chapters 5 and 6, these core regions are losing manufacturing firms and jobs. The good news for Ontario and Québec is that most high-technology firms are located in these provinces; the bad news is the mismatch between the skill level of those losing their jobs in low-end manufacturing and the labour needs of high-tech firms.

Tourism: A Growing Sector in BC's Economy

Tourism is a key and growing element in the service sector and continues to draw more tourists to areas of BC besides Vancouver and Victoria, which are now the main tourist destinations. As noted earlier, the 2010 Winter Olympics will draw millions of tourists to the Games and in the years to come.

The scale of the industry is revealed in the fact that, in 2005, $9.7 billion was spent by 23 million overnight visitors to the province (BC Stats, 2006b). Given BC's scenic landscapes and many parks, tourism has enormous potential for expansion. One example is BC communities tapping into the Alaska cruise ship business. In 2006, cruise ships travelling to Alaska carried nearly a million tourists (VanderKlippe, 2007b). Most cruises begin in Vancouver and sail north through BC's spectacular Inland Passage where whales and grizzly bears are often sighted. From Vancouver, the cruise ships next stop in Juneau, Alaska. A new and distinctive twist for the cruise ships is to stop along the BC coast to learn about local Indian tribes and their cultures. The world's first Indian-themed cruise ship terminal, the Wei Wai Kum Terminal, is located near Campbell River on Vancouver Island. This recently constructed terminal is owned and operated by the Campbell River First Nation. It cost $24.5 million, with principal funding from the federal government and lesser amounts from the province and local governments, including the Campbell River First Nation (Indian and Northern Affairs Canada, 2007). Passengers visit a traditional village complete with totem poles and a Big House, both of which provide a window into Laichwiltach history and culture. The first of four summer visits took place on 5 June 2007 with Indian drummers greeting passengers of the *Seven Seas Mariner* cruise ship (ibid.).

British Columbia's Wealth

The key to British Columbia's past economic prosperity has been the province's natural resources and international trade. In 2005, exports worth $34 billion and imports worth $35 billion passed through BC ports. Four countries, the US, China, Japan, and South Korea, account for over 80 per cent of exports and imports moving through BC ports (Table 7.3).

Trade and the Asian-Pacific Gateway

The **Asian-Pacific Gateway and Corridor** is fundamental to Canada's and BC's future economic well-being. Back in the 1980s, Peter Nemetz (1990) foresaw that BC and the rest of Canada could reap many economic benefits through trade with the growing industrial giants in Asia. Within the geopolitical world, BC has a decided advantage because its ports have the shortest sea links to Asia. However, since increasing amounts of this cargo come in the form of containers, port infrastructures to handle **container** shipment are in short supply. According to the BC government, container traffic to all west coast ports is forecast to rise a staggering 300 per cent by 2020 (British Columbia, 2007). In 2004, the Port of Vancouver had the capacity to handle 1.7 million TEU of container cargo (TEU refers to containers measuring 20 feet), Montréal, 1.2 million TEU, Halifax 0.5 million, and the Fraser River 0.25 million (Padova, 2004). The Asian-Pacific Gateway and Corridor strategy, a largely federal program, aims at expanding BC's transportation system, including its container port facilities. One goal of this strategy is to increase container traffic imports from the Pacific Rim

Artist's rendering of the container facility at Prince Rupert, scheduled to open in the fall of 2007. (CN, 2007)

Table 7.3 Trade through British Columbia, 2005

Exporting Country	$ million	%	Importing Country	$ million	%
US	21,900	64.2	US	14,120	40.1
Japan	4,168	12.2	China	7,501	21.3
China	1,376	4.0	Japan	4,605	13.1
South Korea	1,172	3.4	South Korea	2,291	6.5
Other countries	5,473	16.2	Other countries	6,699	19.0
Total	34,089	100.0	Total	35,216	100.0

Source: BC Stats (2006b). Reprinted by permission of the publisher.

countries from 2 million TEU in 2005 to 9 million TEU in 2020 by expanding port facilities at Vancouver and by constructing a container port at Prince Rupert (British Columbia, 2007; Prince Rupert Port Authority, 2007).

Natural Resources: The Original Basis of BC's Wealth

Natural resources have led the province's growth and continue to contribute to its economic well-being. These resources include: fishing, mining, hydroelectric power, and especially forestry, and they have driven growth in both the primary and secondary sectors. Moreover, a portion of the tertiary sector, the tourism industry, relies largely on the province's natural beauty and resources through ecotourism, hiking, skiing, sports fishing and hunting, and sailing along BC's fjorded coast.

Like Ontario, Québec, Western Canada, and Atlantic Canada, BC can be divided into two parts—a core and a periphery. As with all cores, secondary and tertiary activities are concentrated in the BC core, especially in the Greater Vancouver and Victoria areas. Beyond the BC core, most economic activities focus on primary industrial efforts, especially those associated with the forest and fishing industries. As in other hinterlands, the economic strength of primary industries is declining with the exception of the natural gas activities in northeastern BC.

Fishing Industry

British Columbia, like Atlantic Canada, has an important fishery. The BC fishery, as a renewable resource, ranks fourth in value of production among resource industries behind mining (including natural gas), forestry, and agriculture. Fish processing plants employ 25,000 full-time and part-time employees. More than 80 species of fish and marine animals are harvested from the Pacific Ocean, freshwater bodies, and aquaculture areas. Salmon is the most valuable species, followed by herring, shellfish, groundfish, and halibut. In 2005, the landed value from the sea reached $618 million, while their processed value was over $1 billion (British Columbia, 2006). Most salmon today—ironically, Atlantic salmon—comes from fish farms. In 2004, the wholesale value of farm salmon was $294 million, compared to $219 million for wild salmon (Table 7.4). In recent years, excessive harvesting has reduced fish stocks and, in turn, the number of landings (fish caught). At the same time, production from fish farms has made fresh salmon available to the consumer throughout the year, thus popularizing the product. Environmentalists, however, have long expressed concern about fish farming—estuaries and inlets are subject to pollution from waste; fish meal used as feed in aquaculture results in the disruption of food chains and depletion of species in other parts of the globe; and predators break through the netting where farmed fish are kept, thus permitting their escape to the wild and the unknown effects of breeding with wild species of salmon. In January 2004, fears that salmon from fish farms contain high levels of PCBs and other pollutants that could increase the risk of cancer surfaced in the media and these fears may affect sales of such salmon (Mittelstaedt, 2004).

Unfortunately, BC shares another similarity with Atlantic Canada—overexploitation of fish stocks, particularly the very valuable salmon stocks. Pressure on the fish stocks, especially the salmon stocks, comes from four sources—Canadian commercial fishers, American commercial fishers, the Aboriginal fishery, and the sports fishery. All want larger catches.

Management of the salmon fishery falls under the Pacific Salmon Treaty, which was signed by Canada and the United States in 1985 and sets long-term goals for the sustainability of this resource. The Pacific Salmon Commission, formed by the governments of Canada and the United States to implement the Pacific Salmon Treaty, does not regulate the salmon fisheries but provides regulatory advice and recommendations to the two countries. In Canada, the Department of Fisheries and Oceans (DFO), in consultation with the Pacific Salmon Commission, sets annual quotas for salmon in Canadian waters (www.psc.org/).

West coast fish canneries, like this one at Sechelt, have long played a key role in British Columbia's economy. However, given the presence of American and Canadian fishers, as well as Aboriginal fishers who have special rights, the salmon fishery could suffer the same fate as the Atlantic cod fishery. (©Natalie Forbes/Corbis)

Salmon are migratory fish, so regulating salmon fishing is particularly challenging. Like other fish, they are common property until caught. This principle is based on the 'rule of capture'. Fishers therefore try to maximize their share of a harvest so no one else will take 'their' fish. The problem is complicated further because the Canadian government cannot regulate the 'Canadian' salmon stocks, i.e., those that spawn in Canadian rivers, because they migrate to American waters, where the American fishing fleet harvests them.

The net result is that the salmon stocks are threatened. This problem is commonly referred to as the **tragedy of the commons**. (See Chapter 9 for an account of the overexploitation of the northern cod.) In 2000, Ottawa tried to alleviate the pressure on salmon stocks by reducing the fleet of 4,500 fishing vessels by about one-third. However, this announcement sparked a strong reaction and little was accomplished. At the same time, Ottawa allowed Indian fishers, who have treaty rights to harvest fish for subsistence purposes, a share of the commercial stock (Vignette 7.3). In 1999 Ottawa successfully negotiated a new Pacific Salmon Treaty with the United States, which extends to 2010. Ottawa is able to exert some management of fish stocks in the Pacific Ocean because of its 200-mile fishing zone and because of its role in the Pacific Salmon Commission. Salmon fishing on the Pacific coast is regulated, based on the 1999 treaty, which determines the size of the catch taken by each nation.[5]

Salmon catches by fishing fleets fluctuated during the 1990s and early 2000s, not because of conservation measures that restrict the allowable amount of catches but simply because the number of salmon varied from year to year. Salmon (chinook, sockeye, coho, pink, and chum) spend several years in the Pacific Ocean before returning to spawn in the Fraser and other rivers. The principal Canadian rivers are the Fraser and the Skeena rivers. One of the largest harvests took place in 1997 when 487,000 tonnes of salmon were caught. Two years later, the figure was 17,000 tonnes! In 2003, the harvest was 38,400 tonnes (BC Stats, 2006b).

Ottawa, which is responsible for managing salmon stocks within Canada and for negotiating international fishing agreements, must deal with four perplexing and interrelated issues:

1. the natural cycle (up to five years) of salmon to spawn in rivers, migrate to the sea, and then return to the rivers to spawn again;
2. the harmful effects of the forestry and hydroelectric industries on salmon spawning grounds;
3. the division of the salmon catch among commercial, Native, and sports fishers;
4. the harvesting of 'Canadian' salmon by American fishers in international waters.

Management of fish resources is based on the estimated size of the fish stock. The problem of determining the size of the stock, and therefore the quota to be set for each year's catch, is complicated by the natural population cycle of the fish. For salmon stocks, this problem is complicated by their migratory life cycle, from spawn in Canadian rivers, such as the Fraser and Skeena, then spending most of their adult life in the Pacific Ocean, and finally returning to their original spawning ground to spawn and thereby begin the cycle for another generation. Population fluctuation is illustrated by pink salmon catches in the past. Over one 10-year period, the size of catch varied from a low in 1975 of 38,000 tonnes to a high of 108,000 tonnes in 1985. In 1994, an estimated 2.3 million fish did not return to BC rivers to spawn. What caused this decline is unknown, but a similar drop in 1995 in the number of salmon returning to the Fraser River forced Ottawa to close fishing on the Fraser. Several factors account for this decline:

- *Pollution of fish habitat*. Forestry companies have affected salmon habitat through their logging practices, which have blocked streams and thereby interfered with migration routes to spawning areas. Pulp and paper plants also discharged toxic wastes into rivers and the ocean.
- *Warming ocean temperatures*. With global temperatures increasing, the North Pacific Ocean temperatures may have risen above levels suitable for salmon. For instance, the so-called El Niño effect could provide a natural explanation for the declining salmon

Table 7.4 Fishery Statistics by Species, 2004

Species	Wholesale Value ($ millions)	Percentage
Farm salmon	294	26
Shellfish	221	20
Wild salmon	219	19
Groundfish	191	15
Herring	93	8
Halibut	90	8
Tuna	36	3
Other species	7	1
Total	1,131	100

Source: BC Stats (2006b). Reprinted by permission of the publisher.

stocks, but so far the El Niño effect on ocean temperatures, and then on salmon, is unsubstantiated by scientific evidence.

- *Overfishing*. The most likely explanation for the variation in salmon stocks is overfishing. With a combination of a larger fishing fleet and the application of new technology (radar and sonar equipment), the capacity to track and catch salmon and other fish has improved greatly. Since larger fish are normally caught in the fish nets, too few adult fish are left to reproduce and thus replenish the fish stocks.
- *High fish quotas*. Within Canadian waters, DFO is responsible for estimating the size of the fish stocks and, based on this information, sets fish harvest quotas. Since estimating the size of the fish stocks is a chancy business, being based on historic records and insufficient sampling of fish stocks, DFO estimates tend to be on the high side. Quotas are set from these estimates, though pressure from fishers and the Americans has resulted in some past quotas set too high.
- *The Aboriginal fishery*. First Nations peoples along the BC coast have always harvested fish for subsistence and cultural purposes. More recently, Indian fishers have sold some fish, mainly salmon, as a means of generating cash income. They are seeking a share of the fish quotas in order to participate more openly in the market economy. But how large a share should they receive? This is a difficult question to answer, yet an Aboriginal commercial fishery was created in 1992 (see Vignette 7.3). As modern treaties are concluded with BC's First Nations, the need for the Aboriginal fishery will disappear and be replaced by individual agreements with each First Nation. As part of the first two modern treaty agreements reached in BC, a share of the commercial fishery harvest was defined. Ironically, in regard to the most recent of these two agreements, the members of the Lheidli T'enneh First Nation voted not to ratify it.

The Aboriginal Fisheries Strategy rankled other commercial fishers, who argued that there should be but one commercial fishery. When DFO assigned commercial fishing licences and allocated fixed quotas of salmon to several First Nations along the Fraser River and then opened the annual salmon harvest to First Nations before it permitted fishing by the general commercial fleet, the reaction was swift and soon became a political issue. The commercial fishers argued that the 'advanced' fishing date for First

Vignette 7.3 Aboriginal Fisheries Strategy: A Temporary Arrangement?

First Nations people along the Pacific coast based their economy on the sea. After BC joined Confederation, fishing for subsistence was accepted, but commercial fishing was controlled by Ottawa. While Aboriginal fishers have obtained commercial fishing licences from DFO and participate along with non-Aboriginal fishers in the commercial fishery, First Nations did not have a right to sell fish. However, following the 1987 *Sparrow* decision in which the Supreme Court ruled that First Nations people have a right to fish commercially, Ottawa created an Aboriginal Fisheries Strategy (AFS) for three reasons: to respect the ruling of the Supreme Court of Canada; to develop an interim measure while treaty negotiations were underway; and to recognize and legalize the commercial salmon sales by First Nations fishers.

Nations was a violation of the equality guarantees in the Charter of Rights and Freedoms. In July 2006, Prime Minister Stephen Harper announced plans to do away with 'racially divided fisheries programs'. However, until each First Nation with an interest in commercial fishing achieves a land-claim agreement and therefore specific commercial fishing rights, the Aboriginal Fisheries Strategy is likely to continue. For instance, the Lheidli T'enneh Final Agreement involving the Carrier Indians near and within Prince George, signed on 30 October 2006, included a sub-agreement for the harvesting of 6,000 sockeye salmon for commercial purposes and 12,350 for food and ceremonial purposes (Cernetig, 2006). This treaty might have provided the template for other First Nations who have made a claim for a commercial fishery, so that eventually the Aboriginal Fisheries Strategy would be replaced by individual fishing agreements with each First Nation. As it turned out, however, on 30 March 2007 the Lheidli T'enneh, by a 123–111 vote, refused to ratify the agreement. Reasons for the failure of ratification included the ramifications of extinguishment of Indian title, taxation issues, lack of rank-and-file involvement in the negotiating process, and a multi-million dollar legal bill amassed over more than a decade of negotiations. As for the fish, in this zero-sum game the trick will be to share the salmon stocks 'fairly' among all users and still have a sustainable resource.

Mining Industry

The mineral wealth of British Columbia is found in both of the province's physiographic regions—the Cordillera and the Interior Plains. Each has a distinct geological structure. The Cordillera contains a wide variety of minerals, while the Interior Plains is part of the Western Sedimentary Basin, which contains petroleum deposits. Often, the challenges facing the mining industry are:

- dealing with fluctuating world prices;
- transporting bulky mineral and petroleum products to BC ports for shipment to world markets.

In 2004, the value of total mineral production was $10.3 billion, more than double the 1999 figure of $4.7 billion; estimates for 2005 reach $13 billion, with natural gas accounting for over half of the value of mineral production (BC Stats, 2006b). The explanation for this remarkable rise in value of mineral production is due to the higher prices for most minerals, including coal, copper, and natural gas. By value, natural gas accounted for 56 per cent, followed by coal and copper at 11 and 10 per cent respectively. Mining and mineral processing employ about 3 per cent of the labour force but yield nearly 25 per cent of the value of primary production. Because nearly all of the mineral production is consumed outside of the province, the fortunes of the mining industry are largely determined by external markets and global prices. The outlook for the natural gas and coal industries is bright because of the strong foreign demand. Both commodities are riding the high cycle of energy prices. Some 20 years ago, the sudden drop in world demand saw the closure of the Quintette mine in northeastern BC. In this case, the weak demand was due to reduced shipments to Japanese steel mills. This boom-and-bust pattern of development is common to hinterlands and illustrates both the strength and vulnerability of BC's hinterland economy.

As with other resource development, the capital investment in mining is enormous. For that reason, only very large companies, often multinational firms, can venture into such developments. The Northeast Coal Project is one example. In the late 1970s Denison Mines and Teck Corporation were the major mining companies behind this $4.5 billion construction project (Bone, 2003: 140–1). With a 14-year contract to sell coal to Japanese steel companies, banks and other financial institutions were happy to supply much of the capital (about $2.5 billion) to Denison and Teck. The federal and provincial governments provided $1.5 billion of the $4.5 billion to upgrade the CNR line to Prince Rupert, to extend a BC Rail line to Tumbler Ridge, and to build a coal terminal to load ships that would carry the coal to Japan. The two mining companies contributed about $500 million. Both governments provided another $1 billion to build the coal town of Tumbler Ridge. These agreements were signed in the late 1970s, when the world economy was expanding and the demand for coal was increasing. The Northeast Coal Project comprised Quintette Coal Ltd at Tumbler Ridge, Teck Corporation at Bullmoose, and Gregg River Coal Ltd at Gregg River.

Like all large-scale construction projects, several years were required to complete the undertaking. Production began in 1984. By then, the world demand and price for coal had peaked and a slow but steady decline set in as the global economy slipped into a recession. At that point, the Japanese steel mills no longer needed so much coal and the price set with the Northeast Coal Project mines was no longer attractive. By the time coal was first produced at the Northeast Coal Project, the world spot price (a price charged for goods available for immediate delivery) for coal was 10 per cent lower than the agreed-upon, contract price. Eventually, the price for coal was renegotiated, but Denison Mines went bankrupt during the process. Since then, Teck Corporation has taken over the coal-mining operation. Declining demand and prices forced the company to close its Quintette mine in 2000, leaving the community of Tumbler Ridge with no economic base. Efforts to convert the community into a retirement and 'second-home' town faced an uphill battle because of its remote location (Halseth and Sullivan, 2002). Within a few years, however, a return to strong global demand for coal and a return of high coal prices turned the economic page for Tumbler Ridge. By 2003, coal mining near Tumbler Ridge again became economically feasible. With this positive news and associated construction, the town has seen its population increase from 1,932 in 2001 to 2,526 in 2005 (BC Stats, 2006a). In 2006, coal production began at the nearby Wolverine coal mine.

The importance of the export market for mineral products called for an expanded rail system and rail-to-ship loading facilities. Transportation costs have been reduced by unit trains and bulk-loading facilities. Unit trains consist of a large number of ore cars, sometimes over 100, pulled by one or more locomotives. The terminal of Roberts Bank just south of Vancouver is designed as a large bulk-loading facility where coal in rail cars is dumped and then the coal is moved by conveyor belts to the ship's hold. These ships must be moored in deep water, which necessitated building a long causeway to reach the ships.

Hydroelectric Power

Hydroelectric energy is a renewable energy source dependent on the hydrologic cycle of water, which involves evaporation, precipitation, and the flow of water due to gravity. British Columbia, with abundant water resources and a geography that provides many opportunities to produce low-cost energy, is the second largest producer of hydroelectric power in Canada. In 2005, BC Hydro produced 60.4 gigawatts of electric energy while Hydro-Québec accounted for 160 gigawatts (Natural Resources Canada, 2006a). Within the Cordillera, a combination of elevation, steep-sided valleys, and steady-flowing rivers provides ideal conditions for the construction of hydroelectric dams. The Columbia, Fraser, and Peace rivers offer many excellent hydroelectric sites. In addition, heavy precipitation and meltwater from the mountain snowpack ensure a regular and abundant supply of water. Major hydroelectric developments have taken place on the Columbia and Peace rivers as well as on the Nechako River, a tributary of the Fraser. Minor developments have occurred on Vancouver Island.

In the early days of hydroelectric development, small-scale hydroelectric projects were established near Vancouver and Victoria. These two cities are the major markets in the province for electricity. Hydroelectric dams were also built near smelters that required vast amounts of power to operate. For example, early in the twentieth century, power was generated from privately owned dams on the Kootenay River for the lead-zinc smelter at Trail.

After World War II, public and private companies began to undertake megaprojects. Among these giant industrial construction efforts, three—one on the Columbia River, another on the Peace River, and the third on the Nechako River—had enormous impacts on the economy. They all involved the harnessing of water power from the province's rivers to generate low-cost electrical power, but they also flooded valuable farmland and Indian lands and led to the loss of salmon spawning grounds. Such construction projects were considered major engineering feats. In 1951, Alcan completed the construction of the Kenney Dam across the Nechako River, impounding the water draining from an area more than twice the size of Prince Edward Island. In 1968, BC Hydro built the W.A.C. Bennett hydroelectric dam on the Peace River, creating Williston Lake, the largest freshwater body in the province. From this giant reservoir, vast quantities of water flow through the turbines at the power station, generating much of British Columbia's electrical power and creating a surplus of power for sale in the US.

The most ambitious of these early megaprojects was the one developed on the Columbia River. A combination of rising demand for power in the state of Washington and innovations in the transmission of electricity over long distances enabled the development of water power along the Columbia River in British Columbia. In the 1960s, dams were constructed on the Columbia River to increase the flow of water to the hydroelectric power plant at Grand Coulee in the United States (Vignette 7.4). Initially, BC Hydro sold its share of the electrical power generated at Grand Coulee to its counterpart in the state of Washington. In 1997, this power was no longer needed in the US and British Columbia was faced with a huge surplus of electrical power. Similar to the Québec government's efforts to stimulate industrial development in that province, Victoria has now begun offering this surplus power at a reduced price to firms that locate in British Columbia. So far, no large firms have expressed an interest, but Victoria has used this surplus as part of a settlement it reached with Alcan over the cancellation of one of its hydroelectric projects (see below).

Alcan in Northwest British Columbia

The Alcan hydroelectric development in British Columbia illustrates the advantages and pitfalls of megaprojects. The Aluminum Company of Canada (Alcan) is one of the giants in the world aluminum industry. To keep its production costs low, Alcan is attracted to sites where low-cost hydroelectric power can be developed and then used by its smelting plants to reduce bauxite (a clay-like mineral) into alumina (the chief source of aluminum) and then aluminum. In Canada, water resources are under provincial jurisdiction, so Alcan seeks long-term arrangements

Vignette 7.4 The Columbia River Treaty

In 1961, Canada and the United States agreed to co-operate in the development of the Columbia River for the production of hydroelectric power. Canada undertook to construct three dams for water storage in the Canadian portion of the Columbia River Basin and to operate them to produce maximum flood control and power downstream. In return, the United States would return half the power generated to Canada and pay for half the value of the flood protection for property in the United States. The three dams constructed as a result of this agreement are the Duncan (1967) north of Kootenay Lake, the Hugh Keenleyside (1968) on the Columbia River, and the Mica (1973) north of Revelstoke. The major hydroelectric power station is located in the state of Washington at Grand Coulee. Before the construction of the Grand Coulee Dam, the Columbia River was one of the world's great spawning grounds for salmon.

with provincial governments to develop the water power for its smelting plants. In turn, provincial governments are attracted to megaprojects because they help to develop a region and employ large numbers of people.

In the late 1940s, Alcan proposed to build an aluminum complex in northwest British Columbia. After obtaining the rights to the waters of the Nechako River for 50 years from the provincial government, Alcan began construction of the first phase of this complex in 1951.[5] Three years later, Alcan built a new town called Kitimat, a power plant at Kemano, and a dam on the Nechako River. To produce hydroelectric power, Alcan dammed the Nechako River, thereby reversing its flow westward to the Pacific Ocean near Kemano. To accomplish this task, a tunnel had to be drilled through the Coast Mountains. At Mount DuBose, a 16-km tunnel allowed waters from the Nechako reservoir to flow to Kemano where the power station is located. These waters plunge 860 m from the western end of the tunnel to drive the turbines at Kemano. Electricity is transmitted about 80 km to Alcan's aluminum smelter at Kitimat. Until 1999, Alcan maintained a permanent workforce at Kemano. In that year, Alcan announced that its Kemano powerhouse will be operated and maintained by crews from Kitimat working in Kemano on rotating shifts.

When this project was first planned, concerns about environmental and social impacts of industrial projects were not given much attention. Both the provincial and federal governments bent over backwards to assist Alcan in what was considered one of the great engineering and construction feats of the twentieth century. The flooding of the Cheslatta Indian Reserve, which resulted from the Alcan project, was not considered a significant social impact and the Department of Indian Affairs quickly arranged for a relocation of this tribal settlement. As for the sockeye salmon spawning grounds that were affected, no protest was raised. Three reasons account for the lack of concern: (1) the general public was convinced that industrial growth was 'good' for the province; (2) the public and the fishing industry were not aware that the destruction of the Nechako spawning grounds could have a major impact on salmon stocks; (3) the environmental movement and Aboriginal groups had not yet emerged as forces capable of challenging such projects.

In 1989, Alcan began the second phase of its hydroelectric complex—the Kemano Completion Project. Estimated costs were $1.3 billion. The plan was to enlarge its hydroelectric facility by boring a second tunnel through Mount DuBose, thereby diverting more waters of the Nechako reservoir westward. This time, both the Carrier-Sekani Tribal Council (whose members claim the Nechako River as part of their land claim) and environmental groups, spearheaded by the Rivers Defense Coalition, openly challenged the project, claiming that the social and environmental costs were too great. Even so, the project was approved by the federal government without an environmental review. By 1991, the project was half-completed at a cost of about half a billion dollars, but opposition from environmental groups and Aboriginal leaders had grown so strong that the federal government was forced to order an environmental inquiry. Alcan ceased its construction efforts, hoping to restart work after federal approval.

In 1995, the provincial government cancelled its approval for Alcan's Kemano Completion Project. Alcan threatened to sue. Later, the British Columbia government agreed to compensate Alcan for the money spent on construction by selling the company electricity at a very low price (BC was able to make use of the surplus power it had). Alcan claimed to have spent $500 million towards the construction of the $1.3 billion project. In the arrangement reached in 1997, Alcan will receive electrical power for its proposed smelter in two ways: (1) 115 MW a year from BC Hydro at a price pegged to the price of aluminum on the London Metal Exchange until 1 January 2024; and (2) an additional 60 MW a year from BC Hydro, at 1997 prices, until 1 January 2024 (Cernetig, 1997: 1). At the 1997 price for power, Alcan has an extremely inexpensive source of electricity. By 2006, with electricity prices more than double the 1997 price, Alcan had made a very profitable deal.

With the global economy growing at a rapid rate, the commodity cycle again began to swing upward. By 2003, world demand for aluminum reached record highs. With such encouragement, Alcan dusted off its plans for its BC operations. In August 2006, Alcan decided to invest US$1.8 billion in the modernization of the Kitimat smelter in order to increase production. A key element to this modernization project is the Alcan and BC Hydro agreement for BC Hydro to purchase power from Alcan at the same rate as for new power projects. However, BC municipalities demanded that the rate should be much lower because Alcan's generating station is over 50 years old and therefore Alcan's costs of producing electricity are much lower than those for new plants. The BC government is reviewing the Alcan/BC Hydro agreement.

Key Topic: Forestry

The forest is British Columbia's greatest natural asset. With just over 60 per cent of the province covered by forests, British Columbia contains about half of the nation's softwood timber. With such a dominant position, British Columbia is easily Canada's leading supplier of wood products for construction and finishing. With easy access to ports, forest firms are ideally located to supply the needs of the US and other global markets.

Climate and topography have divided this vast forest into two distinct regions—the rain forest of the coast and the boreal forest of the interior. Size, age, and species vary by these two regions Within the rain forest, the mild, wet climate allows trees to grow to great heights for hundreds, even thousands of years. Old-growth trees that are at least 250 years old are located along the Pacific coast, where, because of heavy precipitation throughout the year, fires are rare. The major species harvested is the western red cedar. In the interior, where the climate is much colder in the winter and hotter in the summer, and certainly drier, trees are smaller and forest stands less compact. Trees have a shorter lifespan (120 to 140 years). Forest fires are a constant threat and, more recently, infestations caused by the pine beetle (discussed above) have become a significant environmental and commercial problem.

British Columbia's forest consists almost entirely of coniferous **softwood forest**. Within Canada, British Columbia accounts for just over half of the total logging output. The main species harvested are lodgepole pine, spruce, hemlock, balsam, Douglas fir, and cedar (Table 7.5). The processing of timber into lumber, pulp, **newsprint**, paper products, shingles, and shakes supports a major manufacturing industry in British Columbia. British Columbia leads in the production of such wood products as lumber and plywood, while Ontario and Québec produce a greater proportion of pulp and paper products. In 2005, the timber harvest was 83.4 million tonnes, with the Interior Forest region accounting for 70 per cent (BC Stats, 2006b).

Forest Regions

British Columbia's forest is far from homogeneous, largely because of varying climatic zones and the varied relief in the Cordillera. The differences in the type of forest have strong implications for forestry. For example, trees along BC's moist west coast grow much more quickly than those in the dry interior. BC's forest lands are divided into two major regions: the Coast Rainforest and the Interior Boreal Forest. Within the Interior Forest, there are four subregions: the Northern Forest, the Nechako Forest, the Fraser Plateau Forest, and the Columbia Forest (Figure 7.6). Within each of these subregions there is great variation due to the differences in local growing conditions, which are affected by precipitation and length of growing season. Other factors are elevation, soil conditions, and topography.

The Coast Rainforest is the most luxuriant coniferous forest in Canada. With its wet and mild marine temperatures and abundant rainfall, this is one of the most densely forested areas of North America. The key species are Douglas fir, western red cedar, and western red hemlock. Under ideal conditions, mature

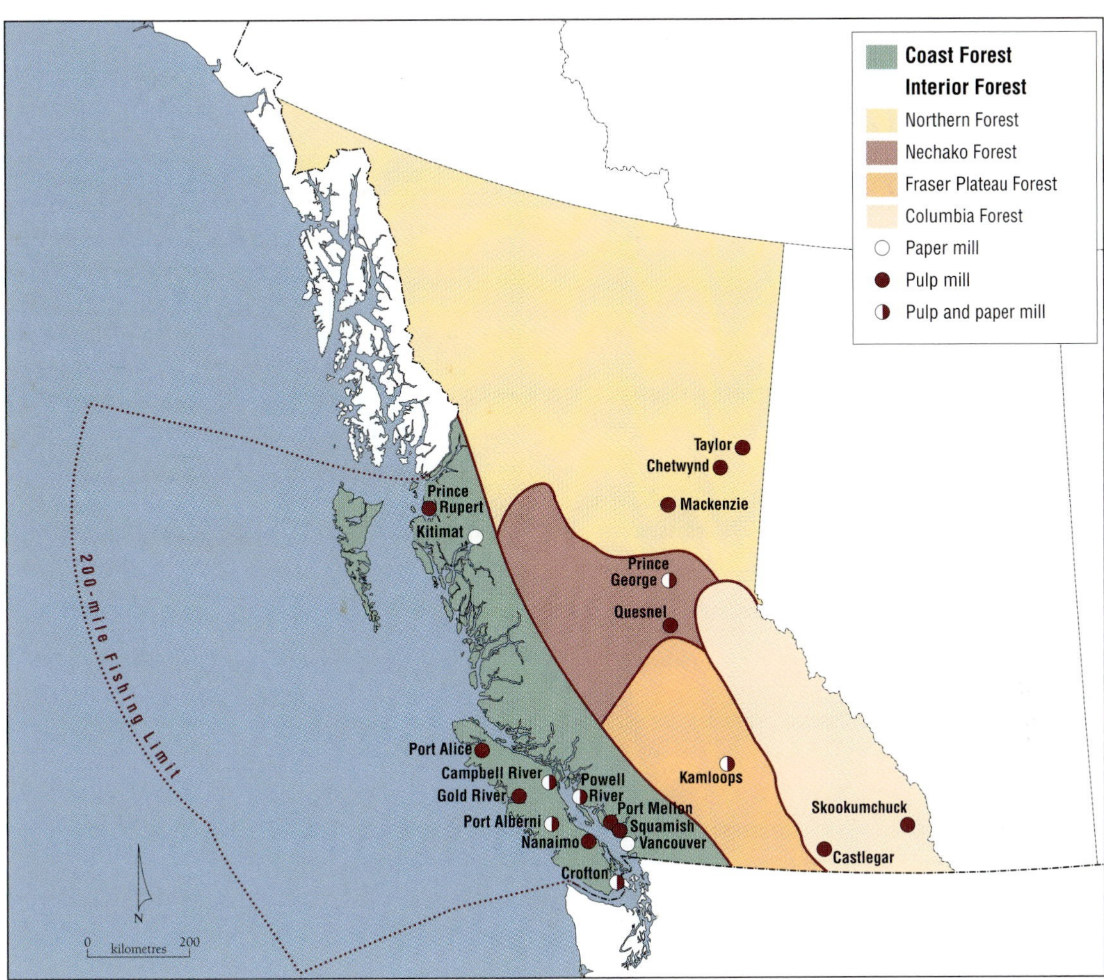

Figure 7.6 Forest Regions in British Columbia. The two principal regions are the Coast Rainforest and the Interior Boreal Forest. The Interior Forest is subdivided into four areas that reflect variations in growing conditions. (Further resources: Student website, National Atlas section, Maps 21, 22, and 23. Website instructions are found on p. xx.)

cedar and hemlock trees reach 45 m in height and about 1 m in diameter. The Douglas fir is an even larger tree. Mature stands may average 60 m in height and over 1.5 m in diameter. Logs from the mature stands (known as old growth) are highly valued for lumber and plywood, while logs from immature stands (known as secondary growth) are used for pulp and paper.

Inland from the Coast Rainforest, the climate changes from a wet marine climate to a semi-arid one. At this point, the Interior Boreal Forest begins. However, differences exist within the Interior Forest. The Fraser Plateau subregion is an open woodland with much smaller trees than those in the Coast Forest. The controlling factor is the dry climate. Trees in the Fraser Plateau subregion must be able to cope with drought conditions. Ponderosa pine and lodgepole pine are the most common species. They often attain heights of 25 m and a diameter of 1 m. These trees are usually converted into lumber.

The Nechako Forest subregion, which lies to the north of the Fraser Plateau, receives more precipitation and experiences lower summer temperatures. The net result is more moisture for tree growth. Consequently, the forest cover is denser than that in the Fraser Plateau. A common species in the Nechako

Forest is the Engelmann spruce, which can attain heights of 40 m and diameters of 60 cm. Timber from this region is processed by pulp and paper mills at Prince George.

The Columbia Forest subregion lies in the easternmost area of southern British Columbia. Again, a dry climate limits tree growth. Because of widely varying terrain, the forest cover is very heterogeneous. Western red cedar and western hemlock are common species, though stands of ponderosa pine and even Douglas fir are also found in this region. Large timber is sent to sawmills, while the smaller logs are used for pulpwood.

The Northern Forest subregion is the most remote of BC's forests. High transportation costs to ship wood to world markets hinder logging. Tree growth is hampered by cool growing conditions and poorly drained land. Large blocks of land that contain muskeg are either devoid of trees or have trees with little commercial value.

Forest Industry

Canada is the world's leading exporter of forest products, accounting for 21 per cent of world trade (Dufour, 2007). Most forest exports—certainly softwood lumber—come from British Columbia. In fact, forest exports are so important to BC's economy that they make up one-quarter by value of all its exports. Thus, fluctuations in total BC exports by value are affected by forest exports. For example, since 2004 softwood lumber exports have declined, mainly due to falling prices. The result is a slight decrease by value in BC exports (Schrier, 2007).

The exploitation of its **commercial forests** remains one of the mainstays of the British Columbia economy. However, the dominant role of forestry has diminished as other sectors of the BC economy have expanded. For instance, the service industry far outranks forestry in number of workers. Still, the value of forest production and exports remains a leading economic force. In 2005, enormous forest reserves supported a major industry with an output valued at over $18 billion (BC Stats, 2006b). Just over 95 per cent of the 60 million hectares of forest lands were owned by the British Columbia government and allocated to forestry companies on long-term leases.

The forest industry has evolved by extending its operations into the interior of BC, by consolidating its operations, and by developing foreign partnerships. At the beginning of the twentieth century, the forest industry consisted of small-scale logging and sawmilling operations and was concentrated along the Pacific coast. Vancouver was a major sawmilling centre. Gradually, logging, sawmilling, and pulp mills began to locate in the interior. In 1956, the completion of the Pacific Great Eastern Railway from North Vancouver to Prince George was a key element in opening the interior to commercial logging, sawmills, and pulp/paper mills. By the 1980s, almost all of British Columbia's vast commercial timberlands were leased to forestry companies. Today, sawmills in the interior account for about three-quarters of the lumber production. Pulp and paper mills are mainly along the coast and in the Prince George area of the interior. Wood manufacturing plants are heavily concentrated in the Lower Mainland.

In today's competitive world, the forest industry has had to consolidate its operations to reduce costs. All firms had to achieve economies of scale to survive. Early casualties were the pulp mills at Gold River and Port Alice. Prince Rupert's pulp mill, Skeena Cellulose, is another marginal operation that has been struggling to survive for a number of years. Twice the provincial government has rushed to the rescue (once in the 1970s and again in the 1990s). As an old, less efficient mill, it was unable to survive. The downturn in the forest industry was particularly severe in 2005–6 when eight mills closed, which had a devastating impact on small communities such as Malakwa, Courtenay, Clearwater, Coquitlam, Ladysmith, and Squamish.

This process of consolidation is not new. MacMillan Bloedel was formed through the merger of three smaller forestry companies (the Powell River Company in 1951, the Bloedel Stewart Welch Company, and the H.R. MacMillan Company in 1959). In 1999, Weyerhaeuser Canada purchased MacMillan

Bloedel, thus becoming the largest producer of softwood lumber and **market pulp** as well as a leader in packaging. Today, a few large forestry companies account for over half of the timber harvest and three-quarters of the processing of wood products—Weyerhaeuser, for example, holds about 15 per cent of the leased forest area and, like other firms, has its headquarters in Vancouver.

Another feature of BC's forest industry is 'foreign partnerships'. Since the 1970s, foreign firms, especially Japanese firms, have invested heavily in the forest industry as minority partners. This shift towards multinational corporations is a common feature of BC's resource industries. The rise in multinational ownership is due to two factors: (1) the huge capital investment required to purchase minority partnership, and (2) foreign owners wishing to secure committed sources of raw materials.

While logging is both highly efficient and mechanized, the practice of clear-cut logging makes the goal of a sustainable industry problematic. According to a report, *Clear-cutting Canada's Rainforests*, produced by the David Suzuki Foundation, clear-cutting remains the dominant system of harvesting the old-growth trees in British Columbia's rain forest. Two impacts of clear-cutting are blowouts (strong winds knock over isolated trees) and landslides (moderate to steep slopes are subject to mud and land slides during periods of heavy rain). (Al Harvey/The Slide Farm)

Global Economic Cycles

As with other export-oriented producers, global economic cycles affect the forest industry. Since the US market is the major destination for Canadian forest products, fluctuations in US demand cause cyclical price swings; in extreme cases, these swings are of a 'boom-bust' nature. In 2005, BC's timber harvest reached a record high of 83.4 million tonnes. This huge harvest had a ready market in the booming US house construction industry. The value of this harvest, including processing, was $18.2 billion. Led by lumber and pulp, forest exports were valued at $14 billion (Table 7.6). In 2006, the US housing industry took a significant downturn, causing prices for Canadian softwood lumber to drop to half of the 2005 prices. Plant closures and staff layoffs swiftly followed. This sudden and unpredictable shift from high demand to low demand illustrates the vulnerability of the forest industry and the concentration of economic hurt and instability on single-industry forest communities. Today, the Canadian industry has to deal with a strong Canadian dollar and rising energy costs (Dufour, 2007).

The forest industry accounts for nearly 100,000 jobs and another 200,000 indirect jobs. Almost 30,000 people are employed in logging operations, 26,000 in wood processing, and 11,000 in pulp and papermaking. These numbers for direct employment do not include the many workers who rely indirectly on forestry, such as tree planters, public officials, and furniture makers.

One recent societal change having an affect on the forest industry is the paperless means of communications that has emerged as the public makes greater use of computers and text-messaging. Newspapers now post their news on the Internet and, in some cases, charge for the privilege. Across BC and the rest of Canada, the impact has led to plant closures and massive layoffs. Since 2003, 80 sawmills, particle board plants, and pulp and paper mills have closed—some temporarily and others permanently. Older, less efficient plants are gone and most of these mills were located in Ontario and Québec. Layoffs have been substantial—6,400 in Ontario, 5,200 in Québec, and 2,800 in British Columbia (Koven, 2006). Analysts have been quoted as saying that more

Table 7.5 Timber Harvest by Species, British Columbia, 2005

Species	Volume (million cubic metres)	Per cent
Lodgepole pine	37.6	45.1
Spruce	11.9	14.3
Douglas fir	10.9	13.1
Hemlock	7.7	9.2
Cedar	6.0	7.1
Balsam	5.1	6.1
Other	4.2	5.1
Total	83.4	100.0

Source: BC Stats (2006b). Reprinted by permission of the publisher.

operations are 'on the bubble' and could also close, and there are increasing calls for governments to take the lead in developing a national strategy (with industry) to mitigate the impacts of the current industry adjustment.

Free Trade and Forest Exports

The major markets for BC forest products lie outside of the province. Like most Canadian resource industries, the economic well-being of BC's forest industry depends on export markets. BC's principal markets are the United States, which absorbs about half of its production, with Japan and the European Union accounting for 25 and 15 per cent, respectively. Given the dominance of the American market, one of the challenges facing the forest industry in BC extends beyond the province's borders and relates to trade relations between

Table 7.6 Forest Product Exports from British Columbia, 2005

Commodity	Value in $ millions	Per cent
Softwood lumber	6,271	44.6
Pulp	2,500	18.3
Paper and paperboard	1,209	8.6
Wood products	1,004	7.1
Panel products	741	5.3
Newsprint	624	4.4
Plywood	460	3.3
Cedar shakes and shingles	235	1.7
Other	936	6.7
Total	14,060	100.0

Source: BC Stats (2006b). Reprinted by permission of the publisher.

Table 7.7 Lumber Production by Provinces/Territories, 2004 (hardwood and softwood lumber)

Province	Production (000 cubic metres)	Production (per cent)
Yukon/Northwest Territories	>100	>0.1
Newfoundland and Labrador	>100	>0.1
Prince Edward Island	>100	>0.1
Manitoba	637	0.7
Saskatchewan	1,184	1.4
Nova Scotia	1,785	2.1
New Brunswick	4,039	4.8
Alberta	8,053	9.6
Ontario	8,727	10.5
Québec	19,883	23.8
British Columbia	39,205	47.0
Canada	83,512	100.0

Source: Natural Resources Canada (2006b).

Canada and the United States. In 2005, the value of Canada–US trade was $629 billion, which is the largest in the world between two countries. The vast majority of this trade takes place without incident, but there are 'hot points' and lumber is one of them. The value of forest exports declined from $40 billion in 2001 to $37 billion in 2005.

NAFTA has reduced trade barriers between the United States and Canada, but it has not prevented trade disputes involving lumber from erupting. Forest exports from British Columbia have already been subjected to the heavy hand of Washington. The US government has imposed a number of countervailing duties against Canadian softwood lumber in 1982, 1986, 1991, and 2002. In each case, Washington's action has been driven by pressure from the American lumber lobby, the Coalition for Fair Lumber Imports. Most Canadians do not understand why NAFTA has not given Canadian products 'free' access to US markets. After all, was not 'free' access to each other's market the purpose of the original Free Trade Agreement and the subsequent NAFTA? The answer to that question is 'freer' but not 'free' trade. For example, when an American producer is adversely affected by imported products, the US company complains to the American government about the lower-priced products, claiming that such lower prices are a form of unfair trade. The company expects Washington to protect it by creating a trade barrier, which it has done on occasion. The trade agreement has a dispute-settlement mechanism, but resolving trade disputes takes time, during which the Canadian exporter loses sales and profits.

Problems over free and fair trade have been particularly frequent in the forest industry. American lumber production, which operates mainly in the Pacific Northwest and Georgia, can produce a maximum of 15 billion board feet per year. Canadian lumber production nearly doubles the American figure. In fact, lumber production from British Columbia (Table 7.7) is roughly equal to that of the entire United States. The main differ-

Softwood lumber exports to the United States are essential to BC's forest industry. (CP/Chuck Stoody)

ence is that US-produced lumber serves its domestic market, while most Canadian-produced lumber is exported to the United States, Japan, and other foreign customers. Trade disputes over lumber exports can therefore significantly affect BC's forest industry.

When American lumber producers lose market share to their Canadian counterparts, they turn to their lobby organization. The US political system is extremely sensitive to lobby efforts because individual members of the House of Representatives and the Senate have the power to intercede on trade matters. In addition, these elected officials rely on lobby organizations for support at election time. Since 1982, the US Coalition for Fair Lumber Imports has managed to convince the American government to reduce Canadian forestry imports into the United States. In 1986, for example, the US launched a trade action, claiming that Canadian softwood lumber production was subsidized by low provincial stumpage fees. In the following year, the US Department of Commerce ruled that Canadian stumpage rates constituted a countervailing subsidy. Ottawa agreed to impose a 15 per cent export tax on lumber exported to the US. British Columbia and several other provinces raised their stumpage rates to counter the American claim of low stumpage rates and to replace the federal export tax. In 1991, Canada terminated the 15 per cent export tax. In response, the US Coalition for Fair Lumber Imports called for another trade action to restrict the flow of Canadian lumber into the US. In 1992, the US government imposed a countervailing duty on Canadian lumber. Over a two-year period, the US government collected more than $800 million from Canadian exporters. In 1994, a bilateral trade panel (the dispute mechanism created by NAFTA) declared the tax invalid under NAFTA and ordered the US government to reimburse the Canadian companies. Not to be outdone, the US Coalition for Fair Lumber Imports called for new restrictions.

In 1996, the US government insisted on a ceiling for Canadian lumber shipments to the US, and Ottawa and Washington reached a new agreement: a limit of 14.7 billion board feet would be allowed into the US in exchange for five years of non-interference with the lumber trade by the US government. The agreed-to figure was an average of the volume of Canadian lumber exported to the United States over the previous three years. The first 650 million board feet in excess of the figure faced a US tax of $50 per 1,000 board feet. Even with this additional cost, Canadian lumber was still competitive in the US market. Once the 650 million board feet figure was exceeded, the tax jumped to $100 per 1,000 board feet. Even at that tax level, Canadian lumber remained competitive in the US market. All of these efforts by the US lumber lobby and the US government have been attempts to bypass NAFTA in order to protect US lumber interests. This US–Canada agreement ended on 31 March 2001. From a Canadian perspective, the softwood lumber trade dispute has never been about subsidies but rather has served as a means by which Washington protects the US lumber producers. In May 2002, as negotiations to extend the previous agreement reached an impasse, Washington announced a duty of 27 per cent on Canadian softwood lumber entering the United States. Four years later, a new agreement finally was signed in October 2006 (Vignette 7.5).

The latest softwood lumber agreement between Ottawa and Washington has ended a bitter trade dispute—at least for the next seven years. The agreement provides protection to the US industry when lumber prices fall. Highlights of the new agreement are:

- The seven-year agreement includes a possible two-year extension.
- Import duties of $4 billion the US charged Canadian companies since 2002 will be returned. The US keeps $1 billion.
- The US is banned from launching new trade actions against lumber imports from Canada.
- Canada is to impose an export tax by Ottawa if softwood lumber prices in the US fall below US $355 per thousand board feet. At that point, Canada

Vignette 7.5 The 2006 Softwood Lumber Agreement

Canadian regions (the BC coast, the BC interior, Alberta, Saskatchewan, Manitoba, Ontario, and Québec) will be responsible for determining which of the following two export charge regimes best serves the interests of their exporters for periods of three years:

Option A: an export charge with the charge varying with price;
Option B: an export charge plus volume restraint, where both the rate and volume restraint vary with the price, as follows:

Price per Thousand Board Feet	Option A: Export Charge	Option B: Export Charge plus Volume Restraint
Over US $355	0	0
US $336–$355	5%	2.5% + regional share of 34% of US consumption
US $316–$335	10%	3% + regional share of 32% of US consumption
US $315 or under	15%	5% + regional share of 30% of US consumption

Source: Canada (2006). Reproduced with the permission of the Minister of Public Works and Government Services Canada, 2007.

must impose a tax on softwood lumber shipped to the United States of at least 5 per cent of the price per thousand board feet. The purpose of the export tax is to protect US lumber producers by increasing the price of Canadian lumber and thus give an advantage to US producers.
- Neutral trade arbitrators are to provide final and binding settlements of disputes.

In the coming years, the forest industry in BC will continue to face a number of challenges. At the time of the signing of the 2006 softwood lumber agreement, the price of lumber had dropped below US $300 per thousand board feet, causing the Canadian government to impose an export tax. In addition to low lumber prices, BC will have to contend with environmental concerns, dwindling forest resources, and the need for more **value-added production**. Industry analysts say that the province has too many pulp mills facing a dwindling supply of wood fibre. The forest industry remains an important part of BC's economy, but in order to protect it, and the economy in general, such issues will have to be addressed. Already the market has caused several pulp and paper mills to close and many more sawmills.

Figure 7.7 Traditional US ball games. Back in 2002, the United States invoked a punitive countervailing duty of 27.2 per cent on Canadian softwood lumber. In 2006, a new softwood lumber agreement was reached. During the time of the duty, larger firms were able to increase their efficiency through economies of scale but many small forest operations were hurt by this duty, causing closures and layoffs. By mid-2006, the US housing market had declined sharply, causing lumber prices to drop dramatically. Unfortunately for Canadian producers, the new softwood lumber agreement requires Ottawa to impose an export tax when the price of softwood lumber reaches US $355 per thousand board feet (see Vignette 7.5). (© Adrian Raeside)

British Columbia's Urban Geography

Canada's third largest city, Vancouver, has a population of over 2 million. Vancouver dominates the urban geography of British Columbia (Table 7.8). Like the province as a whole, Vancouver's population has increased at a rate well above the national average. Until 1996 it was the fastest-growing large metropolitan area in Canada. Since then, Vancouver's spectacular growth rate has continued, but from 1996 to 2005, Calgary was the fastest-growing city in Canada.

Another striking aspect of the population geography of British Columbia is the concentration of people in the southwest corner of the province and the dominant position of Greater Vancouver within that population cluster. Greater Vancouver encompasses a number of cities and towns, including Burnaby, Coquitlam, Delta, Langley, New Westminster, North Vancouver, Richmond, Surrey, and West Vancouver. Besides Greater Vancouver, four other metropolitan centres, Victoria, Abbotsford, Nanaimo, and Chilliwack, are located in this densely populated area of British Columbia. The Okanagan Valley cities of Kelowna, Vernon, and Penticton constitute a modest secondary cluster.

Besides the rapid growth of Vancouver, smaller urban centres in BC are also growing quickly. Since 2001, Kelowna (9.8 per cent), Chilliwack (9.3 per cent), Abbotsford (7.9 per cent), and Nanaimo (7.8 per cent) have exhibited the greatest rates of increase of BC cities with a population of 50,000 or greater (Table 7.8). In comparison, Vancouver and Victoria had increases of 6.5 per cent and 5.8 per cent. At the other end of the scale, the greatest population losses (over 12 per cent) took place in Kitimat and Prince Rupert. Prince Rupert may halt its demographic decline when its container port begins operation in late 2007 and traffic flows along the CN's much touted North West Transportation Corridor. Kitimat's future is less bright, and unless it becomes a transshipment point for Fort McMurray oil going to China, further population losses may occur.

Unlike Ontario and Québec, each of which has more than one population cluster, BC can claim only the Vancouver area as a veritable pop-

Table 7.8 Major Urban Centres in British Columbia, 2001–6

Centre	Population 2001	Population 2006	Percentage Change
Vernon	51,530	55,418	7.5
Chilliwack	74,003	80,892	9.3
Prince George	85,035	83,225	–2.1
Nanaimo	85,664	92,361	7.8
Kamloops	86,951	92,882	4.4
Abbotsford	147,370	159,020	7.9
Kelowna	147,739	162,276	9.8
Victoria	311,902	330,088	5.8
Vancouver	1,986,965	2,116,581	6.5

Source: Statistics Canada (2007). Adapted from Statistics Canada publications *Population and Dwelling Counts, 2001 Census*, Catalogue 93F0050XCB2001013, http://www.statcan.ca/bsolc/english/bsolc?catno =93F0050X2001013 and *Population and Dwelling Counts, 2006 Census*, Catalogue 97-550-XWE2006002 http://www.statcan.ca/bsolc/english/bsolc?catno=97-550-XWE2006002.

Table 7.9 Smaller Urban Centres in British Columbia, 2001–6

Centre	Population 2001	Population 2006	Percentage Change
Kitimat	10,285	8,957	−12.6
Dawson Creek	10,754	10,994	2.2
Prince Rupert	15,302	13,392	−12.5
Squamish	14,435	15,256	5.7
Salmon Arm	15,388	16,205	5.3
Powell River	16,604	16,537	−0.4
Terrace	19,980	18,561	−7.0
Williams Lake	19,768	18,760	−5.1
Quesnel	24,426	22,499	−8.1
Cranbrook	24,275	24,138	−0.6
Fort St John	23,007	25,136	9.3
Port Alberni	25,299	25,297	0
Parksville	24,285	26,518	9.2
Campbell River	35,036	36,461	4.1
Duncan	38,613	41,387	6.6
Penticton	41,564	43,313	4.2
Courtenay	45,205	49,214	8.9

Source: Statistics Canada (2007). Adapted from Statistics Canada publications *Population and Dwelling Counts, 2001 Census*, Catalogue 93F0050XCB2001013, http://www.statcan.ca/bsolc/english/bsolc?catno=93F0050X2001013 and *Population and Dwelling Counts, 2006 Census*, Catalogue 97-550-XWE2006002 http://www.statcan.ca/bsolc/english/bsolc?catno=97-550-XWE2006002.

ulation cluster. Defined in terms of a census metropolitan area, Vancouver's 2006 population was 2.1 million. Victoria, the second largest urban centre, has a much smaller population at 330,088 inhabitants (Table 7.8). As the capital of the province, Victoria is a 'government' town, as well as an important tourist and service centre. Its mild climate has also attracted retired people, especially from the Prairie provinces. Kelowna and Abbotsford, with populations in 2006 of 162,276 and 159,020 respectively, fall into a distant third and fourth place. After Abbotsford, there are four cities with populations between 80,000 and 93,000: Kamloops, Nanaimo, Prince George, and Chilliwack. Below the 80,000 mark are 17 smaller cities with populations of 10,000 or more (Tables 7.8 and 7.9).

Vancouver

Vancouver has a majestic physical setting. The city, located on the shores of Burrard Inlet, lies across the water from the snowcapped peaks of the North Shore Mountains. To the west lies the island-studded Strait of Georgia, while the Fraser River and its deltaic islands (flat, low islands composed of silt and clay near the mouth of a river) mark Vancouver's southern edge. Vancouver has a mild, marine climate, though some find the frequent rain and overcast skies unappealing. Vancouverites are rightfully proud of their city, which has one of the most beautiful settings in the world.

Vancouver emerged from the forest wilderness in a remarkably short time to become one of the great cities of North America. Few cities have grown as quickly. In 1870, the frontier settlement of Granville on Burrard Inlet had 50 residents. Little changed until the arrival of the Canadian Pacific Railway in 1886. At that time, the population of Granville had soared to about 1,000 and the residents incorporated their village as the City of Vancouver. By 1891, the newly incorporated city had reached a population of 15,000. The spectacular growth continued, partly as a result of amalgamation with neighbouring municipalities. By 1931, Vancouver's population had reached a quarter of a million. Twenty years later, Vancouver's population had doubled.

Like Montréal, much of Vancouver's commercial strength stems from its role as a trade centre. Vancouver is the largest port in Canada and one of the largest on the Pacific coast. With most of the world's population located along the Pacific Rim, Vancouver handles $30 billion worth of trade goods each year. In fact, the Port of Vancouver is so busy that some shippers may opt for a longer route through the Panama Canal to Halifax.

Much of Vancouver's economic well-being is closely tied to the United States and Pacific Rim countries. Most of BC's exports go to the United States, China, and Japan. In 2005, 80 per cent of exports from British Columbia and 78 per cent of imports involved those three countries (BC Stats, 2006b: 12). Forest products made up approximately half of all 2005 exports, of which softwood lumber comprised 44 per cent, pulp 20 per cent, newsprint 5 per cent, and paper and paperboard 10 per cent. The remaining 21 per cent consisted of manufactured wood products such as doors and window frames (ibid., 7).

Vancouver is also a transshipment point for resource products from the interior of British Columbia, Alberta, Saskatchewan, and Manitoba; a service centre; and a tourist destination. Unit trains carry coal, grain, lumber, and potash to Vancouver for export. Head offices of private companies (especially fishing, forestry, and mining companies) and federal and provincial public offices are located in Vancouver. While Vancouver (known in 1867 as Gastown) began as a sawmill centre, these air-polluting mills relocated to other centres or were dismantled. False Creek, in downtown Vancouver, had been the prime site for processing logs, but in 1986 False Creek became the site of Expo '86 and then the site of an upscale residential and farmers' market complex known as Granville Island (Vignette 7.6).

Figure 7.8 British Columbia's Lower Mainland. The Lower Mainland of British Columbia, with Vancouver as the focal point, dominates British Columbia's urban geography. (British Columbia Adventure Network © 1996–2007 Interactive Broadcasting Corporation). *Source:* © Davenport Maps Ltd. www.davenportmaps.com.

The face of Vancouverites is changing. Nearly 60 per cent of new arrivals in the census metropolitan area of Vancouver speak Chinese. The impact on Vancouver's schools is profound. In 2001, nearly one in five school-age children were new arrivals while in the City of Richmond, which is part of Vancouver's census metropolitan area, the figure jumped to one in three. (Al Harvey/The Slide Farm)

Vignette 7.6 **Granville Island**

Granville Island, in the heart of downtown Vancouver, is strategically located between two residential areas, the West End and Kitsilano. In 1915, Granville Island was converted from a nondescript sandbar into an industrial core for the growing forest industry. In the 1970s, Granville Island began to change into an upscale residential and specialized commercial area centred on the Granville Island Public Market. Other enterprises have widened its appeal—artists' studios and shops, a wide variety of restaurants, and features like the Kids Market, Maritime Market, and the Coast Salish Houseposts, a joint endeavour between the Emily Carr College and First Nations. Granville Island is unique to Vancouver and has added another dimension to the wide-ranging tourist attractions in the Greater Vancouver region.

Much like British Columbia itself, Vancouver lies at a crossroads. Vancouver is already a major financial, tourist, and cultural centre. The staging of the 2010 Winter Olympics in Vancouver and the ski resort of Whistler will focus world attention on BC, elevating these two sites as world destination for tourists and perhaps attracting business and capital to British Columbia.

British Columbia's Future

British Columbia continues to outperform the national economy. All signs point to more economic expansion and population increase in the coming years. Its future performance will be triggered by ever-increasing trade with Pacific Rim countries, especially with China. The proposed construction of an oil pipeline from Fort McMurray to Kitimat, from where ships would transport the oil to China, indicates the powerful pull of this new economic giant on Canadian resources and the positive spinoffs for BC. While proponents of the scheme to ship Fort McMurray oil to China seem to have lost their initial enthusiasm, the construction of new port facilities at Prince Rupert has provided an outlet for natural resources and agricultural products from Western Canada to reach markets in Asia. Prince Rupert has also become a container port and large flows of cargo are anticipated along CN's North West Transportation Corridor. Potash from Saskatchewan and coal from Alberta and British Columbia plus grain from Western Canada will continue to represent major exports through BC's ports to meet the growing demand from China, Japan, and other Asian countries. Then, too, the 2010 Winter Olympics will have an important economic impact and enhance the province's tourist business. BC's natural beauty, ocean scenery, and mountain terrain offer many attractions as a world-class tourist destination.

The main challenges facing BC are threefold. First, BC needs to ensure that the expanding economy does not harm its natural environment and, by implication, the quality of life. BC citizens enjoy an unprecedented quality of life because of the province's extensive coastal waters, wildlife habitat, and wilderness areas. On the other hand, the very mild winters that some attribute to global warming have allowed the pine beetle to thrive and thus damage the extensive lodgepole pine forest in the interior of BC. Already, the pine beetle has spread to Alberta.

Second, BC needs to foster further development of its high-technology industry and also to expand the processing of its natural resources, especially its wood products. Hayter (1996: 101) argues that the forest industry must change its outlook from that of a **staple producer** to that of a processor for BC to complete its transformation from a resource hinterland to an industrial core. More specifically, forestry companies must shift their efforts from low-value processing of timber, such as lumber, to high-value processing, such as furniture. Given that scenario, Barnes and Hayter (1997: 7) pose two possibilities:

> The pessimistic interpretation is of an industry on the decline, fizzling out as the resource base itself disappears. The optimistic interpretation, though, is of a kind of industrial renaissance, of a newly fashioned forest products industry that emphasizes high-value products, skilled labour, and leading-edge technology.

Yet, evidence strongly suggests that the high-technology sector offers greater prospects for manufacturing in BC and certainly greater economic spinoffs from the supporting research institutes and the high wages earned by high-tech employees.

The final and most contentious challenge will be settling Aboriginal land claims. Land-claim agreements have the potential to benefit both First Nations and BC's economy. These agreements, it is hoped, will also resolve the contentious issue of sharing of BC's resources. The 2006 agreement with the Carrier Indians near Prince George (the Lheidli T'enneh First Nation) seemed to provide an example of sharing resources, particularly fish and forest resources, for commercial purposes, but as we saw the Lheidli T'enneh First Nation voted against accepting the agreement reached by their leaders and the federal and provincial governments. What impact this rejection will have on other negotiating tables, and on the Aboriginal Fisheries Strategy, remains to be seen.

Notes

1. Besides the issue of luring British Columbia into Confederation, there were several other political reasons for constructing a transcontinental railway. First, there was the urgent need to exert political control over the newly acquired but sparsely settled lands in Western Canada. As in British Columbia, the perceived threat to this territory was from the United States. Second, there was the need to create a larger market for manufactured goods produced by the firms in southern Ontario and Québec.
2. The exact number of people in British Columbia in 1871 is not known. How many Aboriginal peoples is a guess because their numbers declined sharply as they came into contact with European diseases. Similarly, the number of Americans and people from other countries who remained in the country after the gold rush is unknown. Certainly most moved on to the next gold rush, but some stayed. The gold rush of 1858 may have attracted about 25,000 Americans who sailed from San Francisco to New Westminster at the mouth of the Fraser River.
3. Professor Jim Miller was Chair of the Department of Geography, University College of the Cariboo, Kamloops, BC. He has since moved to the Maritimes.
4. The provincial government created the Technical University of British Columbia in 1997 and this virtual university began to provide on-line courses in 2000. The objective is to prepare students for the high-tech industry. The university offers certificate programs in electronic commerce and software development. High-technology companies are encouraged to locate offices and research laboratories near the campus. Students will undertake internships and co-operative work sessions with high-technology firms. Its theoretical basis lies in the concept of high-tech clusters around a university.
5. In 1950, Alcan secured the rights to the water in the Nechako River system until 1999. Further hydroelectric development would extend their right to the Nechako waters. Hence, Alcan commenced construction of the Kemano Completion Project. When the British Columbia government signed this long-term agreement, it saw this industrial development as the key to opening up the province's northwest. At that time, Victoria imagined that Kitimat would become an industrial complex with a population quickly reaching 20,000.

Challenge Questions

1. Which physiographic region is associated with BC's natural gas discoveries?
2. Do you agree that BC faces a difficult choice between development and its environment?
3. Is the Aboriginal fishery a treaty right?
4. What are the six stages in land-claim negotiations in BC?
5. What are the anticipated economic impacts of Prince Rupert becoming a major port for container traffic?
6. In the 2006 softwood lumber agreement, what was the intent of the American negotiators when they insisted that Ottawa impose an export tax on Canadian softwood lumber exports if the US price of softwood lumber drops below a certain level?
7. Why is there a lag between the number of seats allocated to BC and its population size?
8. Why have most of the economic problems in the forest industry fallen on small sawmills and on single-industry towns in BC's hinterland?
9. To what extent is the concept of Cascadia rooted in the centralist/decentralist faultline? What other bases do you believe this concept has?
10. Why is nature not controlling the pine beetle as it did in the past?
11. Most sectors of BC's economy are prospering, but the forest industry is struggling. Why is the forest industry in trouble?

Key Terms

Asian-Pacific Gateway and Corridor
A system of transportation infrastructure, including British Columbia's Lower Mainland and Prince Rupert ports, road and rail connections that reach across Western Canada and into the economic heartlands of North America, and major airports and border crossings.

British Columbia–Alberta Trade, Investment, and Labour Mobility Agreement
This agreement gives businesses and workers in both provinces seamless access to a larger range of opportunities across all sectors including energy, transportation, labour mobility, business registration, and government procurement.

commercial forests
Forest lands that are able to grow commercial coniferous (softwoods), deciduous (hardwoods), and mixed woods timber within an acceptable time frame.

container
A sealed steel 'box' of standardized dimensions (measured in 20-foot equivalent units or TEUs) to transport cargo.

market pulp
Pulp sold in competition with other suppliers.

newsprint
A general term used to describe very thin paper used primarily in the publication of newspapers.

producer services
Services that have enabled firms and regions to maintain their specialized roles in marketing, advertising, administration, finance, and insurance industries. Producer services are one of several parts of the growing service sector of the economy.

softwood forest
The predominant forest in Canada. Softwood forests consist mainly of coniferous trees, characterized by needle-like foliage.

staple producer
Resource developer in the primary sector that does relatively little or no actual processing of the resource, and hence the exported product has no value-added costs from manufacturing built into its basic commodity price.

tragedy of the commons
The destruction of renewable resources, such as fisheries or, historically, common pasture, because of the absence of collective control over these resources, which are available to all. The problem lies in individuals maximizing the use of such resources for personal gain; such usage, in total, overwhelms the capacity of renewable resources to maintain and regenerate themselves.

value-added production
Manufacturing that increases the value of primary (staple) goods.

Bibliography

Barman, Jean. 1996. *The West Beyond the West: A History of British Columbia*, 2nd edn. Toronto: University of Toronto Press.

Barnes, Trevor, and Roger Hayter. 1997. *Trouble in the Rainforest: British Columbia's Forest Economy in Transition*. Canadian Western Geographical Series, vol. 33. Victoria: Western Geographical Press.

———, ———, and E. Grass. 1990. 'MacMillan Bloedel: Corporate Restructuring and Employment Change', in M. De Smidt and E. Wever, eds, *The Corporate Firm in a Changing World Economy*. London: Routledge, 145–65.

Boei, Bill, and Amy O'Brian. 2003. '10,000 flee Kelowna, B.C., Blaze', *National Post*, 23 Aug., A1.

Bone, Robert M. 2003. *The Geography of the Canadian North: Issues and Challenges*, 2nd edn. Toronto: Oxford University Press.

Bradbury, John. 1982. 'British Columbia: Metropolis and Hinterland in Microcosm', in L.D. McCann, ed., *Heartland and Hinterland: A Geography of Canada*. Scarborough, Ont.: Prentice-Hall, ch. 10.

BC Stats. 2006a. Regional and Community Facts. Victoria. At: <www.bcstats.gov.bc.ca/data/dd/facsheet/cf267.pdf>.

———. 2006b. Quick Facts about British Columbia. Victoria. At: <www.bcstats.gov.bc.ca/data/bcfacts.asp>.

BC Treaty Commission. 2007. 'Lheidli T'enneh First Nation Vote to Reject the Treaty', Apr. At: <www.bctreaty.net/files/pdf_documents/April07update.pdf>.

British Columbia. 1996. *Forest Practice Code: Timber Supply Analysis*. Victoria: Ministry of Forests.

———. 2000a. British Columbia Home Page: Quick Facts—the Economy. At: <www.bcstats.gov.bc.ca/data/QF_econo.HTM#for>.

———. 2000b. Quick Facts—Statistical Appendix, Victoria. At: <www.bcstats.gov.bc.ca/data/QF_stats.HTM>.

———. 2002. *Profile of the British Columbia High Technology Sector*, 2002 edn. Victoria. At: <www.bcstats.gov.bc.ca/data/bus_stat/hi_tech.htm>.

———. 2003a. British Columbia Home Page: Quick Facts—the Economy. Victoria. At: <www.bcstats.gov.bc.ca/data/QF_econo.HTM#for>.

———. 2003b. *British Columbia Financial and Economic Review*, 63rd edn. Victoria. At: <www.bcstats.gov.bc.ca/tbs/F&Review03.pdf>.

———. 2006. 'Fishery Statistics', Ministry of Agriculture and Lands. At: <www.agf.gov.bc.ca/fish_stats/statistics-commfish.htm>.

———. 2007. 'Pacific Gateway'. Ministry of Transport. At: <www.th.gov.bc.ca/PacificGateway/index.htm>.

Canada. 2006. Office of the Prime Minister. 'Backgrounder—The Canada–US Softwood Lumber Agreement', 1 July. At: <www.pm.gc.ca/eng/media.asp?category=5&id=1234>.

Cassidy, Frank. 1992. 'Aboriginal Land Claims in British Columbia: A Regional Perspective', in Ken Coates, ed., *Aboriginal Land Claims*. Toronto: Copp Clark, 10–43.

Cernetig, Miro. 1997. 'BC Makes Peace with Alcan', *Globe and Mail*, 6 Aug., A1, A5.

———. 2006. 'Native Treaty Worth $76m', *Vancouver Sun*, 30 Oct. At: <www.canada.com/vancouversun/news/story.html?id=b1ce3c49-ca8d-4fbf-a135-7db01a76133c&k=55044>.

Christensen, Bev. 1995. *Too Good to Be True: Alcan's Kemano Completion Project*. Vancouver: Talonbooks.

City of Richmond. 2006. At: <www.richmond.ca/discover/about/profile.htm>.

CN. 2007. 'Port of Prince Rupert Container Terminal'. At: <www.cn.ca/specialized/ports_docks/prince_rupert/fairview/en_KFPortsPrinceRupert_fairview.shtml>.

Davis, H. Craig, and Thomas A. Hutton. 1994. 'Marketing Vancouver's Services to the Asia Pacific', *Canadian Geographer* 38, 1: 18–28.

Department of Fisheries and Oceans (DFO). 1994. *Canadian Fisheries Statistical Highlights 1992*. Ottawa: DFO.

———. 1996. *Fraser River Pink: Report of the Fraser River Action Plan Fisheries Management Group*. Ottawa: DFO.

Dufour, Daniel. 2007. *The Canadian Lumber Industry: Recent Trends*. Statistics Canada Catalogue no. 11–621–MWE2007055. At: <www.statcan.ca/english/research/11-621-MIE/11-621-MIE2007055.htm>.

Edginton, David W. 1994. 'The New Wave: Patterns of Japanese Direct Foreign Investment in Canada during the 1980s', *Canadian Geographer* 38, 1: 28–36.

———. 1996. 'Japanese Real Estate Investment in Canadian Cities and Regions, 1985–1993', *Canadian Geographer* 40, 4: 292–305.

——— and T.A. Hutton. 2000–1. 'Multiculturalism and Local Government in Vancouver', *Western Geography* 10–11: 1–29.

Emmett, Brian. 2005. 'Focus on Fibre for a Forest Fix', *Viewpoint* (Fall). Canadian Forest Service, Natural Resources Canada. At: <www.nrcan.gc.ca/cfs-scf/national/what-quoi/viewpoint/index_e.php?ArticleId=225&IssueId=23>.

Farley, A.L. 1979. *Atlas of British Columbia*. Vancouver: University of British Columbia Press.

Forward, Charles N., ed. 1987. *British Columbia: Its Resources and People*. Victoria: Department of Geography, University of Victoria.

Francis, R. Douglas, Richard Jones, and Donald B. Smith. 1996. *Destinies: Canadian History Since Confederation*, 3rd edn. Toronto: Harcourt Brace.

Gourley, Catherine. 1988. *Island in the Creek: The Granville Island Story*. Madeira Park, BC: Harbour Publishing.

Halseth, Greg. 1999. '"We came for the work": Situating Employment Migration in B.C.'s Small, Resource-based Communities', *Canadian Geographer* 43, 4: 363–81.

——— and Lana Sullivan. 2002. *Building Community in an Instant Town: A Social Geography of Mackenzie and Tumbler Ridge, British Columbia*. Prince George: University of Northern British Columbia Press.

Hanlon, Neil, and Greg Halseth. 2005. 'The Graying of Resource Communities in Northern BC: Implications for Health Care Delivery in Already Underserviced Communities', *Canadian Geographer* 49, 1: 1-24.

Hainsworth, Jeremy. 2007. 'Northern B.C First Nation Rejects Land Claim Settlement in 53 per cent Vote', Canada News Archives, 31 Mar. At: <ca.news.yahoo.com/s/capress/070331/national/first_nations_claims>.

Harris, Cole. 1997. *The Resettlement of British Columbia: Essays on Colonialism and Geographical Change*. Vancouver: University of British Columbia Press.

Hasselback, Drew. 2000. 'Water Shortage Leaves Alcan High and Dry', *National Post*, 9 Dec., D7.

Hayter, Roger. 1992. 'The Little Town That Did: Flexible Accumulation and Community Response in Chemainus, British Columbia', *Regional Studies* 26: 647–63.

———. 1996. 'Technological Imperatives in Resource Sectors: Forest Products', in John N.H. Britton, ed., *Canada and the Global Economy: The Geography of Structural and Technological Change*. Montréal and Kingston: McGill-Queen's University Press, ch. 6.

———. 1997. 'High-Performance Organizations and Employment Flexibility: A Case Study of In Situ Change at the Powell River Paper Mill, 1980–1994', *Canadian Geographer* 41, 1: 26–40.

Howard, Ross. 1996. 'BC Land-Claim Stakes Highest Yet', *Globe and Mail*, 16 Jan., A5.

Hume, Stephen. 2000. 'Did Francis Drake Discover B.C.?', *National Post*, 5 Aug., B1–B2.

Hutchison, Bruce. 1942. *The Unknown Country: Canada and Her People*. Toronto: Longmans, Green and Company.

Hutton, Thomas A. 1998. *The Transformation of Canada's Pacific Metropolis: A Study of Vancouver*. Montréal: Institute for Research on Public Policy.

Indian and Northern Affairs Canada. 2007. 'Backgrounder—Wei Wai Kum Cruise Ship Port'. At: <www.ainc-inac.gc.ca/nr/prs/m-a2007/2-2892-bk-eng.asp>.

Koroscil, Paul, ed. 1991. *British Columbia: Geographical Essays in Honour of A. McPherson*. Burnaby, BC: Department of Geography, Simon Fraser University.

Lheidli T'enneh. 2007a. Lheidli T'enneh Treaty. At: <www.lheidli.ca/ltntreaty/myfile.htm>.

———. 2007b. Home Page. At: <www.lheidli.ca/>.

Liu, Xiao-Feng, and Glen Norcliffe. 1996. 'Closed Windows, Open Doors: Geopolitics and Post-1949 Mainland Chinese Immigration to Canada', *Canadian Geographer* 40, 4: 306–19.

Lush, Patricia. 1998. 'MacBlo Posts Profit, Sells MB Paper', *Globe and Mail*, 24 Apr., B1.

McCullough, Michael. 1998. *Granville Island: An Urban Oasis*. Vancouver: CMHC.

McKee, Christopher. 1996. *Treaty Talks in British Columbia: Negotiating a Mutually Beneficial Future*. Vancouver: University of British Columbia Press.

McVey, Wayne W., and W.E. Kalbach. 1995. *Canadian Population*. Toronto: Nelson Canada.

Marchak, M. Patricia. 1995. *Logging the Globe*. Montréal and Kingston: McGill-Queen's University Press.

Marsh, James H., ed. 1988. *The Canadian Encyclopedia*, 2nd edn. Edmonton: Hurtig.

May, Elizabeth. 1998. *At the Cutting Edge: The Crisis in Canada's Forest*. Toronto: Key Porter Books.

Miller, Jim. 1998. Personal communication, 28 Apr., University College of the Cariboo, Kamloops, British Columbia.

Mittelstaedt, Martin. 2004. 'Farmed Salmon Laced with Toxins, Study Finds', *Globe and Mail*, 9 Jan., A1.

Natural Resources Canada. 2003. Selected Forestry Statistics, 2003. Ottawa. At: <mmsd1.mms.nrcan.gc.ca/forest/section1New/section1New_e.asp>.

———. 2006a. About Hydroelectric Energy. Ottawa. At: <www.canren.gc.ca/tech_appl/index.asp?CaId=4&PgId=26>.

———. 2006b. Table I-4, Lumber Production by Province and Species Group, 1950–2004. At: <mmsdl.mms.nrcan.gc.ca/forest/members/section1/I-4print.asp?>.

Nemetz, Peter N. 1990. *The Pacific Rim Investment, Development and Trade*, 2nd rev. edn. Vancouver: University of British Columbia Press.

Nisga'a. n.d. Nisga'a History. Vancouver. At: <www.schoolnet.ca/aboriginal/nisga1/hist-e.html>.

Office of the Auditor General of British Columbia. 2006. *The 2010 Olympic and Paralympic Winter Games: A Review of Estimates Related to the Provincial Commitments*. Victoria. At: <bcauditor.com/PUBS/2006-07/Report2/Report2%2020062007.pdf>.

Padova, Allison. 2004. *Trends in Containerization at Canadian Ports*. Ottawa: Library of Parliament. At: <www.parl.gc.ca/information/library/PRBpubs/prb0575-e.htm>.

Prince Rupert Port Authority. 2007. 'The New World Port'. At: <www.rupertport.com/about.htm>.

Prudham, W. Scott. 2001. 'Looking to Oregon: Comparative Challenges to Forest Policy Reform and Sustainability in British Columbia and the US Pacific Northwest', *BC Studies* 130: 5–40.

Rajala, Richard A. 1998. *Clearcutting the Pacific Rain Forest: Population, Science, and Regulation*. Vancouver: University of British Columbia Press.

Reed, Maureen G. 1997. 'Seeing Trees: Engendering Environmental and Land Use Planning', *Canadian Geographer* 41, 4: 398–414.

Robinson, J. Lewis, ed. 1972. *British Columbia. Studies in Canadian Geography*. Toronto: University of Toronto Press.

——— and W.G. Hardwick. 1973. *British Columbia: 100 Years of Geographical Change*. Vancouver: Talonbooks.

Saku, James C., Robert M. Bone, and Gérard Duhaime. 1998. 'Toward an Institutional Understanding of Comprehensive Land Claim Agreements in Canada', *Études/Inuit/Studies* 22, 1: 109–21.

Schrier, Dan. 2007. 'Exports: December 2006'. Victoria: BC Stats. At: <www.bcstats.gov.bc.ca/pubs/exp/exp0612.pdf>.

Shidelar, Janet. 2000. *The Geography of Canada Bibliography Series: Vol. 5, British Columbia*. Plattsburgh, NY: Plattsburgh State University, Center for the Study of Canada.

Sparke, Matthew. 1998. 'A Map That Roared and an Original Atlas: Canada, Cartography, and the Narration of Nation', *Annals, Association of American Geographers* 88: 463–95.

Stanford, Quentin H., ed. 1998. *Canadian Oxford World Atlas*, 4th edn. Toronto: Oxford University Press.

Statistics Canada. 1982. *Census Metropolitan Areas and Census Agglomerations with Components*. Catalogue no. 95–903. Ottawa: Ministry of Supply and Services Canada.

———. 1988a. Canadian Statistics—Labour Force, Employed and Unemployed. Ottawa. At: <www.statcan.ca/english/Pgdb/People/Labour/labor07a.htm>.

———. 1988b. Labour Force Characteristics for Both Sexes, Aged 15 and Over. Ottawa. At: <www.statcan.ca/english/econoind/lfsadj.htm>.

———. 1988c. 1981–1996 Census: Labour Force Activity. Ottawa. At: <www.statcan.ca/english/census96/mar17/labour/table6/t6p59s.htm>.

———. 1992. Mother Tongue. Catalogue no. 93–313. Ottawa: Industry Canada.

———. 1996a. *Labour Force Annual Averages 1995*. Catalogue no. 71–220–XPB. Ottawa: Statistics Canada.

———. 1996b. Canadian Statistics: Distribution of Employed People, by Industry, by Province. Ottawa. At: <www.statcan.ca/english/Pgdb/labor21b.htm>.

———. 1997a. *A National Overview: Population and Dwelling Counts*. Catalogue no. 93–357–XPB. Ottawa: Industry Canada.

———. 1997b. 1996 Census: Nation Tables—Population by Mother Tongue, Showing Age Groups, for Canada, Provinces and Territories, 1996 Census—20% Sample Data. Ottawa, 2 Dec. At: <www.statcan.ca/english/census96/>.

———. 1997c. '1996 Census: Mother Tongue, Home Language and Knowledge of Languages', *The Daily*, 2 Dec. Ottawa. At: <www.statcan.ca/Daily/English/>.

———. 1998a. *Canadian Forestry Statistics 1995*. Catalogue no. 25–202–XPB. Ottawa: Ministry of Industry, Science and Technology.

———. 1998b. '1996 Census: Aboriginal Data', *The Daily*, 13 Jan. Ottawa. At: <www.statcan.ca./Daily/English/>.

———. 1998c. '1996 Census: Ethnic Origin, Visible Minorities', *The Daily*, 17 Feb. Ottawa. At: <www.statcan.ca/Daily/English/>.

———. 1998d. *Canadian Economic Observer*. Catalogue no. 11–010–XPB. Ottawa: Statistics Canada.

———. 2002. 2001 Census: Population and Dwelling Counts, for Census Metropolitan Areas and Census Agglomerations, 2001 and 1996 Censuses. Ottawa, 16 July. At: <www.statcan.ca/english/IPS/Data/93F0050XCB2001013.htm>.

———. 2003a. Canadian Statistics: Distribution of Employed People, by Industry, by Province. Ottawa. At: <www.statcan.ca/english/Pgdb/labor21b.htm>.

———. 2003b. *Canada's Ethnocultural Portrait: The Changing Mosaic*. Ottawa. At: <www12.statcan.ca/english/census01/products/analytic/companion/etoimm/contents.cfm>.

———. 2007. Population and Dwelling Counts, for Census Metropolitan Areas and Census Agglomerations, 2006 and 2001 Censuses—100% data. At: <www12.statcan.ca/english/census06/data/popdwell/Table.cfm?T=201&S=3&O=D&RPP=150>.

Stoody, Chuck. 2006. 'Softwood Lumber Dispute,' *CBC News Online*, 23 Aug. At: <www.cbc.ca/news/background/softwood_lumber/>.

Tollerson, Christopher. 1998. *The Wealth of Forests: Markets, Regulations, and Sustainable Forestry*. Vancouver: University of British Columbia Press.

Transport Canada. 2007. 'Gateway-Corridor News, Spring 2007—$1 Billion Commitment to the Gateway and Corridor'. At: <www.tc.gc.ca/majorissues/APGCI/spring2007.htm>.

VanderKlippe, Nathan. 2007a. 'Lumber Waste a Hot Commodity', *National Post*, 6 June, FP1, FP6.

———. 2007b. 'A Town's Ship Comes In', *National Post*, 9 June, FP4.

White, P., M. Michalowski, and P. Cross. 2006. 'The West Coast Boom', *Canadian Economic Observer* (May). Ottawa: Statistics Canada Catalogue no. 11–010.

Wilkinson, Bruce W. 1997. 'Globalization of Canada's Resource Sector: An Innisian Perspective', in Barnes and Hayter (1997: 131–47).

Willems-Braun, Bruce. 1997. 'Buried Epistemologies: The Politics of Nature in (Post)colonial British Columbia', *Annals, Association of American Geographers* 87: 3–31.

Wynn, Graeme, and Timothy Oke, eds. 1992. *Vancouver and Its Region*. Vancouver: University of British Columbia Press.

Yamamoto, Daisaku. 2002. 'Issues of Globalization and Reflexivity in the Japanese Tourism Production System: The Case of Whistler, British Columbia', *Professional Geographer* 54, 1: 83–93.

Further Reading

Molloy, Tom. 2000. *The World Is Our Witness: The Historic Journey of the Nisga'a into Canada.* Calgary: Fifth House.

On 11 May 2000, the Nisga'a treaty passed into law, marking a historic agreement between this small group of Indians and the rest of Canadian society. In *The World Is Our Witness*, Molloy, who was the chief federal negotiator, describes how this agreement ends the centuries-old colonization by the British and, later, Canadians of the Nisga'a lands and people. Furthermore, this treaty begins the search for a place within Canadian society by the Nisga'a. For the other Indian tribes of British Columbia, this treaty has far-reaching implications. Molloy not only provides an insider's view of the struggle to achieve an agreement, but also explains its significance to Canadians. The Nisga'a treaty represents a compromise between the Nisga'a and other Canadians on how to share the lands and resources found in the traditionally occupied lands of the Nisga'a. Originally, there was no sharing. The early English settlers who established British colonies along the east coast of the United States nearly 400 years ago claimed the land by the right of 'discovery' and acknowledged no Aboriginal right to land. Recognition of Aboriginal rights began with the Royal Proclamation of King George III. In 1763, the Royal Proclamation declared that lands to the west of the Appalachian Mountains were Indian lands. While these lands were lost to the Americans in 1783, the legal precedent laid the foundation for treaty-making across British North America and, later, Canada. The Nisga'a treaty is not only the most recent agreement but it has expanded the scope of treaties into the area of resource-sharing.

The Calgary Stampede (Ivy Images)

Overview and Objectives

Western Canada, rich in natural resources, lies in the heart of North America. With exports playing a critical role in its economy, distance to ocean ports remains a challenge. Western Canada is not a homogeneous region in its physical and political geographies. There are two distinct subregions within Western Canada: a northern resource hinterland found mainly in the Canadian Shield and a more densely populated agricultural/industrial core located in the Interior Plains. Western Canada contains three provinces—Alberta, Saskatchewan, and Manitoba. Global prices affect the western economy, with high prices for energy having a positive effect and low prices for agricultural and forest products having a negative impact. Population growth has been greatest in Alberta, while Saskatchewan has seen its population decline slightly. Most growth has taken place in the major cities, especially Calgary and Edmonton, but also Winnipeg and Saskatoon. In sharp contrast, rural areas have witnessed a huge population loss, the disappearance of rural villages, and fewer but larger farms. Is the long-hoped-for transformation of its agricultural economy finally at hand? We explore this question in the Key Topic of 'Agricultural Transition' in this chapter, which will:

- Describe Western Canada's physical and historical geography.
- Present the basic characteristics of its population and economy.
- Examine Western Canada's population and economy within the context of the region's physical setting and dry continental climate.
- Explore Western Canada's changing position within the core/periphery model—from a resource frontier to an upward transitional region.
- Focus on the transition in Western Canada's agricultural sector and in its resource and manufacturing sectors.

CHAPTER 8

Western Canada

INTRODUCTION

Western Canada is at a turning point. Ahead is the prospect of playing a different economic role that involves more processing of the region's agricultural and resource products. As is the case with the British Columbia economy, Western Canada is benefiting from increased trade with Pacific Rim countries as well as with its traditional trading partner, the United States. Higher global prices for natural resources, especially for energy resources, have played a critical role in this transformation process. Even prices for grain show signs of increasing, an ironic result of the conversion of US corn into bioenergy! North of the farmland, the economic outlook for the forest industry remains bleak. This chapter examines the factors underlying the radical departure for Western Canada from its traditional role as an agricultural hinterland to a diversified inland economy. The Key Topic, 'Agricultural Transition', looks at the economic and political forces propelling Western Canada's economy to diversify, shifting more and more into an urban industrial economy and forcing rural land-use changes in the Canadian Prairies. While cities are growing, the exodus from rural villages and towns to cities continues.

Western Canada within Canada

Western Canada is part of the vast western interior of North America (Figure 1.2). This huge area, comprising the three western provinces—Alberta, Saskatchewan, and Manitoba—contains many natural resources, but shipping these products to world markets has proven to be a major stumbling block to the region's economic development. These resources range from the fertile soils of the Canadian Prairies to the commercial timber stands in the northern coniferous forest. As well, there is considerable mineral wealth in the Interior Plains, the Canadian Shield, and the Cordillera.

In terms of economic output, Western Canada falls behind Ontario and Québec. Overall, Western Canada's economy accounts for 22.6 per cent of Canada's GDP (Figure 8.1). An important measure of its economic well-being is Western Canada's low unemployment rate. In 2006, Western Canada had, at 4.0 per cent, the lowest unemployment rate of the six regions, well below the national average of 6.0 per cent (Figure 8.1). Led by Alberta, economic gains in this geographic region are outstripping those of British Columbia, Ontario, and Québec,

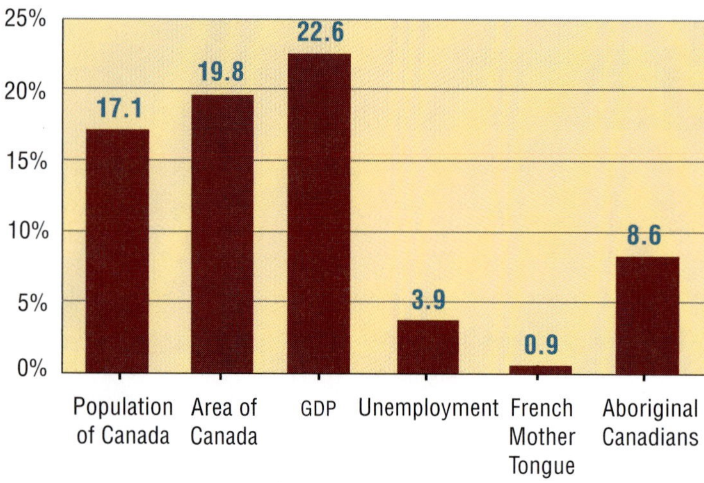

Figure 8.1 Western Canada, 2006. Although Western Canada covers a large area of the country, the region's population strength is more modest than its economic growth. The region has fewer French-speaking Canadians than other parts of the country, but a significant percentage of Aboriginal Canadians.
Sources: Tables 1.1, 1.7, and 4.18.

and percentage population growth in Western Canada has been greater than that in BC or Québec over the past five years.

Since World War II, most of the region's economic and demographic growth has occurred in oil-rich Alberta. While oil and gas are the region's most valuable non-renewable assets, agriculture remains its key renewable resource. A common problem confronting farmers and resource companies in Western Canada is the high cost of transportation, particularly the long rail distance their products must travel to reach ocean ports, from where they are then shipped to world markets. Once products are aboard ocean vessels, the transportation cost per kilometre drops drastically. Overcoming the cost of shipping products to ocean ports has been a persistent struggle in Western Canada's development of its agricultural, forest, and mineral resources. The Hudson Bay Railway was one such effort. In the 1920s, Prairie farmers thought they could reduce their railway costs to Atlantic ports by constructing a rail line from The Pas to Churchill, Manitoba (York Factory), on Hudson Bay and then shipping their grain to their major market in the United Kingdom. The Hudson's Bay Company had used this route for centuries so why not Prairie farmers? Unfortunately, the high cost of marine insurance made this rail/sea route from Churchill to London unattractive.

For years, farmers have struggled with low world prices for their products. Grain prices, for example, have been declining for several decades while costs of production have been increasing. Prices for non-grain crops such as canola have encouraged farmers to switch to these more profitable crops. Unfortunately, non-grain crops also suffer from fluctuating world prices. The continental climate, noted for its annual variations in precipitation, has had an impact on the size of harvests. While the 2006 crop was excellent due to a combination of adequate rainfall in the summer and dry harvest conditions in the fall, the previous five years were troubled by drought, early frost, and wet harvest conditions. With the development of the ethanol industry, a new market for grain is emerging. The spinoff from the rapid expansion of the ethanol industry in the United States and Europe has already increased the demand for bio-raw material from the agricultural sector. In the United States, the heavily subsidized biofuels industry is largely based on corn biomass and this new demand for corn has pushed corn prices up. By June 2007, US future corn prices had nearly doubled in eight months (Macartney and Reid, 2007). At the same time, the growing middle class in China is consuming more and more pork, thus indirectly causing a greater demand for corn as feed for its hogs (Greenwood, 2007a). It remains to be seen, however, whether these global forces will translate into a rise in prices for Canadian agricultural products such as grain and canola.

In 2006, the population of Western Canada totalled over 5.4 million. The vast majority live in the regional core, an area of major cities, towns, and villages located in the southern half of the region. Agricultural, industrial, and service activities are well established within this population core. Calgary, Edmonton, Winnipeg, Saskatoon, and Regina are the leading cities. In the northern half of Western Canada lies its regional hinterland, where the larger centres are linked to resource developments. Western Canada's northern hinterland has less than 5 per cent of the region's population. While most live in resource towns, such as Fort McMurray and Thompson, many Aboriginal settlements exist near former fur trading posts.

Vignette 8.1 Water Deficit and Evapotranspiration

A dry, continental climate prevails over the interior of Western Canada. Due to a combination of low annual precipitation and high summer temperatures, this zone has a 'water deficit'. This zone's low precipitation is primarily the result of the eastward-moving Pacific air masses losing most of their moisture as they rise over the Cordillera's mountain chains. By the time they descend the eastern slopes of the Rocky Mountains, these moist air masses have turned into dry ones. As a result, the Prairies receive little rain or snow from the Pacific Ocean. This water deficit is often measured in terms of potential **evapotranspiration,** which is the amount of water vapour that can potentially be released from an area of the earth's surface through evaporation and transpiration (the loss of moisture through the leaves of plants). There is a water deficit if the evapotranspiration rate is greater than the average annual precipitation. For example, most of the grassland natural vegetation zone receives less than 400 mm annually, while its evapotranspiration rate exceeds 500 mm. The difference indicates a water deficit of over 100 mm. If a water deficit occurs in a dry year, plants can use water found in the soil from previous years when there was a water surplus. Therefore, soil moisture provides a reserve that plants can draw upon, but eventually this reserve is exhausted and plants are damaged or die from a lack of water. The maps of precipitation and natural vegetation in Chapter 2 (Figures 2.5 and 2.7) indicate the geographic location of this water deficit area. If there is a heavy winter snowfall, then the spring melt can add to the soil moisture, thus offsetting the water deficit.

Rich, dark brown soils represent one of the major chernozemic soils in the Canadian Prairies. In this photograph near Bigger, Saskatchewan, the practice of crop rotation is illustrated. (Barrett & MacKay Photography, Inc.)

Western Canada's Physical Geography

Western Canada, which extends over nearly 20 per cent of Canada, has two major physiographic regions—the Interior Plains and the Canadian Shield—and two minor ones—the Hudson Bay Lowland and the Cordillera (Figure 2.1). A thin portion of the Cordillera, the Rocky Mountains, forms a natural and political border between southern Alberta and British Columbia. Each physiographic region has a particular set of geological conditions, physical landscapes, and natural resources. (See also 'The Interior Plains', 'The Canadian Shield', 'The Hudson Bay Lowland', and 'The Cordillera' in Chapter 2.)

The tiny section of the Cordillera, along the eastern flank of the Rocky Mountains in Alberta, provides logging and mining opportunities for Western Canada. However, the main attraction of this slice of the Cordillera is its spectacular mountain landscape. This mountainous terrain has two internationally acclaimed parks, Banff National Park and Jasper National Park, which attract visitors from around the world. Calgarians are especially fortunate in having easy access to the Kananaskis Country Provincial Park, where a number of mountain recreational activities—camping, hiking, and skiing—are available.

In the Interior Plains region the sedimentary rocks contain valuable deposits of fossil fuels. By value, the four leading mineral resources are oil, gas, coal, and potash. Most petroleum production occurs in a geological structure known as the **Western Sedimentary Basin**, which underlies most of Alberta and portions of British Columbia, Saskatchewan, and Manitoba. However, not all mineral extraction takes place deep in these sedimentary rocks. A few deposits are exposed at the surface of the ground. In southeastern

Lake Louise is one of Banff National Park's most stunning natural features. The lake's famous turquoise colour is caused by fine rock particles, called 'glacial flour', that are contained in the stream water from alpine glaciers. (© Derek Trask/Corbis)

Saskatchewan, brown coal is extracted through open-pit mining and then burned to produce thermal electricity. In northeastern Alberta, the huge petroleum reserves in the Athabasca tar sands are exploited by surface mining techniques, such as hydro-transport in which the oil sands are mixed with extremely hot water and transported to an upgrader plant by pipeline.

In the Canadian Shield, which extends into Manitoba and Saskatchewan, rocky terrain makes cultivation virtually impossible. While forestry does take place along the southern edge of the Canadian Shield, particularly in southeastern Manitoba, large-scale commercial enterprises are generally limited to mining and the production of hydroelectricity. In northern Saskatchewan, uranium companies produce most of Canada's uranium from open-pit and underground mines. Large-scale hydroelectric dams and generators are located on Manitoba's northern rivers, particularly the Nelson River.

All of Western Canada has a continental climate characterized by extreme daily and seasonal fluctuations in temperature and low annual precipitation (Figure 2.5). The region is dry not only because of its distance from the ocean, but also because the mountain ranges in the Cordillera block the eastward movement of mild, moist Pacific air masses. During the winter, Arctic air masses often dominate weather conditions in the Prairies, placing the region in an Arctic 'deep freeze' (Figure 2.3). A combination of strong winds and sub-zero temperatures can produce blizzard-like weather. (See 'Prairies Zone' in Chapter 2 for further details.) In southern Alberta, strong winds that become warm and dry as they flow down a mountain slope are known as chinooks. On the other hand, the Alberta clippers with their strong, frigid winds produce true blizzard conditions, due to severe blowing and drifting snow.

Precipitation is a critical climatic element in Western Canada. While southern Manitoba, which receives rainfall from the moist Gulf of Mexico air masses, is rarely troubled by droughty conditions, the remaining prairie lands, but especially the dry lands of southern Saskatchewan and Alberta, are not so fortunate. The annual precipitation in Western Canada is low, varying from a high of 550 mm in southern Manitoba to less than 400 mm in the dry lands of southern Alberta and Saskatchewan (the so-called Palliser's Triangle described in Vignette 8.2). In addition to this spatial variation, precipitation varies from year to year, causing so-called wet and dry years. This annual variation has its greatest effect in the dry lands of Palliser's Triangle. For instance, in the 1930s a series of dry years resulted in the disastrous Dust Bowl, a period of severe drought that devastated Prairie farming and led to extensive wind erosion of the topsoil. The Dust Bowl drove thousands of homesteaders

Too dry and too wet: dust storm southwest of Lakenheath near Assiniboia, Saskatchewan, in 1934. The drought that struck Western Canada in the 1930s was so severe that some grain farmers abandoned their homesteads. Yet within this brown soil zone of southern Saskatchewan, sudden rainstorms can account for much of the local area's annual precipitation. For example, on 30 May 1961, a severe thunderstorm struck Buffalo Gap, some 150 km south of Regina near the United States border. In less than an hour, more than 250 mm of rain fell, ending a long dry spell (Phillips, 1993: 88). (*Saskatchewan Archives Board R-A4665*).

Vignette 8.2 Palliser's Triangle

In 1857, the Palliser Expedition set out from England to assess the potential of Western Canada for settlement. This expedition spanned three years (1857–60). In his report to the British government, John Palliser identified two natural zones in the Canadian Prairies. The first zone was described as a sub-humid area of tall grasses, while the second, located further south, was described as a semi-arid area with short-grass vegetation. Palliser considered the area of tall grasses to be suitable for agricultural settlement. He named this area the Fertile Belt. In Manitoba this belt is south of the Canadian Shield and stretches to the border with the United States. Palliser believed that the semi-arid zone, located in southern Alberta and Saskatchewan, was a northern extension of the Great American Desert. According to Palliser, these semi-arid lands were unsuitable for agricultural settlement. The area described by Captain Palliser became known as **Palliser's Triangle**—its area overlaps with, but is slightly larger than, the agriculture zone known as the Dry Belt (see Figure 8.4).

off the land in the short-grass prairies. Annual variations in precipitation have led Prairie farmers to coin the term 'Next Year Country', referring to the bad harvest years that prompt farmers to hope for a better crop the following year. (See 'Next Year Country' later in this chapter for further information.)

In Western Canada the natural vegetation and the soil, and therefore the success of commercial crops, are largely determined by the evapotranspiration rate, which is a combination of precipitation and temperature (Vignette 8.1). The evapotranspiration rate varies within Western Canada and therefore creates differences in the region's soil and vegetation. In the northern parts of Western Canada, in the Canadian Shield, cooler weather leads to a lower evaporation rate, resulting in sufficient moisture for tree growth. In this area the boreal forest grows in a podzolic soil (Figures 2.7 and 2.8). The only exception is in the Peace River country, a gently rolling plain where degraded black chernozemic soils are found beneath an aspen parkland. Further south, where the evaporation rate is much higher, the podzolic soil gives way to chernozemic soils, which are more favourable for agriculture.

Indeed, the soil and natural vegetation of the southern part of Western Canada have had a significant impact on the region's role as an agricultural heartland. This southern portion, in the Interior Plains, has two natural vegetation zones (parkland and grassland) and three chernozemic soil zones (black, dark brown, and brown) (Figures 2.7 and 8.3). Just south of the boreal forest, podzolic soil gives way to the black chernozemic soil that supports parkland vegetation. This parkland is a transition zone between the boreal forest and the grassland natural vegetation zone, which is located a little further south and is supported by dark brown and brown chernozemic soils. Within the grassland, the evaporation rate increases towards the American border, causing this mid-latitude grassland to change from tall grass to short grass as one moves southward. Tall-grass natural vegetation occurs in a wide arc from southwestern Manitoba to Saskatoon to Edmonton. Here there are dark-brown chernozemic soils, except in southern Manitoba, where black soils occur because of higher annual precipitation. This tall-grass natural vegetation zone and the parkland to its north comprise the area known as the Fertile Belt, where farming has traditionally been more successful. South of this fertile arc is the area of short-grass natural vegetation and brown chernozemic soils known as the Dry Belt (Vignette 8.2). This semi-arid area has presented the highest risk for grain farming, so much of the land is now worked in fallow rotation (a crop is planted every second year) or used for pasture.

Environmental Challenges

Western Canada faces several environmental challenges. One challenge is the urgent need to remove the radioactive wastes surrounding abandoned uranium mines along the north shore of Lake Athabasca. Like a ticking time bomb, these radioactive tailings are slowly seeping into the groundwater and the waters of Lake Athabasca (Bone, 2003: 165). The cost of this cleanup is estimated to be at least $1 billion (Saskatchewan Environment, 2002).

Another challenge is to reduce the release of carbon dioxide and sulphur dioxide gases from the mining and processing operations associated with the Alberta tar sands developments and from coal-burning thermoelectric generation stations in both Alberta and Saskatchewan. So far, little progress has been made in reducing the release of these gases and these emissions are increasing. While the release of carbon dioxide increases the greenhouse gases in the atmosphere and therefore furthers the prospect of global warming, sulphur dioxide emissions cause acid rain, thus making thousands of lakes lying in the Canadian Shield in northern Saskatchewan vulnerable to increased acidity.

Possible pollution of the air, soil, and groundwater from intensive cattle and hog facilities is a serious environmental and health problem. Alberta leads in this shift from the small traditional farms with small herds to large, high-technology operations with thousands of animals. This process of large-scale and highly specialized livestock factory-like operations is part of the trend towards an industrial agricultural economy in Alberta and the rest of North America. The Lethbridge area contains a heavy concentration of beef-cattle feedlots and hog barns. Waste products from cattle and hogs provide the greatest threat to the water resources. The heavy concentration of manure in a small area poses a danger to surface water through runoff and to groundwater through infiltration. Untreated manure may also contain bacteria such as E. coli and salmonella, viruses, and parasites. The potential of fecal bacteria seeping into the limited water supplies in this dry zone of Western Canada is great and yet provincial governments have been slow to react, preferring to encourage agricultural diversification. While world demand for pork is increasing, many European countries, several American states, and Québec have refused to permit additional large-scale hog operations because existing hog farms have caused the contamination of some of their water resources. In the late 1990s, water pollution was so serious that Taiwan banned the construction of factory-like hog barns, forcing Taiwan Sugar Corporation, a major hog operator, to seek permission to build hog barns in Alberta. Strong opposition from concerned citizens in Alberta took the Taiwan hog barn development plan to the Alberta Court of Appeal, where a ruling revoked the company's development permit in January 2001.

Public awareness and pressure from environmental organizations have prompted governments and industry to address the environmental risks associated with intensive livestock operations (Price, 2003: 38). Even so, the so-called 'beneficial management practices' designed to minimize the impacts of animal waste products on water quality are just in the research stage. Even if such research is successful, the likelihood of the livestock industry causing a serious health problem from the contamination of a local water supply along the lines of Walkerton, Ontario, or North Battleford, Saskatchewan, is too great not to invoke the precautionary principle, whereby risk to the environment mandates that certain practices should not be pursued.

Another potentially negative impact of hog barns is the impact on the health of workers and nearby residents. The health of hog barn workers is affected by the air-borne chemicals contained in smelly air generated from the hogs in the confined environment of hog barns. Known as volatile organic compounds, workers are likely to exhibit a range of respiratory problems. The same health issue and the problem of bad odour affecting quality of life may extend beyond the immediate work site and affect people residing near such operations. Johnston and

Weibel (2005–6) investigated the potential impact of the so-called 'neighbourhood effect' of hog farms in the Lethbridge area, that is, the impact on people living near such operations. Johnston and Weibel concluded that the hog-barn neighbourhood effect, which was documented in three locations in the US, seems to exist around Lethbridge, though perhaps at a less intense level.

> Based on available evidence, no one can conclusively identify where that threshold [the point where hog emissions threaten the health and quality of life of nearby residents] is. However, once it has been exceeded, human health will be negatively affected and remedial action will be prohibitively expensive. Development authorities face a difficult and unenviable task. Confronted with much scientific uncertainty, they must strike a balance between the legitimate interests of producers and investors, and local residents who increasingly expect environmental quality to be preserved and enhanced. (Johnston and Weibel, 2005–6: 56–7)

Global warming, caused by the emission of greenhouse gases to the atmosphere, represents a challenge to governments. In 2002, the federal government announced the Pilot Emission Removals, Reductions and Learnings (PERRL) initiative. Grain farmers practising **zero tillage** soon found that they could get paid for doing what they always do—seed their wheat crop without further tillage of the land. Now they receive **carbon credits** from Ottawa for this practice because zero tillage traps in the soil the carbon dioxide absorbed by the wheat. Ottawa leases the credits from the farmers, paying \$2.38 to \$5.43 per acre, depending on the type of soil (CBC News, 2006).

Pincher Creek, Alberta, is famous for its livestock industry. More recently, Pincher Creek has become the site of wind energy production, helping to make Alberta the leading province for wind-produced electricity in Canada. Powerful chinook winds from the Rocky Mountains make Pincher Creek a particularly attractive site for wind turbines. (Barrett & MacKay Photography, Inc.)

Western Canada's Historical Geography

Western Canada's history began long before the three Prairie provinces became part of Canada. In fact, Western Canada's recorded history goes back to the fur-trading days. Beginning in 1670, the Hudson's Bay Company (HBC) administered for 200 years much of Canada's western interior. This area was part of Rupert's Land (all the land draining into Hudson Bay). In 1821, when the company merged with its rival, the North West Company, the HBC acquired control over more land, known as the North-Western Territory (lands draining into the Arctic Ocean). Before 1870, when Canada was ceded these lands by the British government (Figure 3.4), the HBC used these lands exclusively for the fur trade.

The land in Western Canada began to be used for purposes other than fur trading at the beginning of the nineteenth century. In 1810, Lord Selkirk, a Scots nobleman who was concerned with the plight of poor Scottish crofters (tenants) evicted from their small holdings, acquired land in the Red River Valley from the Hudson's Bay Company. The first Scottish settlers arrived in 1812 to form an agricultural settlement near Fort Garry, the principal HBC trading post in the region. This settlement became known as the Red River Colony. Selkirk's settlers, however, faced an unfamiliar and harsh environment and had great trouble establishing an agricultural colony. Over the years, many gave up and left for Upper Canada and the United States.

At the same time, many former officers and servants of the Hudson's Bay Company, along with their Indian wives and children, settled at Fort Garry. In addition to these English-speaking people were the French-speaking Métis who had worked for the North West Company. Because the Métis were Catholic and spoke French, they formed a separate cultural group within the Red River Colony. After the consolidation of the Hudson's Bay Company and the North West Company in 1821, many who worked for the North West Company were no longer employed by the new company. Many Métis, particularly those who were French-speaking, settled in the Red River Colony, where they turned their attention to subsistence farming, freighting, and buffalo hunting.

Agricultural Potential?

Despite the modest settlement near Fort Garry and the thriving fur trade, little was known about the geography of the Canadian West in the mid-nineteenth century. Some men in the British government wondered if the interior of the Canadian West was suitable for agricultural settlement. To the south, in the United States, American settlers had begun to occupy land west of the Mississippi River. By 1854, a railway stretched across the United States from New York through Chicago to St Paul on the Mississippi River. In 1858, Minnesota had a sufficiently large population to warrant statehood. By then, American homesteaders began moving beyond the Mississippi into the northern Great Plains, including the Red River Valley. West of the Red River Valley, however, were the dry lands of the Great Plains, which limited further settlement. In fact, about 30 years earlier American explorers had dubbed Montana part of 'the Great American Desert'. Blocked to the west by these dry lands of North Dakota and Montana, settlers began to look northward to the unoccupied lands of British North America.

In 1857, the British government and the Royal Geographical Society sponsored an expedition into the Canadian West. Their central task was to determine the suitability of the Canadian West for agricultural settlement. John Palliser, an explorer, led the British North American Exploring Expedition. After his party arrived from England, they quickly travelled by rail from New York to St Paul and then by steamboat to the Red River Colony. They travelled by horseback from Fort Garry across the Canadian Prairies to the Rocky Mountains. Palliser reported that fertile land in much of the western interior existed, but that the land in southern Alberta and

Saskatchewan near the border with Montana was far too dry for farming (Vignette 8.2). Palliser believed the Great American Desert that American explorers had already identified extended into the grasslands of southern Alberta and Saskatchewan. Named after him, this semi-arid area is now known as Palliser's Triangle. Another expedition in 1858, organized in Canada West (Ontario) and led by Henry Hind, a geologist and naturalist, confirmed that the parkland (the natural vegetation zone between the grasslands and the boreal forest) offered the best land for agricultural settlement.

Opening the West

By the middle of the nineteenth century, arable land in the Province of Canada was in short supply. Some settlers began to look to the American West for new land. During the negotiations with Britain over Confederation, the subject of the annexation of Rupert's Land into the Dominion of Canada arose, and provision was made in the British North America Act for its admission into Canada. In 1869, the Hudson's Bay Company signed the deed of transfer, surrendering to Great Britain its chartered territory for £300,000—with the exception of the lands surrounding its posts and about 1,133,160 ha of farmland. In 1870, Great Britain transferred Rupert's Land to Canada.

In 1869, Canada's new government sent surveyors into the Red River Colony to prepare a land registry system for the expected influx of settlers. The surveyors employed a grid system known as a township survey. A township consisted of 93 km^2, or 36 sections. Each section was subdivided into four quarter sections of 65 ha each. Each settler would receive a quarter section and would be required to till the land and build a house on that section. The opening of Western Canada for agricultural settlement officially began in 1870 and continued until 1914, when World War I halted the influx of European immigrants into Western Canada. Following World War I, veterans were encouraged to establish homesteads, especially in the Peace River country, where there remained some arable land.

Original Inhabitants

When the Canadian government sent surveyors to the West in 1869, it was not uninhabited. Indians and Métis lived there, but the new agrarian economy would quickly marginalize them and overtake their lands.[1] The forested lands of the Dene and Woodland Cree were outside of the settled areas and these people remained tied to a hunting and trapping economy (Figure 3.2). The Plains Indians (Sarcee, Blood, Peigan, Stoney, Plains Cree, Nakota, Lakota, Blackfoot, and Saulteaux) lost control of the Plains and were driven to sign treaties and live on Indian reserves (Figure 3.10). In 1867, the Red River Colony had a population of nearly 12,000 people, mostly Métis. The arrival of land surveyors and settlers led the Métis, under the leadership of Louis Riel, to mount the Red River Rebellion in 1869. (See 'The Red River Rebellion' in Chapter 3 for more details.) The Métis wanted to negotiate the terms of entry into Canada from a position of strength—that is, as a government. The Métis obtained major concessions from Ottawa—they were guaranteed ownership of land, recognition of the French language, and permission to maintain Roman Catholic schools. In 1870, the fur-trading district of Assiniboia became the province of Manitoba. But the rebels' victory was hollow. As the newcomers poured into Manitoba, the Métis society was overwhelmed.

Many Métis left the colony to search for a new place to settle in the Canadian West. Such a place was Batoche, just north of the site where Saskatoon now is situated. Within 15 years, however, settlers would again encroach on the Métis agricultural settlement. In 1885, as before, the Métis staged a rebellion led by Louis Riel. This time, the Canadian militia defeated the Métis at the Battle of Batoche and Louis Riel was captured, found guilty of treason, and hanged. To this day, Riel remains a controversial figure in Canadian history—a traitor to some, he remains a hero to the Métis.

The experiences of Indian tribes during the early period of western settlement were somewhat different. Indian tribes, such as the Blackfoot, had roamed across the Canadian prairies and the northern Great Plains of the United States long before the arrival of European explorers, fur traders, and settlers. The tribes were nomadic and hunted buffalo. By the 1870s, the buffalo had virtually disappeared from the prairies, leaving the Plains Indians destitute. By then, life for the Blackfoot, Blood, Plains Cree, Peigan, and Saulteaux tribes had become almost unbearable. They had little choice but to sign treaties with the federal government. Between 1873 and 1876, all the tribes (except for three Cree chiefs—Big Bear, Little Pine, and Lucky Man—and their followers) signed numbered treaties in exchange for reserves, cash gratuities, annual payments in perpetuity, the promise of educational and agricultural assistance, and the right to hunt and fish on Crown land until such land was required for other purposes. In 1882, impending starvation for his people also forced the Cree leader, Big Bear, to accept Treaty No. 6. Over the next few years, however, the Cree sought other concessions from the federal government. When these efforts failed, Cree warriors supported the doomed Métis rebellion in 1885 by attacking several settlements, including Fort Pitt, the Hudson's Bay post on the North Saskatchewan River near the Alberta Saskatchewan border.[2]

Treaties with Ottawa offered the Indians prospects for survival and time to find a place in a new economy, but the treaties also made them wards of the Crown. Living on reserves, Indians were isolated from the evolving Canadian society and became increasingly dependent on the federal government. Further north, the Woodland Cree and Dene (Chipewyan) tribes who lived in the boreal forest were not as affected by the encroachment of western settlers. Although they, too, signed treaties, these northern Indians continued their migratory hunting and trapping lifestyle well into the next century. In the 1950s, their dependency on Ottawa grew with the demise of the fur trade and their subsequent relocation to settlements.[3]

The Métis search for a place within Confederation began with rebellions in 1869–70 and 1885. Both rebellions were led by Louis Riel. After losing the Battle of Batoche, Riel surrendered to the Canadian forces. He was tried and convicted of high treason. On 16 November 1885, Riel was hanged as a traitor, though he remains a hero to the Métis to this very day. (*Library and Archives Canada*)

Canadian Pacific Railway

Once treaties ensured the availability of land for homesteading, the Canadian government needed to make this land more accessible for new settlers. The next step in the plan to open Western Canada to agricultural settlement was a transcontinental railway. Macdonald's vision of Canada extending from the Atlantic to the Pacific hinged on a transcontinental railway.

There were already three transcontinental railways in the United States. Without a Canadian counterpart, Ottawa feared the worst, namely, that the West would be lost to the Americans. Even if Canada could retain its western territories in the absence of a Canadian transcontinental railway, the north–south transportation pull exerted by the American railways would prevent Ontario's fledgling industrial core from reaching the market in Western Canada, and western settlers would be unable to ship their products to eastern markets. For instance, in 1870, the new province of Manitoba was linked by steamboat to the rail centre of Fargo, North Dakota. From Fargo, passengers and freight from Manitoba could quickly reach Toronto, Ottawa, and Montréal.

British and Canadian companies were not interested in a risky railway construction project across Canada unless they could obtain substantial financial assistance from Ottawa. Two reasons accounted for their lack of interest: the Canadian Shield and the Cordillera were two formidable (and therefore costly) barriers to overcome in building a railroad. Indeed, physical geography posed a much greater challenge to Canadian railway builders than to their American counterparts. In 1881, Ottawa announced generous terms: the Canadian Pacific Railway Company was awarded a charter, whereby the company received $25 million from the federal government, 1,000 km of existing railway lines in eastern Canada owned by the federal government, and over 10 million ha of prairie land in alternate square-mile sections on both sides of the railway to a maximum depth of 39 km. The terms were successful—the Canadian Pacific Railway was completed in 1885.

Settlement of the Land

The settling of Western Canada marks one of the world's great migrations and the transformation of the Prairies into an agricultural resource frontier. Under the Dominion Land Act that Ottawa had passed in 1872, homesteaders were promised 'cheap' land in Manitoba—by building a house and cultivating some of the land, they could obtain 65 ha of land for only $10. Following 1872, an influx of prospective homesteaders began arriving, most coming from Ontario and, to a lesser degree, from the Maritimes, Québec, and the United States. When the Canadian Pacific Railway was completed, the settlement of Saskatchewan and Alberta began. Many homesteaders now came from Great Britain.

By 1896, the federal government sought to increase immigration by promoting Western Canada in Great Britain and Europe as the last agricultural frontier in North America. The Canadian government initiated an aggressive campaign, administered by Clifford Sifton, Minister of the Interior, to lure more settlers to the Canadian West. Thousands of posters, pamphlets, and advertisements were sent to and distributed in Europe and the United States to promote free homesteads and assisted passages. Prior to 1896, most immigrants came from the British Isles or the United States—these were 'desirable' immigrants. Sifton's campaign, however, cast a wider net to areas of Central and Eastern Europe that were not English-speaking and therefore provided 'less desirable' immigrants. The strategy generated considerable controversy among some English-speaking Canadians who believed in the racial superiority of British people.

Nevertheless, Clifford Sifton's efforts paid off. At the end of the 1880s, the Canadian prairies had few settlers beyond Manitoba, and most of them had taken land near the Canadian Pacific Railway line. Following the recruitment campaign, a flood of settlers arrived and the land was quickly occupied. Thus began the great migration to Western Canada. After 1896, the majority of settlers—about 2 million—were Central or Eastern Europeans from Germany, Russia, and Ukraine. This large influx of primarily non-English-speaking immigrants led to a quite different cultural makeup in Western Canada from that in Central Canada, where the French and English dominated. Within a remarkably short span of time, cultural and linguistic acculturation had forged a non-British but English-speaking society from the sons and daughters of these immigrants. Some, however, kept separate. For instance, Doukhobor settlers were Russian-speaking peasants who adhered to a sect-like religion and who preferred to live in a communal setting (see 'The Doukhobors' in Chapter 3). By 1905, Alberta and Saskatchewan had sufficient numbers to warrant provincial status. By the outbreak of World War I, the region of Western Canada was settled.

The decision to build the CPR along a southern route (from Winnipeg to Regina to Calgary) meant that much of the land opened to homesteaders lay in the driest part of Western Canada.

Life was not easy for homesteaders. Many were ill-prepared for farming, let alone farming in a dry continental environment. Securing supplies of wood and water often posed a problem. Those settlers who could not afford to import lumber were forced to live in sod houses and burn buffalo chips and cow dung for heat. While many members of ethnic and religious groups settled together in the same area, forming communities, isolation still posed a problem for many. The land survey system encouraged a dispersed rural population, with individual farmsteads rather than rural villages. As a result, farm families sometimes did not visit the nearest town or see their neighbours for weeks or even months. Such isolation was particularly hard on farmwives. In spite of these difficulties, the land was settled, towns sprang up, and institutions were created to meet the local and regional needs. In short, a new society was in the making.

By 1921, there were over 250,000 farms in Western Canada. Homesteaders now had to turn to the Peace River country for arable land. The Peace River country, part of the high Alberta Plain, is much further north and its short growing season makes agriculture risky. Even so, the Peace River country has a climate and soils that allow for mixed farming (a combination of grain and hay crops with livestock). The section of the Rocky Mountains located to the west of the Peace River country is considerably lower and thus allows more rainfall from Pacific air masses to reach the area. The final settlement of the Peace River country took place after World War I, when returning soldiers were encouraged to settle there.

Emergence of an Agricultural Economy

By the early twentieth century a Prairie agricultural economy had emerged. The market was Great Britain and other Western European countries. Based on a single staple—grain—this economy had to contend with fundamental geographic weaknesses. The first of these was its geographic position within the interior of Canada. Western Canada's geographic isolation from Europe was partly overcome by the building of a railway system across the Prairies. The purpose of the CPR was twofold: (1) to bring settlers to the West, and (2) to allow these settlers to export their farm products to markets in Central Canada, Britain, and other European countries. However, once the railway system was in place, farmers faced a new problem—the high cost of shipping grain long distances by rail and low grain prices. World War I saw prices rise to record levels and then decline (See Figure 8.2 and Vignette 8.4 and 'Grain Transportation Subsidy' later in this chapter.)

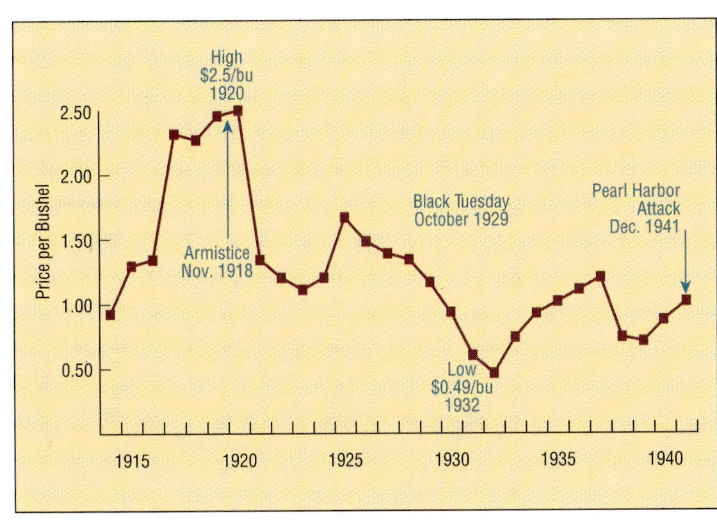

Figure 8.2 Declining world grain prices
Source: US Bureau of the Census, *Historical Statistics of the United States, Colonial Times to 1957*.

Railway expansion within the Prairies solved the second geographic problem, namely, the difficulty of transporting grain from the farm to the grain elevator. In the late nineteenth century most farmers transported their grain by horse-drawn wagons to loading points (grain elevators) along the two east–west railway lines. Under the best conditions, farmers operating 15 km from a grain elevator were fortunate if they could haul their grain to an elevator within one day. For that reason, farmers did not cultivate much land beyond 15 km of a grain elevator. By building many branch lines to the main east–west railway, railway companies expanded their rail systems and made grain farming commercially viable in almost all areas of the Prairies.

A third geographic challenge was the region's short growing season. In 1910, a new strain of wheat, Marquis wheat, was developed. Its shorter maturation period overcame the threat of frost.[4] Marquis wheat also extended the growing area for wheat in the Prairies. By 1920, Marquis wheat was the most popular spring wheat in Western Canada and the adjoining Great Plains states.

Drought posed a fourth geographic problem for farmers, especially those farming in the semi-arid Dry Belt of southern Alberta and Saskatchewan. While drought remained a threat, the technique of dry-land farming reduced the risk of crop failure. In dry-land farming, part of the land is left in **summer fallow** each year. In this way, sufficient soil moisture is accumulated over several years, allowing for the seeding of the land every other year.[5]

From Intensive to Extensive Agriculture

Over the last century, agriculture in Western Canada underwent significant changes. The first change was the shift in the farm economy from a labour-intensive operation to a capital-intensive one. At the end of the nineteenth century, many hands were required to successfully deal with the sowing, growing, and harvesting of grain. The introduction of machinery, such as self-propelled steam tractors and threshing machines, changed the way farms were run and reduced the need for labour. Further technological changes continued to affect the size of the farm labour force and the size of the farms. For instance, the development of the combine harvester, which can cut and harvest a swath of grain as wide as 15 metres, allowed farmers to harvest hundreds of hectares in a single day. The days of the quarter section of land, the land size obtained by homesteaders, were over. By 1921, farmers required a half section or more to make a profit.

After the mechanization of the farm economy, a trend of consolidating farms into larger and larger units occurred. The number of farms declined, while the size of farms increased (Tables 8.1 and 8.2). In the Canadian West, this shift began before World War I but accelerated during the Depression (1930s) because of extremely low prices and poor harvests (Figure 8.2). Many farm households fled the land penniless, the bulk from the semi-arid zone known as Palliser's Triangle. Depopulation of rural Western Canada continues. In the 35-year period from 1971 to 2006, the number of farms declined by 35 per cent, while the average farm size doubled from 726 acres in 1971 to 1,425 acres in 2006.. Nearly 62,000 farms disappeared, with Saskatchewan alone losing more than half of these farms. Some lost heart and sold their lands while others were less efficient farmers or lacked sufficient capital. For the surviving grain farmers, larger farms increased productivity and thereby lowered per unit costs.

The move from intensive to extensive agriculture transformed the grain economy of Western Canada. With fewer people engaged in agriculture, the rural landscape lost much of its population to towns and cities. The ripple effect was that many villages, which had been small service centres, were abandoned and disappeared from the landscape.

Economic Diversification

While Western Canada remains an important agricultural region, development of its forests, minerals, and petroleum deposits has diversified the region's economy, particularly in Alberta. Following World War II, rising world prices for primary products led to a resource

Table 8.1 Number of Farms in Western Canada, 1971–2006

Year	Alberta	Saskatchewan	Manitoba	Western Canada
1971	62,702	76,970	34,981	174,653
1981	58,056	67,318	29,442	154,816
1991	57,245	60,840	25,706	143,791
2001	53,652	50,598	21,031	125,281
2006	49,431	44,329	19,054	112,814
Change 1971–2006	−13,271	−32,641	−15,927	−61,839
Percentage Change	−21.2	−42.4	−45.5	−35.4

Sources: Statistics Canada (1992b, 2003a, 2007a).

boom in Western Canada. American demand for oil and gas from Alberta rose sharply, while Saskatchewan saw major resource developments in potash and uranium. At the same time, Manitoba became a major producer of nickel and hydroelectric power. Furthermore, the forest industry in all three provinces expanded with the rising demand for lumber, pulp, and paper in the United States and other industrial countries.

While these resource developments helped diversify the economy of Western Canada, rising oil prices in the 1970s triggered a major resource boom that had a most dramatic impact on the economy of Alberta. This boom had important spinoffs for the Albertan economy, including jobs and royalties, technological advances that allowed for the mining of the vast tar sands in northern Alberta, pipeline construction projects to supply the large markets of Ontario and the United States, and the emergence of Calgary as the headquarters for the offices of major oil companies.

By the 1980s, Alberta's economy had not only grown rapidly, it had also diversified, transforming Alberta into a 'have' province. While Alberta's economy and population expanded rapidly during the 1970s and 1980s, Manitoba and Saskatchewan followed at a much slower pace. In fact, during the

Table 8.2 Average Size of Farms in Western Canada, 1971–2006 (acres)

Year	Alberta	Saskatchewan	Manitoba	Western Canada
1971	790	845	543	726
1981	813	952	639	801
1991	898	1,091	743	911
2001	970	1,283	891	1,048
2006	1,824	1,451	1,000	1,425
Change 1971–2006	1,034	606	457	699
Percentage Change	131	72	84	96

Sources: Statistics Canada, 1992b, 2003a, 2007a

decade that followed, low prices for primary products had a greater impact on Manitoba's and Saskatchewan's economies because they continued to rely heavily on agriculture and resources. By 2006, however, prices for energy and minerals had reached new heights, driven by the expanding economies of Pacific Rim countries, especially China. Even agricultural prices show signs of improving, leaving only forest products with low prices. However, a downturn in the global economy could see prices drop sharply, though such a downturn seems unlikely for the next few years.

The Prairie Psyche and Western Alienation

Western Canada's geography but particularly its unpredictable weather for agriculture forged a 'Prairie psyche'. Homesteaders, confronted with a harsh environment and long shipping distances to ocean ports, developed a sense of place based on the lack of control over their environment and economy. Unlike southern Ontario and even southern Quebec, arable land in Western Canada lies much further north. Windsor at 42° N and Québec City at 47° N represent Central Canada's latitudinal range of arable land while Winnipeg, 50° N, and Edmonton, 54° N, represent the western latitudinal range. Hence, climatic conditions in the Canadian Prairies are much less favourable for a successful harvest. In fact, drought, grasshoppers, hail, and untimely frosts or rain at harvest time can turn a bumper crop into crop failure.

Over half the wheat grown in Canada comes from Saskatchewan. Wheat remains the principal crop in Saskatchewan because of its ability to grow with little moisture, and after a 'million-dollar' rain in June 2007 a bumper crop was anticipated. Demand for grain crops—especially corn—from the growing biofuel industry and from China for hog production has meant increased opportunity for farmers in the Prairie West. (Al Harvey/The Slide Farm)

Before World War I, farmers felt frustrated by external forces—world prices for grain remained low, while prices for agricultural machinery made in Ontario increased. Furthermore, dealers at the Winnipeg Grain Exchange bought low from the farmers and sold high to the grain buyers, and operators of private grain elevators along the railways assigned the farmers a low grade for their wheat (resulting in a low price). These conditions led farmers to form new institutions or movements. For example, in 1913 farmers banded together to create the United Grain Growers, to provide an alternative means of selling their grain and thus protect themselves from the low grades and prices offered by private grain companies.

The deep sense of alienation expressed by farmers has been an ongoing theme in the history of Western Canada and can now be found in all sectors of society in Western Canada. This negative feeling stems from the peripheral position of Western Canada within Canada and the global economy. Ottawa is often seen as either an uncaring government that ignores western grievances or a manipulative state power that places the interests of Central Canada over those of Western Canada. For instance, at the time that Alberta and Saskatchewan joined Confederation in 1905, the two western provinces were denied control over natural resources while Ontario,

Québec, Nova Scotia, and New Brunswick had obtained this control (and taxing power) when they united to form the Dominion of Canada in 1867. A more recent example is the federal initiative of the early 1980s known as the National Energy Program. In this case, Ottawa exerted its control over oil prices and levied taxes on oil production. In Alberta's eyes, the National Energy Program was both a 'tax grab' and a political means by which Ottawa favoured energy-deficient Ontario over Alberta, securing low oil prices for Central Canada's manufacturing industry while interfering with Western Canada's resource revenue.

Sometimes western alienation has led to the formation of new political movements, particularly political parties, as a means of combatting the apparent inequities of centralist governments. One such example was the Co-operative Commonwealth Federation (CCF), a political party formed in 1932 by a coalition of labour and farming interests in an effort to combat the destitution people were experiencing during the Great Depression. (The CCF later became the NDP.) It met with considerable success, and is attributed with leading the way for the creation of a variety of social programs in Canada. Social Credit, another Western-based political party, came into existence at about the same time. The Reform Party of Canada, which became the Canadian Reform Conservative Alliance in 1999 and then amalgamated with the old-line Progressive Conservative Party to form the Conservative Party in 2003, provides the most recent example of a western protest party.

The political fragmentation of Canada's federal parties is strongest in the West (British Columbia and Western Canada) and Québec. In the 1960s, Howard Richards, who founded the Department of Geography at the University of Saskatchewan in 1960, argued that isolation from other populated regions and from the political decision-making in Ottawa is at the root of western alienation. This isolation is worsened by the physical barriers of the Cordillera and the Canadian Shield. As a consequence, westerners feel exploited by Central Canada's businesses and politicians, who wield the economic and political power that ultimately affects those living in the western hinterland. For farmers, the targets of their resentment were the banks and railways.[6] Later, for oil producers and provincial governments, the target would become Ottawa's centralist policies. Western alienation remains rooted in the centralist/decentralist faultline where western interests are seemingly ignored by Ottawa in favour of national (i.e., Central Canadian) interests. Following the Liberal re-election in November 2000, Prime Minister Jean Chrétien attributed western alienation to voters focusing too much on local issues. The Prime Minister went on to say: 'It's possible [that we will try to have a better approach] but we have to ask them [Westerners] to look at the national scene from a national perspective' (Bellavance and Fife, 2000: A1). In 2002, Prime Minister Chrétien signed the Kyoto Protocol, which he believed was in the national interest. Alberta, however, saw the Kyoto Protocol as a threat to its heavy oil industry. In 2007, Prime Minister Harper declared that the Kyoto commitments were unachievable, but that his government would introduce 'hard' caps for industrial polluters, including the automobile producers in Ontario and the oil sands companies in Alberta. Another form of western alienation reared its ugly head over the 2007 revisions to the equalization formulation (Vignette 3.5). The Saskatchewan government was particularly concerned about keeping non-renewable resources out of the formula. Furthermore, Saskatchewan argues that what is good for Atlantic Canada is good for Saskatchewan. The federal government, which has total responsibility for the equalization program, decided that non-renewable resources would be included in the formula and that a cap was necessary to ensure that have-not provinces' fiscal capacity—the capacity of a province to raise it own revenues—will not exceed the fiscal capacity of Ontario and/or Alberta, both of which have higher fiscal capacities than the other eight provinces (Vignette 8.3).

Vignette 8.3 **Atlantic Goose, Prairie Gander**

Provinces with non-renewable resources but especially oil and gas have seen their revenues rise sharply with the increase in the price for these energy products. Newfoundland and Labrador and Nova Scotia complained that the new revenue was being clawed back through reduced equalization payments. The federal government under Prime Minister Paul Martin negotiated the Atlantic Energy Accord, which basically corrected this clawback issue. Saskatchewan then asked for a similar arrangement. The Premier of Saskatchewan, Lorne Calvert, stated that 'Saskatchewan should receive the same opportunity to fully retain our non-renewable resource revenues as Atlantic Canada, through the negotiations of a Saskatchewan Energy Accord' (Calvert, 2005). By June 2007, Ottawa had given no indication of an interest in such negotiations and the Saskatchewan government was considering legal action against Ottawa.

Western Canada Today

Western Canada has evolved from a narrowly based agrarian economy to a more diversified one. Increased exploitation of its natural resources, a growing trend towards processing these resources, and a fundamental shift in its agricultural sector are the factors contributing to the region's transformation. Alberta's economy, driven by the oil and gas industry, has become the most diversified of the three provinces. Saskatchewan and Manitoba have also diversified but, lacking Alberta's huge petroleum reserves, their economic growth is occurring at a much slower rate. The 1995 cancellation of the Crow Benefit, a transportation subsidy that allowed farmers to ship their products by rail at a reduced cost, initiated massive changes in the agricultural sector (Vignette 8.4). In 1995, prices for grain were relatively high and farmers could manage the rail charges. But in the following years, grain prices dropped while rail charges rose, pushing many farmers into bankruptcy. The challenge remains to process more agricultural products within Western Canada and to ship higher valued process products in containers by rail and then by ship to world markets..

Alberta is the economic giant of the three provinces. It has over half of the population in Western Canada and accounts for about 63 per cent of the region's GDP (Table 8.3). In Alberta the 2000 GDP per capita amounted to just over $40,000. Both Saskatchewan and Manitoba fall well below this level. What accounts for this variation? Each province has much natural wealth. Saskatchewan, for example, has most of the cropland and is the leading producer of potash and uranium. In addition to having the richest agricultural land in the West, Manitoba produces vast amounts of hydroelectric power from the Nelson River. Even so, Alberta holds the trump resource card—oil and gas.[7] Indeed, by 2015 Alberta is forecast to rank as the fifth largest producer of oil in the world! (Grauman, 2006).

Back in 1972, when a barrel of oil was worth $2 on the world spot market, Alberta oil was not valuable enough to dominate the western economy. Now that oil prices have increased by 40 times (in mid-September 2007 a barrel reached the $80 mark), the same oil reserves have greatly appreciated in value and thus in economic importance. When the 2002 price was around $30/barrel, the annual value of natural gas and oil production in Alberta was just over $40 billion. By 2005, the price was closer to $50/barrel and the total value had increased to $81.5 billion. Saskatchewan, too, benefited from higher prices. Its 2002 value of petroleum production was $5.6 billion, and this climbed to $8.7 billion in 2005 (Table 8.7).

Vignette 8.4 The Origin and End of the Crow Benefit

When the Canadian Pacific Railway was built, the railway company required public financial assistance to cover the construction costs. In exchange for that assistance, the CPR agreed to lower its shipping rates to Thunder Bay, thus reducing the cost of marketing grain in Canada's major market, Great Britain. Signed in 1897, the Crow's Nest Pass Agreement between the Canadian Pacific Railway and the federal government called for the CPR to lower the rate of moving grain to Thunder Bay by 3 cents per hundredweight in return for a $3 million federal subsidy to extend the rail line from Lethbridge to the Kootenay Valley in British Columbia. The purpose of this subsidy was to ensure that the rail rates for grain were low, that is, below actual shipping costs. This would allow the agricultural sector in Western Canada to thrive despite geographic distance from markets.

Over time, the transportation subsidy for Canada's two national railways increased, reaching $550 million in 1994. It increased for two reasons: shipping costs increased, and the subsidy was extended to include west coast ports. On 1 August 1995 the Western Grain Transportation Act ended and so did the railway subsidy known as the Crow Benefit. Now farmers would bear the full burden of shipping grain by rail to Canadian ports.

Though Western Canada exports both primary and processed products to other countries—recently increasing its exports to Pacific Rim countries—the petroleum industry is more closely tied to the North American market. Most of the natural gas and oil goes to markets in Ontario and the United States. After World War II, a network of oil and gas pipelines was constructed to serve both the Canadian and American markets. Energy, unlike manufactured goods, had easy access to American markets before the Free Trade Agreement.

World prices hold the key to economic growth for the resource industries. Imagine the impact on the Prairie economy if other **primary prices** (prices for primary resources) followed the upward path of oil prices! Imagine, too, if Europeans and Americans dropped

Table 8.3 Basic Statistics for Western Canada by Province, 2006

Province	Population (000s)	Unemployment Rate (%)	GDP (%)	Aboriginal Population (2001) (%)	Canada's Cropland (%)
Alberta	3,290	3.4	16.4	16.0	26.8
Saskatchewan	968	4.7	3.1	13.3	41.7
Manitoba	1,148	4.3	3.1	15.4	13.1
Western Canada	5,407	3.9	22.6	44.7	81.6
Canada	31,613	6.1	100.0	100.0	100.0

Sources: Statistics Canada (2003d, 2007a, 2007b, 2007c, 2007d).

Alberta's economic success stems from its huge petroleum deposits and high world prices for oil. With the continued development of the huge oil sands deposits near Fort McMurray, Alberta's place as the leading oil producer in Canada seems assured. (Al Harvey/The Slide Farm)

their wheat subsidies to Canadian levels! Unfortunately, Prairie farmers, miners, and loggers are all too well acquainted with commodity price cycles, which are controlled by global demand for such products. For a long time, prices, except for oil, were depressed. In the first years of the twenty-first century, an upward trend in many commodity prices emerged—all caused by changes in world demand but especially demand from China.

Industrial Structure

Employment by industrial sector in Western Canada reveals the prominence of primary economic activities. While Table 8.4 presents only a generalized picture of the western economy, it illustrates two important aspects. One is the importance of the primary sector. The percentage of people employed in this sector (10 per cent) is over five times the figure for Canada's principal industrial core region—Ontario. The second is the relatively smaller size of the secondary and tertiary sectors. The proportion of employment in the secondary sector is well below that of Ontario while the tertiary sector is just slightly lower (Table 8.4). For instance, percentage of employment in the manufacturing sector is less than half that of Ontario. However, the percentage of workers in the construction industries, a subset of the secondary sector, is higher in Western Canada. The explanation lies in the rapidly expanding resource industries, especially the oil sands in Alberta. In turn, construction of houses and office buildings is taking place at a record pace in Calgary, Edmonton, and Saskatoon.

Manufacturing

The manufacturing sector in Western Canada, while growing, is still relatively small. The four leading manufacturing centres are Edmonton, Calgary, Winnipeg, and Saskatoon. The explanation for the relatively weak state of manufacturing in the Prairies has been attributed to its small market. Two factors have changed the equation. First, the specialized demands of the expanding resource industries in Alberta and Saskatchewan have created local manufacturers of mining equipment. Second, market size has increased. From 1981 to 2005, a significant increase in Western Canada's domestic market took place as the region's population went from 4.2 million to 5.5 million. Western Canada's portion of total Canadian manufacturing increased to 10 per cent, with most of this occurring in Alberta. Much of the increase stems from the petrochemical, agriculture, and mining industries. In a few cases, global manufacturing companies are turning to Western Canada. Case New Holland recently announced its 2006 global consolidation plan, which called for the closure of several plants in the US Midwest and the relocation of those activities to Saskatoon. The Saskatoon plant will add the production of Case New Holland air-seeders to its planter and seeder production lines for the North America market. Winnipeg, already a centre for the aerospace industry, has high hopes to receive a substantial portion of the $3.4 billion targeted for four C-17 Globemaster transport planes, which can carry tanks, soldiers, and large equipment around the world without stopping. While the airplanes would be built in the US, Boeing has agreed to spend an amount equal to the $3.4 billion purchase price on projects in Canada.

Table 8.4 Employment by Industrial Sector in Western Canada, 2005

Industrial Sector	WC Workers (000)	WC Workers (%)	Ontario Workers (%)	% Difference
Primary	284.3	10.0	2.0	8.0
Agriculture	132.8	4.7	1.5	3.2
Secondary	468.6	16.5	23.6	−7.1
Construction	214.2	7.5	6.2	1.3
Manufacturing	229.7	8.1	16.6	−8.5
Tertiary	2,095.3	73.5	74.4	−0.9
Total	2,848.2	100.0	100.0	

Source: Statistics Canada (2006c). Adapted from Statistics Canada website http://www40.statcan.ca/l01/cst01/labor21c.htm, Searched 12 September 2006.

Agriculture

Agriculture was the driving force behind the settlement and development of Western Canada in the late nineteenth and early twentieth centuries. At that time, most homesteaders grew spring wheat for export to Great Britain. Today, while spring wheat remains the principal crop, other crops, but particularly durum wheat, oilseed, and specialty crops, are strengthening their place in Prairie agriculture. Cattle and hog production is on the rise, too. These changes are due primarily to low prices for grain, rising costs of shipping grain by rail, the loss of the grain rail subsidy (the Crow Benefit), and much higher grain subsidies in the European Union and the United States.[8] The combination of these four factors has forced some farmers into bankruptcy while others have sought alternative crops or have turned to livestock and hog production. As well, approximately 30 per cent of farmers seek off-farm employment to supplement their farm incomes. In the process of change, the number of farmers has decreased sharply, leaving few people on the land and in the rural communities. Signs of decline, such as decaying communities, abandoned branch rail lines, and the closure of many rural grain elevators, are everywhere in the rural landscape. By the middle of the twentieth century, agriculture had lost its central place in the economy of Western Canada. The decline of rural towns and villages and the loss of farm population illustrate the weak position of the rural economy. The struggle to find its footing has seen a shift from grain to other crops and farm activities. By 2006, prices for grain and canola showed signs of increasing because of high oil prices. What is the connection between oil and agricultural prices? Oddly enough, higher prices for canola and spring wheat have been triggered by the indirect effect of the expanding demand on the part of European and US ethanol plants for biomass, which has taken much land out of cultivation for human food. Whether these higher prices will continue depends on the growth of the ethanol industry in North America and the demand for agricultural products from Pacific Rim countries, especially China.

Western Canada is blessed with rich chernozemic soils that are well-suited for growing cereal crops (Figure 8.3). Other crops, such as peas and lentils, require more moisture and are therefore restricted to the black soil zone. Unlike canola, durum wheat thrives in hot, dry weather commonly occurring in the brown soil zone. It also requires a longer growing season than spring wheat and therefore is usually grown south of the fifty-first parallel. Consequently, the agricultural land use varies, forming three distinct agricultural regions. They are: (1) the **Fertile Belt** (black soil associated with parkland and long-grass natural vegetation, (2) the **Dry Belt** (brown soils with short-grass natural vegetation), and (3) the **agricultural fringe** (southern end of the boreal forest) and the Peace River country (Figure 8.4). These subregions each have very different growing conditions. The major factors controlling those conditions are the number of frost-free days and the soil moisture. The Fertile Belt provides the best environment for crop agriculture. The Dry Belt occupies semi-arid lands and has become a grain/livestock area. The agricultural fringe is along the southern edge of the boreal forest, and the Peace River country forms a pocket in the northwestern area of this forest. In the agricultural fringe, the short growing season encourages farmers to grow feed grains and raise livestock, while in the Peace River country, farmers grow both grain and feed grain for raising livestock.

The Canadian Prairies are on the northern margin of crop agriculture where the length of the growing season is short. Slight variations in weather conditions have either a positive or negative impact on crops. Wheat, for example, is one of the more resilient crops that can grow in areas of low precipitation, but wheat harvests can vary widely from year to year in both size and quality. Because of its continental climate, temperature and precipitation vary from year to year so that the risk to the Prairie farmer is high compared to the southern Ontario farmer. Other weather conditions affecting crop farming include late spring seeding due to cold or wet weather, summer frosts before the crop is mature, and wet weather in the fall that impedes harvesting. All of these weather conditions will reduce the quality of the harvest. Beyond weather conditions, pests, such as grasshoppers, have greatly reduced the size of the harvest in

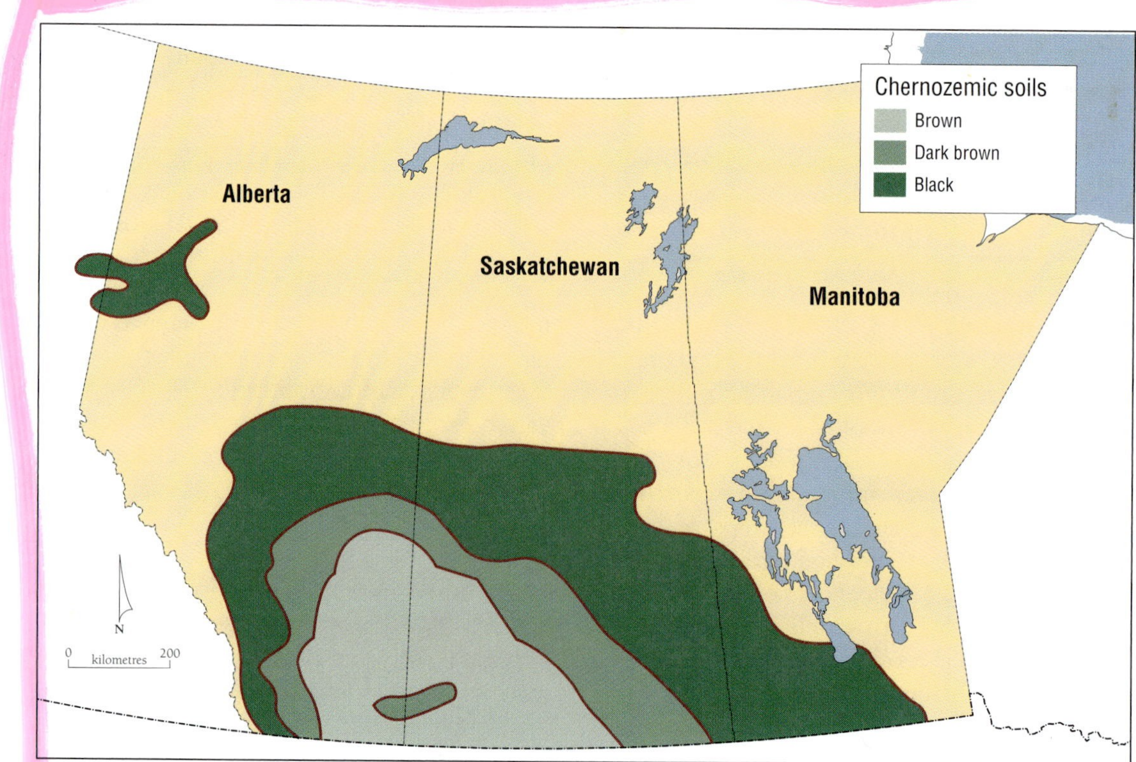

Figure 8.3 Chernozemic soils in Western Canada. There are three types of chernozemic soils in the Canadian Prairies: black, dark brown, and brown. The differences in colour are due to varying amounts of humus in the soil. The Peace River country's soil, formed under an aspen forest, is 'degraded' black soil.

Vignette 8.5 The Death of a Grain Elevator

The destruction of Saskatchewan Wheat Pool No. 8 near Melfort, Saskatchewan, in November 2000 reflects the economic transition occurring in the Prairie provinces. Until recently, the rural landscape of Western Canada was dominated by local grain elevators such as the one operated by United Grain Growers shown in the photograph. Each small rural community had at least one grain elevator, which was connected by a branch line to the main railway line to Vancouver and Thunder Bay. These elevators were the economic basis of rural communities but are no longer economically viable. The grain companies are replacing them with a smaller number of very large cement terminals at strategic locations along mainline railroads. The loss of these smaller elevators not only symbolizes the restructuring of Western Canada's agriculture but also signals the demise of many rural communities. The consolidation of the grain transportation network calls for the destruction of over 2,000 wooden grain elevators and the abandonment of many branch lines. Farmers are now faced with hauling their grain by truck to large grain terminals located on the CPR and CN main lines. As well, the heavy truck traffic on rural roads has caused their rapid deterioration, forcing provincial governments to repair and upgrade these country roads.

A grain elevator is demolished near Melfort, Saskatchewan. (*Peter Wilson*, Saskatoon Star Phoenix, *9 December 2000, E1.*)

some years. How would global warming affect farming? A longer growing season would be an advantage. The key question is precipitation. Would a warmer climate have the same, less, or more precipitation? If the precipitation is the same or less, farming would be more difficult because of the threat of droughts. On the other hand, if precipitation increased, perhaps because of more open water in Hudson Bay and the Arctic Ocean, then conditions for farming would be more favourable than at the present.

The Fertile Belt

The Fertile Belt extends from southern Manitoba to the foothills of the Rocky Mountains west of Edmonton (Figure 8.4). The higher levels of soil moisture, an adequate frost-free period, and rich soils make this belt ideal for a variety of crops and livestock. The most popular crop, since farmers first arrived in the West, has been wheat. In recent years, however, the acreage in grain has declined, while the planting of oilseed and specialty crops, such as beans, field peas, and sunflowers, has increased. This change was fuelled by rising prices for these crops and declining prices for wheat. Between 1996 and 2006, farmers moved away from traditional crops to alternatives that would reduce their input costs or increase their per acre revenues. Wheat is still the largest single crop. In Saskatchewan, wheat makes up close to 40 per cent of the area under field crops, but just six years earlier the figure was over 50 per cent.

The predominance of wheat in the Fertile Belt is most evident in western Saskatchewan and eastern Alberta. In southern Manitoba and the adjacent parts of eastern Saskatchewan, there is more annual precipitation so farmers there can grow a wider variety of crops as well as keep livestock. As a result, mixed farming is common. Grain and specialty crops (canola, flax, sunflowers, and lentils) are combined with beef, pork, and poultry production. Near Winnipeg, for instance, a livestock industry has developed where cattle are fattened before shipment to meat-packing plants in Winnipeg and Ontario. Feedlots and nearby meat-packing plants are located in other parts of Western Canada, particularly in Brandon, Calgary, Edmonton, Lethbridge, and Saskatoon. Since NAFTA, meat production has increased, especially pork, and most meat products are exported to markets in the United States.

As the urban markets grow, market gardens, dairy farms, and other specialized forms of intensive agriculture are developing near major cities in the Fertile Belt. The demand for specialty-crop products has spawned a number of smaller but very intense production units, including nurseries and greenhouses, to produce flowers and vegetables for local sale. For that reason, farm sizes are much smaller around the major cities.

The Dry Belt

The Dry Belt contains both cattle ranches and large grain farms. It extends from the Saskatchewan–Manitoba boundary to the southern foothills of the Rockies and north nearly to Saskatoon (Figure 8.4). However, the driest area, or heart of the Dry Belt, occupies a much smaller area, stretching southward from the South Saskatchewan River to the US border. The arid nature of the Dry Belt is not due to low annual precipitation (it is comparable to

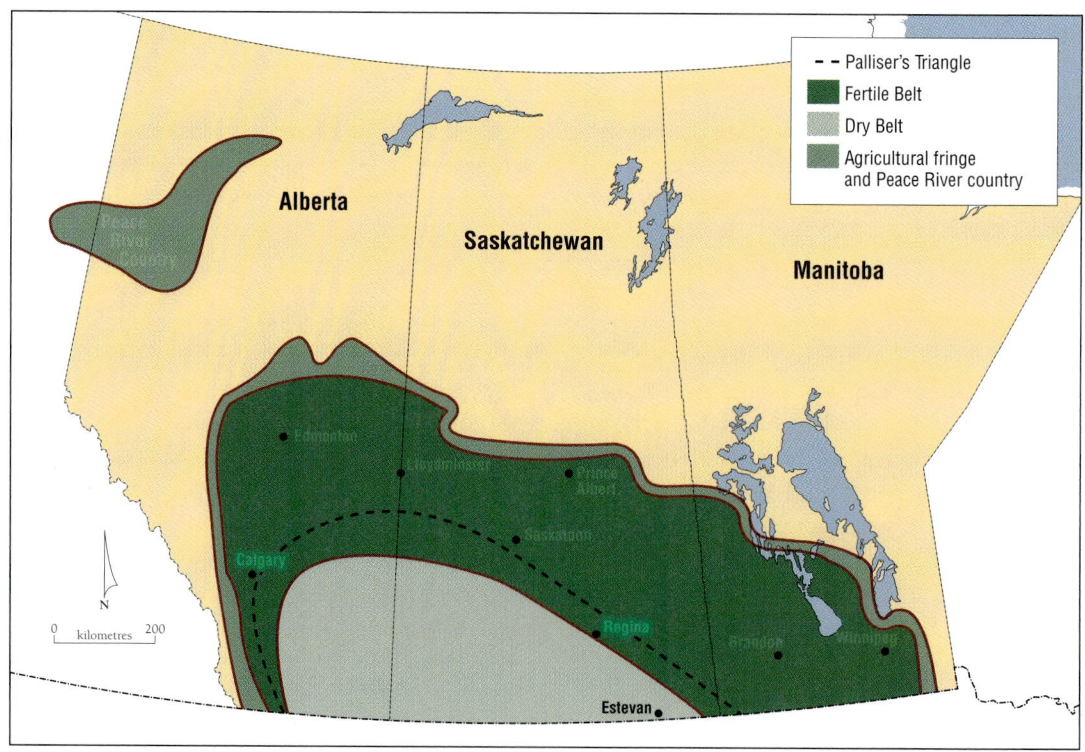

Figure 8.4 Agricultural regions in Western Canada. Farming in the Prairies can be divided into three areas: the Fertile Belt, the Dry Belt, and the agricultural fringe and Peace River country. Each region has a different type of agriculture because of variations in physical geography.

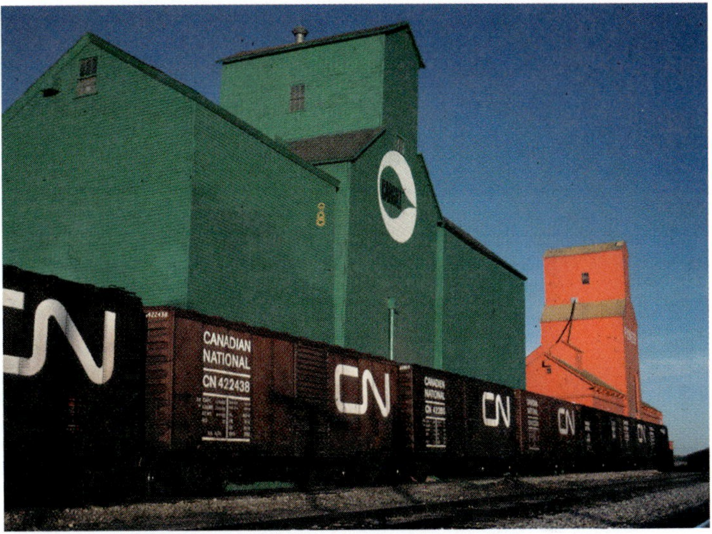

The export of grain to foreign markets requires an elaborate transportation and handling system. Grain elevators play a key role by serving as a collection point along railway lines where the grain is graded, cleaned, and weighed before being transported by railway cars to Canada's major ports. (Al Harvey/The Slide Farm)

other areas of Western Canada) but to longer summers and higher evaporation rates. Within the Dry Belt, feed grain and hay crops are grown to supply winter feed for the cattle ranching that dominates in this area. Prior to settlement, the Dry Belt's short-grass vegetation provided a natural grazing area for buffalo. Later, cattle replaced the buffalo. Cattle ranching began in this area in the 1880s. Today, ranches are large, often many times the size of grain farms, because of the lower productivity of the dry land and the need for huge grazing areas to support a rotational grazing system.

Along the northern edge of the Dry Belt, grain farming is pursued, but the risk of crop failure is high. To conserve soil moisture and control weeds, summer fallowing remains a widespread practice, though continuous cropping is gaining favour. The purpose of summer fallowing is that two years of precipitation will accumulate sufficient soil moisture to germinate the seed and sustain the young wheat plant. The crop will still require summer rainfall to reach maturity. On average, one-third of the arable land in the Dry Belt is kept in summer fallow each year. The drawback to summer fallowing is that the ploughed (fallow) land is exposed to water and wind erosion. In a dry spring, windy weather can result in extensive loss of topsoil with huge clouds of dust stretching for many kilometres across the Prairies. Continuous cropping keeps short, stiff stalks of grain or hay remaining on a field after harvesting, thus protecting topsoil from water and wind erosion. Then, too, advances in technology have allowed for 'one-pass' seeding, spraying, and fertilizing. The expense and time not spent on repeated tilling of a field reduce a farmer's costs and conserve moisture.

Irrigation on these semi-arid lands has provided another solution to dry conditions. The most extensive irrigation systems are in southern Alberta. In fact, nearly two-thirds of the 750,000 ha of irrigated land in Canada are located in Alberta. In the 1950s and 1960s, two major irrigation projects were developed in the dry lands of Alberta and Saskatchewan: the St Mary River Irrigation District is based on the internal storage reservoirs of the St Mary and Waterton dams in southern Alberta, and Lake Diefenbaker serves as a massive reservoir on the South Saskatchewan River. In 1992, the Oldman River Dam was completed. At first, development of irrigated land in Saskatchewan was hampered by the problem of finding crops that would provide sufficient revenue to offset the high cost of irrigation. Farmers grew cereals, alfalfa, hay, oilseeds, sugar beets, and canning vegetables. Until the location of three large french-fry-processing plants near Lethbridge, few potatoes were grown. The demand for potatoes by Lamb-Weston, McCain Foods, and Maple Leaf Foods resulted in potato acreage jumping from 28,000 acres in 1998 to nearly 40,000 acres in 1999 (Robertson, 1999: D3). After the rapid expansion of the potato industry from 1998 to 2003, potato acreage has levelled out to 51,000 acres (Alberta, 2006). Another problem, faced by Saskatchewan farmers in particular, is that with a shorter growing season their selection of crops is more limited compared to the corn, sugar beets, and other specialty crops that farmers in southern Alberta are able to grow. However, the major problem facing Saskatchewan is the longer distance to major markets compared to southern Alberta.

The Great Sand Hills, situated approximately 100 km northwest of Swift Current, Saskatchewan, is a fragile ecosystem that is unique within the larger environment of the short-grass vegetation zone. Located in the centre of the Palliser Triangle, some sand dunes remain void of natural vegetation, making them subject to wind erosion, while others have been stabilized by a covering of native prairie grasses. Cacti, creeping juniper, and small shrubs like rose, saskatoon, chokecherry, and silver sagebrush grow in the Great Sand Hills. (Al Harvey/The Slide Farm)

The Agricultural Fringe and Peace River Country

The agricultural fringe refers to the narrow strip of forested land located just to the north of the Fertile Belt while the Peace River country lies in the Interior Plateau where the deeply entrenched Peace River flows from the Rocky Mountains and across the High Plains. Grain, livestock, and hay crops are the principal agricultural activities. Specialty crops include legumes and grass seeds.

As the last agricultural frontier, settlers were not attracted to the agricultural fringe or the Peace River country until the best lands in the Prairies were settled. At the beginning of the twentieth century, pioneers first moved into these areas. For the Peace River country two other events were necessary to draw settlers. First, the signing of Treaty No. 8 in 1899 opened large areas to agricultural settlement. Second, accessibility to export markets improved with the construction of a railway to Peace River in 1915, thus allowing wheat and other agriculture products to be shipped to world markets. While homesteaders first arrived before World War I, most land in both the agricultural fringe and the Peace River country was brought under cultivation between 1920 and 1940. In these higher latitudes (55° to 57°N), crop agriculture is challenged by a short growing season and the threat of frost. These two natural factors plus the high cost of shipping agricultural products to world markets make farming even more risky and marginal than in the rest of Western Canada. In 1955, BC farmers in the Peace River country benefited from the extension of BC Rail to Dawson Creek. This shorter link to grain terminals in North Vancouver and then to Pacific Rim customers reduced shipping costs for these farmers.

Next Year Country

The dry continental climate in Western Canada makes farming a risky business. The threat of crop failure is greatest along the margins—the northern agricultural fringe is subject to frost, while grain farmers in the Dry Belt constantly face the possibility of drought. Prairie farmers often describe the land as 'Next Year Country'. The meaning behind these words is simple: Our crops did poorly this year, but we hope that they will do better next year. Often the reason for a bad year with low yields is insufficient precipitation during the growing period. However, farmers face a whole range of natural hazards: summer frosts, which occur when a cold air mass slides unimpeded from the Canadian Arctic to Western Canada; hail and early snowfall; pests, such as grasshoppers; and diseases, such as stem rust. All of these hazards can have devastating effects—from delaying a harvest or lowering the grade of wheat, to reducing a yield or destroying an entire crop. Nevertheless, when compared to the days of the pioneer farmers, who did not have the advantages of modern farm technology and improved strains of wheat, farmers now have a better chance of dealing with adverse weather conditions.

Since 2000, half of the crop years have suffered from a lack of moisture. Yet, dry spells are no surprise because Western Canada lies in a continental climatic zone. Long-term dry spells have occurred. Some fear that global warming could create such arid conditions again. Others argue that since weather records indicate that Prairie weather follows a cycle of wet and dry years, it is only a matter of time before another dry cycle occurs (Akinremi et al., 2001). Successive hot, dry summers can quickly dry up soil moisture, bringing back the spectre of drought, dust storms, and crop failure in the semi-arid areas. David Jones's *Empire of Dust* (1987) describes the settling and abandonment of homesteads in the 1930s in the Dry Belt of Saskatchewan and Alberta. Perhaps Palliser's original assessment of the limited agricultural potential of the semi-arid areas of Western Canada was correct? In any case, a rise in temperature, without a corresponding increase in precipitation, would threaten grain-growing in the Dry Belt by increasing the risk of crop failure.

The shortage of water in the arid lands of southern Alberta has been partly solved by damming rivers and using the water for irrigating sugar beet, hay, and potato crops. Fodder crops, including hay, supply the feeder cattle and hog farms with necessary animal feed. The Oldman River Dam, shown above, added significant amounts of water for irrigation. Alberta accounts for 60 per cent of Canada's irrigated farmland. (Bayne Stanley/Viewpoints West)

The Canadian Wheat Board

The Canadian Wheat Board is under great pressure. In 2006, the federal government removed barley from its monopoly. Friends of the Canadian Wheat Board believe that a monopoly is necessary for its success and therefore the success of farmers. The federal government, with its free-market perspective, appears to favour the abolishment of the Canadian Wheat Board.

Back in 1935, the federal government established the Canadian Wheat Board (CWB). Without this assistance, grain farmers felt helpless because they had no control over the prices for their commodities. Private grain buyers, it seemed, made money, but not the farmers. Since 1943, Prairie wheat farmers have been compelled by law to sell their crops only to the Board. While the CWB ceased to be a Crown corporation in 1998, it remains a quasi-federal government agency. The CWB remains the grain-handling and marketing agency for wheat, barley, and oats destined for export or use by the food industry in Canada. The twin goals of the CWB are to sell as much grain as possible at the best possible prices and to ensure that each producer gets a 'fair' share of the market.

Today, some farmers want to sell their grain on the open market because they believe that they can obtain better prices than the Canadian Wheat Board. In recent years, some Prairie farmers living near the border have challenged the authority of the Canadian Wheat Board by trucking their grain across the border to the US market, where prices have been higher for durum. In 2000, the Wheat Board retaliated, taking the issue to court, and in 2002 the court ruled in favour of the CWB. Grain farmers in Saskatchewan paid a $500 fine while a handful of Alberta farmers refused to pay the fine and, as a protest against this monopoly, went to jail (CBC, 1999, 2002).

The Canadian Wheat Board has also been under attack from Washington. In 2003, the United States served notice that it intends to try to dismantle all state trading enterprises because Washington believes such enterprises subsidize exports. For that reason, Washington has gradually raised the duties on durum and spring wheat. By August 2003, the duties were raised again, reaching a 13.55 per cent duty on high-quality durum wheat and a 14.6 per cent duty on spring wheat. Following a NAFTA decision, both tariffs were removed by the US government in February 2006. The Wheat Board sells about $400 million worth of wheat a year in the US market, representing about 5 per cent of the American market. Millers and bakeries in the United States prefer Canadian wheat because of its high quality, which is ideally suited for pasta.

Key Topic: Agricultural Transition

Prairie agriculture is in a state of transition. In the late twentieth century, the liberalization of international trade opened up the US and other foreign markets for Western Canadian agricultural products and, at the same time, exposed farmers more directly to global forces of demand and supply. For Prairie agriculture, NAFTA provided unfettered access to the US market, thus encouraging farmers and ranchers to expand their production. The agricultural transition also saw a cutback in federal subsidies, including the Grain Transportation Subsidy. The initial impact on Prairie farmers was unfavourable for two reasons. First, their input and grain shipping costs increased while prices for their products remained low. Second, their access to the US market was unexpectedly restricted and therefore their expanded production designed for the US market became an unwanted surplus. The second phase of the agricultural transition, which is just unfolding, sees a brighter future for western farmers because of world price increases for agricultural products and the removal of US trade barriers for grain and livestock.

First Phase: Time of Low Prices and Rising Input Costs

The first phase of this agricultural transition was marked by the end of federal subsidies for grain transportation (see the subsection 'Grain Transportation Subsidy') and by rising prices for fertilizers, fuel, pesticides, and other farm inputs. This transportation subsidy was known as the Crow Benefit. With its passing, grain farmers had to pay all the freight charges for grain shipments. Worse yet, the price of grain declined from a high in 1995 while freight rates increased each year. Therefore, the first phase of this agricultural transition represented hard times for farmers.

While the shipment of feed grain to eastern cattle and hog farms dropped dramatically because of the higher freight rates, the geographic pattern of foreign exports did not alter, with the US remaining the main market for most agricultural exports, including spring wheat, canola, hogs, and cattle. The reason was simple: even with higher freight rates, the railways represented the only route for the export of grain to world markets. Farmers (except those close to the US border who could truck their product to US markets) had three choices: continue to ship to world markets and absorb the additional transportation costs; switch production to another crop or other crops; or sell the farm. Switching to another crop had its drawbacks. Crop rotation might be affected; new crops could require new equipment and certainly new farming techniques; and prices for new crops might be as low as those for wheat.

For Western Canadian farmers, then, trade liberalization was and is a double-edged sword. This new world of so-called open markets has provided export opportunities, but as exports to the United States increased, duties were imposed. With prices remaining low and cost of production increasing, many farmers went bankrupt or sold their farms. From 2001 to 2006, the number of farms declined by nearly 13,000 (Table 8.1). Trade liberalization has also led to the elimination of farm subsidies, such as the transportation subsidy for grain, and it threatens farm marketing boards and farm organizations, such as the Canadian Wheat Board.[9] Canada, being so dependent on exports to the United States, is vulnerable to US trade barriers. While Washington aims at protecting its domestic producers, the cost of this protection is borne by Canadian farmers and ranchers. Perhaps the worst aspect of trade with the United States is the uncertainty—if Canadian producers expand production to service the US market, then their operations are extremely vulnerable to barriers to that market.

In 1995, the federal grain transportation subsidies that cushion grain farmers from high rail costs ended. This single policy shift unleashed economic forces that are reshaping Western Canada's agricultural economy. While this transformation affects all sectors of the western economy, the changes are most visible in the agricultural sector. The end of the grain transportation subsidy, the continued use of agricultural subsidies by the United States and Europe, and the persistent use of trade barriers by Washington to impede the flow of Canadian grain and livestock products to the US market have made this transition to a more open market difficult. Another problem is low prices for wheat, making it difficult for farmers to cover the cost of shipping their grain to ocean ports.

The dry climate has always been a challenge for farmers. Since 2000, Western Canada has experienced three years of hot and dry weather and four years of hot and wet weather. In the dry years, drought affected both farmers and ranchers—farmers saw their yields drop and ranchers had little hay for winter feed. In the three wet years, two other natural problems occurred. A frost in August 2004 caused the quality and price of grain to drop while a wet fall made harvesting difficult in 2005. Added to difficult climatic conditions, in 2003, the discovery of BSE (bovine spongiform encephalopathy) in Alberta had devastating implications for the livestock industry. Almost immediately, the United States and Japan (the chief export markets) halted the importation of Canadian livestock and meat products. The Canadian livestock industry is concentrated in Western Canada, with Alberta having almost half of Canada's 13.4 million cattle. Beef cattle comprise Alberta's largest agricultural sector, providing about half of farm revenue. Exports are extremely important, making up around 40 per cent of total sales. The largest customer is the United States, followed by Japan and Mexico. With the halt of exports, Canadian prices to the ranchers dropped dramatically, though retail prices did not follow. The US lifted the ban of Canadian cattle in 2005.

The Second Phase: Higher Prices

Has a second phase of the agriculture transition begun? By June 2007, prices for most agricultural products were up substantially over the previous year and the price for wheat had recovered to the highest point (1995) in the last 20 years. Two questions are: (1) why have prices risen and (2) are these higher prices just a brief anomaly? Professor Brian Oleson, an agribusiness professor at the University of Manitoba, believes that these higher prices are not anomalous but represent a dramatic shift to permanently higher prices for wheat and other agricultural products. Oleson states: 'We are entering a new era. It's almost as if the platform [for wheat prices] has been raised and we're all dancing on a new dance floor' (Greenwood, 2007b).

The second phase, if real, has been triggered by two international developments. Both spell good news for Prairie farmers. Higher prices for grain, canola, and pork can be attributed to (1) demand for more meat, in particular pork, from China's growing middle class and (2) the need of the biofuel industry for corn and other crops. China has become a global economic power and, as a result, a substantial middle class has emerged with money to pay for a better diet, which often includes pork. Consequently, Chinese farmers are increasing their production of hogs and more feed (corn) is needed. In the first six months of 2006, prices for both hogs and corn have risen sharply in China. And there is no turning back. As Keith Bradsher (2007) reported:

> Few things are as essential to the Chinese as their pigs. From pork spare ribs and mu shu pork to char siu bao—barbecued pork buns—pork is a staple of the Chinese diet. So in this Year of the Pig, an acute shortage of pork has been national news, as butchers raise prices almost daily and politicians scramble to respond.
>
> Chinese officials offer several reasons for the high pig prices. The cost of animal feed has risen by one-quarter in the last year, partly because more corn is being made into ethanol and partly because more prosperous workers are eating more meat, which requires more animal feed.

As well, the biofuel industry has become a major industry in Europe and the United States. For the European Union, the assumption is that ethanol-blended fuel produces cleaner emissions than regular gasoline and thus is an important part of Europe's attempt to reduce the emission of greenhouse gases to the atmosphere. In the case of the United States, the political rationale supporting the ethanol industry is to reduce dependency on foreign oil. In the US, ethanol is made from corn. But corn is also used to feed chickens, hogs, and cattle. As demand for corn increases, its price will also increase, which means a rise in prices for meat, eggs, and dairy products and for other farm produce. While the federal Conservative government has committed $2 billion in incentives for the construction of ethanol and biodiesel plants in its June 2007 budget, at least a year will pass before these plants are consuming vast quantities of grain and other forms of biomass. At that time, the Canadian demand will affect prices in the Canadian Prairies. However, US ethanol and biodeisel plants are already operating and their consumption of corn has pushed the US price up. Signs indicate that US prices are translating into higher prices for Canadian grain and canola. The large subsidies provided by governments to the biofuel industry represent the only dark cloud on the horizon, i.e., if the subsidies end, then the biofuel industry will likely collapse and grain prices will fall.

A third factor suggesting a permanent turnaround in grain prices is that as cities sprawl into agricultural land, the amount of high-quality land available for food production decreases. Of course, the counter-argument to a smaller arable land base is that technology will allow farmers to produce more on a smaller amount of land. Such an argument, however, is itself countered by the realization that additional technological inputs can and do have negative environmental consequences.

Grain Transportation Subsidy

The greatest change affecting agriculture in Western Canada is the loss of the Crow Benefit in 1995. Since then, farmers have had to pay the full cost of rail transportation to ship their grain to port.[10] The magnitude of the new transportation costs is illustrated by the size of the former subsidy. In 1994, the Crow Benefit subsidy of $550 million amounted to approximately half the cost of shipping grain. This federal subsidy went directly to the railway companies, which were compelled to keep their rates for grain below cost. Now that farmers face the full transportation costs, the burden of farming in the heart of North America makes shipping to ocean ports costly. As a result, Prairie farmers are searching for crops and livestock that can be sold and processed locally, thus reducing their transportation costs. The shift in land use and the movement to local processing varies across the Prairies. In 2006, spring wheat was still the dominant crop in the Prairie provinces, but the land seeded to spring wheat had declined sharply between 1996 and 2001. Since then, farmers have seeded around 20 million acres in spring wheat each year (Statistics Canada, 2006a). In 2006, Prairie farmers harvested 19.6 million tonnes of spring wheat, up 6.7 per cent from 18.3 million tonnes in 2005 and well above the five-year average of 15.7 million tonnes. Saskatchewan, Alberta, and Manitoba accounted for 9.1 million tonnes, 7.0, and 3.5 million tonnes respectively (Statistics Canada, 2000d).

Farmers in eastern Saskatchewan and in Manitoba are growing less grain but more feed grains and specialty crops. Since the climate in the eastern Prairies provides more moisture and a longer growing season than in most other areas of the Canadian West, farmers have a wider variety of crops to choose from. Feed grains are popular because they can be sold locally to beef and hog producers in these provinces. Similarly, specialty crops, such as canary seed, grown under contract to larger firms, are also sold locally. Risk does exist in shifting away from wheat. For instance, the expansion of livestock and pig herds and slaughtering plants, such as the new hog slaughtering plant at Brandon, is dependent on sales to the United States. Recently, sales to the US have declined. Still, the announcement of two canola crushing plants at Yorkton, Saskatchewan, and the expansion of existing Saskatchewan crushing plants at Nipawin and Clavet will encourage more canola production (probably at the expense of wheat). Added to the canola expansion, the expected demand for industrial rapeseed for biodiesel and other non-food uses provides more evidence of a new land use and local processing.

Farmers in western Saskatchewan and in Alberta did not decrease their wheat acreage for two reasons. First, wheat is an ideal crop for semi-arid growing conditions. Second, western Saskatchewan and Alberta have the lowest rail costs for shipping spring wheat to Vancouver. Ranchers, too, benefit from short distances to major markets, with a relatively short rail haul to Vancouver to reach the Asian market. Southern Alberta has greatly expanded its cattle and hog slaughtering plants, which has resulted in an expansion of the livestock industry and the growth of more hay and fodder crops for winter feed.

The Prairie Staple

Grain production, particularly spring wheat, has been the prairie staple for over 100 years. Grains, especially spring wheat, do well in dry conditions whereas other crops would fail. Recent developments, however, such as low world prices for spring wheat and the loss of the rail transportation subsidy for grain, have led grain farmers to seek alternative crops.[11] From 1986–7 to 2005–6, wheat cultivation has dropped by 41 per cent while canola has increased by 152 per cent (Table 8.5). The concern about global warming has opened a new opportunity for Prairie farmers. The most promising prospects lie in the emergence of ethanol and biodiesel production, which is creating a new demand for canola. Already, Europeans are importing Canadian canola to fuel their ethanol and biodiesel plants. The net result is higher prices for canola and an anticipated switch from wheat to canola in the 2006–7 crop year.

Access to the US market has meant:

- larger herds of cattle and hogs with more beef and pork processing in Western Canada,
- increased demand for feed grains;
- an expanding canola industry with local crushing plants shipping the value-added oil product to the US, Japan, and other Asian countries;
- the growth of specialty crops, such as sunflowers, lentils, chick peas, and field peas;
- the production of alfalfa pellets from hay for shipment to the US and Japan for feed in their dairy industries.

Farmers are growing more oilseed crops, such as canola and flax, and more specialty

Table 8.5 Shift in Acreage: Wheat, Durum, and Canola, 1986–7 to 2005–6 (million acres)

Crop	1986–7	2005–6	Change in Acreage	% Change
Spring wheat	17.8	10.5	(7.3)	41.0
Canola	2.5	6.3	3.8	152.0
Durum	3.5	3.7	0.2	5.7

Sources: Saskatchewan Agriculture and Food (1996); Saskatchewan Agriculture, Food, and Rural Revitalization (2007).

crops. The primary reason for the viability of these alternative crops is that their prices, although fluctuating, have risen over the last 20 years, while the price for spring wheat has not changed appreciably. In fact, the price of spring wheat dropped so low that from the crop year 1998–9 to 2001–2 many farmers were losing money. However, price fluctuations in alternate crops do occur. The price for canola, for example, dropped by 30 per cent from 1994–5 to 2001–2 but rose again by the 2005–6 crop year. During that seven-year period, however, the price for canola did not decline steadily but varied widely from year to year, making choices about what crops to seed difficult and risky. Another factor is the lower transportation costs for the farmer because oilseed and specialty crops only need to be trucked to local processing plants. Major food companies have built large canola-crushing plants in Western Canada over the past 10 years, encouraging farmers to grow more canola. The proposed ethanol crushing plants would consume more canola. By shipping their products to local plants by truck, farmers' transportation costs are minimized. However, canola has its limitations. Canola cannot be grown in the more arid areas of Western Canada, and the crop is prone to disease and weed infestations if cultivated repeatedly.

With the end of the Crow Benefit in 1995, livestock producers in southern Ontario can no longer afford to import feed grains from Western Canada. However, a growing livestock industry in Western Canada now supplies beef and pork products to Ontario and Québec. As well, the demand for pork is increasing, opening more markets for Western Canadian producers in both the United States and Pacific Rim countries. To meet the demand, huge hog farms, similar to those in the United States, have sprung up across Western Canada that take advantage of modern agro-technology, mass production techniques, and economies of scale. Because of the smell and the risk of contaminating groundwater, large hog barns are

Table 8.6 Shift in Price: Wheat, Durum, and Canola, 1986–7 to 2005–6

Crop	1986–7 ($ tonne)	2005–6 ($ tonne)	% Change
Spring wheat	105	170	61
Canola	199	350	78
Durum	122	180	58

Sources: Saskatchewan Agriculture and Food (1996); Saskatchewan Agriculture, Food, and Rural Revitalization (2007).

located in rural settings far from settlements. Here, the hog barns are providing employment for residents of rural communities, though their waste poses a threat of fecal bacteria seeping into groundwater.

The livestock industry is also undergoing change—a combination of restructuring and consolidation of its processing plants. As a result of the Free Trade Agreement in 1989, competition within the North America market has become fierce, forcing Canadian operators to build larger hog-processing plants, to specialize in a single product in each plant, and to demand lower wages from employees. In Western Canada, new hog-slaughtering plants have been built at Brandon, Red Deer, and Lethbridge and another one is planned for Winnipeg. Plant specialization is driven by cost of production—larger plants gain economies of scale, and specialized production lines achieve higher productivity. While it does cost to ship meat from the slaughterhouse to a number of processing plants, the saving gained from efficiencies in processing plants dedicated to a single product offsets the shipping costs. For example, Maple Leaf Foods' new hog-slaughtering operation at Brandon supplies carcasses to its Winnipeg 'ham' plant and its North Battleford 'bacon' plant. Michael McCain, president of Maple Leaf Foods, describes competing for a place in the North American market for pork products as 'akin to dancing with elephants. We have to be a little more nimble, a little more agile than our dancing partners' (Bell, 1999: 60).

Western Canada now accounts for 40 per cent of hog production in Canada. Manitoba leads with 3 million pigs, Alberta has 2.1 million, and Saskatchewan follows with 1.2 million. Large-scale hog barns have led the way to this remarkable expansion. Hog barns, like cattle feedlots, produce a great deal of waste that could threaten the local water supply. At a community meeting in 2003 in Foam Lake, Saskatchewan, the issue of economic benefits from a proposed large hog barn was challenged by the possible environmental costs. The proposal included a breeder farrow barn for 5,000 sows, a nursery barn, and three finisher barns. Pig excretion would be placed in lagoons and, in time, would be removed as manure that could be placed on fields as fertilizer. Grain farmer Harry Abtosway had heard enough. He complained that 'Pollution, environment, smell, you name it. I think we are going to ruin our land, our water supply' (Hall, 2003: C8). Large hog barns are not popular with local people, but then, as Terry Markusson, the proponent of the proposed hog barn, put it: 'There is nothing that smells as bad as a dying community.' In March 2003, the Foam Lake municipality council rejected the proposal after residents submitted a petition with over 600 signatures opposing the hog barn plan.

Western Canada's Resource Base

Western Canada's resource economy began to diversify in the 1970s, when oil and gas developments in Alberta benefited from rising prices, uranium mining gained a foothold in Saskatchewan, and Manitoba expanded its hydroelectric facilities on the Nelson River. The dramatic rise in oil prices, resulting from the action of the Organization of Petroleum Exporting Countries in 1973 to control supply, gave an extra boost to Alberta's economy in the 1970s. By 2005, oil, natural gas, electricity, potash, and uranium all benefited from global demand and high prices. All signs suggest that global demand will continue. By value, these exports amounted to $65 billion or about 15 per cent of Canada's total exports (Statistics Canada, 2006c). On the other hand, agricultural products continue to suffer from low prices. Spring wheat, for instance, accounted for less than $2 billion in 2005. While signs are positive for global demand for agricultural products to increase, reality tells Western Canada that with record prices for energy and potash deposits, the region's wealth remains with these resources (Figure 8.5). Manitoba Hydro has three new projects on the drawing board—Wuskwatim (200 MW), Gull/Keeyask (620 MW), and Conawapa (1,380 MW). Power from these projects will flow to Ontario and neighbouring American states.

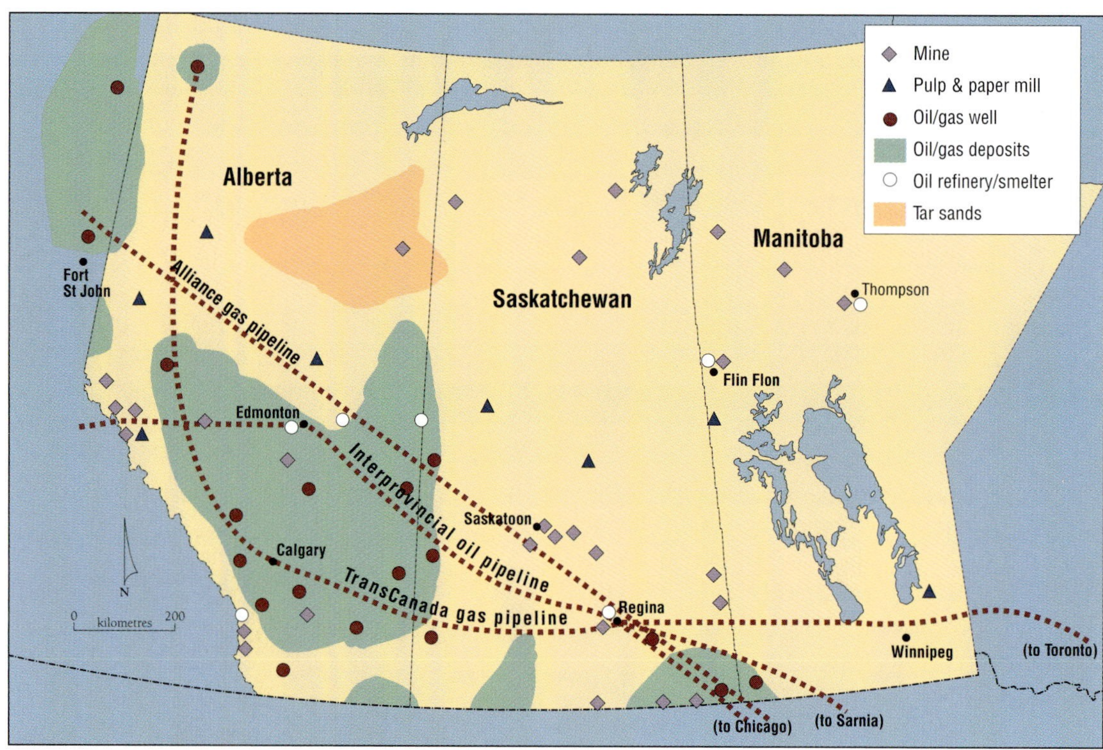

Figure 8.5 Western Canada's resource base, 2007. The number of mines, pulp and paper mills, oil and gas wells, and oil refineries and smelters that dot Western Canada's landscape attest to the region's range of natural wealth. The Alliance Pipeline, completed in 2000, is a natural gas pipeline extending 3,000 kilometres from just north of Fort St John, BC, to Chicago, Illinois. Huge petroleum resources are found in Alberta and, to a lesser degree, in Saskatchewan. Saskatchewan has the world's largest reserves of potash and uranium.

Energy: Oil and Natural Gas

Energy is Western Canada's most valuable natural resource. Western Canada contains over 70 per cent of Canada's oil reserves. These oil deposits are located in the southern half of the Western Sedimentary Basin. Alberta has the largest oil reserves (55 per cent of Canada's reserves), followed by Saskatchewan (17 per cent). Manitoba has less than 1 per cent.

Oil production from Alberta's vast reserves of oil and natural gas drives its economy (Vignette 8.6). In 2005, Western Canada's production of oil and natural gas was valued at $90.5 billion, with Alberta accounting for 90 per cent and Saskatchewan 10 per cent (Table 8.7).

In addition to oil and natural gas deposits, vast amounts of oil are contained in the tar sands—oil mixed with sand, known as **bitumen**—of northern Alberta. The two major oil sands mining sites are situated near Fort McMurray and Cold Lake. Major oil sands projects planned or underway in the Fort McMurray area are (1) Syncrude expansion; (2) Suncor Energy expansion; (3) Canadian Natural Resources planned project; (4) Mobile Canada planned project; (5) Petro-Canada planned project; (6) Gulf Canada planned project; (7) PanCanadian Petroleum planned project; (8) Shell Canada underway; and (9) True North Energy planned project.

The economic impact of petroleum on Western Canada, especially Alberta, has been remarkable, and prices, as shown in Figure 8.6, have continued to climb. A combination of investments, demand for labour, and royalties has transformed Alberta into the fastest-growing province in Canada as well as the richest province. In terms of the core/periphery model, the export of oil reversed a century-old economic relationship—Central Canada was dependent on the western provinces. Ottawa interfered with this new relationship by introducing the National Energy Program in 1981 (which was later jettisoned when Brian Mulroney and the Progressive Conservatives took office in 1984). The federal government was seen by Alberta as intruding on provincial rights and attempting to prevent the transfer of wealth. The results of this federal program were to widen the political gap between Alberta and Ottawa and to deepen the sense of western alienation and mistrust towards Ottawa and Central Canada.

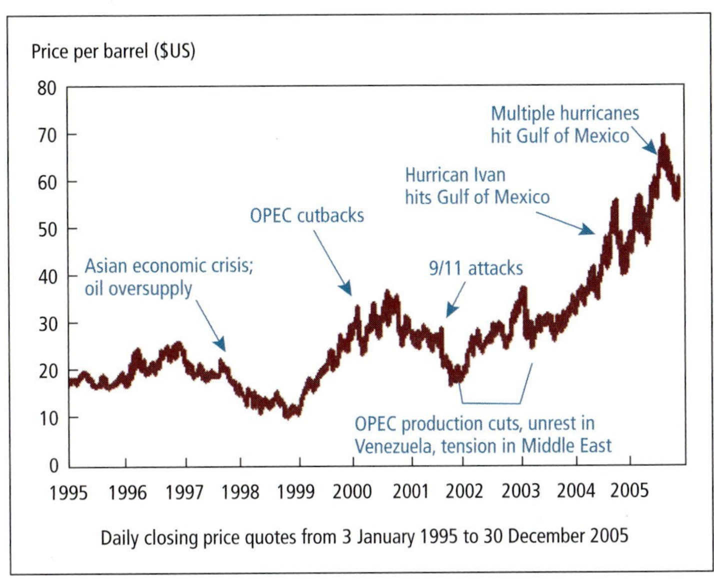

Figure 8.6 Crude oil prices, 1995–2005
Source: US Department of Energy and Energy Information Administration, West Texas Intermediate.

Mining Industry

The mining industry has helped to diversify the economy of Western Canada. The variety and value of mineral production in Western

Vignette 8.6 Alberta Oil Sands

Canada's oil sands deposits represent one of the world's largest oil reserves. Because of these vast oil sands, Canada ranks second in the world in terms of proven oil reserves. Oil sands are found in three areas of northern Alberta—Fort McMurray, Cold Lake, and Peace River (Figure 8.7). Alberta oil sands production reached 966,000 barrels per day in 2005 and was expected to reach 1.2 million barrels per day in 2006, more than double the production levels of the year 2000.

Development of the oil sands is not without environmental consequences. For instance, the mining process requires huge amounts of water to extract the tar-like bitumen and turn it into synthetic crude oil. According to the Pembina Institute (2007), the ratio of water to bitumen is 2 to 4.5 cubic metres of water to 1 cubic metre of bitumen. During this process, the water taken from the Athabasca River becomes highly toxic and, thus, cannot be returned to the river but is stored in tailing ponds. Production is expected to double again in the next five years, which means that the demand for water will grow, as will the size of the tailing ponds, while water levels downstream and in Lake Athabasca will continue to fall.

To accomplish the increase in production, companies will be investing at the rate of $10 billion or more a year. Recent investments have grown. For example, from 1998 to 2005, the annual capital investment has increased from $1.5 billion to $10.4 billion. The impact of such investment has created an enormous demand for construction and operational workers. By 2006, just over 200,000 workers were involved in the oil sands, and many of these workers came from other parts of Canada and from throughout the world. In fact, Newfoundlanders make up an important part of the population of Fort McMurray.

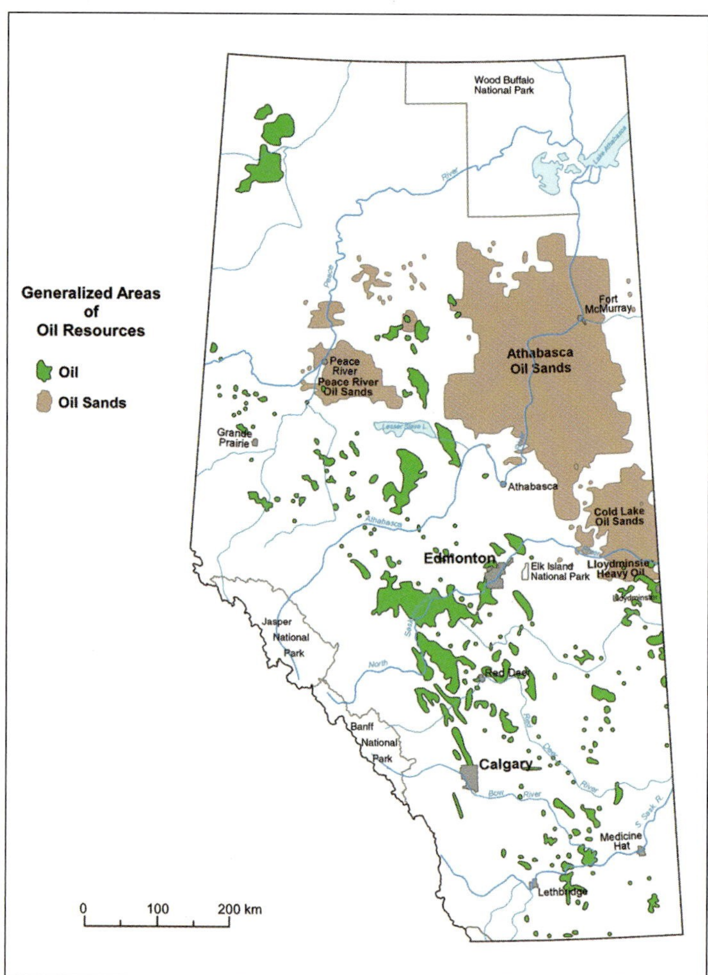

Figure 8.7 Alberta's hydrocarbon resources: Oil sands and oil fields
Source: Alberta (2007). Alberta Energy and Utilities Board/Alberta Geological Survey.

Canada are enormous. Mining companies produce a full range of mineral products, including metals, non-metals, structural materials, and fuels. In 2005, the three Prairie provinces produced $90.5 billion of petroleum, with Alberta accounting for 90 per cent (Table 8.7). The production in Western Canada amounted to 67 per cent of the national output (Statistics Canada, 2003b).

In Western Canada the geology of each province differs sufficiently to produce three distinct types of mining. Alberta contains rich coal reserves along the eastern slopes of the Rocky Mountains. Coal mining began in 1872 just west of Lethbridge. By the 1990s, the province of Alberta produced more coal than any other province, which contributed to Canada's position as the fourth-largest coal exporter in the world. Today, coal mining takes place in the East Kootenay and Peace River coalfields. Like other resource industries, these Alberta mines depend on world demand and prices. In 1998, the Asian economic slump resulted in a drop in demand and price for Alberta coal.

Potash and uranium are the major mineral deposits in Saskatchewan. The potash deposit lies approximately 1 km below the surface of the earth, reaching its thickest extent around Saskatoon, where six of the nine potash mines are located. Canada is the world's largest producer and exporter of potash. Saskatchewan has the largest and highest-quality deposit in the world, and except for a potash mine in New Brunswick, all of Canada's potash production comes from Saskatchewan. Potash is used to produce potassium-based fertilizers and is sold to firms in the United States and Asia, especially China and Japan. Even though China and Japan have been suffering from the Asian economic downturn, they continue to import Canadian potash to ensure high yields for their crops. Still, world demand varies, forcing potash mines to adjust the volume of their production.

Uranium mining also takes place in Saskatchewan. Production began in 1953 on the northern shores of Lake Athabasca near Uranium City. Since the late 1970s uranium mining has shifted south to the geological area of the Canadian Shield known as the Athabasca Basin. Two mines are currently operating (McArthur River and Rabbit Lake). Cluff Lake mine ceased operations in 2002. Subject to the approval of the federal environmental agency, three new mines (Cigar Lake, McClean Lake, and Midwest) are expected to begin production. Saskatoon is the major supply centre for both potash and uranium mines as well as their corporate headquarters.

Manitoba has two major mineral deposits—copper-zinc and nickel—both located in the Canadian Shield. Mining for

Table 8.7 Petroleum Production by Value in Western Canada, 2005

	Alberta	Saskatchewan	Manitoba	Western Canada Total
Production ($ billions)	81.5	8.7	0.3	90.5
Per cent	90	9.6	0.4	100

Source: Canadian Association of Petroleum Producers (2007). Reprinted by permission of the publisher.

copper and nickel in Manitoba is relatively expensive. As well, the high cost of shipping the processed product to distant markets adds to their economic disadvantage. Copper and zinc ore bodies are found near Flin Flon, a mining and smelter town in northern Manitoba that began production in 1930, shortly after a rail link to The Pas was completed in 1928. At one time, Flin Flon produced most of Canada's copper and zinc but today it is an aging resource town with a declining population. Thompson, located some 740 km north of Winnipeg, is a nickel-mining town. In 1957, after a rail link to the Hudson Bay Railway was completed, the mine facility, smelter, and town were constructed. Unlike Flin Flon, Thompson was a specially designed resource town with a complete array of urban amenities. The first nickel was produced in 1961. During the 1960s, Thompson's population soared to 20,000. Since then, two events have reduced the population size of Thompson and many other resource towns: (1) productivity of nickel mining, enhanced by greater mechanization, has increased, resulting in a smaller labour force, and (2) the service sector of Thompson has contracted. In 1991, Thompson had a population of 14,977, but by 2006 it had fallen to 13,446. For similar reasons, Flin Flon's population dropped from 7,119 in 1991 to 5,594 in 2006.

Forest Industry

The boreal forest stretches across the northern part of Western Canada. Within the boreal forest, a few coniferous species—spruce, fir, pine, and tamarack—predominate. The commercial forest zone lies in the Interior Plains where soil conditions are more favourable for tree growth than in the thin, rocky soils of the Canadian Shield. In 2005, logging in Western Canada amounted to 15 per cent of Canada's total timber harvest (Table 8.8). Approximately 75 per cent of this production takes place in Alberta.[12] Alberta's advantage comes from its large northern forest stands. The forest-covered Interior Plains region, which extends beyond the North Saskatchewan River to the border with the Northwest Territories, is the largest forest stand in Western Canada. This, together with the timber in the foothills of the Rocky Mountains, means that Alberta has by far the largest commercial forest stands of the three Prairie provinces.

Each of Western Canada's three provinces has a diversified timber operation, ranging from sawmills to pulp and paper plants. Most pulp and paper mills are located on the southern edge of the forest. In Saskatchewan these mills are located at Prince Albert and Meadow Lake. In Alberta they are in Athabasca, Grande Prairie, Hinton, and Peace River. In Manitoba The Pas is the site of a major pulp and paper mill.

By the end of the 1980s, virtually all the commercial forest stands had been leased, signalling the end of available virgin timber. In fact, several mills were unable to obtain additional Crown timber leases and have suffered from a shortage of wood fibre. Now that the original boreal forest has been harvested, the second growth is best suited for the pulp and paper industry. These trees, small and large, known as pulpwood, are cut and then trucked to a pulp mill. Each mill has a forest lease, giving it the right to log on Crown land. While a

Table 8.8 Forest Area and Production in Western Canada, 2005

Province	Area (millions of ha)	Productive Forest Area (millions of ha)	Industrial Roundwood (millions of m^3)
Manitoba	26.3	13.2	2.1
Saskatchewan	28.8	10.6	6.1
Alberta	38.2	20.9	23.5
Western Canada	93.3	44.7	31.7
Per cent Canada	22.3	20.9	15.4

Source: Natural Resources Canada (2007).

small amount of lumber production does occur, pulp and paper products predominate. Distance from markets also affects the focus of the forest industry. Pulp and paper products, which have a much higher value per tonne than lumber, are better able to withstand rail transportation costs from Western Canada to distant markets.

In the 1980s, it was discovered that aspen forests in Alberta, Saskatchewan, and Manitoba provided an excellent source of pulpwood for high-quality paper production. With considerable provincial financial support, mills were established at Hinton, Grande Prairie, Prince Albert, and The Pas. The Alberta government, seeking to diversify its economy by encouraging more processing of resources, tried to lure pulp and paper companies to northern Alberta by offering them vast northern hardwood timber leases. Two Japanese firms, Daishowa Canada Company and Alberta-Pacific Forest Industries, established mills at Peace River and Athabasca in northern Alberta. By the end of the 1980s, virtually all Alberta timber had been leased to five pulp and paper companies.

Similar developments occurred in Saskatchewan and Manitoba. One Alberta-based firm, Millar-Western, turned to Saskatchewan where it arranged to purchase pulp logs from Norsask, a local company that holds the timber lease to lands north of Meadow Lake. With an assured supply of pulpwood, Millar-Western built a mill at Meadow Lake. The Millar-Western pulp mill is a unique industrial development because: (1) the mill's internal circulating system means that no toxic wastes are released into the local rivers and lakes, and (2) it has a business arrangement with Norsask Forest Products, which is jointly owned by the Meadow Lake Tribal Council and Techfor Services, a company owned by the employees of the local sawmill (Anderson and Bone, 1995: 127). This business venture represents a new approach for resource development, combining the financing and expertise of an established pulp and paper company with the commitment by Aboriginal peoples to supply the raw materials.

The twenty-first century has not started well for the forest industry. Massive layoffs and plant closures have been the order of the day. This important Canadian forest industry remains challenged by three factors. First, pulp and paper mills faced declining demand for their products. In 2006, the Prince Albert pulp and paper mill closed along with several sawmills and other wood-processing plants in various places in Western Canada. Second, prices for lumber dropped sharply in 2006 because of the collapse of house construction in the United States. Third, the rising value of the Canadian dollar has cut the profit margins of all exporters, including those in the forest industry. The future of the softwood lumber industry in Canada depends heavily on a recovery of the

US housing market and a decline in the Canadian dollar.

Population

Western Canada's population has undergone significant changes since settlers first came to this region, and these changes have affected and been affected by many aspects of Prairie life. In 1921, Western Canada had a population of nearly 2 million. At that time, it comprised 22 per cent of Canada's population. By 2006, its population had increased to 5.4 million but accounted for only 17 per cent of the national population.

Two migrations transformed Western Canada. Just 100 years ago, land-hungry homesteaders poured into the Prairies, creating a rural landscape of small farms and villages. No one could have foreseen that this rural landscape would lose most of its population in a remarkably short time. In the second migration, the so-called rural-to-urban migration, rural people moved from the countryside to towns and cities in Western Canada and to urban centres in British Columbia, Ontario, and the United States. The push factors of this migration were the mechanization of the agricultural sector, the consolidation of farms, and the shrinking need for farm labour that resulted. Pull factors included the employment opportunities and greater amenities that existed in towns and cities. The rural-to-urban migration resulted in a redistribution of people within Western Canada. Today, most people live in the five major cities of the region: Calgary, Edmonton, Winnipeg, Saskatoon, and Regina (Figure 8.4).

Within Western Canada, Alberta has the largest population with 3.3 million residents. Manitoba has 1.1 million and Saskatchewan just under 1 million. Over the last 25 years, the rate of population increase has been highest in Alberta, followed by Manitoba. Recent population changes between the 2001 and 2006 censuses confirm that this trend continues, with Alberta growing at a rate of 10.6 per cent over this five-year period followed by

Saskatoon, situated on the South Saskatchewan River, is Saskatchewan's largest city with a population of nearly 234,000 (2006). Known as the 'Bridge City', Saskatoon has witnessed rapid population growth over the last 25 years. During that period, many Indians and Métis have settled in Saskatoon. By 2006, the Aboriginal population comprised over 10 per cent of Saskatoon's residents. (Al Harvey/The Slide Farm)

Manitoba at 2.6 per cent. Saskatchewan's population decreased by 1.1 per cent (Statistics Canada, 2007d).

Another demographic feature of Western Canada is the size of the Aboriginal population. By 2001, approximately 440,000 Aboriginal people lived in Western Canada, forming nearly 9 per cent of Western Canada's population. A growing number reside in urban centres, largely because Indian reserves and Métis communities have insufficient employment opportunities and limited urban amenities.

Western Canada's Urban Geography

The process of urbanization in Western Canada has lagged behind that of Ontario, British Columbia, and Québec. Even so, nearly three-quarters of Western Canada's residents live in urban centres, and most of these reside in the five major cities—Regina, Saskatoon, Winnipeg, Edmonton, and Calgary. A second order of urban centres includes Lethbridge, Red Deer, Grande Prairie, Medicine Hat, Wood Buffalo (Fort McMurray), Brandon, Prince Albert, and Moose Jaw. In 2006, these cities had a total population of 3.7 million (Table 8.9).

An important urban corridor is emerging within Alberta, linking Edmonton with Calgary. Some 2.5 million people live and work within this corridor, so it not only forms a major market within Western Canada but also has the potential for the establishment of high-tech industries. While the three Prairie provinces have not drawn high-tech industry in the same concentration as Ontario, Québec, and British Columbia, all five major cities have a cluster of high-tech firms. Alberta leads with 66,000 high-tech workers, followed by Manitoba with 16,000 and Saskatchewan with 12,000 (Howes and McKinnon, 2000: C7). Two anchors—Calgary and Edmonton—contribute special functions to the corridor. Calgary (1,079,310 in 2006) is the headquarters of the oil and gas industry, while Edmonton (1,034,945 in 2006), besides being the provincial capital, is known as the 'Gateway to the North'. Over the past five years, this urban corridor has had a high rate of population growth, exceeding 10 per cent.

From 2001 to 2006, the rate of urban growth varied considerably for the towns and cities of Western Canada. This variation reflects differences in local economic growth and in the pace of consolidating populations into regional centres. From 2001 to 2006, the fastest-growing cities in Western Canada were in Alberta: Wood Buffalo at 23.6 per cent, Grande Prairie at 22.3 per cent, and Red Deer at 22 per cent. The fastest-growing cities in Saskatchewan and Manitoba were Saskatoon at 3.5 per cent and Brandon at 4.3 per cent (Table 8.9).

Western Canada has, in effect, two types of urban settlement pattern. These two patterns reveal a north–south split based on a **regional core** and a hinterland. In the south, centres are aligned close to the southern border with the US, whereas an oasis-like pattern exists in the northern hinterland with settlements far apart. Such a pattern is common in resource hinterlands where a single-industry town is located near an ore body. Examples include Flin Flon (copper), Fort McMurray (heavy oil), and Thompson (nickel).

Across Western Canada, the trend for over 60 years has been a rural-to-urban migration with many of the smallest centres disappearing. These small communities formed the heart of the local farming area during the first half of the twentieth century, but today they have lost their function. Professors Stabler and Olfert, who have studied the migration from rural areas in Western Canada for many years, see the rural decline as inexorable, causing many small centres to disappear. They also predict that this migration in Western Canada, but in Saskatchewan in particular, will escalate (Stabler and Olfert, 2002). The main reasons why these rural centres are on a downward slope are complex and interconnected:

- Low prices for grain caused farmers to declare bankruptcy and to exit the rural economy.
- Farms consolidated, thereby reducing the size of the rural population.
- Highways were built, facilitating shopping in larger centres, where the variety and prices of goods were superior to those in smaller centres; eventually, stores in smaller communities closed.
- Schools, hospitals, and other public services were concentrated in larger centres, attracting the people who use and need these services.
- Railways abandoned branch lines, causing grain elevator companies to close their operations; for smaller towns, the closure of the grain elevator and the loss of rail service meant the loss of the town's last function.
- Grain companies built larger grain elevators on main lines, expecting that the lower freight rates offered at the larger grain elevators would attract farmers, drawing business away from elevators in small centres.
- Farmers moved to larger centres, where more urban amenities were available, and commuted to their farms.

Table 8.9 Major Urban Centres in Western Canada, 2001–6

Centre	Population 2001	Population 2006	% Change
Moose Jaw	33,519	33,360	–0.5
Prince Albert	41,460	40,766	–1.7
Brandon	46,273	48,256	4.3
Wood Buffalo	42,581	52,643	23.6
Medicine Hat	61,735	68,822	11.5
Grande Prairie	58,787	71,868	23.3
Red Deer	67,829	82,772	22.0
Lethbridge	87,388	95,196	8.9
Regina	192,800	194,971	1.1
Saskatoon	225,927	233,923	3.5
Winnipeg	676,594	694,668	2.7
Edmonton	937,845	1,034,945	10.4
Calgary	951,494	1,079,310	13.4
Total	3,424,232	3,731,500	9.0

Source: Statistics Canada (2007d). Adapted from Statistics Canada publication *Population and Dwelling Counts, 2006 Census*, Catalogue 97-550-XWE2006002 http://www.statcan.ca/bsolc/english/bsolc?catno=97-550-XWE2006002.

Western Canada's Future

Western Canada occupies the western interior of Canada. Its geographic location poses two challenges for its future development: distance to ocean ports for its exports and a dry climate for agriculture. Its economic position within the global economy makes it dependent on global prices for its primary products. For years, Alberta has benefited from high prices for its oil production. Recent increases in the demand for other forms of energy—uranium and hydroelectric power—as well as for mineral products such as potash has created boom-like conditions in Western Canada, especially in Alberta. Alberta's oil sands, which contain the second largest deposit of oil in the world, continue to drive this resource boom. Saskatchewan has most of the potash and uranium reserves in the world, and Manitoba is blessed with vast hydroelectric potential. Record high prices for potash and uranium have created a mini-boom in Saskatchewan, which is especially noticeable in Saskatoon. Given events in China and in the biofuel industry, even farmers are finally enjoying higher prices for their products. Crop conditions for 2007 looked very favourable and the 'million-dollar' June rain greatly benefited the crops. The only question is whether these prices and the weather will hold.

The direction of future change in Western Canada's economy is clear—more processing of agricultural products, a larger service sector, and a growing urban population. Higher prices may have signalled a more robust agricultural sector. If so, prosperity may have finally arrived for rural Western Canada. But the final shape of this new economy will not be revealed until the next decade of the twenty-first century.

Vignette 8.7 Winnipeg

Winnipeg, the capital and the largest city in Manitoba, is located at the confluence of the Red River and the Assiniboine River. With the building of the Canadian Pacific Railway, Winnipeg became known as the 'Gateway to the West'. By the turn of the twentieth century, Winnipeg was the principal city in Western Canada, controlling the grain trade and also acting as the administrative, financial, and wholesale hub for Western Canada. Today, Winnipeg dominates the economy of Manitoba, but its role within Western Canada has been considerably diminished by the growth of Calgary and Edmonton and, to a lesser degree, by Saskatoon and Regina.

Winnipeg's role in Western Canada has gradually weakened over time. After the completion of the Panama Canal in 1914, Alberta farmers and some Saskatchewan farmers began to ship their wheat through Vancouver instead of through Winnipeg. Next, service industries began to emerge in Edmonton, Calgary, Saskatoon, and Regina. Each city captured some of the trade previously held by Winnipeg merchants. Winnipeg's stranglehold over the sale and distribution of agricultural machinery and products to the farmers of Western Canada was broken.

After World War II, Winnipeg still remained the largest city in Western Canada. However, Calgary and Edmonton grew rapidly in the following years, partly because of developments in the oil and gas industries. Still, Winnipeg was the largest city in Western Canada in 1951 with a population of 357,000 compared to 177,000 for Edmonton and 142,000 for Calgary. By 1981, both of these cities surpassed Winnipeg in population size. This demographic trend continued. By 2006, Calgary and Edmonton's populations had jumped to just over one million. Winnipeg, in sharp contrast, saw its population slowly expand to 706,900.

Notes

1. Two novels illustrate the powerful impact of settlement on the Indians and Métis. Rudy Wiebe's *The Temptations of Big Bear* (1973) focuses on the Cree; Guy Vanderhaegh's *The Englishman's Boy* (1996) looks at the Cypress Hills Massacre.
2. The Cree attacked the outpost at Frog Lake, laid siege to Fort Battleford, and defeating the North West Mounted Police at Fort Pitt and Cut Knife Hill. With the arrival of the Canadian militia from eastern Canada, the Cree were defeated.
3. It was only in 1960 that the federal government extended the rights and privileges accorded to Canadian citizens to its Aboriginal peoples. Until then, Indians could not vote in provincial and federal elections without losing their status. The right to vote marked the start of a long journey to find a place in Canadian society. This journey is far from over.
4. The search for a quicker-maturing wheat began in 1892 when a cross was made between Red Fife, a popular wheat grown on the Prairies, and an earlier maturing wheat. After a decade of trials, Marquis wheat was tested at the Dominion Experimental Farms at Indian Head, Saskatchewan. By 1910, this variety of wheat made Canada famous for producing an exceptionally high-quality, hardy spring wheat.
5. Summer fallowing accomplished two goals: it conserved moisture and controlled weeds. In the 1990s, farmers began to abandon this technique because advances in technology enabled them to accomplish these goals without having to resort to summer fallowing.
6. Western alienation is based on past experience of real or perceived 'abuse' by big business, such as the CPR, and big government, meaning

government controlled by eastern interests. Such feelings were deeply felt by homesteaders who settled the West. By early in the twentieth century, western alienation was particularly strong and the subsequent resentment was often aimed at the Canadian Pacific Railway. In fact, farmers regarded this railway company as the most rapacious agent of eastern Canadian interests. Not only had the company obtained millions of acres of fertile land, but its real estate offices often manipulated station sites to ensure their location on Canadian Pacific Railway property. Since farmland increased in value with proximity to rail-loading sites, such Canadian Pacific Railway land increased in value and could be sold to settlers at a higher price. Even more galling to westerners was the fact that CPR landholdings could not be taxed for 20 years. To extend this tax-free period, the CPR sometimes delayed selecting land, which meant that large areas were not available for homesteading because the railway could opt to select such lands as part of its grant.

7. Alberta, Manitoba, and Saskatchewan did not gain full control over Crown lands, and therefore over natural resources, until 1930. At that time, Ottawa transferred federal property to these provinces, giving them access to lucrative sources of taxation associated with natural resource developments. Until then, taxes from these lands went to Ottawa, supposedly to pay for railway building in the West. At first, the revenue from natural resources was small, but with the discovery of oil at Leduc, Alberta, in 1947, royalties from the production of petroleum provided most of the revenue for Alberta. Equally important, the exploitation of the vast oil and gas deposits in Alberta grew rapidly and eventually resulted in a variety of petroleum-processing plants and pipeline-construction firms in Alberta. Later, more modest energy and mineral developments were found in Saskatchewan, British Columbia, and Manitoba.
8. In June 1999, the Organization for Economic Co-operation and Development announced that 1998 wheat subsidies in American dollars were $141 a tonne in the European Union, $61 a tonne in the US, and $8 a tonne in Canada (Hursh, 1999: C9).
9. Changes in the structure of farming are also anticipated. With expansion of the livestock and meat-processing industries, the number of farm workers may also increase. Large-scale hog farms, for instance, require farm workers. Also, the increase in specialty crops is causing more farmers to undertake contract farming; for example, they grow a certain quantity of canary seed for a set price. Some of these crops are now processed locally, a trend that is expected to intensify.
10. Farmers are becoming less sheltered from economic forces. Through contract farming, agribusiness—the sector of the economy that provides inputs to farms, such as chemical fertilizers, and procures agricultural products from farms for processing and distribution to consumers—is increasingly affecting the daily lives of farmers. Some fear that in the next severe economic downturn or drought, family farming operations will be replaced by corporate farms. In fact, the trend towards contract farming by agribusiness may be the first sign of such an ownership shift. While not yet apparent in the grain industry, agribusiness has established itself in specialty crops, poultry, and hog farming.
11. Farmers' decisions are affected by a complex set of variables, but ultimately the cost of doing business is the determining factor. For instance, the increased production of canola may affect the price of livestock feed. A protein by-product derived from the crushing of canola seeds is suitable for livestock. Instead of importing a similar protein supplement and paying for the built-in freight costs, using a local canola-based product to feed livestock may reduce costs.
12. Since the early 1990s, sawmills in the interior of British Columbia have exhausted local supplies of timber. They have had to purchase logs from Alberta ranchers who own timber lands in the foothills around Calgary.

Challenge Questions

1. Why does the annual precipitation on the Prairies vary so much from year to year and season to season?
2 Long distance to ocean ports is hurtful for wheat exports from Western Canada to world markets. The Hudson Bay Railway was one solution. Was it successful? Explain.
3. Spring wheat and canola prices have remained low for many years. Some suggest that ethanol production will cause prices for canola and, indirectly, wheat to rise. Why?
4. Why did Lorne Calvert, the Premier of Saskatchewan, argue so strongly for a Saskatchewan Energy Accord?
5. Wheat harvests vary from year to year. Besides low precipitation, what other weather conditions can affect size and quality of the harvest?
6. What was the impact of OPEC on the price of oil?
7. In terms of the core/periphery model, Alberta's oil and natural gas industry has reversed a century-old economic relationship with Central Canada. Alberta is no longer economically dependent on Ottawa. In 1981, Ottawa tipped the balance in favour of Central Canada by introducing the National Energy Program. How did this federal program work and why did it favour Ontario? See Chapter 3 for more on the National Energy Program.
8. Why do farmers practising zero tillage receive payments known as carbon credits?
9. Since Western Canada has 'shallow historic roots', what effect could this have on the region's sense of place?

Key Terms

agricultural fringe
Agriculture at its physical limits. Along the southern edge of the boreal forest and in the Peace River country, farmers clear the land, but the short growing season prevents most crops from maturing, so many farmers turn to cattle.

bitumen
A tar-like mixture of sand and oil.

carbon credit
Federal payment to farmers for storing carbon dioxide in the soil, part of the federal initiative (the Pilot Emission Removals, Reductions and Learnings) to combat the release of greenhouse gases into the atmosphere.

Dry Belt
An agricultural area in the semi-arid parts of Alberta and Saskatchewan primarily devoted to grain farms and cattle ranches. Crop failures due to drought are more common.

evapotranspiration
An important part of the water cycle, the sum of evaporation of water from the soil to the air and the transpiration of water from plants and its subsequent loss as vapour.

Fertile Belt
A mixed farming area where crop failures due to drought are less common. This area of long-grass and parkland natural vegetation is associated with black and dark-brown chernozemic soils.

Palliser's Triangle
Captain John Palliser led an expedition organized by the British Colonial Office and the Royal Geographical Society to survey the Canadian West in 1857–60. He concluded that this short-grass natural vegetation area in southern Alberta and Saskatchewan was a northern extension of the Great American Desert and was therefore unsuitable for agricultural settlement.

primary prices
The prices for commodities such as foodstuffs, raw materials, and other primary products.

regional core
Within the core/periphery model, cores can occur at different geographic levels. A regional core is an area (often a large city) that dominates trade and stimulates economic growth in the region.

summer fallow
The farming practice of leaving land idle for a year or more to accumulate sufficient soil moisture to produce a crop; also practised to restore soil fertility; being replaced by continuous cropping.

Western Sedimentary Basin
Within the geological structure of the Interior Plains, the normally flat sedimentary strata are bent into a basin-like shape. These basins often contain petroleum deposits.

zero tillage
Leaving the land undisturbed except for the planting of seed, a practice currently followed by grain farmers to replace strip farming, which often resulted in wind erosion of the topsoil.

Bibliography

Akinremi, O.O., S.M. McGinn, and H.W. Cutforth. 2001. 'Seasonal and Spatial Patterns of Rainfall on the Canadian Prairies', *Journal of Climate* 14, 9: 2177–82.

Alberta. 2006. 'Annual Report—Potato Agronomy', Agriculture, Food and Rural Development. At: <www1.agric.gov.ab.ca/$department/deptdocs.nsf/all/opp10699>.

———. 2007. 'Oil Reserves'. At: <www.energy.gov.ab.ca/docs/oil/pdfs/AB_OilReserves.pdf>.

Alberta Food and Agriculture. 2007. 'Alberta 2006 Crop Season in Review with Feed Availability Report'. At: <www1.agric.gov.ab.ca/$department/deptdocs.nsf/all/sdd11361>.

Anderson, Robert B., and Robert M. Bone. 1995. 'First Nations Economic Development: A Contingency Perspective', *Canadian Geographer* 39, 2: 120–30.

Artibise, Alan. 1975. *Winnipeg: A Social History of Urban Growth, 1874–1914*. Montréal and Kingston: McGill-Queen's University Press.

Barr, Brenton M., and John C. Lehr. 1982. 'The Western Interior: Transformation of a Hinterland Region', in L.D. McCann, ed., *A Geography of Canada: Heartland and Hinterland*. Scarborough, Ont.: Prentice-Hall, ch. 8.

Barron, F. Laurie, and Joseph Garcea, eds. 1999. *Urban Indian Reserves: Forging New Relationships in Saskatchewan*. Saskatoon: Purich Publishing.

Bell, Ian. 1999. 'Dancing with Elephants', *Western Producer*, 2 Sept., 60–1.

Bellavance, Joël-Denis, and Robert Fife. 2000. 'PM May Try "tough love" on West', *National Post*, 23 Dec., A1.

Blackbourn, Anthony, and Robert G. Putnam. 1984. *The Industrial Geography of Canada*. London: Croom Helm.

Bone, Robert M. 2003. *The Geography of the Canadian North: Issues and Challenges*, 2nd edn. Toronto: Oxford University Press.

Bonsal, B.R., X. Zhang, and W.D. Hogg. 1999. 'Canadian Prairie Growing Season Precipitation Variability and Associated Atmospheric Circulation', *Climate Research* 11: 191–208.

Bradsher, Keith. 2007. 'Rise in China's Pork Prices Signals End to Cheap Output', *New York Times*, 8 June. At: <www.nytimes.com/2007/06/08/business/worldbusiness/08prices.html>.

Broadway, Michael J. 1998. 'Where's the Beef? The Integration of the Canadian and American Beef-packing Industries', *Prairie Forum* 23, 1: 19–30.

Budhia, Narendru. 1995a. 'Manitoba—Outlook for 1995 Improves', *Provincial Outlook: Economic Forecast* 10, 2 (Spring): 38–41.

———. 1995b. 'Saskatchewan—As the Crow Flies', *Provincial Outlook: Economic Forecast* 10, 2 (Spring): 42–5.

Calvert, Lorne. 2005. 'Atlantic Goose, Prairie Gander', *National Post*, 24 Feb., A16.

Canadian Association of Petroleum Producers. 2007. At: <www.capp.ca/default.asp?V_DOC_ID=1072&SectionID=3&SortString=TableNo>.

Cattaneo, Claudia. 2002. 'Hedging over Kyoto Begins in Oilpatch', *National Post*, 11 Dec., FP1.

CBC. 1999. 'Farmers Fined $500 for Each Illegal Border Crossing', 22 Nov. At: <www.cbc.ca/news/features/wheat_board.html>.

———. 2002. 'Prairie Farmers Would Rather Go to Jail', 1 Oct. At: <www.cbc.ca/news/features/wheat_board.html>.

CBC News. 2006. 'Farmers Learn to Grow Carbon Credits'. At: <www.cbc.ca/canada/saskatchewan/story/2006/09/12/carbon-farming.html>.

Evans, Simon M. 1979. 'American Cattlemen on the Canadian Range, 1874–1914', *Prairie Forum* 4: 121–35.

Eyton, J. Ronald. 1997. 'Spatial Externalities and Edmonton Dwelling Values', *Canadian Geographer* 41, 2: 202–6.

Found, William C. 1996. 'Agriculture in a World of Subsidies', in John N.H. Britton, ed., *Canada and the Global Economy: The Geography of Structural and Technological Change*. Montréal and Kingston: McGill-Queen's University Press, ch. 9.

Garrod, Stan. 1984. *Economics in Society: Canadian Case Studies*. Don Mills, Ont.: Addison-Wesley.

Grauman, Meny. 2006. 'Alberta Profile—Economic View', *Provincial Pulse* (Scotiabank), 3 Feb., 1.

Greenwood, John. 2007a. 'Pork Price Causes Stir', *National Post*, 13 June, FP3.

———. 2007b. 'Surging Prices Separate Wheat from the Chaff', *National Post*, 15 June, FP1.

Hall, Angela. 2003. 'Hogs Touchy Issue in Rural Sask.', *Saskatoon Star-Phoenix*, 25 Feb., C8.

Hamburg, Karen T., C. Emdad Haque, and John C. Everitt. 1997. 'Municipal Waste Recycling in Brandon: Determinants of Participatory Behavior', *Canadian Geographer* 41, 2: 149–65.

Haque, C.E. 1998. 'Coping Responses to the 1997 Red River Valley Flood: Research Issues and Agenda', *Prairie Perspectives* 1: 47–62.

Hildebrandt, Walter, and Brian Hubner. 1994. *The Cypress Hills: The Land and Its People*. Saskatoon: Purich Publishing.

Howes, Carol, and Ian McKinnon. 2000. 'Calgary's Critical Mass', *National Post*, 15 Nov., C7.

Hursh, Kevin. 1999. 'Subsidy Complaints Don't Hold Water', *Saskatoon Star-Phoenix*, 28 July, C9.

Jones, David, C. 1987. *Empire of Dust: Settling and Abandoning the Prairie Dry Belt*. Edmonton: University of Alberta Press.

Johnston, Tom, and Amber Weibel. 2005–6. 'Industrial Hog Production and the Hog-barn Neighbourhood Effect in Lethbridge County, Alberta', *Western Geography*, 15 and 16: 53–7.

Kariel, H.G. 1996. 'Land Use in Alberta's Foothill Country', *Western Geography* 7: 20–46.

Macartney, Jane, and Tim Reid. 2007. 'Food Price Rises Force a Cut in Biofuels', *The Times*, 12 June. At: <www.timesonline.co.uk/tol/news/world/asia/china/article1917927.ece>.

MacLachlan, Ian. 2001. *Kill and Chill: Restructuring Canada's Beef Commodity Chain*. Toronto: University of Toronto Press.

———. 2005. 'A Neo-Fordist Interpretation', in Andrew Gilg, Richard Yarwood, Stephen Essex, John Smithers, and Randall Wilson, eds, *Rural Change and Sustainability: Agriculture, the Environment and Communities* (London: CABI Publishing, 28–47.

McNicol, B.J. 1997. 'Views about Industrial Tourist Pressures in Canmore, Alberta', *Western Geography* 7: 47–72.

Marsh, James H., ed. 1988. *The Canadian Encyclopedia*, 2nd edn. Edmonton: Hurtig.

National Farmers' Union. 1999. 'Farm Subsidies', *National Post*, 27 Oct., C6.

National Forestry Database Program. 2000. Forest Products: Industrial Roundwood by Category, 1990–8. At: <nfdp.ccfm.org/frames2_e.htm>.

Natural Resources Canada. 2003a. Mineral Production of Canada, by Provinces and Territory. At: <mmsd1.mms.nrcan.gc.ca/mmsd/production/production_e.asp>.

———. 2003b. The State of Canada's Forests 2001/02. At: <www.nrcan.gc.ca/cfs-scf/national/what-quoi/sof/sof02/profiles_e.html>

———. 2007. The State of Canada's Forests 2005/06. At: <cfs.nrcan.gc.ca/sof06/profiles_e.html>.

Neatby, L.H. 1979. *Chronicle of a Pioneer Prairie Family*. Saskatoon: Western Producer Prairie Books.

Paul, Alec H. 1997. 'Shortlines, Mainlines, Branchlines, Dead Lines: Rural Railways in Southwestern Saskatchewan in the 1990s', in John Welsted and John Everitt, eds, *The Yorkton Papers: Research by Prairie Geographers*. Brandon Geographical Studies No. 2. Brandon, Man.: University of Brandon.

Pembina Institute. 2007. 'Athabasca River Expedition: Connecting the Drops'. At: <www.connectingthedrops.ca/river/stresses>.

Phillips, David. 1993. *The Day Niagara Falls Ran Dry!* Toronto: Canadian Geographic and Key Porter Books.

Potyondi, Barry. 1995. *In Palliser's Triangle: Living in the Grasslands, 1850–1930*. Saskatoon: Purich Publishing.

Price, Jacqueline D. 2003. 'Are Factory Farms Fouling Our Water?', *Alberta Views* (May–June): 34–9.

Rannie, W.F. 1998. 'The Red River Flood in Manitoba, Canada', *Prairie Perspectives* 1: 1–24.

Rice, Murray D. 1996. 'Functional Dynamics and a Peripheral Quaternary Place: The Case of Calgary', *Canadian Journal of Regional Science* 19, 1: 65–82.

Richards, J. Howard. 1968. 'The Prairie Region', in John Warkentin, ed., *Canada: A Geographical Interpretation*. Toronto: Methuen, ch. 12.

Robertson, Grant. 1999. 'Big Business Harvests Potato Profit', *Calgary Herald*, 20 July, D1, D3.

Rowat, Miles Ryan. 2006. *Boom Times: Canada's Crude Petroleum Industry*. Ottawa: Statistics Canada. At: <www.statcan.ca/english/research/11-621-MIE/11-621-MIE2006047.htm>.

Rumney, Thomas. 1999. *The Geography of Canada Bibliography Series: Vol. 2, The Prairies*. Plattsburgh, NY: Plattsburgh State University, Center for the Study of Canada.

Russell, Bob. 1999. *More with Less: Work Reorganization in the Canadian Mining Industry*. Toronto: University of Toronto Press.

Saskatchewan Agriculture and Food. 1997. *Agricultural Statistics 1996*. Regina: Statistical Branch.

Saskatchewan Agriculture, Food, and Rural Revitalization. 2002. Agricultural Statistics 2001. At: <www.agr.gov.sk.ca/DOCS/statistics/finance/other/Handbook01.pdf>.

———. 2007. Table 2-2: Winter Wheat Production and Value. At: <www.agr.gov.sk.ca/apps/agriculture_statistics/HBV5_Result.asp>.

———. 2007. Market Trends. At: <www.agr.gov.sk.ca/apps/MarketTrends/>.

Saskatchewan Environment. 2003. *An Assessment of Abandoned Mines in Northern Saskatchewan*. A report prepared by Clifton Associates Ltd. File R3160. Regina: Clifton Associates Ltd.

Sauchyn, David J., ed. 1993. *Quaternary and Late Tertiary Landscapes of Southwestern Saskatchewan and Adjacent Areas*. Canadian Plains Research Centre, University of Regina. Winnipeg: Hignell Printing.

Schmitz, A., and H. Furtan. 2000. *The Canadian Wheat Board*. Regina: Canadian Plains Research Centre.

Selwood, J., and S. Kohm. 1988. 'Location, Location, Location: Selling Sex in the Suburbs', *Prairie Perspectives* 1: 161–71.

Semple, R.K. 1994. 'The Western Canadian Quaternary Place System', *Prairie Forum* 12: 81–100.

Simpson, J., and S. Hathout. 1998. 'Optimum Route Location Model for an All-Weather Road on the East Side of Lake Winnipeg', *Prairie Perspectives* 1: 113–24.

Smith, P.J., ed. 1972. *The Prairie Provinces: Studies in Canadian Geography*. Toronto: University of Toronto Press.

Spry, Irene M. 1963. *The Palliser Expedition: An Account of John Palliser's British North American Expedition 1857–1860*. Toronto: Macmillan.

———, M.R. Olfert, and Murray Fulton. 1992. *The Changing Role of Rural Communities in an Urbanizing World: Saskatchewan 1961–1990*. Canadian Plains Report no. 8. Regina: Canadian Plains Institute.

Stabler, Jack C. 2002. *Saskatchewan Communities in the 21st Century: From Places to Regions*. Regina: Canadian Plains Research Centre.

——— and M.R. Olfert. 1992. *Restructuring Rural Saskatchewan: The Challenge of the 1990s*. Regina: Canadian Plains Institute.

Stanford, Quentin H., ed. 1998. *Canadian Oxford World Atlas*, 4th edn. Toronto: Oxford University Press.

Statistics Canada. 1982. *Census Metropolitan Areas and Census Agglomerations with Components*. Catalogue no. 95–903. Ottawa: Minister of Supply and Services Canada.

———. 1992a. *Census Divisions and Subdivisions*. Catalogue no. 93–304. Ottawa: Minister of Supply and Services.

———. 1992b. *Census Overview of Canadian Agriculture 1971–1991*. Catalogue no. 93–348. Ottawa: Ministry of Industry, Science and Technology.

———. 1993. *Canada Year Book 1994*. Ottawa: Ministry of Industry, Science and Technology.

———. 1994a. *Canada's Aboriginal Population by Census Subdivision and Census Metropolitan Area*. Catalogue no. 94–326. Ottawa: Ministry of Industry, Science and Technology.

———. 1994b. *Canadian Forestry Statistics 1993*. Catalogue no. 25–202–XPB. Ottawa: Ministry |of Industry, Science and Technology.

———. 1996a. *Canada's Mineral Production*. Catalogue no. 26–202–XPB. Ottawa: Mining Sector, Natural Resources Canada.

———. 1996b. *Labour Force Annual Averages 1995*. Catalogue no. 71–220–XPB. Ottawa: Statistics Canada.

———. 1997a. *A National Overview: Population and Dwelling Counts*. Catalogue no. 93–357–XPB. Ottawa: Industry Canada.

———. 1997b. 1996 Census: Nation Tables—Population by Mother Tongue, Showing Age Groups, for Canada, Provinces and Territories, 1996 Census—20% Sample Data, 2 Dec. At: <www.statcan.ca/english/census96/>.

———. 1997c. '1996 Census: Mother Tongue, Home Language and Knowledge of Languages', *The Daily*, 2 Dec. At: <www.statcan.ca/Daily/English/>.

———. 1998a. *Canadian Economic Observer*. Catalogue no. 11–010–XPB. Ottawa: Statistics Canada.

———. 1998b. '1996 Census: Aboriginal Data', *The Daily*, 13 Jan. At: <www.statcan.ca/Daily/English/>.

———. 1998c. '1996 Census: Ethnic Origin, Visible Minorities', *The Daily*, 17 Feb. At: <www.statcan.Daily/English/>.

———. 1998d. Census of Agriculture—Canada Highlights. At: <www.statcan.ca/english/censusag/can.htm>.

———. 1998e. *Canadian Forestry Statistics 1995*. Catalogue no. 25–202–XPB. Ottawa: Ministry of Industry, Science and Technology.

———. 2000a. Table 2: Revised Statistics of the Mineral Production of Canada, by Province, 1996. At: <www.nrcan.gc.ca/mms/efab/mmsd/production>.

———. 2000b. Canadian Statistics—Export of Goods on a Balance-of-Payments Basis. At: <www.statcan.ca:80/english/Pgdb/Economy/International/>.

———. 2002a. Canadian Statistics: Census of Agriculture Farms: Land tenure, provinces. At: <www.statcan.ca/english/Pgdb/econ113a.htm>.

———. 2002b. 2001 Census: Population and Dwelling Counts, for Census Metropolitan Areas and Census Agglomerations, 2001 and 1996 Censuses, 16 July. At: <www.statcan.ca/english/IPS/Data/93F0050XCB2001013.htm>.

———. 2002c. Census of Canada 2001—Census Geography. Highlights and Analysis: Canada's 2001 Population. At: <www12.statcan.ca/English/census01/>.

———. 2003a. Agriculture 2001 Census Farm Operations: provincial/regional trends. At: <www.statcan.ca/english/agcensus2001/first/regions/contents.htm>.

———. 2003b Canadian Statistics—Export of Goods on a Balance-of-Payments Basis. At: <www.statcan.ca/english/Pgdb/gblec04.htm>.

———. 2003c. Canadian Statistics: Distribution of Employed People, by Industry, by Province. At: <www.statcan.ca/english/Pgdb/labor21b.htm>.

_____. 2003d. *Census of Canada 2001—Aboriginal Peoples of Canada: A Demographic Profile*. Analysis series 96F0030XIE2001007, 21 Jan. Ottawa. At: <www12.statcan.ca/english/census01/products/analytic/companion/abor/contents.cfm>.

———. 2006a. Field Crop Reporting Series No. 7. At: <www.statcan.ca/english/freepub/22-002-XIB/22-002-XIB2006007.pdf>.

———. 2006b. 'Estimates of Production of Principal Field Crops', *The Daily*, Sept. At: <www.statcan.ca/Daily/English/061005/d061005a.htm>.

———. 2006c. Exports of Goods on a Balance-of-Payments Basis, by Product. At: <www40.statcan.ca/l01/cst01/gblec05.htm>.

———. 2006d. 'Production of Principal Field Crops', *The Daily*, 7 Dec. At: <www.statcan.ca/Daily/English/061207/d061207a.htm>.

———. 2007a. 'Total Farm Area, Land Tenure and Land in Crops, by Province', *2006 Census of Agriculture*. At: <www40.statcan.ca/l01/cst01/agrc25j.htm>.

———. 2007b. Labour Force, Employed and Unemployed, Numbers and Rates. At: <www40.statcan.ca/l01/cst01/labor07c.htm>.

———. 2007c. Gross Domestic Product, Expenditure-based, by Province and Territory. At: <www40.statcan.ca/l01/cst01/econ15.htm>.

———. 2007d. Population and Dwelling Counts, for Canada, Provinces and Territories, 2006 and 2001 Censuses—100% Data. At: <www12.statcan.ca/english/census06/data/popdwell/Table.cfm?T=101>.

Thraves, B.D., and G. Barriault. 1998. 'Change in the Size and Functions of Regina's Central Business District, 1964–1997', *Prairie Perspectives* 1: 141–60.

Vanderhaeghe, Guy. 1996. *The Englishman's Boy*. Toronto: McClelland & Stewart.

Wiebe, Rudy. 1973. *The Temptations of Big Bear*. Toronto: McClelland & Stewart.

Zakreski, Dan. 2000. 'Change in the Wind: Destruction of Elevators Signals Symbolic Shift in Saskatchewan Agriculture', *Saskatoon Star-Phoenix*, 9 Dec., E1–E2.

Further Reading

Fung, Ka-iu, ed. 1999. *Atlas of Saskatchewan*. Saskatoon: PrintWest. (Also available as a CD.)

The *Atlas of Saskatchewan* adds greatly to our understanding of Western Canada. The section on agriculture (pp. 217–58) is particularly pertinent to the Key Topic in this chapter. Maps, graphs, and text provide the reader with a detailed account of agriculture in Saskatchewan. For example, approximately half of Saskatchewan is classified as farmland. Of this land, natural pasture forms some 20 per cent while another 20 per cent is placed in summer fallow. Cropland, including that for hay and alfalfa, comprises the remaining 60 per cent. From 1966 to 1996, cropland increased from 18.6 million ha to 20 million ha. Spring wheat remains the dominant crop, grown on 30 per cent of the cropland in 1996, followed by barley at 9 per cent, durum wheat at 8 per cent, and canola at 7 per cent. Trends in agricultural production provide the reader with a clear picture of the agricultural shift underway in Saskatchewan. From 1966 to 1996, six major trends were identified: (1) spring wheat cropland declined 25.4 per cent, from 7.5 million ha to 5.6 million ha; (2) durum wheat cropland increased nearly five times, from 366,000 ha to 1.7 million ha; (3) barley nearly doubled in acreage, from 913,000 ha to 1.8 million ha; (4) canola increased threefold, from 296,000 ha to 1.6 million ha; (5) specialty crops, such as mustard, lentils, peas, and canary seed, jumped from 36,000 ha to 1 million ha; and (6) the livestock industry is expanding. The removal of the freight subsidy on grains in 1995 was a major impetus to expand hog production. Large-scale hog producers within Saskatchewan purchase feed grain from local farmers. Saskatchewan farmers account for 38 per cent of Canada's primary agricultural exports, 44 per cent of its cultivated land, and 21 per cent of its farms.

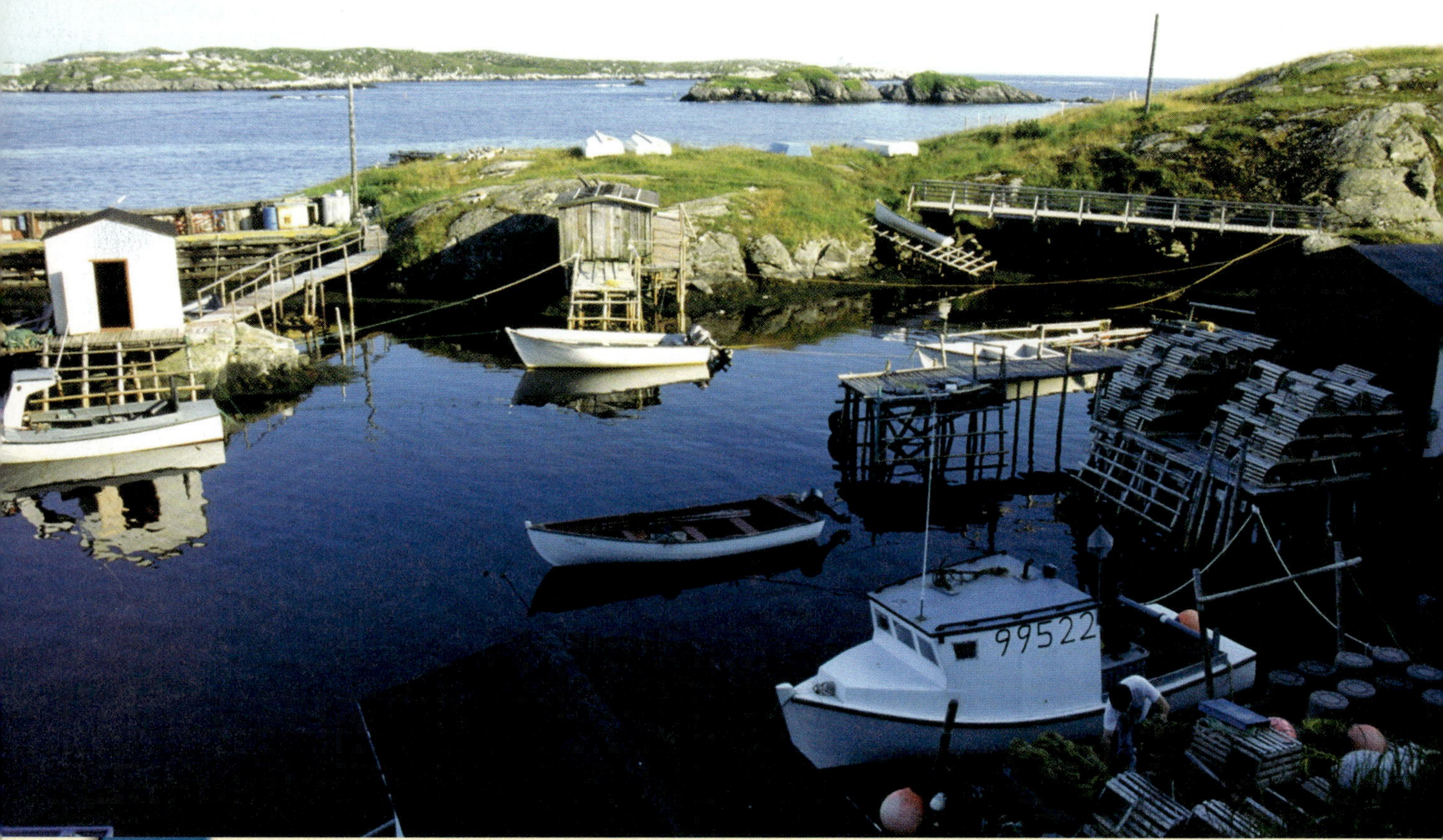

Diamond Cove, Newfoundland
(Victor Last, Geographical Visual Aids)

Overview and Objectives

Atlantic Canada, located on the eastern margins of the country, is far from the centres of economic and political power. As an 'old' resource hinterland, Atlantic Canada lags far behind the other geographic regions in economic growth and population growth. As well, it suffers from high out-migration with many relocating to Alberta's boom economy. While the North American Free Trade Agreement has provided access to the larger New England market, Atlantic Canada has few resources suitable for export apart from energy. The demise of the northern cod stocks has seriously hurt the region's economy, especially in Newfoundland and Labrador, which has the highest unemployment rate in Canada. Still, there are signs of economic rejuvenation. Offshore oil and gas developments have quickened the region's economic pace and are having a ripple effect on manufacturing and service sectors associated with the petroleum industry. Then, too, the age of supertankers and the widespread use of containers are allowing Halifax to regain its former prominent roles in ocean shipping and as a transshipment point for North American markets. In spite of these bright spots in Atlantic Canada's economy, the challenge remains to revitalize Atlantic Canada's economy so that the region can break its dependency on Ottawa and halt the outflow of its younger population to the faster-growing areas in Canada. This chapter will:

- Describe Atlantic Canada's physical and historical geography.
- Present the basic characteristics of Atlantic Canada's population and economy from the perspective of an 'old' resource hinterland, but, at the rejuvenation.
- Examine the population and economy within the context of Atlantic Canada's physical setting, especially its fragmented geography.
- Explore Atlantic Canada's role as a resource hinterland.
- Focus on the cod fishery as an example of the 'tragedy of the commons'.

CHAPTER 9

Atlantic Canada

INTRODUCTION

Stretching along the country's eastern coast, the region of Atlantic Canada consists of two parts: the Maritimes (Nova Scotia, Prince Edward Island, and New Brunswick) and Newfoundland and Labrador (Figure 1.2).[1] Despite the differences this physical separation has created between the Maritimes and Newfoundland and Labrador, Atlantic Canada is united by a rich sense of place that has grown out of the region's history and geographic location. The Atlantic Ocean has dominated the history of this region from the early days of the fishery to the 'Golden Age of Wooden Ships and Iron Men'. Today, Atlantic Canadians continue to rely heavily on the sea, fishing on the continental shelf and conducting energy exploration below it. Most important of all, the sea provides a link between Atlantic Canada, the markets found in the interior of North America, and countries on the Atlantic Rim. Because of its geographic position, could Atlantic Canada emulate BC as a trade corridor? The Atlantic Gateway remains a dream but two signs of this geographic link are the growing traffic for container cargo and the emergence of **liquefied natural gas** (LNG) facilities, both of which are designed to serve the larger markets in eastern North America but especially the New England market. Like the Pacific Gateway, the goal is to combine ocean-based transportation between continents with a rail and trucking network that penetrates into major markets in North America (McMillan, 2006). As in the region's early history, the Atlantic Ocean asserts itself in the affairs of Atlantic Canada. In this chapter, the Key Topic is the fishing industry.

Atlantic Canada within Canada

As Canada's oldest hinterland, Atlantic Canada has experienced both growth and decline over the last 200 years. Today, Atlantic Canada is, in Friedmann's regional version of the core/periphery model, a downward transitional region. The region is troubled by past exploitation of its renewable resources and the exhaustion of its most accessible and richest non-renewable resources. Atlantic Canada's troubles are summed up by the subpar performance of its economy relative to the rest of Canada and the seemingly unstoppable out-migration to faster-growing regions of Canada. Yet, Atlantic Canada has received a second chance with three developments. First, the production from the huge nickel deposit at Voisey's Bay and from the vast offshore hydrocarbon deposits is having an impact on the economy of Atlantic Canada and the revenues of Newfoundland and Labrador and Nova Scotia. Also, the construction of two LNG facilities, one near Saint John, New Brunswick, and a similar project planned for Goldboro, Nova Scotia, will add strength to the goal of making the region an energy hub for eastern North America. Second, major cities are benefiting from an ever-widening tertiary base to their urban economies. Halifax, with its superb harbour, is participating in the growth of oceanic container trade. As well, Halifax effectively serves as the regional capital of the Maritimes. The offshore oil industry represents a critical element of the economy of St John's, Newfoundland, while the environmental assessment of the planned $7 billion oil refinery for Saint John, New Brunswick, is proceeding. Finally, the Atlantic Accord, signed in 2004, offers a greater share of resource royalties and equalization payments to the governments of Nova Scotia and Newfoundland and Labrador.

As a resource hinterland, the rural part of Atlantic Canada has not been faring well. Communities dependent on fishing and forestry are in trouble. While the rural residents of Atlantic Canada have a strong sense of place and are proud of their land and heritage, jobs are in short supply. For that reason, an increasing number are either relocating to Alberta or commuting from villages and towns in Atlantic Canada to construction sites around Fort McMurray. At the same time, people are increasingly determined to overcome their position as poor cousins within the Canadian federation. In each province of Atlantic Canada, people are struggling to find better ways to manage the region's resources in a more sustainable fashion; to achieve a higher degree of processing of its non-renewable resources; and to expand the base of high-technology industries. Some changes are already taking place. Offshore oil and gas developments have stimulated the growth of highly specialized manufacturing firms in both Newfoundland and Nova Scotia, while high-technology firms are locating in the major cities of Atlantic Canada. Information technology is also growing in the Maritimes. For instance, Bathurst, Fredericton, Halifax, Moncton, Saint John, and St Stephen are telephone call centres for firms operating in Canada (call centres enable customers to phone a 1–800 number to obtain the services that companies provide). However, these information service jobs are at the low end of the labour market in terms of pay and skills and, unlike the innovative high-technology industries, provide limited spinoffs.

Atlantic Canada's economy continues to struggle. While Halifax leads the region in economic growth, most other communities are not faring so well and many rural communities are on the edge of collapse. This urban/rural split in economic well-being is due to several factors. First, jobs in the primary sector, such as logging and fishing, are disappearing as firms replace workers with more advanced machinery. Second, the collapse of the northern cod fishery resulted in great hardships in fishing communities, especially the outports of Newfoundland. Third, the need to share the fishing resource (and possibly other natural resources) with the Mi'kmaq will give the Indian communities a much needed boost but, as a zero-sum game, non-Indian communities will see some of their fishers leave this industry. Economic decline has created another problem—out-migration, which makes the task of rebuilding the region's economy even more difficult.

One measure of Atlantic Canada's overall economic performance is reflected in the region's per capita gross domestic production figures and its level of unemployment. In 2006, Atlantic Canada's GDP per capita was the lowest in Canada, while the region's 2006 unemployment rate was the highest (Figure 9.1 and Table 9.1). The primary reasons for Atlantic Canada's weak economic performance include the following:

- The geographic division of Atlantic Canada into subregions adds to companies' transportation costs and discourages the emergence of an

integrated economy in which economies of scale might occur.
- Atlantic Canada has been exploiting its resources for a long time and some of these, such as coal and iron, have been exhausted, while its renewable resources, such as the northern cod, have been overexploited.
- Atlantic Canada has a series of small internal markets.
- Atlantic Canada, but especially Newfoundland and Labrador, is far from Canada's major markets.

All these factors have made it extremely difficult for economic development to flourish in the region. Furthermore, Atlantic Canada has, over time, become heavily dependent on Ottawa for economic support through equalization payments and social programs. For both economic and political reasons, this cycle of dependence seems unlikely to end.

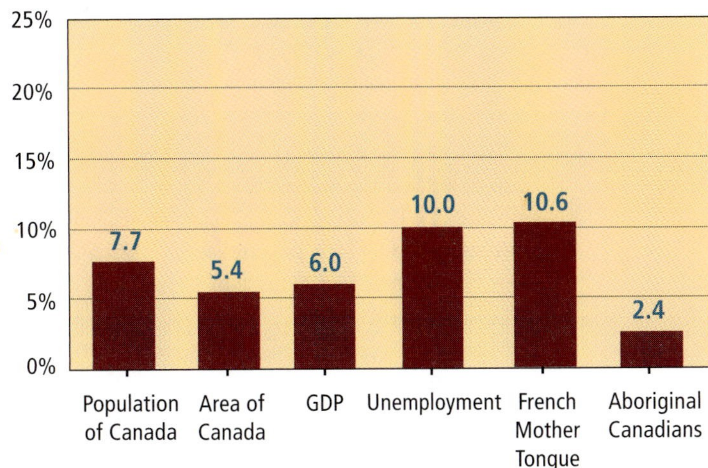

Figure 9.1 Atlantic Canada, 2006. Except for its oil and gas sector, Atlantic Canada has a weak economy. All four provinces are considered have-not provinces and they receive equalization payments. The region has 7.2 per cent of Canada's population but produces only 6.0 percent of the country's GDP. The high rate of unemployment is another indication of its poor economic performance. Acadians account for the high number of French-speaking Canadians in Atlantic Canada, which has the second highest percentage by region in Canada. In 2001, Aboriginal peoples, however, formed only 2.4 per cent of the region's population.

Atlantic Canada's Physical Geography

Atlantic Canada is not a homogeneous region but 'a region of geographical fragmentation' (Macpherson, 1972: xi). It consists of three subregions: the Maritimes, Newfoundland, and Labrador. Macpherson argues that this physical division has impacted on all aspects of life in Atlantic Canada, hindering the emergence of a common political will, a common regional consciousness, and an economic union among the four Atlantic provinces. Whether or not an economic union is a desirable political arrangement for Atlantic Canada

Table 9.1 Basic Statistics for Atlantic Canada by Province, 2006

Province	Population (000s)	Population Density	% GDP	Unemployment Rate	% Aboriginal (2001)
Prince Edward Island	136	23.9	0.3	11.0	0.1
Newfoundland/Labrador	506	1.4	1.9	14.8	3.7
New Brunswick	730	10.2	1.7	8.8	2.4
Nova Scotia	914	17.3	2.1	7.9	1.9
Atlantic Canada	2,285	4.9	6.0	10.0	2.4
Canada	31,613	3.5	100.0	7.7	3.3

Sources: Statistics Canada (2003d, 2007a, 2007b, 2007c).

is uncertain, though the topic of union has long been debated.

Each subregion has its distinct set of physical characteristics. In this section, climate and physiography are emphasized. The climate in the Atlantic region is quite varied. As shown in Figure 2.6, Atlantic Canada has three climatic zones—Atlantic, Subarctic, and Arctic zones. Why do three zones exist in Atlantic Canada? The great north–south extent of this region is one reason. For example, the distance from the southern tip of Nova Scotia (44° N) to the northern extremity of Newfoundland and Labrador (60° N) is over 2,000 km. In addition, the Labrador Current chills the Labrador coastline, creating an arctic climate, while inland a subarctic climate is found. Milder air masses affect the Atlantic climatic zone, including the hot, moist air masses from the Gulf of Mexico.

The general atmospheric circulation system brings several air masses to the Maritimes and Newfoundland, where they interact and result in unsettled weather conditions. The main air masses originate in the interior of North America, the Gulf of Mexico, and the North Atlantic Ocean. Consequently, summers are usually cool and wet, while winters are short and mild but often associated with heavy snow and rainfall. Most precipitation falls in the winter. Temperature differences between inland and coastal locations are striking. Coastal locations are usually several degrees warmer in the winter than inland locations. During the summer, the reverse is true, with coastal areas usually several degrees cooler.

Along the narrow coastal zone of Atlantic Canada, the climate is strongly influenced by the Atlantic Ocean. The summer temperatures of coastal settlements along the shores of Newfoundland are markedly cooled by the very cold water of the Labrador Current (a cold ocean current in the North Atlantic Ocean), which brings with it icebergs. Another effect of the sea occurs in the spring and summer—fog and mist result when the warm Gulf Stream waters (which originate in the Gulf of Mexico) mix with the cold Labrador Current. In the winter, the clash of warm and cold air masses sometimes results in severe winter storms characterized by heavy snowfall. For example, St John's had a record 648.2 cm of

Newfoundland still has hundreds of small outports along its rocky and indented coastline. Since the loss of the cod fishery, most outports no longer have an economic base and their future is in doubt. (Al Harvey/ The Slide Farm)

snow (almost 22 feet) in the winter of 2000–1 (see Vignette 9.1). Such storms are usually short-lived and more normal weather conditions return quickly. Annual precipitation is abundant throughout Atlantic Canada. It averages around 100 cm in New Brunswick and 140 cm in Newfoundland. Much precipitation comes from nor'easters—strong winds off the ocean from the northeast—that draw their moisture from the Atlantic Ocean.

Atlantic Canada is situated in two of Canada's physiographic regions: the Maritimes and Newfoundland are part of the Appalachian Uplands, and Labrador is part of the Canadian Shield. Within the Appalachian Uplands, the region's landforms are dominated by a rugged coastline and an upland interior consisting of hilly to mountainous terrain. In the Canadian Shield, where arable land is even more scarce than in the Appalachian Uplands, the geography consists of a coastline with fjords, a rugged interior, and the lofty Torngat Mountains.

Atlantic Canada's physiography lends itself to a resource hinterland based on fish, forests, and minerals. The region has little arable land, except on Prince Edward Island, in the Saint John River Valley, and in the Annapolis Valley (Vignette 9.2). Beyond these few favoured lowlands, the land is too rough and rugged for agricultural settlement. As a consequence, most settlements in Atlantic Canada are clustered along the coastline, where for years fishers made their living harvesting the riches of the ocean. Further inland, settlements evolved where mineral and forest resources were profitable. For instance, in the Appalachian Uplands, a dense forest bolstered the economies of the Maritime provinces, especially New Brunswick. Similarly, mining has long been a way of life for some Atlantic Canadians—from the famous coal mines of Cape Breton Island, Nova Scotia, to the recently developed rich nickel-copper-cobalt deposit at Voisey's Bay, Labrador. Furthermore,

Vignette 9.1 Weather in St John's

St John's has one of the most varied but mild climates in Canada. Weather is characterized by foggy days, overcast skies, and stormy, wet weather. Throughout the year, a salty smell is in the air. While most of the year has mild temperatures, strong northwest winds result in heavy winter snowfalls. In the winter of 2000–1, St John's had a record amount of snow—just under 6.5 metres! Jutting into the Atlantic Ocean at the eastern tip of the island of Newfoundland, the marine climate of St John's is affected by the westerly circulation of the atmosphere and by the waters of the North Atlantic. As David Phillips (1993: 155), Canada's well-known weather pundit, put it, 'Of all major Canadian cities, St John's is the foggiest, snowiest, wettest, windiest, and cloudiest.' These weather characteristics are due to four major climatic factors:

- The proximity of St John's to the Atlantic Ocean, which keeps summer temperatures cool and winter temperatures mild. Its marine location also brings abundant precipitation, with just over 1,500 mm falling each year.
- The flow of continental air masses from west to east brings warm weather in the summer and cold weather in the winter. These continental air masses intersect with marine air masses, resulting in highly unstable weather.
- The cold waters of the Labrador Current bring a taste of Arctic weather by filling St John's harbour with pack ice each winter. Icebergs are often sighted just offshore sometimes as late as June.
- The warm, moisture-laden air masses from the Gulf of Mexico make their way to St John's and, by mixing with the cold air just above the waters of North Atlantic, result in thick, cool fog. Fog is common from April to September.

the region has long benefited from its extensive continental shelf, where fishers pursued codfish and recent discoveries of petroleum and natural gas have led to the development of major oil-field sites, such as Hibernia on the Grand Banks.

The natural division of Atlantic Canada has led to different settlement patterns and a diverse cultural geography. In the early seventeenth century, French colonists arrived to begin the process of European settlement. Indeed, Nova Scotia was settled first by French colonists—the Acadians. In 1755 in Nova Scotia and then in 1758 from Prince Edward Island, the Acadians were rounded up and expelled by the British because they initially refused to sign an oath of allegiance to the British Crown. German Protestants arrived at Lunenburg on Nova Scotia's southeastern Atlantic coast beginning in 1753; the Loyalists, including about 3,000 black refugees, settled in Halifax and other parts of Nova Scotia following the American Revolution; and Gaelic-speaking Highlands Scots, both Catholic and Protestant, made Cape Breton and the Northumberland shore their home beginning in the early 1800s. The British-run merchant fishery saw many English and Irish workers settle along the coast of Newfoundland. In addition to these different settlement patterns, the physical obstacles between the subregions and the separation of the region from the rest of Canada fostered and maintained cultural differences:

- The Maritimes, though a relatively cohesive economic unit, has closer ties to New England than to the rest of Canada.
- The island of Newfoundland stands alone, separated from the rest of Atlantic Canada by the Cabot Strait and from Labrador by the Strait of Belle Isle. These water barriers hamper Newfoundland from having close economic links with the Maritimes and Labrador.
- Labrador, while part of the province of Newfoundland and Labrador, has been drawn into Québec's economic orbit because of the land connection between Québec and Labrador. Examples include: (1) the transmission of hydroelectric energy to markets across southern Québec; (2) the shipment of Labrador's iron ore to the port of Sept-Îles, Québec; and (3) the Labrador–Québec highway that runs from Labrador's largest town, Happy Valley–Goose Bay, to Baie-Comeau and Québec's provincial highway system.

Vignette 9.2 The Annapolis Valley

The Annapolis Valley is a low-lying area in Nova Scotia. At its western and eastern edges the land is at sea level, but it rises to about 35 m in the centre. The area is surrounded by a rugged, rocky upland that reaches heights of 200 m and more. Beneath the valley's fertile soils are soft sedimentary rocks. The sandy soils of the Annapolis Lowlands originate from marine deposits that settled there about 13,000 years ago. After glacial ice retreated from the area, sea waters flooded the land, depositing marine sediments that consisted of minute sand and clay particles. Isostatic rebound then caused the land to lift and slowly these lowlands emerged from the sea. In the seventeenth century, the favourable soil of the Annapolis Valley attracted early French settlers, who later became known as Acadians. Today, the Annapolis Valley's stone-free, well-drained soils and its gently rolling landforms provide the best agricultural lands in Nova Scotia.

Despite the differences between Atlantic Canada's provinces, many similarities exist. In addition to a common geographic location—the waters of the Atlantic Ocean wash up on the shores of all four provinces—the Atlantic provinces have all undergone a similar economic process of growth, decline, and dependence. As the region of Canada where Europeans first settled, Atlantic Canada has undergone a long period of resource exploitation. Its role as a resource hinterland brought early prosperity, which was followed by decline as its non-renewable resources were exhausted and some of its renewable resources were overexploited. This economic malaise drew Ottawa into the affairs of Atlantic Canada, often at the request of the Atlantic provinces, which turned to Ottawa for financial assistance in times of crisis. Ottawa continues to have a strong presence in the region and, some say, an unduly strong influence. Ottawa's dominating role through the federal Department of Fisheries and Oceans in the management of the Atlantic fisheries is particularly resented.

Environmental Challenges

Without a doubt, the major environmental disaster to strike Atlantic Canada has been the collapse of the cod fishery. But this disaster was due more to technological advances that enable much larger catches, coupled with federal mismanagement, than to natural factors such as the warming of the Atlantic waters or the expanded seal population consuming vast quantities of cod. Clearly a case of a 'tragedy of the commons', this ecological collapse of one of the world's greatest fish stocks is documented in the later pages of this chapter. Here, our attention is focused on an unwanted remnant of Cape Breton's glory days as a major iron and steel centre in Canada—the so-called tar ponds.

By the end of the twentieth century, Sydney, Nova Scotia, not only had lost its industrial sector but the community was saddled with an environmental disaster—the Sydney Tar Ponds. But what are the tar ponds and how did this industrial disaster occur? The tar ponds are composed of tar and a wide variety of chemicals that are dangerous to human beings. Arsenic, lead, and other chemicals are found in the tar ponds. These toxic wastes came from the operation of the Sydney Steel Co., or Sysco, especially its coke ovens. A coke oven is a large chamber where coal is heated at extremely high temperatures. Waste products from coke ovens include tar and toxic gases. These toxic wastes included benzene, kerosene, and naphthalene.

In the case of Sysco, these wastes were discharged into a nearby stream and gradually seeped into Muggah Creek. Since the 1980s, residents living near the tar ponds complained of an orange goo seeping into their basements, while others found that dust from the tar ponds was blowing into their yards and houses. When Health Canada stated that those living closest to the tar ponds were in greater danger than those living further away, Ottawa and Halifax began to pay attention. Action began after a cancer specialist concluded that there was a higher risk of dying from cancer in Whitney Pier and Ashby, communities closest to the tar ponds, than the national average. At that point, the federal and provincial governments joined forces with the local government to fund a cleanup of the tar ponds.

The Sydney Tar Ponds are the site of the biggest environmental cleanup project in Canadian history. In 1998, Ottawa and Halifax agreed to a $62 million cost-sharing agreement to clean up the tar ponds and the site of the coke ovens. Since 2001, the tar ponds, coke ovens site, and other areas within the Muggah Creek watershed have been the scene of intense activity to repair the environmental damage. Since the cleanup began, the community is seeing improvements on a number of fronts: the construction of a sewer interceptor that will divert the tonnes of raw sewage that now flow daily into the tar ponds; the demolition and removal of derelict structures on the coke ovens site; and the closure and capping of the old Sydney landfill. Still, much remains to be done.

Atlantic Canada's Historical Geography

Atlantic Canada was the first part of North America to be discovered by Europeans. While the Vikings arrived first, establishing for a brief time around AD 1000 a settlement at the northern extreme of the Northern Peninsula in Newfoundland, the first documented discovery of land in North America was made by John Cabot. The Italian navigator, employed by the King of England, reached the rocky shores of Atlantic Canada (the exact location, either Cape Bonavista, Newfoundland, or Cape Breton Island, Nova Scotia, is in dispute) on 24 June 1497. Like Columbus far to the south in 1492, Cabot was searching for a sea route to Asia. In England, Cabot's report of the abundance of **groundfish**—cod, grey sole, flounder, redfish, and turbot—in the waters off Newfoundland lured European fishers to make the perilous voyage across the Atlantic to these rich fishing grounds. Canada's Atlantic coast quickly became a popular area for European fishers and though landings on shore took place—for drying the fish and establishing temporary habitation during the fishing season—permanent settlements were slow to take hold in this part of North America.

In 1605, a handful of French settlers, at Port-Royal on the Bay of Fundy coast of present-day Nova Scotia, established the first permanent European settlement in North

Vignette 9.3 Strangers in Their Own Land

Four historic events changed the Mi'kmaq into strangers in their own land. First, the defeat in 1713 of their European ally, the French, by the British, placed the Mi'kmaq in a vulnerable position. Second, the expulsion of the Acadians in 1755 eliminated another ally of the Mi'kmaq. Third, following the founding of Halifax in 1749, relations between the British and the Mi'kmaq deteriorated. British rangers were unleashed to harass the Mi'kmaq, to destroy their villages, and to drive them far beyond the British settlements. Fourth, by 1783 the British Loyalists from the new republic of the United States relocated to Nova Scotia and New Brunswick, where they occupied the fertile lands of the recently departed Acadians and the prime hunting lands and fishing areas of the Mi'kmaq. No longer regarded as a military threat, the Mi'kmaq were pushed into more marginal lands and were surrounded by a growing British population. By then, the ancient world of the Mi'kmaq had collapsed. Denied a place in the new economy, the Mi'kmaq were forced into a state of dependence in which they soon became outcasts in their own land.

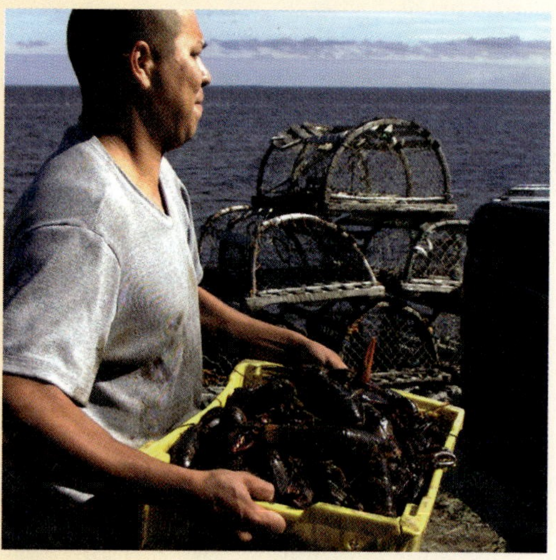

In 1999, the Supreme Court of Canada ruled that the Mi'kmaq should have a share of the commercial fishery (see Further Readings for a summary of the *Marshall* decision). Five years later, the Mi'kmaq had gained access to about 3 per cent of the federal lobster allocation by receiving about 320 commercial fishing licences. When operating at full capacity, these licences should allow the Mi'kmaq fishers to generate around $25 million worth of lobster each year. (CP/Jaques Boissinot)

America north of Florida. During the seventeenth and part of the eighteenth centuries, French settlers spread into the Annapolis Valley and other lowlands in the Maritimes. These French settlements, united by culture, language, and a common economy, became known as Acadia.

While French settlers had established themselves in the Maritimes by the early seventeenth century, English settlers began arriving in 1610 in another part of Atlantic Canada—at Conception Bay, Newfoundland. Over the next 100 years, small English, French, and Irish coastal settlements developed because fishers preferred to remain in Newfoundland at the close of the fishing season rather than make the long journey back to Europe. By the 1750s, there were over 7,000 permanent residents living in hundreds of small fishing communities along the shores of the island of Newfoundland. The French Shore, for example, is a coastal area of southwest Newfoundland where French fishers enjoyed treaty rights to fish. These rights were granted by the British in 1713 and did not end until 1904. Today, many communities along the French Shore are populated by descendants of these French settlers.

Over the first half of the eighteenth century, war between the two European colonial powers in North America—England and France—was almost continuous. During that time, the French forged an alliance with the Mi'kmaq and Maliseet, drawing them into the conflict with the English and their Iroquois allies.[2] Under the terms of the 1713 Treaty of Utrecht, France surrendered Acadia to the British. However, many French-speaking settlers, the Acadians, remained in this newly won British territory, which was renamed Nova Scotia. During the previous century, the Acadians had established a strong presence in the Maritimes with settlements and forts. Most Acadians lived along the Bay of Fundy coast, tilling the soil. Until the mid-1700s, Britain made little effort to colonize these lands, leaving the Acadians to till the land in this British-held territory. By 1750, French-speaking Acadians numbered over 12,000.

British settlement began to spread to the Maritimes in the mid-eighteenth century, with the first serious attempt occurring in 1749 when the British government began to recruit settlers for its newly acquired lands in Nova Scotia. Halifax, founded in 1749, would become the centre of British influence and military power in Nova Scotia and would provide a counterbalance to the well-fortified French military base of Louisbourg on Cape Breton Island. Through a series of wars culminating with the Seven Years' War (1756–63), the British would gradually gain control of Atlantic Canada.

In 1756, the final struggle for North America began. Britain and France engaged in an all-out battle for supremacy around the world. In North America, the British commanders began preparing for war in 1755. When the Acadians refused to swear an oath of unconditional loyalty to the British Crown, the British deported 6,000 Acadians from Nova Scotia in 1755 to more secure British colonies to the south, from Massachusetts to South Carolina and Georgia. Then, in 1758, in Prince Edward Island, a few thousand more Acadians were rounded up and expelled by the British; this time they were transported to France, though perhaps as many as half of these deportees died en route from sickness aboard ship or by drowning. Historically, such deportations have been common (Lockerby, 2004), yet today such instances might be called 'ethnic cleansing'. A few thousand other Acadians escaped further north from Nova Scotia and PEI, into northern New Brunswick. At the same time, to establish a British presence in the newly conquered land, the British recruited yeoman farmers from New England, the New England Planters, to settle on what had been Acadian lands.

After the British defeated the French, all French territories in North America were under British control, ending the dream of a French North America. Under the terms of the Treaty of Paris in 1763, France ceded all its territories to Britain, except for the islands of Saint-Pierre and Miquelon near the southern coast of Newfoundland. Today, most Acadians live in northern New Brunswick, where they constitute just under 40 per cent of the province's population.

The next event to influence the evolution of Atlantic Canada was the American Revolution. Following victory by the American colonies, approximately 40,000 Loyalists made their way to Nova Scotia and New Brunswick (see 'The Loyalists' in Chapter 3 for more details). Most settled along the shores of the Bay of Fundy, in the Annapolis Valley, on Cape Breton Island, and around the British stronghold of Halifax. Halifax became known as the 'Warden of the North' because it provided a superb port for the ships of the British navy. Over the next 50 years, the arable land was cleared and occupied. During that time, more British settlers came to the Maritimes. Nova Scotia alone received 55,000 Scots, Irish, English, and Welsh. Most Scots went to Cape Breton and the Northumberland shore. While many newcomers tried their hand at farming, the fishery still dominated life in the Maritimes and Newfoundland.

In the early nineteenth century, the harvesting of Atlantic Canada's natural wealth increased. This frontier hinterland of the British Empire exploited its rich natural resources—the cod off the Newfoundland coast and the virgin forests in the Maritimes—and became heavily dependent on transatlantic trade of these resources. Overseas trade with other British colonies in the Caribbean, as well as with Britain, formed the cornerstone of Atlantic Canada's trade and prosperous economy. In the first half of the nineteenth century, trade with New England was relatively limited because both the Maritimes and New England had almost identical resource-oriented economies. However, after the American Civil War, New England would industrialize, leading to greater trade between the two regions. In addition, Britain's move to free trade in 1849 would mean the loss of Atlantic Canada's protected markets for its primary products, resulting in even greater interest in the American market, especially in the Maritimes.

Log boom on the Miramichi River, Blackville, New Brunswick, c. 1915. The cribwork at mid-river was used by raftsmen and log drivers for sorting logs, breaking up jams, and keeping a channel clear. In the nineteenth and early twentieth centuries the spring log drive on the Miramichi was one of the largest in Canada. Most logs went to sawmills. Today, logs are transported by truck and logging in Atlantic Canada primarily feeds pulp mills in the region. (View-8155. Notman Photographic Archives, McCord Museum, Montréal)

Historic Head Start

With the fishery and the timber trade well underway in the early nineteenth century, Atlantic Canada had a head start in developing a commercial economy. Furthermore, the availability of timber and the region's favourable seaside location provided the ideal conditions for shipbuilding in Atlantic Canada. By 1840, Atlantic Canada entered the 'Golden Age of Sail', with Nova Scotia and New Brunswick becoming the leading shipbuilding centres in the British Empire. Exports from Atlantic Canada were primarily cod and timber, while imports were manufactured goods from England and sugar and rum from the British West Indies. Several world events in the mid-nineteenth century added strength to this economic boom in the Maritimes. Demand for its exports rose, especially for timber, which was used to build British merchant ships and warships. The Crimean War (1853–6), the Reciprocity Treaty with the United States (1854–66), and the American Civil War (1861–5) opened new markets for fish, minerals, and lumber from the Maritimes and cod from Newfoundland. International trade was increasing and Atlantic Canada was looking outward to the rest of the world, while the Province of Canada was more consumed with internal developmental issues at that time.

Just before Confederation, however, external events dampened Atlantic Canada's resource-based economy. Iron was replacing wood in shipbuilding; the three-way trade with Britain and its Caribbean possessions collapsed, largely because a world glut of sugar caused prices to drop; and the end of the Reciprocity Treaty in 1866 cut off access to the Maritimes' natural trading partner, New England. These events resulted in the deterioration of Atlantic Canada's economic position among the Atlantic trading nations and colonies.

Confederation

The provinces of Atlantic Canada joined Canada at different times. Nova Scotia and New Brunswick joined at the time of Confederation (1867); Prince Edward Island followed in 1873; while Newfoundland did not join Canada until 1949. At the time of Confederation, Newfoundlanders, being so isolated, saw little advantage to joining Canada. Then, too, Newfoundlanders cherished their independence, albeit as a colony at the edge of the British Empire, and they feared that Newfoundland would be dominated by the larger provinces. Besides, Newfoundland's trade relations at the time were tied to Britain, not Canada. For the Maritime colonies, the decision to join Canada was not an easy one. In spite of their historic head start and their excellent access to sea and world markets, they were on the margins of the new country and their once-booming economy lost its momentum.

Confederation favoured economic growth in Central Canada. The federal government's National Policy of 1879 reinforced Central Canada's advantage, placing the Maritimes' fledgling manufacturing base in danger. In order to stimulate Canadian manufacturing, Canada imposed duties on imported goods. In retaliation, Americans increased their trade barriers on Canadian manufactured goods. Canada even placed a tariff on foreign-produced coal to protect the high-cost coal mines of Nova Scotia. In effect, Canada had created a 'closed' economy that facilitated the industrialization of Central Canada by relegating the rest of Canada to the dual roles of domestic market for Canadian-manufactured goods and resource hinterland for Central Canada, the United States, and the rest of the world (see Chapters 1, 3, and 4).

At Confederation, the manufacturing base in the Maritimes suffered from small local markets, and the region's natural market, New England, was now less accessible because of high US tariffs. What Atlantic Canada needed was access to the market of Central Canada. Ottawa's answer was the Intercolonial Railway (completed in 1876). Built to fulfill the terms of Confederation, the Intercolonial provided rail access to Central Canada. The railway was never a commercial success, but it did stimulate economic growth in the Maritimes. The Intercolonial was operated by the federal government: freight rates were kept low to promote trade, and substantial annual deficits were paid by Ottawa. With low-cost transportation to the national market and the possibility of firms achieving economies of scale, a general economic surge occurred in the Maritimes. Cotton mills, sugar refineries, rope works, and iron and steel manufacturing plants were established or expanded to serve the much larger national market and to grab a share of the expanding western market. The railway boom in the Canadian prairies had a positive impact on Nova Scotia in the form of heavy investment in the province's steel mills. With the construction of new railway lines, steel production was focused on manufacturing steel rails and locomotives. By taking advantage of Cape Breton's coalfields and iron ore from Bell Island, Newfoundland, the steel industry in Sydney prospered, accelerating Nova Scotia's economic growth well above the national average.

The Maritime economy, however, suffered a deadly blow when, in 1919, freight rates were increased to levels common in Central Canada. Immediate access to the national market became more difficult and, with the loss of sales, many firms had to lay off workers, while others were forced to shut down. Even before these troubled times, the Maritimes economy was unable to absorb its entire workforce, leading many to migrate to the industrial towns of New England and southern Ontario. After World War I, the out-migration increased.

After World War II, the Maritimes grew but at a slower pace than other parts of Canada. Newfoundland joined Canada in 1949, mainly for economic reasons. Newfoundland stood to gain from the social programs available to Canadians—for instance, unemployment insurance assisted seasonal workers like fishers and family allowance payments provided a monthly cheque to each mother. Above all, Newfoundlanders hoped that Canada's booming economy would pull Newfoundland from its poverty and its dependence on the fishery. Even during periods of national affluence—for instance, the 1950s and 1960s—Atlantic Canada experienced limited economic growth; numerous federal loans and grants to firms in Atlantic Canada made little difference. No doubt this region languished because of its scattered and relatively small population, its narrow resource base, and the high cost of transporting goods to the national market. In 1968, Professor David Erskine at the University of Ottawa drew a rather dismal picture of Atlantic Canada as a hinterland:

> The region is, in the Canadian context, one of 'effort' rather than of 'increment'. Small scale resources once encouraged small scale development, but only large scale resources encourage modernization. The small scale and lack of concentration of its resources makes the region one in which government investment is easily dispersed without bringing about growth. Low levels of professional services and low levels of education result from the high taxes from low incomes; thus, still further retardation of economic growth occurs. (Erskine, 1968: 233)

Until the Free Trade Agreement, Atlantic Canada's limited access to external markets hindered its economic development. Manufacturing firms in the region have had to overcome high transportation costs to deliver their products to Canada's major markets in Central Canada. Competition from firms in Central Canada often drove Atlantic-based firms from the marketplace for two reasons:

1. Manufacturing firms in Ontario and Québec could achieve economies of scale, making their production costs lower and resulting in lower prices for consumers.
2. Manufacturing firms in Ontario and Québec had much lower transportation costs in reaching their western customers than similar firms in Atlantic Canada.

Unable to compete within the national marketplace, most manufacturers in Atlantic Canada had to limit their production to the local market and kept wages low in order to compete with imported products from Central Canada.

Steel, Iron, and Coal—The Rise and Fall of Nova Scotia's Industrial Base

For over 100 years, coal and iron mining provided the basis for the Nova Scotia iron and steel industry. Based in Cape Breton Island, Nova Scotia's iron and steel industry was a mainstay of the province's industrial sector. The story of the rise and fall of this industry provides insights into the challenges facing manufacturers in Nova Scotia and the rest of Atlantic Canada.

The iron and steel industry in Cape Breton Island near Sydney, Nova Scotia, was for a long time the heavy industrial heartland of Atlantic Canada. During that time, Sydney was the principal city on Cape Breton Island and the second-largest city in Nova Scotia.[3] Sydney's fate was closely tied to its major industrial firm, the Sydney steel mill, and to local coal mines. As the mill and mines prospered in the early part of the twentieth century, the town of Sydney expanded, but since the end of World War II, both the mines and mill suffered financial losses and the size of their labour forces were reduced. This process of deindustrialization, while delayed by federal and provincial subsidies, has taken its toll on the people of

Sydney and Cape Breton. In 2001, the steel mill was closed, ending this long history of heavy industry. The future of Cape Breton, without the steel and coal industry to fuel its growth, is uncertain.

The rise and fall of heavy industry in Cape Breton spans a period of 100 years. In 1899, the Dominion Iron and Steel Company began to produce steel from Cape Breton coal and Newfoundland iron ore. Demand for its steel proved so great that plans were made to expand steel production. In 1901, a large, more integrated steel complex was built to supply steel rails to firms constructing new rail lines in Canada and other countries. Located at a fine harbour on the Atlantic coast, the iron and steel mill at Sydney on Cape Breton Island had ready ocean access to raw materials and world markets. It was able to use local coal and iron ore from Bell Island, Newfoundland. It supplied markets in Canada (via the St Lawrence River and the Intercolonial Railway) and in other Commonwealth countries. Now called the Dominion Steel and Coal Co. (Dosco), the firm benefited from the strong demand for steel during World War I. Under these prosperous conditions, Dosco became Canada's largest private employer.

In the 1920s, the Sydney steel mill began to fall on hard times. By 1925, Canada's railway boom had ended, weakening the demand for steel, and the demand for coal slumped as many coal customers turned to another source of energy, petroleum. A modest recovery for coal, however, occurred during World War II. After the war effort, it became evident that the Sydney steel mill was no longer viable without public subsidies. Sydney simply could not compete with offshore steel producers in Brazil and South Korea. The basic problems facing the Sydney steel mill were threefold: little local demand in the Maritimes, stiff competition in Canada from low-cost producers in southern Ontario and even stiffer competition in external markets, and increasing costs for its raw materials and labour.

Even with the assistance provided by government subsidies, Dosco continued to lose money. As a result, in 1965, Dosco announced plans to close its coal mines, which would eliminate 6,500 jobs, and in 1966 the Bell Island iron mine, which had been a Dosco operation, was shut down. In 1967, the federal government took over the operation of the coal mines through a Crown corporation, the Cape Breton Development Corporation (Devco). The steel mill was on the brink of bankruptcy when it was sold to the Nova Scotia government in 1967. From that time on, a provincial corporation, the Sydney Steel Corporation (Sysco), not only ran the mill but, with help from the federal government, also paid for the annual shortfall. The two governments intervened because they did not wish to add to the unemployment problem in Cape Breton, an already economically depressed area with a high unemployment rate. As Blackbourn and Putnam observed:

> The continued existence of an uneconomic and obsolete steel mill in an unsatisfactory location is a classic illustration of the importance of inertia in industrial location. The closure of the steel mill would create unemployment in a depressed province. Not only would jobs be lost in steel but the coal mines which supply the mill would probably close leading to further unemployment. (Blackbourn and Putnam, 1984: 112)

In the late 1960s, neither the federal government nor the Nova Scotia government, for social and political reasons, could accept closure of the Sydney steel mill. At that time, the two governments' treasuries could absorb the annual losses and even advance more funds for plant improvements. However, by the early 1990s, governments had to deal with their enormous debts, and cutbacks were the order of the day. Even after the Nova Scotia and federal governments stepped forward with financial support, the industry continued to limp along. In 1999, the Crown corporation owned by Sydney Steel Corporation, which had received more than $2 billion in federal and provincial subsidies over the previous 20 years, tried to sell the company. By 1999, only 700 workers were employed at Sydney Steel Corporation, a far cry from the peak of 4,000 in the late 1960s. By 1987, Sydney's steel mill had been reduced to an electric-arc furnace using scrap metal to manufacture steel rails. Sales to the United States were crucial but the US imposed countervailing duties because the Nova Scotia government was subsidizing Sysco. Unable to find a buyer, the Nova Scotia government closed the plant and, in 2001, announced plans for its demolition.

Since the 1980s, demand for Cape Breton coal declined. With the switch at Sysco from a coal-burning furnace to an electric-arc furnace, a major customer for Cape Breton coal was lost. Cape Breton bituminous coal mines were operated by the federal agency, Cape Breton Development Corporation (Devco). Most coal was sold under contract to Nova Scotia Power. At first glance, coal from the Cape Breton coalfield was ideally suited for the nearby thermoelectric stations operated by Nova Scotia Power. While the cost of transporting coal to these generating stations is low, the cost of mining Cape Breton coal is high due to its underground mining operations, which extend several kilometres. As a result, coal from deep-tunnel mining at Glace Bay, Cape Breton, cost more to deliver to Nova Scotia Power than coal from Pennsylvania. With NAFTA, American coal is no longer barred from Canada by high tariffs. For that reason, Nova Scotia Power, now a privatized corporation, purchased lower-priced American coal to fire its thermoelectric generating plants in order to keep its electricity rates at their current level. In February 1999, the government announced it was ending subsidization of Devco, which employed about 1,700 people at its two Cape Breton mines. One mine (Phalen) was closed in 2000 and the other one (Prince) closed in 2001.

Atlantic Canada Today

Atlantic Canada contains a natural beauty and captivating cultural roots that continue to foster a quality of life for Atlantic Canadians that is very special in Canada. Yet, Atlantic Canada continues to see its young people seek economic opportunities in other parts of the country. The prospect for strong economic growth remains elusive and the challenge facing Atlantic Canada is to translate its energy, minerals and metals, forests, and marine and freshwater resources into a strong sustainable economy.

Vignette 9.4 Coal Mining—A Dangerous Occupation

Coal mining is a very dangerous occupation—nowhere was this more evident than in Nova Scotia where coal miners worked deep underground. The closure of the Prince mine in 2001 marked the end of a long history of coal mining in Nova Scotia. While the pay was good, danger went with the job and, for some, the cost was their lives. In 1956 and again in 1958, the Springhill mine near the head of the Bay of Fundy was the site of two underground explosions that took the lives of nearly 100 miners. In 1992, the Westray mine in Pictou County, Nova Scotia, was the site of another mine disaster. A buildup of methane gas caused an underground explosion. Twenty-six men lost their lives. The mine had failed safety inspections and yet continued to operate. No doubt, the marginal nature of this coal-mining operation contributed to the substandard conditions of the mine. Westray was under pressure to keep its costs low to remain competitive, so work stoppages to correct the problems identified in the safety inspections would not only have been costly but might have resulted in Westray's failure to deliver sufficient coal to Nova Scotia Power. Unfortunately, the Westray mine was closed permanently only after the loss of the 26 miners.

The Confederation Bridge, completed in 1997, connects Prince Edward Island with New Brunswick. As an integral part of the Trans-Canada Highway network, the 12.9 km Confederation Bridge is the longest bridge over ice-covered waters in the world. (Ron Garnett/AirScapes.ca)

Like many other regions of Canada, Atlantic Canada has two economies—a weak rural economy and a much more vigorous urban one. While Atlantic Canada is witnessing economic growth in its cities, its rural areas are struggling. Cities are the focus of the tertiary sector of the economy while rural areas are home to the primary sector, especially the fishery. Within Atlantic Canada, the internal migration pattern sees more and more rural residents moving to the six largest cities in search of employment and business opportunities. The role of its traditional economy, the fisheries, has positive and negative elements. The shell fishery has done well, largely because of increase in prices. On the other hand, the failure of the northern cod stocks to recover remains a hurtful economic reality, especially to small coastal villages but also to the fish-processing sector. The demise of Great Harbour Deep represents a trend towards depopulation of parts of rural Atlantic Canada (Vignette 9.6). Major resource projects, such as the offshore oil and gas and the Voisey's Bay nickel mine, provide a spark to Atlantic Canada's economy, but the future economic success of Atlantic Canada depends on the prosperity of its major cities, especially Halifax, St John's, Moncton, Saint John, Fredericton, and Charlottetown.

Relationships with Ottawa: The Atlantic Accord/ Equalization Dispute

Relationships with Ottawa remain extremely important. Direct federal assistance takes place through the Atlantic Canada Opportunities Agency (ACOA). Formed in 1987, ACOA provides federal funds to foster economic growth in Atlantic Canada. The level of funding for 2007–8 was $1.1 billion (Atlantic Canada Opportunities Agency, 2007).

Equalization payments from Ottawa are much more important than funds from ACOA. Equalization payments go directly to the provincial governments, which then decide how to spend the money. Atlantic Canada receives about half of its revenue from equalization payments (see Vignette 3.5 and Table 3.7).

Equalization is the responsibility of the federal government. The challenge facing Ottawa is to balance the interests of the have provinces and those of the have-not provinces. Because each province has different interests, finding a solution to calculating equalization payments that pleases all provincial governments has eluded many federal governments. Equalization, then, is one of the hot-button issues in the centralist/decentralist faultline. In the 2007 budget, the federal position was to include half of the oil and/or gas revenues in the equalization formula as well as to add a fiscal capacity cap. Both of these measures pleased Ontario, but they raised a storm of protest from three oil/gas producing provinces—Newfoundland and Labrador, Nova Scotia, and Saskatchewan. While the idea of a cap on fiscal capacity (the ability of provinces to raise revenue) seems straightforward and fair-minded, for Newfoundland and Labrador and Nova Scotia the matter is complicated by three factors:

- Both Atlantic provinces strongly believe that Ottawa has failed to honour the Atlantic Accord in its 2007 budget by adding a fiscal capacity cap.
- Both Atlantic provinces interpret the Atlantic Accord as a means to turn these provinces from have-not to have provinces and, therefore, see Ottawa's new measures as hampering that process of economic development.
- The 'curse of declining populations' in such calculations means that, as population decreases in these provinces, per capita fiscal capacity increases even if their revenues remain constant.

There is a saving grace: Nova Scotia and Newfoundland and Labrador can maintain the benefits of the Atlantic Accord by opting to remain under the previous equalization system.

If equalization was not a sufficiently hot-button item to begin with, adding the Atlantic Accord and the 2007 federal budget into the mix certainly made it more complex and heated. The Atlantic Accord is by far the most significant recent agreement between Ottawa and the provinces of Newfoundland and Labrador and Nova Scotia. This agreement ensures that offshore oil and natural gas revenues provide an economic boost for these two provinces (Vignette 9.5 and Figure 9.2; see also Vignette 3.4). As important, the Accord protects their equalization payments, that is, these payments, according to the agreement, will not be affected by rising oil/gas royalties. In the 2007 federal budget, the funds allocated to equalization payments increased to just over $12.7 billion, an increase of $1.5 billion over 2006–7, thereby providing a larger flow of payments to these two have-not provinces. However, the 2007–8 equalization program includes a fiscal capacity cap to ensure that equalization payments do not raise a province's total per capita fiscal capacity above that of any non-receiving province, namely Ontario. As Ottawa likes to point out, equalization payments for 2007–8 will reach record levels because more funds were allocated to the program. In the case of Newfoundland and Labrador, equalization payments increase from $477 million to $733 million and for Nova Scotia from $1.3 billion to $1.53 billion for the fiscal year 2007–8.

So what is the problem? For Newfoundland and Labrador and Nova Scotia, the rub is that the projected 2008–9 equalization figures show the impact of the fiscal capacity cap for the next fiscal year (Table 9.2). Accordingly, Newfoundland and Labrador would see its equalization payments drop by nearly $500 million in the new system, while Nova Scotia would see a decline of $92 million. Saskatchewan drops to zero. Keep in mind that the value of Newfoundland and Labrador's oil production in 2005 was $7 billion while gas was $1.4 billion (Table 9.7). Not surprisingly, Newfoundland and Labrador opted to stay with the 2006–7 equalization calculations, thereby avoiding the cap on fiscal capacity. This means that the province keeps all the benefits from the Atlantic Accord but loses out on the increased equalization payments—oil revenues are expected to increase in the coming years for Newfoundland and

Table 9.2 Equalization Payments by Have-Not Province, 2006–7 to 2008–9 ($ millions)

	NL	PEI	NS	NB	Que.	Man.	Sask.	BC	Total
2006–7	632	291	1,386	1,451	5,539	1,709	13	260	11,281
2007–8 Renewed Equalization	477	294	1,308	1,477	7,160	1,826	226	0	12,768
2008–9 (Projection)	197	310	1,294	1,492	7,622	2,003	0	0	12,918
Difference 2006–7 & 2008–9	–435	19	–92	41	2,083	294	–13	–260	1,637

Source: Adapted from Department of Finance Canada (2007).

Labrador. Nova Scotia, on the other hand, chose the new equalization program because the province believes that it will gain more with the increased equalization payments than from gas royalties, which, if high enough, would be subjected to the cap on fiscal capacity—continued gas production in the offshore waters of Nova Scotia is not as assured as oil production on the Grand Banks of Newfoundland and Labrador.

Population

Since Confederation, Atlantic Canada's population has increased very slowly, well below the national average. In recent years, its population growth has stalled and then declined. From 1996 to 2001, Atlantic Canada experienced a population decline of 2.1 per cent, which amounted to a loss of 48,000 residents. The next five-year saw considerably less pop-

Vignette 9.5 The Atlantic Energy Accord

Since 1982, revenues from non-renewable resources have been included in the calculations for equalization payments. By the beginning of the twenty-first century, the net result for three oil- and gas-producing provinces—Newfoundland and Labrador, Nova Scotia, and Saskatchewan—was the clawback of their equalization payments. In short, as their energy royalties increased, their equalization payments decreased. Newfoundland and Labrador, which had wallowed in a weak economy for years, complained that the clawback should end. In 2004, the government of Canada reached an arrangement with Newfoundland and Labrador and Nova Scotia that provides 100 per cent protection from equalization reductions resulting from the inclusion of offshore resource revenues for eight years. As a result of the Accord, the two provinces are estimated to receive, respectively, \$2.6 billion and \$1.1 billion more in equalization payments over the next eight years plus higher resource revenues because of higher prices for oil and natural gas. With the 2007 budget, however, funds for equalization payments have increased to \$12.8 billion for 2007–8 up from \$11.3 billion in 2006–7. To obtain these enriched future equalization payments, these two provinces must opt for the new equalization program, thereby losing gains made in their Atlantic Energy Accord agreement.

Figure 9.2. Jackpot! In 2004, Premiers Danny Williams and John Hamm signed the Atlantic Accord with Prime Minister Paul Martin, thus allowing both Newfoundland and Labrador and Nova Scotia to receive all of their offshore revenues without reducing their equalization payments. In 2005, a dramatic increase in oil and natural gas prices resulted in much higher resource revenues for the two provinces, allowing Newfoundland and Labrador to declare a surplus. The province of Saskatchewan wants a similar agreement, but the Ontario government was strongly opposed to the removal of a clawback of non-renewable resource revenue from the calculation of equalization payments because it feared that more federal money would be allocated to equalization payments and therefore less federal money would be available for Ontario from other federal transfer payments. (Michael de Adder/*The Halifax Daily News*)

ulation decline in the region: from 2001 to 2006, the population figure for Atlantic Canada dropped by less than 1,000.

The population change varied among the four provinces, largely reflecting differences in economic conditions. From 2001 to 2006, only Newfoundland and Labrador lost population—nearly 7,500—but in the previous five-year period the province's population fell by nearly 39,000 (Table 9.3). The implication of this population loss on the regional economy is quite serious. First, the loss leads to smaller markets within Atlantic Canada and smaller markets will affect local businesses. Second, the large number of people leaving Atlantic Canada to search for jobs in more prosperous regions of Canada, especially southern Ontario, Alberta, and British Columbia, represents a loss of the younger, more ambitious members of the labour force. Third, there is a psychological cost associated with a declining population.

Since the 1920s, the economic circumstances in Atlantic Canada have been on a downward cycle, driving hundreds of thou-

Table 9.3 Population Change in Atlantic Canada, 1996–2006

Province	1996	2001	2006	Difference 2001–6	% Difference
Prince Edward Island	134,557	135,294	135,851	557	0.4
Newfoundland/Labrador	551,792	512,930	505,469	–7,461	–1.5
New Brunswick	738,133	729,498	729,997	499	0.1
Nova Scotia	909,282	908,007	913,462	5,455	0.6
Atlantic Canada	2,333,764	2,285,729	2,284,779	–950	–0.1

Sources: Statistics Canada (2002b, 2007a). Adapted from Statistics Canada *Population and Dwelling Counts (On-line) – Highlight Tables, 2001 Census*, Catalogue 93F0051XIE, http://www.statcan.ca/bsolc/english/bsolc?catno=93F0051XIE.

sands to seek their fortunes elsewhere. Over the years, Atlantic Canada has lost more people through out-migration than it has gained through in-migration. In fact, few immigrants to Canada choose to settle in Atlantic Canada because they see few opportunities. The resulting loss of people, while boosting economies elsewhere, has hindered economic growth in Atlantic Canada. An indication of the serious nature of this net outflow is shown in the proportion of Canadians living in this region. In 1949, when Newfoundland joined Canada, Atlantic Canada constituted 11.6 per cent of the nation's population. By 2006, the region's population had fallen to 7.2 per cent of the national total.

Until the United States placed restrictions on the movement of Canadians into the United States in the 1930s, New England was the favourite destination of migrants from the Maritimes. Many went to the Boston area. After the 1930s, most Maritime migrants settled in Ontario. After Newfoundland joined Confederation in 1949, its migrants went mainly to Ontario. From 1986 to 2001, about 200,000 people left Atlantic Canada for Ontario, Alberta, and British Columbia. During that 15-year period, Ontario was the destination of just over half of the migrants from Newfoundland, nearly half from Nova Scotia, about 40 per cent from New Brunswick, and one-third from Prince Edward Island. Within Ontario, Toronto attracts most of the migrants from Atlantic Canada. In the 2006 census, the figures may show that Alberta is the leading destination for Atlantic Canadians.

Dispersed Population

Atlantic Canada's population of over 2 million is widely dispersed along its coastline in both small and large centres. Geography explains why this settlement pattern developed. The original selection of sites for fishing villages, particularly in Newfoundland and Nova Scotia, was largely determined by small, sheltered harbours suitable for mooring boats and affording quick access to good fishing grounds. While many of these coastal villages still remain, they are anachronisms in the twenty-first century. Delivering basic public services, such as education and health care, in these coastal places is expensive because populations are small and are located in isolated, hard-to-reach areas.

From an economic perspective, this dispersed pattern of population distribution is a major weakness for three reasons:

- The highways and rail lines required to connect the urban centres in Atlantic Canada are more extensive than in regions with a more compact population geography, and are therefore more expensive to build and maintain.

- High construction and maintenance costs for transportation networks result in higher transportation costs for firms operating in Atlantic Canada, thus discouraging companies from investing and locating in the region.
- Because of the absence of a large population/market, firms are hampered from achieving economies of scale and **economies of association** (having parts and service firms close to the manufacturing plant).

The rural-to-urban migration, common to other regions of Canada, is occurring in Atlantic Canada. Migration of younger members of outport families is accelerating, while older members may prefer to live in their communities of origin where the cost of living is much lower than in a regional centre. For the province, the cost of supplying public services is extremely high. Yet 'planned' relocation is fraught with two dangers: social disruption and economic transfer of rural poverty to urban poverty. The so-called remedy for isolated communities in Newfoundland was the 'planned' resettlement program.

How did Atlantic Canada develop such a dispersed coastal pattern of settlement? Back in the seventeenth and eighteenth centuries, when the first British and French families arrived in Newfoundland, the newcomers formed isolated coastal fishing outports (often with only three or four families each). As isolated self-sufficient outports, these communities were ideal for a small-scale coastal fishery. By 1945, nearly 1,400 tiny outports with populations under 200 had been established along the coast of Newfoundland (Staveley, 1987: 257). But by 2001, there were less than 400 outports with populations under 200. This decline was the result of two forms of relocation—voluntary and government-assisted. In short, people, particularly young people, no longer wanted to live in these tiny places where life revolved around the fishery. Outport living could not survive a depressed fishing industry and the growing desire/need for urban amenities. The abandonment of outport villages began in 1946 and by 1954, 49 small communities were abandoned without government assistance.

After Confederation, both Ottawa and St John's became concerned about this unusual rural-to-urban migration. First, the Newfoundland government made financial help available under its Centralization Program. Under this scheme, all households in the community had to agree to relocate for public assistance to be made available. From 1955 to 1966, 110 communities were abandoned and the residents resettled in larger centres of their choice. In 1967, the federal government joined forces with Newfoundland. The federal–provincial Newfoundland Resettlement Program added new conditions to relocation. Migrants had to relocate to one of 77 designated 'growth centres'—selected communities that offered more social and economic opportunities. Another stipulation was that at least 75 per cent of the households must agree to move for them to receive public relocation funds. Under the joint program, another 150 communities were abandoned from 1967 to 1975. After this program ended in 1976, few communities were abandoned until the late 1990s when the cod fishery was closed. Following the 1992 cod moratorium, many communities saw their fish plants close, causing many to seek work elsewhere. Such was the fate of Great Harbour Deep, a tiny outport of 180 inhabitants on the Northern Peninsula that was founded some 400 years ago. In April 2003, the residents accepted the provincial government's offer of financial help if all agreed to leave their community. Each household received up to $100,000 to help cover the costs of their relocating (Vignette 9.6).

Access to External Markets

The FTA and freer trade of the GATT/WTO regime greatly improved Atlantic Canada's access to foreign markets. Sales of fish, pulp, natural gas, and potatoes to consumers in New England rose sharply. During the 1990s, a low Canadian dollar compared to the US dollar and an expanding American economy created ideal conditions for an increase in exports to the United States and in the number of US

Vignette 9.6 The End of Great Harbour Deep

For 400 years, people fished cod in the waters off Great Harbour Deep. Located on the Northern Peninsula, Great Harbour Deep was one of the most isolated communities in Newfoundland. Access to the outside world was by a three-hour boat ride in the summer and by a ski-equipped aircraft in the winter. While residents cherished their independence, the need for outside amenities, including health services, had become more and more pressing. In the 1990s, the loss of the inshore fishery and the closure of the town's fish plant led to the economic collapse of this outport. Some left, while others worked on the large trawlers for months at a time. By 2003, the population had dropped to 180 residents. After a series of public meetings, community residents decided to accept the government's $5 million package and leave their community for good. In the summer of 2003, the 50 families of Great Harbour Deep left their community, leaving behind memories and empty homes but hoping to start a new life in a new community.

(CP/Mike McDonald)

tourists visiting Atlantic Canada. By the twenty-first century, however, the difference between the two dollars narrowed, making exports to the US more expensive. US tourists found prices in Canada less attractive than in the 1990s; in addition, post 9/11, the American government has made border crossings more complicated for its citizens.

Provincial Barriers to Trade

While interprovincial trade barriers have come down between Alberta and British Columbia, the rest of the provinces, including the four Atlantic Canada provinces, have shown little serious interest in taking similar action. In addition to the physical obstacles that Atlantic Canada has faced in trading with external markets—Canadian and foreign—the region has also grappled with trade barriers between its provinces. Provincial barriers divide the region's population into four provincial markets, often impeding interprovincial trade and labour mobility. The purchasing policies of provincial governments and their various agencies frequently favour local producers and act as non-tariff barriers to producers from other provinces. Other barriers take the form of provincial marketing boards, provincially controlled liquor sales, and provincial-preference hiring practices in the construction industry. Such internal barriers are a major impediment to stimulating economic growth within Atlantic Canada. The very division of the region into four provinces results in higher administrative costs and thus higher tax rates, which exert an equally negative impact on economic development. Simply having four provincial governments for a relatively small population costs taxpayers in Atlantic Canada about $400 million annually, making them bear a higher tax burden than residents of other provinces.

From time to time, there has been talk of an Atlantic economic union or the creation of a single province to overcome this regional fragmentation. Such an amalgamation would not overcome the differences within the region's physical geography, but it might consolidate Atlantic Canada's economic and political power. It might even reduce the cost of government, but it would raise a host of issues including the thorny question of a capital city for Atlantic Canada. As early as 1949, Joey Smallwood, Premier of Newfoundland, proposed a union of the four eastern provinces to enable them to increase their bargaining power in Ottawa. In the 1970s, the Deutsch Commission on Maritime Union recommended a 'full political union as a definite goal'. However, there was no support within Atlantic Canada, and before the 1970s had ended Nova Scotia Premier Gerald Regan pronounced the proposal for a Maritime Union 'as dead as a doornail'. For the residents of Atlantic Canada, the advantages of a union are not readily apparent, and Atlantic premiers have not shown much enthusiasm for this proposal.

How much the taxpayers of Atlantic Canada would save with a single government is difficult to estimate, but political union could achieve considerable savings by eliminating overlapping public agencies and provincial trade barriers. Unfortunately, an Atlantic union would likely lead to massive layoffs in the public sector. For a region with high unemployment, such a prospect makes a union much less attractive.

Industrial Structure

Atlantic Canada's three primary resources are its fish, forests, and minerals, but the region's economic future lies in the development of its tertiary sector, especially high-technology industries. Employment provides an overall picture of the basic economic structure of Atlantic Canada and of each of its provinces. In 2005, employment in primary activities accounted for 5.8 per cent of the labour force in Atlantic Canada; secondary employment constituted 16.0 per cent; and tertiary employment 78.2 per cent (Table 9.4). Compared with the figures for 2002, the greatest percentage point decline took place in the secondary sector, largely because of employment losses in manufacturing. Modest increases took place in the primary and tertiary sectors, the latter in the major cities of Atlantic Canada (Table 9.5).

Significant variations among the provinces' employment figures exist. Newfoundland, as expected, ranks first in employment in the fisheries, but it also ranks first in mining, primarily because of the large iron-mining operations in Labrador and the offshore oil extraction. Employment in manufacturing is much stronger in Nova Scotia and New Brunswick than in Newfoundland and Prince Edward Island, mostly because there is greater processing of local raw materials in these two provinces—from the processing of agricultural, fish, and wood products to the processing of mineral products. However, manufacturing has also suffered setbacks. For instance, heavy industry in Cape Breton used to employ many workers, but closure of the coal mines and steel factory has hurt the economy of Nova Scotia and made Cape Breton's unemployment figures among the highest in Atlantic Canada. Commuting or relocating to labour-short Alberta has provided one option for workers in Cape Breton and elsewhere in Atlantic Canada.

Table 9.4 Employment by Industrial Sector in Atlantic Canada, 2005

Industrial Sector	AC Workers (000s)	AC Workers (%)	Ontario Workers (%)	% Difference
Primary	62.5	5.8	2.0	3.8
Agriculture	18.2	1.7	1.5	0.2
Secondary	171.7	16.0	23.6	(7.6)
Manufacturing	99.8	9.3	16.6	(7.3)
Tertiary	841.6	78.2	74.4	3.8
Total	1,075.9		100.0	

Source: Statistics Canada (2006c). Adapted from Statistics Canada website http://www40.statcan.ca/l01/cst01/labor21c.htm, Searched 12 September 2006.

Atlantic Canada's Economy

As a result of its weak economy compared to other regions of Canada, Atlantic Canada has become heavily dependent on Ottawa, which sends money to the region through federal programs and transfer payments. The most important subsidies are equalization payments and employment insurance payments. Unemployment insurance in Atlantic Canada is extremely important, particularly to those engaged in seasonal industries, such as self-employed fishers and those employed in fish plants and logging operations. Ottawa first extended this insurance to self-employed fishers in 1947. Before the cod moratorium in 1992, inshore fishers caught fish during the summer and drew unemployment insurance payments in the winter. The combined loss of fish income and less employment in fish processing plants has resulted in fewer workers able to claim unemployment payments. For those residing in small fishing communities, their earned and unearned income dropped significantly.

For decades, Ottawa tried to overcome the economic malaise in Atlantic Canada by pro-

Table 9.5 Employment Changes by Industrial Sector in Atlantic Canada, 2002–5

Industrial Sector	AC Workers (% 2002)	AC Workers (% 2005)	% Change
Primary	5.6	5.8	0.2
Agriculture	1.6	1.7	0.1
Secondary	16.9	16.0	(0.9)
Manufacturing	10.3	9.3	(1.0)
Tertiary	77.5	78.2	0.7
Total		100.0	

Source: Statistics Canada (2003c, 2006c).

viding funds for new businesses and for improving the region's transportation system. Such initiatives were meant to create jobs in areas of high unemployment. Since the 1960s, federal grants for businesses have totalled over $6 billion. Most federal funds have been administered by several key federal agencies. In 1962, the Atlantic Development Board was established to stimulate businesses and strengthen the region's public infrastructure. In 1969, the federal Department of Regional Economic Expansion (DREE) began to offer incentives to encourage companies to locate in Atlantic Canada and other less-favoured areas of the country. DREE also spent money on highway construction, schools, and municipal services. In the late 1980s, the Atlantic Canada Opportunities Agency was formed to develop the economy through business and job opportunities. By the mid-1990s, however, Ottawa began to reduce its regional subsidies, including more stringent qualifications for employment insurance, as part of its effort to combat the federal debt, thereby threatening the financial aid upon which Atlantic Canada relied. The November 2000 election saw the Liberals reinstate regional subsidies and relax the qualifications for employment insurance. In July 2000, Jean Chrétien announced the Atlantic Investment Partnership, a new (but old) system of federal grants and loans to firms and business people in Atlantic Canada.

Ottawa's ongoing influence on Atlantic Canada's economy was evident after the closure of the cod fisheries in 1992. After it announced a moratorium on cod fishing because of the dismal state of the stocks, the federal government promised additional financial assistance for the short term. The Atlantic Groundfish Strategy (TAGS) began in May 1994. TAGS made nearly $2 billion in monthly payments available to about 40,000 fishers and fish-plant workers displaced by the moratorium. While cod fishers are found in all parts of Atlantic Canada, the greatest impact of the cod moratorium occurred in the isolated fishing communities totally dependent on the fishing industry. TAGS was intended to shift fishers into other types of work through retraining and education programs, but most fishers considered the program to be another form of unemployment insurance. In their hearts, the fishers hoped that the cod would return before the payments from TAGS ran out. The success of TAGS was hindered by the following factors: alternative employment opportunities in remote fishing communities are virtually non-existent; few fishers choose to leave their homes and villages for an uncertain future in larger towns or cities; and given the age and education level of most fishers, retraining becomes an almost impossible task.

By early 1998, TAGS was coming to a close but neither had the cod returned to their former numbers nor had the number of cod fishers declined. Ottawa was faced with a dilemma. Pressure from Atlantic Canada, particularly from Newfoundland, was intense. In August 1998, Ottawa announced that an additional $730 million would be available to cod fishers under the retraining program initiated by TAGS (MacAfee, 1998: A3). Such funding would extend payments to cod fishers for another year but Ottawa made it clear that this would be 'the end of the line' (Greenspon, 1998: A4). The cod fishery still has not reopened, as the depleted cod stocks have not returned to sustainable levels, and prospects for the foreseeable future are not encouraging.

As an older resource hinterland, Atlantic Canada is suffering from a problem that is all too common to such regions. Over the years, Atlantic Canada has depended heavily on the exploitation of its natural wealth. Except for the iron and steel industry based in Cape Breton Island, the region's natural wealth was exported to the United States, Britain, and other European countries where the resources underwent further processing. With the demise of the Cape Breton coal and steel complex, a strong industrial base no longer exists and the prospects for such a base in Atlantic Canada are dim. Other types of development are possible.

One of the promising alternatives is the vast wealth of energy lying below its continental shelf. However, the pattern of resource exploitation, processing, and export is the same as before. The problem is that little processing takes place. For example, the oil from

Hibernia, the $6.2 billion offshore oil project east of St John's, is destined for refineries along the eastern seaboard of the United States, while a gas pipeline sends the natural gas from the Sable Island deposit off the Nova Scotia coast to New England. To counter the pattern of exploitation and export, provincial governments have tried to intervene in the marketplace to obtain more benefits from large-scale resource developments, but they have had only limited success. For example, the Newfoundland and Labrador government has obtained some concessions from the oil industry—the key components for the Terra Nova oil-production vessel were manufactured within the province at Bull Arm, the construction site for the Hibernia platform (Vignette 9.7). With rising prices for oil but especially for natural gas, both Newfoundland and Labrador (the province of Newfoundland officially changed its name to Newfoundland and Labrador in 2001) and Nova Scotia should receive a sharp economic boost in coming decades.

As Atlantic Canada searches for its place in the highly competitive global economy, diversification of its resource hinterland role is essential. Economic diversification lies in two directions: greater emphasis on value-added activities (i.e., manufacturing and processing of its natural resources, especially fish stocks) in the primary sector, and an expansion of high technology in the tertiary sector.

Key Topic: The Fishing Industry

Nature has given Atlantic Canada a vast continental shelf that provides an excellent physical environment for fish: the warm ocean currents from the Gulf of Mexico and the cold Labrador Current create ideal conditions for fish reproduction and growth. The continental shelf extends almost 400 km offshore (Figure 9.3). In some places, where the continental shelf is raised, the water is relatively shallow. Such areas are known as banks. The largest banks are the Grand Banks off Newfoundland's east coast and Georges Bank off the south coast of Nova Scotia (Vignette 9.8). Scallops, one of the most profitable shellfish, is normally harvested in beds on Georges Bank. However, in 2006, new beds of mature scallops were discovered on Browns Bank, which will allow fishers to leave the young and small scallops on Georges Bank to mature.

The diversity that characterizes Atlantic Canada extends to the fishery as well. Although each province relies on the fishery, there are

Vignette 9.7 Bull Arm Complex: Success or Failure?

In 1990–1 the Bull Arm complex, an elaborate and highly specialized construction site, was built by the provincial government at a cost of nearly $500 million. The complex was designed to build the Hibernia offshore oil-drilling platform, which was towed out to the Grand Banks and began production in 1997. The Newfoundland government hoped that the Bull Arm complex could continue to build offshore components for the oil and gas industry's future energy developments. While the Bull Arm complex did acquire work from two other offshore projects—Terra Nova and White Rose—this industrial site never regained the level of construction work achieved during the Hibernia project. Worse yet, the complex was idle for several years and, after its White Rose contract, the future of the complex appears uncertain. The spotty performance of the Bull Arm complex represents a classic problem faced by hinterlands—how to diversify the economy through public investment.

Source: Adapted from Jang (1998).

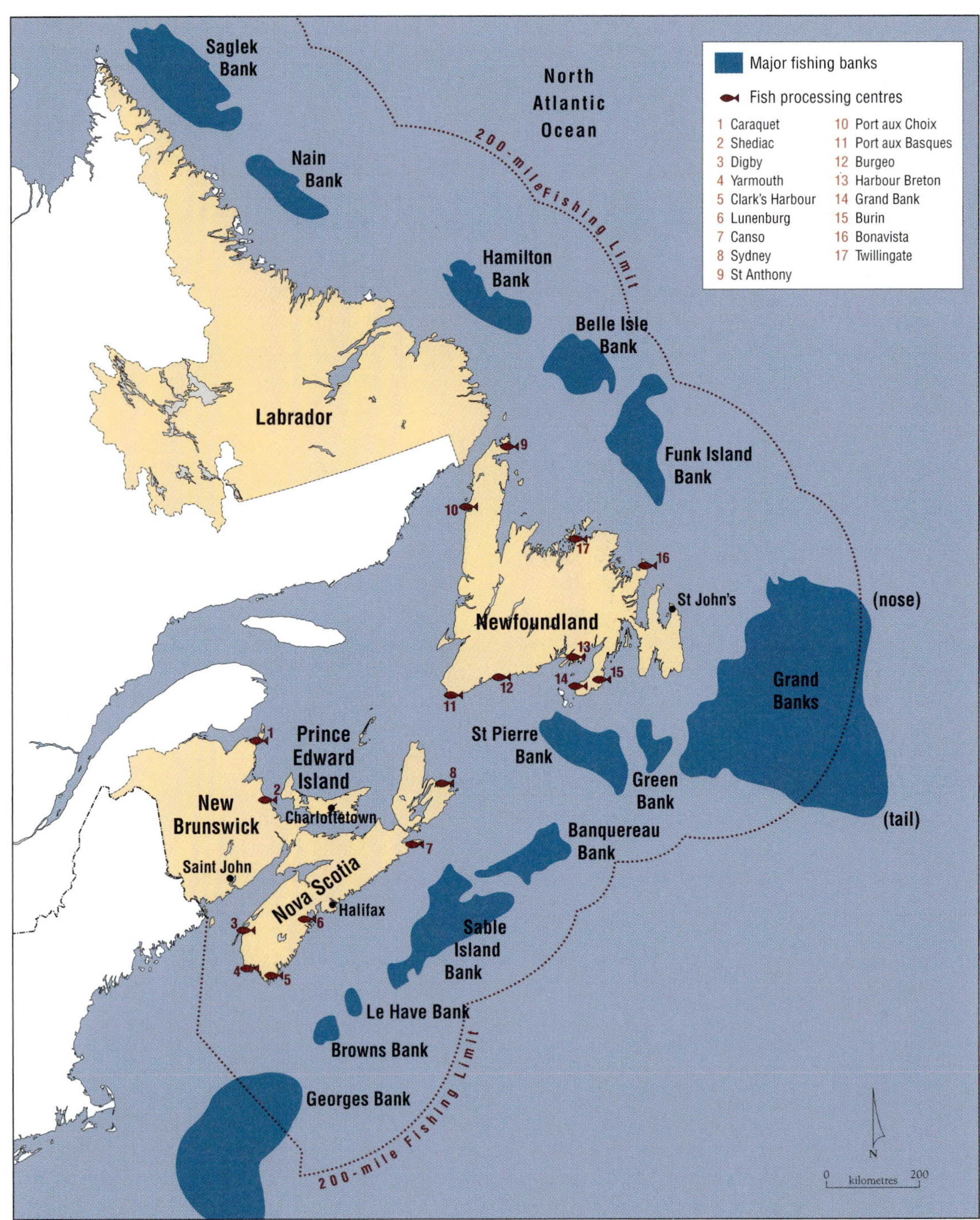

Figure 9.3 Major fishing banks in Atlantic Canada. The Atlantic coast fishery operates within a vast continental shelf that extends some 400 km eastward into the Atlantic Ocean, southward to Georges Bank, and north to Saglek Bank. Within these waters are at least a dozen areas of shallow water known as 'banks'. The Grand Banks of Newfoundland is the most famous fishing ground while Georges Bank contains the widest variety of fish stocks. (Further resources: Student website, National Atlas section, Map 26. Website instructions are found on p. xx.)

Vignette 9.8 Georges Bank

As part of the Atlantic continental shelf, Georges Bank is a large, shallow-water area that extends over nearly 4,000 km^2. Water depth usually ranges from 50 to 80 m, but in some areas the water is 10 m or less. Georges Bank is one of the most biologically productive regions in the world's oceans because of the tidal mixing that occurs in its shallow waters. This brings to the surface a continuous supply of regenerated nutrients from the ocean sediments. These nutrients support vast quantities of minute sea life called plankton. In turn, large stocks of fish (such as herring and haddock), as well as scallops, feed on plankton. Cod also feed in these waters, but their numbers have dwindled due to overfishing, especially in international waters. Properly managed, Georges Bank can sustain annual fishery yields of about 400,000 tonnes.

striking differences (particularly between Newfoundland and Labrador and the Maritime provinces) in the type of catch and the location. Before the ban on cod fishing, Newfoundlanders depended mostly on cod fishing, while Maritimers harvest a variety of sea life, including cod, lobsters, and scallops. Fishers from the Maritimes ply the waters of Georges Bank and smaller banks just offshore of Prince Edward Island and Nova Scotia. In these waters, Maritimers find cod, grey sole, flounder, redfish, and shellfish. Newfoundlanders, on the other hand, historically fished in the waters of the Grand Banks, around the island of Newfoundland, and along the shore of Labrador, where cod and other groundfish congregate in the shallow areas of the continental shelf. Since the collapse of the cod fishery, the Newfoundland fishery industry has diversified. In 2006, for example, of a total landed value for the Newfoundland and Labrador fishery of $445.3 million, shellfish (chiefly snow crab and shrimp) accounted for $315 million, with the balance including the annual seal hunt as well as both groundfish (e.g., flounder, sole) and **pelagics** (e.g., mackerel, capelin) (Newfoundland and Labrador, 2007).

The fishing industry consists of an inshore fishery, an offshore fishery, and fish-processing plants. Traditionally, Newfoundland and Maritime fishers used a flat-bottomed skiff called a dory and stayed close to the shore. Only the larger skiffs ventured to the Grand Banks and other offshore banks. Today, inshore fishers still use smaller boats to fish close to the shore. During the winter, rough waters and ice prevent the smaller craft from leaving port. Offshore fishers travel in steel stern trawlers to the Grand Banks and other offshore fishing banks where they fish for weeks. Upon returning to port, they unload hundreds of tonnes of iced groundfish at a fish-processing plant. Offshore vessels, often much larger than 25 tonnes, are equipped with radar, radios, sonar detectors, and other sophisticated equipment that enable them to operate in winter as well as summer.

The modernization of the Canadian fishing industry, with its larger trawlers and more efficient nets and gear, was accompanied by huge investments by fish companies and individuals and an expansion of the fishing fleet in number and size. Atlantic Canada's fishery, particularly in Newfoundland and Nova Scotia, was able to increase its catch and its income. Now, fishing companies usually own the ships, so fishers no longer share in the value of the catch but receive a salary. However, modernization had many downsides:

- Fewer fishers were required by this more efficient fishing system.
- The new fishing system based on technological advances and huge dragger nets created the circumstances for overfishing.

- Overfishing and the collapse of the northern cod stocks forced inshore fishers to turn to government assistance to survive.

Management of Fish Stocks

The expansion and modernization of the world fishing fleets have threatened world fish stocks. With no management of fish stocks in international waters, pressure on selected fish stocks grew. By the 1970s, the world fishing fleets had greatly increased their take of groundfish on the Grand Banks. As is the case in British Columbia with the salmon, the issue of fishing limits in Atlantic Canada was crucial in trying to manage the fish stocks in a sustainable way. Management of the international fish stocks in Atlantic Canada was so important that, in 1977, Ottawa claimed the right to manage the fisheries within a 200-nautical-mile zone off the east coast of Atlantic Canada. Washington also extended its fishing zone to the 200-mile limit, making Georges Bank an area of dispute between the two countries.

Georges Bank is one of the richest fishing banks, where fish and shellfish, such as scallops and mussels, are found. Fishers from both the Maritimes and New England harvest the fish stocks on this bank, which is located at the edge of the Atlantic continental shelf between Cape Cod and Nova Scotia. In the 1970s, fish stocks, especially herring, dropped to dangerously low levels due to overfishing. A

Table 9.6 Cod Landings for the Atlantic and Newfoundland Fisheries, 1990–2005 (metric tonnes live weight)

Year	Newfoundland	Atlantic Canada	% Newfoundland
1990	245,896	395,266	62.2
1991	178,687	309,031	57.8
1992	75,138	187,804	40.0
1993	37,068	76,644	48.4
1994	2,292	22,719	9.8
1995	863	12,438	6.9
1996	1,147	15,541	7.4
1997	12,317	29,899	41.2
1998	22,764	37,894	60.0
1999	38,663	55,527	69.6
2000	30,216	46,177	65.4
2001	23,774	40,440	58.8
2002	21,083	35,741	59.0
2003	14,290	22,768	62.8
2004	14,534	24,729	58.8
2005	16,257	26,156	62.2

Note: Since 1992, a modest cod catch was allocated to inland fishers.
Source: Fisheries and Oceans Canada (2006). Reproduced by permission of the Minister of Public Works and Government Services Canada, 2007.

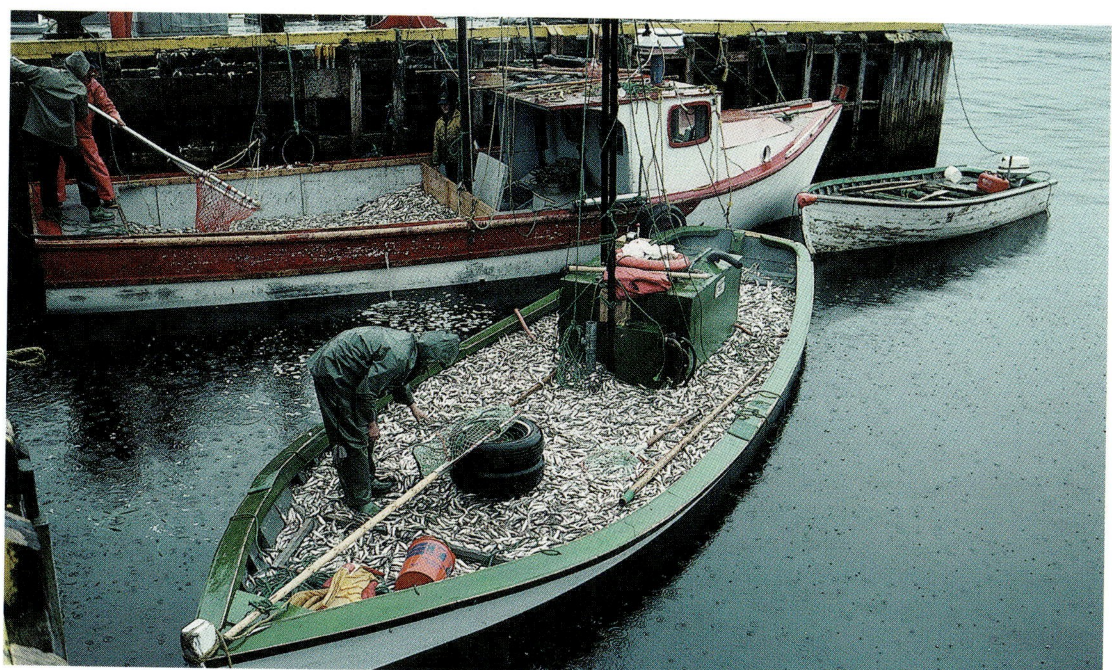

In spite of difficulties, fishing remains a way of life in Atlantic Canada. As shown in this photograph, inshore fishers operate close to the shore by using small boats and, because of their small-scale operations, pose little threat to the fishery. On the other hand, the Canadian and foreign fishing fleets with their drag nets endanger the Atlantic fishery. (Barrett & MacKay Photography, Inc.)

boundary decision made by the International Court at The Hague in October 1984 allocated five-sixths of Georges Bank to the United States. However, the easternmost one-sixth that was awarded to Canada is rich in groundfish and scallops. The Nova Scotia fishers have greatly benefited from this decision. With the scallops now under Canadian management, sustainable harvesting of the scallops and groundfish has brought a measure of stability to this sector of the fishing industry.

However, overexploitation of international fish stocks remains a worldwide problem. The root of the problem is that public resources are decimated by the selfish actions of individuals who have no regard for the well-being of the resource. Such mismanagement is known as the 'tragedy of the commons'. Such ecological disasters occur when no one is responsible for ensuring the proper management of resources. Northern cod stocks fell victim to this tragedy twice: in the 1970s when there was no management of fishing on the Grand Banks, and again in the 1990s when the resource was mismanaged (Vignette 9.9). Mismanagement was based on three factors: (1) the federal government's estimate of cod stocks was overly optimistic and Ottawa consequently set the quota too high; (2) strong pressure for a high cod quota came from Newfoundland politicians, various fisher groups, and outport communities; and (3) Canada had no control over foreign fishing along the nose and tail of the Grand Banks. The circumstances and consequences of the demise of cod stocks in Atlantic Canada can lead to a greater appreciation of the need for sustainable harvesting in the fishing industry.

Overfishing and the Cod Stocks

The Atlantic cod (*Gadus morhua*) is the major natural resource in Atlantic Canada. While the Atlantic cod (sometimes called the northern cod) is found as far north as the southern tip of Baffin Island and as far south as Cape Hatteras, cod stocks are concentrated off

Vignette 9.9 An Ecological Crisis

The decline in groundfish stocks in Atlantic Canada is an ecological crisis. Much of this crisis is due to the use of draggers by the Canadian and foreign fishing fleets. Ottawa does limit draggers in certain areas within Canada's 200-km fish management zone, but supports dragging operations because they are so cost-efficient. Most scallops, for example, are harvested by Canadian draggers that operate in the waters of Georges and Browns banks.

Cod stocks suffered badly from draggers. The Canadian Atlantic Fisheries Scientific Advisory Committee revealed in mid-1992 that stocks had declined sharply since the 1960s, when the **biomass** (the volume of fish aged three years or older) was over 3 million tonnes. By 1991, the biomass had dropped to between 530,000 and 700,000 tonnes, possibly the lowest level ever observed. **Spawning biomass** (generally seven years or older) represents the reproductive part of the fish stock. In 1991 it amounted to between 50,000 and 110,000 tonnes. Thirty years before, the spawning biomass had been as high as 1.6 million tonnes, and even in 1987 it was estimated at 400,000 tonnes. Faced with such a serious crisis, the Canadian government announced in 1992 a moratorium on cod fishing in the waters of Atlantic Canada. By the summer of 2006, cod stocks had not yet recovered sufficiently to allow the reopening of the cod fishery. As shown in Table 9.6, the last large catch occurred in 1991. Since then, cod fishing has been restricted to inland fishers.

Newfoundland and along the Labrador coast. The largest single concentration of northern cod, historically, was on the Grand Banks.

The Atlantic cod has been one of the world's leading food fishes for centuries. Cod is sold fresh, frozen, salted, and smoked. Salted cod was very popular in the past, but now most of the cod sold is frozen. During the sixteenth century, the catch was small, probably averaging less than 25,000 metric tonnes a year. By the end of the seventeenth century, however, the annual catch may have reached 100,000 metric tonnes. Annual cod landings during the nineteenth century were about 150,000 to 400,000 metric tonnes. As fishing technology advanced, catches of cod rose to nearly 1 million metric tonnes in the 1950s. By the 1960s, the annual catch reached a peak of almost 2 million metric tonnes. European trawlers accounted for most of this catch.

With the emergence of a modern fishing fleet, fish stocks around the world were overfished. Several fish stocks, including the California sardine, Peruvian anchovy, and Namibian pilchard, collapsed in the 1980s from overfishing—not so much from local fishing operations but from the highly modernized fishing fleets that scan the world's oceans for fish. The simple hook-and-line fishing system did not have the capacity to overfish, but foreign and Canadian fishing fleets that employ sophisticated fishing equipment and that drag huge, weighted nets across the ocean floor indiscriminately trapping all groundfish regardless of age, species, and value do have that capability. Known as draggers, these fleets create enormous waste because 'non-commercial' fish—the bycatch—are simply discarded. In addition, the draggers destroy fragile ocean-floor ecosystems, including coral reefs and breeding habitat. In spite of the many excuses for the collapse of the northern cod stocks, including changing ocean temperatures and an increase in seal population, the primary cause has been the multinational seafood products companies that ran the technologically advanced fishing fleets with factory ships that could stay at sea for extended periods of time. These fleets ruthlessly exploited the cod and other fish of the sea, to the point where a study by 14 international scholars published in late 2006, which analyzed all existing datasets, forecast that all of the world's seafood fisheries will collapse by the year 2050 (physorg.com).

In the case of cod, Ottawa's ability to manage the fish stocks was complicated because, as noted above, the nose and tail of the Grand Banks extend into international waters. This made it difficult for Canadian fisheries officials to monitor and enforce catches in these areas. An agreement was reached with the countries belonging to the European Common Market, but monitoring the actual catches proved difficult. Some European fishing vessels, particularly those from Spanish ports, were exceeding their fish quotas and were catching undersized cod by using nets with a small mesh, which is illegal. During the 1980s, groundfish stocks, especially the northern cod stocks, dropped significantly—perhaps by as much as 60 per cent (Cashin, 1993: 19). This collapse translated into a decline in fish landings and processing. Cod landings were reduced by over half from 1991 to 1992. In 1992, Ottawa was forced to impose a moratorium on the cod fishery. The fear was that the northern cod stocks had collapsed and further fishing would make their recovery more difficult. While fisheries landings have remained small, the lowest point occurred in 1995 (Table 9.6). By 2007, scientists were still unsure as to how long it will take for the northern cod stocks to recover, if indeed they ever will. Cod must reach the age of seven before the fish can reproduce, so recovery to a point to make the cod fishery even marginally viable is a long-term proposition at best.

The impact of the cod moratorium on Atlantic Canada, particularly Newfoundland, was severe. The cod has been the mainstay of Newfoundland's economy. Quite justifiably, the Atlantic cod was called 'Newfoundland currency'. In the past, the total catch by Newfoundland fishers was close to four-fifths groundfish. By comparison, the groundfish would constitute only about half of the catch in Nova Scotia and about one-third of the catch in Prince Edward Island and New Brunswick.

Before the moratorium, most jobs were in the fish-processing plants. Often these jobs lasted for only a short period, sometimes less than a month. However, employment at fish-processing plants provided two highly valued items—much-needed cash and enough weeks of work to qualify for unemployment insurance. In May 1994, fish-processing employees became eligible for financial support from The Atlantic Groundfish Strategy and were included in the extension to TAGS announced in 1998. In Newfoundland, the fish-processing sector was dominated by two firms, Fisheries Products International and National Sea Products. Prior to the moratorium on cod fishing, these two firms accounted for about half of the fish-processing jobs or nearly 8 per cent of the total number of employed people in Newfoundland.

The fishing industry in Atlantic Canada has suffered under the July 1992 moratorium on cod fishing. Since 1995, however, Newfoundland's fishing industry has recorded higher and higher production values because of high prices for crab, shrimp, lobster, clams, turbot, and other stocks. In this highly managed industry, only a few fishers have licences for these special kinds of catches. For example, only about 1,200 licences to catch crab are issued to Newfoundland fishers each year. On the other hand, before the cod moratorium, approximately 7,000 licences were issued for groundfish. Without cod, wealth from the fishery is concentrated in fewer and fewer hands. By 1997, small catches of cod by inshore fishers were permitted, but, in spite of strong pressure from fishers, Ottawa is still not prepared to allow the cod fishery to reopen. In Atlantic Canada, fishing is not only an important economic industry, it is a way of life. By 2006, cod stocks show no signs of recovery and the inshore fishing way of life based on outports is disappearing.

Atlantic Canada's Land Wealth

Though Atlantic Canada's early settlement and subsequent development revolved very much around the harvest of the sea, the land mass of

Figure 9.4 Natural resources in Atlantic Canada. Besides fish, Atlantic Canada contains important natural resources. Enormous iron and nickel deposits are located in Labrador, huge oil and gas reserves are found below the continental shelf, and one of Canada's largest hydroelectric facilities is situated on the Churchill River. Especially important for Newfoundland and Labrador is that, in late August 2007, Premier Danny Williams negotiated a stake in the Hebron-Ben Nevis offshore project, including a 4.9 per cent equity share that could amount to $16 billion over the 25-year life of the project (McCarthy, 2007).

the four provinces has also provided the region's economy with important and profitable natural resources for forestry, mining, and agriculture (Figure 9.4).

Forest Industry

Beyond the settled areas of Atlantic Canada, the land is covered by forest. Atlantic Canada has 22.9 million ha of forest land. Within the region, the forest industry plays an important role in the economy, particularly in the economy of New Brunswick. In 2001, the value of forest products was approximately $4 billion. Almost half of this production originated in New Brunswick. Most trees harvested, including logs from private woodlots and especially spruce, are sold to local pulp mills. In 2002, nine pulp plants operated in New Brunswick, three in Nova Scotia, and three in Newfoundland. Since then, the forest industry has faced US duties on softwood lumber and declining demand for newsprint. Pulp and paper mills have closed across Canada. In Newfoundland and Labrador, the newsprint mill at Stephenville closed in 2005. As well, Edmundston (New Brunswick) lost its paperboard mill and Weymouth (Nova Scotia) saw its sawmill close.

Besides employment losses due to plant closures, the long-term trend in the forest industry is for few workers due to the introduction of efficient logging equipment and the popularity of clear-cutting operations. For example, a single-grip timber harvester with one operator can cut as many trees in a day as 20 men using chainsaws.[4] Capital investment is also replacing high-priced labour in wood processing. The forestry companies' goal is to maximize profits by keeping their products competitive in the global marketplace. Unfortunately, this process adversely affects resource hinterlands by reducing the size of their labour force, leaving many unemployed. Mechanization has also made logging difficult for the small logging operator, who cannot afford the high cost of current machinery and who often owns a small woodlot where such machinery is unsuitable. Small operators, including those with woodlots, are finding it difficult to survive in the face of such stiff competition.

Because Atlantic Canada was settled first, its hardwood and softwood forests have been harvested for a very long time—over 300 years. In Nova Scotia and New Brunswick, some areas have been logged three or four times. Logging is an important occupation in Atlantic Canada. Unlike in the rest of Canada, where forest land is usually Crown land (public), the proportion of private timberlands to Crown lands in the Maritimes is extremely high. Private timberlands comprise 92 per cent of the commercial forest in Prince Edward Island, 70 per cent in Nova Scotia, and 50 per cent in New Brunswick. In Newfoundland and Labrador, like the rest of Canada, private ownership makes up only 2 per cent of the forested area. On average, the rate of logging on private lands is very high, sometimes exceeding the annual allowable cut estimated by the province. Exceeding the allowable cut impedes sustainable harvesting practices in the region. The high rate of logging often takes place on farms where timber sales are an important source of income.

A new issue in the forest industry in Atlantic Canada relates to Aboriginal people's right to timberlands. Aboriginal Canadians are demanding greater access to timber. Most Crown land in Atlantic Canada has been leased to large companies, mainly to the Irving Forest Corporation. In 1998, a dispute erupted over cutting rights on Crown land. The dispute went to court. Following a provincial court ruling in 2003, the right of Mi'kmaq Indians to harvest timber on Crown lands claimed by these Indians was confirmed. However, in 2005, the Supreme Court of Canada, in a decision described as a setback for Native rights, ruled that the Mi'kmaq of Nova Scotia and New Brunswick need permits if they want to cut down trees on Crown land for commercial gain. The Mi'kmaq are still seeking a compromise with the provincial governments whereby some Crown land currently leased to large companies would be reallocated to Indians. Just how this ruling of the Supreme Court will translate into a negotiated settlement remains unclear at this time.

Mining Industry

Atlantic Canada is endowed with a number of minerals with commercial value. In 2005, mineral production in this region included metallics (such as iron), non-metallics (such as potash), and fuels (such as oil and natural gas). In 2005, the sharp rise in oil and natural gas prices accounted for most of the increase in value of production from $5.2 billion in 2002 to $8.4 billion in 2005 (Table 9.7). Vast iron ore and nickel deposits exist in the Canadian Shield of Labrador, and coal, lead, potash, and zinc deposits are found in the Appalachian Uplands. Historically, coal mining on Cape Breton Island was extremely important, but its last coal mine was closed in 2001.[5]

In 2005, the annual value of mineral production in Atlantic Canada was $10.7 billion (Table 9.7). In 1999, the figure was $3.2 billion. This dramatic increase can be attributed to offshore petroleum developments. In 2005, Newfoundland and Labrador produced oil valued at $7 billion while natural gas from Nova Scotia's Sable Island field contributed $1.4 billion (Table 9.7). The leading mining operations in Atlantic Canada in 2005 were based on the Newfoundland offshore oil deposits, lead-zinc and potash deposits in New Brunswick, iron ore in Labrador, and coal in Nova Scotia. In 2005, production of nickel from Voisey's Bay added $78 million to the value of Newfoundland and Labrador's mineral output.

Mining is an important employer in Atlantic Canada. For instance, the largest single employer in Labrador, the Iron Ore Company of Canada, employs nearly 400 workers. In 2005, these workers mined, processed, and shipped nearly $1 billion worth of iron in the form of pellets and concentrates to steel mills in Canada and the United States. Another 400 workers are employed at the lead-zinc mine operated by Brunswick Mining and Smelting Corp. near Bathurst, New Brunswick. Similar numbers of workers are employed in the oil and gas industry.

Megaprojects

Large-scale resource projects do generate an immediate construction boom by creating a high demand for workers, especially skilled tradesmen, and for a variety of products and services. Once operational, the economic benefits slow and the number of permanent workers is relatively small. In the first decade of the twenty-first century, the major megaprojects affecting Atlantic Canada were offshore oil and gas developments led by the Hibernia oil project and the Sable Island gas development. Terra Nova began to produce oil in 2002 while production from the White Rose oil field and the Voisey's Bay nickel mine began in 2005. Discovered in 1984, the White Rose offshore oil field is located in the Jeanne d'Arc Basin 350 km east of St John's. The field consists of

Table 9.7 Mineral Production in Atlantic Canada, 2005 ($ millions)

Product	NL	PEI	NS	NB	Atlantic Canada 2005	Atlantic Canada 2002
Metallic	1,093	0	0	565	1,658	1,360
Non-metallic	46	3	286	310	645	489
Fuel	7,000	0	1,400	0	8,400	5,171
Total	8,139	3	1,686	875	10,703	7,020

Sources: Canadian Association of Petroleum Producers (2006); Natural Resources Canada (2006).

both oil and gas pools, including the South White Rose oil pool. The oil pool covers approximately 40 km^2 and contains an estimated 200–50 million barrels of recoverable oil. Unlike the fixed platform used by Hibernia, the Terra Nova oil field uses a Floating Production Storage and Offloading (FPSO) vessel. This facility is a ship-shaped vessel with integrated oil storage from which oil is offloaded onto a shuttle tanker. Husky Oil uses an FPSO vessel for its White Rose oil field.

These megaprojects have added a new dimension to Atlantic Canada's economy. Other potential projects—the Hebron Ben Nevis oil field, the Lower Churchill hydroelectric site, and a mega thermoelectric plant that will use gas from a liquefied natural gas terminal near Saint John, New Brunswick—could make a fundamental difference to the economies of Newfoundland and Labrador and New Brunswick.

The Hibernia oil project is located 315 km east of St John's, Newfoundland, on the Grand Banks above the site of a huge deposit of oil and natural gas. To tap the estimated 615 million barrels of oil from the Hibernia deposit, an innovative offshore stationary platform was needed (Vignette 9.10). About 4,000 workers

Vignette 9.10 The Hibernia Platform

The Hibernia platform is a gravity-based structure positioned on the ocean floor on the southeast corner of the Grand Banks. The 111-m-high oil platform and oil-storage units weigh over 650,000 tonnes. In the summer of 1997 the platform was placed on the ocean floor just above the oil deposits. The depth of the water at this point is about 80 m, leaving the oil platform approximately 30 m above the surface of the ocean. This structure is designed to be a platform for the oil derricks, to house pumping equipment and living quarters for the workers, as well as to provide storage for the crude oil. The rig extracts huge amounts of oil from the Avalon reservoir (2.4 km under the seabed) and from the Hibernia reservoir (3.7 km deep). The crude oil is then pumped from the Hibernia storage tanks to an underwater pumping station and then through loading hoses to three 900,000-barrel supertankers for shipment to foreign refineries.

Source: Adapted from Cox (1994).

The Hibernia platform has a massive concrete base which supports its drilling and production facilities as well as the workers' accommodations. Since the platform was positioned on the ocean floor in 1997, the province has joined the ranks of other oil-producing provinces. The government of Newfoundland and Labrador at first received half of the royalties generated by these offshore developments. By 2006, the province received all of the oil revenues. (CP/*Toronto Star*/Peter Power)

built a specially designed offshore oil platform that can withstand the pounding storms of the North Atlantic and crushing blows from huge icebergs. The massive concrete and steel construction sits on the ocean floor, with 16 'teeth' in its exterior wall designed to absorb the impact of icebergs.

The Hibernia drilling site has an annual output of about 30 million barrels and adds greatly to Newfoundland's mineral output. Based on an average price of $60 per barrel, the annual value of production is $1.8 billion. Over the next 20 years with the same price and production, the total value of production is estimated at $36 billion. Oil production began in 1998. By 2000, Hibernia accounted for 12 per cent of Canadian oil production. Oil is exported to American and other foreign refineries. According to the owners, the Hibernia Consortium, production should continue for approximately 20 years.

A new megaproject at Voisey's Bay in Labrador is the site of a huge nickel deposit, the Ovoid deposit, which lies close to the surface. It consists of 32 million tonnes of relatively rich ore bodies—2.8 per cent nickel and 1.7 per cent copper. Another deposit, Eastern Deeps, though larger (about 70 million tonnes), has a lower grade of nickel and copper (similar grade to that found at the Sudbury mines in Ontario).

The Voisey's Bay deposit has three attractive features: high-grade nickel, a surface deposit that has the potential for open-pit mining, and proximity to ocean shipping. These characteristics may make it the lowest-cost nickel mine in the world. In fact, there is concern that production from this mine might force higher-cost producers, such as the nickel mines in Sudbury, to close their operations. For Newfoundland and Labrador, the Voisey's Bay mining and smelting operation could employ over 1,000 workers. In late 1996, after Inco Ltd had purchased the Voisey's Bay deposit for $4.3 billion, the company announced that it would build a $1 billion nickel smelter and refinery at Argentia on the southeastern coast of Newfoundland with approximately 750 workers employed at the smelter and refinery. Later, Inco proposed to ship the nickel ore to its Sudbury smelter. Newfoundlanders were outraged and their

Discovered in 1993, the Voisey's Bay nickel deposit lies along the coast of Labrador approximately 350 km north of Happy Valley–Goose Bay. In 1996, Inco Ltd acquired the rights to the Voisey's Bay property and its subsidiary, Voisey's Bay Nickel Company, is responsible for constructing and operating a mine and concentrator. Once in operation, the mine and concentrator are expected to employ 400 people. Since the deposit lies within the land claims of both the Labrador Innu and Inuit, Inco was obliged to reach an Impacts and Benefits Agreement with each of the Aboriginal organizations. (CP/Andrew Vaughan)

government refused to sign the surface lease that would allow Inco to move forward with its plans for a mine. In 2002, the government of Newfoundland and Labrador concluded an agreement with Inco that included the possibility of some of the ore being processed in Newfoundland at a yet-to-be-built experimental hydromet smelter. In the meantime, the provincial government has approved the shipment of ore to Inco's Sudbury smelter.

Despite the lure of megaprojects, they present problems for regional development. First, they are capital-intensive undertakings. During the construction phase a large labour force is required, but in the operational phase relatively few employees are needed. Second, megaprojects in resource hinterlands lose much of their spinoff effects to industrial areas. As a consequence, economic benefits related to the manufacture of the essential parts of a megaproject go outside the hinterland. Efforts by hinterland governments to address this classic problem of economic leakage from resource hinterlands to industrial cores have had mixed results, as is perhaps demonstrated by the indefinite nature of Inco's intent to build a smelter/refinery at Argentia, Newfoundland, and the difficulties faced by the Bull Arm complex (Vignette 9.7). Third, megaprojects sometimes take place on Crown lands where Aboriginal groups often have land claims based on traditional occupancy of the land. The Voisey's Bay nickel mine development is situated on such lands. Labrador, therefore, presents a special version of the Aboriginal/non-Aboriginal faultline.

Resource Development and the Aboriginal/ Non-Aboriginal Faultline

The Voisey's Bay mining development has sparked a renewed interest in land-claims settlements in Labrador. With such agreements, resource developments are more assured of a trouble-free working environment. This point was made back in 1973 when the courts forced Hydro-Québec to reach an agreement with the Cree and Inuit of Québec who claimed that northern Québec was part of their traditional lands (Crown lands that they or their ancestors used for hunting and trapping). Such lands have given rise to a legal claim known as Aboriginal title. With an unresolved claim of Aboriginal title, there is an uncertainty over ownership of the land in dispute. The Québec government (as did all other provincial governments of the day) considered that Aboriginal people could use Crown lands for hunting and trapping at the pleasure of the Crown and, therefore, no special arrangement with Aboriginal residents was necessary when the provincial government granted a lease or sold the land to a developer. How wrong they were! Without such an agreement, construction of the James Bay Hydroelectric Project could have been halted. By the twenty-first century, however, companies can proceed with their developments before a land-claim agreement is achieved if they reach a special arrangement known as an Impact and Benefits Agreement with the concerned Aboriginal group(s).

By 2003, Inco had gained Impact and Benefits Agreements with both the Labrador Inuit and the Labrador Innu. Yet in all of the negotiations, the Innu—whose traditional homeland includes the site of the Voisey's Bay mining project, who are without treaty, who have never achieved a comprehensive land-claim agreement, and who continue to suffer from incredibly high suicide rates and substance abuse problems—are drawn into the vortex of a huge construction project and forced to answer the question, 'What do you want in exchange for development?' (Samson, 2003: 96–102). While the Innu, like the Labrador Inuit, have reached an Impact and Benefits Agreement with Inco that provides specific benefits such as employment opportunities, some Innu (as well as many Inuit) have only a limited command of the English language, and most lack the basic education level (high school diploma) or employment experience to take advantage of opportunities provided by the construction of the Voisey's Bay nickel mine.

Then, too, the issue of a comprehensive land claim is more complicated for the Labrador Innu than for the Labrador Inuit because the Innu are considered Indians by the federal government and therefore they must be dealt with under the Indian Act. The Labrador Inuit began negotiations in the 1990s and, in August 2003, reached a land-claim agreement.[6] In 2005, the three levels of government (federal, provincial, and Inuit) signed this agreement, thus bringing it into law. This agreement, which includes provision for self-government, is the first of its kind in Atlantic Canada. On the other hand, progress on the Innu land claim is proceeding very slowly because Ottawa insists that the Labrador Innu first register as Indians under the Indian Act, the benefits of which they had been denied for more than 50 years since Newfoundland joined Canada. This has meant that they must first negotiate for and be given small reserve lands, and only then seek a comprehensive land-claim settlement that would involve their traditional lands (lands that would extend far beyond their reserves) and that may then remove them from the Indian Act they had to adhere to before seeking a comprehensive agreement in the first place (Canadian Human Rights Commission, 2003).

Agriculture

Agriculture is limited by the physical geography in Atlantic Canada. Arable land constitutes less than 5 per cent of the Maritimes. Arable land is even rarer in Newfoundland and Labrador, making up less than 0.1 per cent of its territory. Though limited in size, agricultural production significantly contributes to the economy of Atlantic Canada. In 2005, the value of agricultural production in the region was about $1 billion. Specialty crops, especially potatoes, contributed heavily to the value of this production.

Atlantic Canada has nearly 400,000 ha in cropland and pasture. Almost all of this farmland is concentrated in three main agricultural areas—Prince Edward Island, the Saint John River Valley in New Brunswick, and the Annapolis Valley in Nova Scotia. Specialty crops, especially potatoes and tree fruit, are extremely important cash crops. In all three agricultural areas, dairy cattle graze on pasture land. The dairy industry in Atlantic Canada has benefited from the orderly mar-

The rich, red soils of Prince Edward Island are famous for growing potatoes, which are the primary cash crop in the province. Prince Edward Island is Canada's leading potato province, responsible for almost one-third of Canadian production. Its potatoes are grown for three specific markets: seed, table potatoes, and processing. Seed potatoes are sold to commercial potato growers and home gardeners to produce next year's crop; table potatoes go to the retail and food service sectors; and processing potatoes are manufactured into french fries, potato chips, and other processed potato products. (Barrett & MacKay Photography, Inc.)

keting of fluid-milk products through marketing boards.

Prince Edward Island is the leading agricultural area in Atlantic Canada. It has almost half of the arable land in the region. Most of Prince Edward Island's 155,000 ha of farmland is devoted to potatoes, hay, and pasture, with the principal cash crop being potatoes. Since the 1980s, most potato growers have had contracts with the island's major potato-processing plants—Irving's processing plant near Summerside and McCain's plant at Borden–Carleton now dominate the potato industry on the island. The second major agricultural area, the Saint John River Valley, is in New Brunswick. Its 120,000 ha of arable land comprise about one-third of Atlantic Canada's farmland. The Saint John River Valley has the best farmland in New Brunswick. Nova Scotia has nearly one-quarter of Atlantic Canada's farmland, with 105,000 ha. Nova Scotia's famous Annapolis Valley, the region's third agricultural area, is the site of fruit orchards and market gardens. The valley's close proximity to Halifax, the major urban market in Atlantic Canada, has encouraged vegetable gardening. In both New Brunswick and Nova Scotia, potatoes are a major cash crop. Almost all potato farmers in these two provinces seed their potatoes under contract to McCain Foods, a multinational food-processing corporation based in New Brunswick. The company has benefited from NAFTA after the removal of tariffs on its food products, especially french fries, to the United States. Newfoundland has the least amount of farmland—just over 6,000 ha.

When the British founded Halifax in 1749, the high hill overlooking the harbour offered a perfect place to construct a fortress to defend the new town and its naval base. Named the Halifax Citadel, this fortress is an impressive star-shaped masonry structure complete with defensive ditch, earthen ramparts, musketry gallery, powder magazine, garrison cells, guard room, barracks, and school room. The Citadel is now a National Historic Site. (Ron Garnett/AirScapes.ca)

Atlantic Canada's Urban Geography

The urban geography of Atlantic Canada is characterized by few large cities and many small, isolated coastal villages. These features are attributed to the region's geography and natural resources. It is not surprising, then, that a relatively small percentage of Atlantic Canada's population lives in urban places. In 2001 Atlantic Canada was the least urbanized region of Canada with less than half of its population living in urban centres (Figure 9.5).

Atlantic Canada has a coastal settlement pattern. Major cities are situated along its coastline. Trade and fishing have played a key role in the development of this pattern. The region's interior is relatively 'empty'. The region's physical geography has shaped this pattern because interior lands have few resources to attract settlement. The main exception is mining towns. The iron-mining town of Labrador City is an example of a single-industry community in the interior of Labrador's mining hinterland. Labrador City's population has been declining due to cost reductions and the substitution of machinery for labour. Between 1996 and 2006, Labrador City's population declined from 10,473 to 7,240.

Halifax, the largest city in Atlantic Canada, ranks only thirteenth in size among Canada's major cities (Vignette 9.11). Halifax's relatively small size is due to three factors: (1) a less well-developed urban system and regional economy compared to other regions of Canada; (2) the fragmented nature of Atlantic Canada's popu-

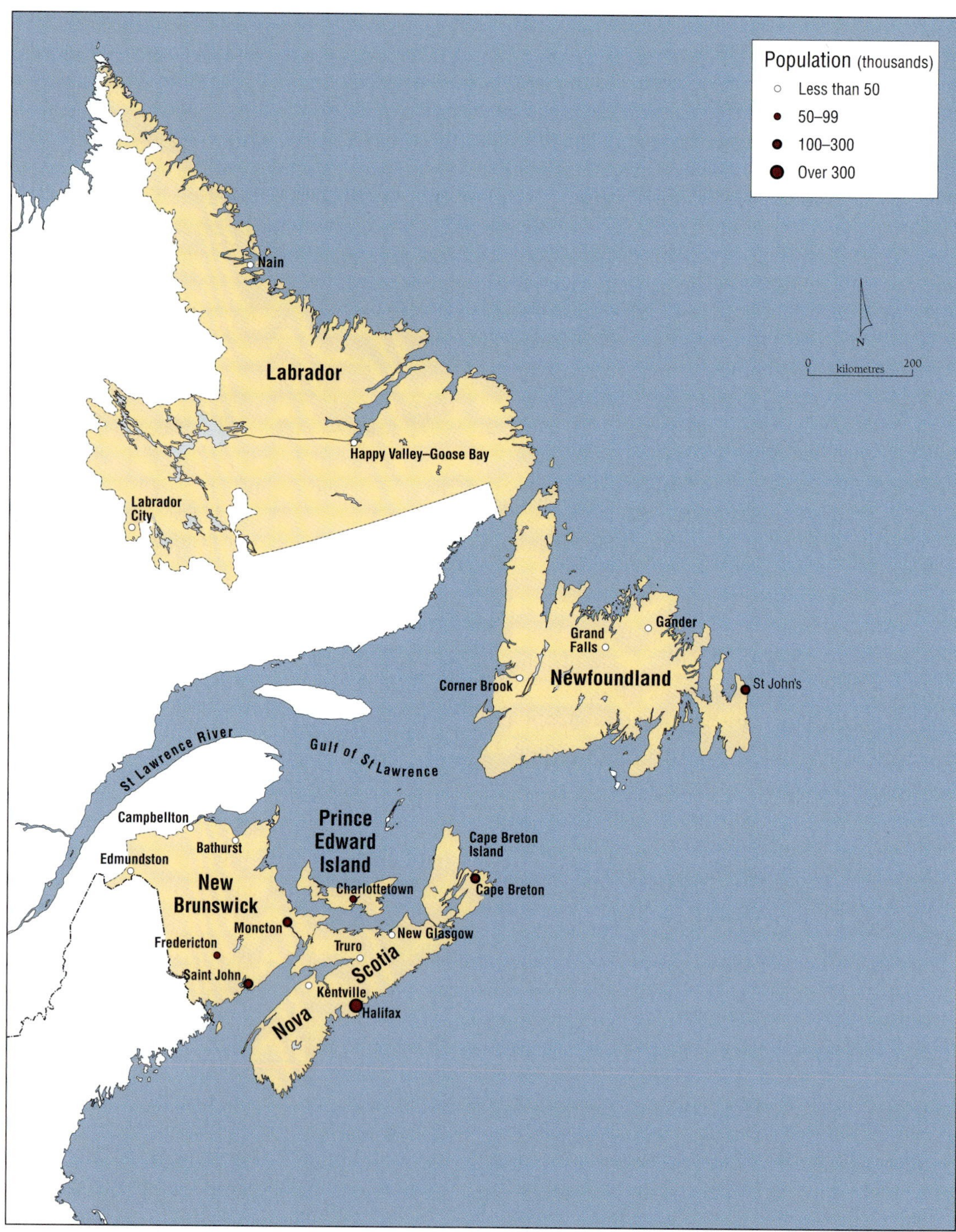

Figure 9.5 Major urban centres in Atlantic Canada. Atlantic Canada has few large cities and most urban centres are scattered across the region. Only five centres—Halifax, St John's, Moncton, Saint John, and Cape Breton—have populations exceeding 100,000, and Cape Breton is a political amalgamation of all the communities on Cape Breton Island.

lation and resulting small markets; and (3) Atlantic Canada's relatively small economy. However, Halifax, with the best natural harbour on the east coast of North America, has the potential to become a superport for the eastern seaboard of North America. The need for such a port has arisen from two recent developments. First, the construction of container ships that are too large to pass through the Panama Canal. These ships, known as post-Panamax, are longer than three football fields. In the coming years, post-Panamax ships are anticipated to dominate the container business. Second, the exploding volume of trade from China to North America has created unloading delays at west coast ports, including Vancouver. Canadian Tire, Wal-Mart, and other major retail companies have faced long delays in receiving their goods from China and are now encouraging shipping firms to redirect their container ships to Halifax. With trade increasing every year, Halifax dreams of becoming a superport for North America. In spite of Halifax having the deepest harbour along the east coast of North America, New York City remains the primary destination of container ships because of its huge market and its extensive rail/truck network that connects it to other major markets in North America. From the perspective of container shipping companies, these two advantages have trumped Halifax's superior harbour. For the world's shipping companies, Halifax remains a secondary port.

Halifax, St John's, Moncton, and Saint John, the four largest urban centres in Atlantic Canada, have experienced different population increases over the past 25 years. Saint John has seen virtually no change over 25 years with its population increasing from 121,000 to 122,000. However, the construction of the $750 million regasification plant (LNG) near Saint John will be completed in late 2007, giving a boost to the city's economy. This facility is one element of the Atlantic Gateway concept, which sees Atlantic Canada as a link between the gas-producing countries of North Africa and the Middle East and the energy-deficient market of New England. Most of the natural gas from this facility will be shipped by pipeline from Saint John to markets in New England.

Vignette 9.11 Halifax

Halifax, the capital of Nova Scotia and the largest city in Atlantic Canada, was founded in 1749. On 1 April 1996 Halifax ceased to exist as an incorporated city and became the Halifax Regional Municipality, including the former cities of Halifax and Dartmouth. By 2006, Halifax had a population of almost 373,000. As in the past, its strategic location allowed Halifax to play a major role on the Atlantic coast as a naval centre, an international port, and as a key element in the Atlantic Gateway concept. Along the east coast of North America, its deep, ice-free harbour is ideally suited for huge post-Panamax ships. Yet, because of its relative distance from the major markets in North America and its reliance on indirect transferring of goods between ships, trains, and trucks, Halifax cannot provide lower transportation costs than New York. The economic strength of the Halifax Regional Municipality rests on its defence and port functions, its service function for smaller cities and towns in Nova Scotia, and its provincial administrative functions. Halifax also has a small manufacturing base and a growing service sector, as well as a small but growing high-technology industry. In 2005, two high-tech firms, SupportSoft and Versata, expanded their Halifax operations, with Versata relocating its software development operations from India.

Artist's rendition (top) and the construction site (bottom) for the LNG plant near Saint John. Canaport LNG is contructing a state-of-the-art LNG receiving and regasification terminal in Saint John, New Brunswick, that will begin operations in late 2008. The terminal's send-out capacity will be 1 billion cubic feet (28 million cubic meters) of natural gas per day to markets in Canada and the US. Canaport LNG will be the first LNG terminal in Canada. (Canaport LNG, 2007)

The population of St John's increased by 26,000 over the period 1981–2006, from 155,000 to 181,000. Unlike Newfoundland and Labrador as a whole, St John's is growing, with much of this population increase attributable to the offshore oil developments. Halifax witnessed the greatest increase over this 25-year period when its population grew by 95,000, from 278,000 to 373,000 residents. Much of this increase, however, is related to the creation in 1996 of the Halifax Regional Municipality, which encompasses other, smaller communities such as Dartmouth, Sackville, and Bedford. Halifax's economic strength stems from its role as the regional centre of the Maritimes, which makes it the logical centre for tertiary activities. Economic conditions have attracted few new immigrants to the cities of Atlantic Canada, and many of the region's residents have left for cities in other provinces. From 2001 to 2006, the three largest cities in Atlantic Canada, Halifax, Moncton, and St John's, grew by 3.8, 6.5, and 4.7 per cent respectively (Table 9.8). The next two largest centres, Saint John and Cape Breton, fared less well. Saint John saw a slight decline in its population from 2001 to 2006 while Cape Breton's population dropped by 3.1 per cent. On the other hand, New Brunswick's capital, Fredericton, experienced growth of 5.3 per cent, while Charlottetown and Truro had population increases of around 2 per cent. Beyond these major urban centres, 10 smaller cities and towns form a lower level in the urban hierarchy of Atlantic Canada. Their populations range from 7,000 to 36,000. Of these 10 towns, six saw their populations decline from 2001 to 2006. The iron-mining town of Labrador City suffered the largest percentage decline, at 6.5 per cent, followed by three other resource towns, Campbellton, Bathurst, and Edmundston (Table 9.8).

Atlantic Canada's urban geography remains highly dispersed, broken into provincial units, and lacking a single dominant metropolitan centre. Instead, two centres, Halifax and St. John's, serve as the urban focal points for the Maritimes and Newfoundland and Labrador, respectively. Halifax, with its deep, ice-free harbour and existing container facilities, has the potential of becoming the hub of the Atlantic Gateway concept. In spite of the geographic dichotomy that separates Newfoundland and Labrador from the Maritimes, one demographic trend remains common—a rural-to-urban movement within Atlantic Canada coupled with a strong out-migration from Newfoundland and Labrador to other parts of Canada.

Atlantic Canada's Future

Atlantic Canada lies on the eastern rim of Canada. The region's future is handicapped by its physical fragmentation, population dispersion, and failed cod fishery. Physical fragmentation works against a strong Atlantic Canada.

Table 9.8 Urban Centres in Atlantic Canada, 2001–6

Urban Centre	Population 2001	Population 2006	% Change
Labrador City	7,744	7,240	–6.5
Gander	9,651	9,951	3.1
Bay Roberts	10,531	10,507	–0.2
Grand Falls–Windsor	13,340	13,558	1.6
Campbellton	18,820	17,888	–5.0
Edmundston	22,173	21,442	–3.3
Kentville	25,172	25,969	3.2
Corner Brook	26,153	26,623	1.8
Bathurst	32,523	31,424	–3.4
New Glasgow	36,735	36,288	–1.2
Truro	44,276	45,077	1.8
Charlottetown	57,234	58,625	2.0
Fredericton	81,346	86,688	5.3
Cape Breton*	109,330	105,928	–3.1
Saint John	122,678	122,389	–0.2
Moncton	118,678	126,424	6.5
St John's	172,918	181,113	4.7
Halifax	359,183	372,858	3.8

*Cape Breton includes Sydney, North Sydney, Glace Bay, and other Cape Breton Island municipalities.
Source: Statistics Canada (2007a). Adapted from Statistics Canada publication *Population and Dwelling Counts, 2001 Census*, Catalogue 93F0050XCB2001013, Released 16 July 2002, http://www.statcan.ca/bsolc/english/bsolc?catno=93F0050X2001013.

Geography has divided Atlantic Canada into three subregions: the strongest economy and largest market are in the Maritimes, followed by the island of Newfoundland, and then Labrador. The powerful attraction of high-paying jobs in Alberta's boom economy has drawn many workers, but some of these workers have kept their homes in Atlantic Canada and, by sending money home, are helping their families and the local economy. Within Atlantic Canada, future prospects seem brightest for the more populated Nova Scotia, which has a vigorous metropolitan centre in Halifax.

The future for Atlantic Canada in the twenty-first century is uncertain, but the prospects for sustainable economic growth are brighter than ever before. Why? Mega-resource developments, especially offshore oil, have the potential to transform the regional economy, and the possibility of an Atlantic Gateway—a trade corridor connecting Atlantic Canada with North America and the rest of the world—if realized, would be a tremendous boon. For Newfoundland and Labrador and Nova Scotia, the expansion of offshore oil and gas is stimulating their economies. No such hope yet exists for New Brunswick and Prince Edward Island. Still a nagging question persists—does energy production hold the key to economic rejuvenation? The answer seems more positive today than it did in the past. First, the two provinces producing fossil fuels now receive 100 per cent of the offshore oil and gas revenues compared to 30 per cent. Second, prior to the Atlantic Energy Accord of 2004 the link between provincial revenues and the size of equalization payments reduced the actual gains by Newfoundland and Labrador and Nova Scotia. The reason is simple—as provincial oil and/or gas revenues increased, the amount of equalization payments declined. While not quite a zero-sum game, the impact of oil royalties was diminished by reduced equalization payments, thus slowing real economic gains in these provinces. The Atlantic Accord protects the two provinces from a clawback in their equalization payments, although the 2007 federal budget has reintroduced the potential of a clawback. Third, the price of oil and natural gas has reached record levels in the past few years, thus greatly increasing royalties to the two provinces. The unknown question is future prices, though it is probably safe to assume, given estimated global supplies and expected increases in demand, that the trend will be towards higher prices.

In the next decade, several events could have a positive effect on Atlantic Canada. These events include:

- A reversal of the out-migration of skilled workers. Currently, Atlantic Canada suffers a shortage of skilled workers, and the proponents of major industrial construction projects, such as the LNG projects near Saint John, New Brunswick, and Goldboro, Nova Scotia, have expressed concern about the availability of skilled labour.
- Additional oil and gas discoveries off Newfoundland and Labrador and Nova Scotia.
- Development of the Lower Churchill hydroelectric site remains a potential project, but a political accord between Québec and Newfoundland and Labrador is required because the power would be transmitted across Québec to reach markets in eastern Canada and the US.
- Recovery of the codfish stocks, which would provide work for the inland fishers and employment for fish-plant workers.
- Inco going ahead and building a nickel smelter at Argentia, Newfoundland, which would become a major employer.
- Expansion of the tourism industry.
- The continued growth of high-technology and call-centre industries.
- Halifax becoming the major container port for the east coast of North America, which would be a pivotal part of the Atlantic Gateway concept.

Because of these economic opportunities, Atlantic Canada's future for a more prosperous economy seems more plausible than ever before. Indeed, Atlantic Canada's economy may be at a tipping point with offshore oil and gas production at Hibernia, Terra Nova, White Rose, and Sable Island well underway, and the potential for three major resource projects within the next five years: the hydroelectric dam on the Lower Churchill River; an oil well at the Hebron Ben Nevis oil field near Hibernia; and the LNG plants at Saint John, New Brunswick, and at Goldboro, Nova Scotia. Continuing high world prices for these resources could propel Atlantic Canada's economy into a more prosperous state—one in which unemployment rates decline, business opportunities increase, and out-migration slows and even reverses. Such an economic state may create that elusive balance between economic growth and a 'down East' way of life.

Notes

1. On 30 April 1999, the Newfoundland House of Assembly gave unanimous consent to a constitutional amendment that would officially change the name of the province to Newfoundland and Labrador.
2. The Maritime region was a focal point for the struggle between France and Britain. Indian allies were critical for both European powers. 'Involvement with the French brought benefits—and immediate consequences. Before they realized the full scope of the French–British rivalry, the Mi'kmaq and the Maliseet discovered they had already chosen sides' (Coates, 2000: 31).
3. As with Fort McMurray (Wood Buffalo), Alberta, Statistics Canada has created a geographically sprawling census metropolitan area—Cape Breton—out of Sydney and other Cape Breton Island municipalities.
4. The FMG Timberjack 990 requires only one person—an operator—to fell a tree in a very short time. Priced at $600,000, this giant machine is one example of capital substitution for labour. The Timberjack grips the base of a tree, slices its steel blade straight across the trunk, and strips the tree of its branches. The operator then uses the Timberjack to cut the tree into measured lengths before utilizing a boom arm and clamp grip to load the cut logs onto a truck.
5. Coal, a combustible sedimentary rock formed from the remains of plant life during the Carboniferous Age (a geological period in the Paleozoic era), is classified into four types: anthracite, bituminous, sub-bituminous, and lignite. Anthracite is the highest grade of coal, while bituminous is used in the iron and steel industry and for generating thermoelectric power. Most coal mined in Nova Scotia has been bituminous coal.
6. On 26 May 2004 the Labrador Inuit voted 76 per cent in support of the agreement, with an 86.5 per cent voter turnout. The provincial and federal governments passed legislation in 2004 and 2005, respectively, giving legal effect to the Labrador Inuit Land Claims Agreement Act. This agreement contains the provision that the Labrador Inuit will receive 25 per cent of the revenue from mining and petroleum production on their settlement land, as well as 5 per cent of provincial royalties from the Voisey's Bay project. The federal environmental review panel examining the possible environmental and social impacts of the Voisey's Bay Nickel Company's mine proposal expressed concern about the disposal of the 15,000 tonnes of mine tailings to be produced each day. The mining company proposes to deposit the toxic tailings in a pond and prevent this from draining into surrounding streams and rivers by building two dams. While federal and provincial environmental officials are satisfied with the company's solution to the tailings problem, local people, especially the Inuit and Innu, are skeptical and worried about the effects of these toxins on the wildlife they depend on for food.

Challenge Questions

1. What recent developments provide Atlantic Canada with an opportunity to break the downward economic spiral described by Friedmann's 'downward transitional region'?
2. From 2002 to 2005, which of the three industrial sectors has increased its percentage of Atlantic Canada's industrial labour force and which has seen a decrease?
3. Is it true that geography explains why Atlantic Canada has such a dispersed coastal settlement pattern?
4. Why does Georges Bank contain such large and rich fish stocks?
5. Why is Saskatchewan supportive of the Atlantic Accord and opposed to the inclusion of the fiscal cap in the 2007 budget, while Ontario takes the opposite position?
6. The value of oil production in Newfoundland and Labrador increased sharply in 2005. Why?
7. Why are draggers considered a danger to marine life and why does Ottawa resist outlawing the use of draggers in Canadian waters?
8. What developments suggest that the dream of an Atlantic Gateway is coming true?
9. Compared to the other regions, how would you rank Atlantic Canada's sense of place?

Key Terms

biomass
The total quantity or weight of an organism (e.g., codfish) in a given area.

economies of association
Manufacturing plants located near their suppliers obtain lower prices because of the transportation savings associated with proximity to suppliers; localization economies.

groundfish
Fish that live on or near the bottom of the sea. The most valuable groundfish are cod, halibut, and sole.

liquefied natural gas (LNG)
A liquid form of natural gas chilled to –162 degrees Celsius. The cooling process, called liquefaction, reduces the volume to one six-hundredth of its original volume. At regasification terminals, the LNG is warmed until it returns to a gaseous state.

pelagics
Marine life, especially fish species such as salmon, mackerel, and capelin, found in the upper levels of the sea, as opposed to the bottom-dwelling groundfish.

spawning biomass
The total quantity or weight of a sexually mature organism in a given area that can reproduce. Cod, for example, reach sexual maturity around the age of seven.

Bibliography

Atlantic Canada Opportunities Agency. 2007. 'Report on Plans and Priorities: 2007–2008 Estimates'. At: <www.acoa.ca/e/library/reports/rpp2008/section1.shtml#_Toc158716790>.

Bailey, Alfred G. 1969. *The Conflict of European and Eastern Algonkian Cultures, 1504–1700*. Toronto: University of Toronto Press.

Bennett, Margaret. 1989. *The Last Stronghold: Scottish Gaelic Traditions in Newfoundland*. Edinburgh: Canongate.

Blackbourn, Anthony, and Robert G. Putnam. 1984. *The Industrial Geography of Canada*. London: Croom Helm.

Blades, Kent. 1995. *Net Destruction: The Death of Atlantic Canada's Fishery*. Halifax: Nimbus Publishing.

Bradfield, Michael. 1991. *Maritime Economic Union: Sounding Brass and Tinkling Symbolism*. Halifax: Canadian Centre for Policy Alternatives.

Canada. 1993. *Canada Year Book 1994*. Ottawa: Ministry of Industry, Science and Technology.

———, Natural Resources. 1994. 1993 *Canadian Minerals Yearbook: Review and Outlook*. Ottawa: Minister of Supply and Services.

———. 1995. *The State of Canada's Forests: A Balancing Act*. Ottawa: Natural Resources.

Canadian Human Rights Commission. 2003. *Report to the Canadian Human Rights Commission on the Treatment of the Innu of Canada by the Government of Canada*, by Professors Constance Backhouse and Donald M. McRae. At: <www.chrc-ccdp.ca/publications/Rapport_Innu_Report/RapportInnuReport_Page3.asp?l=e>.

Canaport LNG. 2007. '2007 Project Update'. At: <www.canaportlng.com/project_journal_2007.php>.

Cashin, Richard. 1993. *Charting a New Course: Towards the Fishery of the Future*. Ottawa: Department of Fisheries and Oceans.

Choyce, Lesley. 1996. *Nova Scotia: Shaped by the Sea*. Toronto: Viking.

Clapp, R.A. 1998. 'The Resource Cycle in Forestry and Fishing', *Canadian Geographer* 42, 2: 129–44.

Conrad, Margaret, and James Hiller. 2001. *Atlantic Canada—A Region in the Making*. Toronto: Oxford University Press.

Coward, Harold, Rosemary Ommer, and Tony Pitcher, eds. 2000. *Just Fish: Ethics and Canadian Marine Fisheries*. St John's: ISER Books.

Cox, Kevin. 1994. 'How Hibernia Will Cast Off', *Globe and Mail*, 12 Nov., D8.

Department of Fisheries and Oceans. 1994. *Canadian Fisheries Statistical Highlights 1992*. Ottawa: Department of Fisheries and Oceans.

Department of Finance Canada. 2007. Federal Transfers to Provinces and Territories. At: <www.fin.gc.ca/FEDPROV/eqpe.html>.

Doeringer, Peter B., and David G. Terkla. 1995. *Troubled Waters: Economic Structure, Regulatory Reform, and Fisheries Trade*. Toronto: University of Toronto Press.

DRI Canada. 1994. *Atlantic Canada: Facing the Challenge of Change*. Moncton: Atlantic Canada Opportunities Agency.

Eaton, Peter B., Alan G. Gray, Peter W. Johnson, and Eric Handert. 1994. *State of the Environment in the Atlantic Region*. Halifax: Environment Canada.

Erskine, David. 1968. 'The Atlantic Region', in John Warkentin, ed., *Canada: A Geographical Interpretation*. Toronto: Methuen, 231–80.

Fife, Robert. 2001. 'N.S. Recruits Albertans in Cash Grab', *National Post*, 7 June, A4.

Fisheries and Oceans Canada. 2006. Statistical Services: Landings. At: <www.dfo-mpo.gc.ca/communic/statistics/commercial/landings/seafisheries/s2005aq_e.htm>.

Forbes, E.R., and D.A. Muise, eds. 1993. *The Atlantic Provinces in Confederation*. Toronto: University of Toronto Press.

Foster, Gilbert. 1988. *Language and Poverty: The Persistence of Scottish Gaelic in Eastern Canada*. St John's: Institute of Social and Economic Research, University of Newfoundland.

Gordon, Daniel V., and Gordon R. Munro. 1996. *Fisheries and Uncertainty*. Calgary: University of Calgary Press.

Greenspon, Edward. 1998. 'Ottawa Approves New Aid for Fishery', *Globe and Mail*, 12 June, A1, A4.

Hardin, Garrett. 1968. 'The Tragedy of the Commons', *Science* 162: 1243–8.

Hutchings, Jeffrey A., and Ransom A. Myers. 1994. 'What Can Be Learned from the Collapse of a Renewable Resource: Atlantic Cod, *Gadus morhua*, of Newfoundland and Labrador', *Canadian Journal of Fisheries and Aquatic Science* 51: 2126–46.

Isaac, Thomas. 2001. *Aboriginal and Treaty Rights in the Maritimes: The Marshall Decision and Beyond*. Saskatoon: Purich Publishing.

Jang, Brent. 1998. 'Hibernia Halves Output', *Globe and Mail*, 5 Mar., B1.

Kirby, J.L. 1982. *Navigating Troubled Waters: A New Policy for the Atlantic Fisheries: Report of the Task Force on Atlantic Fisheries*. Ottawa: Department of Fisheries and Oceans.

Lockerby, Earle. 2004 [1998]. 'The Deportation of the Acadians from Ile St.-Jean, 1758', in J.M. Bumsted and Len Kuffert, eds, *Interpreting Canada's Past: A Pre-Confederation Reader*. Toronto: Oxford University Press, 88–93.

MacAfee, Michelle. 1998. 'Diversity in Fisheries Paying Off: Minister', *Globe and Mail*, 14 Aug., A3.

McCalla, Robert J. 1991. *The Maritime Provinces Atlas*. Halifax: Maritext.

McCarthy, Shawn. 2007. 'Nfld. Grabs a Seat at Oil Table', *Globe and Mail*, 23 Aug. At: <www.reportonbusiness.com/servlet/story/RTGAM.20070823.whebron0823/BNStory/robNews/home>.

McManus, Gary E., and Clifford H. Wood. 1991. *Atlas of Newfoundland and Labrador*. St John's: Breakwater.

McMillan, Charles. 2006. *Embracing the Future: The Atlantic Gateway and Canada's Trade Corridor*. A study prepared for the Asia-Pacific Foundation of Canada, 30 Nov. At: <www.acoa.ca/e/library/reports/sectorFocussed/embracingFuture.shtml>.

MacKenzie, A.A. 1979. *The Irish in Cape Breton*. Antigonish, NS: Formac.

Macpherson, Alan G., ed. 1972. *The Atlantic Provinces: Studies in Canadian Geography*. Toronto: University of Toronto Press.

Marsh, James H., ed. 1988. *The Canadian Encyclopedia*, 2nd edn. Edmonton: Hurtig.

Matthews, Ralph. 1983. *The Creation of Regional Dependency*. Toronto: University of Toronto Press.

———. 1993. *Controlling Common Property: Regulating Canada's East Coast Fishery*. Toronto: University of Toronto Press.

Millward, Hugh, and Lorna Winsor. 1997. 'Twentieth-Century Retail Change in the Halifax Central Business District', *Canadian Geographer* 41, 2: 194–201.

Natural Resources Canada. 2003a. Mineral Production of Canada, by Provinces and Territory. At: <mmsd1.mms.nrcan.gc.ca/mmsd/production/production_e.asp>.

———. 2003b. The State of Canada's Forests 2001/02. At: <www.nrcan.gc.ca/cfs-scf/national/whatquoi/sof/sof02/profiles_e.html>.

Neis, Barbara, and Lawrence Felt, eds. 2000. *Finding Our Sea Legs: Linking Fishery People and Their Knowledge with Science and Management*. St John's: ISER Books.

Newfoundland and Labrador, Economic Research and Analysis Division. 2007. 'Fishery'. At: <www.economics.gov.nl.ca/E2007/fishery.asp>.

Norcliffe, Glen, and Judy Bates. 1997. 'Implementing Lean Production in an Old Industrial Space: Restructuring at Corner Brook, Newfoundland, 1984–1994', *Canadian Geographer* 41, 1: 41–60.

Palmer, Craig T., and Peter Sinclair. 1997. *When the Fish Are Gone: Ecological Disasters and Fishers in Northwest Newfoundland*. Halifax: Fernwood.

Phillips, David. 1993. *The Day Niagara Falls Ran Dry!* Toronto: Canadian Geographic and Key Porter Books.

Physorg.com. 2006. 'Ocean Study Predicts the Collapse of All Seafood Fisheries by 2050'. At: <www.physorg.com/news81778444.html>.

Ponting, J. Rick. 1997. *First Nations in Canada: Perspectives on Opportunity, Empowerment, and Self-Determination*. Toronto: McGraw-Hill Ryerson.

Power, Thomas P., ed. 1991. *The Irish in Atlantic Canada, 1780–1900*. Fredericton: New Ireland Press.

Rumney, Thomas. 1998. *The Geography of Canada Bibliography Series: Vol. 3, Atlantic Canada*. Plattsburgh, NY: Plattsburgh State University, Center for the Study of Canada.

Samson, Colin. 2003. *A Way of Life That Does Not Exist: Canada and the Extinguishment of the Innu*. St John's: ISER Books.

Savoie, Donald. 2000. *Aboriginal Economic Development in New Brunswick*. Moncton: Canadian Institute for Research on Regional Development.

——— and Ralph Winter. 1993. *The Maritime Provinces: Looking to the Future*. Canadian Institute for Research on Regional Development. Sackville, NB: Tribune Press.

Stanford, Quentin H., ed. 1998. *Canadian Oxford World Atlas*, 4th edn. Toronto: Oxford University Press.

Statistics Canada. 1982. *Census Metropolitan Areas and Census Agglomerations with Components*. Catalogue no. 95–903. Ottawa: Minister of Supply and Services Canada.

———. 1994. *Agricultural Economic Statistics*. Catalogue no. 21–603E. Ottawa: Statistics Canada.

———. 1995. *Annual Demographic Statistics, 1994*. Catalogue no. 91–213. Ottawa: Statistics Canada.

———. 1996. *Labour Force Annual Averages 1995*. Catalogue no. 71–220–XPB. Ottawa: Statistics Canada.

———. 1997a. *A National Overview: Population and Dwelling Counts*. Catalogue no. 93–357–XPB. Ottawa: Industry Canada.

———. 1997b. 1996 Census: National Tables—Population by Mother Tongue, Showing Age Groups, for Canada, Provinces and Territories, 1996 Census—20% Sample Data, 2 Dec. At: <www.statcan.ca/english/census96/>.

———. 1997c. '1996 Census: Mother Tongue, Home Language and Knowledge of Languages', *The Daily*, 2 Dec. At: <www. statcan.ca/Daily/English/>.

———. 1998a. Canada Highlights: Census of Agriculture. At: <http://www.statcan.ca/english/censuag>.

———. 1998b. *Canadian Economic Observer*. Catalogue no. 11–010–XPB. Ottawa: Statistics Canada.

———. 1998c. '1996 Census: Aboriginal Data', *The Daily*, 13 Jan. At: <www. statcan.ca/Daily/English/>.

———. 1998d. '1996 Census: Ethnic Origin, Visible Minorities', *The Daily*, 17 Feb. At: <www.statcan.ca/Daily/English/>.

———. 1998e. Table 2: Revised Statistics of the Mineral Production of Canada, by Province, 1996. At: <www.nrcan.gc.ca/mms/efab/mmsd/production>.

———. 1998f. 1981–1996 Census: Labour Force Activity. At:<www.statcan.ca/english/census96/mar17/Labour/table6/>.

———. 2000a. Population. At: <www.statcan.ca/english/Pgdb/People/Population/demo 02.htm>.

———. 2000b. Labour Force, Employed and Unemployed, Numbers and Rates. At: <www.statcan.ca/english/Pgdb/People/Labour/labor07a.htm>.

———. 2000c. Gross Domestic Product. At: <www.statcan.ca/english/Pgdb/Economy/Economic/econ15.htm>.

———. 2000d. Preliminary Estimates of the Mineral Production of Canada, by Province, 1999. At: <nrcan.gc.ca/mms/efab/mmad/production/1999/99p.pdf>.

———. 2002a. 2001 Census: Population and Dwelling Counts, for Census Metropolitan Areas and Census Agglomerations, 2001 and 1996 Censuses, 16 July. At: <www.statcan.ca/english/IPS/Data/93F0050XCB2001013.htm>.

———. 2002b. Census of Canada 2001—Census Geography. Highlights and Analysis: Canada's 2001 Population. At: <www12.statcan.ca/English/census01/>.

———. 2003a Agriculture 2001 Census Farm Operations: Provincial/Regional Trends. At: <www.statcan.ca/english/agcensus2001/first/regions/contents.htm>.

———. 2003b Canadian Statistics—Export of Goods on a Balance-of-Payments Basis. At: <www.statcan.ca/english/Pgdb/gblec04.htm>.

———. 2003c. Canadian Statistics: Distribution of Employed People, by Industry, by Province. At: <www.statcan.ca/english/Pgdb/labor21b.htm>.

———. 2003d. *Aboriginal Peoples of Canada: A Demographic Profile*. <www12.statcan.ca/english/census01/products/analytic/companion/abor/contents.cfm>.

———. 2007a. Population and Dwelling Counts, for Census Metropolitan Areas and Census Agglomerations, 2006 and 2001 Censuses—100% Data. At: <www12.statcan.ca/english/census06/data/popdwell/Table.cfm?T=201&S=3&O=D&RPP=150>.

———. 2007b. Gross Domestic Product, Expenditure-based, by Province and Territory. At: <www40.statcan.ca/l01/cst01/econ15.htm>.

———. 2007c. Canadian Statistics: Distribution of Employed People, by Industry, by Province. At: <www40.statcan.ca/l01/cst01/labor21c.htm>.

Stavely, Michael. 1987. 'Newfoundland: Economy and Society at the Margin', in L.D. McCann, ed., *Heartland and Hinterland: A Geography of Canada*, 2nd edn. Scarborough, Ont.: Prentice-Hall, 247–85.

Wynn, Graeme. 1987. 'The Maritimes: The Geography of Fragmentation and Underdevelopment', in L.D. McCann, ed., *Heartland and Hinterland: A Geography of Canada*, 2nd edn. Scarborough, Ont.: Prentice-Hall, 175–245.

Further Reading

Coates, Ken S. 2000. *The Marshall Decision and Native Rights*. Montréal and Kingston: McGill-Queen's University Press.

On 7 September 1999 the Supreme Court of Canada ruled in a case involving Donald Marshall Jr (who 16 years earlier had gained national attention when he was exonerated from a murder conviction after spending more than a decade in prison) that the Mi'kmaq could earn a 'modest income' from the fishery in the Maritimes. In one swoop, Atlantic Canada woke up to the Aboriginal desire and right to participate in resource development. While the Supreme Court limited its decision to the fishery, further legal action could see Aboriginal peoples gaining a share of other resources. Such events are taking place across Canada, as witnessed by the James Bay and Northern Québec Agreement, a series of comprehensive land-claim agreements in the North, and the Nisga'a Agreement in British Columbia. The Mi'kmaq, like other Indian, Inuit, and Métis peoples, want a share of Canada's natural resources, a place in the global market economy, and respect for their culture.

Coates places the *Marshall* decision, which changes the relationship between the Mi'kmaq and other Maritimers, within the larger Canadian context where a search for a new relationship between Aboriginal Canadians and other Canadians is underway. The tone and message of Coates's book is aptly captured by his call 'to restructure the relationship between First Nations and Other Canadians'. In Chapter 1, 'Of Eels, Judges, and Lobsters: The Marshall Challenge and the Supreme Court Decision', students will find an easy-to-read account of the complexity and urgency of revisiting historic treaties (in this case, a 1760–1 treaty) to ensure that they 'work' in our contemporary world.

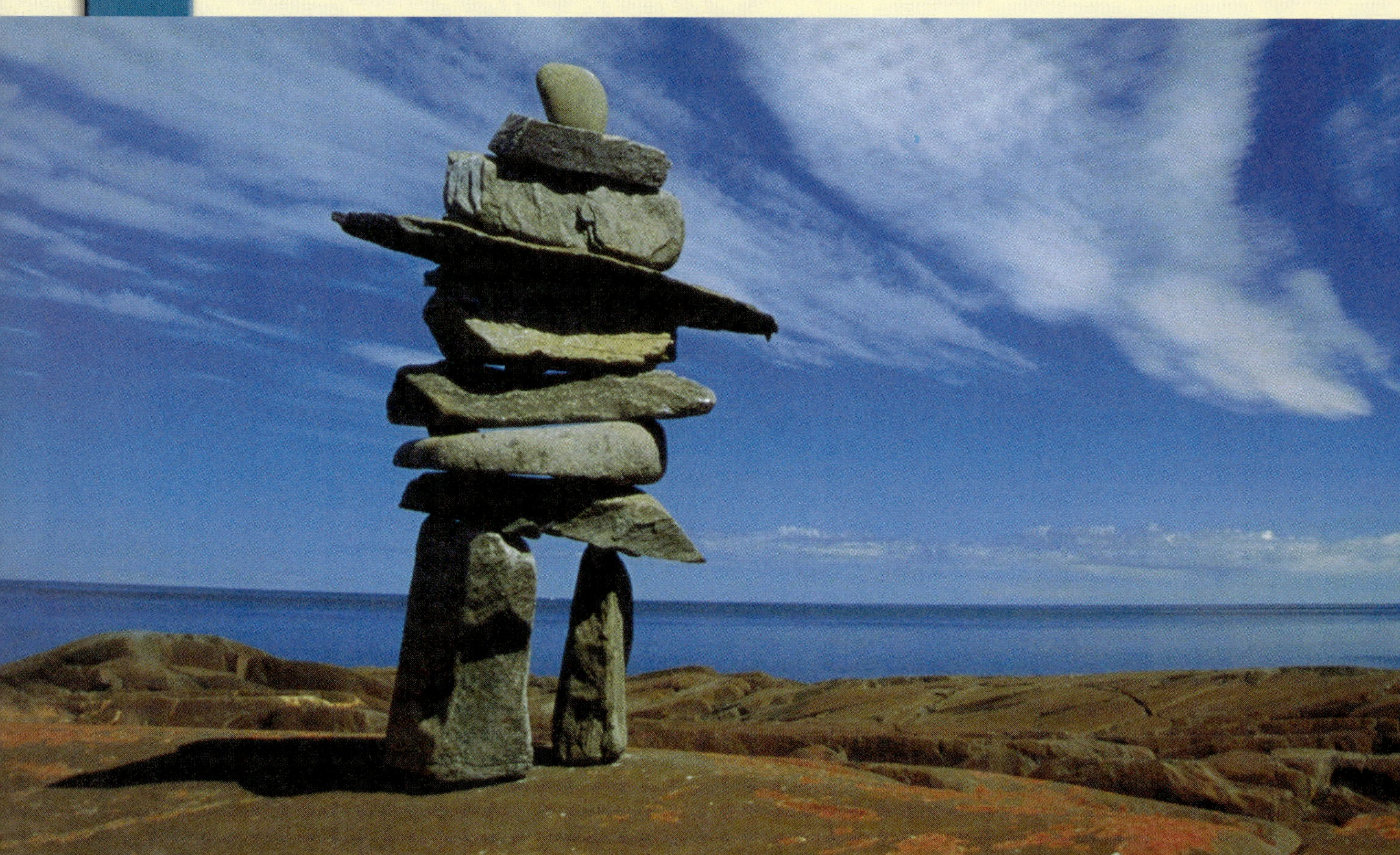

Cape Merry, Churchill, Manitoba
(Mike Grandmaison/Firstlight.ca)

Overview and Objectives

While the Territorial North is the largest geographic region in Canada, it has by far the smallest economy and population. Its cold climate and remote location greatly limit economic development and settlement. The region is a resource frontier for both Canada and the world, with energy and mineral products accounting for most of its output. While hunting remains an important source of food for Aboriginal peoples, trapping for commercial purposes has lost its former prominence. With the settlement of land claims and the formation of the new territory of Nunavut, more and more Aboriginal peoples are involved in governance and the market economy. Because Aboriginal peoples constitute half of the region's population, their participation in governance and economic development is crucial. Economic development in the Territorial North involves huge capital investments, so large-scale projects are common while social development calls for large transfer payments by Ottawa. The Key Topic in this chapter is megaprojects. The chapter will:

- Describe the Territorial North's physical and historical geography, including the birth of Nunavut and modern land claims.
- Present the dualistic nature of its population and economy.
- Examine the population and economy of the Territorial North's in the context of its physical setting and resource potential.
- Explore the region's changing economic and political position within Canada, North America, and the emerging global economy.
- Focus on the role of megaprojects in the North's development.

CHAPTER 10

The Territorial North

INTRODUCTION

Consisting of Yukon, the Northwest Territories, and Nunavut, the Territorial North is Canada's largest region (Figure 1.1). In Friedmann's regional scheme of the core/periphery model, the Territorial North would be described as a resource frontier (Table 1.3). Like other resource frontiers, the Territorial North has four principal characteristics:

- It is far from world markets.
- Resource development is dependent on external demand.
- Resource development is hampered by a cold environment.
- Its economy is sensitive to fluctuations in world prices for its resources.

The Territorial North is also a homeland for Aboriginal peoples. It is the only geographic region in Canada in which Aboriginal peoples constitute a majority of the total population (Figure 10.1). In 2001, just over half of the population of the Territorial North was classified as Aboriginal. Nunavut provides a more revealing example because 85 per cent of its population have declared themselves as Inuit. Nunavut, the most recently created territory, is a political expression of the homeland concept for the Inuit. (Nunavut is discussed later in this chapter.) The Territorial North, then, as both a frontier and a homeland, has a dual character. This duality is a theme that runs through this chapter. The Key Topic explored is megaprojects in the Territorial North.

The Territorial North within Canada

Of the six geographic regions, the Territorial North has the smallest population. In 2006, it had 101,000 residents spread over nearly 4 million km^2. With a population density of only 2.5 people per 100 km^2, this region is one of the world's most sparsely populated areas. The Territorial North has the smallest economic output of the six geographic regions in Canada (Figure 10.1).

The Territorial North's demographic features have been shaped by two factors. First, its small population is due to its small economy, which can support relatively few people. Second, the Aboriginal population—especially the Inuit—has a very high birth rate that accounts for the population growth in the Territorial North. Third, the non-Aboriginal population tends to move to job opportunities so that when the North's economy stalls, non-Aboriginal residents are more likely to move to southern Canada than Aboriginal residents.

As a **resource frontier**, the economy of the Territorial North is based on the exploitation of its energy and mineral resources. Such industrial activities greatly disturb the natural environment and wildlife. Exploitation of non-renewable resources has spurred economic growth in the region, but this type of economic development lacks stability because it is entirely dependent on exports to national and global markets. Variation in demand makes the Territorial North's economy subject to a boom-and-bust cycle. This cycle is caused by the finite nature of non-renewable resources, which results in mine closures, and by fluctuations in world prices. Downward price movements have caused temporary shutdowns of mining operations. Two examples of price-induced shutdowns involved the Faro lead-zinc mine in Yukon and the Giant gold mine in the Northwest Territories.

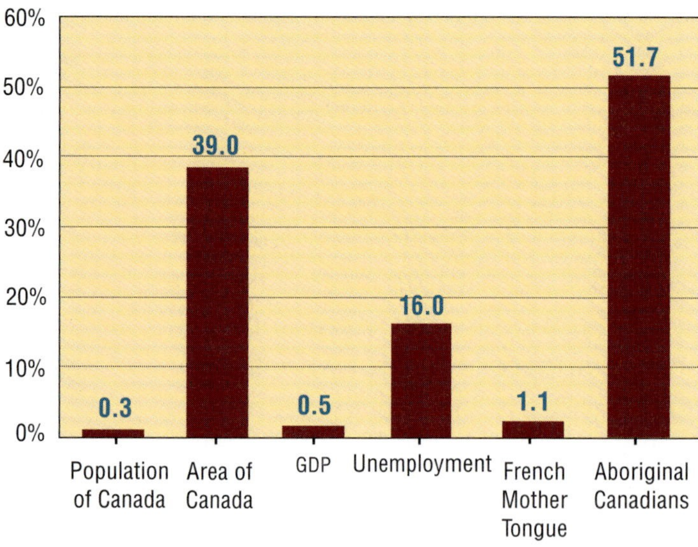

Figure 10.1 The Territorial North, 2006. Though the region is the largest in Canada, its population and economy are the smallest. For Aboriginal peoples and French-speaking Canadians residing in the North, the most recent census data are from 2001. At that time, over half of the population was Aboriginal and, by 2006, the figure is probably slightly higher. The Territorial North suffers from high **unemployment** as well as **underemployment**, i.e., when no jobs are available, individuals do not seek employment and hence are not classified as unemployed.

Note: The 2001 unemployment figure is based on territorial data. Aboriginal unemployment may be as high as 30 and, when combined with underemployment, perhaps as much as 60 per cent.

Sources: Tables 1.1 and 1.7.

Two Images

The Territorial North has two powerful and seemingly contradictory images—one is as a northern frontier, while the other is as a homeland. The traditional image of the northern frontier is one of great wealth just waiting to be discovered. For example, during the Klondike gold rush (1897–8), prospectors flooded the Yukon to pan for gold along the Klondike River. A more contemporary version of this image is one of large multinational corporations with their vast capital and advanced technology undertaking megaprojects—mining for gold, diamonds, lead, and zinc, and drilling for oil and gas. **Megaprojects** are large-scale resource developments financed and managed by multinational corporations designed to meet global needs for primary products. Such projects create an economic boom during the construction period, but in their operational phase fewer employment opportunities are available and economic spinoffs for local businesses are limited. Because of the risks associated with developing resources in a frontier—from overcoming physical barriers unique to the Territorial North to coping with downturns in world prices for resources—such projects are usually undertaken by large corporations. In return, these corporations reap large profits and supply the industrial cores of the world with raw materials and energy.

Northerners, however, particularly Aboriginal peoples, see the North as a **homeland**. This perception is based on a special, deep commitment to the North, which cultural geographers often attribute to a sense of place. Local people have a strong appreciation

Vignette 10.1 Northern Lights

Northern lights (aurora borealis) are dazzling displays of rapidly moving light that appear mostly as white-like flashes extending across the sky. Sometimes these flashes display various colours. Northern lights occur most frequently in higher latitudes. At Yellowknife (62° 30′ N), for example, northern lights occur, on average, nearly 250 nights a year; at Toronto (43° 42′ N), northern lights may occur 20 nights a year. Since Japanese culture associates the aurora borealis with good luck for newly married couples, many Japanese couples travel to Yellowknife to see the northern lights. But what are northern lights? They are the visible portions of the dissipation of solar energy carried to the earth's magnetosphere by solar winds. Solar winds are the stream of electrically charged particles emitted by the sun. When these particles near the earth, they are drawn to the earth's two magnetic poles where they form a magnetic field known as the magnetosphere. While the aurora borealis causes spectacular displays of light in northern skies, northern lights are also associated with geomagnetic disturbances that can disrupt communication and electrical systems.

The South Nahanni is one of the world's great wild rivers. Located in boreal wilderness of Nahanni National Park Reserve in the southwestern part of the Northwest Territories, this untamed river is seen surging through the steep-walled First Canyon. Later, its waters rush past hot springs, plunge over a waterfall twice the height of Niagara, and cut through canyons more than one kilometre deep. (Stephen J. Krasemann/Valan Photos)

Offshore drilling in the Beaufort Sea. Megaprojects are common on resource frontiers, especially in the North, where adverse conditions require that developers have the investment power and experience to manage risks. (CP/Bob Weber)

for natural features, cultural traits, and the political and economic issues affecting their homeland. A sense of place evokes a sense of belonging to a particular place. (discussed in Chapter 1).

The concepts of homeland and frontier are not always compatible. For instance, the interests of multinational companies relate to the profitability of the northern frontier's untapped natural wealth. The interests of northerners, however, especially Aboriginal northerners, are best served when the long-term environmental and social well-being of their communities is considered. 'Frontier' economic activities often bring jobs and investment, but the Aboriginal community has had minimal involvement, partly because of a mismatch between the needs of the companies and the skill sets of Aboriginal workers.

In the Territorial North, the concepts of homeland and regional consciousness have resulted in the devolution of political power from the federal government to the territorial governments. Elected governments exist in the three territories and six land-claim agreements with First Nations have been concluded, with the most recent being the Tlicho Final Agreement (2003). Significantly, the Tlicho Final Agreement included provisions for self-government for the Dogrib who live between Great Bear and Great Slave lakes in the Northwest Territories. Future comprehensive land-claim agreements are likely to include a section on self-government. Aboriginal self-government received an enormous boost in 1992 when the provision for the territory of Nunavut was placed in the Nunavut Land Claims Agreement under the Nunavut Political Accord. Unlike Yukon and the Northwest Territories, Nunavut is an expression of 'ethnic' regional consciousness and yet Nunavut is a public government and therefore is different from First Nations' 'ethnic' self-governments. The next challenge facing the Territorial North, but especially Nunavut, is to generate sufficient economic growth to break its economic dependency on Ottawa, while still preserving the homeland to which many northerners are committed.

The Territorial North's Physical Geography

The Territorial North includes the Yukon Territory, the Northwest Territories, and the new territory of Nunavut. The region extends over four of Canada's physiographic regions: the Canadian Shield, the Interior Plains, the Cordillera, and Arctic Lands (including the Arctic Archipelago) (Figure 2.1). As a result, the region encompasses a varied topography, from mountainous terrain and forest stands in the west, to barren plains, ice-covered islands, and Arctic seas further east and north. The vegetation in this region is also quite varied and includes a small portion of the boreal forest in the southwest, the tundra with its mosses and lichens further north, and a polar desert in the high reaches of the Territorial North (Figure 2.7). The region is also known for the several rivers that wind through it, the many lakes that dot its landscape, and the Arctic Ocean that supports a range of aquatic wildlife.

However, the physical geography of the Territorial North is governed not so much by physiography as by a cold environment. Cold persists throughout most of the year and, in many ways, it affects human activities in the Territorial North. The Territorial North's cold environment includes permafrost (Figure 2.9) and long winters with sub-zero temperatures. Except for the northern half of Yukon, the Territorial North was subjected to glaciation. For that reason the rich Klondike placer deposits were not affected by the scouring effects of the Cordillera ice sheets that affected southern Yukon. The region's main climate zones, the Arctic and the Subarctic (Figure 2.6), are characterized by very short summers.[1] In the case of the Arctic climate, summer is limited to a few warm days interspersed with more autumn-like weather, including freezing temperatures and snow flurries. The Subarctic climate has a longer summer that lasts at least one month. During the short but warm summer, the daily maximum temperature often exceeds 20° C and sometimes reaches 30° C.

Arctic air masses dominate the weather patterns in the Territorial North. They are characterized by dry, cold weather and originate over the ice-covered Arctic Ocean, moving southward in the winter. The Arctic zone has an extremely cold and dry climate. Distinguished by long winters and a brief summer, the Arctic climate is normally associated with high latitudes and lower levels of solar energy. The ice-covered Arctic Ocean and continuous permafrost keep summer temperatures cool even though the sun remains above the horizon for most of the summer. These cool summer temperatures, which Köppen defined as an average mean of less than 10° C in the warmest month, prevent normal tree growth. For that reason, the Arctic climate region has tundra vegetation, which includes lichens, mosses, grasses, and low shrubs. In the very cold Arctic Archipelago, the ground is often bare, exposing the surface material. As there is little precipitation in the Arctic Archipelago (often less than 20 cm per year), this area is sometimes described as a 'polar desert'.

Beyond 70° N, growing conditions become very limited. With lower temperatures and less precipitation than in the lower latitudes of this climatic zone, the tundra vegetation cannot sustain itself. In these high latitudes, much of the land is barren. The Arctic climate, however, does extend into lower latitudes in two areas: along the coasts of Hudson Bay and the Labrador Sea. In both cases, these cold bodies of water chill the summer air along the adjacent land mass. In this way, the Arctic climate extends along the coasts of Ontario, Québec, and Labrador well below 60° N, sometimes extending as far south as 55° N.

The Arctic Ocean was called the 'Frozen Sea' by early explorers. This extensive ice cover, known as the **Arctic ice pack** or polar pack ice, drifts in a clockwise motion in the Beaufort Sea. Because of the extent and thickness of the Arctic ice pack, few ships can navigate these waters without the assistance of icebreakers. Two former mining operations in the High Arctic—the Nanisivik mine on northern Baffin Island and the Polaris mine on Little Cornwallis Island—stored their ore until the summer navigation season, when ships reinforced against ice, sometimes aided by icebreakers, transported the ore to European and other world markets.

The geology of the Territorial North provides much of its wealth. For example, the vast sedimentary basins within the Interior Plains and Arctic Lands sometimes contain large deposits of oil and natural gas (Vignette 10.2). Similarly, the Cordillera and Canadian Shield have already yielded some of their mineral wealth to prospectors and geologists. These minerals include diamonds, gold, lead, uranium, and zinc.

Environmental Challenges

The Territorial North is faced with two major environmental challenges. First, global warming is expected to reach its maximum temperature increase in the Arctic because of the

Vignette 10.2 Petroleum Basins

The Territorial North has many sedimentary basins, some of which contain petroleum deposits. The three main petroleum-containing basins are the Western Sedimentary Basin, the Mackenzie Basin, and the Canadian Arctic Basin. The Canadian Arctic Basin contains several smaller basins, including the Sverdrup Basin. According to Indian and Northern Affairs Canada, oil reserves in the Territorial North total 1,663 million barrels and gas reserves amount to 30.9 trillion cubic feet. In the same report, INAC states that 'The Northwest Territories has an ultimate potential for crude oil estimated at 0.9 x 109 m^3 (5.7 billion barrels) and Nunavut has a potential of 0.43 x 109 m^3 (2.7 billion barrels).' The market value of discovered oil and gas at current prices (2006) is approximately $263 billion.

Discovered Resource Inventory

Region	Crude Oil 106 m^3	Million Barrels	Natural Gas 109 m^3	Trillion Cubic Feet
Northwest Territories	70.5	443	178.3	6.3
Nunavut	0.9	6	190.7	6.7
Arctic Offshore	193.0	1,214	506.5	17.9
Total	264.4	1,663	875.5	30.9

Note: Compiled from several published sources, which may underestimate or overestimate actual field resources.
Source: Indian and Northern Affairs Canada (2007a). Reproduced with the permission of the Minister of Public Works and Government Services Canada, 2007.

Total reserves in the Mackenzie Delta and the Beaufort Sea are estimated at 12 trillion cubic feet of natural gas and 1.7 billion barrels of oil.

albedo effect, which involves greater solar warming of the land and water because of the reduction of ice and snow cover. Second, since the economy in the region is relying more and more on large-scale industrial projects, the environment will necessarily be subjected to an ever-increasing negative impact from these industrial projects, and the fragile environment of the North, because of the cold temperatures, has great difficulty recovering from industrial degradation.

As discussed in Chapter 2, global warming is the increase over time of the earth's average surface temperature. Now our attention is focused on why most temperature increase is expected to take place in the Arctic. Several factors are involved, such as heat transfer from lower latitudes to higher ones, but the primary factor is the albedo effect, i.e., a shift from a high albedo in the Arctic to a low one. Albedo is a technical term that refers to the ratio of light from the sun that is reflected by the earth's surface to the light absorbed by the earth's surface. Light from the sun takes the form of short-wave radiation while energy emitted from the earth's surface takes the form of long-wave radiation. Long-wave radiation is more readily absorbed by the atmosphere and thus warms the atmosphere while short-wave radiation escapes into outer space without warming the atmosphere. Currently, the Arctic has a high albedo because of its cover of snow and ice, which means most solar energy is reflected back into outer space without warming the atmosphere. However, as ice and snow cover decreases in the Arctic, its albedo will shift from high to low, meaning that the solar energy reaching the Arctic will be more effective in warming its atmosphere and then raising temperatures. For this reason, temperatures are expected to increase more in the Arctic than in other regions in the world.

The impact of global warming in the Arctic is expected to have both positive and negative impacts on the wildlife and northern peoples. Global warming may melt the sea ice covering the Arctic Ocean and Hudson Bay, thus allowing for ocean transportation and the reduction of transportation costs in the Arctic. Wildlife may be affected. Already scientists have noted negative impacts on whale and polar bear populations. Another example is the huge migrating caribou herds that have their calving grounds in the Arctic. The reduction of the size of calving grounds may have a negative impact on the natural process of reproduction of caribou. The Dene and Inuit communities that rely on these herds for much of their country food may have to purchase more of their food from local stores. The cultural and nutrition implications are large. Caribou meat forms an important part of the culture of Dene and Inuit while the superior nutritional value of caribou meat compared to store-bought meat is well recognized.

Given that the North has a fragile environment, the growing presence of resource industries presents a challenge because of long-lasting environmental damage caused by such industries. Then, too, the northern lands, largely because of the cold environment, which limits the biological capacity of the environment to recover from the effects of industrial impacts, are more vulnerable to industrial impacts than temperate and tropical lands. Past exploitation, prior to environmental regulations, took a heavy toll on the landscape and such landscapes are still scarred. The Yukon gold rush provides such an example. Forests were ravaged for fuel while the river landscape was mutilated by hydraulic

The Yukon River Valley near Dawson City. From Aboriginal peoples to fur traders and prospectors, this long and winding river has been an important transportation route in the history of the Territorial North. (Bill Terry/ Take Stock Inc.)

mining that was employed as placer gold mining declined. To extract the gold, water and steam were forced by high-pressure hoses into the frozen sands and gravels of the valley **terraces** on the tributaries of the Yukon River. Gold was recovered after thawing and sorting these deposits, but in the process, trees were cut down for firewood and building materials, riverbanks were destabilized, and tailings (the huge piles of discarded sand and gravel) were scattered across the landscape. Environmentally, gold mining had an enormously destructive impact on the Yukon landscape and those scars are still visible today.

Gold mining around Yellowknife provides another example. The local refining of gold at the Giant gold mine at Yellowknife involved the use of arsenic. Now closed, this mine has left a toxic time bomb of 270,000 tonnes of arsenic trioxide, a by-product of gold refining. Since the last mining company (Royal Oak Mines) declared bankruptcy, the cost of the cleanup, estimated at a quarter of a billion dollars, is left to the federal government (Bone, 2003a: 160).

Historical Geography of the Territorial North

European Contact

At the time of contact with Europeans, seven Inuit groups and seven Indian groups belonging to the Athapaskan language family (also known as Dene) occupied the Territorial North. The Inuit stretched across the Arctic: the Mackenzie Delta Inuit lived in the west; further east were the Copper Inuit, Netsilik Inuit, Iglulik Inuit, Baffinland Inuit, Caribou Inuit, and Sadlermiut Inuit. Inuit also lived in northern Québec and Labrador. By the early twentieth century, two groups (most of the Mackenzie Delta Inuit and all of the Sadlermiut Inuit) would succumb to diseases that European whalers brought to the Arctic. The Indian tribes that resided in the Subarctic were the Kutchin, Hare, Tutchone, Dogrib, Tahltan, Slavey, and Chipewyan.

These Inuit and Indians had developed hunting techniques that were well adapted to these two cold but different environments. The Inuit employed the kayak and harpoon to hunt seals, whales, and other marine mammals, which enabled them to occupy the Arctic coast from Yukon to Labrador. As a result, the Inuit depended extensively on marine mammals and fish. The Indians hunted and fished in the northern coniferous forest, where the birchbark canoe, the bow and arrow, and snowshoes enabled them to hunt in summer and winter. The Dene tribes relied heavily on big game like the caribou, the Chipewyans often following the caribou to their calving grounds in the northern barrens of the Arctic. Both the Inuit and the Dene moved across the land in a seasonal rhythm, following the migratory patterns of animals. Operating in small and highly mobile groups, these hunting societies depended on game for their survival. Cultural traits, such as the ethic of sharing, developed from this dependency on the land and sea for food.[2]

The Time of European Exploration

Though the Vikings were the first to make contact with northern Aboriginal peoples around 1000, little is known of those encounters.[3] About five centuries later, in 1576, Martin Frobisher, in searching for a Northwest Passage to the Far East, sailed to the northern Arctic. He reached Baffin Island where he encountered a group of Inuit, some on land, others in their kayaks. A skirmish between Frobisher's men and the Inuit took place in which Frobisher was wounded by an arrow. In the exchange, five of his men were lost and three of the Inuit were captured. The Inuit and one kayak were taken back to England, as often was done in the early years of European exploration, as proof of Frobisher's discovery. All three of the captives soon succumbed to illness. Following Frobisher, various European explorers ventured into Arctic waters in search of the Northwest Passage, including John Franklin, whose famous last expedition ended in disaster

(Vignette 10.3). However, cultural exchange between Europeans and the original inhabitants of these lands remained limited until the nineteenth century, when the trade in fur pelts and whaling peaked in North America.

Whaling and the Fur Trade

Whaling, which was the first commercial venture in the Arctic, began in the late sixteenth century in the waters off Baffin Island. During those early years of whaling, whalers had little opportunity or desire to make contact with the Inuit living along the Arctic coast. The Inuit probably felt the same, particularly those who had heard stories of the nasty encounter with Frobisher's men. During early summer, whaling ships set sail from British, Dutch, and German ports to Baffin Bay, where they hunted whales for several months. By September, all ships would return home. In the early nineteenth century, the expeditions of John Ross (1817) and William Parry (1819) sailed farther north and west into Lancaster Sound. Their search for the Northwest Passage had limited success but opened virgin whaling grounds for whalers. These new grounds were of great interest as improved whaling technology had reduced the whale population in the eastern Arctic. In fact, the period from 1820 to 1840 is regarded as the peak of whaling activity in this area. At that time, up to 100 vessels were whaling in Davis Strait and Baffin Bay.

As whaling ships went further to find better whaling grounds, it became impossible to return to their home ports within one season. By the 1850s, the practice of 'wintering over' (that is, allowing ships to freeze in sea ice along the coast) was adopted by English, Scottish, and American whalers. This allowed whalers to get an early start in the spring, providing for a long whaling season before the return trip home at the onset of the next winter. Wintering over took place along the indented coastlines of Baffin Island, Hudson Bay, and the northern shores of Québec and Yukon. Permanent shore stations were established at Kekerton and Blacklead Island in Cumberland Sound, at

Vignette 10.3 The Northwest Passage and the Franklin Search

In 1845, Sir John Franklin headed a British naval expedition to search for the elusive Northwest Passage through the Arctic waters of North America. This British naval expedition took place at the end of the Little Ice Age, meaning that ice conditions would have been much more challenging than those occurring today. He and his crew never returned. Their disappearance in the Canadian Arctic set off one of the world's greatest rescue operations, which was conducted both on land and by sea and stretched over a decade. The British Admiralty organized the first search party in 1848. Lady Franklin sent the last expedition to look for her husband in 1857. These expeditions accomplished three things: (1) they found evidence confirming the loss of Franklin's ships (the *Erebus* and *Terror*) and the death of his crew; (2) one rescue ship under the command of Robert McClure almost completed the Northwest Passage; and (3) the massive rescue effort resulted in a greater knowledge of the numerous islands and various routes to the north and west of Baffin Island in the Arctic Ocean. The exact sequence of events that led to the Franklin disaster is not known. However, archaeological work, conducted in the early 1980s on the remains of members of the expedition, revealed that lead poisoning, caused by the tin cans in the ships' food supplies, probably contributed to the tragic demise of the Franklin expedition.

Cape Fullerton in Hudson Bay, and at Herschel Island in the Beaufort Sea. Life aboard whaling ships was dirty, rough, and dangerous, and many sailors died when their ships were caught in the ice and crushed.

The Inuit welcomed the whaling ships because of the opportunity for trade. The Inuit were attracted to shore stations and often worked for the whalers by securing game, sewing clothes, and piloting the whaling ships through difficult waters to promising sites for whale hunting. Some Inuit men signed on as boat crew and harpooners. In exchange for this work, the Inuit obtained useful goods, including knives, needles, and rifles, which made domestic life and hunting easier. While this relationship brought many advantages for the Inuit, there were also negative social and health aspects, including the rise in alcoholism and the spread of European diseases among the Inuit (Vignette 10.4). Perhaps the most devastating result of this trade relationship for the Inuit was the unexpected end of commercial whaling, which, for the Inuit, represented the loss of access to highly valued trade goods. Just as the twentieth century began, demand for products made from whales decreased sharply, halting the flow of whalers, and thus trade goods, that were sailing into Arctic waters. By now, the Inuit were dependent on trade goods for their hunting activities. Somehow, they had to find another means of obtaining these useful goods.

Fortunately for the Inuit, the fur trade had been expanding northward into the Arctic, thereby providing a replacement for whaling. The fur trade had already been successfully operating in the Subarctic for some time—a relationship between European traders and the Indian tribes in this part of the Territorial North was established through the trade of fur pelts, especially beaver. However, by the beginning of the twentieth century, the fashion world in Europe had discovered the attractive features of the Arctic fox pelt. Demand for Arctic fox pelts rose, which led the Hudson's Bay Company to establish trading posts in the Arctic. Soon the Inuit were deeply involved in the fur trade. The working relationship between the Hudson's Bay Company and the Inuit was based on barter trade: white fox pelts could be traded for goods.

Until the 1950s, the fur trade dominated the Aboriginal land-based economy. It lasted

Vignette 10.4 European Diseases

Whalers, fur traders, and missionaries introduced new diseases to the Arctic. As the Inuit had little immunity to measles, smallpox, and other communicable diseases such as tuberculosis, many of them died. In the late nineteenth century, the Sadlermiut and the Mackenzie Delta Inuit were exposed to these diseases. As a result, all the Sadlermiut died. The Mackenzie Delta Inuit, whose numbers were as high as 2,000, almost suffered the same fate but managed to survive.

The Mackenzie Delta Inuit occupied the northwestern Arctic coast, in present-day Yukon, the Northwest Territories, and part of Alaska. Herschel Island, lying just off the Yukon coast, was an important wintering station for American whaling ships. Whalers often traded their manufactured goods with the local Mackenzie Delta Inuit, who became involved with the commercial whaling operations. Through contact with the whalers, the Inuit were infected by European diseases. By 1910, only about 100 Mackenzie Delta Inuit were left. Gradually, Inupiat Inuit from nearby Alaska and white trappers who settled in the Mackenzie Delta area intermarried with the local Mackenzie Delta Inuit, which secured the survival of these people. Today, their descendants are called Inuvialuit.

for less than 100 years in the Arctic and for over three centuries in the Subarctic. Did the fur trade, as well as Arctic whaling, create a form of dependency whereby Indians and Inuit could not survive without trade goods? The answer is a qualified yes. At first, Aboriginal peoples had a form of partnership with European traders and whalers. Each side had power—for instance, the European traders needed the Indians to trap beaver and the Indians needed the traders to obtain European goods and technology. Gradually, however, the power relationship shifted in favour of the European traders. By the nineteenth century, the fur companies controlled the fur economy. Fur-trading posts dotted the northern landscape. Indians, who had long ago integrated trade goods into their traditional way of life—including their hunting techniques and their migration patterns—were therefore heavily dependent on trade. In fact, when game was scarce, tribes relied on the fur trader for food. Ironically, by securing game for the traders, Indians reduced the number of animals that would be available for their own sustenance. In the Territorial North, game became scarce around fur-trading posts not from natural causes but from overexploitation.

The problems of a growing dependency on European goods and a changing way of life for northern Aboriginal peoples were compounded after the arrival of Anglican and Catholic missionaries in the 1860s and the North West Mounted Police (NWMP) in the 1890s. The Indians and Métis were confronted with the full force of Western culture in the late nineteenth century and the Inuit in the early twentieth century. The Western ideas and rules introduced by the missionaries and police who now lived near the trading posts had a profound impact on Aboriginal culture. The NWMP (which added 'Royal' to its name in 1904 and, in 1920, was renamed the Royal Canadian Mounted Police) imposed Canada's system of law and order on Aboriginal peoples, while the missionaries challenged their spiritual values and encouraged the Inuit, Indians, and Métis to remain in the settlements. Also, on behalf of the Canadian government, both Anglican and Catholic missionaries placed young Aboriginal children in church-run residential schools, where they were taught in either English or French. In this attempted assimilation, most children learned to read and write in English or French, but they were inadequately prepared for northern life. As they lost the opportunity to learn from their parents about how to live on the land, they became trapped between the two very different worlds of their Aboriginal communities and the Euro-Canadians. Under these circumstances, many lost their indigenous language, animistic beliefs, and cultural customs. Fur traders opposed many of these induced Western cultural adaptations because they needed the Aboriginal peoples on the land to trap. Nevertheless, the influence of the churches, the power of the state, and the number of non-Aboriginal residents in the North increased in the twentieth century, placing Aboriginal cultures under siege and crippling their land-based economy. However, political and social changes were occurring at this time that would first lead to territorial governments, then to relocation of Aboriginal peoples to settlements, and finally to land-claim agreements and self-government.

Political and Territorial Evolution

When Canada was formed in 1867, the Territorial North remained a British possession. Britain had claimed British North America on the basis of settlement, trade, and exploration. In the Territorial North, the British declared their ownership of the Arctic islands, basing their claim on the British Navy's efforts to find the Northwest Passage, including the search for the missing Franklin expedition. In the rest of the Territorial North, the Hudson's Bay Company had established a number of fur-trading posts in the forested lands of the Mackenzie Basin and the Yukon. By extending its fur-trading economy over this area, the British government claimed these Subarctic lands. Canada came into possession of the Territorial North with the purchase of

Rupert's Land in 1869 and its transfer to Canada in 1870, and in 1880 Britain transferred the Arctic islands to Canada.

At first, these northern territories were governed from Ottawa. The first territorial government was established in Yukon. The demand for self-government came not from the Aboriginal population but from those who had been lured north by the Klondike gold rush. In 1898, Yukon became a separate territory—a territorial government was formed that consisted of a federally appointed commissioner and council located in Dawson City.[4] By 2006, Yukon's government consisted of 18 elected members.

The formation of the government of the Northwest Territories was much different. From 1905 until after World War II, the NWT was governed by an appointed commissioner and council, which was composed entirely of senior civil servants based in Ottawa. When its population reached 16,004 in 1951, elected members were gradually added to the previously appointed council until it became a fully elected body in 1975. Until 1963, the commissioner was a deputy minister in the federal department in charge of the administration of the Northwest Territories. In 1967, the seat of territorial government was moved to Yellowknife and the commissioner was relocated there with the nucleus of what has since become a territorial public service. By 2005, the government of the Northwest Territories consisted of 19 elected members while the territories' population had reached 43,000.

Nunavut, formed in 1999, was carved out of the Northwest Territories. The territories, including Nunavut (discussed later in the chapter), include most islands in Hudson Bay. The 2007 Nunavik Offshore Claim Agreement assigned the coastal islands adjacent to arctic Québec to Nunavik (this new boundary is shown in Figure 10.4). Nunavut does not have all the powers given to the provinces through the Canadian Constitution. In fact, territorial powers do not stem from the Constitution but are assigned to the territories by the federal government. Territorial powers include education, social services, tax collection, highways, and community services. Given the importance of wildlife to Aboriginal peoples, the territorial governments are also responsible for wildlife. But the territories do not control tax revenues from their natural resources, which go to Ottawa. To offset this tax loss, the territories receive most of their revenue as transfer payments from the federal government. Without these transfer payments, the territorial governments would not be able to afford their existing programs.

With the closing of its two lead/zinc mines, Polaris and Nanisivik, Nunavut had almost no mining industry in 2005 (Table 10.6). Fortunately, the Jericho diamond mine opened late in 2006. Still, Nunavut faces a dilemma—it has the smallest mining industry in the Territorial North but the fastest-growing population (Table 10.3). These two factors mean that Nunavut faces a challenging economic future.

Forgotten Frontier: Confederation to 1939

From the time the Territorial North in its entirety had been transferred to Canada in 1880 until World War II, the region was a forgotten part of Canada. The Territorial North was forgotten because it had little agricultural land and few commercial resources. For the federal government, the Territorial North was not a 'priority' region and thus received little attention from Ottawa. With the exception of Yukon, the economy was dominated by hunting and trapping. Ottawa had adopted a laissez-faire policy towards the Territorial North to minimize federal expenditures, leaving the fur traders and missionaries to deal with the needs of a hunting society.

Strategic Frontier

With the outbreak of World War II, the Territorial North became a strategic frontier. Military investments and activities took the form of military bases, highways, landing fields, and radar stations. While the nature of its strategic role changed over time, the Territorial North served as a buffer zone between North America and the Soviet Union for over 50 years. This role ceased with the collapse of the Soviet Union and the end of the Cold War in 1991.

For the Americans, Canada's North provided a secure transportation link to the European theatre of war and, in 1942, to Alaska, where the threat of Japanese attack was real. The air routes consisted of the Northwest Staging Route and Project Crimson. Each consisted of a series of northern landing strips that would enable American and Canadian warplanes to refuel and then continue their journey to either Europe or Alaska. In the Northeast, Project Crimson involved constructing landing fields at strategic intervals to allow Canadian and American airplanes to fly from Montréal to Frobisher Bay (now Iqaluit) and then to Greenland, Iceland, and, finally, England. In Canada's Northwest, American aircraft came to Edmonton and then flew along the Northwest Staging Route to Fairbanks, Alaska, where their major military base was located. The Alaska Highway, which was built at the same time, provided road access to the various landing fields and to Alaska. The US Army command had decided that the oil needed by the American armed forces in Alaska must be made secure by increasing the oil production at Norman Wells in the NWT and sending the oil by pipeline across several mountain ranges to Whitehorse and then northward to the military facilities at Fairbanks. Known as the Canol Project, the oil pipeline was completed in 1944, but with the disappearance of the Japanese threat it was closed within a year.

After World War II, the geopolitical importance of northern Canada changed. The North's new strategic role was to warn of a surprise Soviet air attack. The defence against such an attack was a series of radar stations that would detect Soviet bombers and allow sufficient response time for American fighter planes and (later) American missiles to destroy the Soviet bombers. In the 1950s, 22 radar stations, called the Distant Early Warning line, were constructed in the Territorial North along 70° N. Before the end of the Cold War, these radar stations were abandoned and replaced with more sophisticated methods of detecting incoming Soviet planes or missiles. With the collapse of the Soviet Union, Ottawa withdrew its military establishment at Inuvik, did not proceed with its plans for a military base at Nanisivik, and downsized its operation at Alert.

American military investment did expand the Norman Wells oil fields, along with a pipeline to Whitehorse, but most of their investment went into improving the transportation system, such as the construction of the Alaska Highway. Private resource development moved along much more slowly as it responded to world demand.

Arctic Sovereignty

With the exception of Hans Island, the international community recognizes Canada's ownership of the islands in the Arctic Ocean. However, the ownership of Arctic waters between the islands in Canada's archipelago remains unclear. The United States considers the Northwest Passage as an international sea route. While this route is not used for commercial purposes because of difficulty of navigating through ice, a number of 'unknown' countries have sent their nuclear submarines under the ice cover, and the Americans have made several trips through the Northwest Passage on the surface. In 1969, an American tanker, the S.S. *Manhattan*, made a voyage through the Northwest Passage without asking Canada's permission. It was an attempt to prove the passage was a viable route for shipping oil from Prudhoe Bay oil fields. Ottawa did provide a Canadian icebreaker to escort the *Manhattan*. In 1970, the *Manhattan* made another trip through the passage. In 1985, the US Coast Guard icebreaker *Polar Sea* transited the passage—once again, without asking the Canadian government for permission. The political fallout over what was considered the most direct challenge to Canada's sovereignty in the Arctic led to the signing of the Arctic Co-operation Agreement in 1988 by Prime Minister Brian Mulroney and US President Ronald Reagan. The document states that the

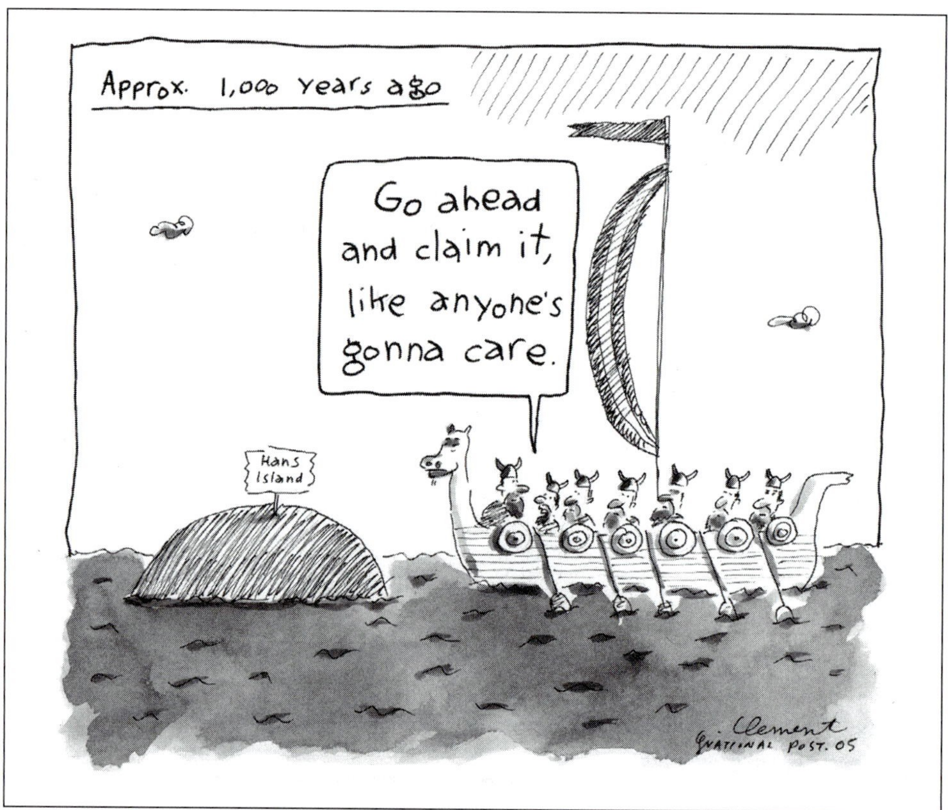

Figure 10.2 Whose island? Hans Island is a small, barren rocky knoll located along the border between Canada and Greenland (Denmark). While Hans Island has no known value and is uninhabited, its importance lies in two areas. First, the underlying continental shelf may contain oil and gas deposits. Second, its location in the centre of Kennedy Channel of Nares Strait marks the beginning of the Northwest Passage and therefore the ownership of this island may have implications for the Northwest Passage and offshore petroleum deposits. (Gary Clement, *National Post*, 28 July 2005 © Gary Clement)

US is to refrain from sending icebreakers through the Northwest Passage without Canada's consent; in turn, Canada will always give consent. However, the issue of whether the waters are international or internal was again left unresolved.

Canada has sought to legalize its sovereignty over the Arctic. In 1907, Canada first announced the 'Sector Principle', which divided the Arctic Ocean among those countries with territory adjacent to the Arctic Ocean. Accordingly, Canada could claim title over a wedge of the Arctic Ocean north of its territory between 60° W and 141° W longitudes and extending to the North Pole. While the Soviet Union supported this principle, the United States and other countries do not.

More recently, Canada has looked to environmental legislation as a means of exercising its sovereignty over Arctic waters. In the age of supertankers and container ships, the threat of toxic spills is more likely than ever before. Specifically, Ottawa has two concerns. First, how would Canada ensure its sovereignty over Arctic waters in order to manage such ocean traffic? Second, how would Canada protect its waters and adjacent islands from toxic wastes discharged from foreign ships passing through the Northwest Passage. In 1985, Ottawa's response took the form of the Arctic Waters Pollution Prevention Act, which gives Canada the right to control navigation in its sector of the Arctic Ocean and to manage its ocean environment.

In 2003, Canada ratified the UN Convention on the Law of the Sea, which specifies that coastal countries have the right to control access to their coasts. This access zone is 12 nautical miles (22.2 kilometres) wide. While the Convention on the Law of the Sea supports Canada's claim to Arctic waters, the convention leaves some grey areas, including the fact that some of the islands in Canada's archipelago are separated by more than 90 kilometres of Arctic waters. For that reason, the greatest threat to Canada's sovereignty claims remains the practice of foreign countries passing through the Northwest Passage without obtaining permission from the federal government.

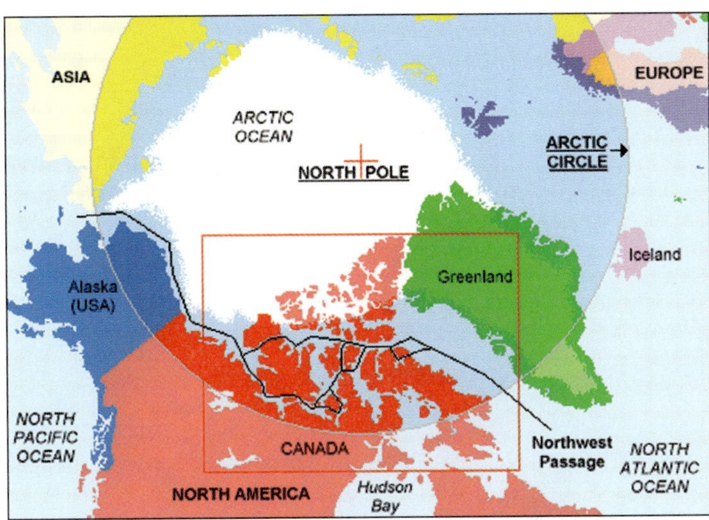

Figure 10.3 The Circumpolar North and Northwest Passage
Source: Athropolis Productions Limited, at: <www.athropolis.com/map9.htm>.

Early Resource Development

Other than a few gold mines, the Territorial North remained a wilderness area where Aboriginal peoples continued to live on the land. After World War II, the US demand for raw materials and energy exceeded its domestic production. American companies began to look to Canada for additional supplies to fill the demand. By the 1960s, multinational corporations had turned their attention to the region's mineral wealth, spending millions of dollars exploring to find profitable mineral deposits. By the early 1970s, three types of major resource development were underway: (1) major oil companies, such as Gulf Canada and Dome Petroleum, were drilling for oil and gas in the Mackenzie Delta and the Beaufort

Vignette 10.5 The Arctic Ocean and the Beaufort Sea

The Beaufort Sea is part of the Arctic Ocean. Named after the British hydrographer, Sir Francis Beaufort, this 'frozen' sea occupies an area of approximately 450,000 km^2. Most of its surface is permanently covered by floating sea ice known as the Arctic pack ice. One-year ice is about 1 m thick while older ice can reach 5 m thick. Only large icebreakers can penetrate this formidable ice barrier. In August 1994, two icebreakers, the American *Polar Sea* and the Canadian *Louis S. St Laurent*, ploughed through this ice on a scientific voyage to the North Pole. During the short summer, the shore ice melts, leaving a narrow stretch of open water between the shore and the polar pack ice. Small ships take advantage of this open water to bring supplies to communities located along the coast of the Beaufort Sea. Most of these supplies are transported by barge northward along the Mackenzie River.

The shallow waters of the Beaufort Sea are often less than 50 m, so offshore drilling for oil and gas is possible. Sparked by high oil prices in the 1970s, oil companies drilled hundreds of wells in the Beaufort Sea over the subsequent 20 years. Significant deposits of oil and gas were discovered, but the high cost of production and transportation to southern markets has prevented further offshore oil and gas development.

Sea; (2) Cominco, a large mining corporation, began extracting lead and zinc from a mine at Pine Point in the Northwest Territories (by 1988 this mine closed); (3) Cyprus Anvil, now a defunct mining company, began mining lead and zinc at Faro, Yukon. Production at Faro was halted several times because of low prices. The mine was permanently closed in 1993.

From the Land to Native Settlements

Until the 1950s, Aboriginal peoples continued to live a semi-nomadic life as hunters, fishers, and trappers. They would visit fur-trading posts several times a year to trade furs for supplies and to socialize, and some Aboriginal people, the so-called 'Homeguards', resided year-round near the trading posts from early in the history of the Subarctic and Arctic posts, supplying the European traders with country food. By early in the twentieth century, many trading posts consisted of three organizations: the fur trade company, the RCMP, and the church(es). The relocation from the land to the trading posts marks the beginning of a new way of life. Government facilities and services soon came to these posts, transforming them into Native settlements.

Relocation is a controversial subject in northern history. Was it a badly planned attempt at social engineering? Was it a voluntary or forced relocation? Federal officials saw relocation from two perspectives. First, relocation was seen as a necessary step in protecting northern peoples from the hardships of living on the land. Second, concentrating Aboriginal peoples in settlements allowed Ottawa to provide a variety of services, including schooling for their children. These two perspectives were part of the 'modernization' of northern Aboriginal peoples. Aboriginal peoples, while satisfied with their hunting lifestyle, did desire greater access to the trading store for food and supplies as well as to the nursing station for health care.

Yet, how serious was the hardship of living on the land? First, the search for game (food) might not always be successful and so hardship and hunger were part of their ancient culture. For most of the time, however, Aboriginal peoples found their hunting, fishing, and trapping lifestyle most satisfying. However, when game was scarce, hardships and hunger were an unpleasant reality. Some areas were more risky for hunting peoples. The **Barren Grounds** of the central Arctic were a particularly challenging place to live off the land because of the heavy dependence for sustenance on the migrating caribou herds. Failure to find the caribou translated into hunger and even starvation. Reports of deprivation and even cases of death by starvation among the Caribou Inuit had reached Ottawa before, but no action was taken. In the early 1950s, the Canadian media reported that about 60 Caribou Inuit starved to death. How could people living in a modern country like Canada starve to death? This sad event pushed the government into action, leading to a relocation policy. By 1958, Ottawa made the decision to relocate the Caribou Inuit to settlements, such as Baker Lake and Eskimo Point, but by then starvation had taken its toll—the Caribou Inuit population had dropped from 'about one hundred and twenty in 1950 to about sixty in 1959' (Williamson, 1974: 90). At the same time, Ottawa extended this relocation program to coastal Inuit and to Indians and Métis in the Subarctic who also lived off the land. However, Ottawa was unprepared for the economic, psychological, and social consequences of settlement life for hunting peoples.

Life in settlements had a number of social consequences. For example, access to food and medical services helped reduce the infant mortality rate, which caused the birth rate to rise. Since the 1940s, the Aboriginal birth rate has remained well above the national average, often more than twice as high. As a consequence, most northern settlements have tripled their population since 1951.[5] Even so, most are small by southern standards (under 1,000 people). For example, the largest centre in Nunavut is Iqaluit, with 6,184 residents in 2006, an 18 per cent increase over 2001.

Life in Native settlements is strongly influenced by Western ideas and values. Residents

At 62° N, the Inuit community of Salliut lies within the treeless Arctic. Just offshore in the waters of Hudson Strait, a Canadian Coast Guard vessel brings supplies to this isolated settlement. Modern housing is one benefit coming from the James Bay and Northern Québec Agreement. (© Isabelle Dubois/ArcticNet)

of Native settlements have had to accept many of the cultural values and ways of the dominant society. Aboriginal children have learned English through the school system. The old way of life has become more costly and less attractive to young people. For example, snowmobiles rather than a dog team supply the transportation to the hunting and trapping grounds.[6] In Native settlements, income from trapping plays a very minor role while the two principal sources of income are wages and various forms of government payments, including social assistance. The major employer is the government. As well, food security, or hunger, among lower-income households remains a problem due to limited access to country food and insufficient income to purchase store food. Another alarming trend reported by Chan (2006) is the increase in the consumption of store food rich in carbohydrates, particularly by younger generations, which has serious long-term health implications, such as obesity and diabetes.

The Territorial North Today

The Territorial North consists of three territorial governments: Yukon, the Northwest Territories, and Nunavut. The capital cities of these three territories are Whitehorse, Yellowknife, and Iqaluit. Territorial governments have fewer powers than provincial governments and, in this sense, they are political hinterlands. For example, the federal government retains power over natural resources and collects a substantial amount of tax revenue from companies using natural resources, so this important source of revenue is not available to territorial governments. Instead, they depend on Ottawa for transfer payments. Without these transfer payments, the territorial governments could not offer the basic services available in southern Canada. For example, Nunavut receives 90 per cent of its revenue from Ottawa. The level of fiscal dependency is somewhat lower for Yukon and the Northwest Territories. The Northwest Territories is challenging Ottawa over the issue of resource royalties. With more diamond and natural gas projects on the drawing board, the Northwest Territories is demanding a share of resource revenues.

The Territorial North is changing, especially politically. One recent political change was the division of the Northwest Territories to create a new territory, Nunavut. The second political change is being triggered by the resolution of outstanding land claims between Aboriginal peoples and Ottawa through comprehensive land-claim agreements that transfer land, cash, and administrative powers to northern Aboriginal organizations. The most recent comprehensive land-claim agreement—finalized in 2003 with the Dogrib First Nation and known as the Tlicho Final Agreement—includes provisions for self-government.

The Territory of Nunavut

The new territory of Nunavut was made possible through a land settlement agreement between Canada and the Inuit of the Eastern Arctic in 1993. The terms of the agreement included the use of Crown lands for the Inuit to hunt, fish, and trap, and the transfer of part of the land to the Inuit, with a portion of this area involving rights to subsurface minerals. The same year the land agreement was reached, the federal government made the commitment to create a new territory by passing the Nunavut Act. This Act, which provided the legal basis for the creation of a distinct territory and territorial government, also allowed for a six-year transition period, giving the Inuit time to form their government, recruit their civil servants, and select a capital city. By means of a plebiscite, Iqaluit was selected as the capital of the new territory. Following an election in February 1999, the 19 members of the Nunavut Assembly took office on 1 April 1999. Unlike First Nations, the Inuit created a public form of government, meaning that every resident has the same political rights.

The word 'Nunavut' means 'our land' in Inuktitut. The vast majority of the approximately 30,000 people who reside in Nunavut are Inuit. A primary goal of the Nunavut government is to reflect the people's aims and aspirations. In addition to expanding economic development and increasing the number of jobs in the new public service for residents of the highly decentralized administration of Nunavut, the new government strives to promote Inuit culture and the Inuktitut language. For many Inuit, Nunavut is a dream—one that ensures their cultural survival within Canadian society. The dream also brings with it the challenge of building a northern economy that can provide jobs for its growing labour force and thereby reduce its financial dependency on Ottawa.

Land Claims

Aboriginal land claims are based on the peoples' long-time use of the land for hunting, fishing, and trapping. In Canada, Aboriginal land claims are settled by treaty: the Aboriginal tribe surrenders its claim to all the land in exchange for title to a smaller amount of land, a cash settlement, and, in most instances, usufructuary right to a larger territory owned by the Crown for hunting and fishing. Other issues involved in land-claim negotiations are self-government, management of wildlife resources, and preservation of language and culture. In simple terms, land claims are an attempt by Aboriginal peoples and the federal government to resolve the issue of Aboriginal rights. (See Chapter 3 for more information on these rights and the types of land claims.)

During the latter part of the twentieth century, Indian, Inuit, and Métis organizations in the Territorial North made land claims on behalf of their members. In the 1970s, three major land claims were made in the region: Yukon First Nations (Yukon), Dene/Métis (Denendeh), and Inuit (Arctic). Over the next two decades, the Inuit land claim was split into the Western Arctic (Inuvialuit) and the Eastern Arctic (Inuit/Nunavut), and the Dene/Métis land claim for Denendeh was divided into five separate land claims (Gwich'in, Sahtu/Métis, North Slavey (also referred to as Dogrib or Tlicho), South Slavey, and Deh Cho).[7]

Until the late 1970s, Ottawa did not recognize traditional claims to Crown land. The evolution of claims policy and objectives has been greatly influenced by jurisprudence. In the *Calder* case in 1973, the Supreme Court of Canada acknowledged the possible existence of Aboriginal title in Canadian law, that is, some claims to landownership might be valid but such claims had to prove earlier occupancy. The Supreme Court ruling caused Ottawa to reconsider its position. Between 1973 and 1985, the rethinking of the federal government's position on land claims resulted in the comprehensive and specific land-claims policy, which was formally announced in 1986. However, negotiations with the Inuvialuit followed this new policy and, in 1984, the Inuvialuit reached a final agreement with Ottawa. They were followed by the Gwich'in in 1992, and, in the next year, the Sahtu/Métis, Inuit, and Yukon First Nations. In 2003, the Dogrib or Tlicho Agreement was completed, leaving only two outstanding claims—Deh Cho and South Slavey (Figure 10.4).

As with all comprehensive land-claim agreements, the Inuvialuit Final Agreement (IFA) was based on the Inuvialuit's traditional use and occupancy of lands in the Western Arctic, that is, the land that they and their ancestors used for hunting and trapping. The Inuvialuit's land claim was originally part of the Inuit land claim to the entire Arctic. However, the Inuvialuit broke away from the Inuit and, as a result, there are two large **settlement areas** in the Arctic, the Inuvialuit and the Nunavut settlement areas.[8] The goals of the Inuvialuit were to preserve their cultural identity and values within a changing northern society; to enable themselves to be equal and meaningful participants in the northern and national economy and society; and to protect and preserve the Arctic wildlife, environment, and biological productivity (Canada, 1985: 1). Canada's goal, on the other hand, was to extinguish the Inuvialuit claim to the Western Arctic.

Though the JBNQA (1975) represents the first modern treaty, Ottawa, as pointed out above, instituted a new land-claims negotiating system composed of specific and comprehensive agreements some 10 years later. As the first comprehensive land-claim agreement in Canada, the IFA served as a model for subsequent ones. Since 1984, as shown in Table 10.1, five more comprehensive land-claim agreements have been reached. Each agreement is very similar to the IFA. One common feature is that each agreement has created economic and environmental administrative sectors. In the case of the IFA, the economic sector is the Inuvialuit Regional Corporation, which manages and invests the cash settlement received as part of the agreement.[9] The second sector, the Inuvialuit Game Council, is responsible for environmental issues that affect their hunting economy. The creation of these two sectors was the Inuvialuit's attempt to straddle two worlds;

Table 10.1 Comprehensive Land-Claim Agreements in the Territorial North

Aboriginal Peoples	Date of Agreement	Cash Values*	Land (km^2)
Inuvialuit	1984	$45 million (1977)	90,650
Gwich'in	1992	$75 million (1990)	22,378
Sahtu/Métis	1993	$75 million (1990)	41,000
Inuit (Nunavut)	1993	$580 million (1989)	350,000
Yukon First Nations	1993	$243 million (1989)	41,440
Dogrib (Tlicho)	2003	$152 million (1997)	39,000

*By dollar for the stated year.
Sources: Canada (1985: 6, 31; 1991: 3; 1993a: 3; 1993b: 81, 215; 2004); *Globe and Mail* (1993: A1).

their old world was based on harvesting game from the land, while their new world is part of the global industrial economy. However, the IFA was silent on one important element so necessary for Aboriginal peoples—self-government. Over time, the issue of self-government became part of comprehensive land-claim agreements. The Nunavut and Tlicho final agreements do spell out the specific nature and unique structures of self-government for the Inuit and the Dogrib.

The Yukon First Nation agreement, signed in 1993, was different in that it only established the basic elements of final agreements for each of the 14 Yukon First Nations. Known as the Umbrella Final Agreement, this 1993 arrangement provided the basic framework within which each of the 14 Yukon First Nations (Carcross/Tagish; Champagne and Aishihik; Dawson; Kluane; Kwanlin Dun; Liard; Little Salmon/Carmacks; Nacho Nyak Dun; Ross River Dena; Selkirk; Ta'an Kwäch'än Council; Teslin Tlingit Council; Vuntut Gwitchin; and White River) can conclude a final claim settlement agreement. So far, five First Nations have achieved a final agreement: Teslin Tlingit Council (1993), Selkirk First Nation (1997), Little Salmon/Carmacks (1997), Ta'an Kwäch'än Council (2002), and the Kluane First Nation (2003).

For the Inuvialuit, the agreement was an important step towards defining their place within Canadian society. In exchange for cash and land, the Inuvialuit gave up their

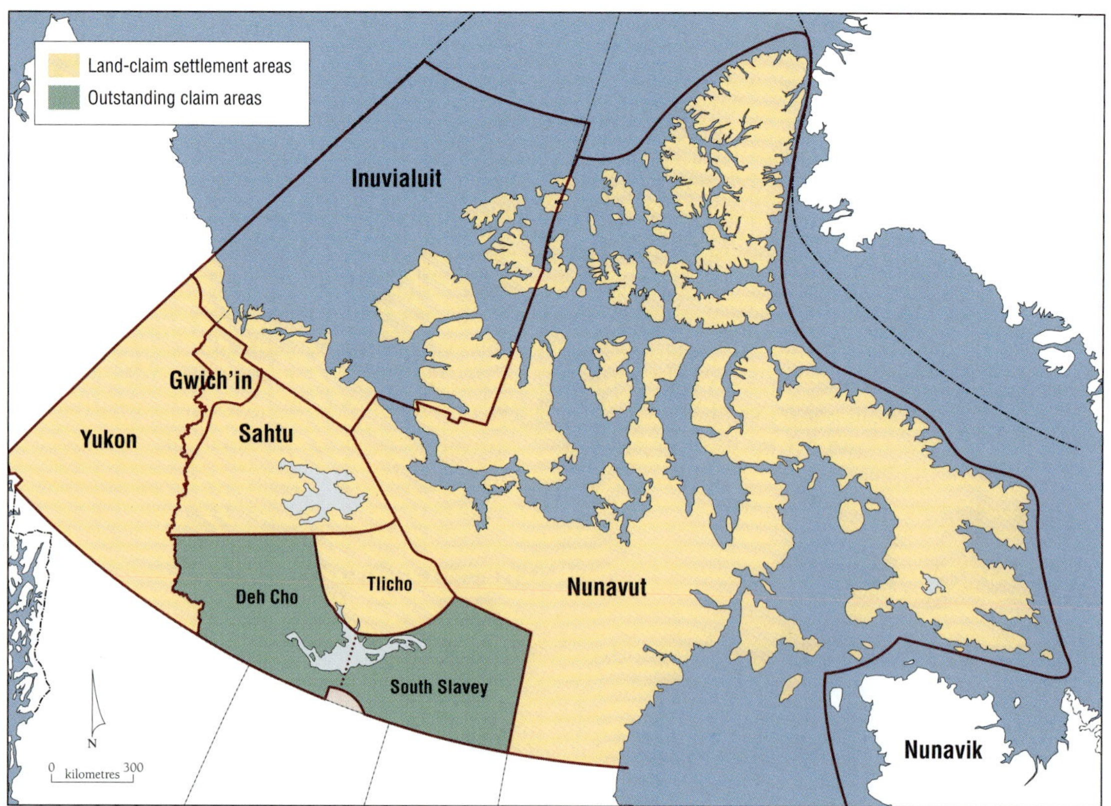

Figure 10.4 Land-claim agreements in the Territorial North. Six comprehensive land-claim agreements have been finalized while two claims, the Deh Cho and South Slavey, remain outstanding. The new coastal boundary between Nunavut and Nunavik (northern Québec) places islands close to Nunavik within the Québec boundary. Prior to the Nunavik Offshore Claim Agreement (2007), all islands in Hudson Bay belonged to Nunavut. Nunavut gained ownership of all the islands in Hudson and James bays because the 1670 Charter of the Hudson's Bay Company assigned these islands to the company and, in 1870, Canada gained ownership of the Hudson's Bay Company lands.

Aboriginal rights, including their claim to the vast lands of the Western Arctic. In exchange, they received legal title to about 90,650 km^2 of land, which is slightly less than 20 per cent of the settlement area (Canada, 1985: 6). The land to which the Inuvialuit gained title lies within the Inuvialuit settlement area, which extends from the Arctic mainland into the Arctic Ocean (Figure 10.4). A number of islands, including Banks Island, Prince Patrick Island, and the western parts of Melville and Victoria islands, are situated here. Of the total settlement land, 12,950 km^2 include surface and subsurface mining rights for the Inuvialuit. The Inuvialuit enjoy exclusive rights to hunting, trapping, and fishing over the remaining 77,700 km^2.

Population

The population of the Territorial North is small and concentrated in settlements. In 2006, the Territorial North had a population of 101,310. At that time, the Northwest Territories had 41,464 residents, Yukon 30,372, and Nunavut 29,474 (Statistics Canada, 2007a). Overall, Aboriginal peoples make up close to 52 per cent of the northern population according to the 2001 Census of Canada. However, the percentage of Aboriginal peoples varies widely between the three territories. Nunavut has the highest percentage of Aboriginal peoples at 85 per cent, followed by the Northwest Territories at 50 per cent and Yukon at 23 per cent (Table 10.2).

Almost everyone in the Territorial North lives in a settlement, town, or city. Most settlements are very small (Figure 10.5). In 2006, for example, approximately three-quarters of these have populations under 1,000, and more than 40 per cent of the Territorial North's population were in three cities: Whitehorse (22,898), Yellowknife (18,700), and Iqaluit (6,184). By function, centres in this region fall into three categories: Native settlements, resource towns, and regional service centres. Most Aboriginal peoples reside in Native settlements, where they form more than half the population. A growing number, however, have moved to regional centres, particularly the three capital cities where employment opportunities in the public sector are greatest. As well, a growing number of Aboriginal families have relocated to cities in southern Canada.

In 2004–5, the rate of natural increase in the Territorial North was approximately 1.5 per cent. Within the three territories, the highest rate (2.2 per cent) occurred in Nunavut. In comparison, the national figure was only 0.3 per cent. The reason for this significant difference is the much higher birth rate in the Territorial North (18 births per 1,000 compared to the national figure of 11 births per 1,000) and a lower death rate, (4.0 deaths per 1,000 compared to the national figure of 7.3 deaths per 1,000), the result of a younger population. Within the Territorial North, the highest birth rates and lowest death rates are among Aboriginal peoples, especially among the Inuit in Nunavut (Table 10.3).

Table 10.2 Population and Aboriginal Population, Territorial North

Territory	Population 2006	% Change 2001–6	Aboriginal Population 2001	% Aboriginal of Total 2001 Population
Yukon	30,372	5.9	6,540	22.8
Northwest Territories	41,464	11.0	18,725	50.1
Nunavut	29,474	10.2	22,720	85.0
Territorial North	101,310	9.2	47,985	51.7

Source: Bone (2003: 86); Statistics Canada (2007a).

Table 10.3 Components of Population Growth for the Territories, 1980 and 2004–5

Demographic Event	Canada 1980	Yukon 1980	NWT* 1980	Canada 2004–5	Yukon 2004–5	NWT 2004–5	Nunavut 2004–5
Births/1,000 persons	15.1	19.4	28.2	10.5	11.2	16.6	26.3
Deaths/1,000 persons	7.0	5.2	5.2	7.3	5.3	4.0	4.6
Natural rate of increase (%)	0.8	1.4	2.3	0.3	0.6	1.3	2.2
Net interprovincial migrants		400	–900		–6	–427	–296

*Includes Nunavut.
Sources: Bone (2003a); Statistics Canada (2006d).

The population of the Territorial North is also affected by migration. Migration to the North normally occurs when economic expansion creates jobs. Since World War II, many southerners have moved north to take jobs in the mining industry, the public service, and the business sector. However, since many newcomers remain for only a few years, a rapid turnover in the non-Aboriginal population occurs. Since the end of the resource boom in the early 1980s, there has been a net flow of people moving to southern Canada. From 1996 to 2005, there was a net loss of nearly 6,900 people (Table 10.4). There are three basic reasons for this out-migration. First, these migrants have an education and job skills that allow them to find employment in many areas of Canada, so they are economically mobile. Second, with friends and relatives living in southern Canada, they consider southern Canada their 'homeland'. Third, families tend to move south when their children reach school age.

The Aboriginal Economy

Aboriginal peoples participate in the wage economy. However, some continue to harvest land and sea resources. This harvesting economy is most prevalent in the smaller, more isolated Native settlements. Trapping, hunting, and other land-based activities persist among Aboriginal peoples in the North not so much because of their commercial value but largely

Table 10.4 Net Number of Migrants, Territorial North, 1986–2005

Territory	1986–91	1991–6	1996–2001	2001–5
Yukon	790	665	–2,760	51
Northwest Territories	–1,700	–400	–3,170	–206
Nunavut			–330	–605
Territorial North	–910	265	–6,260	–862

Sources: Statistics Canada (1998d, 2002b, 2006d).

Figure 10.5 Major urban centres in the Territorial North. The major cities are the territorial capitals, Whitehorse, Yellowknife, and Iqaluit. Numerous smaller settlements have populations under 1,000. Alert is a military base.

because of their cultural importance. For instance, hunting produces food for the table and country food remains a core cultural feature among northern Aboriginal families.

Declining prices for furs and efforts by animal rights groups to ban the sale of seal pelts in Europe made the situation for trappers and hunters very difficult. As shown in Figure 10.6, the value of fur production in the Northwest Territories dropped drastically from the mid-1980s to 2002. In 1992–3, for example, the value of fur sales dropped below $1 million for the first time. In the subsequent 10 years, trapping exceeded $1 million in sales only once (1996–7). In 2003–4, the value of fur production in the Northwest Territories was $812,000. With low prices and rising costs, trapping does not generate enough cash income, and individual trappers must supplement their revenue from trapping with other income to make ends meet.

Country Food

Though cultural change has occurred among Aboriginal peoples, such as the growth of the wage economy, some core cultural elements have remained. For Aboriginal peoples, the core elements are a strong attachment to the land, to country food, and to the ethic of sharing. **Country food** is food obtained from the land, a preferred source of meat and fish. As equivalent store-bought foods are expensive, most Aboriginal northerners keep their food costs low by consuming country food. While it is true that there are substantial costs expended in harvesting country food, to some degree this cost is offset by the pleasure and spiritual rewards of being on the land and participating in hunting and fishing. Sharing also remains an important component in the harvesting and distribution of country food among family members, relatives, and close

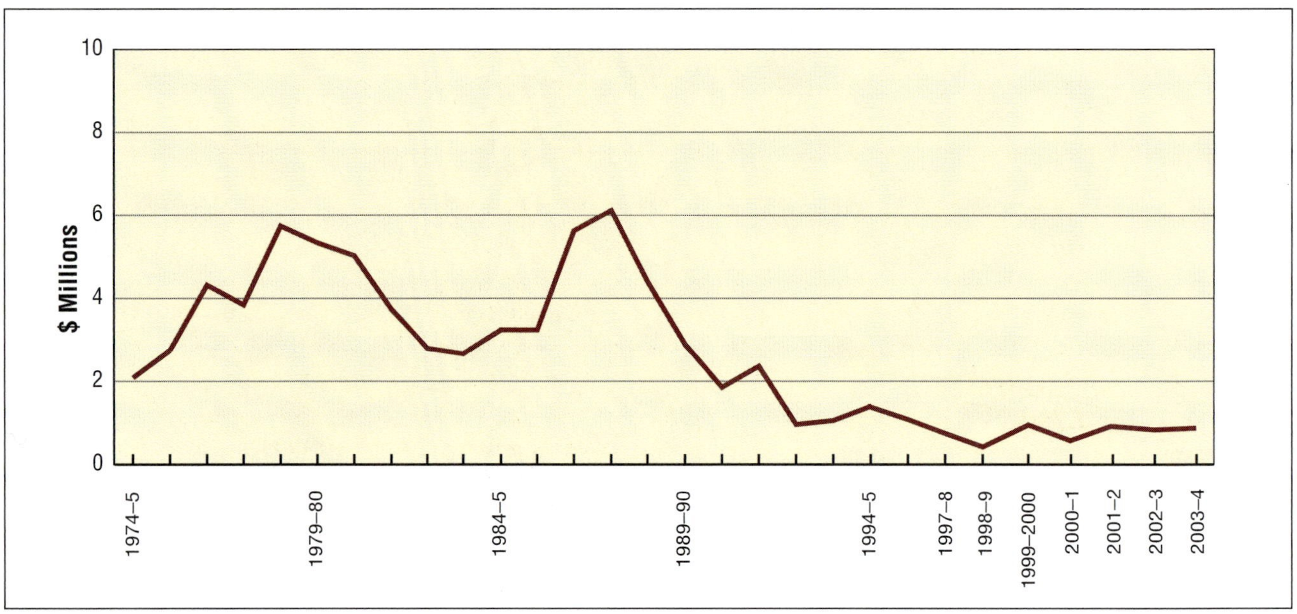

Figure 10.6 Value of fur production in the Northwest Territories. Since peaking in the late 1980s, the value of fur production has been steadily declining, signalling the shrinking importance of the commercial trapping economy in the North. *Source:* Government of the Northwest Territories (2003, 2004).

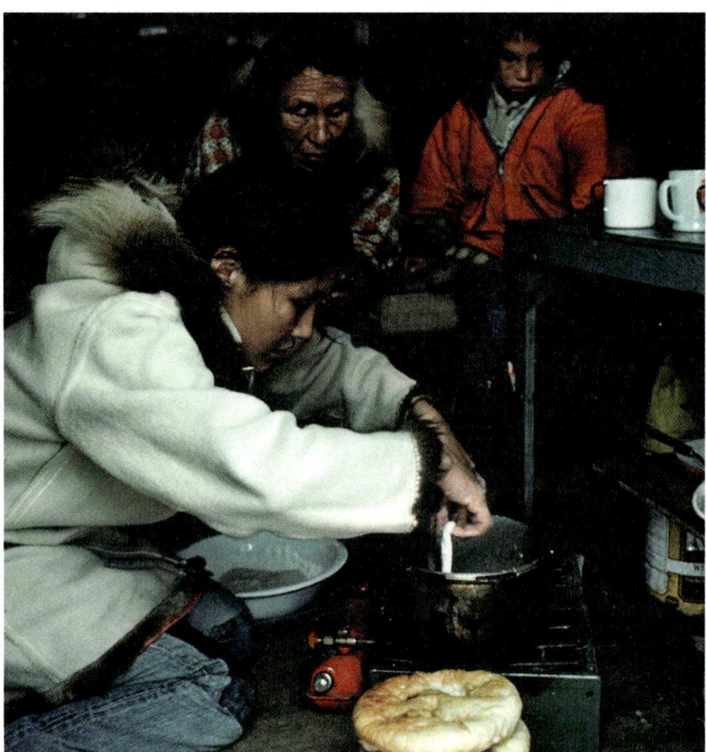

Country food among Aboriginal peoples in the Territorial North is not only an alternative to costly store-bought food but also an intrinsic feature of their culture. (John Eastcott/Ya Momatiuk/Valan Photos)

friends. However, this ethic does not apply to the wage economy—wages, for instance, are not shared.

Figures on the cost, size, and value of harvesting country food are not available. However, estimates suggest that the 'substitute value' of country food (the value of equivalent store-bought food) was about $40–$50 million per year in the mid-1980s (Usher and Wenzel, 1989). More recent estimates are not available, but a visit to an Aboriginal community quickly reveals the importance of wildlife harvesting to the daily diet. While some store-bought foods can be found on the tables of Aboriginal families, most of the meat and fish they consume is from the land and waters.

Country food is so important to Aboriginal northerners that they have made wildlife an essential issue in land-claim negotiations. The agreements establish co-management committees that administer the environment and wildlife. For example, these committees approve (or reject) proposals for new industrial projects and determine the total allowable harvest of wildlife based on biological principles and **traditional ecological**

knowledge. In the case of the Inuvialuit Final Agreement, the members of the co-management committees are named by the Inuvialuit Game Council and the three governments (Canada, Northwest Territories, and Yukon). The members of the Inuvialuit Game Council are selected by the Hunters and Trappers Associations found in each of the six Inuvialuit communities. Under the direction of these committees, professional scientists record and monitor the state of the environment and wildlife. As a result of comprehensive land-claim agreements, power to control the use of the environment and wildlife now rests, in part, with most Aboriginal peoples in the Territorial North.

Industrial Structure

As a northern frontier, the Territorial North's economy depends heavily on its primary industries. Energy and mining are the principal elements of this economy. In terms of employment, the primary sector in the Territorial North is much larger than the same sector in other geographic regions. For example, approximately 15 per cent of the workers fall into the primary sector in the Territorial North compared to only 2.7 per cent in Ontario. Similarly, the tertiary sector is much larger than in the other geographic regions. The Territorial North has approximately 83 per cent of its workforce allocated to the service sector compared to about 73 per cent for Ontario. A large tertiary sector is characteristic of resource frontiers in modern industrial nation-states. In Canada, the large transfer payments from Ottawa to the three territorial governments play a major role in their budgets, allowing them to have similar levels of education, health, and social services to those found in the provinces. As a result, the contemporary economic structure of the Territorial North mirrors these two forces—resource extraction and transfer payments—making the tertiary and primary sectors of greatest significance. The secondary sector, manufacturing, is almost non-existent in the Territorial North (Table 10.5).

The northern industrial structure is dominated by non-renewable resource industries. These energy and mineral companies export their production to world markets. The northern transportation system is geared for that role. The value of production from the primary sector accounts for 90 per cent of all production in the Territorial North. While the value of mineral output varies from year to year, it hovered around \$2.3 billion for 2005. Production also varies by territory, with 96 per cent coming from the Northwest Territories. However, this highly efficient and capital-intensive industry employs relatively few people and generates few indirect benefits for the Territorial North because companies purchase their equipment and supplies from firms in southern Canada or in foreign countries. Thus, the question of northern benefits

Table 10.5 Estimated Employment by Economic Sector, Territorial North, 2006

Economic Sector	North Workers (%)	Ontario Workers (%)	Difference (percentage points)
Primary	13	2.7	10.3
Secondary	2	24.6	−22.6
Tertiary	85	72.7	12.3
Total	100.0	100.0	

Source: Author's estimate.

remains a contentious issue because most benefits, whether the purchase of supplies and equipment or the employment of skilled trades workers, flow to the south.

Resource companies struggle with high transportation costs. Air transportation of bulky commodities is out of the question. Winter or ice roads provide a solution. The long, cold winters permit the construction of winter or ice roads, which allow the trucking of bulky commodities great distances, thus reducing the companies' transportation costs (Vignette 10.6).

Northern Benefits

Northern benefits are a key political concern to northerners. The basic question is: What do northerners get from resource development? The market economy works poorly for the Territorial North for several reasons. First, the region has a small and dispersed labour force with relatively few trades workers. Second, the business community in the Territorial North has limited capacity to respond to the needs of megaprojects. Given this situation, special efforts are needed to ensure a minimum of economic benefits to the Territorial North. In the early 1980s, the construction of the Norman Wells Pipeline saw two new benefits specially designed for northerners. One was the northern air commuting system that provided access to the project for northerners from communities in the Mackenzie Basin. Another was the special effort to ensure some contracts went to local businesses, especially to businesses owned by Dene or by their First Nation government. Both measures significantly increased economic benefits to the North. Some 20 years later, First Nations and their communities are exploring an innovative way to secure an annual flow of cash. One way is to participate in the ownership of the pipeline, which would generate annual revenues. The Aboriginal Pipeline Group has obtained a one-third interest in the pipeline for an initial investment of $4 million. Another proposal was for communities to seek 'tax revenue' from the company for access to the Mackenzie Delta gas pipeline corridor. So far, Imperial Oil has rejected this demand and plans to pay a one-time payment for access, which is the normal easement arrangement made with landowners south of 60°.

Air commuting, while an effective way to obtain skilled southern workers for remote resource projects, does have a drawback for the North. For the Territorial North, southern-based air commuting systems hurt the North's economy because: (1) workers spend their wages in their home communities in southern Canada, thereby stimulating provincial, not territorial, economies; and (2) workers who reside in a province but work in the territories pay personal income tax to provincial rather than territorial governments, thereby depriving territorial governments of valuable personal income tax.

Most of the 85 per cent of the Territorial North's workforce in the tertiary sector are employed by one of three governments: federal, territorial, or local. Local governments are

Vignette 10.6 Air Commuting and Winter Roads

Lying far from Canada's highway system, mining companies operating in remote areas of the Northwest Territories and Nunavut employ air commuting systems to transport their employees and winter roads for hauling bulky goods and diesel fuel. Nunavut's first diamond mine—the Jericho mine, located in the Barren Lands—combats high transportation costs by trucking bulky commodities along a winter road and using an air commuting system for employees. Workers from Inuit communities along the Arctic coast, such as Kugluktuk (formerly Coppermine), and southern workers from Edmonton use the company-operated air commuting service.

settlement councils, band councils, and other organizations funded by a higher level of government. The reason the tertiary sector is so large in the Territorial North is that geography demands such an investment of people and capital to ensure the delivery of public services. Territorial governments must spend more money per resident than do provincial governments to provide basic services. Much of the cost differential is attributed to overcoming distance in the Territorial North and hiring staff for small communities. The drawback is that economies of scale are difficult to achieve in small communities where teachers and nurses often have a relatively small number of students and patients compared to those working in larger centres.

However, the social importance of the public service sector goes beyond the number of employees. The wide geographic distribution of public jobs across the North is a major social benefit to those in small communities. As a result, employment opportunities, while concentrated in the three capital cities, exist in every community. In small, remote communities, where unemployment rates are often as high as 30 per cent, virtually all jobs are associated with one or more levels of government. A second social benefit is the governments' ability to implement social policies in their hiring practices. For example, Nunavut's government seeks to employ mainly Inuit in its civil service.

Resource Development in the Territorial North

Since the sixteenth century, explorers—now replaced by multinational companies—sought to exploit the North's natural wealth, particularly its minerals. The search for precious metals began soon after explorers reached North America. In Mexico and South America, the Spanish and Portuguese were particularly successful in obtaining vast quantities of gold and silver. In North America, English and French explorers searched for similar wealth. Northern Indians trading at the Hudson's Bay posts often wore copper ornaments and had copper tools. Richard Norton, the factor or chief trader at Fort Prince of Wales, ordered one of his men, Samuel Hearne, to find the source of this copper and assess its commercial worth to the company. In one of the great overland journeys of all time, Hearne travelled with Chipewyan leader Matonabbee's tribe from Fort Prince of Wales on the west coast of Hudson Bay (present-day Churchill, Manitoba) to the mouth of the Coppermine River at Coronation Gulf on the Arctic Ocean. In 1771, Hearne found the source of copper, but this natural copper only occurred at the surface in small amounts. Others seeking mineral wealth in the North were more fortunate. In 1896, George Carmack, an American prospector, and his Indian brothers-in-law, Skookum Jim and Tagish Charley, found gold nuggets in a small tributary (later renamed Bonanza Creek) of the Yukon River. When word of their find reached the outside world, the Klondike gold rush began.

The Klondike Gold Rush

The discovery of gold in Yukon sparked the greatest gold rush of all time. It caught the imagination of the world and gave credence to the myth of a northern Eldorado. While the discovery of gold took place in 1896, the Klondike gold rush did not begin until the following year. Dawson City was the centre of this frenzy. Thousands of prospectors flooded the area to scour the sandbars of the rivers and pan for gold. Within two summers, the more accessible placer gold was gone. When word of a new gold find in Alaska reached Dawson City, most prospectors moved down the Yukon River into Alaska.

By 1900, gold mining in the Klondike had changed from an individual pursuit to a large-scale capital-intensive operation. With the accessible placer gold effectively gone, the only remaining gold was deep in the frozen terraces found along the sides of the Klondike River. Permafrost is widespread at latitudes above 60° N. Mining in permafrost was beyond the abilities of a prospector, so a second phase of mining began, organized by companies with both capital and technology. Mining companies soon purchased the leases to large sections of the land surrounding the rivers. Gold mining became highly organized, capital-intensive, and technically advanced. To expose the gold locked in the frozen terraces, hydraulic and steam-mining techniques were used to thaw the ground and force the sand, gravel, and fine gold particles into separating devices commonly known as sluice boxes (Vignette 10.7). The emergence of commercial, large-scale gold mining marked the birth of the mining industry in the Territorial North.

Environmental and Social Impacts

In the late nineteenth century, Western society regarded the North as a source of wealth. Little regard was given to the natural environment or to the Aboriginal peoples. The Klondike gold rush exemplifies this attitude. The North, popularized by the writings of Jack London and Robert Service, and later by Charlie Chaplin's portrayal of a prospector in the motion picture *The Gold Rush*, was seen as a frontier, a place of adventure and excitement. In those days, Yukon's environment was at the mercy of prospectors and mining companies. Wood was in great demand for power by steamboats and the hydraulic operations in the goldfields, and for heat by companies, government offices, and individuals. As a result, the forests adjacent to the Yukon River and its tributaries were logged, leaving the riverbanks and adjacent lands bare. Wildlife habitat suffered and so did the Aboriginal peoples who depended on wildlife for their food supply. In the post-placer gold-mining period, the gold was washed out of the frozen terraces, leaving behind a scarred landscaped (see the earlier section, 'Environmental Challenges').

The Klondike gold rush also had social impacts on the Territorial North, but these were more subtle. The influx of miners, administrators, and, eventually, the North West Mounted Police transformed southern Yukon into an organized territory within Canada. The impact on Aboriginal tribes in Yukon was overwhelming—now they were not only a minority in their own country but were also living within a Western culture with values often very different from their own. Aboriginal peoples found themselves on the margins in this new world.

The disregard for the environmental and social impacts of resource development in the North continued until after World War II. Following the Berger Report (1977) on the effects of building a gas pipeline in the North, the public began demanding that Ottawa take action to deal with the hidden costs of resource development.[10] The federal government responded by enacting legislation that requires an environmental and social assess-

Vignette 10.7 Gold Rushes

Placer gold was found in commercial quantities mainly in the western Cordillera region from California to Alaska where a series of gold rushes occurred in the mid- to late nineteenth century. The discovery of placer gold in some remote places triggered a sudden influx of prospectors, traders, gamblers, merchants, bankers, prostitutes, and various hangers-on. All sought to increase their wealth either by finding gold or by 'extracting' gold from the miners.

Placer gold can be worked cheaply by amateurs. It involves the separation of gold nuggets and particles from sand and gravel. The common practice is to pan for gold, that is, to jiggle the sand, gravel, and gold in a pan, causing the heavier gold to separate from the other material. Sluicing is another more elaborate method of separating the gold from other loose material obtained from the stream bed. This requires sluice boxes—channels or troughs fitted with grooves to separate gold from gravel.

ment of development projects before they can be approved. The federal agency responsible for enforcing these regulations is the Canadian Environmental Assessment Agency. By their very nature, industrial projects alter the natural and social environments. This federal agency's task is to review the proposals of planned developments, to assess the degree of their impact on the natural and social environments, and, if necessary, to recommend measures that will mitigate any identified impacts.

However, this process does not always manage to reconcile the interests of all the parties affected by developments. For instance, in 1994 the NWT Diamond Project was submitted to Ottawa for review under the Federal Environmental Assessment and Review Process. BHP Diamonds Inc. and the Blackwater Group proposed to construct open-pit and underground diamond mines in a remote area about 300 km northeast of Yellowknife near Lac de Gras. The ensuing review process did not satisfy everyone. Aboriginal people complained that they did not have enough funding and time to examine and understand the project. They also complained that the public review process did not hold enough meetings in their communities. (Public hearings were held in eight Native communities but most took place in Yellowknife.) The Northern Environmental Coalition expressed concern that the panel did not pay enough attention to the cumulative effects of the project on the environment (O'Reilly, 1996). Nevertheless, in 1996, the federal environmental assessment panel filed its report, recommending approval subject to a series of restrictions and qualifications (MacLachlan, 1996: 1).

BHP Diamonds' Ekati diamond mine. The value of mineral production from this open-pit mine exceeds that of all other mines in the Territorial North. Located in the remote Arctic, the mine represents the backbone of the Northwest Territories economy. (Copyright © 2007 Tim Atherton)

Though resource developers often promise long-term economic development, non-renewable resource projects frequently have a short lifespan and therefore fail to generate lasting economic benefits. Because of their boom-and-bust nature, mining operations cause social dislocations when they cease to operate. The question of the long-term economic and social value of resource developments is further complicated by the mismatch between the North's labour force and the workforce needs of megaprojects. The net result is that the more skilled workers for megaprojects are drawn from southern labour markets.

Development in Remote Locations

In the Territorial North, the development of known mineral deposits in remote areas is often prevented by high transportation costs. The pattern of mining developments in the North clearly indicates the importance of transportation, especially water transportation. In 1935, the first major gold discovery in the Northwest Territories took place near Yellowknife. The Con-Rycon gold mine was a viable proposition because the mine was located on the shores of Great Slave Lake and therefore had excellent access to the Mackenzie River transportation system. Mines located far from water transportation, such as the Lupin gold mine (about 250 km northeast of Yellowknife), required the construction of **winter roads**. The need for roads in mining is not always so much for transporting minerals to external markets but for bringing equipment and supplies to the mine site. Mines along the Arctic coast can take advantage of ocean transportation. Three examples are the former mines at Rankin Inlet on the shores of Hudson Bay, the lead-zinc mine at Nanisivik near the northern tip of Baffin Island, and the Polaris lead-zinc mine on Little Cornwallis Island in the Arctic Archipelago (see Figure 10.7).

In assessing the cost of transportation, the value of the mineral is taken into account. For example, copper, lead, nickel, and zinc are all low-grade ores (much of the ore has no commercial value). Even after the first-stage separation of some of the waste material from the valuable mineral, the enriched ore still remains a bulky product, with significant waste material remaining that can only be removed through smelting. Shipping such a low-value commodity is very expensive. Ideally, such ore is transported to a smelter by ship. Railways are the second most effective transportation carrier for low-grade ore. The critical nature of transportation for such mines is clear in the example of the lead-zinc deposit at Pine Point in the Northwest Territories. Discovered in 1898 by prospectors heading overland to the Klondike gold rush, the Pine Point deposit was not developed until 1965, when a railway was extended to the mine site. With a means of transporting the ore to a smelter, the company, Cominco, could begin sending massive amounts of ore by rail to its smelter at Trail in British Columbia. Unfortunately, the mine closed in 1983, leaving Pine Point a ghost town.

Few Aboriginal workers are employed in the mining industry. The Rankin Inlet nickel-mining operation proved to be an exception to this. The majority of the employees at this mine were from the local community. The local Inuit had a chance at these jobs because it was difficult to recruit outside workers to work in such a remote place, the mining operation was small, and mine officials were willing to hire entire Inuit crews. This meant that most underground communications were in Inuktitut.

Key Topic: Megaprojects

The Territorial North has entered a new phase of resource development characterized by megaprojects controlled by multinational companies. Megaprojects usually cost more than $1 billion and require several years to complete. This phase of development has integrated the Territorial North's economy into the global economy, thereby firmly locking the North into a resource hinterland role in the world economic system. The most recent megaprojects in the Territorial North have involved diamond

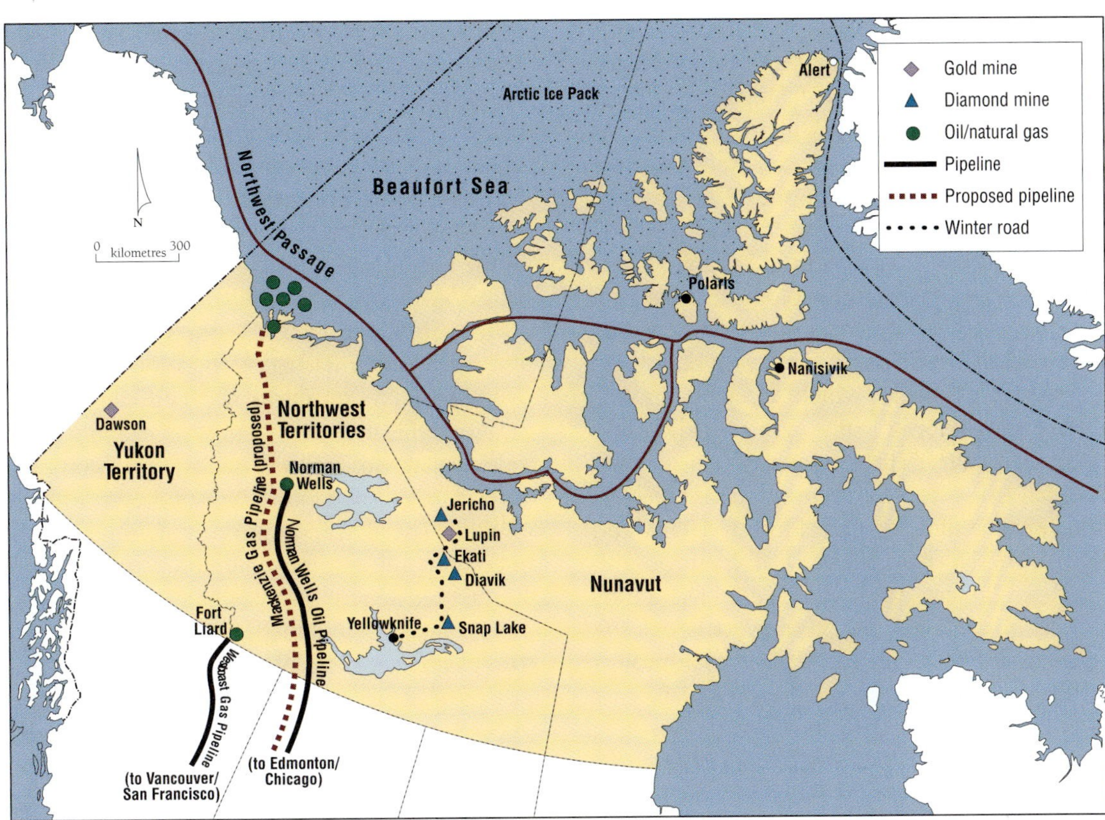

Figure 10.7 Resource development in the Territorial North. The mineral wealth of the Territorial North lies mainly in the Northwest Territories. Diamonds, gold, natural gas, and oil drive this territory's resource economy. In contrast, the resource economies of Yukon and Nunavut are much smaller (Further resources: Student website, National Atlas section, Map 29. Website instructions are found on p. xx.)

mining. The first diamond mine to come into production was the Ekati mine. As part of the NWT Diamonds Project (MacLachlan, 1996), preliminary construction of buildings, roads, and mines began in the early 1990s, but the Ekati mine was not completed until 1998. Two other mines, Diavik and Jericho, are now in production, while the Snap Lake mine will begin production in 2008. Supplies for these remote mines are trucked along a winter road (Figure 10.7). All are located in the Slave geological region that straddles the border between the Northwest Territories and Nunavut.

Proponents of resource development describe megaprojects as the economic engine of northern development, though others challenge this assumption, claiming that they offer few benefits to the region and, more particularly, to the Aboriginal communities (Bone, 1992: 143–5). These large-scale ventures are designed for the export market. By injecting massive capital investment into the construction of giant engineering projects, megaprojects create a short-term economic boom. However, most construction monies are spent outside hinterlands because the manufactured equipment and supplies are produced not in hinterlands but in core industrial areas. This reduces the benefits of megaprojects to the hinterland economy and virtually eliminates any opportunity for economic diversification. As well, since all megaprojects in the Territorial North are based on non-renewable resources, these developments have a limited lifespan. At the end of these projects, the local economy suffers a collapse. Three examples are the closure of mines at Faro, Yukon; Pine Point, Northwest Territories; and Rankin Inlet, Nunavut.

However, megaprojects can more readily inject much-needed capital and create development in the region. Megaprojects in resource hinterlands are high-risk ventures. Multinational companies can reduce their risks in three ways. First, they can create a consortium of companies and thereby spread the investment risk among several firms. Second, they can arrange for long-term sales of the product at a fixed price before proceeding with construction. Third, they can obtain government assistance, which often takes the form of low-interest loans, cash subsidies, and tax concessions. Three megaprojects are discussed in the following sections: the Mackenzie Valley Pipeline Project: 1970s and 2000s, the Norman Wells Oil Expansion and Pipeline Project, and the NWT Diamonds Project. Except for the 2000 version of the Mackenzie Valley Pipeline Project, each project was approved after an environmental impact assessment. The Norman Wells Project received its approval in 1980, while the NWT Diamonds Project was approved in 1996.

The Mackenzie Valley Pipeline Project: 1970s

By 1970, the plans formulated by oil and pipeline companies to bring natural gas from Prudhoe Bay, Alaska, to major markets in the United States, primarily in the Midwest, were submitted to Ottawa.[11] Economic conditions at the time included a high demand for natural gas in the continental United States and the fact that natural gas was a by-product of oil production at Prudhoe Bay. One plan was to build a pipeline, known as the Mackenzie Valley Pipeline Project, which would transport natural gas from the Arctic coast of Alaska to the Mackenzie Delta and then south along the Mackenzie Valley, eventually reaching Alberta and then the United States. Shortly after these plans were announced, world oil prices increased sharply as a result of joint action to limit supply by the Organization of Petroleum Exporting Countries (OPEC). OPEC's control of world oil production changed the economic landscape for energy and added another economic reason for the Mackenzie Valley Pipeline Project. At these higher prices, the Mackenzie Delta deposits were no longer too expensive to develop, providing its natural gas could be added to the proposed pipeline by means of a short spur pipeline.

In Canada, this project was subjected to the Mackenzie Valley Pipeline Inquiry (also known as the Berger Inquiry). The Berger Inquiry began in 1974 and was concluded in 1976. Led by British Columbia Justice Thomas Berger, who a few years earlier had been chief counsel for the Nisga'a in their landmark land-claim case (*Calder*), this federal investigation examined the potential environmental, social, and economic impacts of the proposed gigantic construction project. The Berger Inquiry established a new precedent by holding community hearings in numerous remote Native settlements, and its final report, *Northern Frontier, Northern Homeland*, issued in 1977, recommended that a natural gas pipeline from the Mackenzie Delta to Alberta was feasible but should only proceed after further study and after settlement of the Dene land claims. To complete these two tasks, Berger recommended a 10-year delay in the construction of such a pipeline. At the same time, he rejected the construction of a pipeline from Prudhoe Bay to the Mackenzie Delta because of the danger of harming the delicate Arctic environment and wildlife found along the North Slope of Alaska and Yukon.

Without the link to Prudhoe Bay and faced with rising construction costs, the proposal lost its economic attraction and the pipeline was never built. With new supplies of Mexican and Albertan natural gas reaching the United States in the 1980s, prices declined in the North American market. However, by 1998, this surplus had disappeared and prices began to rise. From 1998 to 2003, natural gas prices more than quadrupled. Consequently, deposits in the Territorial North again commanded the attention of the major oil and gas companies servicing the North American market. In 2000, for instance, natural gas from northeastern British Columbia (just south of the Northwest Territories) began to flow through the Alliance Pipeline to the Chicago market, underlining the continental system of energy production and marketing in North America.

The Mackenzie Valley Pipeline Project Revisited

The anticipated development of the natural gas deposits in the Mackenzie Delta has brought about a revisiting of the old Mackenzie Valley Pipeline Project. Imperial Oil, Gulf Canada Resources, Shell Canada, and Exxon Mobile, all of which hold considerable natural gas reserves in the Western Arctic, are poised to begin the construction of a natural gas line along the Mackenzie River Valley. This proposed megaproject includes the development of an estimated six trillion cubic feet of natural gas in the three largest onshore fields discovered in the Mackenzie Delta and the building of a gas and liquids pipeline system. The Mackenzie Valley gas pipeline would have a capacity of 1.2 billion cubic feet per day and could be expanded to accommodate gas from other fields in the future. The first step in obtaining Ottawa's approval is to complete a statement of the project called a Preliminary Information Package (PIP), which is then reviewed by the National Energy Board (now completed) and by the Joint Review Panel, which, as of October 2007, was still holding community meetings into environmental and social issues.[12] Thirty years ago, Aboriginal peoples strongly opposed the project for two reasons. First, the Dene insisted that their land claims be settled before the construction of a Mackenzie Valley pipeline. Second, the Dene had concerns about the impact of this construction project on their lands and people. More specifically, they feared that wildlife would become less plentiful, causing great harm to their hunting/trapping life. Other concerns also were identified, such as the Dene fear that their communities would not gain a fair share of jobs and business opportunities, and that the expected increase in non-Aboriginal population would generate negative impacts on Dene culture and social well-being.

By 2003, these concerns were no longer seen as major obstacles. Several land-claim agreements have been concluded, and the Dene workforce and business community are more developed and better prepared to take advantage of such a mega-construction project. The Aboriginal Pipeline Group has obtained a one-third interest in the pipeline. When project details were submitted to Ottawa in 2003, the cost of such a pipeline was approximately $5 billion. Nellie Cournoyea, chairwoman and chief executive officer of the Inuvialuit Regional Corporation, believes that 'it signals the recognition we are building the capacity to participate in the economy of the twenty-first century.'

The attention focused on the Mackenzie Delta gas fields is clearly related to the rising price of natural gas, which in turn is a result of the growing demand for natural gas in the United States. By 2015, demand for natural gas is projected to soar to 31.3 trillion cubic feet a year in North America, well above the current production of 24 trillion cubic feet, possibly pushing prices well over current rates. By then, existing production in southern Canada and the United States is expected to diminish—perhaps to 20 trillion cubic feet—leaving a possible gap of 11 trillion cubic feet. One of the two proposed Arctic pipelines should be completed by that time, possibly making up as much as half of this expected shortfall.

The race to build the first Arctic pipeline to North American markets pits Prudhoe Bay against the Mackenzie Delta. At the moment, the Mackenzie Delta is favoured because of two main advantages: (1) a much lower construction cost because of a shorter distance to market and because the diameter of its pipe is much smaller (and therefore less costly) than that proposed for the Alaskan pipeline; and (2) a much shorter construction time (six years) compared to the proposed Alaskan pipeline (10–12 years). However, the US government may subsidize the construction of the much more expensive Alaskan pipeline (estimated in 2007 at over $30 billion compared to nearly $8 billion for the all-Canadian pipeline). The stakes are high, causing much lobbying in Washington. Alaska has two strong points: (1) Alaska's reserves are much larger than those in the Mackenzie Delta (Alaska holds known reserves of 35 trillion cubic feet and an estimate of an additional 100 trillion cubic feet, compared to about 9 trillion cubic feet of known reserves in the Mackenzie Delta, plus estimated reserves similar in size to those in

Alaska); and (2) American security concerns call for the development of American petroleum deposits, though the Prudhoe Bay pipeline would follow the Alaska Highway and therefore must pass through Canadian territory (Yukon, British Columbia, and Alberta) to reach the main markets in the United States.

The Mackenzie Gas Project proposes to build a 1,220-kilometre pipeline system along the Mackenzie Valley. It would link northern natural gas sources to southern markets. The main Mackenzie Valley pipeline would connect to an existing natural gas pipeline system in northwestern Alberta. The proposed project crosses four Aboriginal regions in Canada's Northwest Territories (NWT): the Inuvialuit Settlement Region, the Gwich'in Settlement Area, the Sahtu Settlement Area, and the Deh Cho Territory. A short segment will be in northwestern Alberta near the NWT border. The Deh Cho, the only Aboriginal group without a land-claim agreement, have been adamant that no pipeline can cross their territory in the Fort Simpson area until they have a settlement, but the Conservative government in Ottawa has pressed ahead with the pipeline project regardless. Indeed, in the spring of 2007 a small deployment of Canadian military personnel called Operation Narwhal came to the region with the ostensible purpose of stopping possible 'terrorist' sabotage of the pipeline (*The Dominion*, 2007).

The natural gas exploration and development companies involved in the Mackenzie Gas Project have interests in three discovered natural gas fields in the Mackenzie Delta—Taglu, Parsons Lake, and Niglintgak. Together, they can supply about 800 million cubic feet per day of natural gas over the life of the project. Other companies exploring for natural gas in the North are also interested in using the pipeline. In total, as much as 1.2 billion cubic feet per day of natural gas could be available initially to move through the Mackenzie Valley Pipeline.

But will the project go ahead? Natural gas prices have increased since 2003, when the project's costs were estimated at $5 billion. In May 2007, however, Imperial Oil announced revised figures for the costs to build the Mackenzie Gas Project. The total Imperial estimate is now $16.2 billion, with $7.8 billion for the construction of the pipeline, $4.9 billion for the development of the gas fields, and $4.9 for the gas gathering system. Could such a jump in estimated costs be a ploy to obtain concessions from Ottawa?

Imperial Oil, a company that made record profits in 2006, and the Northwest Territories government, which seeks economic development, see federal financial involvement as crucial. For Imperial, federal support would reduce the risk of low gas prices and help ensure a 'double-digit' rate of return (CBC News, 2007). The NWT government sees the project as in the 'national interest' because, like the CPR, it will open up the last major frontier in Canada's North.

The Norman Wells Project, 1982–5

The Norman Wells Project, which cost almost $1 billion in 1996 dollars, was hailed by industry as a model megaproject for the North. Both the federal government and Esso Resources Canada believed that the project would not harm the natural and social environments in the construction impact zone, which stretched from Norman Wells to Fort Simpson and then to the Alberta border. Aboriginal organizations, such as the Dene Nation, and environment groups, such as the Canadian Arctic Resources Committee, did not agree.

Plans for this major oil and pipeline construction effort by Esso Resources Canada began in the late 1970s, when the price of oil was rising sharply. Once the proposal was approved by the federal government, construction began, running from 1982 to 1985. The petroleum pipeline route begins at the resource town of Norman Wells in the Northwest Territories and ends at Zama in northern Alberta, the northern terminus for the national oil pipeline system. Much of the terrain north of 60° N lies in the zone of discontinuous permafrost. As a result, special construction measures were necessary. Limiting construction to the winter months prevented excessive damage to the terrain and vegetation by trucks and other heavy equipment, and efforts to re-establish vegetation along the

A drilling rig near Norman Wells (second island from the left). The Norman Wells project allowed for a significant increase in oil production and supply to southern markets. (Terry A. Parker/Viewpoints West)

pipeline route following construction were designed to minimize soil erosion and the warming of ground temperatures by solar radiation. The danger from higher ground temperatures in the summer is that this could accelerate the melting of ground ice, leading to the subsidence (sinking) of the ground and possible rupturing of the buried pipeline.

The local oil fields were first discovered in 1920. By 1982, geological investigations had revealed the quantity of proven reserves to be as high as 100 million m^3 (Bone, 1992: 145). The purpose of the Norman Wells Project was to increase the oil produced and shipped to southern markets by means of a pipeline. Prior to 1985, the output from the Norman Wells oil fields served only the local Mackenzie Valley market. At that time, annual output was less than 180,000 m^3. With the completion of the pipeline and the expansion of oil production by a factor of 10, the Norman Wells oil fields had the capacity to send vast quantities of oil to southern customers. Since 1985, annual output has exceeded 1 million m^3 and almost reached 2 million m^3 several times.

The Norman Wells Project represents a successful and profitable operation. For 25 years, Esso has produced and transported crude oil to markets in southern Canada and the United States. During that time, while some subsidence of the ground along the pipeline route has occurred, the pipeline has not been damaged and there have been no major oil spills. Equally important, the operation of the Norman Wells Project has not triggered social problems in the communities along the pipeline route. From the oil industry's perspective, the Norman Wells Project demonstrated that pipelines can be built in areas with permafrost. However, from the Dene Nation's perspective, land claims should

Figure 10.8 Diamond and gold mines in the Territorial North. While gold mining has declined, diamonds are now the leading non-renewable resource by value in the Territorial North. The producing mines are Ekati, Diavik, and Jericho. Snap Lake should be operating by 2008; Gahcho is some years from production.
Source: Canadian Arctic Diamonds, at: <www.canadianarcticdiamond.com/03_history/Map.html>. Image courtesy Diavik Diamond Mines, Inc.

have been settled before the project was approved. Under such circumstances, the Dene would have become business partners in the construction and ensuing operation.

The NWT Diamonds Project

Canada is now the third largest producer of diamonds in the world, behind Botswana and Russia, and accounts for 15 per cent of the world supply, thanks to the two diamond mines in the Northwest Territories and one in Nunavut. In 2005, the Northwest Territories accounted for nearly all of the $1.7 billion of diamond production. In 2008, another diamond mine, the Snap Lake mine in the Northwest Territories, is expected to come into production. How did this remarkable development come about? In 1991, two prospectors, Charles Fipke and Stewart Blusson, defied conventional opinion that the Canadian Shield was not a geological structure where diamond could be formed. In their search, they discovered diamond-bearing kimberlite near Lac de Gras in the Northwest Territories. Following extensive drilling, their find proved to have commercial viability, and BHP Diamonds Inc. decided to develop a mine. The company applied to Ottawa to secure approval, and in 1996 the federal government approved the NWT Diamonds Project, which entails the operation of the Ekati diamond mine in the Lac de Gras area of the Northwest Territories for about 25 years. Five diamond-bearing **kimberlite pipes** are being mined, four of them located within a few kilometres of each other. All five kimberlite pipes lay under lakes that had to be drained to facilitate mining operations.

During the operation of the Ekati diamond mine, between 35 and 40 million tonnes of waste rock are removed from the kimberlite pipes each year. Since only a small proportion of that rock contains diamonds, the rest of the rock is placed in piles near the mines. Recovery of diamonds from the ore takes place in a processing plant where the ore is crushed and diamonds are separated. Final sorting is done using X-rays to separate the diamonds from the remaining waste material. BHP then sells rough-cut diamonds at international diamond markets. The mining operation employs about 800 workers with an estimated annual wage bill of $40 million. BHP uses an air-commuting system to bring workers from Yellowknife to its mine site on a two-weeks-in and two-weeks-out rotation.

Diamond mining has quickly become the backbone of the mining industry in the Territorial North. By value of production, number of employees, and spinoff effects, BHP Diamonds is the leading mining company in the North. In 2006, the two mines, Ekati and Diavik, produced $1.6 billion worth of diamonds (Table 10.6). Their effect on the economy of the Northwest Territories has been profound. One spinoff was the designation of Yellowknife as the pickup point for miners who commute by air on a rotational scheme to the mine site. This reverses the usual trend of cash and tax flowing out of the territories. A second spinoff was the location of three diamond-cutting and -polishing businesses in Yellowknife.

Another diamond mine, Jericho (Nunavut), began production in 2006 and Snap Lake is to begin production in late 2007 (Vignette 10.8). A deposit at Kennedy Lake, a De Beers diamond property, remains on hold.

How important is the diamond industry to the Northwest Territories? The figures are dazzling—diamond production has gone from zero in 1997 to $1.6 billion in 2006 and the total value of the diamond reserves at the three mines is around $23.3 billion (Statistics Canada, 2004). The newly opened mine in Nunavut accounted for $400,000 worth of diamonds in 2006. According to Randy Turner, president of Winspear Resources Ltd, the Ekati diamond field contains about 71 million carats with an average value of $133 a carat or a total value of $9.4 billion; the Diavik field has about 101 million carats with an average value of $80 a carat or a total value of $8 billion; and Snap Lake has 42 million carats with an average value of $140 a carat or a total value of $5.9 billion (Belford, 2000: C9).

Megaprojects: Achilles Heel

Megaprojects in the Territorial North are based on non-renewable resources, which consist of petroleum and mineral deposits. Forests, water, and wildlife are classified as renewable resources. The value of non-renewable resources in the North far outweighs the economic value derived from renewable resources. In 2006, mineral production in the Territorial North was valued at $2.25 billion, which was almost 2.5 per cent of Canada's mineral output (Table 10.6). Several mines have closed: the Nanisivik and Polaris lead/zinc mines closed in 2002; Con gold mine in 2003; and Giant gold mine in 2005. That same year, approximately 66 per cent of the Territorial North's production came from the Northwest Territories. The leading minerals by value were diamonds, tungsten, oil, and natural gas. The major mines were located at Ekati, Diavik, and Jericho (diamonds); and Fort Liard (natural gas). Placer gold, valued at $50 million per year, is obtained from the valleys of the Yukon River and its tributaries. Norman Wells is the only oil-producing site in the Territorial North, but there are much larger oil and gas deposits in the Mackenzie Delta, Sverdrup Basin, and the Beaufort Sea (Vignette 10.2).

Non-renewable resource development, because of the limited lifespan of such projects, subjects the Territorial North to a boom-and-bust economic cycle. Since mineral and petroleum deposits are in finite quantities, they do not offer long-term economic stability for the region. This economic cycle is the Achilles heel of northern development. In addition, because resource development is based on world demand, production fluctuates with this demand, creating great instability in resource

Table 10.6 Mineral Production in the Territorial North, 2006 ($ millions)

Mineral Product	Yukon	Northwest Territories	Nunavut	Territorial North
Metals	37.7	55.6	0	93.3
Non-metals	55.7	1,573.3	29.2	1,658.2
Fuels	0*	497.0*	0	497.0*
Total	93.4	2,125.9	29.2	2,248.5

*Estimate.

Sources: Natural Resources Canada (2007); Indian and Northern Affairs Canada (2007b).

Vignette 10.8 Nunavut's First Diamond Mine

The construction of the Jericho diamond mine, located south of the **Arctic Circle** near the Lupin gold mine, marks the beginning of a diamond industry in Nunavut. Diamond production from this small but rich ore deposit is expected to extend over eight years from 2006 to 2014. Mine and plant construction was completed in 2006. Work at the open-pit mine requires 100 employees and another 40 workers are employed at the processing plant. About one-quarter of the workers commute from the Inuit community of Kugluktuk. With a two week rotation system (two weeks at site followed by two weeks at home), only half of the labour force is on site at any one time. Open-pit mining will take place for the first four years and then be replaced by underground mining. While other kimberlite pipes exist in the area, the short lifespan of this mine demonstrates the difficulty of developing a sustainable economy based on non-renewable resources.

communities. Fluctuations in mineral production provide a measure of this instability, and although production has been on the upswing in recent years, especially with the development of diamond deposits, individual communities in the past have been devastated by mine closings and the same, inevitably, will occur in the future.

The negative economic and social effects are deeply felt in resource towns. Mines even suspend their operations during periods of low prices. In 1997, for example, the Faro mine was closed for the fourth time in its short history. Again, low metal prices made the zinc, lead, and silver-mining operation unprofitable. The Faro mine had been the principal mining activity for 30 years in Yukon. Each time the mine closed, the value of mineral production in Yukon dropped drastically. For example, in 1992 Yukon accounted for nearly half a billion dollars worth of minerals. After the Faro mine closed in 1993, production dropped to $81 million in 1994.

Megaprojects have made important contributions to the northern economy. These contributions include generating the bulk of the North's GDP, expanding the northern transportation infrastructure, and providing high wages for employees. The diamond industry holds out some hope of even greater spinoffs such as the processing of diamond gems in Yellowknife. Megaprojects, however, though touted as the engine of northern development in the 1970s, have so far failed to transform the Territorial North's economy. While generating profits for the corporations, they have failed to diversify the northern economy, have not solved the massive unemployment problem, and have increased the region's economic vulnerability due to sudden changes in world demand for its primary products. Worse yet, all northern megaprojects are based on non-renewable resources, which, having a fixed lifespan, cannot provide the basis for a stable, long-term economy. From this perspective, megaprojects are not the engine, but rather are the Achilles heel of the northern economy.

The Territorial North's Future

As the only geographic region in Canada where Aboriginal peoples form a majority of the population, the Territorial North is destined to remain both a resource frontier and a homeland for Aboriginal peoples. Nunavut, for instance, is both a political expression of this homeland concept for the Inuit of the Eastern Arctic and a region dependent on non-renewable resource development for its economic growth.

Comprehensive land-claim agreements have provided Aboriginal peoples with capital, land, and control over wildlife to chart a new future. Although they are involved in the market economy and residing in communities, Aboriginal peoples, but especially the elders, retain a strong attachment to the land, country food, and their culture. However, Aboriginal youth appear to have less interest in the land, country food, and their language than their parents.

As a resource hinterland, the Territorial North has three principal characteristics: it is far from world markets; resource development depends on external demand; and the economy is sensitive to fluctuations in world prices for resources. While the three governments are seeking sustainable and robust economies, the geographic realities make those goals difficult to achieve. For instance, the economy of the Territorial North has two serious flaws, namely, that it is based on non-renewable resources and that economic benefits from resource development accrue mainly to firms and workers in southern Canada and foreign countries. Combined, these two flaws lead to a state of underdevelopment known as a staple trap.

In the next decade, megaprojects will continue to draw the Territorial North more closely into the global economy as a resource frontier. While diamond production is now the leading industry, if the Mackenzie Gas Project proceeds, the natural gas will become the leading resource by value. The implication of such resource development, based on its non-renewable nature, is that stable economic growth and economic diversification remain elusive goals. Without a more diversified economy, the Territorial North will continue to suffer from boom-and-bust economic cycles and the economic instability thus created. Even with ownership of natural resources transferred to territorial governments, these governments will remain financially dependent on Ottawa because of the 'have-not' character of the region's geography.

Notes

1. There is a third climatic zone in the Territorial North—the Cordillera. However, the Cordillera climate, often described as a mountain climate, is affected by elevation (as elevation increases, temperature drops). North of 60° N, the Cordillera climate is also affected by latitude so that boreal natural vegetation is found at lower elevations and tundra natural vegetation at high elevations.
2. Sharing food was essential for small hunting groups to be able to live together harmoniously. By sharing in all hunters' successes, they could adjust for the vagaries of individual luck and reduce the threat of starvation. Today, the sharing of food remains a pivotal component of Aboriginal culture and the Native economy.
3. Sometime around the year 1000 the Vikings made contact with the ancestors of the Inuit, the Thule (*c.* AD 1000 to 1600). The Thule originated in Alaska where they hunted bowhead whales and other large sea mammals. They quickly spread their whaling technology across the Arctic, travelling in skin boats and dogsleds. With the onset of the Little Ice Age in the fifteenth century, climate conditions affected the distribution of animals and the Thule who were dependent on them. An increased amount of sea ice blocked the large whales from their former feeding grounds, resulting in the collapse of the Thule whale hunt. With the loss of their main source of food, the Thule had to rely more and more on locally available foods, usually some combination of seal, caribou, and fish. By the eighteenth century, the Thule culture had disappeared and was replaced by the Inuit hunting culture.
4. Territorial governments, unlike provinces, are governed under delegated powers from the federal government. In that sense, they are dependent on Ottawa for their political powers. Under the Canadian Constitution, Ottawa has the power to govern its territories. In the past, responsible government occurred when a territory had a sufficiently large population and tax base to warrant having an elected council. Until that time, the governing members were appointed by the federal government.
5. Social programs may have encouraged large families. For example, the family allowance

program begun in 1945 provided a payment for each child. For a family of five, the annual cash derived from a family allowance was often greater than the cash a father might earn from the sale of fur pelts.

6. Today, capital and operating costs are still serious problems for hunters. Transportation from a settlement to hunting/trapping areas is a major expenditure. Snowmobiles, for example, may cost as much as $20,000 in a northern retail store (in the 1990s, they cost $5,000 to $10,000), and the price of gasoline to fuel them is significantly higher in the North. With a continuing rise in the prices of manufactured goods, the cost of a snowmobile 10 years from now may double again. Since a snowmobile used for long-distance travel has a short lifespan, the hunter/trapper must replace it about every three or four years.
7. The first comprehensive land-claim submission came from the Dene/Métis in 1974. The land claimed by the Dene/Métis extended over most of the Mackenzie Basin north of 60° N. This land was called Denendeh. In 1988, an agreement-in-principle was signed by the two negotiating parties (the federal government and Dene Nation). This agreement called for a cash payment of $500 million over 15 years plus title to 181,230 km^2 of land. Nearly 6 per cent of this land (10,000 km^2) was to include subsurface rights for the Dene/Métis. As well, the Dene/Métis would obtain a share of federal resource royalties, including those generated by the Norman Wells oil field. Chiefs and elders from the Great Slave Lake area refused to approve this agreement for two main reasons—the agreement-in-principle contained no reference to self-government and it called for the surrender of Aboriginal rights. With this rejection, the Gwich'in and Sahtu/Métis subsequently negotiated their own comprehensive land-claim agreements with Ottawa.
8. The Inuvialuit broke away from the other Inuit who were seeking a common land-claim settlement that would have stretched from the Arctic Coast of the Yukon to Baffin Island. The Inuvialuit wanted to advance their own land claim before the huge oil and gas deposits in the Beaufort Sea were developed. The Inuvialuit, with their separate agreement, hoped to obtain land with oil and gas deposits and achieve a taxation arrangement similar to that of the Arctic Slope Regional Corporation of the Inupiat in Alaska near Prudhoe Bay. This municipality received substantial tax revenues from the oil companies.
9. Under the IFA, Canada agreed to transfer $45 million in 1977 dollars. By 1984, these funds were valued at $152 million. The payments to the Inuvialuit Regional Council, the business corporation of the IFA, began on 31 December 1984 with $12 million. By 1997, there had been three annual payments of $1 million beginning 31 December 1985; five annual payments of $5 million beginning 31 December 1988; four annual payments of $20 million beginning 31 December 1993; and a final payment on 31 December 1997 of $32 million (Canada, 1985: 107).
10. Until the 1970s, Canadians did not question the merit of northern industrial projects. The Mackenzie Valley Pipeline Inquiry headed by Thomas Berger made Canadians aware of the hidden costs of building a gas pipeline from Prudhoe Bay along the Arctic coast to the Mackenzie River and then south along the Mackenzie Valley to markets in southern Canada and the United States. The hidden costs identified by Berger were: the potential threat to wildlife, especially the Porcupine caribou herd; the ground subsidence that could result from a buried pipeline; and the feared social impacts on Aboriginal peoples living in the construction zone. For more on this subject, see Bone (2003a: 169–70).
11. This pipeline would be constructed by Canadian Arctic Gas Pipeline Ltd, which represented 27 Canadian and American oil producers, such as Exxon, Gulf, and Shell, and Trans-Canada Pipelines. Later, a rival bid was made by Foothills Pipeline Ltd, which consisted of Alberta Gas Trunk Line and Westcoast Transmission. Prudhoe Bay is located along the North Slope of Alaska. Alaskan oil accounts for 20 per cent of US oil production. An oil pipeline, known as the Trans-Alaska Pipeline System (TAPS), transports Alaskan oil from Prudhoe Bay to Valdez and then by tanker to markets along the west coast of the United States.
12. In 1977, a rival proposal by Foothills Pipeline called for a natural gas pipeline to follow the route of the Alaska Highway. In 1982, Ottawa approved the Alaska Highway Gas Pipeline Project. However, this proposed project was never built because of declining natural gas prices in the 1980s. Now that gas prices have soared to historically high levels, the Yukon government is calling for the construction of the Alaska Highway Gas Pipeline.

Challenge Questions

1. Why is global warming expected to reach its maximum temperature increase in the Arctic?
2. Why would the British naval expedition led by Sir John Franklin stand a better change of navigating across the Northwest Passage today than in 1845?
3. What are the key weaknesses in Canada's case for claiming the Arctic waters separating the islands in the Arctic Archipelago?
4. Table 10.3 indicates that Nunavut has the highest birth rate while Table 10.6 shows that its mineral production ranks well below both Yukon and the Northwest Territories. What are the implications for Nunavut's economic future?
5. On the one hand, megaprojects are the driving force behind the economy of the Territorial North. On the other hand, these projects have an Achilles Heel that prohibits sustainable growth. Why?
6. How have comprehensive land-claim agreements equipped Aboriginal peoples to chart a new future?
7. What is the difference between 'ethnic' First Nation self-governments and the Nunavut form of self-government?
8. Should the federal government financially support the proposed Mackenzie Gas Project because it is in the 'national interest' to open the North to development?

Key Terms

air commuting
Travel to a work site, such as a mine, by aircraft owned or hired by the company. Commuting to work is a common theme in geographic literature. However, air commuting is a relatively new phenomenon. Until the 1970s, companies built resource towns to house workers and their families. Since then, companies have opted to employ air transportation to take their workers to and from the work site. These employees remain at the work site for a week or two, working long shifts (often 12 hours per day). Workers then have a week or two at home. The company pays for the air transportation and for the food and lodgings at the work site.

Arctic Circle
An imaginary line that signifies the northward limit of the sun's rays at the time of the winter solstice (21 December). At a latitude of 66° 32' N, the sun does not rise above the horizon for one day of the year (the winter solstice). Except for a short period of twilight, darkness prevails for 24 hours. At the summer solstice (21 June), the sun's rays do not fall below the horizon, providing constant daylight for 24 hours.

Arctic ice pack
Floating sea ice in the Arctic Ocean that has consolidated into an ice pack, with an extent of over 10 million km^2. New sea ice (less than one year in age) is often about 1 m thick; old sea ice can reach 5 m in thickness. Ice ridges are formed, reaching 20 m in thickness. Scientists believe that higher temperatures are reducing the geographic extent and thickness of the Arctic ice pack.

Barren Grounds
The vast area of tundra stretching from the west coast of Hudson Bay to the Great Slave and Great Bear lakes and northward to the Arctic Ocean. The Barren-Ground Caribou use this region for calving each summer before migrating to the boreal forest.

country food
Food, primarily game, such as caribou, fish, and sea mammals, obtained by Aboriginal peoples from the land and sea. Although Indians, Inuit, and Métis now live in settlements, they still fish and hunt for cultural and economic reasons.

homeland
People who live in a region develop a strong attachment to that place; a sense of place.

kimberlite pipes
Intrusions of igneous rocks in the earth's crust that take a funnel-like shape. Diamonds are sometimes found in these rocks.

megaprojects
Large-scale construction projects, often related to resource extraction, that exceed $1 billion and take more than two years to complete.

resource frontier
The perception of the Territorial North as a place of great mineral wealth that awaits development by outsiders.

settlement areas
Each comprehensive land claim has a geographic extent that is known as a settlement

area. While less than 25 per cent of this land is allocated to the Aboriginal beneficiaries as a collective (not individual) landholding, the entire area is subject to the environmental and wildlife regulations exercised by the settlement area's co-management boards.

terraces
Streams and rivers form terraces when they cut downward into the flood plain and form a new and lower flood plain. The old flood plain (now a terrace) is found along the sides of the stream or river.

traditional ecological knowledge
Familiarity with and knowledge of the natural surroundings. This knowledge of the environment and wildlife in a particular area or ecosystem has accumulated over the centuries among a group of people who live close to the land/sea; also called local ecological knowledge.

underemployment
Several definitions exist: a classical one refers to workers who are employed, but not in the desired capacity, whether in terms of compensation, hours, or level of education, skill, and experience. In this text, the term refers to persons in small communities where the very few jobs are already filled, and consequently, because these potential workers are aware that no job opportunities exist, they therefore do not seek jobs.

unemployment
Lack of paid work, but this term and statistics based on it measure only those who are seeking such paid employment.

winter roads
Temporary ice roads over muskeg, lakes, and rivers built during the winter to provide ground transportation for freight and travel to remote communities.

Bibliography

Anderson, Robert B., and Robert M. Bone. 1995. 'First Nations Economic Development: A Contingency Perspective', *Canadian Geographer* 39, 2: 120–30.

——— and ———. 2003. *Natural Resources and Aboriginal People in Canada: Readings, Cases and Commentary*. Concord, Ont.: Captus Press.

Belford, Terrence. 2000. 'Moiling for Diamonds in Canada's North', *National Post*, 8 May, C9.

Berger, Thomas R. 1977. *Northern Frontier, Northern Homeland: The Report of the Mackenzie Valley Pipeline Inquiry*, 2 vols. Ottawa: Minister of Supply and Services.

Bockstoce, John R. 1994. *Whales, Ice and Men: The History of Whaling in the Western Arctic*. Seattle: University of Washington Press.

Boden, Jürgen F., and Elke Boden, eds. 1991. *Canada North of Sixty*. Toronto/Osteinbek: McClelland & Stewart/Alouette Verlag.

Bone, Robert M. 1992. *The Geography of the Canadian North: Issues and Challenges*. Toronto: Oxford University Press.

———. 1998. 'Resource Towns in the Mackenzie Basin', *Cahiers de Géographie du Québec* 42, 116: 249–59.

———. 2003a. *The Geography of the Canadian North: Issues and Challenges*, 2nd edn. Toronto: Oxford University Press.

———. 2003b. 'Power Shifts in the Canadian North: A Case Study of the Inuvialuit Final Agreement', in Anderson and Bone (2003: 382–93).

——— and Shane Long. 1995. 'Population Change in Aboriginal Settlements in the Mackenzie Basin North of 60°', *Proceedings of the 1995 Symposium of the Federation of Canadian Demographers*. Ottawa: St Paul University.

Canada. 1985. *The Western Arctic Claim: The Inuvialuit Final Agreement*. Ottawa: Department of Indian Affairs and Northern Development.

———. 1991. 'Comprehensive Land Claim Agreement Initialled with Gwich'in of the Mackenzie Delta in the Northwest Territories', Communiqué 1–9171. Ottawa: Department of Indian Affairs and Northern Development.

———. 1993a. 'Formal Signing of Tungavik Federation of Nunavut Final Agreement', Communiqué 1–9324. Ottawa: Department of Indian Affairs and Northern Development.

———. 1993b. *Umbrella Final Agreement between the Government of Canada, Council for Yukon Indians and the Government of the Yukon*. Ottawa: Department of Indian Affairs and Northern Development.

———. 2004. Agreements. Ottawa: Department of Indian and Northern Affairs. At: <www.ainc-inac.gc.ca/pr/agr/index_e.html#Comprehensive%20Claims%20Agreements>.

Canadian Press. 2003. 'Ottawa to Phase Out Yukon Placer Mining', *National Post*, 28 Jan., FP10.

Cattaneo, Claudia. 2000. 'Inuvialuit Press PM for Pipeline Support', *National Post*, 24 Nov., C5.

CBC News. 2007. 'Mackenzie Gas Line Still "Leading Case" Despite Bloating $16.2B Cost Outlook', 12 Mar. At: <www.cbc.ca/cp/business/070312/b031292A.html#skip300x250>.

Chan, Laurie H.M. 2006. 'Food Safety and Food Security in the Canadian Arctic', *Meridian* (Publication of the Canadian Polar Commission): 1–3.

Coates, Ken, and Bill Morrison. 1992. *The Forgotten North: A History of Canada's Provincial Norths*. Toronto: Copp Clark.

Crowe, Keith J. 1991. *A History of the Original Peoples of Northern Canada*, 2nd edn. Montréal and Kingston: McGill-Queen's University Press.

Dahl, Jens, Jack Hicks, and Peter Jull, eds. 2000. *Nunavut: Inuit Regain Control of Their Lands and Their Lives*. Copenhagen: International Work Group for Indigenous Affairs.

Dickason, Olive Patricia. 2002. *Canada's First Nations: A History of Founding Peoples from Earliest Times*, 3rd edn. Toronto: Oxford University Press.

Difrancesco, R.J. 2000. 'A Diamond in the Rough? An Examination of the Issues Surrounding the Development of the Northwest Territories', *Canadian Geographer* 44, 2: 114–34.

Dominion, The. 2007. 'Events in April', no. 45 (2 May). At: <www.dominionpaper.ca/articles/1151>.

Elias, Peter Douglas. 1995. *Northern Aboriginal Communities: Economies and Development*. North York, Ont.: Captus Press.

French, Hugh M., and Olav Slaymaker. 1993. *Canada's Cold Environments*. Montréal and Kingston: McGill-Queen's University Press.

Frideres, James. 1993. *Native Peoples in Canada*. Scarborough, Ont.: Prentice-Hall.

Globe and Mail. 1993. 'Mackenzie Land Claim Settled', 7 Sept., A1.

Government of the Northwest Territories, Bureau of Statistics. 2003. *Statistics Quarterly* 25, 2. At: <www.stats.gov.nt.ca/Statinfo/Generalstats/statsquarterly/_statq.html>.

———. 2004. Northwest Territories—2003 . . . by the numbers. At: <www.stats.gov.nt.ca/Statinfo/Generalstats/bythenumbers/BTNhome(dvo).html>.

———. 2006. 2006 NWT Socio-Economic Scan. At: <www.stats.gov.nt.ca/Statinfo/Generalstats/Scan/Scan_2006.pdf >.

Grant, Shelagh D. 1988. *Sovereignty or Security? Government Policy in the Canadian North, 1936–1950*. Vancouver: University of British Columbia Press.

Hamelin, Louis-Edmond. 1978. *Canadian Nordicity: It's Your North, Too*, trans. William Barr. Montréal: Harvest House.

———. 1982. 'Originalité culturelle et régionalisation politique: Le project Nunavut des Territoires-du-Nord-Ouest (Canada)', *Recherches Amérindiennes au Québec* 12, 4: 251–62.

Hamilton, John David. 1994. *Arctic Revolution: Social Change in the Northwest Territories, 1935–1994*. Toronto: Dundern Press.

Hesketh, Bob, ed. 1996. *Three Northern Wartime Projects: The Alaska Highway, the Northwest Staging Route, and Canol*. Edmonton: Canadian Circumpolar Institute.

Indian and Northern Affairs Canada. 2007a. 'Northern Oil and Gas Activities'. At: <www.ainc-inac.gc.ca/oil/ann/ann2006/intro_e.html>.

———. 2007b. *Northern Oil and Gas Annual Report 2006*. At: <www.ainc-inac.gc.ca/oil/ann/ann2006/dev_e.html>.

Ironside, R.G. 2000. 'Canadian Northern Settlements: Top-down and Bottom-up Influences', *Geografiska Annaler* 82 B, 2: 103–14.

Johnston, Margaret E. 1994. *Geographic Perspectives on the Provincial Norths*. Northern and Regional Studies Series, vol. 3. Centre for Northern Studies, Lakehead University. Mississauga, Ont.: Copp Clark Longman.

Keith, Robbie. 1998. 'Arctic Contaminants: An Unfinished Agenda', *Northern Perspectives* 25, 2: 1–3.

Kenney, Gerard. 1994. *Arctic Smoke and Mirrors*. Prescott, Ont.: Voyageur Publishing.

Légaré, André. 1993. 'Le project Nunavut: Bilan des revendications des Inuit des Territoires-du-Nord-Ouest', *Études/Inuit/Studies* 17, 2: 29–62.

———. 1996. 'Le gouvernement du Territoire du Nunavut', *Études/Inuit/Studies* 20, 1: 7–43.

McGhee, Robert. 1996. *Ancient People of the Arctic*. Vancouver: University of British Columbia Press.

MacLachlan, Letha. 1996. *NWT Diamonds Project: Report of the Environmental Assessment Panel*. Ottawa: Canadian Environmental Assessment Agency.

McPherson, R. 2003. *New Owners in Their Own Land: Minerals and Inuit Land Claims*. Calgary: University of Calgary Press.

Marcus, Alan R. 1995. *Relocating Eden: The Image and Politics of Inuit Exile in the Canadian Arctic*. Hanover, NH: University Press of New England.

Marsh, James H., ed. 1988. *The Canadian Encyclopedia*, 2nd edn. Edmonton: Hurtig.

Myers, H. 2001. 'Changing Environment, Changing Times: Environmental Issues and Political Action in the Canadian North', *Environment* 43, 6: 32–44.

Natural Resources Canada. 2005. Mineral Production of Canada, by Provinces and Territories. At: <mmsd1.mms.nrcan.gc.ca/mmsd/production/2005/WEB05P.pdf>.

———. 2007. 'High Metal Prices Spur Mineral Production to a Record $34 billion in 2006', *Information Bulletin: Mineral Production*, Mar. At: <www.nrcan.gc.ca/mms/pdf/minprod-07_e.pdf>.

Notzke, Claudia. 1994. *Aboriginal Peoples and Natural Resources in Canada*. Toronto: Captus Press.

O'Reilly, Kevin. 1996. 'Diamond Mining and the Demise of Environmental Assessment in the North', *Northern Perspectives* 24, 1–4: 1–6.

Page, Robert. 1986. *Northern Development: The Canadian Dilemma*. Toronto: McClelland & Stewart.

Poelzer, Greg. 2001. 'Aboriginal Peoples and Environmental Policy in Canada: No Longer at the Margins', in Debora L. VanNijnatten and Robert Boardman, eds, *Canadian Environmental Policy: Context and Cases*, 2nd edn. Toronto: Oxford University Press, 87–106.

Purich, Donald. 1992. *The Inuit and Their Land: The Story of Nunavut*. Toronto: Dormer.

Ray, Arthur J. 1990. *The Canadian Fur Trade in the Industrial Age*. Toronto: University of Toronto Press.

Rowley, Graham W. 1996. *Cold Comfort: My Love Affair with the Arctic*. Montréal and Kingston: McGill-Queen's University Press.

Royal Commission on Aboriginal Peoples. 1996. *Report of the Royal Commission on Aboriginal Peoples*, 5 vols. Ottawa: Minister of Supply and Services.

Saku, James A., and Robert M. Bone. 2000. 'Looking for Solutions in the Canadian North: Modern Treaties as a New Strategy', *Canadian Geographer* 44, 3: 259–70.

———, ———, and Gérard Duhaime. 1998. 'Towards an Institutional Understanding of Comprehensive Land Claim Agreements in Canada', *Etudes/Inuit/Studies* 22, 1: 109–21.

Shidelar, Janet. 2000. *The Geography of Canada Bibliography Series: Vol. 6, The North*. Plattsburgh, NY: Plattsburgh State University, Center for the Study of Canada.

Stanford, Quentin H., ed. 1998. *Canadian Oxford World Atlas*, 4th edn. Toronto: Oxford University Press.

Statistics Canada. 1997a. *A National Overview: Population and Dwelling Counts*. Catalogue no. 93–357–XPB. Ottawa: Industry Canada.

———. 1997b. 1996 Census: Nation Tables—Population by Mother Tongue, Showing Age Groups, for Canada, Provinces and Territories, 1996 Census—20% Sample Data, 2 Dec. At: <www.statcan.ca/english/census96/>.

———. 1997c. '1996 Census: Mother Tongue, Home Language and Knowledge of Languages', *The Daily*, 2 Dec. At: <www. statcan.ca/Daily/English/>.

———. 1998a. *Canadian Economic Observer*. Catalogue no. 11–010–XPB. Ottawa: Statistics Canada.

———. 1998b. '1996 Census: Aboriginal Data', *The Daily*, 13 Jan. At: <www.statcan.ca/Daily/English/>.

———. 1998c. '1996 Census: Ethnic Origin, Visible Minorities', *The Daily*, 17 Feb. At: <www.statcan.Daily/English/>.

———. 1998d. '1996 Census: Education, Mobility and Migration', *The Daily*, 14 Apr. At: <www.statcan.Daily/English/>.

———. 2002a. 2001 Census: Population and Dwelling Counts, for Census Metropolitan Areas and Census Agglomerations, 2001 and 1996 Censuses, 16 July. At: <www.statcan.ca/english/IPS/Data/93F0050XCB2001013.htm>.

———. 2002b. '2001 Census: Language, Mobility and Migration', *The Daily*, 10 Dec. At: <www.statcan.Daily/English/>.

———. 2003a. Canadian Statistics, International Trade: Exports of Goods on a Balance-of-Payments Basis. At: <www.statcan.ca/english/Pgdb/gblec04.htm>.

———. 2003b. Canadian Statistics: Population of Census Metropolitan Areas. At: <www.statcan.ca/english/Pgdb/demo05.htm>.

———. 2003c. Census of Canada 2001—Labour, Employment and Unemployment, by Industry, by Province. At: <www.statca.ca/english/Pgdb/labor21b.htm>.

———. 2003d. Census of Canada 2001—Canada's Ethnocultural Portrait: The Changing Mosaic. Analytical Series 96F0030XIE2001008, 21 Jan. At: <www12.statcan.ca/english/census01/products/analytic/companion/etoimm/subprovs.cfm>.

———. 2003e. Canadian Statistics: Distribution of

Employed People, by Industry, by Province. At: <www.statcan.ca/english/Pgdb/labor21b.htm>.

———. 2004. 'Study: Diamonds Are Adding Lustre to the Canadian Economy', *The Daily*, 13 Jan. At: <www.statcan.ca/Daily/English/040113/d040113a.htm>.

———. 2006a. Deaths and Death Rate, by Province and Territory. At: <www40.statcan.ca/l01/cst01/demo07b.htm>.

———. 2006b. Births and Birth Rate, by Province and Territory. At: <www40.statcan.ca/l01/cst01/demo04b.htm>.

———. 2006c. Components of Population Growth, by Province and Territory. At: <www40.statcan.ca/l01/cst01/demo04b.htm>.

———. 2006d. Annual Demographic Statistics 2005. At: <www.statcan.ca/english/freepub/91-213-XIB/0000591-213-XIB.pdf>.

———. 2007a. Population and Dwelling Counts, for Census Metropolitan Areas and Census Agglomerations, 2006 and 2001 Censuses—100% Data. At: <www12.statcan.ca/english/census06/data/popdwell/Table.cfm?T=201&S=3&O=D&RPP=150>.

———. 2007b. Gross Domestic Product, Expenditure-based, by Province and Territory. At: <www40.statcan.ca/l01/cst01/econ15.htm>.

Stevenson, Marc, Culley Schweger, Valerie Raey, Elaine L. Maloney, and Susan Macara. 1996. *Environmental and Economic Issues in Fur Trapping*. Edmonton: Canadian Circumpolar Institute.

Toronto Star. 2004. 'Canada Climbs to Third in World Diamond Output', 14 Jan., C1.

Usher, Peter J. 1982. 'The North: Metropolitan Frontier, Native Homeland?', in L.D. McCann, ed., *Heartland and Hinterland: A Geography of Canada*. Scarborough, Ont.: Prentice-Hall, 411–56.

——— and George Wenzel. 1987. 'Native Harvest Surveys and Statistics: A Critique of Their Construction and Use', *Arctic* 40, 2: 145–60.

——— and ———. 1989. *A Strategy for Supporting the Domestic Economy of the Northwest Territories*. Report for the Legislative Assembly's Special Committee on the Northern Economy. Ottawa: P.J. Usher Consulting Services.

——— and ———. 1989. 'Socio-Economic Aspects of Harvesting', in Randy Ames, Don Axford, Peter Usher, Ed Weick, and George Wenzel, *Keeping on the Land: A Study of the Feasibility of a Comprehensive Wildlife Harvest Support Program in the Northwest Territories*. Ottawa: Canadian Arctic Resources Committee, ch. 1.

Wenzel, George. 1991. *Aboriginal Rights, Animal Rights: Ecology, Economy and Ideology in the Canadian Arctic*. Toronto: University of Toronto Press.

Williamson, Robert G. 1974. *Eskimo Underground: Socio-Cultural Change in the Canadian Central Arctic*. Occasional Papers II. Uppsala, Sweden: Almqvist & Wiksell.

Further Reading

Morrison, William R. 1998. *True North: The Yukon and Northwest Territories*. Toronto: Oxford University Press.

This is a finely illustrated history of Canada's North. Some 200 photographs and five maps accompany the lively narrative that describes the main elements of the North's history. From the coming of the First Peoples to our contemporary period, Professor Morrison has etched out an entertaining story of the dual nature of the Canadian North—both a northern frontier and a northern homeland. In Chapter Eight, 'Invasion: War and the Militarization of the North', construction of the Alaska Highway is described both in words and photographic images. Black American troops cleared the right-of-way for the highway and horses hauled supplies to the work camps in the first year of this mega-construction project. The difficulty of constructing a highway across the rugged northern terrain is well illustrated by a photograph of the highway turning into a quagmire after a heavy rain. The completed military highway was not suited for faint-of-heart drivers—places like 'Suicide Hill' are discussed and illustrated. The final chapter, 'Search for a Future: the Modern North, 1970–1995', depicts the North as a place where 'its future seems to be in a state of permanent uncertainty', but Morrison goes on to identify one constant, the people, particularly the Native people. Students will find this illustrated book both easy to read and very informative.

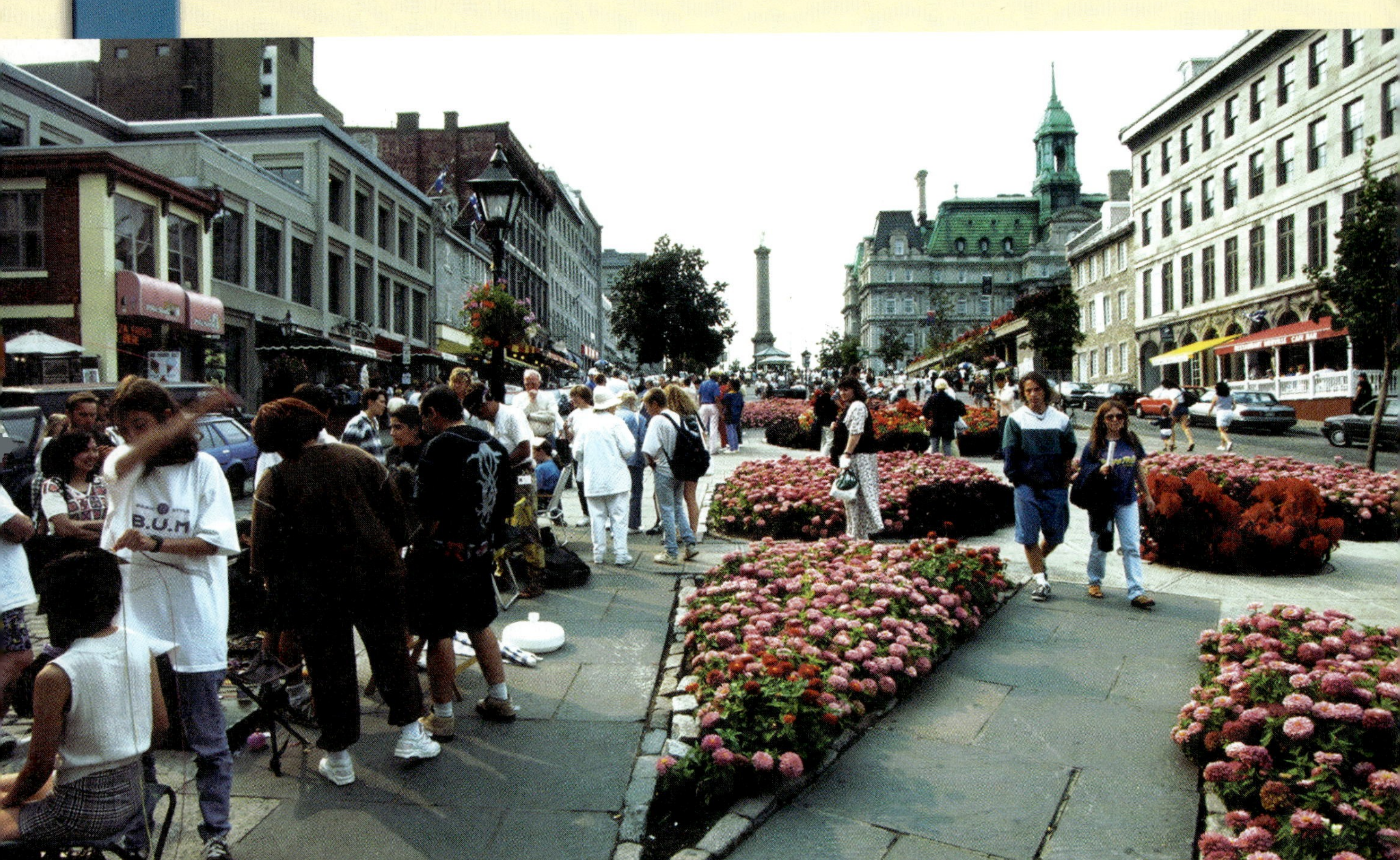

Montréal, Québec (© David G. Hauser/Corbis)

Overview and Objectives

Canada is a country composed of six major geographic regions, each with its own sense of place. This final chapter summarizes the complex regional differences caused by Canada's physical and historical geography, its core/periphery economic structure, and the tensions existing in its social and political structures. The principal tensions or faultlines, as we have seen, are between centralists and decentralists, Aboriginal and non-Aboriginal Canadians, French-speaking and English-speaking Canadians, and new immigrants and those born in this country. The future, while full of uncertainty, contains one truth: Canada will remain a country of cultural and regional diversity. The four faultlines are but a recognition of the process of adjustment and change so necessary in a 'soft' democratic country. Though problems have arisen because of this diversity in the past, and will continue to arise in the future, Canadians have recognized that differences cannot be obliterated. Instead, compromises must and can be found.

Canada's geography and the consequent political arrangements dictate that the nation and its six geographic regions are forever searching for negotiated solutions to vexing issues, whether equalization payments or the number of members of Parliament. What is important—beyond the heated words—is the existence of a political process that addresses these regional differences even though the agreement is often temporary and incomplete. Final solutions are not the stuff of democracies. This chapter will:

- Demonstrate that Canada is and will remain a country of regions.
- Re-examine the core/periphery model in the light of globalization.
- Re-evaluate Canada's four faultlines, not as centrifugal forces but as centripetal ones.
- Stress that Canada's future lies not in division but in the 'process' of seeking compromises.

CHAPTER 11

Canada: A Country of Regions

INTRODUCTION

'Canada is big—preposterously so', wrote geographer Kenneth Hare (1968: 31). The sheer size of Canada has meant that the country spans a diverse physical geography, varying climate conditions, and a wide range of vegetation. These geographic variations, combined with regional histories, have transformed the nation into a country of regions. Geography and history have woven Canada's regions into a federated nation and, in so doing, have shaped and then reshaped Canada's identity and the identities of its regions.

During the historical evolution of the country, tensions between certain groups have arisen. These tensions, known as faultlines, are manifested in four different contexts: between centralists and decentralists, between French-speaking and English-speaking Canadians, between Aboriginal Canadians and non-Aboriginal Canadians, and between new immigrants and those born in Canada. However, these cracks, rather than fragmenting the country, have forced Canadians to build bridges—between regions and governments and between different peoples. Canada, by incorporating cultural diversity and regional differences within its socio-political structure, provides an alternative to monolithic nation-states, where the majority settles differences—usually in its favour. Canada, on the other hand, has accepted minority positions, whether same-sex marriages or self-government for Aboriginal groups. In doing so, Canada is prepared to live with contradictions—and these contradictions make Canada an exciting and different place to live. In this concluding chapter, five issues will be addressed:

- Canada's regional character;
- Canada's core/periphery spatial structure;
- Canada's faultlines;
- Canada's past; and
- Canada's future.

Regional Character

The geographic complexity of Canada begins with its six geographic regions—Ontario, Québec, British Columbia, Western Canada, Atlantic Canada, and the Territorial North. These regions, though primarily created by geography, were shaped further by Canada's historical development. For instance, southern

Ontario presented some of the most favourable geographic conditions for economic growth and industrialization. It had fertile soils and a moderate climate, which supported agriculture and a growing population, as well as geographic proximity to water transportation routes and to US markets. Complementing these favourable natural features, historically Ontario was supported by economic conditions that allowed it to prosper, such as the Reciprocity Treaty and then protective tariffs—policies that ensured a stable market for the region's industries.

All of Canada's regions have been similarly shaped by a combination of geography and history. For instance, Québec has been shaped by the St Lawrence River and by its French-speaking inhabitants. British Columbia has been shaped by the Cordillera, which separates it from the rest of Canada, and by the railway that first connected it to the country. Western Canada has been shaped by its fertile soils and by a sense of alienation from the country's centre of power. Atlantic Canada has been shaped by its coastal geography and by economic downturns in its resource-oriented industries. The Territorial North has been shaped by its cold and harsh climate and by Aboriginal land-claim agreements. This is to name but a few of the geographic and historical features that have defined Canada's regions.

From each region's history and geography, a sense of place has developed. Regional geography is very much determined by a sense of place, the power of which has been elegantly expressed by Canadian writer John Ralston Saul (1997: 69):

> Because, in spite of intellectual claims to the contrary, not religion, not language, not race but place is the dominant feature of civilizations. It decides what people can do and how they will live.

Within each region of Canada, the collective experiences of people, both past and present, evoke a sense of belonging to a place. Like other geographic concepts, this psychological sentiment operates at several geographic levels—national, regional, local. In this book, attention has been directed to regions and the common experiences within those regions that have fuelled a sense of place—for instance, the struggle to farm in a dry continental climate in Western Canada.

Within a federation of regions, sense of place can take on political connotations as divisions between regions feed geographic 'separatism'. Though Québec is identified with this sense of separation, almost every region experiences it to some degree. For instance, historian Margaret Ormsby (1958: 257) claimed that 'British Columbia was in, but not of, Canada', meaning that union with Canada for British Columbia was not based on sentiment but on material advantage, such as the promise of a railway and the absorption of BC's debt by Ottawa. In British Columbia, history created a sense of independence while geography formulated a sense of isolation from the rest of Canada. That Canada 'ends' at the Rocky Mountains is part of this region's sense of identity. Another part of this identity, or sense of place, comes from British Columbians' pride in their natural surroundings—this feeling is expressed in the provincial licence-plate motto, 'Beautiful British Columbia'.[1]

If Canada is so clearly made up of very different regions that are separate from the whole, how did it ever overcome its geographic and historical differences to evolve from regional clusters into a modern nation-state? Why is Canada more than a collection of economic regions dominated by self-interest? Canada was able to overcome its regional differences through a willingness to share power and thereby create a balance between central interests and those of regions and groups. In fact, the sharing of power was embedded in the Canadian Constitution right from the start. With the creation of a federation, the BNA Act divided governing powers between the provinces and the federal government. This initial sharing of powers set the stage for a trend to develop in which different Canadian regions and peoples would share power and economic wealth. In fact, Canadian history illustrates this trend. A number of struggles between those holding power and those wanting a share of that power have occurred. Often, the result has been compromise, with power being shared between parties. Equalization payments (1957), official bilingualism (1969), multiculturalism (1971), and modern land-claim agreements (the first in 1975) are examples of how the federation has been able to share wealth and cultural recognition between various regions and peoples. Spatial and cultural diversity did not create barriers to national unity that needed to be obliterated, but, rather, served as the very building blocks necessary to construct a humanitarian society.

Canada's strength—you might even say what makes it interesting—is its regional diversity. Physical geography laid the groundwork for Canada's regional structure while Canada's history unfolded within these natural settings. Economic, political, and social change has occurred—some regions have grown swiftly while others have lagged behind and in the process lost some of their best people—but Canada's regions have proven remarkably stable. Furthermore, compromises have been reached to reconcile the inequities between regions as well as the peoples across these regions. Equalization payments play a key role in keeping regional economic differences from getting out of balance. Calculating equalization payments is a complex process and some provinces are demanding changes. Change is possible, and the ultimate responsibility lies with the federal government. Each have-not region of the country sees these payments as an opportunity to correct regional fiscal imbalances so that they don't have 'second-class' education and social services systems. Atlantic Canada, for instance, gained relief from the 'negative' or clawback link between oil royalties and equalization payments with the 2004 Atlantic Accord (but then, the 2007 federal budget indicated that these gains may be diminished). The Territorial North but especially the Northwest Territories would like to obtain control over natural resources and thereby receive the royalties from resource developments. British Columbia, normally a province that does not receive equalization paymetns, went through a rough economic patch beginning in the late 1990s when trade with Asia declined sharply. Natural disasters, such as forest fires in 2003 that cost over half a billion dollars, the pine beetle invasion that began in earnest in 2004, and flooding in late 2006, added to BC's economic difficulties. Not surprising, then, the BC government was happy that it received equalization payments for seven of the last eight years (1999–2000 and 2001–2 to 2006–7). With its economy back on track, BC is once again a have province (Table 3.7).

The Core/Periphery Model

Canada's regional geography has significantly affected the country's national and regional economies. Physiography, climate, permafrost, and other natural elements have in many ways determined what economic activities could be pursued in each region and, therefore, how regional economies would develop. A way to examine this phenomenon is through the core/periphery model—cores represent the more prosperous and industrialized regions, which draw on the peripheries, the more resource-oriented, less economically diversified regions. Using Friedmann's version of this model (see Chapter 1), Canada's six geographic regions fall into: core regions (Ontario and Québec); upward transitional regions (British Columbia and, to a lesser degree, Western Canada); a downward transitional region (Atlantic Canada); and a resource frontier (the Territorial North). Furthermore, within each of Canada's geographic regions, natural variations have created internal core/periphery structures. For instance, within Ontario, southern Ontario represents an industrial core, while northern Ontario is a resource periphery.

For years, the core/periphery structure of Canada remained relatively stable. Central Canada benefited from geography and favourable federal policies that allowed its industrial sector to grow by guaranteeing a domestic market for its goods in other regions. The rest of Canada's regions acted as peripheries, supplying raw materials and purchasing the core's manufactured items. However, during the latter part of the twentieth century a new global economic order emerged that is affecting and will continue to affect Canada's core/periphery structure. A remarkable shift of demographic and economic power has already reached British Columbia and Alberta.

Since World War II, Canada has taken several steps towards a new economic goal of integration into larger economies. First, Canada became a participant in the GATT (more recently, the WTO) regime in order to be involved more fully in the global post-war economic order. Then came the Auto Pact, an agreement that integrated the North American automobile industry. Finally, the FTA and NAFTA were signed in order to achieve uninhibited access to the American economy. The net effect of trade liberalization has been to strengthen Canada's economic ties with the United States. For example, from 1992 to 2006 the percentage of Canada's exports to the United States increased from 75.5 per cent to nearly 79 per cent (Table 1.6).

The effects of trade liberalization on Canada's regional economies have been mixed. All the regions have increased their service sectors, but, except for Ontario, exports are predominantly primary products, with most destined for the American market. Ontario, thanks to the Auto Pact, saw its automobile industry thrive, almost doubling its share of the North American market. But following the Free Trade Agreement in 1989, the rest of the manufacturing industry in Central Canada was either subject to fierce competition from American firms or to closure of American branch plants. As a result, the manufacturing sectors of both Ontario and Québec went through difficult times.

For Canada's peripheries, the process of globalization accomplished two, long-sought-after goals—more local processing of primary products and access to foreign manufactured goods. New trade links with the United States and, to a lesser degree, with Japan and Europe quickly took shape. British Columbia, for example, benefited from the rapid economic expansion of Japan and other Pacific Rim countries. In the early 1990s, BC exported around 37 per cent of its goods to Pacific Rim countries, with Japan accounting for 25 per cent of this trade (IDAC, 1998: 6). However, by 1997, the economies of Pacific Rim countries faltered, chilling BC's export-dependent economy. By 2001, British Columbia was receiving equalization payments and had therefore slid into the position of a have-not province. Other hinterlands gained greater access to particular parts of the American market through trade liberalization. For instance, Atlantic Canada has access to the New England market. McCain Foods and the various Irving enterprises have taken advantage of the American market, but smaller producers were less able to achieve economies of scale and therefore have had more difficulty selling their products in the US. In 2004, New Brunswick's Irving family opened a diaper manufacturing plant (known as Irving Personal Care) in Moncton to specialize in diaper production for the North American market. Certainly, trade liberalization has allowed various regions of Canada to tap into the much larger American market and high production levels allow economies of scale to take place, keeping cost per unit low. In this way, Canadian firms can compete in the North American marketplace. Greater trade with the United States has brought economic benefits, but it has made Canada more dependent on access to American markets—and when that access is disrupted, Canadian firms and workers suffer.

In becoming a member of NAFTA and the GATT/WTO, Canada surrendered certain controls over its economy. No longer can Ottawa or the provinces impose policies reflecting economic nationalism, such as having one set of rules for foreign companies and another set for Canadian companies, or maintaining regional development programs that lend support to businesses competing with foreign firms. For instance, subsidies to sustain regional industries, such as the Cape Breton Development Corporation (Devco), or to reduce grain transportation costs for Prairie farmers by means of the Crow Benefit are no longer an option for the federal government. In fact, Canada, being the smaller partner, has had to bend on sensitive trade issues that affect American producers. One example was the Canada–US softwood lumber dispute that ended in 2006. Even the new agreement has its testy elements. The key issue was access of Canadian lumber to US markets. Washington, through the powerful influence of the American lumber lobby, was and remains determined to limit the amount of lumber Canada can export to the United States. The 2006 softwood lumber agreement not only limits exports, but it also forces lumber-producing provinces to limit their exports accordingly—or suffer a penalty. This agreement has caused friction among the major provincial producers—BC, Ontario, and Québec. Each claims that its allocation is too low. The matter is so complicated and contentious that the agreement's 'surge' mechanism is in dispute among the producer provinces. Ottawa is to increase the tax on softwood lumber exports to the US when the volume of exports hits a certain trigger point, yet each province disagrees on a 'trigger point' and so far Ottawa has not imposed one. According to Montana Senator Max Baucus, chairman of the US Senate Finance Committee:

> Canada struck a deal on lumber and they need to abide by it. Canada can stick by our deal on lumber, or expect to be taken to the woodshed—in this case, international arbitration—over this dispute. (Godfrey, 2007)

In spite of shortcomings in trade with the US and the belligerent attitude of some American political leaders, approximately four-fifths of Canada's exports flow to the United States. Such a trade pattern affects Canada in many ways. Given the changes that have already resulted from trade liberalization, what future transformations await Canada in these times of US concern over security and the resulting increased costs of doing business with the US.? Three questions arise, all having implications for Canada's core/periphery structure:

1. Will the economic regional alignment of Canada shift from an east–west alignment to a north–south one?
2. Will the national core/periphery structure be supplanted by a continental or global core/periphery structure?
3. Will Washington continue to stress internal security over trade with Canada?

The north–south 'grooves' of geography naturally encourage north–south regional trade flows. For instance, trade moves more naturally between the provinces of Western Canada and the American states to their south than between these three provinces and the more distant provinces to the east. With the erosion of trade barriers, north–south trade has increased rapidly. All along the US–Canada border, ever-increasing quantities of products move between Canadian regions and American markets according to the just-in-time principle. Ontario trades with New York and the Midwest, Québec with New England and New York, British Columbia with the US Pacific Northwest, Atlantic Canada with New England, and Western Canada with the American Great Plains. The Territorial North is an exception. While its oil and gas flow southward to American markets, mineral production has a more widespread distribution, including Europe and the Far East.

Perhaps because of the long-standing trade links between provinces, east–west trade, while growing more slowly, remains an essential element of Canada's economic structure. Economic studies by McCallum (1995), Helliwell (1996), and Anderson and Smith (1999) support the notion that the flow of trade between Canadian regions and provinces remains much more powerful than north–south trade between Canada and the United States, perhaps by a factor of 10 to 20 times. The explanation lies in the 'national border effect'. This effect is likely caused by a number of factors: the east–west Canadian transportation and communications system that encourages interprovincial trade; trade inertia based on pre-liberalization trading patterns between Canadian firms that remained strong in spite of slightly lower US prices; surviving trade barriers such as the milk marketing boards that shut out American milk imports and that, through the assignment of a large milk quota to Québec, ensure a high level of Québec milk sales to Ontario customers; and finally, Canadian taste preferences, such as Canadians consuming more domestic beer rather than lower-priced American beer. Then, too, the unexpected happened—security trumped trade, making crossing the border into the United States more difficult, more time-consuming, and more expensive. The just-in-time principle did not take into account the harsh security crackdown at the Canada–US border that followed the terrorist attacks on New York and Washington on 11 September 2001. Now an older principle rules—the just-in-case principle. Inventories are growing in size on both sides of the border because deliveries are no longer a matter of hours.

Still, integration into the North American economy has realigned Canada's industrial core and resource peripheries from within a national structure to within continental and global structures. While the cost of doing business has increased because of higher transportation and inventory costs, the Canadian core—Ontario and Québec—serves both the North American market and the domestic one while the Canadian peripheries are receiving more of their manufactured goods from adjacent parts of the United States and from other industrial cores, and are shipping more of their primary products to the United States and world markets.

However, the core/periphery structure of Canada's regions has so far remained intact. Central Canada is still Canada's core. The resilience of the Canadian core/periphery spatial structure under trade liberalization is remarkable, given the noticeable increase in north–south trade flows. This persistency represents a paradox: as regional economic growth occurs, the regional economic structure remains relatively stable. Geography, by limiting the economic opportunities available to each region, has promoted a particular type of economic growth in each region, even while foreign countries have provided new markets. At the same time, regional economies have undergone major modifications in response to trade liberalization through finding new markets and processing more primary products. For instance, livestock production and alternative crops in Western Canada have expanded because of increased trade with the United States and Japan. But economic integration of Western Canada with the outside world makes this region's prosperity dependent on continued exports to these countries. Trade dependency brings with it a measure of vulnerability to unexpected events. In May 2003, Canadian beef exports were banned by other countries, including the United States, because one cow was found to have mad cow disease (BSE). With about one-third of beef production exported, the impact on ranchers and the beef industry was immense. In 2004, the US lifted import restrictions on ground beef, bone-in cuts of beef, and offal from animals younger than 30 months. The import of live cows and meat from older animals from Canada, however, was still banned. By 2007, the US lifted its ban and Canadian live cattle could again enter the US.

In re-examining the core/periphery model in the light of trade liberalization, much change has taken place and more is expected. Canada's core now forms a subset of the American industrial core and, as a result, competes for customers in all regions of North America. Canadian peripheries are changing, too. With new trade opportunities and higher prices for resources, regional peripheries are realigning their economic ties along continental and global lines. Among the peripheries, however, there are significant differences. Those peripheries with a rich resource base and a highly skilled labour force are accelerating their economic diversification and are moving towards a more independent economic status. Those with a weak economy, high unemployment rates, and an outflow of young people are becoming more and more dependent on Ottawa and on world market prices for their resources.

The global economy has taken Canada and other countries into unexplored territory. Economic structures, such as trade agreements, are in some ways overcoming political structures, such as national borders. Furthermore, multinational corporations, growing ever larger and more powerful, hold more and more of the world's investment capital and therefore have a significant impact on global economic structures. Capital, now the key element in the new market economy, is so mobile and multinational firms are so powerful that a new political order will have to emerge to regulate this world economy. Of particular concern is the need to ensure global environmental standards and labour practices, as well as to maintain employment levels in Western economies, while bringing greater development to global peripheries. Varying opinions exist on what form a new political order should take, but most agree that political globalization, which can match the economic globalization already underway, is greatly needed (Wallerstein, 1997: 141, 158; Dicken, 1998: 462–3).

What will be the consequences for Canada if an efficient political global order is not established? Stripped of more and more of its economic powers, Canada will, at best, retain control over its domestic affairs. Even domestically, however, international market forces, unless kept in check by a new political order, can exert exceptional pressure on governments. Take Canada's tax system, for instance. Ottawa is under pressure to lower Canadian taxes to the level of US taxes. While such a policy would keep more money in the pockets of Canadians, it could lead to the underfunding of our public health-care system, the slashing of social programs, and the widening of the income gap between Canada's regions and between the rich and poor within the regions. Consider the unexpected impact of international terrorism on world trade. For Canada, exports to the United States are critical for our economic well-being and any delays at the border are costly.

Canada's Faultlines

Canada is a complex and dynamic country. Part of its complexity is rooted in the four faultlines that represent the main tensions within the country. The first faultline, between non-Aboriginals and Aboriginal peoples, has existed since Europeans arrived on what is now Canadian soil: for centuries, Aboriginal Canadians have struggled with non-Aboriginals over such issues as land, rights, and the environment. The second faultline, between English-speaking and French-speaking Canada, represents the relationship between the two founding European nations that first settled Canada. The third faultline, between centralists and decentralists, developed as governing and economic structures began to create 'have' and 'have-not' regions. The last faultline—between 'newcomers' and 'old-timers'—reflects a gap in values and expectations between recent immigrants to the country and those who were born and raised in Canada. While Canada's early history is characterized by imposed solutions to these faultlines, the country's wisest leaders, acknowledging the strength of Canada's complexity, sought compromises instead. Compromise usually took the form of shared power rather than power enforced by the dominant party. Sharing power allowed Canada to evolve into a pluralistic society where both individual and collective differences are accepted and respected.

The sharing of power, however, has not been easy. For instance, in the past, Aboriginal peoples were marginalized by such 'imposed solutions' as reserves and residential schools. Today the place of Aboriginal peoples within Canada has changed. A shift in power is embodied in modern land-claim agreements (Sparke, 1998: 463–95). Yet, the scale and complexity of the land settlement process is huge, and involves both comprehensive claims by First Nations not covered at present by a treaty (e.g., the Nisga'a Final Agreement) and specific claims in which First Nations seek to have their original treaty rights to land and/or funding met (e.g., Ipperwash and Caledonia in Ontario; Oka in Québec). At the present rate, the last current claim would be implemented in the next century!

Since specific land-claim negotiations began in 1973, nearly 1,400 have been filed by First Nations and less than 300 have been settled, with 123 claims in negotiations (Canada, Department of Indian and Northern Affairs, 2007). The primary reason for the slowness of the settlement process lies in the complex nature of most claims. Each claim must be documented by the claimant (First Nation) and then reviewed by the Specific Claims Branch of the Department of Indian Affairs and Northern Development. The next step is negotiations between Ottawa and the First Nation. If an agreement is reached, the First Nation community must approve it by vote and then Canada ratifies the agreement by legislation. The final step is implementation, that is, the transfer of funds and/or land to the First Nation by Ottawa. The cost of settling existing specific claims is estimated to fall between $6 billion to $12 billion (*National Post*, 2007: A14).

The attraction for an agreement, whether specific or comprehensive, is that First Nations can move on from their political agenda to a developmental one and hopefully play a more productive role in Canada's economy. The Nisga'a, for example, reached a compromise with British Columbia and the federal government that secured them territory, guaranteed them a share of resources, and established a locally designed form of self-government. The Nisga'a gained full title to some of their lands in northern BC and then began the process of re-establishing their place within Canadian society. This journey of reconciliation between non-Aboriginals and Aboriginal peoples through land-claim agreements, begun in the 1970s, will continue well into the twenty-first century and perhaps beyond. Ottawa's decision in 2007 to accelerate specific land-claim settlements by creating an independent commission and by boosting funding may ease the backlog of claims. Such a commission, which may be created by Parliament in 2008, will apply to treaties and therefore will likely exclude the Métis, non-status Indians, and Inuit.[2]

Canada's French/English faultline is perhaps the one that is the most deeply felt. Since Confederation, these two linguistic groups have struggled for power. Over time, French Canadians tried to win compromises from the more dominant power of English Canada. Despite the cultural divide between the two groups, the fact that both societies have contributed to Canada's unique diversity is recognized. However, the rise of Québec nationalism and a series of failed attempts at constitutional reform have made the need for further reconciliation even more urgent. Hollow gestures, such as the Calgary Declaration following the 1995 Québec referendum, were replaced by a greater sharing of power between Ottawa and Québec. Political reconciliation and 'solving the fiscal imbalance' in the 2007 federal budget have muted the separatist voices calling for Québec's separation from the rest of the country, so much so that a survey carried out by Leger Marketing in June 2007 found that '83 per cent of Quebecers believe Quebec will be part of Canada when the country celebrates its 150th anniversary in 2017' (CanWest News Service, 2007).

Canada's third faultline, between centralists and decentralists, stems from one of Canada's most prominent features—regional diversity. Regional diversity has led to regional disparity—some regions are endowed with both political and economic power while other regions lack both. This disparity has generated a tension between those regions having power, as well as the federal government, and those regions struggling for more power. Though federalism already affords a sharing of power between the central government and regional (provincial and territorial) governments, some argue that power is not shared equally. Because of this sentiment as well as other issues, such as the unilateral cuts in transfer payments to the provinces by Ottawa in the 1990s, provincial governments have been demanding a greater share of power, including more taxing power. With trade liberalization and the federal government's reduced ability to inject money into depressed regions of the country, centralist/decentralist tensions will only intensify. Compromises will have to be reached if provinces are to meet their socio-economic responsibilities, while still maintaining a level of power for the central government that ensures the preservation of national standards among the services available to Canadians in all regions.

The equalization program of the federal government only too often reveals the difficulty of reaching an agreement acceptable to all provinces. Certainly that was the case when the 2007 federal budget not only indicated how much each province would receive, but its fiscal cap affected the Atlantic Accord side deal and thus created unseemly and, with every day, escalating political reactions in St John's, Halifax, and Ottawa. Perhaps these political shenanigans prompted a prominent Canadian commentator, Andrew Coyne, a strong defender of centralist policies, to argue that the more Ottawa gives to provinces, the more they complain. As Coyne bluntly remarked in regard to the ever-increasing flow of federal dollars to provinces:

> With what result? The provinces are more bitter, more resentful, than ever, every one of them doing its best to convince its citizens they are getting done out by the rest. And if the feds should ever dare to resist, they are accused of that worst of crimes, offences against 'co-operative federalism'. If the feds cannot win this game, they should stop playing. (Coyne, 2007)

As Canada entered the twenty-first century, western alienation loomed as a major political problem. The National Atlas map of the voting pattern in the 2000 federal election presents a spatial version of western alienation (see Student website, National Atlas section, Map 31). One of the driving forces behind western alienation and such voting patterns is the desire for more political power or at least recognition in Ottawa. Population growth has fuelled this desire in British Columbia and Western Canada. Alberta is the current sparkplug for such growth. Demographic growth rates, but particularly net interprovincial migration (Figure 11.1), demonstrate this point.

Canada's last faultline, between newcomers and old-timers, has always been a factor in Canada's history. In the last decade, immigration has not only added to Canada's labour force but also has kept Canada's population increasing. Immigrants come from all corners of the world, but the vast majority have settled

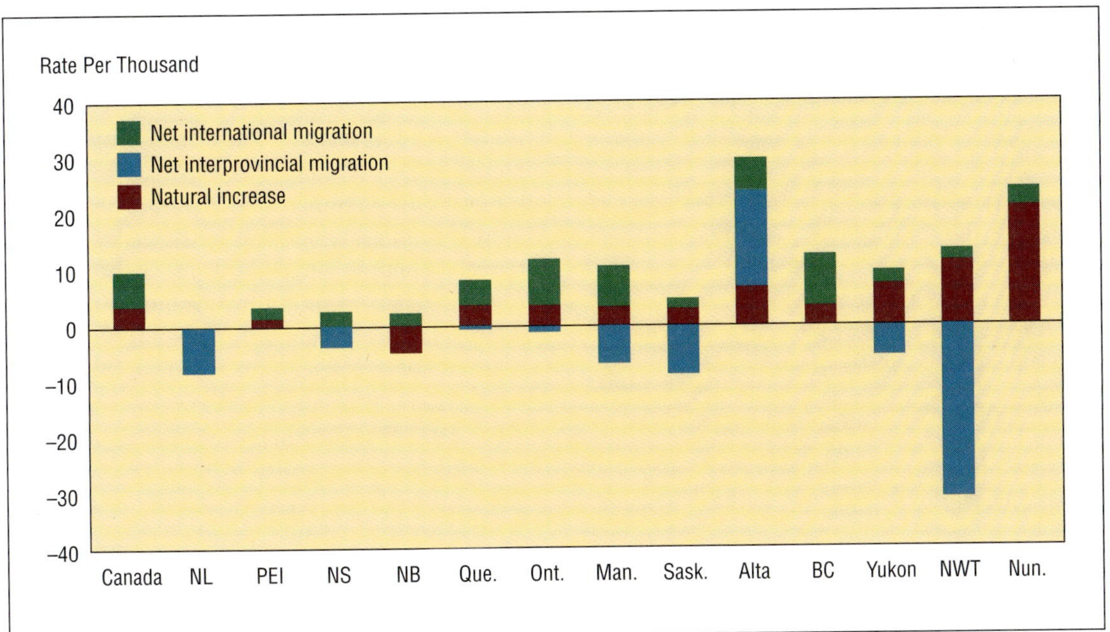

Figure 11.1 Demographic growth rates by component, Canada, provinces, and territories, 2005–6
Source: Statistics Canada (2007). Adapted from Statistics Canada publication *Annual Demographic Estimates: Canada, Provinces and Territories 2005–2006*, Catalogue 91-215-XWE, http://www.statcan.ca/bsolc/english/bsolc?catno=91-215-XWE.

in the three largest cities—Toronto, Vancouver, and Montréal. As a result, two Canadas have been created—one with a high proportion of newcomers and the other with relatively few newcomers. Ottawa has toyed with the idea of requiring new immigrants to locate outside of the three main cities, partly to ensure a source of labour and partly to disperse the newcomers more broadly across Canada.

Canada's faultlines continually challenge the country and its regions, but these dynamic struggles also strengthen us. One of Canada's leading thinkers, John Ralston Saul, sees these faultlines as pillars. In his book *Reflections of a Siamese Twin: Canada at the End of the Twentieth Century*, Saul argues that the Aboriginal, the francophone, and the anglophone are dependent on each other, forming a triangular reality. 'Each of their independent beings has been interwoven with the other two for over 450 years of continuous existence on the northern margins of the continent' (Saul, 1997: 81). He further argues that this interdependence has, over time, allowed Canada to avoid the rigidity of nation-states that compel their citizens to conform. In a sense, then, Canada's faultlines, rather than dividing the country, have managed to unite its different regions and peoples.

Northern lights near the Victor diamond mining camp in the Hudson Bay Lowland of Ontario. (Photo: Trish Buttineau. From Debeers Canada, at: www.debeerscanada.com/files_2/photogallery-f.html)

The Last Century: An Overview

Since 1867, Canada's society and economy have undergone a remarkable transformation. At Confederation, Canada was a small rural society of some 3 million people clustered into four regions of Canada (New Brunswick, Nova Scotia, southern Ontario, and southern Québec). Canada's population was composed of Aboriginal peoples and settlers from two main ethnic groups—English and French. Eventually, the country's territory expanded and new provinces joined or came into being. The country was rich in resources but small in population and therefore seemed destined to produce primary goods for export markets. Ottawa, recognizing the problem of an economy entirely devoted to primary production, set in motion an economic policy that imposed high tariffs on imported manufactured goods. That policy stimulated industrial growth in southern Ontario and southern Québec and encouraged foreign firms to build branch plants in the emerging industrial core of Canada. The same policy forced the rest of Canada to buy Central Canada's manufactured goods but to sell their primary goods on world markets. The settling of Western Canada added a third dimension to Canada's population—people from Northern, Central, and Eastern Europe added to the diversity of Canada's population and culture. As Canada grew more independent from Britain, it had to face the harsh realities of federalism, including Aboriginal issues, the concept of two founding peoples, and regional self-interests.

Canada has evolved into an independent nation consisting of 10 provinces and three territories. Technological advances have transformed Canada. Natural barriers like the Canadian Shield and the Rocky Mountains have been overcome by innovations in transportation; advances in plant and seed genetics have shortened the growing time for crops and provided more crop alternatives in Western Canada; new techniques such as insulated buildings, winter roads, and utilidors (above-ground insulated corridors for water, sewerage, and electricity) have helped overcome the cold environment found in the Territorial North.

Political decisions have also transformed Canada. The liberalization of immigration and trade has had a profound impact on Canada's society and economy. Immigration from a broader range of countries is creating a more cosmopolitan society with strong international links. Free trade has moved Canada into a North American regional economy and strengthened its economic links with other countries of the world. Canada is now a more urbane society composed of Canadians whose origins are from all corners of the globe. While progress has been made in building compromise among Canada's four faultlines, more work remains. In fact, Canada's faultlines will continue to demand attention because absolute solutions are impossible within such a diverse but democratic country—but then that combination of democracy and diversity has become Canada's true strength.

The Twenty-First Century

First and foremost, the feature that characterizes Canada and makes it strong—diversity—has evolved through compromise. But compromise is never fully achieved: it takes ongoing and committed efforts to maintain. Already in the twenty-first century, Canadians are facing new challenges. For instance, with more and more newcomers bringing their cultures and religions to their adopted country, the newcomers/old-timers faultline seems destined to have its rumblings. So far, most are minor, such as the issue of veiled voters revealing their faces before voting. Much more serious is the dispute over public funding for religious schools. Why, argue supporters of universal funding for all religious schools, should only Catholic schools benefit from public funding? Still, the magnitude and geopolitical consequences of this potentially defining issue for Canadian society remain dwarfed by the French/English faultline, which could literally split the country. In the aftermath of the 1995 referendum, the sharper edges of conflict were blunted, and in 2003 the separatist Parti Québécois government was pushed from political power in Québec and then, in 2007, fell to third place in the provincial election. Except for hard-core separatists, few Canadians can contemplate their country without Québec and some fear that Québec separation would lead to Canada's disintegration. The potential outcomes of Québec separating are sobering to consider. Many argue that, once split, the durability of 'the rest of Canada' is uncertain. For instance, history and geography have shown that discontiguous states have short lifespans.[3]

Canada's regionalism, if not fostered with ongoing compromise, also has the potential to fragment the country. Some, like journalist Joel Garreau, have argued that Canada's regions could be 'picked off', one by one, and absorbed into the United States. Robert Kaplan, an American writer, argues that economic integration with the United States will inevitably lead to regional self-interest fracturing Canada into new political amalgamations. For instance, British Columbia, Kaplan speculates, could become part of the Pacific Northwest, which might be renamed 'Cascadia' (Kaplan, 1998: 322–4). Economists Courchene and Telmer (1998: 1–3) support this economic integration thesis by declaring that Ontario is no longer Canada's heartland but is now a North American region state. They also base their thesis on 'regional self-interest'.

The faultline between British Columbia and Western Canada and Ottawa has deepened because these two regions are growing more rapidly than the rest of the country. As this demographic and economic shift of power continues, these western regions are likely to challenge Ottawa and the core regions of Ontario and Québec on such issues as equalization payments and the number of seats in the House of Commons.

Geographic regions and their senses of place are enduring. In the twenty-first century, the six major regions of Canada will continue to exist. But are Canada's regions likely to become part of a new spatial arrangement (i.e., North America), as predicted by some experts? Or do Canada and its regions comprise an idea driven by more than just economic concerns? Clearly, to ensure that Canada is enduring, internal tensions will have to be addressed. This demands a political will and a clear recognition that Canada is not a homogeneous nation-state but a pluralistic country where regional differences not only exist but are cherished by those living in these regions. Canada's survival will depend on our ability to accept the country's geographic structure and, as a result, to recognize that Canada's identity is a regional matter (Frye, 1971: i–iii). In the twenty-first century, Canada will need to build cultural bridges between its regions and, in so doing, strengthen national identity and unity.

Acceptance of Canada's cultural diversity, then, is the linchpin holding Canada and its six regions together. The future, while full of uncertainty, contains two certainties: Canada will remain a heterogeneous country and, as such, will continue to experience regional differences. However, by engaging in open and vigorous debate, compromises—not solutions—will follow. Compromise works at the heart of the exercise of democratic power in a diverse society by demanding a balance between individual and collective rights, between regional and national interests, and between those with power and those in need of it. The ability to strike this balance is critical to Canada's well-being and to regional harmony.

As pointed out at the beginning of this chapter, the future, while full of uncertainty, contains one truth: Canada will remain a country of ever-changing cultural and regional diversity. Another truth is that tensions will continue to flare up and the well-tested Canadian 'ad hoc' process of seeking adjustments leading to compromises so necessary in a 'soft' democratic country will need to keep coming into play.

Notes

1. Geographers view cultural landscapes, particularly cityscapes, as creating a human imprint that evokes a sense of place. However, this human imprint has been affected in recent years by a sense of urban placelessness wherein all cities seem alike due to a spatial standardization by global-based firms like McDonald's and Shell, and, especially in North America, the ubiquity of shopping malls, plazas, and big-box stores. Uniformity is erasing the uniqueness of place (Relph, 1976: 79–80). While the impact of standardization of cultural landscapes is most apparent in urban areas, it does exist elsewhere, especially in international tourist centres. Some cities have opposed such imposed architectural regularity in an effort to conserve their unique cultural landscapes. For instance, Québec City forced McDonald's to construct a fast-food restaurant that matched the architectural style of surrounding buildings in old Québec City.
2. The ability to negotiate modern land-claim agreements varies among Aboriginal peoples. Those Aboriginal peoples who did not sign treaties are best positioned to negotiate comprehensive land-claim agreements, which generally include rights to land, resources, and self-government. First Nations in British Columbia, the Territorial North, and Labrador fall into this category. Aboriginal peoples who have signed treaties but have not received all the benefits entitled to them under their treaties can seek redress under specific land-claim agreements. Aboriginals who are classified as non-status Indians by the federal government have lost their legal right to land claims.
3. In today's world, discontiguous states are rare. Of the more than 200 nations in the world, less than 10 states are divided. Most are small enclaves, such as Kaliningrad, which is separated from Russia by Lithuania and Belarus. Others have evolved into separate states, such as Pakistan, which existed as two parts for 25 years but saw its eastern lands form a new state, Bangladesh, in 1972. If Québec separated, Atlantic Canada would be cut off from the rest of the country, much like Alaska is from the rest of the United States. Alaska, like

Atlantic Canada, is a large, remote region that relies on a resource economy. However, several differences exist between Alaska and Atlantic Canada. Alaska has less than 1 per cent of the US population while Atlantic Canada represents 8 per cent of Canada's population. Furthermore, unlike Alaska, Atlantic Canada relies more heavily on road transportation and road links to keep its economy functioning. Finally, while both regions receive large subsidies, the cost of supporting an isolated Atlantic Canada would place a much greater financial burden on the rest of Canada than Alaska now places on the United States.

Challenge Questions

1. What arguments can you put forward to defend the position that 'Geographic regions and their senses of place are enduring'? Do you agree with the notin that those regions with the longest history of European settlement have the strongest sense of place? Can you identify some aspect of physical geography that has promoted a sense of place in each region? Finally, how do these three thoughts apply to the region you live in?
2. Regarding Canada's future, what are the two so-called 'truths'. Why do you agree or disagree with these truths? Can you think of more 'truths'?
3. 'Canada is big—preposterously so', wrote geographer Kenneth Hare back in 1968. In your opinion, which of the four faultlines were more prominent in the 1960s and which were less prominent? What events/factors caused these changes? You may wish to refer to 'Further Reading' in Chapter 1.
4. What geographic concept(s) supports John Ralston Saul's statement that 'in spite of intellectual claims to the contrary, not religion, not language, not race but place is the dominant feature of civilizations. It decides what people can do and how they will live'?
5. Do you think that Andrew Coyne wants to end equalization payments? If these payments were ended, what would the regional consequences be, i.e., who would gain and who would lose?
6. With increasing immigration, it seems inevitable that the demand for public funding for all religious schools will become a major political issue. How would you resolve this religious/political issue and thus prevent rumblings from becoming an earthquake on the faultline between newcomers and old-timers?

Bibliography

Anderson, Michael A., and Stephen L.S. Smith. 1999. 'Canadian Provinces in World Trade: Engagement and Detachment', *Canadian Journal of Economics* 32, 1: 22–38.

Barnes, Trevor J. 1993. 'A Geographical Appreciation of Harold A. Innis', *Canadian Geographer* 37, 4: 352–64.

Britton, John N.H., ed. 1996. *Canada and the Global Economy: The Geography of Structural and Technological Change*. Montréal and Kingston: McGill-Queen's University Press.

Canada, Department of Indian Affairs and Northern Development. 2007. Specific Claims Branch. At: <www.ainc-inac.gc.ca/ps/clm/scb_e.html>.

CanWest News Service. 2007. 'Quebec Will Still Be Part of Canada in 10 Years: Poll', *National Post*, 23 June, A10.

Clark, Joe. 1994. *A Nation Too Good to Lose: Renewing the Purpose of Canada*. Toronto: Key Porter Books.

Courchene, Thomas J., and Colin R. Telmer. 1998. *From Heartland to North American Region State: The Social, Fiscal and Federal Evolution of Ontario*. Monograph Series on Public Policy, Centre for Public Management. Toronto: University of Toronto Press.

Coyne, Andrew. 2007. 'Sooner or Later, Ottawa Caves', *National Post*, 20 June, A12.

Dicken, Peter. 1998. *Global Shift: Transforming the World Economy*. London: Guilford Press.

Frye, Northrop. 1971. *The Bush Garden: Essays on the Canadian Imagination*. Toronto: Anansi Press.

Garreau, Joel. 1981. *The Nine Nations of North America*. Boston: Houghton Mifflin.

Godfrey, Mike. 2007. 'US Accuses Canada of Reneging on Lumber Deal', News.com, 5 Apr. At: <www.tax-news.com/asp/story/US_Accuses_Canada_Of_Reneging_On_Lumber_Deal_xxxx26882.html>.

Hare, F. Kenneth. 1968. 'Canada', in John Warkentin, ed., *Canada: A Geographical Interpretation*. Toronto: Methuen.

Haynes, David M., ed. 1997. *Can Canada Survive? Under What Terms and Conditions?* Toronto: University of Toronto Press.

Helliwell, John F. 1996. 'Do National Boundaries Matter for Quebec's Trade?', *Canadian Journal of Economics* 29, 3: 507–22.

Innis, Harold A. 1930. *The Fur Trade in Canada: An Introduction to Canadian Economic History*. New Haven: Yale University Press.

Investment Dealers Association of Canada (IDAC). 1998. *British Columbia: Economic & Fiscal Outlook*. Vancouver: Investment Dealers Association of Canada.

Joyce, Greg. 1999. 'Treaty with the Nisga'a', *Saskatoon Star-Phoenix*, 13 Apr., B8.

Kaplan, Robert D. 1998. *An Empire Wilderness: Travels into America's Future*. New York: Random House.

McCallum, John. 1995. 'National Borders Matter: Canada–US Regional Trade Patterns', *American Economic Review* 85, 3: 615–23.

National Post. 2007. 'By the Numbers', 2 June, A14.

Ormsby, Margaret A. 1958. *British Columbia: A History*. Toronto: Macmillan.

Pally, Thomas. 1998. 'Building Prosperity from the Bottom Up', *Challenge* 51, 5: 59–71.

Penrose, Jan. 1997. 'Construction, De(con)struction and Reconstruction: The Impact of Globalization and Fragmentation on the Canadian Nation-State', *International Journal of Canadian Studies* 16: 15–50.

Relph, Edward. 1976. *Place and Placelessness*. London: Pion.

Saul, John Ralston. 1997. *Reflections of a Siamese Twin: Canada at the End of the Twentieth Century*. Toronto: Viking.

Sparke, Matthew. 1998. 'A Map that Roared and an Original Atlas: Canada, Cartography, and the Narration of Nation', *Annals, Association of American Geographers* 88, 3: 463–95.

Statistics Canada. 1997. 'Provincial GDP and Interprovincial Trade 1996', *The Daily*, 16 May. At: <www.statcan.ca/Daily/English/ 970516/d970516.htm>.

———. 2003. Imports and Exports of Goods on a Balance-of-Payments Basis. At: <www.statcan.ca/english/Pgdb/gblec02a.htm>.

———. 2007. Annual Demographic Estimates: Canada, Provinces and Territories 2005-2006. At: <www.statcan.ca/english/freepub/91-215-XIE/2006000/ct003_en.htm>.

Wallerstein, Immanuel. 1998. 'Contemporary Capitalist Dilemmas, the Social Sciences, and the Geopolitics of the Twenty-first Century', *Canadian Journal of Sociology* 23, 2–3: 141–58.

Further Reading

Saul, John Ralston. 1997. *Reflections of a Siamese Twin: Canada at the End of the Twentieth Century*. Toronto: Viking.

John Ralston Saul is one of Canada's outstanding thinkers. In *Reflections of a Siamese Twin* the reader is offered an intellectual challenge, for knowledge of Canada's past is necessary to interpret Saul's concept of complexity and his many metaphors. Canada's complexity is linked by Saul to his concept of '*un grand pays mou*' (a big soft country). 'Soft' does not mean weak but flexible enough to seek compromise rather than confrontation, and the idea is related to the diversity of our society—Canada as 'a country of minorities'. Saul's vision of Canada hinges on relations between anglophone, francophone, and Aboriginal Canadians, and these three pillars of Canada form the basis of what the author calls our 'triangular reality'. The regional fault-line (the centralist/decentralist argument) discussed in this textbook adds another dimension to Saul's triangle and reinforces the importance of Canada's regional nature.

Index